NATO ASI Series

Advanced Science Institutes Series

A Series presenting the results of activities sponsored by the NATO Science Committee, which aims at the dissemination of advanced scientific and technological knowledge, with a view to strengthening links between scientific communities.

The Series is published by an international board of publishers in conjunction with the NATO Scientific Affairs Division

A Life Sciences
B Physics

Plenum Publishing Corporation
London and New York

C Mathematical and Physical Sciences
D Behavioural and Social Sciences
E Applied Sciences

Kluwer Academic Publishers
Dordrecht, Boston and London

F Computer and Systems Sciences
G Ecological Sciences
H Cell Biology

Springer-Verlag
Berlin, Heidelberg, New York, London, Paris and Tokyo

Series C: Mathematical and Physical Sciences - Vol. 285

Climate and Geo-Sciences

A Challenge for Science and Society in the 21st Century

Climate and Geo-Sciences

A Challenge for Science and Society in the 21st Century

edited by

A. Berger

Institut d'Astronomie et de Géophysique Georges Lemaître,
Université Catholique de Louvain-la-Neuve, Belgium

S. Schneider

National Center for Atmospheric Research,
Boulder, Colorado, U.S.A.

and

J. Cl. Duplessy

Centre des Faibles Radioactivités,
Laboratoire Mixte CNRS-CEA,
Gif-sur-Yvette, France

NATO Advanced Research Workshop on...

Kluwer Academic Publishers

Dordrecht / Boston / London

Published in cooperation with NATO Scientific Affairs Division

Proceedings of the NATO Advanced Research Workshop on
Climate and Geo-Sciences
A Challenge for Science and Society in the 21st Century
Louvain-la-Neuve, Belgium
May 22–27, 1988

Library of Congress Cataloging in Publication Data

Climate and geo-sciences : a challenge for science and society in the
21st century / A. Berger, S. Schneider, J. Cl. Duplessy, editors.
p. cm. -- (NATO ASI series. Series C, Mathematical and
physical sciences ; no. 285)
Includes index.
ISBN 0-7923-0404-7
1. Climatology--Congresses. 2. Climatology--Social aspects--Congresses. 3. Climatic changes--Congresses. 4. Paleoclimatology--Congresses. 5. Greenhouse effect, Atmospheric--Congresses.
I. Berger, A. (André), 1942- . II. Schneider, Stephen Henry.
III. Duplessy, J. Cl., 1942- . IV. Series.
QC980.C53 1989
551.6--dc20 89-15404

ISBN 0–7923–0404–7

Published by Kluwer Academic Publishers,
P.O. Box 17, 3300 AA Dordrecht, The Netherlands.

Kluwer Academic Publishers incorporates the publishing programmes of
D. Reidel, Martinus Nijhoff, Dr W. Junk and MTP Press.

Sold and distributed in the U.S.A. and Canada
by Kluwer Academic Publishers,
101 Philip Drive, Norwell, MA 02061, U.S.A.

In all other countries, sold and distributed
by Kluwer Academic Publishers Group,
P.O. Box 322, 3300 AH Dordrecht, The Netherlands.

Printed on acid free paper

Printed in The Netherlands

This book contains the proceedings of a NATO Advanced Research Workshop held within the programme of activities of the NATO Special Programme on Global Transport Mechanisms in the Geo-Sciences running from 1983 to 1988 as part of the activities of the NATO Science Committee.

Other books previously published as a result of the activities of the Special Programme are:

BUAT-MENARD, P. (Ed.) – *The Role of Air-Sea Exchange in Geochemical Cycling* (C185) 1986

CAZENAVE, A. (Ed.) – *Earth Rotation: Solved and Unsolved Problems* (C187) 1986

WILLEBRAND, J. and ANDERSON, D.L.T. (Eds.) – *Large-Scale Transport Processes in Oceans and Atmosphere* (C190) 1986

NICOLIS, C. and NICOLIS, G. (Eds.) – *Irreversible Phenomena and Dynamical Systems Analysis in Geosciences* (C192) 1986

PARSONS, I. (Ed.) – *Origins of Igneous Layering* (C196) 1987

LOPER, E. (Ed.) – *Structure and Dynamics of Partially Solidified Systems* (E125) 1987

VAUGHAN, R. A. (Ed.) – *Remote Sensing Applications in Meteorology and Climatology* (C201) 1987

BERGER, W. H. and LABEYRIE, L. D. (Eds.) – *Abrupt Climatic Change – Evidence and Implications* (C216) 1987

VISCONTI, G. and GARCIA, R. (Eds.) – *Transport Processes in the Middle Atmosphere* (C213) 1987

SIMMERS, I. (Ed.) – *Estimation of Natural Recharge of Groundwater* (C222) 1987

HELGESON, H. C. (Ed.) – *Chemical Transport in Metasomatic Processes* (C218) 1987

CUSTODIO, E., GURGUI, A. and LOBO FERREIRA, J. P. (Eds.) – *Groundwater Flow and Quality Modelling* (C224) 1987

ISAKSEN, I. S. A. (Ed.) – *Tropospheric Ozone* (C227) 1988

SCHLESINGER, M.E. (Ed.) – *Physically-Based Modelling and Simulation of Climate and Climatic Change 2 vols.* (C243) 1988

UNSWORTH, M. H. and FOWLER, D. (Eds.) – *Acid Deposition at High Elevation Sites* (C252) 1988

KISSEL, C. and LAY, C. (Eds.) – *Paleomagnetic Rotations and Continental Deformation* (C254) 1988

HART, S. R. and GULEN, L. (Eds.) – *Crust/Mantle Recycling at Subduction Zones* (C258) 1989

GREGERSEN, S. and BASHAM, P. (Eds.) – *Earthquakes at North-Atlantic Passive Margins: Neotectonics and Postglacial Rebound* (C266) 1989

MOREL-SEYTOUX, H. J. (Ed.) – *Unsaturated Flow in Hydrologic Modeling* (C275) 1989

BRIDGWATER, D. (Ed.) – *Fluid Movements – Element Transport and the Composition of the Crust* (C281) 1989

ANDERSON, D.L.T. and WILLEBRAND, J. (Eds.) – *Ocean Circulation Models: Combining Data and Dynamics* (C284) 1989

BERGER, A., SCHNEIDER, S. and DUPLESSY, J. Cl. (Eds.) – *Climate and Geo-Sciences* (C285) 1989

TABLE OF CONTENTS

SECTION 1 - CLIMATE RECORD / WHAT CAN WE LEARN FROM DATA ?

SECTION 2 - UNDERSTANDING CLIMATE AND CLIMATIC CHANGES

SECTION 3 - SIMULATION OF THE GREENHOUSE WARMING

SECTION 4 - TECHNOLOGY FOR CLIMATE STUDIES

SECTION 5 - CLIMATE ENVIRONMENT AND INTERNATIONAL SECURITY

SECTION 6 - CONFERENCE OVERVIEW AND RECOMMENDATIONS

INTRODUCTION

It has been widely recognized recently that in order to make scientific progress on large and important problems (eg, carbon dioxide effects on climate, viability of various sites for nuclear waste disposal etc.), it is necessary to integrate knowledge from wide ranging sets of disciplines. This is certainly true in the climate sciences, for progress in understanding the cause of the ice ages or the effects of industrial pollution on the future climate or even the likelihood of severe climatic consequences in the aftermath of nuclear war. All require state-of-the-art input from many geoscience disciplines : climatology, oceanography, meteorology, chemistry, ecology, glaciology, geology, astronomy, space technology, computer technology, mathematics etc.

Major international meetings have called for interaction of such geo-science disciplines to solve real world problems. To move beyond the rhetorical level, the NATO Special Programme on Global Transport Mechanisms in the Geo-Sciences whose activities started in 1983, decided to organise his closing symposium on such a topic which focus on the relationship between climate and geo-sciences. This symposium was held at the end of May 1988 at the Université Catholique de Louvain, Louvain-la-Neuve, Belgium.

One hundred-and-thirty participants from the 16 NATO countries and a number of non-NATO countries assembled for the Symposium. Another feature was the attendance by special invitation of 16 promising young scientists who might well become leading scientists on climate and geo-sciences in their respective countries in the next century.

The symposium has reviewed recent scientific progress in many climate-related disciplines, inviting all specialists to contribute in their own field and asking world leading scientists to prepare and present the reviews. The following main topics were thus covered :

. Climate Record : What Can We Learn from Data ?
. Climate System and Simulation of Climate Variations
. Understanding Climatic Change
. Technology for Climate Studies
. Climate, Environment and International Security.

But to make real progress at interdisciplinary research, the symposium has to be forward looking. Therefore, conclusions sessions were discussing emerging technologies, such as space observing instruments or super computers, and a panel was charged with drafting <u>specific</u> recommendations on how individual disciplines might work together to make accelerated progress on important climate-related problems on the global change agenda. The following brief summary of the main results is based on the closing remarks made by the panel and reported by Dr. John Perry in section 6 of the present volume.

"The presentations made during the Symposium revealed how greatly the understanding of climate has enhanced in recent years. A great deal is known about the climates of the past, from the Archaean through the brief moments of our species' time on Earth. The roles of astronomical forcing and of life processes in shaping climate seem even more firmly established than a decade ago. In modelling the behaviour of the climate system, it still appears that the ocean, clouds and the land surface pose major challenges. Symposium participants learned much about influences of chemistry, the problems of coupling media, the information gleaned from proxy records, climate modelling techniques, atmospheric chemical processes, and the details of modelled response and atmosphere-biosphere interactions. Discussions such as this between specialists tend to focus on details, uncertainties, and the many things, it is wished were known better. They tend to obscure the common core of consensus on what is known or believed. So it is perhaps worthwhile to state explicitely this consensus that implicitly underlies all the work of scientists in the field :

. The atmosphere, and indeed the entire global environment, are changing;

. Humans are today the principal agents of change, and the rich nations of NATO are among the most prominent of these agents provocateurs;

. In particular, the climates of all the world's nations will change, rich and poor alike.

. Climate is an important resource, constraint, and perhaps hazard that must be taken into account in the prudent management of our little planet condominium.

The question facing Symposium participants at the terminal session was **"Where do we go from here ?"**. This question could be answered in at least three ways, according to the definition of "we".

First, if "we" are the **physical climate system researchers**, the papers presented during the week laid out a challenging of scientific problems. Some of these problems are being analysed under the aegis of large programmes, as the World Climate Research Programme, which momentum should be maintained and further developed.

If the "we" is the **broader array of scientists** concerned with all the interlinked changes going on in our complicated planet, then a companion international effort - the International Geosphere Biosphere Programme (IGBP) - promises to play a similar catalytic and coordinating role. This Programme seeks to describe and understand the interactive physical, chemical, and biological processes that regulate the total Earth system, the changes that are occurring in this system, and the manner in which they are influenced by human actions. Although its scope in principle encompasses that of the World Climate Research Programme, the IGBP will be complementary to that Programme, in particular by providing a context for the involvement of geochemists, biologists, and ecologists in the effort to understand what is happening to our planet.

Finally, one can suppose that the "we" is **who participate in the NATO Science Programmes**. The NATO effort seeks to advance the scientific and technological capabilities of the Alliance by promoting international scientific cooperation in scientific disciplines relevant to the challenges facing modern society. The discussions during the Symposium indicated strongly that the problems posed by man-made influences on climate comprise a prime example of exactly the type of inescapably international interdisciplinary issue in which activities such as NATO's have great and beneficial leverage.

The number of scientists actively contributing to this research is but a trace constituent in the body of science. In order to work effectively, they require access to resources of data and computing power that are available in only a few places, and they must be able to gather together from time to time in a critical intellectual mass. Furthermore, as Paul Ehrlich so forcefully held during the Symposium, the ultimate goal of our research is to influence policies in the economic and political realms. Results of our research will influence policies only if politicians understand them and believe in them. To build this broad base of confidence and credence we must therefore involve the scientific estabishments, government bureaucracies, and diverse constituencies of many countries."

As is customary, all the papers were reviewed by the authors'peers in order to ensure their strict adherence to the high standards of international research publishing. The editors express their appreciation to all the scientists who gave time and energy as referees for this expression of the important work done throughout NATO Scientific Programmes.

In addition to the financial support of NATO Special Programme Panel on Global Transport Mechanisms in Geo-Sciences, the co-sponsorship of the following institutions is acknowledged : Université Catholique de Louvain; Belgian National Funds of Scientific Research; Belgian Ministry of National Education; Commission of European Communities (Climatology and Natural Hazards Research Programme); International Commission on Climate of the International Union of Geodesy and Geophysics; International Union For Quaternary Research Paleoclimate Commission; United Nations Educational, Scientific and Cultural Organization - International Geological Correlation Program, sub-programme "Quaternary Geosciences and Human Survival"; and the Committee on Science and Technology in Developing Countries.

We thank also the following international organizations who have sent an observer to participate in the discussions : The International Geosphere-Biosphere Programme (IGBP), International Geographical Union (IGU), International Association of Hydrological Sciences (IAHS) and the European Science Foundation (ESF).

Warm thanks are due to all the members of the Local Committee who took care of all the practical organization : MMrs N. Materne-Depoorter, L. Moens-Haulotte, M.F. Loutre, I. Marsiat, I. van der Mersch, and MMr. J. Dormaels, Th. Fichefet, J.F. Lamarque, F. Mercier, V. Ntezimana.

Special thanks are due to Mrs N. Materne-Depoorter and Miss Fr. Bouvy who were involved in the technical production of this book.

The Editors, April 1989.

1. Ehrlich A., 2. Verstraete M., 3. Shackleton N., 4. Liverman D., 5. Street-Perrott A.,
6. Crowley T., 7. Bradley R.S., 8. Teixeira A.G., 9. Juillet-Leclerc A., 10. Seret G.,
11. Marsiat I., 12. Ruiz de Elvira A., 13. Schneider S., 14. Chen R., 15. Brasseur G.,
16. Duplessy J.Cl., 17. Rebeyrol Y., 18. Delvaux A., 19. Materne N., 20. Siegenthaler U.,
21. Lorius Cl., 22. Moens L., 23. Muszinski I., 24. Sveinbjornsdottir A., 25. Robock A.,
26. Muller A., 27. Holcombe T., 28. Sorensen H., 29. Gleick P.H., 30. Kattenberg A.,
31. van Ypersele J.P., 32. Charlson R.J., 33. Diamantopoulos J., 34. Henderson-Sellers A.,
35. Henderson-Sellers B., 36. Souchez R., 37. Loutre M.F., 38. Gullentops A., 39. Boulton G.S.,
40. Bryan K., 41. Gordon C., 42. Blanchet J.P., 43. Gobbi G.P., 44. Alexandre P.,
45. Fichefet Th., 46. Wigley T., 47. Harvey D., 48. Chanin M.L., 49. Ghil M., 50. Jones Ph.,
51. Vlachos E., 52. Mitchell J., 53. Berger A., 54. Sundquist E., 55. Pantic N.,
56. Dalfes N., 57. Ghan S., 58. Olafsson J., 59. Mélice J.L., 60. Fearnside Ph., 61. Morel P.,
62. Jenne R., 63. Perry J.S., 64. Nicolis C., 65. Schayes G., 66. Malingreau J.P., 67. Mix A.,
68. Ntezimana V., 69. Goudriaan J., 70. Durkee P.A., 71. Thill G., 72. van der Mersch I.,
73. Ronday F., 74. Dooge J., 75. Cess R.D., 76. Bjorge D., 77. Lamarque J.F., 78. Rasmussen M.S.,
79. Thompson S., 80. Beran M., 81. Gregory S., 82. Tricot Ch., 83. Gallée H., 84. Kukla G.,
85. Fosberg M.A., 86. Schlesinger M., 87. Oerlemans J., 88. Kassimidis G., 89. Bolle H.

LIST OF PARTICIPANTS

ACKERMAN M.
DIRECTEUR DE L'I.A.S.
3 AV. CIRCULAIRE
1180 BRUXELLES
BELGIUM

ALEXANDRE P.
RUE J. MUSCH 23/54
4920 EMBOURG-CHAUDFONTAINE
BELGIUM

ANTOINE J.P.
PRESIDENT DU DEPT. DE PHYSIQUE
U.C.L.
2 CHEMIN DU CYCLOTRON
1348 LOUVAIN-LA-NEUVE
BELGIUM

AUDENAERT W.
CHEF DE CORPS WING METEO
RUE DU LONG CHENE 86
1970 WEZEMBEEK-OPPEM
BELGIUM

BARRETT E.C.
DIRECTOR RSU
UNIVERSITY OF BRISTOL
REMOTE SENSING UNIT
DEPT. OF GEOGRAPHY
UNIVERSITY ROAD
BRISTOL BS8 1SS
UNITED KINGDOM

BARRON E.J.
EARTH SYST. SCIENCE CENTER
537 DEIKE BUILDING
PENSYLVANIA STATE UNIVERSITY
UNIVERSITY PARK, PA 16802
U.S.A.

BERAN M.
INSTITUTE OF HYDROLOGY
MACLEAN BUILDING
CROWMARSH GIFFORD
WALLINGFORD OXON OX10 8BB
UNITED KINGDOM

BERGER A.
INST. ASTR. GEOPHYSIQUE
G. LEMAITRE, U.C.L.
2 CHEMIN DU CYCLOTRON
1348 LOUVAIN-LA-NEUVE
BELGIUM

BJORGE D.
INSTITUTE OF GEOPHYSICS
UNIVERSITY OF OSLO
P.O. BOX 1022
BLINDERN
N-0315 OSLO
NORWAY

BLANCHET J.P.
ATMOSPHERIC ENVIRON. SERVICE
OF CANADA
4905 DUFFERIN STREET
DOWNSVIEW, ONT. M3H 5T4
CANADA

BOLLE H.
INSTITUUT FUR METEOROLOGIE
WE 07 IM FB GEOWISS
FREIE UNIVERSITAT BERLIN
DIETRIECH-SCHAFER WEG 6-8
D-1000 BERLIN 41
FED. REPUBLIC OF GERMANY

BOSSY L.
I.A.S.
3 AV. CIRCULAIRE
1180 BRUXELLES
BELGIUM

BOULTON G.S.
GRANT INST. OF GEOLOGY
EDIMBURGH UNIV.
WEST MAINS ROAD
EDIMBURGH EH9 3JW
UNITED KINGDOM

BOUVIER J.M.
INSTITUT DU QUATERNAIRE
UNIVERSITE BORDEAUX 1
F-33405 TALENCE
FRANCE

BRADLEY R.S.
DEPT. OF GEOL. AND GEOG.
MORRILL SCIENCE CENTER
QUATERNARY STUDIES GROUP
UNIV. OF MASSACHUSETTS
AMHERST, MASS 01003-0026
U.S.A.

BRASSEUR G.
INST. D'AERONOMIE SPATIALE
3 AV. CIRCULAIRE
1180 BRUXELLES
BELGIUM

BRYAN K.
G.F.D.L.
PRINCETON UNIVERSITY
P.O. BOX 308
PRINCETON, NJ 08540
U.S.A.

BUNGUM H.
ROYAL NORWEGIAN COUNCIL FOR
SCIENT. AND INDUST. RESEARCH
NORWEGIAN SEISMIC ARRAY -
NORSAR
P.O. BOX 51
N-2007 KJELLER
NORWAY

BUYSE H.
CONSEILLER SCIENTIFIQUE
HALLES UNIVERSITAIRES
PLACE DE L'UNIVERSITE 1
1348 LOUVAIN-LA-NEUVE
BELGIUM

BUZBEE W.
DIRECTOR SCIENTIFIC
COMPUTING DIVISION
NCAR COMPUTER CENTER
P.O. BOX 3000
BOULDER, CO 80307
U.S.A.

CESS R.D.
DEPT. OF MECHANICAL ENG.
LAB. FOR PLANET. ATMOS. RES.
STATE UNIV. OF NEW YORK AT
STONY BROOK L.I., NY 11794
U.S.A.

CHANIN M.L.
MEMBRE DU SPECIAL COMMITTE
FOR IGBP
DIRECTEUR AU CNRS
SERVICE AERONOMIE DU CNRS
B.P. 3
F-91370 VERRIERES-LE-BUISSON
FRANCE

CHARLSON R.J.
UNIVERSITY OF WASHINGTON
INST. FOR ENVIRON. STUDIES
AND ATMOSPHERIC SCIENCES
ENGINEERING ANNEX, FM-12
SEATTLE, WA 98195
U.S.A.

CHEN R.
WORLD HUNGER PROGRAM
BROWN UNIVERSITY
P.O. BOX 1831
PROVIDENCE, RI 02912
U.S.A

COLENBRANDER H.J.
SECRETARY GENERAL OF IAHS
COMMITTEE ON HYDROLOGICAL
RESEARCH T.N.O.
P.O. BOX 297
NL-2501 BD THE HAGUE
THE NETHERLANDS

CROWLEY T.
ARC TECHNOLOGIES
305 ARGUELLO DRIVE
COLLEGE STATION, TEXAS 77840
U.S.A.

DA CUNHA L.
NATO
SCIENT. AFFAIRS DIVISION
1110 BRUSSELS
BELGIUM

DAILY G.C.
C/O PROF. DR. WERNER KOCH
LEHRSTUHL FUR FORSTBOTANIK
AMALIENSTRASSE 52
D-8000 MUNCHEN 40
FED. REPUBLIC OF GERMANY

DALFES N.
TUBITAK
RES. INST. FOR BASIC SC.
P.O. BOX 74
41401 GEBZE - KOCAELI
TURKEY

DELVAUX A.
UNITE ASTR
U.C.L.
2 CHEMIN DU CYCLOTRON
1348 LOUVAIN-LA-NEUVE
BELGIUM

DIAMANTOPOULOS J.
ARISTOTELIAN UNIVERSITY
SECTOR OF ECOLOGY
UP BOX 119
BIOLOGY DEPT.
GR-54006 THESSALONIKI
GREECE

DOOGE J.
DEPT. OF CIVIL ENGINEERING
UNIVERSITY COLLEGE
EARLSFORT TERRACE
DUBLIN 2
IRELAND

DRICOT E.
UNITE PAGE
BATIMENT MERCATOR
PL. L. PASTEUR 3
1348 LOUVAIN-LA-NEUVE
BELGIUM

DUCUING J.
ASSISTANT SECRETARY GENERAL
OF THE NATO SCIENTIFIC
AND ENVIRONMENTAL AFFAIRS
NATO
1110 BRUXELLES

DUPLESSY J.C.
CENTRE DES FAIBLES
RADIOACTIVITES
LAB. MIXTE CNRS-CEA
DOMAINE DU CNRS
AV. DE LA TERRASSE
F-91190 GIF-SUR-YVETTE
FRANCE

DURKEE P.A.
NAVAL POSTGRADUATE SCHOOL
ROOT HALL 231, CODE 63DE
MONTEREY. CA 93943-5100
U.S.A.

EHRLICH A.
C/O PROF. P. EHRLICH
DEPT. OF BIOLOGICAL SCIENCE
STANFORD UNIVERSITY
STANFORD, CA 94305
U.S.A.

EHRLICH P.
DEPT. OF BIOLOGICAL SCIENCE
STANFORD UNIVERSITY
STANFORD, CA 94305
U.S.A.

FANTECHI R.
C.E.C. - CLIMATE PROGRAM
DRG XII
RUE DE LA LOI 200
1049 BRUXELLES
BELGIUM

FEARNSIDE PH.M.
INPA, DEPT. OF ECOLOGY
CX POSTAL 478
69.000 MANAUS-AMAZONAS
BRAZIL

FICHEFET T.
UNITE ASTR
U.C.L.
2 CHEMIN DU CYCLOTRON
1348 LOUVAIN-LA-NEUVE
BELGIUM

FOSBERG M.A.
FOREST FIRE AND ATMOS.
SC. AND RES.
US DEPT. OF AGRICULTURE
12TH & INDEPENDENCE SW
P.O. BOX 96090
WASHINGTON, DC 20090-6090
U.S.A.

FRATTA M.
ESF - PALEOCLIMATOLOGY
1 QUAI LEZAY-MARNESIA
F-67000 STRASBOURG
FRANCE

GALLEE H.
UNITE ASTR
U.C.L.
2 CHEMIN DU CYCLOTRON
1348 LOUVAIN-LA-NEUVE
BELGIUM

GHAN S.
LAWRENCE LIVERMORE
NATIONAL LABORATORY.
LIVERMORE, CA 94550
U.S.A.

GHAZI A.
C.E.E.
DIRECTION GENERALE XII
PROGRAMME CLIMATOLOGIE
RUE DE LA LOI 200
1049 BRUXELLES
BELGIUM

GHIL M.
UNIVERSITY OF CALIFORNIA
DEPT. OF ATMOS. SCIENCES
405 HILGARD AV.
LOS ANGELES, CA 90024
U.S.A.

GLEICK P.H.
PACIFIC INST. FOR SIDES
1681 SHATTUCK AV., SUITE H
BERKELEY, CA 94709
U.S.A.

GOBBI G.P.
ISTITUTO DI FISICA
DELL'ATMOSFERA
C.N.R. - CP 27
I-00044 FRASCATI
ITALY

GORDON C.
DYNAMICAL CLIMAT. BRANCH
METEOROLOGICAL OFFICE
LONDON ROAD
BRACKNELL, BERK. RG12 2SZ
UNITED KINGDOM

GOUDRIAAN J.
DEPT. OF THEOR. PROD. ECOL.
AGRICULTURAL UNIVERSITY
BORNSESTEEG 65
NL-6708 PD WAGINENGEN
THE NETHERLANDS

GREGORY S.
CHAIRMAN OF CLIMATIC CHANGE
STUDY GROUP OF I.G.U.
DEPT. OF GEOGRAPHY
UNIVERSITY OF SHEFFIELD
SHEFFIELD,
YORKSHIRE S10 2TN
UNITED KINGDOM

HARVEY D.
DEPT. OF GEOGRAPHY
UNIVERSITY OF TORONTO
100 ST-GEORGE STREET
TORONTO, ONTARIO M5S 1A1
CANADA

HEKSTRA G.
MINIST. VOLKSHUISVESTING
RUIMTELYKE ORDENING EN
MILIEUBEHEER
DIRECTORAAT.-GEN. VOOR
DE MILIEUHYGIENE
P.O. BOX 450
BM-2260 LEIDSCHENDAM
THE NETHERLANDS

HENDERSON-SELLERS A.
MACQUARIE UNIVERSITY
SCHOOL OF EARTH SCIENCES
NORTH RYDE
NEW SOUTH WALES 2109
AUSTRALIA

HENDERSON-SELLERS B.
MACQUARIE UNIVERSITY
SCHOOL OF EARTH SCIENCES
NORTH RYDE
NEW SOUTH WALES 2109
AUSTRALIA

HOLCOMBE T.
PROG. MANAGER FOR PALEOCLIM.

NAT. GEOPHYS. DATA CENTER
325 BROADWAY
BOULDER, CO 80303-3328
U.S.A.

HOLLINGWORTH A.
E.C.M.R.W.F.
SHIEFIELD PARK
READING, BERKS. RG2 9AX
UNITED KINGDOM

JENNE R.
NCAR
P.O. BOX 3000
BOULDER, CO 80307
U.S.A.

JONES P.D.
SECRETARY OF ICCL
CLIMATIC RESEARCH UNIT
UNIVERSITY OF EAST ANGLIA
NORWICH, NR4 7TJ
UNITED KINGDOM

JUILLET-LECLERC A.
CENTRE DES FAIBLES RADIOACT.
LABORATOIRE MIXTE CEA-CNRS
DOMAINE DU CNRS
F-91198 GIF-SUR-YVETTE
FRANCE

KANDEL R.
ECOLE POLYTECHNIQUE
LABO. DE METEOR. DYNAMIQUE
ROUTE DEPARTEMENTALE 36
F-91128 PALAISEAU CEDEX
FRANCE

KASSIMIDIS G.
HELLENIC NAT. METEO. SERV.
P.O. BOX 73 502
GR-16603 ATHENS
GREECE

KATTENBERG A.
K.N.M.I.
P.O. BOX 201
NL-3730 AE DE BILT
THE NETHERLANDS

KUKLA G.
LAMONT-DOHERTY GEOL. OBS.
OF COLUMBIA UNIVERSITY
PALISADES, NY 10964
U.S.A.

LADURON D.
DOYEN DE LA
FAC. DES SCIENCES
PLACE DES SCIENCES 2
1348 LOUVAIN-LA-NEUVE
BELGIUM

LAMARQUE J.F.
UNITE ASTR
U.C.L.
2 CHEMIN DU CYCLOTRON
1348 LOUVAIN-LA-NEUVE
BELGIUM

LIVERMAN D.
DEPT. OF GEOGRAPHY
UNIVERSITY OF WISCONSIN
MADISON, WI 53706
U.S.A.

LORIUS C.
LAB. DE GLACIOLOGIE
RUE MOLIERE
DOMAINE UNIVERSITAIRE
B.P. 96
F-38402 ST MARTIN
D'HERES CEDEX
FRANCE

LOUTRE M.F.
UNITE ASTR
U.C.L.
2 CHEMIN DU CYCLOTRON
1348 LOUVAIN-LA-NEUVE
BELGIUM

MALINGREAU J.P.
CENTRE COMMUN DE RECHERCHE
CCE
ETABLISSEMENT D'ISPRA
I-21020 ISPRA (VARESE)
ITALY

MARSIAT I.
UNITE ASTR
U.C.L.

2 CHEMIN DU CYCLOTRON
1348 LOUVAIN-LA-NEUVE
BELGIUM

MELICE J.L.
UNITE ASTR
U.C.L.
2 CHEMIN DU CYCLOTRON
1348 LOUVAIN-LA-NEUVE
BELGIUM

MILANKOVIC V.
157 MOUNTAINVIEW RD.
NORTH BALWYN 3104 (VIC.)
AUSTRALIA

MITCHELL J.
METEOROLOGICAL OFFICE
LONDON ROAD
BRACKNELL, BERKS RG12 2SZ
UNITED KINGDOM

MIX A.
COLLEGE OF OCEANOGRAPHY
OREGON STATE UNIVERSITY
OCEANOGR. ADMIN. BLDG 104
CORVALLIS, OR 97331-5503
U.S.A.

MOREL P.
JOINT PLANNING STAFF
W.M.O.
C.P. N 5
1211 GENEVA 20
SWITZERLAND

MULLER A.
GEOLOGISCH INSTITUT
FACHABT. GEOWISSENSCHAFT
TEMPLERGRABEN 55
D-5100 AACHEN
FED. REPUBLIC OF GERMANY

MUSZYNSKI I.
UNITE ASTR
U.C.L.
2 CHEMIN DU CYCLOTRON
1348 LOUVAIN-LA-NEUVE
BELGIUM

MYERS N.
CONSULTANT IN ENVIRON.
AND DEVELOPMENT
UPPER MEADOW
OLD ROAD
HEADINGTON
OXFORD, OX3 8SZ
UNITED KINGDOM

NICOLIS C.
I.A.S.
3 AVENUE CIRCULAIRE
1180 BRUXELLES
BELGIUM

NICOLIS G.
INST. INTERN. DE PHYS.
ET DE CHIMIE
CODE POSTAL 231
CAMPUS PLAINE ULB
BD. DU TRIOMPHE
1050 BRUXELLES
BELGIUM

NTEZIMANA V.
UNITE ASTR
U.C.L.
2 CHEMIN DU CYCLOTRON
1348 LOUVAIN-LA-NEUVE
BELGIUM

NUNES CORREIA FR.
DIRECTOR-GERAL DOS
RECURSOS NATURAIS
AV. ALMIRANTE GAGO
COUNTINHO 30
1000 LISBON
PORTUGAL

OERLEMANS J.
INST. FOR METEO.
AND OCEANOGRAPHY
PRINCETONPLEIN 5
POSTBUS 80.005
NL-3508 TA UTRECHT
THE NETHERLANDS

OLAFSSON J.
MARINE RESEARCH INSTITUTE
SKULGATA 4
P.O. BOX 390

REYKJAVIK
ICELAND

PAEPE R.
DIRECTEUR DU
SERVICE GEOLOGIQUE
RUE JENNER 13
1040 BRUXELLES
BELGIUM

PANTIC N.
SERBIAN ACADEMY OF SC.
AND ARTS
35 KNEZ-MIHAILOVA
YU-11000 BEOGRAD
YOUGOSLAVIA

PARRY M.
DEPT. OF GEOGRAPHY
UNIVERSITY OF BIRMINGHAM
P.O. BOX 363
BIRMINGHAM, B15 2TT
UNITED KINGDOM

PERRY J.S.
CLIMATE RESEARCH BOARD
JOSEPH HENRY BUILDING
NATIONAL ACADEMY OF SC.
2101 CONSTITUTION AV., NW
WASHINGTON, DC 20418
U.S.A.

PORTER J.
THE UNIV. OF BIRMINGHAM
DEPT. OF GEOGRAPHY
ATMOSPHERIC IMPACTS
RESEARCH GROUP
P.O. BOX 363
BIRMINGHAM B15 2TT
UNITED KINGDOM

RASCHKE E.
INST. FUR GEOPHYSIK
UND METEO.
UNIVERSITAT ZUR KOLN
KERPERNER STRASSE 13
D-5000 KOLN 41
FED. REPUBLIC OF GERMANY

RASMUSSEN M.S.
INSTITUTE OF GEOGRAPHY
THE UNIV. OF COPENHAGEN
OSTER VOLDGADE 10
DK-1350 COPENHAGEN K
DENMARK

REBEYROL Y.
LE MONDE
5 RUE DES ITALIENS
F-75427 PARIS CEDEX 09
FRANCE

ROBOCK A.
THE UNIVERSITY OF MARYLAND
DEPT. OF METEOROLOGY
SPACE SCIENCE BUILDING
COLLEGE PARK, MARYLAND 20742
U.S.A.

ROELANTS DU VIVIER FR.
DEPUTE EUROPEEN
RUE BELLIART 97-113
1040 BRUXELLES
BELGIUM

RONDAY FR.
UNIV. DE LIEGE
SERV. MEC. DES FLUIDES
GEOPHYSIQUES
SART TILMAN B5
4000 LIEGE
BELGIUM

RUIZ DE ELVIRA A.
UNIV. DE ALCALA DE HENARES
DEPARTAMENTO DE FISICA
APDO. 20
SP-28880 ALCALA DE HENARES
SPAIN

SCHAYES G.
U.C.L.
UNITE ASTR
2 CHEMIN DU CYCLOTRON
1348 LOUVAIN-LA-NEUVE
BELGIUM

SCHLESINGER M.
DEPT. OF ATMOS. SC.
OREGON STATE UNIVERSITY
CORVALLIS, OREGON 97331
U.S.A.

SCHNEIDER S.
NCAR
P.O. BOX 3000
BOULDER, CO 80307
U.S.A.

SCHUURMANS C.J.E.
K.N.M.I.
P.O. BOX 201
WILHELMINALAAN 10
NL-3730 AE DE BILT
THE NETHERLANDS

SERET G.
UNITE PAGE
BATIMENT MERCATOR
PL. L. PASTEUR 3
1348 LOUVAIN-LA-NEUVE
BELGIUM

SHACKLETON N.
GODWIN LAB.
FREE SCHOOL LANE
CAMBRIDGE, CB2 3RS
UNITED KINGDOM

SIEGENTHALER U.
PHYSIKALISCHES INSTITUT
UNIVERSITAT BERN
SIDLERSTRASSE 5
CH-3012 BERN
SWITZERLAND

SORENSEN H.
INSTITUT FOR PETROLOGY
UNIVERSITY OF COPENHAGEN
OSTER VOLGADE 10
DK-1350 COPENHAGEN
DENMARK

SOUCHEZ R.
U.L.B.
LAB. GEOMORPHOLOGIE
C.P. 160
AV. F.D. ROOSEVELT 50
1050 BRUXELLES
BELGIUM

STREET-PERROTT A.
SCHOOL OF GEOGRAPHY
MANSFIELD ROAD
OXFORD, OX1 3TB
UNITED KINGDOM

STUIVER M.
DEPT. OF GEOLOGICAL
SCIENCES AND ZOOLOGY
QUATERNARY ISOTOPE LAB.
UNIVERSITY OF WASHINGTON
SEATTLE, WA 98195
U.S.A

SUNDQUIST E.
U.S. GEOLOGICAL SURVEY
431 NATIONAL CENTER
WESTON, VA 22092
U.S.A.

SVEINBJORNSDOTTIR A.E.
SCIENCE INSTITUTE
UNIVERSITY OF ICELAND
DUNHAGA 3
107 REYKJAVIK
ICELAND

TEIXEIRA A.G.
INSTITUTO GEOFISICA
DO INFANTE D. LUIS
FACULDADE DE CIENCIAS
R. DA ESCOLA POLITECNICA
1200 LISBOA
PORTUGAL

THILL G.
149 BD. J.F. KENNEDY
L-4930 BASCHARAGE
LUXEMBOURG

THOMPSON S.
NATURAL SYSTEMS GROUP
NCAR
BOULDER, CO 80307-3000
U.S.A.

TICKELL C.
UNITED KINGDOM MISSION
TO THE UNITED NATIONS
845 THIRD AVENUE
NEW YORK, NY 10022
U.S.A.

TRICOT CH.
U.C.L.
UNITE ASTR
2 CHEMIN DU CYCLOTRON
1348 LOUVAIN-LA-NEUVE
BELGIUM

USHER P.
UNITED NATIONS
ENVIRONMENT PROGRAMME
P.O. BOX 30552
NAIROBI
KENYA

VAN DER MERSCH I.
U.C.L.
UNITE ASTR
2 CHEMIN DU CYCLOTRON
1348 LOUVAIN-LA-NEUVE
BELGIUM

VAN YPERSELE J.P.
UNITE ASTR
U.C.L.
2 CHEMIN DU CYCLOTRON
1348 LOUVAIN-LA-NEUVE
BELGIUM

VERSTRAETE M.
NCAR
P.O. BOX 3000
BOULDER, CO 80307
U.S.A.

VLACHOS E.
COLORADO STATE UNIVERSITY
SOCIOLOGY DEPARTMENT
FORT COLLINS, CO 80523
U.S.A.

WHELPDALE D.M.
ATMOS. DISPERSION DIVISION
DEPT. OF THE ENVIRONMENT
4905 DUFFERIN STREET
DOWNSVIEW, ONTARIO M3H 5T4
CANADA

WIGLEY T.
CLIMATIC RESEARCH UNIT
UNIVERSITY OF EAST ANGLIA
NORWICH, NR4 7TJ
UNITED KINGDOM

WOLLAST R.
LAB. D OCEANOGAPHIE
U.L.B.
CAMPUS DE LA PLAINE
CP 208
1050 BRUXELLES
BELGIUM

SECTION 1 – CLIMATE RECORD

WHAT CAN WE LEARN FROM DATA ?

PRE-PLEISTOCENE CLIMATES: DATA AND MODELS

Eric J. Barron
Earth System Science Center
Pennsylvania State University
512 Deike Building
University Park, PA 16802

ABSTRACT. Six major problems in paleoclimatology are utilized to illustrate how critical observations from the pre-Pleistocene are leading focused, interdisciplinary investigations which are guiding future research: (1) the faint early sun paradox, (2) the Proterozoic "ice house", (3) Cretaceous warmth, (4) Eocene equator-to-pole temperature gradients, (5) gateways and the Oligocene to Miocene transition to glacial climates, and (6) Milankovitch sedimentary rhythms during non-glacial climates. The research in these areas have produced a variety of advances and have fostered new areas of research including (1) consideration of atmospheric chemistry as a major regulator of the Earth's climate, (2) illustration of critical limits to Radiative Convective Models and General Circulation Models, (3) consideration of feedbacks between temperature, continental weathering, and CO_2, (4) evaluation of the mechanisms of glaciation, (5) illustration of the importance of tropical temperature history, (6) illustration of the importance of climates in continental interiors, (7) consideration of the importance of the role of the oceans in heat transport, (8) consideration of the role of barriers and gateways to ocean circulation and (9) indication of the potential of sedimentary rhythms as opportunities to investigate climate sensitivity.

INTRODUCTION

The data available to reconstruct any past climate are far from comprehensive. Three major problems define the limitations of the observations. First, the record cannot be fully sampled either spatially or temporally. In fact, the geologic record only rarely yields much information from above the surface of the Earth. Second, climatic variables such as temperature, pressure or wind velocity are not measured directly. The indices, or paleothermometers and paleobarometers, are recorded as physical, chemical or biologic signatures. The evaluation of these signatures requires an interpretation of a variety of physical, chemical or biologic processes. Third, a comprehensive reconstruction requires an accurate sense of time and a capability for global

A. Berger et al. (eds.), Climate and Geo-Sciences, 3–19.

correlation. Global stratigraphic correlation is problematic for any interval of geologic time, and becomes increasingly problematic with the age of the rock units.

Certainly, all three of the major problems which define limits in climatic reconstruction increase in degree as a function of age. The record becomes increasingly sparse. The physical, chemical and biologic signatures are increasingly difficult to interpret because of diagenesis and evolution. Finally, correlation becomes tenable only if the stratigraphic limits become increasingly broad. However, the "great" problems in paleoclimatology, those which attract attention across disciplinary boundaries and lead advances in understanding the climate system, are apparently defined more by the availability of critical data than by the capabilities of comprehensive reconstruction and interpretation.

The great problems in paleoclimatology are defined by the degree to which data fail to fit preconceived notions or the level of physical understanding of the climate system, and the degree to which the data indicate major differences in the factors which influence this system. Clearly, even a single point in space and time can challenge our level of physical understanding of the climate system. Observations may also demand a specific type of explanation. A prime example is the close association of deep-sea faunal and isotopic data with Milankovitch orbital variations discovered by Hays et al. (1976). The explanation of these data required mechanisms to promote climatic sensitivity to orbitally-induced insolation changes. This challenge catalyzed research efforts across the disciplines of the Earth Sciences. Alternatively, the observations may be sufficiently different from other times or from the present day to provide a challenge even to the development of a plausible hypotheses. A diversity of problems in earth history fulfill the criteria of challenge across disciplinary boundaries and the potential to lead advances in understanding in the climate system. In these cases, the data from the record define a major problem, and hypothesis development and models of the physical system focus attention on the critical limitations in theory and observations. These limitations guide future research.

Six major problems in paleoclimatology illustrate how critical data are leading focused, interdisciplinary investigations which are guiding future research: (1) the faint early sun paradox, (2) the Proterozoic "ice house", (3) Cretaceous warmth, (4) Eocene equator-to-pole temperature gradients, (5) gateways and the Oligocene to Miocene transition to glacial climates, and (6) Milankovitch order sedimentary rhythms during non-glacial climates.

(1) Faint Early Sun Paradox

A major, apparently robust, result of stellar models is that the solar luminosity during the Archean (4.6 - 2.4 billion years ago) was 70-80% of the present value (Gough, 1981). The specification of this reduction in solar luminosity in simple climate models, based on an energy balance between incoming visible and near-infrared solar energy and outgoing middle-infrared thermal emission, with current assumptions on the role of an atmospheric greenhouse yield an ice-covered earth (Sagan and Mullen, 1972).

The nature of the rock record defines the paradox. The oldest rock record (~3.8 billion years ago) from Isua in western Greenland contains evidence for waterborne processes of erosion, transport and sedimentation (Lowe, 1980). The earliest record of glaciation is 2.7 billion years old. The record of evaporite minerals extends to 3.5 billion years, and because these minerals were originally gypsum, the evaporites set an upper limit of 58°C (Holland, 1978). Given the current temperature sensitivities of Prokaryota, Eukaryota and Metazoa, Archean life would be difficult to imagine on a frozen Earth. This life also sets upper temperature limits near 50-70°C. Oxygen isotope ratios for hydrogen and oxygen in cherts from the pre-Cambrian have been interpreted as indicating temperatures as high as 70°C (Knauth and Epstein, 1976; Karhu and Epstein, 1986), although the interpretation of this evidence has been questioned by Walker (1982).

These data (Figure 1), literally points in space and time, define broad temperature limits (0-70°C), imprecisely dated and with little knowledge of latitudinal position or representativeness of global average conditions. Despite the limitations in the Archean data (the absence of ice and evidence for liquid water), these data, along with a condition of substantially reduced luminosity, present a well-posed problem for climate modelling.

The faint early sun paradox has fostered a variety of research advances and focused attention on new research directions:

(1) Archean climate research focused attention on early Earth chemistry, specifically questioning hypotheses of a methane-ammonia atmosphere (Kuhn and Atreya, 1979). The atmospheric composition has implications for hypotheses of the origin of life.

(2) The pioneering work by Sagan and Mullen (1972) placed the "greenhouse" character of the atmosphere as potentially the most important of all long term regulators of the Earth's climate, one which is far from fully studied.

(3) The presentation and explanation of the faint early sun paradox by Sagan and Mullen (1972) discounted higher CO_2 because the fundamental CO_2 bands rapidly saturate, weakening the prospect that higher levels would explain a drastically reduced solar luminosity. The Archean problem was one of the factors which focused attention on the weaker CO_2 bands which do not saturate (Ramanathan, 1976). The potential for very high levels of CO_2 in Earth history became a matter of debate (Augustsson and Ramanathan, 1977; Owen et al., 1979).

(4) The faint early sun paradox certainly draws attention to the limits of Radiative Convective Models (RCM's), particularly in terms of cloud-feedbacks (Schneider and Thompson, 1980; Wang et al., 1981; Henderson-Sellers, 1979) and water vapor feedback (Kuhn and Kasting, 1983). Walker (1982) illustrates how significant the wide range of cloud interactions, humidity and albedo assumptions are for estimating the amount of CO_2 required to solve the faint-early sun paradox. For different assumptions, models require in the range of 5 to 1000 times

present day levels in order to maintain a non-ice planet.

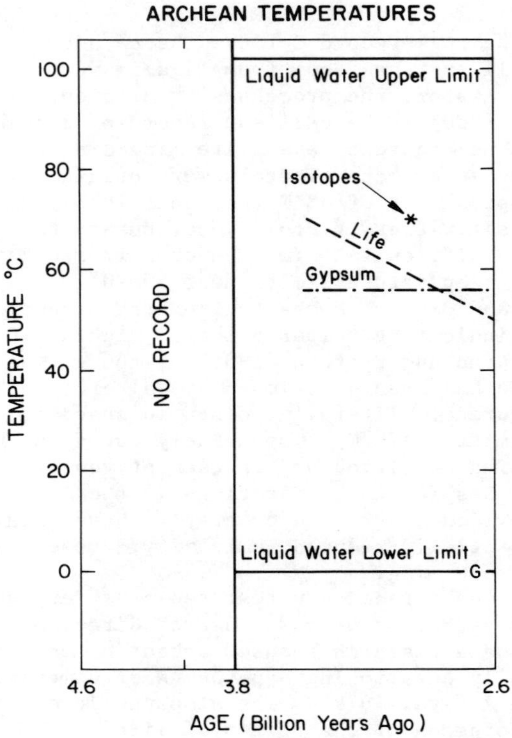

Figure 1. Archean temperature limits based on the presence of liquid water, life, gypsum, oxygen isotopes and the occurrence of glacial sediments (G) after Knauth and Epstein (1976) and Walker (1982).

(5) The Archean problem was a building block for coupling climate and geochemical cycles. For example, Walker et al. (1981) described the significant feedback between weathering rates, temperature and CO_2 which might limit large temperature excursions because higher temperatures promote increased CO_2 removal from the atmosphere through consumption in the increased weathering of silicate rocks. This feedback is one suggestion for how the Earth has maintained conditions suitable for life despite a variety of significant climatic forcing factors.

Remarkably many of these directions and advances stems from a few data sites, poorly constrained in time, which indicate the presence of liquid water and from the theory of stellar evolution.

(2) A Proterozoic Ice-House?

The implications of the climatic record from the upper Proterozoic are also dramatic, especially following the arguments of how Archean warmth might be maintained. The Upper Proterozoic (approximately 650 to 950 m.y. ago) is in marked contrast with the Archean. The record of continental glaciation is extensive, and is now described from every continent except Antarctica (Hambrey and Harland, 1981). The broad extent of glaciation is in itself evidence for an extreme condition in the Earth's climatic history.

The extent of glaciation appears to be even more dramatic when glacial deposits are placed in a paleolatitudinal framework using paleomagnetism. Well described tillites from Scotland have been placed by Tarling (1974) within 10° of the equator. The reconstruction of latitudes by Irving and Park (1972) orients North America entirely between $\pm$ 20° latitude from 800 to 600 m.y.b.p. North America is characterized by extensive glaciation during the Late Proterozoic (Figure 2). Piper (1973) presents data from Africa indicating a low latitude position from 1,000 m.y. ago to the Early Cambrian. A diverse group of paleomagnetic studies supports an interpretation of widespread low-latitude glaciation.

Throughout most of Earth history continental glaciation occurs at high latitudes or polar regions when data are reconstructed from paleomagnetism. The Upper Proterozoic is a major exception. One interpretation of these data is a Proterozoic "ice-house", and the most extreme glacial climate in Earth history.

The Upper Proterozoic tillites have other unusual characteristics. Williams (1975) emphasized the association of tillites with indicators of relative warmth and indicators of extreme climatic variability. The evidence includes tillites interbedded with limestones, dolomites and evaporites which are usually associated with warmth, evidence for permafrost structures, rhythmic or laminated sedimentation and glacial 'dropstones' within carbonate sequences. Williams argues not for an "ice-house", but for a climate characterized by large and rapid excursions.

The data from the Upper Proterozoic present a considerable challenge: (1) How would the global climate recover from a high albedo ice-house scenario? (2) Was the temperature-weathering rate-CO_2 feedback of Walker et al. (1981) inoperable?

Despite the potential of the data to catalyze investigation, climate models have not been applied to understand the Upper Proterozoic record. The explanation may in part be due to the lack of credible forcing factors or mechanisms which can be utilized to guide model simulations. The hypotheses put forth to explain the record of glaciation range from rapid polar wander, the shadow effect of earth-orbiting rings, rapid greenhouse excursions and an Earth obliquity in excess of 53° (e.g. Williams, 1975; Sheldon, 1986). Almost certainly the problem is sufficiently defined by data that major advances can be derived by modeling experiments designed either to discard hypotheses or to set limits to the conditions or plausibility of equatorial glaciation.

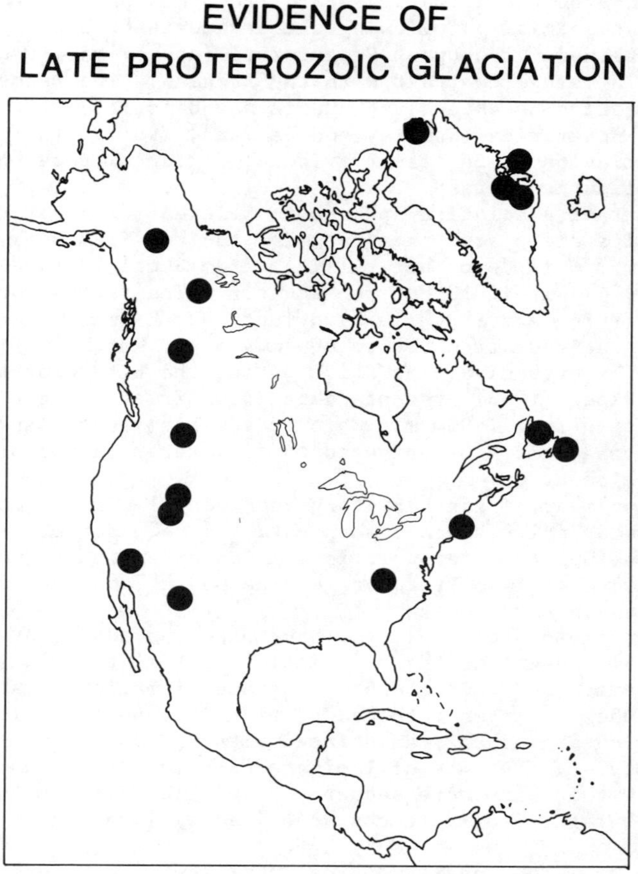

Figure 2. The record of Proterozoic North America glaciations after Hambrey and Harland (1981). North America was within $\pm$ 20° latitude of the equator.

(3) Cretaceous Warmth

The Cretaceous Period (140 m.y. to 65 m.y. ago) has been frequently cited as the largest contrast from the present day climate which can be well-documented. Barron (1983) summarizes the evidence for a globally-averaged surface temperature 6-12°C higher than the present day (Figure 3) and specific evidence for abundant high latitude floras, ocean deep water temperatures near 15°C, lack of cold water faunas and floras and even evidence for warmth and lack of seasonality in continental interiors.

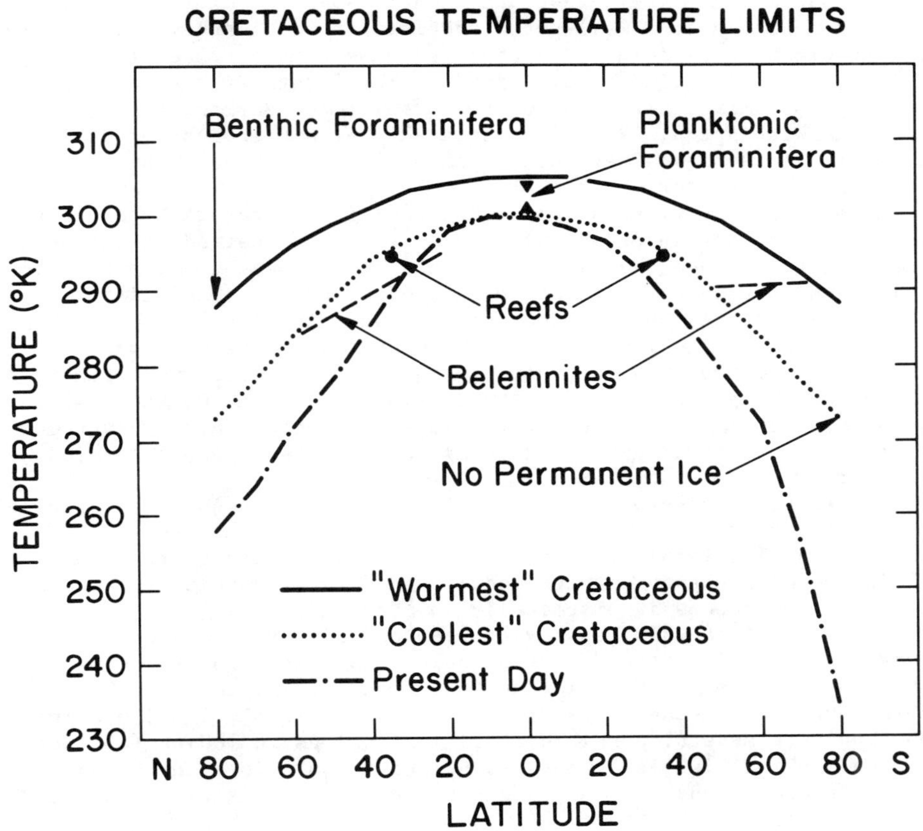

Figure 3. Cretaceous mean annual surface temperature limits in comparison with modern values (Barron, 1983) including the sources of data. A "warmest" (line) and "coolest" (dots) estimate is given with respect to the present day (dot-dash line).

In particular, the concept of plate tectonics revolutionized views on the causes of "ice-free" and glacial episodes during the Phanerozoic. Suggestions that Cretaceous warmth resulted from different continental positions (e.g., Crowell and Frakes, 1970; Frakes and Kemp, 1972, and summaries by Tarling, 1978 and Beaty, 1978) followed directly after the major breakthroughs in plate tectonics. Each focused on the role of high latitude land area as the major difference between ice-free and ice age climates because land at the poles provides a surface for the accumulation of high albedo snow, and high latitude land area blocks poleward heat transport by the ocean.

The large contrast in geography and in apparent climatic conditions again presented a well-posed problem for model study: Is geography sufficient to explain the contrasts between major glacial episodes and warm, apparently ice-free climates. The first efforts to explore geography as an explanation of Cretaceous warmth (Barron et al., 1981; Barron and Washington, 1984) demonstrated that a hierarchy of climate models fail to achieve Cretaceous warmth given paleogeography as the primary forcing factor.

More importantly, these contributions did not refute hypotheses about geography as the principal forcing factor. Instead, the results focused on three open questions: Are the differences between the model results and Cretaceous observations a product of (a) model limitations, (b) data limitations, or (c) the need for additional climatic forcing factors. These three limitations are ubiquitous in paleoclimatic studies, but they provide a useful framework for further investigation.

The data supporting the interpretation of extreme Cretaceous warmth and inability of models to achieve this warmth have fostered a variety of research directions and have established areas of needed new research:

(1) The most severe model limitation is the inadequate coupling of the ocean-atmosphere system. Although a few experiments (e.g., Schneider et al., 1985) have cast doubt on the ability of the oceans to explain the discrepancy between models and data, the role of the oceans during the Cretaceous is practically uninvestigated.

(2) Much of the discrepancy between model results and Cretaceous observations involves the degree of polar amplification of the warming. Discussions of General Circulation Model sensitivity to CO_2 changes (Schlesinger and Mitchell, 1987; Washington and Meehl, 1986; Dickinson, 1985) indicate several possible model limitations including (a) lapse rate feedback at high latitudes, (b) cloud formulations and cloud-climate feedbacks and (c) snow and sea-ice parameterizations and ice-albedo feedback.

(3) The model-data discrepancies focus attention on three critical types of observations: (a) Climatic reconstructions indicate tropical sea surface temperatures similar to or warmer than present day. Tropical temperatures are critical to deciphering the role of atmospheric heat transport and the role of the oceans. Tropical sea surface temperatures have received far too little attention in pre-Pleistocene paleoclimatology. (b) The Antarctic is assumed to be ice-free based on the data from the continental margin. Is this evidence for Cretaceous warmth really incompatible with the occurrence of some permanent ice at high latitudes in continental interiors? The constraint that interior Antarctica is ice-free forces consideration of extreme climatic solutions. (c) Data from continental interiors are critical to the verification of climate model experiments. The classic interpretation of the Cretaceous includes equable climates in continental interiors. Again such a condition forces consideration of extreme climatic solutions. With the condition of winter sunlight, continental interiors at high latitudes should experience freezing temperatures even under conditions of global warmth. Data from continental interiors are essential and have received far too little attention. The role of seasonality may

also play an essential role in explaining climate evolution as a function of geography (Crowley et al., 1986).

(4) The model-data discrepancies also focus attention on alternative or additional climatic forcing factors. Geochemical models (Berner et al., 1983; Lasaga et al., 1985) suggest that major variations in atmospheric carbon dioxide levels through the Phanerozoic are plausible (Figure 4). A four-fold increase in carbon dioxide level during the Cretaceous might be sufficient to explain Cretaceous warmth (Barron and Washington, 1985). Again, paleoclimatic studies focus attention on the "greenhouse" character of the atmosphere as potentially the most important of all long term regulators of the Earth's climate. Currently, many assumptions and simplifications limit firm conclusions.

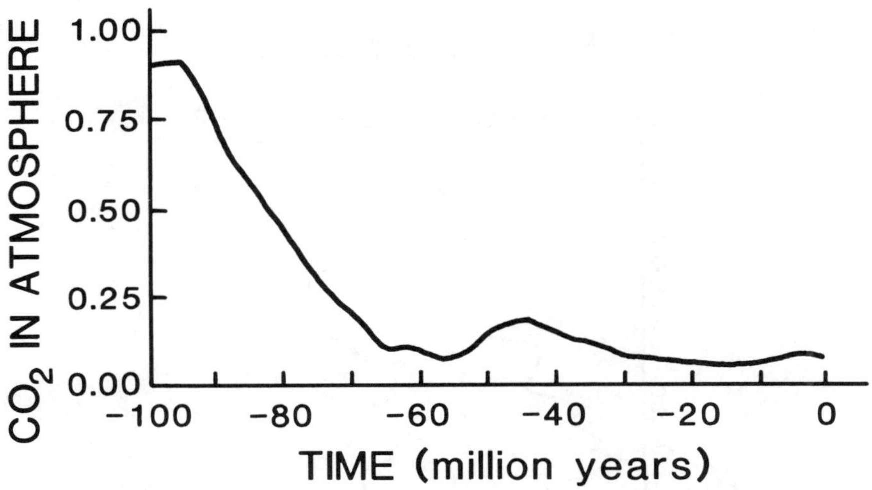

Figure 4. Estimated level of atmospheric carbon dioxide for the last one hundred million years after Lasaga et al. (1985) in units of 10^{18} moles.

(4) Eocene Equator-to-Pole Temperature Gradients

The evidence for Eocene (~50 m.y. ago) warmth is substantial (e.g., Wolfe, 1978; Estes and Hutchison, 1980; Shackleton and Kennett, 1975). The record of salamanders, lizards, turtles and alligators at high latitudes within the Arctic circle imply warmth with essentially frostless winters. The interpretation of latitudinal temperatures from oxygen isotopes are particularly interesting. Both Shackleton and Boersma (1981) and Keigwin and Corliss (1986) estimate a latitudinal temperature gradient less than half its present value with high latitudes near 10°C and, interestingly, tropical temperatures several degrees cooler than at present.

The condition of cooler tropics and warmer poles would seem to imply a greater poleward heat transport. Yet, an active role by the atmosphere given a reduced surface temperature gradient and cooler tropics is certainly problematic. Among others, Shackleton and Boersma (1981) proposed a greater role by the oceans in poleward heat transport, in other words, a climate dominated by the role of the oceans rather than the atmosphere. Barron (1987) considered the climatic conditions which would result from such a situation: (a) a weak atmospheric circulation and (b) given the condition of low winter insolation and a weaker atmospheric circulation, extreme seasonal contrasts within continental interiors in middle and high latitudes.

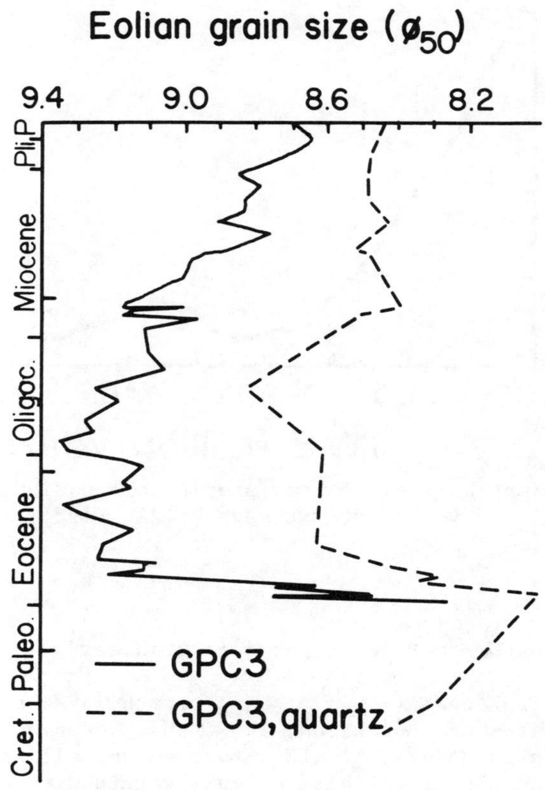

Figure 5. Eolian grain size distribution for quartz and total sediment for core GPC3 in the central Pacific Ocean from the Pleistocene to the Late Cretaceous; data from Janecek and Rea (1983).

Eolian sediment size from the deep-sea record (Rea et al., 1985) supports the interpretation of a substantially weaker Eocene atmospheric circulation (Figure 5). However, the record of Eocene floras from continental interiors (MacGinite, 1969, 1974; Wolfe, 1978) suggest continental interiors were equable, with a much reduced annual cycle of temperature. The Eocene climate is clearly problematic, as the model results and theory concerning the atmospheric circulation with a reduced temperature gradient, the oxygen isotopic data, the eolian sediment data and the paleobotanical data are not consistent.

The hypotheses related to Eocene climate are largely untested with physical models, but the stage is set to consider the plausibility of a climate dominated by the ocean. Certainly, the potential for climatic change through the role of the ocean, and even interaction and coupling of the ocean-atmosphere system, are largely unknown. The Eocene data, and the need to achieve a consistent interpretation of diverse sources of information, requires a major reevaluation of the potential role of the ocean and/or one or more major sources of paleoclimatic data.

(5) Gateways and the Oligocene to Miocene Transition to Glacial Climates

Berggren and Hollister (1974) correlated the occurrence of ocean circulation barriers and gateways with a large and diverse number of paleoceanographic events. The events of the last thirty million years are particularly noteworthy.

The opening of the Drake Passage occurred nearly 30 million years ago. The first formation of the Circum-Antarctic Current is associated with climatic cooling and increased bottom water flow. By 22 million years ago Antarctica was becoming increasingly glaciated. The frequently cited interpretation of the cause of glaciation is the thermal isolation of Antarctica due to the full development of the Circum-Antarctic Current. Concurrently, oceanic data indicate a substantial expansion of the high latitude siliceous sediment province associated with the Antarctic divergence, increased circulation vigor and consequently a larger number of deep-sea hiatuses. Opening of the Drake Passage, the formation of the Circum-Antarctic Current and Antarctic glaciation represent a critical stage in the climatic evolution of the time period preceeding the present day (Kennett, 1977). Kennett (1977) suggests three critical factors explain this major climatic step: (a) opening of the Drake Passage to deep-water flow, (b) a north Atlantic gateway to the Arctic and (c) closure of the Tethys gateways between North and South America and Europe-Africa.

For more than a decade, the importance of gateways and barriers to ocean circulation has been proposed as a major causative factor in climatic change based on correlation between paleoceanographic "events" and plate tectonic reconstruction of continents and morphologic features. The potential for insight into the climate system are substantial, yet very few physically-based experiments (e.g., Gill and Bryan, 1971) have been performed.

(6) Milankovitch-Order Sedimentary Rhythms During Non-Glacial Climates

Rhythmic variations in the stratigraphic record within a 10^4- to 10^6-year range prior to the Pleistocene glacial cycles are quite common. Rhythmic sedimentary variations in the Cretaceous record are noted in a variety of environments including epicontinental and deep-sea sediments (Fischer, 1980; deBoer, 1982; Schwarzacher and Fischer, 1982; Arthur et al., 1984; Dean et al., 1977). Rhythmic sediments of the Cretaceous have been tied to Milankovitch-type orbital periodicities by a large number of authors (Kauffman, 1977; Fischer, 1982; Pratt, 1981; Herbert and Fischer, 1986).

Initially, the link between orbital variations and rhythmic sediments were tied to ice volume and the role of ice as an amplifier of the insolation signal related to orbital variations. More recently, Kutzbach and Otto-Bliesner (1982) and Kutzbach and Guetter (1984) have demonstrated the importance of low latitude solar insolation changes as an important modulator of the monsoon circulation during the last 9,000 years. Barron et al. (1985) have suggested that any region of geographically-influenced precipitation, such as the monsoons may be influenced by changes in the amplitude of the annual cycle of insolation. Changes in the amplitude of the annual cycle of insolation are a primary effect associated with Pleistocene orbital variations.

Currently, a large numberof mechanisms have been proposed to link Milankovitch orbital variations to rhythmic sedimentation throughout Earth history, including times with and without the occurrence of polar ice. These include: (a) modulation of monsoonal circulations, (b) modulation of coastal wind systems which influence upwelling and primary productivity, (c) modulation of winter storms and jet stream tracks as a function of land-sea thermal contrasts, (d) regional participation changes in conjunction with the location of continental divides, (e) modulation of deep water formation associated with climatic change in marginal seas which are sites of water mass formation, and (f) climatic sensitive "zones" related to small shifts in the areas of major climatic belts or circulation systems. Many of these hypotheses can be tested or investigated.

The geologic record offers multiple cases to map rhythmic sedimentation for variety of global climates and paleogeographies. If Milankovitch type orbital variations are established as the primary explanation of these sedimentary periodicities, we have a remarkable opportunity to examine climatic sensitivity.

SUMMARY AND RECOMMENDATIONS

Although our ability to define and characterize climate change in Earth history becomes increasingly difficult as the age of the record increases, critical observations back to the earliest record are sufficient to lead advances in our understanding of the climate system. Six examples have been utilized to illustrate how such data define major climate problems, guide hypothesis development and how models of the physical system focus attention on the critical limitations in theory and observations. These limitations guide future research.

A number of recommendations for disciplinary and interdisciplinary research follow from the discussion of the examples given below:

(1) Spatial and temporal resolution and well-defined climatic indices are the key criteria for interpretation of past climates. Within the inherent limitations of the geologic record several sources of data are essential and have received too little attention: (a) a history of tropical temperatures and knowledge of their extent of variation in Earth history, (b) a history of the climatic record in continental interiors, typically the poorest area of preservation of the rock record, (c) evidence for the climatic state above the surface of the Earth (e.g. from aeolian sediments records), (d) the relationships which govern continental weathering and control geochemical cycles, (e) the distribution in space and time of rhythmic sediments and (f) the correlations of oceanic events with global climate.

(2) Many of the major problems in paleoclimatology focus attention on critical questions concerning our understanding of the physical climate system. These include (a) ocean-atmosphere feedbacks in climate change, (b) cloud-climate feedbacks, (c) snow and sea-ice processes and ice-albedo feedbacks, and (d) lapse rate feedbacks at high latitudes.

(3) Problems in paleoclimatology focus attention on critical questions of the mechanisms which influence climate change. Several questions have been posed concerning (a) solar evolution and solar variability, (b) evolution of atmospheric composition, (c) the mechanisms which might explain equatorial glaciation, (d) the role of geography in climate change, (e) the mechanisms which would amplify the role of the oceans in poleward heat transport, (f) the role of barriers and gateways in governing the ocean circulation, (g) the links between climate and geochemical cycles, and (h) climate sensitivity to orbital variations throughout Earth history.

This list is far from exhaustive and tends to focus on areas where major new initiatives or extended efforts are required.

REFERENCES

Arthur, M. A., W. E. Dean, D. J. Bottjer and P. A. Scholle, 1984: Rhythmic bedding in Mesozoic-Cenozoic pelagic carbonate sequences: the primary and diagenetic origin of Milankovitch-like cycles. Milankovitch and Climate, J. Imbrie, J. Hays, G. Kukla and B. Saltzman, Eds., D. Reidel, Dordrecht, 191-222.

Augustsson, T., and V. Ramanathan, 1977: A radiative-convective model study of the CO_2 climate problem. J. Atmos. Sci., 34, 448-451.

Barron, E. J., 1987: Eocene equator-to-pole surface ocean temperatures: A significant climate problem? Paleoceanography, 2, 729-739.

Barron, E. J., 1983: A warm, equable Cretaceous: The nature of the problem. Earth Sci. Rev., 19, 309-338

Barron, E. J., M. A. Arthur and E. G. Kauffman, 1985: Cretaceous rhythmic bedding sequences: a plausible link between orbital variations and climate. Earth and Planetary Science Letters, 72, 327-340.

Barron, E. J., S. L. Thompson and S. H. Schneider, 1981: An ice-free Cretaceous? Results from climate model simulations. Science, 212, 501-508.

Barron, E. J., and W. M. Washington, 1985: Warm Cretaceous climates: high atmospheric CO_2 as a plausible mechanism. The Carbon Cycle and Atmospheric CO_2: Natural Variations Archean to Present. E. T. Sundquist and W. S. Broecker, Eds., American Geophysical Union, Washington, DC, 546-553.

Barron, E. J., and W. M. Washington, 1984: The role of geographic variables in explaining paleoclimates: Results from Cretaceous climate model sensitivity studies. J. Geophys. Res., 89, 1267-1279.

Beaty, C., 1978: The causes of glaciation. Am. J. Sci., 66, 452-459.

Berggren, W., and C. Hollister, 1974: Paleogeography, paleobiogeography and the history of circulation in the Atlantic Oceans. Studies in Paleo-oceanography, W. W. Hay, Ed., Society of Economic Paleontologists and Mineralogists, Spec. Publ. 20, 126-186.

Berner, R. A., A. C. Lasaga and R. M. Garrels, 1983: The carbonate-silicate geochemical cycle and its effect on atmospheric carbon dioxide over the past 100 million years. Am. J. Sci., 283, 641-683.

Crowell, J., and L. A. Frakes, 1970: Phanerozoic glaciation and the causes of the Ice Ages. Am. J. Sci., 268, 193-224.

Crowley, T. J., D. A. Short, J. G. Mengel and G. R. North, 1986: Role of seasonality in the evolution of climate during the last 100 million years. Science, 231, 579-584.

Dean, W. E., J. V. Gardner, L. F. Jansa, P. Cepek and E. Seibold, 1977: Cyclic sedimentation along the continental margin of northwest Africa. Initial Reports of the Deep Sea Drilling Project, Y. Lancelot, E. Seibold et al., U.S. Government Printing Office, Washington, 41, 965-986.

de Boer, P. L., 1982: Cyclicity and the storage of organic matter in Cretaceous pelagic sediments. Cyclic and Event Stratification, G. Einsele and A. Seilacher, Eds., Springer, New York, 456-475.

Dickinson, R. E., 1985: Climate sensitivity. Advances Geophys., 28A, 99-129.

Estes, R., and J. H. Hutchison, 1980: Eocene lower vertebrates from Ellesmere Island, Canadian Arctic Archipelago. Palaeogeo. Palaeoclim. Palaeoecol., 30, 325-347.

Fischer, A. G., 1980: Gilbert-bedding rhythms and geochronology, in: The Scientific Ideas of G. K. Gilbert, E. I. Yockelson, Ed., Geol. Soc. Am. Spec. Pap., 183, 93-104.

Fischer, A. G., 1982: Long-term climatic oscillations recorded in stratigraphy. Climate in Earth History, W. H. Berger and J. C. Crowell, Eds., National Academy Press, Washington, DC, 97-104.

Frakes, L. A., and E. Kemp, 1972: Influence of continental positions on Early Tertiary climates. Nature, 240, 97-100.

Gill, A. E., and K. Bryan, 1971: Effects of geometry on the circulation of a three-dimensional southern hemisphere ocean model. Deep-Sea Res., 18, 685-721.

Gough, D. O., 1981: Solar interior structure and luminosity variations. Solar Physics, 74, 21-34.

Hambrey, M. J., and Harland, W. B., 1981: Earth's Pre-Pleistocene glacial record. Cambridge Univ. Press, Cambridge, 1004 pp.

Hays, J., J. Imbrie and N. J. Shackleton, 1976: Variations in the Earth's orbit: Pacemaker of the Ice Ages. Science, 194, 1121-1132.

Henderson-Sellers, A., 1979: Clouds and the long term stability of the Earth's atmosphere and climate. Nature, 279, 786-788.

Herbert, T. D., and A. G. Fischer, 1986: Milankovitch climatic origin of mid-Cretaceous black shale rhythms in central Italy. Nature, 321, 739-743.

Holland, H. D., 1978: The Chemistry of the Atmosphere and Oceans. Wiley Interscience, New York, 351 pp.

Irving, E., and J. K. Park, 1972: Hairpins and superintervals. Can. Jour. of Earth Sci., 9, 1318-1324.

Janecek, T. R., and D. K. Rea, 1983: Eolian deposition in the northwest Pacific Ocean: Cenozoic history of atmospheric circulation. Geol. Soc. Am. Bull., 94, 730-738.

Karhu, J., and S. Epstein, 1986: The implication of the oxygen isotope records in coexisting cherts and phosphates. Geochim. Cosmochim. Acta, 50, 1745-1756.

Kauffman, E. G., 1977: Upper Cretaceous cyclothems, biotas, and environments, Rock Canyon Anticline, Pueblo, Colorado. Cretaceous Facies, Faunas, and Paleoenvironments across the Western Interior Basin, E. G. Kauffman, Ed., Rocky Mtn. Assoc. Geol., 14, 129-152.

Keigwin, L. D., and B. H. Corliss, 1986: Stable isotopes in late middle Eocene to Oligocene foraminifera. Geol. Soc. Am. Bull., 97, 335-345.

Kennett, J. P., 1977: Cenozoic evolution of Antarctic glaciation, the circum-Antarctic Ocean, and their impact on global paleoceanography. J. Geophys. Res., 82, 3843-3860.

Knauth, L. P., and S. Epstein, 1976: Hydrogen and oxygen isotope ratios in nodular and bedded cherts. Geochim. Cosmochim. Acta, 40, 1095-1108.

Kuhn, W. R., and S. K. Atreya, 1979: Amonia photolysis and the greenhouse effect in the primordial atmosphere of the Earth. Icarus, 37, 207-213.

Kuhn, W. R., and J. F. Kasting, 1983: The effects of increased CO_2 concentration on surface temperature of the early Earth. Nature, 301, 53-55.

Kutzbach, J. E., and P. J. Guetter, 1984: The sensitivity of monsoon climates to orbital parameter changes for 9000 years B.P.: experiments with the NCAR general circulation model. Milankovitch and Climate, A. Berger, J. Imbrie, J. Hays, G. Kukla and B. Saltzman, Eds., D. Reidel, Dordrecht, 801-820.

Kutzbach, J. E., and B. L. Otto-Bliesner, 1982: The sensitivity of the African-Asian monsoonal climate to orbital parameter changes for 9000 years B.P. in a low resolution general circulation model. J. Atmos. Sci., 39, 1177-1188.

Lasaga, A. C., R. A. Berner and R. M. Garrels, 1985: An improved geochemical model for atmospheric CO_2 fluctuations over the past 100 million years. in The Carbon Cycle and Atmospheric CO_2: Natural Variations Archean to Present, E. T. Sundquist and W. S. Broecker, Eds., American Geophysical Union, Washington, DC, 397-411.

Lowe, D. R., 1980: Archean sedimentation. Annual Reviews in Earth and Planetary Science, 8, 145-167.

MacGinitie, H.D., 1969: The Eocene Green River flora of northwestern Colorado and northeastern Utah. Calif. Univ. Pubs. Geol. Sci., 83, 1-143.

MacGinitie, H. D., 1974: An early middle Eocene flora from the Yellowstone-Absaroka volcanic province, northwestern Wind River Basin, Wyoming. Calif. Univ. Pubs. Geol. Sci., 108, 1-103.

Owen, T., R. D. Cess and V. Ramanathan, 1979: An enhanced carbon dioxide greenhouse to compensate for reduced solar luminosity on Early Earth. Nature, 277, 640-642.

Piper, J. D. A., 1973: Latitudinal extent of Late Precambrian glaciations. Nature, 244, 342-344.

Pratt, L. M., 1981: A paleo-oceanographic interpretation of the sedimentary structures, clay minerals, and organic matter in a core of the Middle Cretaceous Greenhorn Formation near Pueblo, Colorado, Princeton University, Dissertation, 176 pp.

Ramanathan, V., 1976: Radiative transfer within the Earth's troposphere and stratosphere: A simplified radiative-convective model. J. Atmos. Sci., 33, 1330-1346.

Rea, D. K., M. Leinen and T. R. Janecek, 1985: Geologic approach to the long-term history of atmospheric circulation. Science, 277, 721-725.

Sagan, C., and G. Mullen, 1972: Earth and Mars: Evolution of atmospheres and surface temperatures. Science, 177, 52-56.

Schlesinger, M. E., and J. F. B. Mitchell, 1987: Climate model simulations of the equilibrium climatic response to increased carbon dioxide. Rev. Geophysics, 25, 760-798.

Schneider, S. H., and S. L. Thompson, 1980: Cosmic conclusions from climatic models: Can they be justified? Icarus, 41, 456-469.

Schneider, S. H., S. L. Thompson and E. J. Barron, 1985: Mid-Cretaceous continental surface temperatures: Are high CO_2 concentrations needed to simulate above freezing winter conditions?. The Carbon Cycle and Atmospheric CO_2: Natural Variations Archean to Present. Geophys. Monogr. Ser. 32, E. T. Sundquist and W. S. Broecker, Eds., AGU, Washington, DC, 554-560.

Schwarzacher, W., and A. G. Fisher, 1982: Limestone-shale bedding and perturbations of the Earth's orbit. Cyclic and Event Stratification, G. Einsele and A. Seilacher, Eds., Springer, New York, 72-95.

Shackleton, N., and A. Boersma, 1981: The climate of the Eocene ocean. J. Geol. Soc. London, **138**, 153-157.

Shackleton, N., and J. P. Kennett, 1975: Paleotemperature history of the Cenozoic and the initiation of Antarctic glaciation: oxygen and carbon isotope analyses in DSDP sites 277, 279, 281. Initial Reports of the Deep-Sea Drilling Project, J. P. Kennett, R. E. Houtz, et al., U. S. Government Printing Office, Washington, DC, **29**, 801-807.

Sheldon, R. P., 1986: Evidence for ring systems orbiting Earth in the geologic past. Earth and the Human Future, K. R. Smith, F. Fesharaki and J. P. Holdren, Eds., Westview Press, Boulder and London, 45-70.

Tarling, D., H., 1978: The geological-geophysical framework of ice ages. Climatic Change, J. Gribben, Ed., Cambridge Univ. Press, 3-24.

Tarling, D. H., 1974: A paleomagnetic study of Eocambrian tillites in Scotland. J. Geol. Soc. of London, **130**, 163-177.

Walker, J. C. G., 1982: Climatic factors on the Archean Earth. Palaeogeogr. Palaeoclimat. Palaeoecol., **40**, 1-11.

Walker, J. C. G., P. B. Hays and J. F. Kasting, 1981: A negative feedback mechanism for the long-term stabilization of Earth's surface temperature. J. Geophys. Res., **86**, 9776-9782.

Wang, W.-C., W. B. Rossow, M.-S. Yao and M. Wolfsen, 1981: Climate sensitivity of a one-dimensional radiative-convective model with cloud feedback. J. Atmos. Sci., **38**, 1167-1178.

Washington, W. M., and G. A. Meehl, 1986: General circulation model CO_2 sensitivity experiments: snow-sea ice albedo parameterizations and globally averaged surface air temperature. Climatic Change, **8**, 231-241.

Williams, G. E., 1975: Late Precambrian glacial climate and the Earth's obliquity. Geological Magazine, **112**, 441-465.

Wolfe, J. A., 1978: A paleobotanical interpretation of Tertiary climates in the northern hemisphere. Amer. Sci., **66**, 691-703.

ARCHAEAN ATMOSPHERE–BIOSPHERE INTERACTIONS

A. HENDERSON-SELLERS
Department of Geography,
University of Liverpool,
P.O. Box 147,
Liverpool, L69 3BX.
U.K.

ABSTRACT. The Archaean, which comprises about 46% of the history of the Earth, has a fundamentally important characteristic in common with the present-day: both are periods in which the biota modified the atmosphere so much that climate changed. The future of mankind depends upon our ability to understand and predict the climatic consequences we ourselves induce through increasing atmospheric carbon dioxide and other trace gas amounts and, perhaps, by massive injections of aerosols into the atmosphere. Changes like these, but much greater in magnitude, occurred throughout the Archaean and yet the climate remained equable and the biota, a diverse and versatile collection of prokaryotes, flourished.

1. Introduction: When, Where and How

1.1 THE ARCHAEAN: HALF EARTH'S HISTORY

The Archaean era terminated about two and a half thousand million years before present (2.5 Ga, Figure 1). By that time the majority of the continental crust had formed, thereby differentiating tectonic processes in the Proterozoic and the Phanerozoic, which have recycled this continental crust, from those of the Archaean that formed it. The recent 2500 million years of the Earth's history have been characterized by a plate tectonic regime. During the Archaean however, vulcanism, driven by much steeper geothermal gradients, was the predominant geological activity and, at least in the early part, impact bombardment was commonplace.

The transition from the Archaean to the Proterozoic is not as clear-cut as the mark on the time axis in Figure 1 makes it appear or, more correctly, it's the mark itself that is ambiguous since it spans a million or more years. By comparison, the Quaternary period, which encompasses the Pleistocene and the Holocene and all the history of mankind, began only 1.6 million years ago and this length of time, almost the full history of man's evolution and civilization, is merely the thickness of a tick mark on the axis of time in Figure 1. The transition from the Archaean to the Proterozoic cannot be much better defined, as yet.

The start of the Archaean is also somewhat ambiguous. Geological history began when the Earth was formed approximately 4.6 Ga but there is no direct evidence from the earliest 700–800 million years. The oldest yet discovered rocks of the Earth, which form part of the Isua outcrop in Greenland, have been dated at 3.83 Ga. This first and mysterious period of Earth history, the Hadean, is sometimes, but not always considered as part of the Archaean.

A. Berger et al. (eds.), Climate and Geo-Sciences, 21–38.

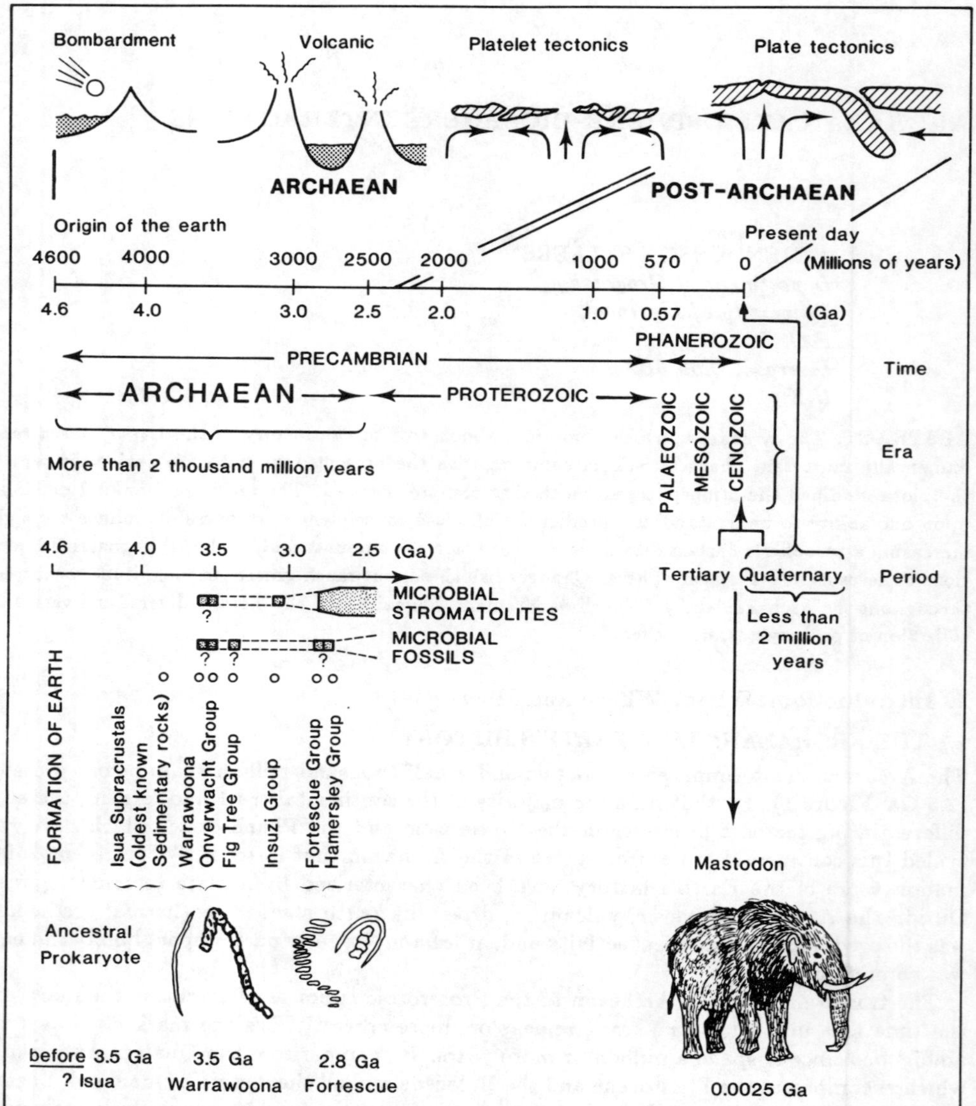

Figure 1 Geological timescale for the Earth. The Archaean encompasses approximately half Earth's history from its origin 4600 million years ago to 2500 million years ago. Tectonic processes during the Archaean differed from the plate tectonics characterizing the post-Archaean. Fossil evidence for microbial life extends from the latter half of the Archaean and up to the present day. For comparison mastodons flourished for only 2.5 million years and Man has existed for the latter part of the Quaternary period.

As we hope that the "unknown" era of Earth history will be reduced by the discovery of still older rocks we will include the Hadean in our definition of the Archaean. Thus the Archaean extends from the origin of the Earth to the beginning of the Proterozoic, about 2,100 million years. It is the single most extensive geological era, beating the Proterozoic (1,930 million years) and the Phanerozoic (570 million years), and comprises about 46% of the Earth's history (e.g. Smith, 1981).

The Archaean environment was not very similar to the present-day. Solar luminosity was lower, between 70% and 80% of present-day values (Newman and Rood, 1980). Oceans covered a greater proportion (perhaps as much as 90%) of the planetary surface primarily because the continents had not formed (Windley, 1977; Cogley and Henderson-Sellers, 1984). The land surface along with the rest of the crust was recycled rapidly. Day length was shorter, perhaps as short as 15 hours, and tidal amplitudes were larger, up to 1.5 times greater than the present day (Walker, 1977). Carbon dioxide was more abundant in the atmosphere; values of a few hundred times present atmospheric levels are not inconsistent with the few data available. The atmosphere and oceans were oxygen deficient (Schopf, 1983). Despite these, apparently, hostile conditions life was abundant during, at least, the latter half of the Archaean; microbial fossils and stromatolites have been identified in rocks as old as 3.5 Ga. Moreover, there is persuasive evidence that a highly diverse prokaryotic biota was an important agent of change causing gross modifications to the Archaean environment.

1.2 THE BIOSPHERE: MORE THAN THE BIOTA

The biosphere incorporates life and the place where it lives. Here the term "biosphere" will be taken to mean life and its essential nutrients (cf. Lovelock, 1979).

The Archaean biosphere contained purely prokaryotic biota, composed of much simpler cells than most of the life that we recognize today. Plants and animals contain eukaryotic cells which have a nucleus containing the genetic material as well as specialized organelles. In contrast (Figure 2) the prokaryotic cell is much smaller and simpler; does not contain a nucleus, the genetic material being spread throughout the protoplasm, and biochemical functions are performed on complex cell wall membranes. The oldest eukaryotes (biota composed of eukaryotic cells) yet identified are dated at 1.4 Ga while fossils of the oldest prokaryotes have been found in rocks dated at 3.5 Ga.

The Archaean fossil record is diverse and as yet relatively unexplored. Stromatolites are the best known Archaean fossils and, until the late 1970s, stromatolites were regarded as the only undisputed fossils in Archaean rocks (Rothpletz, 1916; Walter, 1977). They are organosedimentary calcareous structures which are usually laminated (Walter *et al.*, 1980). The microbial fossil record for the Archaean is, at present, very sparse (Niklas and Brown, 1981; Schopf and Oehler, 1976) and tends to parallel the microbial stromatolite record (Figure 1). An urgent research requirement is increased understanding of the few Archaean microbiotas now known and identification of assemblages which bridge the paleontologic gap that exists between the Warrawoona fossils (3.5 Ga) and the relatively well-known microbiotas of the later Precambrian (2.0 Ga).

A chemical fossil is a well-preserved biosynthetic product such as a nucleic acid or carbohydrate and, although these products disappear too rapidly for Archaean investigations, there have been successful retrievals of lipids and amino acids from Archaean sediments. As well as these extractable molecules there is often an insoluble macro-molecular material generically termed ***"kerogen"*** which can be classified according to its ecological source (e.g. Tissot and Welte, 1978; Hayes *et al.*, 1983).

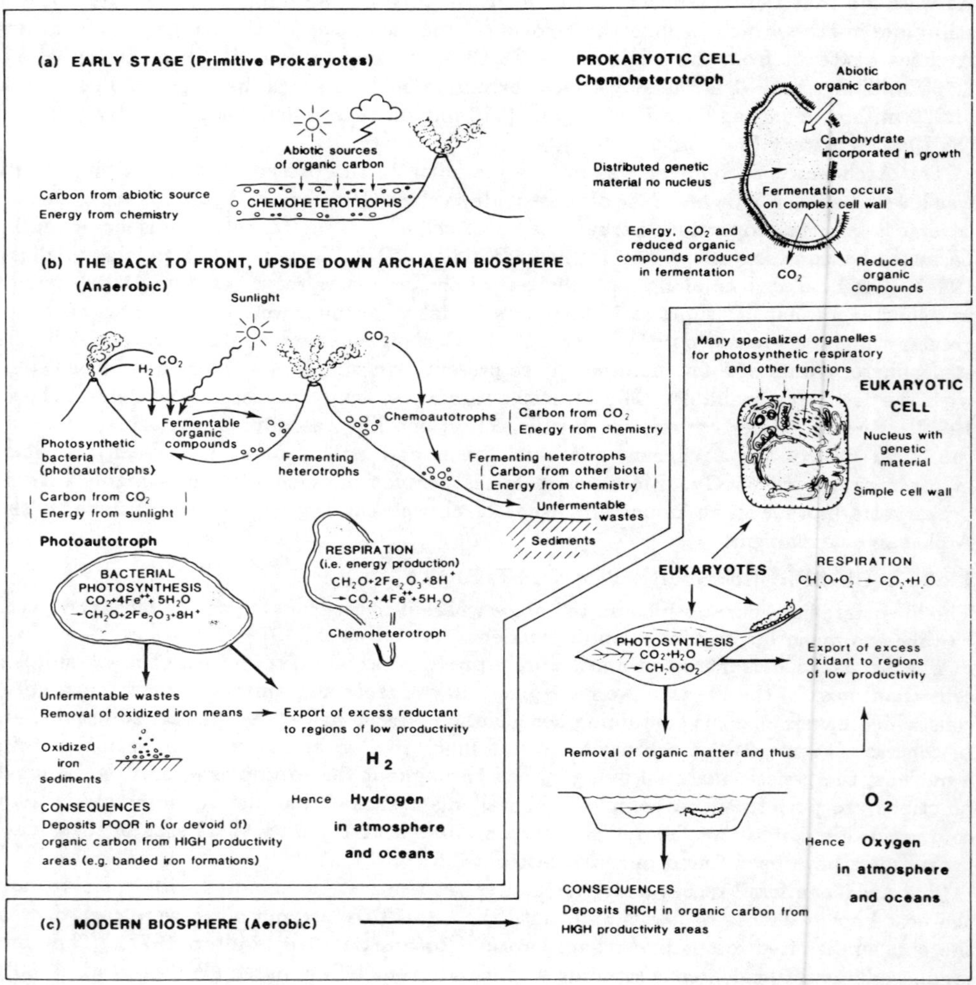

Figure 2 Stages in the evolution of the biosphere.

(a) The earliest life was probably chemoheterotrophic (i.e. it relied on abiotic sources of carbon and derived energy from chemical reactions). Certainly it was prokaryotic. The prokaryotic cell has no nucleus but possesses a complex cell wall on which processes such as fermentation and photosynthesis take place.

(b) The Archaean biota were all prokaryotic but diverse forms existed. Cycling between producers and fermenters must have been common. Deposits from high productivity areas might have been poor in organic carbon and, as oxidized iron was precipitated out, hydrogen would be exported.

(c) By contrast, the modern biosphere is composed of eurokaryotes possessing complex cells. Cycling produces deposits rich in organic carbon from highly productive areas and export of oxygen.

Although the very old (3.8 Ga) Isua metasediments seem to be unfossiliferous, they do exhibit stable-isotope systematics (Schidlowski *et al.*, 1979; Schidlowski, 1982) which have been interpreted as exhibiting well-advanced photosynthesis (McKirdy and Hahn, 1982) which is consistent with activity of contemporary organisms. Eleven occurrences of stromatolites (and possible stromatolites) are known from the Archaean dating back to 3.5 Ga (Figure 3(a,b)). Three of these are too poorly known to be definitively identified as stromatolitic and some of the Late Archaean units are very imprecisely dated, and may prove to be earliest Proterozoic. Although over half of the eleven occurrences were discovered or first described during the last decade (cf. Figure 3(c)) (Walter, 1983; Schopf, 1983) the studies have already proved fruitful and fascinating. In particular the Warrawoona microbial community (Schopf, 1983) including both heterotrophs and autotrophs, occupied a surficial sediment-water interface environment that was intermittently exposed as long ago as 3.5 Ga. The diversity and cellular complexity of the preserved microbes indicate that the beginnings of life on Earth must have occurred substantially earlier.

1.3 THE ATMOSPHERE: A BROAD DEFINITION

In order to try to understand the interactions between the Archaean biota and their environment, a very wide definition of the atmosphere is most useful. Here the Rubey inventory (Rubey, 1951) of excess volatiles which today compose the atmosphere, oceans, ice sheets, fresh water and hydrated and carbonated sediments, also termed the Earth's early hydrosphere (Cogley and Henderson-Sellers, 1984), will be used.

Although most sediments are uninformative as to the climate under which they were deposited, orders-of-magnitude excursions in temperature, pressure and atmospheric composition (for at least some constituents) are ruled out by the occurrence throughout the geological record of sediments with modern aspect; at the least, temperatures above 0°C and pressure above 6 hPa have always prevailed. Archaean turbidites (Eriksson, 1982) suggest water depths of 1 km to a few kilometres, and therefore substantial oceans, at 3.0–3.5 Ga. The continuous presence of carbonate rocks shows that the atmosphere has contained CO_2 throughout geological history. Undoubted tillites dating from 2.6–2.2 Ga, but not from other times, may suggest broad climatic change. On the other hand, in common with all such interpretations of a temporally patchy record, the absence of Archaean tillites may simply be a preservation effect.

The least well understood feature of the hydrospheric volatiles is the apparent sharp contrast between their oxidation state and that of the underlying reduced mantle volatiles (Anderson, 1975; Wyllie, 1979; Davies, 1981). On the other hand, if the volatiles were evolved or acquired during accretion, it is likely that the near-surface volatiles were never in equilibrium with those of the mantle (Anders and Owen, 1977; Arculus and Delano, 1980 cf. Haggerty and Tompkins, 1983).

The idea that degassing was substantially completed at or near the end of the Earth's formation depends upon (i) the enrichment in ^{129}Xe of earth-interior gases with respect to the present-day atmosphere and to Archaean shales (Frick and Chang, 1977; Phinney, 1972), requiring outgassing of the atmosphere within a few half-lives (17 Ma) of ^{129}I, with negligible net additions since 3.3 Ga (Thomsen, 1980) and (ii) the recognition that geothermal energy sources were much larger during accretion than times since so that models with outgassing due to radioactive warm-up (Li, 1972; Melton and Giardini, 1982), or similar gradual processes, seem mechanically and thermally improbable (Cogley and Henderson-Sellers, 1984).

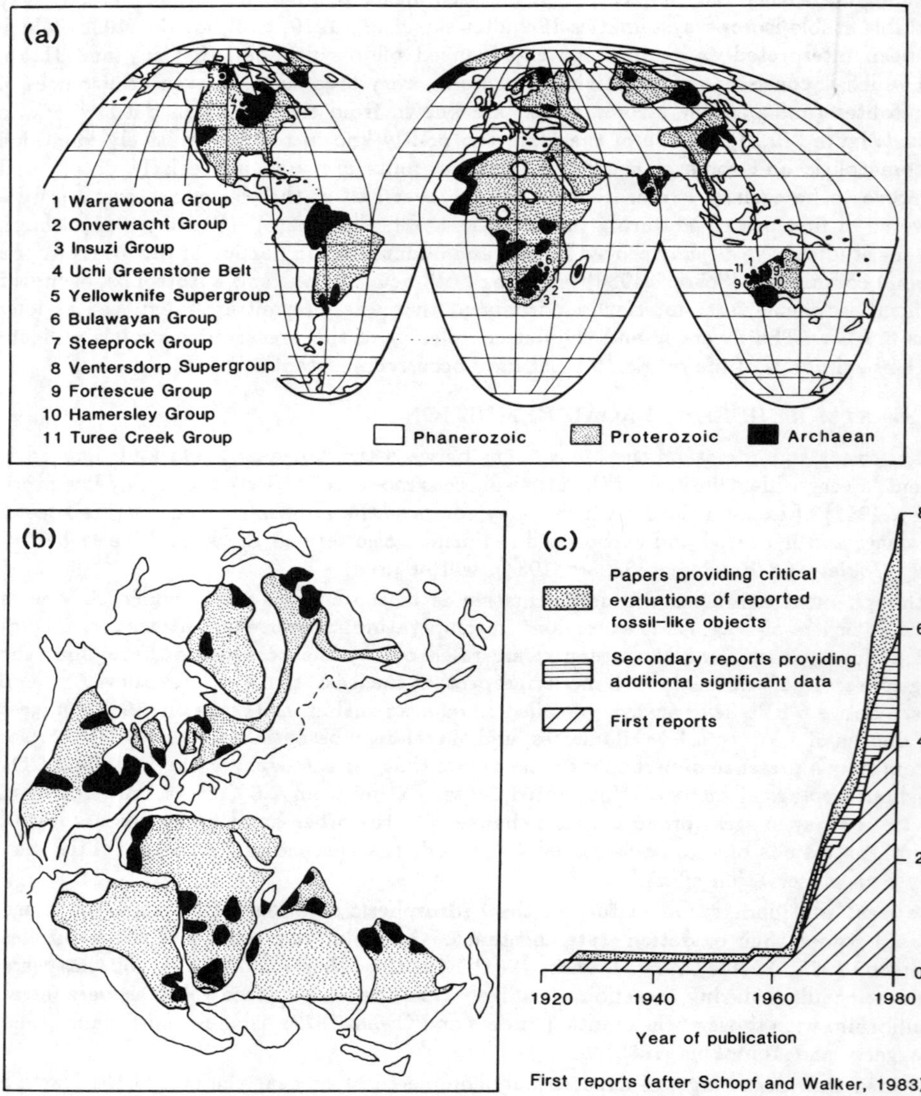

Figure 3 (a) Present-day distribution of Archaean rocks with sites of microfossils and stromatolites identified.

(b) The identified remains of Archaean protocontinents occur in small areas in what were contiguous land masses.

(c) During the last twenty years there has been a fairly steady increase in reports of Archaean microfossils. The rock record undoubtedly offers a wealth of further information. [Redrawn from Windley, 1977 and Schopf, 1983].

Detrital Uraninite, UO_2, and pyrite, FeS_2, found in some Archaean (3.0–2.3 Ga) sediments (e.g. Simpson and Bowles, 1977; Utter, 1980; Clemmey, 1981; Clemmey and Badham, 1982) have long been taken as evidence for a "strongly" reducing Precambrian atmosphere, but oxygen-abhorring minerals in apparently well-ventilated settings do not necessarily imply an atmosphere without oxygen (Grandstaff, 1976, 1980). Most Precambrian sedimentary iron is found (Garrels *et al.*, 1973) in banded iron formation comprising oxidized (dominated by haematite, Fe_2O_3, and magnetite Fe_3O_4), reduced (siderite, $FeCO_3$) and strongly reduced (pyrite) assemblages. Moreover, Walker (1987) has shown that oxidized iron in sediments poor in organic carbon may well be the result of highly productive anaerobic biotic regions (Figure 2) which exported H_2.

Several lines of evidence suggest that substantial excursions of atmospheric CO_2 concentration have occurred. There is much more dolomite, Ca Mg $(CO_3)_2$, in Precambrian than in Phanerozoic rocks, and although it may be that conversion of calcite to dolomite becomes more likely with the age of a sediment, it seems that a primary origin must be hypothesized for at least some Precambrian dolomite (e.g. Baker and Kastner, 1981; Tucker, 1982). Models of more recent excursions in atmospheric CO_2 (Budyko and Ronov, 1979; Berner *et al.*, 1983) include a negative feedback in which CO_2 consumption by weathering increases as atmospheric CO_2 levels increase (Walker *et al.*, 1981) and a two-way reaction between carbonate and silicate reservoirs.

Thus, at a time prior to 3.3 Ga and possibly prior to 3.8 Ga, life became established on Earth. If the formation of the oceans concurrently with accretion, perhaps in its later stages, is accepted the geography of the Earth's surface throughout the Archaean is likely to have consisted of a global archipelago rather than a series of extensive landmasses (Cogley and Henderson-Sellers, 1984). The existing record of microfossils shows that habitable, shallow water environments have been present since at least 3.5 Ga. They were probably restricted in area during the Archaean and possibly short-lived in terms of geological time scales (Schopf *et al.*, 1983). The growth of extensive areas of stable continental shelf around the end of the Archaean may have had a major impact on life, permitting a significant increase in biological productivity. On the other hand it is equally likely that the rapid tectonic recycling has removed the majority of the evidence of earlier communities.

2. The Archaean Atmosphere and Biosphere: A Story of Traumatic Co-evolution

2.1 LIFE

It is likely, but by no means certain, that the "origin of life" occurred many times although it is possible that living organisms, once formed, operated so efficiently that only one genesis was required. When this origin occurred, or where, or how many times, is unknown but the early Archaean atmosphere contained no free oxygen and therefore life originated and evolved in anaerobic conditions. The first organisms could have exploited a supply of abiotically synthesised organic material. Thus first prokaryotes were heterotrophs and used carbon compounds as their source of cellular carbon. They were also probably chemoheterotrophs (i.e. they obtained their energy chemically through a reaction of one compound with another) perhaps by fermentation, which is a widespread, and relatively simple, biochemical process (Bernal, 1967; Oparin, 1968; Schopf, 1981; Schopf *et al.*, 1983).

If the atmosphere was only very weakly reducing or neutral, it is unlikely that abiotic synthesis could produce an adequate organic carbon supply for this ecosystem to persist for long. The very early biosphere was, therefore, carbon-limited.

The next organisms to evolve may well have been autotrophic which use CO_2 (often from the environment but sometimes generated from another compound) as its source of cellular carbon. A photoautotroph has the additional facility of capturing energy directly from sunlight rather than needing to obtain chemical energy, say by fermentation (cf. Margulis and Lovelock, 1978). Bacterial (prokaryotic) photosynthesis is a simpler process than green plant photosynthesis.

$$H_2X + CO_2 \Rightarrow CH_2O + X \tag{1}$$

where X may be sulphur, nothing or small carbon radicals. This can continue if a more reduced compound is available to donate electrons but does not result in the release of oxygen to the environment. Rapid growth of photosynthetic bacteria formed a large source of organic material. Heterotrophs could have tapped this supply of organic material so that an interdependency may have developed.

An alternative scenario for this later stage (cf. Figure 2), depicts the first autotrophs as deriving their energy from chemical interactions not directly from sunlight (i.e. chemoautotrophs). There are microbes in existence today, called methanogenic bacteria, that derive energy from a reaction between hydrogen and carbon dioxide to produce methane and water as waste products. These autotrophs would have flourished initially as they converted abundant supplies of carbon dioxide and hydrogen into methane and water. This may have initiated a massive and biotically generated impact on the atmosphere as CH_4 increased at the expense of CO_2. It has been proposed that photochemical reactions in the atmosphere converted some methane back into hydrogen and carbon dioxide but quantitative estimates of the net effect have not been made. The heterotrophs would have flourished utilizing the organic material synthesized by the autotrophs.

The climatic impact of the addition of a source of CH_4 can be estimated using the recent results of Kiehl and Dickinson (1987). They used a new narrow-band radiative convective model to compute the volume mixing ratios of CO_2 and CH_4 required to just compensate (i.e. resulting in present day temperatures) for a range of reductions in solar luminosity (Figure 4). They find that about 428 times the present day atmospheric mixing ratio of CO_2 would balance a 24% reduction in solar luminosity but that this amount is halved (240 times) if there is an atmospheric mixing ratio of 10^{-4} of CH_4.

The order of evolution may not be important because prokaryotes are remarkably versatile. For example *Rhodopseudomonas capsulata* which is one of the most versatile prokaryotes known, exhibits at least five growth modes (Madigan and Gest, 1979). In any case, the geological record may ultimately reveal little that can definitively date the evolutionary developments since the prokaryotes suffered from a "Volkswagen syndrome": a lack of change in the external body form masking evolutionary and even revolutionary developments in internal biochemistry (Schopf *et al.*, 1983).

The environmental impact of the evolution of life through the stages described above, in whatever order, is the reduction of the amount of H_2 in the atmosphere. This is an important, and inevitable, result. A secondary impact may have been an increase in atmospheric methane. Kasting *et al.* (1979) demonstrate that free oxygen can be produced on an abiological Earth in only very small quantity and even after the origin of oxygenic photosynthesis there must have been a prolonged period of fluctuating imbalance between oxygen sources and sinks prior to the establishment of widespread aerobic conditions. Theoretical calculations, based on consideration of the numerous parameters involved, suggest a probable duration for this transition on the order of "a few, or perhaps several, hun-

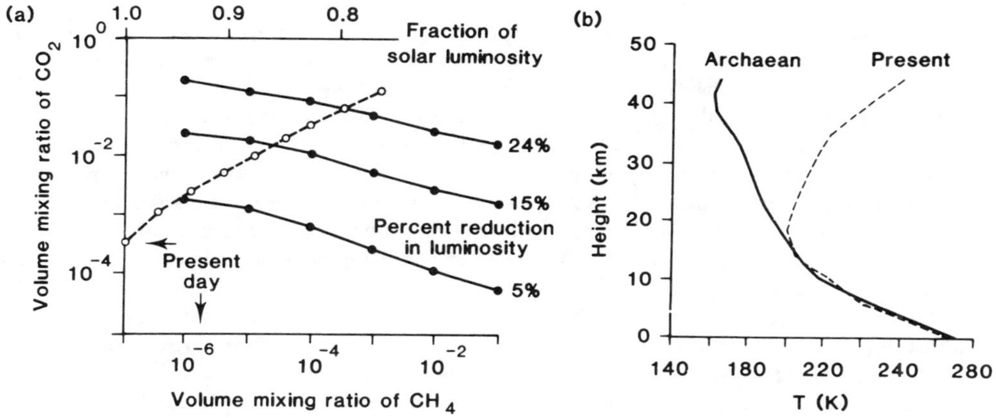

Figure 4 The impact on computed values of temperatures in the Archaean of lower values of solar luminosity and higher mixing ratios of greenhouse gases.
(a) Solid lines show that plausible combinations of CO_2 and CH_4 can compensate for a range of lowered values of solar luminosity. (The present-day mixing ratios are arrowed). Dashed line shows that CO_2 alone can also balance lowered solar luminosity (top axis).
(b) Temperature profiles computed for Archaean conditions (24% reduction in solar luminosity, 969 PAL of CO_2 and no ozone) and the present-day differ only in the higher atmosphere; surface temperatures are quite similar (see also Rossow *et al.*, 1982). [All redrawn from Kiehl and Dickinson, 1987].

dred millions of years" (Schopf, 1980). Interpretation of the geological record too must be undertaken with care (Walker, 1978a,b; Holland, 1978; Walker, 1987).

Over the Archaean the important biotic atmospheric phases are: the origin of life (earlier than ~3.5 Ga), anaerobic chemoheterotrophy and anaerobic photoautotrophy (as early as ~3.5 Ga), anaerobic chemoautotrophy (at least since ~2.8 Ga); aerobic photoautotrophy and amphiaerobes (organisms that can use O_2 like anaerobes or exploit some other energy conversion process) (probably ~2.8 Ga); and obligately aerobic microorganisms (originating earlier than at least ~2.0 Ga but abundant and globally widespread between ~1.7 and ~1.5 Ga).

There remains considerable uncertainty about the nature of the oldest known rocks from the Isua outcrop. The available data do not allow a decisive choice among three plausible sources for the Isua kerogen: (i) biologically produced organic matter, (ii) abiotically synthesized organic matter and (iii) metamorphically altered carbonate minerals but it seems likely that some portion of this carbonaceous material is post-depositional and, possibly, post-metamorphic. Schopf *et al.* (1983, p.367) conclude: "like many others in the field, it is our hunch that life may well have existed in Isua time, but rigorously considered, the established facts do not seem to us even "permissive" of such a statement. With regard to the timing of the origin of life, we conclude that living systems must have been extant earlier than ~3.5 Ga ago, but how much earlier is at present a matter of speculation."

2.2 CLIMATE AND ENVIRONMENT

Amongst the wide range of Archaean environmental disturbances three stand out: the lowered solar luminosity, the changing chemistry of the atmosphere and oceans and the faster rotation rate of the Earth. These factors may not have had the largest effects but they are, at least, generally agreed upon and their impacts have been discussed.

The impact on the early Earth of 20%–35% reduced solar luminosity (e.g. Newkirk, 1980) has been a topic of discussion for almost thirty years. The most common argument is that the gaseous composition of the Archaean atmosphere was very different from today's, including larger mixing ratios of gases which absorb strongly in the thermal infrared. Sagan and Mullen (1972) suggested that the Archaean atmosphere contained considerable amounts of NH_3, CH_4 and CO_2. Of these, CO_2 is the longest lived, hence most likely, while NH_3 is too short lived to have any notable effect, and CH_4 is possible if a source can be established and of importance for radiative balance in the early atmosphere. Hart (1978) developed a scenario for CO_2 and CH_4 over time, using a simple geochemical model, the CO_2 component of which Owen *et al.* (1979) used in a radiative-convective model to calculate the early Earth's surface temperatures. The results of Owen *et al.* (1979) have been disputed by others using different climate models of similar basic type (Kuhn and Kasting, 1983; Kasting *et al.*, 1984).

Recently Kiehl and Dickinson (1987) have identified the reason for the discrepancy: Owen *et al.* (1979) inadvertently neglected the effects of pressure scaling on two CO_2 absorption bands and the effects of absorption band overlap. With these errors corrected, the differences are much smaller. Kiehl and Dickinson (1987) undertook two simulations of the surface temperature through time as a function of the changing solar luminosity and Hart's (1978) CO_2 partial pressures, and with and without present-day ozone. Although these O_3 differences were to facilitate intercomparison with Owen *et al.* (1979) who oddly included present-day O_3 amounts in their calculation and Kasting *et al.* (1984) who excluded O_3 entirely, they also show that the impact of removing O_3 from the early Archaean atmosphere is to increase the surface temperature by $\sim$4K. All models now seem to agree that plausible increases in CO_2 could have compensated the decreased solar luminosity perhaps even causing higher temperatures than today. Figure 4(b) compares the temperature profiles for the present-day and 4.25 Ga (24% reduction in solar luminosity, 969 times the CO_2 partial pressure and no ozone) as computed by Kiehl and Dickinson (1987). Surface temperatures are similar although the upper atmosphere is much colder. Similar vertical temperature distributions are shown by Rossow *et al.* (1982).

Other workers have evaluated the compensatory effects of different Archaean characteristics. Henderson-Sellers and Meadows (1975) showed that the lower planetary albedo, due primarily to the domination of the surface by oceans, decreased the impact of reduced solar luminosity somewhat. The same authors also investigated the evolution of the Earth's surface temperature as a function of the infrared opacity of the atmosphere and the planetary rotation rate (Henderson-Sellers and Meadows, 1977). The shorter day and night lengths are likely to result in weaker day/night temperature contrasts as the period of darkness becomes closer to the radiative relaxation time. Hunt (1979a,b) examined the effects of faster rotation rate on the dynamics of a simplified hemispheric model. The poleward transports of sensible and latent heat were much reduced and the equator-to-pole temperature difference was about 75 K for a 6 hour day, more than twice that predicted by the control model.

In summary, it is possible to make estimates of the concentrations of the species likely to have dominated the atmosphere of the early Earth. Carbon dioxide was present in concentrations at least as great as the present-day value, probably in the range one to a few hundred times the present atmospheric level (Owen *et al.*, 1979; Walker *et al.*, 1981; Kuhn and Kasting, 1983) and can, alone, compensate for astrophysically predicted reductions in solar luminosity (Figure 4). The water vapour concentration was determined by the saturation vapour pressure and the relative humidity (i.e. as a function of the surface temperature). The N_2 concentration was probably similar to today's (Walker, 1977). H_2 mixing ratios were determined by the balance between diffusion-limited escape (Walker (1977) but see also Visconti (1982)) and surface vulcanism. Kasting and Walker (1981) estimate the H_2 mixing ratio to have been between 10^{-5} and 10^{-3}. Oxygen may have been very scarce; however, high-energy UV fluxes (Canuto *et al.*, 1982, 1983) could have increased oxygen concentrations substantially, and there is geological evidence from later times which suggests concentrations higher than the low value of Kasting and Walker (1981). Probably the most important result from photochemical modelling studies is that an abiotic atmosphere composed of reduced components is extremely unlikely to have persisted but some of the prokaryotes, the methanogens, could have been a very significant source of methane (Kasting and Walker, 1981; Levine, 1982; Schopf, 1983; Zahnle, 1986).

The single most striking feature of the Archaean environment was the strong and regular interaction between all the major nonnoble gases in the atmosphere and the prokaryotic biota (Margulis and Lovelock, 1978). These interactions must have modified the environment and were certainly a primary, if not *the* primary, characteristic of the Archaean.

3. Sensitivity and Stability of the Archaean Environment: an Analogue of the Future?

3.1 ARCHAEAN CLIMATIC SENSITIVITY

Environmental disturbing agents were more prevalent and more extreme during the Archaean than at any other time in geological history. For example, bombardment by bodies of all sizes from meteoritic to cometary must have been commonplace during at least the first half of the Archaean; massive fluctuations in atmospheric chemistry occurred at least twice: (i) H_2 was reduced significantly by the early biota and, much later, (ii) O_2 began to be released by other biota. Other chemical excursions resulting from biospheric activity such as methanogenic bacteria (early chemoautotrophs), which obtain energy from a reaction between hydrogen and carbon dioxide, release methane as a waste product. This CH_4 could have built up to substantial levels in the atmosphere.

The justification for studying the Archaean which prefaced this narrative was that the Archaean shares an important characteristic with the present-day (and the future): the biota demonstrably influence the atmosphere. There is another characteristic of the Archaean which it seems to share with the rest of the geological history of the Earth: there are no catastrophic climatic excursions. The evidence from the Archaean does not show large-scale glaciation, and certainly not an ice-covered Earth, nor does it suggest extremely warm conditions in which a "runaway greenhouse" might have occurred. The massive injections of aerosols caused by cometary and meteoritic bombardment neither froze nor cooked the planet and the simple, prokaryotic life seems to have persisted throughout.

This apparent environmental stability demands careful examination. Before tackling the causes and responses to perturbations it is important to identify and clarify the possible reasons for this behaviour. Figure 5 is an attempt to illustrate the differences and complementary nature of stability and sensitivity. The more sensitive a system, the more readily it responds to a perturbation (the sphere moves more readily than the cube). The degree of system sensitivity is unimportant if, for example, the system is in one of the lower two stability states shown since perturbations which induce a response have no effect on the state. Progress towards greater instability in Figure 5 increases the importance of the system's sensitivity and, more critically, the timescale of the response. For example, if equilibrium is restored rapidly in the effectively stable case the impact on the biota may be much less than if restoration takes very much longer than the time span of the perturbation. The Archaean Earth cannot be represented by any of the cartoons on the top row nor any on the two lower rows since it was not catastrophically unstable but environmental excursions are detectable in the geological record.

The surprising outcome, at least to me, of the piecemeal research to date about the Archaean is not so much the considerable stability of the climate system but the remarkably large number of ways of explaining this stability. The simplest example is the discussions of what is probably the largest perturbation: the 20%–30% lowering of solar luminosity. The most plausible climatic "fixes" to this disturbance include greatly increased levels of CO_2 (Owen *et al.*, 1979; Pollack and Yung, 1980; Kasting *et al.* 1984), lower surface albedo (Henderson-Sellers *et al.*, 1980), more cirrus clouds (Rossow *et al.*, 1982) and increased CH_4 (Kasting *et al.*, 1983; Kiehl and Dickinson, 1987); although others, such as greatly increased levels of NH_3, have also been discussed (Kuhn and Atreya, 1979; Kasting, 1982).

Each of these "solutions" can apparently "solve the problem" so that there are a number of plausible ways to compensate for the early lowered solar luminosity. An obvious, but so far unasked, question is, "did these compensatory changes and feedbacks occur simultaneously and, if so, what are the climatic and environmental consequences?"

The effect on the computed surface temperature of external forcing and internal feedbacks has been described by Schlesinger (1985). He shows that the change in surface temperature, ΔT, in response to the sum of external forcing effects Q is

$$\Delta T \propto \frac{Q}{1 - \sum F_i} \tag{2}$$

where the impact of adding one further feedback effect, F_i, is to increase ΔT if the additional F_i is positive and decrease ΔT if F_i is negative. If the hypothesized explanations are external changes then the relationship is straightforward. On the other hand, as can be seen, the impact of positive internal responses (i.e. feedback factors) becomes increasingly large as $\sum F_i$ approaches 1. Thus the computed climatic effect depends upon other internal responses to temperature changes and if a large number of positive feedbacks are already operating the system may be highly sensitive to further changes (cf. Figure 5). Since, to-date, the literature has described, almost exclusively, positive additions to Q and positive feedback effects for the Archaean Earth we have the odd situation of an over abundance of climatic fixes perhaps suggesting a "runaway greenhouse". There must have been compensatory negative feedbacks which kept the Archaean climate in check but how this "balancing act" worked is unclear.

3.2 ARCHAEAN (AND FUTURE) HOMOEOSTASIS

Homoeostasis is the relative stability of a collection of interdependent elements of an organism or large system. This implies the regulation of system parameters within a window of variability. It is now well established (e.g. Tregonning and Roberts, 1979; Holden, 1986), that some multi-component mathematical systems evolve towards homoeostasis. On the

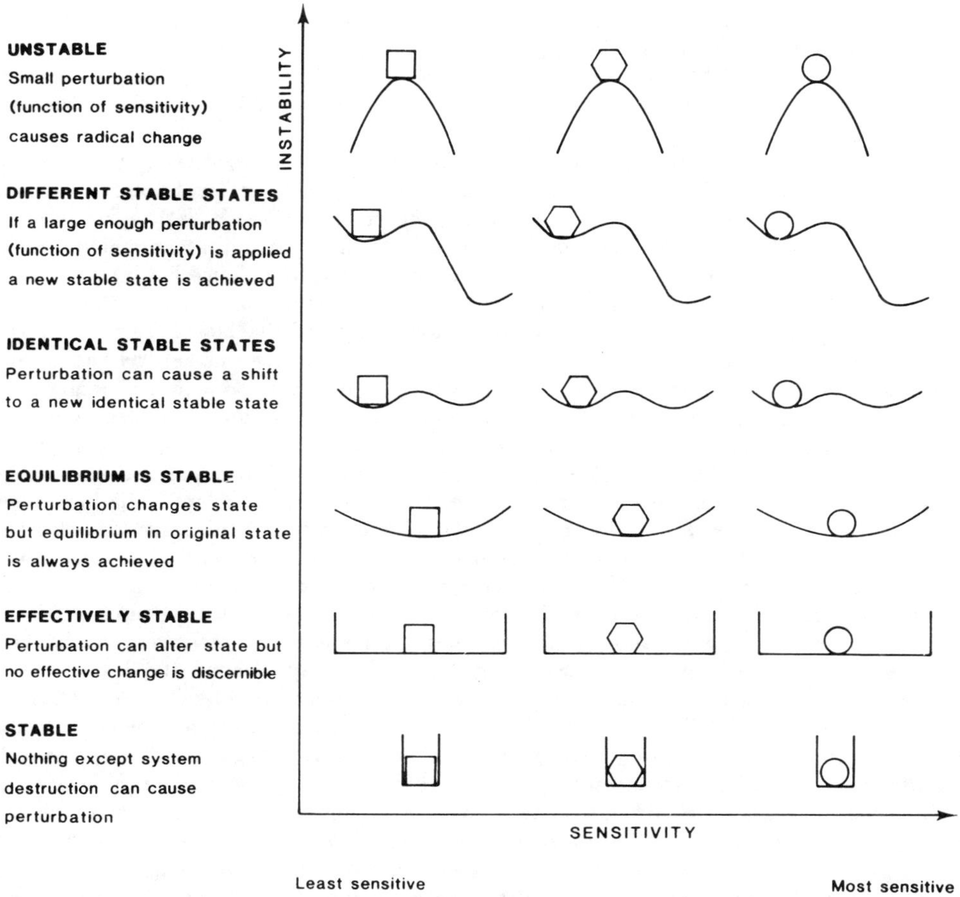

Figure 5 Cartoons illustrating the relationship between sensitivity and stability of a system. Sensitivity is related to the degree of response whereas stability describes the ultimate result. Some aspects of the Earth's hydrosphere influence the sensitivity of the climate system. In particular, the large specific heat and latent heats of water seem to decrease sensitivity. The biota may influence the stability by homoeostasis.

other hand biological systems also exhibit this tendency (e.g. Lovelock and Margulis, 1974; Watson and Lovelock, 1983; Charlson *et al.*, 1987).

The Archaean environment, which here incorporates the atmosphere, oceans, cryosphere and the biota (i.e. the climate system, see GARP, 1975) was remarkably robust. The first half of Earth's history included traumatic changes in external and internal features which were, apparently, withstood with ease. How could the environment be so robust? One possible explanation, and the one I have personally always favoured, is that once the hydrosphere was established it offered the necessary stability. Indeed, I and my co-author (Cogley and Henderson-Sellers, 1984, p168) state this hypothesis in our concluding paragraph:

"a substantial early hydrosphere, then, is both consistent with results from a variety of disciplines and, as our results show, powerful in solving the paradox of the faint young sun. The most plausible date for its origin is at the very beginning. Most of its properties, and especially its stability against large, irreversible changes of composition and climate, seem to follow once the hydrosphere is established."

It is implausible to propose that the hydrosphere acts as an active regulator of the environment. Thus either the hydrosphere offers the biota large feedbacks with which to modify the environment, or the original formation of the Earth with a ready-made hydrosphere was just a lucky coincidence but once formed the multi-component system has tended towards homoeostasis. Although, in my opinion, the choice between the Gaia hypothesis (that life, together with the physical characteristics, operates as the positive regulator of processes so as to optimize the maintenance of the biosphere overall) and the Goldilocks principle (that the Earth was coincidentally formed at exactly the correct distance from the Sun that it was neither "just too hot nor just too cold") cannot yet be made, there are a number of important conclusions that can, and must, be identified.

First, although the Archaean system was as complex as the modern environment and occupies about half of Earth's history, surprisingly little research has focussed upon it to-date. Over three-quarters of the reports or analyses of Archaean microfossils have been published in the last twenty years (Figure 3(c)). The considerable disparity between the Archaean and the Proterozoic fossil records may be due, at least in part, to neglect of the apparently less fruitful Archaean record rather than any lack of fossil evidence. The first lesson to be drawn from the Archaean is that positive encouragement of research produces results, one outstanding example being the careful collection and interpretation of data undertaken by the Precambrian Palaeobiology Research Group (Schopf, 1983).

The Archaean biogeochemical cycles may well have differed considerably from those of the present. Walker (1987) has proposed that careful reconsideration should be given to the interpretation of the sedimentary rock record. Sediments poor in organic material may well identify locations of high Archaean productivity (Figure 2). Moreover re-interpretation of the biogeochemical activity also gives rise to different conclusions about the impact on the atmosphere and oceans. A global geosphere/biosphere perspective is as urgently required for the Archaean as it is for the present. Large alterations in single components can no longer be accepted as views of the planet and model results must be carefully tested (cf. Kiehl and Dickinson, 1987).

The final lesson the Archaean teaches us may be the most crucial: not only was the Archaean climate relatively stable but life flourished. However the tolerable width of the "window of stability" for microorganisms is very much greater than that for multicellular organisms. The Archaean prokaryotes flourished because they responded rapidly to environmental change and exploited disturbances. Multicellular organisms find this difficult or impossible as the threat of a nuclear winter demonstrates. If the maintenance of the non-equilibrium chemical and thermodynamical system which we call the biosphere is the result of the positive regulation of the Gaia hypothesis only the microorganisms have yet demonstrated the ability to survive environmental trauma.

4. References

Anders, E. and Owen, T., 1977, Mars and Earth: Origin and abundance of volatiles, *Science*, **198**, 453–465.

Anderson, A.T., 1975, Some basaltic and andesitic gases, *Rev. Geophys. Space Phys.*, **13**, 37–55.

Arculus, R.J. and Delano, J.W., 1980, Implications for the primitive atmosphere of the oxidation state of the earth's upper mantle, *Nature,* **288**, 72–74.

Baker, P.A. and Kastner, M., 1981, Constraints on the formation of sedimentary dolomite, *Science,* **213**, 214–216.

Bernal, J.D., 1967, *The Origin of Life*, World, New York, 345pp.

Berner, R.A., Lasaga, A.C. and Garrels, R.M., 1983, The carbonate-silicate geochemical cycle and its effect on atmospheric carbon dioxide over the past 100 million years, *Am. J. Sci.,* **283**, 641–683.

Budyko, M.I. and Ronov, A.B., 1979, Chemical evolution of the atmosphere in the Phanerozoic, *Geochem. Int. 1979*, Engl. Transl., 1–9.

Canuto, V.M., Levine, J.S., Augustsson, T.R. and Imhoff, C.L., 1982, UV radiation from the young sun and oxygen and ozone levels in the prebiological palaeoatmosphere, *Nature,* **296**, 816–820.

Canuto, V.M., Levine, J.S., Augustsson, T.R., Imhoff, C.L. and Giampapa, M.S., 1983, The young sun and the atmosphere and photochemistry of the early earth, *Nature,* **305**, 281–286.

Charlson, R.J., Lovelock, J.E., Andrae, M.O. and Warren, S.G., 1987, Oceanic phytoplankton, atmospheric sulphur, cloud albedo and climate, *Nature,* **326**, 655–661.

Cogley, J.G. and Henderson-Sellers, A., 1984, The origin and earliest state of the earth's hydrosphere, *Rev. Geophys.,* **22**, 131–175.

Clemmey, H., 1981, Some aspects of the heavy mineral assemblages in lower Proterozoic uranium-gold conglomerates, *Mineral. Mag.,* **44**, 388–408.

Clemmey, H. and Badham, N., 1982, Oxygen in the Precambrian atmosphere: An evaluation of the geological evidence, *Geology,* **10**, 141–146.

Davies, G.F., 1981, Earth's neodymium budget and structure and evolution of the mantle, *Nature,* **290**, 208–213.

Eriksson, K.A., 1982, Sedimentation patterns in the Barberton Mountain Land, South Africa, and the Pilbara Block, Australia: Evidence for Archaean rifted continental margins, *Tectonophysics,* **81**, 179–193.

Frick, U. and Chang, S., 1977, Ancient carbon and noble gas fractionation, *Proc. Lunar Sci. Conf. 8th*, 263–272.

GARP, 1975, The physical basis of climate and climate modelling, *GARP Publications Number 16*, World Meteorological Organization, Geneva, 265pp.

Garrels, R.M., Perry, E.A. and Mackenzie, F.T., 1973, Genesis of Precambrian iron-formations and the development of atmospheric oxygen, *Econ. Geol.,* **68**, 1173–1179.

Grandstaff, D.E., 1976, A kinetic study of the dissolution of uraninite, *Econ. Geol.,* **71**, 1493–1506.

Grandstaff, D.E., 1980, Origins of uraniferous conglomerates at Elliott Lake, Canada and Witwatersrand, South Africa: Implications for oxygen in the Precambrian atmosphere, *Precambrian Res.,* **13**, 1–26.

Haggerty, S.E. and Tompkins, L.A., 1983, Redox state of earth's upper mantle from kimberlitic ilmenites, *Nature,* **303**, 295–300.

Hart, M.H., 1978, The evolution of the atmosphere of the earth, *Icarus,* **33**, 23–29.

Hayes, J.M., Kaplan, I.R. and Wedeking, K.W., 1983, Precambrian organic geochemistry, preservation of the record, Chapter 5 in *Earth's Earliest Biosphere*, (ed. J.W. Schopf), Princeton University Press, Princeton, 93–134.

Henderson-Sellers, A. and Meadows, A.J., 1975, Longterm variations in the albedo and surface temperature of the Earth, *Nature,* **254**, 44.

Henderson-Sellers, A. and Meadows, A.J., 1977, Surface temperature of early earth, *Nature,* **287**, 589–591.

Henderson-Sellers, A., Benlow, A. and Meadows, A.J., 1980, The early atmospheres of the terrestrial planets, *Q. J. R. Astron. Soc.*, **21**, 74–81.

Holden, A.V. (ed.), 1986, *Chaos*, Princeton University Press, Princeton, N.J., 311pp.

Holland, H.D., 1978, The CO_2 – HCO_3^- – CO_3^{2-} system through geologic time and its implications for the evolution of the atmosphere, in H.U. Dütsch (ed.), Influence of the biosphere on the atmosphere, *Pure and Applied Geophysics,* **116**, 232–233.

Hunt, B.G., 1979a, The influence of the earth's rotation rate on the general circulation of the atmosphere, *J. Atmos. Sci.,* **36**, 1342–1408.

Hunt, B.G., 1979b, The effects of past variations of the earth's rotation rate on climate, *Nature,* **281**, 188-191.

Kasting, J.F., 1982, Stability of ammonia in the primitive terrestrial atmosphere, *J. Geophys. Res.,* **87**, 3091–3098.

Kasting, J.F. and Walker, J.C.G., 1981, Limits on oxygen concentration in the prebiological atmosphere and the rate of abiotic fixation of nitrogen, *J. Geophys. Res.,* **86**, 1147–1158.

Kasting, J.F., Liu, S.C. and Donahue, T.M., 1979, Oxygen levels in the prebiological atmosphere, *J. Geophys. Res.,* **84**, 3097–3107.

Kasting, J.F., Zahnle, K.J. and Walker, J.C.G., 1983, Photochemistry of methane in the earth's early atmosphere, *Precambrian Res.,* **20**, 121–148.

Kasting, J.F., Pollack, J.B. and Crisp, D., 1984, Effects of high CO_2 levels on surface temperature and atmospheric oxidation state of the early earth, *J. Atmos. Chem.,* **1**, 403–428.

Kiehl, J.T. and Dickinson, R.E., 1987, A study of the radiative effects of enhanced atmospheric CO_2 on early Earth surface temperatures, *J. Geophys. Res.,* **92**, 2991–2998.

Kuhn, W.R. and Atreya, S.K., 1979, Ammonia photolysis and the greenhouse effect in the primordial atmosphere of the earth, *Icarus,* **37**, 207–213.

Kuhn, W.R. and Kasting, J.F., 1983, Effects on increased CO_2 concentrations on surface temperature of the early earth, *Nature,* **301**, 53–55.

Levine, J.S., 1982, The photochemistry of the paleoatmosphere, *J. Mol. Evol.,* **18**, 161–172.

Li, Y.-H., 1972, Geochemical mass balance among lithosphere, hydrosphere and atmosphere, *Am. J. Sci.,* **272**, 119–137.

Lovelock, J.E., 1979, *Gaia*, Oxford University Press, New York, 154pp.

Lovelock, J.E. and Margulis, L., 1974, Atmospheric homeostatis, by and for the biosphere: The Gaia hypothesis, *Tellus,* **26**, 1–10.

Madigan, M.T. and Gest, H., 1979, Growth of the photosynthetic bacterium *Rhodopseudomonas Capsulata* chemoautotrophically in darkness with H_2 as the energy source, *J. Bacteriol.,* **137**, 524–530.

McKirdy, D.M. and Hahn, J.H., 1982, The composition of kerogen and hydrocarbons in Precambrian rocks, in *Mineral Deposits and the Evolution of the Bisophere*, edited by H.D. Holland and M. Schidlowski, pp. 123–154, Springer–Verlag, New York.

Margulis, L. and Lovelock, J.E., 1978, The biota as ancient and modern modulator of the Earth's atmosphere, *Pageoph.,* **116**, 239–243.

Melton, C.E. and Giardini, A.A., 1982, The evolution of the earth's atmosphere and oceans, *Geophys. Res. Lett.,* **9**, 579–582.

Newkirk, G., 1980, Solar variability on time scales of 10^5 years to $10^{9.6}$ years, in *Proceedings of the Conference on the Ancient Sun: Fossil Record in the Earth, Moon, and Meteorites*, edited by R.O. Pepin, J.A. Eddy and R.B. Merrill, pp. 293–320, Pergamon, New York.

Newman, M.J. and Rood, R.T., 1980, Implications of solar evolution for the earth's early atmosphere, *Science,* **198**, 1035–1037.

Niklas, K.H. and Brown, R.M. Jr., 1981, Ultrastructural and paleobiochemistry correlations among fossil leaf tissues from the St. Maries River (Clarkia) area, northern Idaho, USA, *Am. J. Bot.,* **68**, 332–341.

Oparin, A.I., 1968, *Genesis and Evolutionary Development of Life*, Academic Press, New York, 203pp.

Owen, T., Cess, R.D. and Ramanathan, V., 1979, Enhanced CO_2 greenhouse to compensate for reduced solar luminosity on early earth, *Nature,* **277**, 640–642.

Phinney, D., 1972, ^{36}Ar, Kr and Xe in terrestrial materials, *Earth Planet. Sc. Lett.,* **16**, 413–420.

Pollack, J.B. and Yung, Y.-L., 1980, Origin and evolution of planetary atmospheres, *Annu. Rev. Earth Planet. Sci.,* **8**, 425–487.

Rossow, W.B., Henderson-Sellers, A. and Weinreich, S.K., 1982, Cloud feedback: A stabilizing effect for the early earth?, *Science,* **217**, 1245–1247.

Rothpletz, A., 1916, Über die systematische Deutung und die stratigraphische Stellung der ältesten Versteinerungen Europas und Nordamerikas mit besonderer Berücksichtigung der Cryoptozoon und Oolithe. Teil II. Über *Cryoptozoon, Eozoon* und *Atikokania, Abh. Bayer. Akad. Wiss.,* **28(4)**, 91pp.

Rubey, W.W., 1951, Geologic history of sea water: An attempt to state the problem, *Geol. Soc. Am. Bull.,* **62**, 1111-1147.

Sagan, C. and Mullen, G., 1972, Earth and Mars: Evolution of atmospheres and surface temperatures, *Science,* **177**, 52–56.

Schidlowski, M., 1982, Content and isotopic composition of reduced carbon in sediments, in *Mineral Deposits and the Evolution of the Biosphere*, edited by H.D. Holland and M. Schidlowski, pp. 103–122, Springer–Verlag, New York.

Schidlowski, M., Appel, P.W.U., Eichmann, R. and Junge, C.E., 1979, Carbon isotope geochemistry of the 3.7 $\times 10^9$-yr-old Isua sediments, West Greenland: Implications for the Archaean carbon and oxygen cycles, *Geochim. Cosmochim. Acta,* **43**, 189–199.

Schlesinger, M.E., 1985, Analysis of results from energy balance and radiative convective models, in M.C. MacCracken and F.M. Luther (eds.), *The potential climatic effects of increasing carbon dioxide*, U.S. Dept. of Energy DOE/ER–0237, Washington, 281–319.

Schopf, J.W., 1980, The origin and Archean evolution of life, in M.H. Hickman (ed.), *The Primitive Earth Revisited*, Oxford, Ohio, Dept. Geol. Miami University, 94–103.

Schopf, J.W., 1981, The Precambrian development of an oxygenic atmosphere, in F.C. Armstrong (ed.), *Genesis of Uranium- and Gold-Bearing Precambrian Quartz-Pebble Conglomerates*, U.S. Geol. Surv. Prof. Pap., **1161**, B1–B11.

Schopf, J.F., (ed.), 1983, *The Earth's Earliest Biosphere*, Princeton University Press, Princeton, N.J., 543pp.

Schopf, J.F. and Oehler, D.Z., 1976, How old are the eurokaryotes?, *Science,* **193**, 47–49.

Schopf, J.W., Hayes, J.M. and Walter, M.R., 1983, Evolution of Earth's earliest ecosystems: recent progress and unsolved problems, Chapter 15 in *Earth's Earliest Biosphere*, (ed. J.W. Schopf), Princeton University Press, Princeton, 361–384.

Simpson, P.R. and Bowles, J.F.W., 1977, Uranium mineralizations of the Witwatersrand and Dominion Reef Systems, *Philos. Trans. R. Soc. London, Ser. A,* **286**, 527–548.

Smith, D.G., 1981, *The Cambridge Encyclopædia of Earth Sciences*, Cambridge University Press, Cambridge, 496pp.

Thomsen, L., 1980, ^{129}Xe on the outgassing of the atmosphere, *J. Geophys. Res.,* **85**, 4374–4378.

Tissot, B.P. and Welte, D.H., 1978, *Petroleum Formation and Occurrence*, Springer, New York, 538pp.

Tregonning, K. and Roberts, A., 1979, Complex systems which evolve towards homoeostatis, *Nature,* **281**, 563–564.

Tucker, M.E., 1982, Precambrian dolomites: Petrographic and isotopic evidence that they differ from Phanerozoic dolomites, *Geology,* **10**, 7–12.

Utter, T., 1980, Rounding of ore particles from the Witwatersrand gold and uranium deposit (South Africa) as an indicator of their detrital origin, *J. Sediment. Petrol.,* **50**, 71–76.

Visconti, G., 1982, Radiative-photochemical models of the primitive terrestrial atmosphere, *Planet. Space Sci.,* **30**, 785–794.

Walker, J.C.G., 1977, *Evolution of the Atmosphere*, Macmillan, New York, 318pp.

Walker, J.C.G., 1978a, Oxygen and hydrogen in the primitive atmosphere, *Pure Appl. Geophys.,* **116**, 222–231.

Walker, J.C.G., 1978b, The early history of oxygen and ozone in the atmosphere, *Pure Appl. Geophys.,* **117**, 498–512.

Walker, J.C.G., 1987, Was the Archaean biosphere upside down?, *Nature,* **329**, 710–712.

Walker, J.C.G., Hays, P.B. and Kasting, J.F., 1981, A negative feedback mechanism for the long-term stabilization of earth's surface temperature, *J. Geophys. Res.,* **86**, 9776–9783.

Walter, M.R., 1977, Interpreting stromatolites, *Am. Scient.,* **65**, 563–571.

Walter, M.R., 1983, Archean stromatolites: evidence of the Earth's earliest benthos, Chapter 8 in *Earth's Earliest Biosphere*, (ed. J.W. Schopf), Princeton University Press, Princeton, 187–213.

Walter, M.R., Buick, R. and Dunlop, J.S.R., 1980, Stromatolites 3,400–3,500 Myr-old from the north pole area, Western Australia, *Nature,* **284**, 443–445.

Watson, A.J. and Lovelock, J.E., 1983, Biological homoeostatis of the global environment: the parable of Daisyworld, *Tellus,* **35B**, 284–289.

Windley, B.F., 1977, *The Evolving Continents*, John Wiley and Sons, London, 385pp.

Wyllie, P.J., 1979, Magmas and volatile components, *Am. Mineral.,* **64**, 469–500.

Zahnle, K.J., 1986, Photochemistry of methane and the formation of hydrocyanic acid (HCN) in the earth's early atmosphere, *J. Geophys. Res.,* **91**, 2819–2834.

DATING PROXY DATA

M. STUIVER
Department of Geological Sciences and
Quaternary Research Center
University of Washington
SEATTLE, WA 98195
U.S.A.

Instrumental records span only a small fraction of the Earth's history. To extend the knowledge of past events, i.e. climate, one resorts to the study of natural processes which are climate dependent. A proxy record of climate can be obtained if a measurable parameter (e.g. an isotope ratio) that records the climate dependence has been preserved.

There are many types of proxy data available, each with its assumptions and measuring limitations. The methods employed have recently been summarized by R.S. Bradley in Quaternary Paleoclimatology (Bradley, 1985). Dating methods are discussed (dating by radiocarbon, potassium-argon, uranium series, thermoluminescence, fission track, amino acid, obsidian hydration, paleomagnetism, lichenometry and dendrochronology) as well as proxy records obtained for ice cores, marine sediments, non-marine geological environments and biological evidence. As it is an impossible task to improve upon Bradley's excellent treatment of the diverse field of dating proxy data, the current discussion is limited to details of the radiocarbon dating method.

The most recent improvements in radiocarbon dating are twofold. A major breakthrough has been the direct detection of ^{14}C in mg samples through accelerator mass spectrometry. In addition, the conventional decay counting technique now delivers an enhanced level of exactness (Stuiver, 1978) that has resulted in a "high precision" radiocarbon age terminology. It is especially the latter improvement that is important for radiocarbon timescale calibration and thus the dating of proxy data.

A conventional radiocarbon age (Stuiver and Polach, 1977) of a sample (expressed in radiocarbon years BP, with 0 yr BP = A.D. 1950) is derived from the ratio of the currently remaining ^{14}C activity to its ^{14}C activity when part of a living organism. The latter activity is estimated by assuming identical ^{14}C specific activity (activity per gram C), after correction for isotope fractionation, for all living organisms over the entire interval for which radiocarbon dating is applicable. Thus the correctness of a radiocarbon age not only depends on the precision and accuracy of the laboratory measurement but also on the extent to which the assumption of constancy of past ^{14}C levels is met. Both factors impart a certain degree of uncertainty to the radiocarbon age,

A. Berger et al. (eds.), Climate and Geo-Sciences, 39–45.

and their relative importance needs to be explored. It will be shown that the search for improved laboratory precision (Stuiver and Pearson, 1986) is partially contained by a natural nemesis producing a fluctuating ^{14}C environment.

The error (one standard deviation) in a radiocarbon activity measurement can be derived from the data obtained for repeat analyses of identical samples. The histogram of repeat analyses gives the precision of the measurement. A portion of the scatter of the repeat analyses is attributable to the statistical uncertainty in the number of decaying ^{14}C atoms (Poisson distribution). Even the laboratory with the most sophisticated instrumentation cannot escape the realities of Poisson counting statistics. Thus the Poisson error in the count rates, when coverted into radiocarbon years, provides a minimum error.

Although most laboratories derive a radiocarbon age standard deviation from Poisson counting statistics, it is clear that this is an unwise choice as it often leads to substantial underreporting of radiocarbon age errors (International Study Group, 1982, Stuiver, 1982). However, even when the entire spectrum of uncertainties is taken into account, it is currently possible to determine ^{14}C activity of young samples with a precision of circa 2 parts per thousand (per mil). This capability leads to the so-called high precision dates with conventional radiocarbon age errors of about ± 20 years for samples less than 5000 years old (Stuiver and Pearson, 1986; Pearson and Stuiver, 1986). These high precision measurements are made on large samples (containing 7 g C or more) counted for several days in order to generate a large number of counts n (the relative Poisson error is $n^{-1/2}$). For smaller samples a two per mil precision is not attainable, and here the radiocarbon age errors will usually be in the 40 to 100 yr range. Accuracy problems (offset between laboratories) can be limited to about ten radiocarbon years for high precision work.

Given the levels of precision in one component of the ratio used in age determinations, we now explore the uncertainty caused by a fluctuating ^{14}C environment.

Because the ^{14}C production in the atmosphere has been variable during the past (Stuiver and Quay, 1980; Sternberg and Damon, 1983; Stuiver et al., 1986), and because the distribution of ^{14}C between the atmosphere and the ocean depends on ocean circulation and climatic change, a variable atmospheric ^{14}C level results. With such changing levels of the $^{14}CO_2$ utilized by organisms, a conventional radiocarbon age is only an approximation of the actual age in calendar years. For a conversion of a radiocarbon age to a more realistic single or plural calendar date, a calibration curve is needed where radiocarbon ages (in radiocarbon yr BP) are plotted versus calibrated ages (cal AD, cal BC or cal BP).

Such calibration curves are obtained by determining the radiocarbon ages of wood samples dated independently by dendrochronological means. High precision curves covering the last 4500 calendar years, with radiocarbon age errors averaging 12 yr, are now available for bi-decadal wood samples (Stuiver and Pearson, 1986; Pearson and Stuiver, 1986).

When converting a radiocarbon age with the aid of a calibration curve, a substantial uncertainty remains in the cal AD or cal BC age because the wiggly nature of the calibration curve often results in multiple intercepts, each of which yields a calibrated date. A continuous spectrum is even possible when atmospheric $^{14}CO_2$ levels decline at the same rate as ^{14}C declines through radioactive decay. All samples formed during such an interval will have the same radiocarbon content at the end of the interval, and thus yield identical radiocarbon ages.

The conversion procedure not only yields single or multiple calibrated AD/BC ages that are compatible with a single radiocarbon age, but also range(s) of calibrated ages that correspond to the standard deviation σ_s in the radiocarbon age. The probability that a certain calibrated age is the actual sample age will vary within the calibrated age range, with higher probabilities to be expected around the intercept ages. Because for most scientific analyses the spread of calibrated AD/BC age groupings is the parameter of overriding interest, the following discussion focuses on calibrated age ranges only.

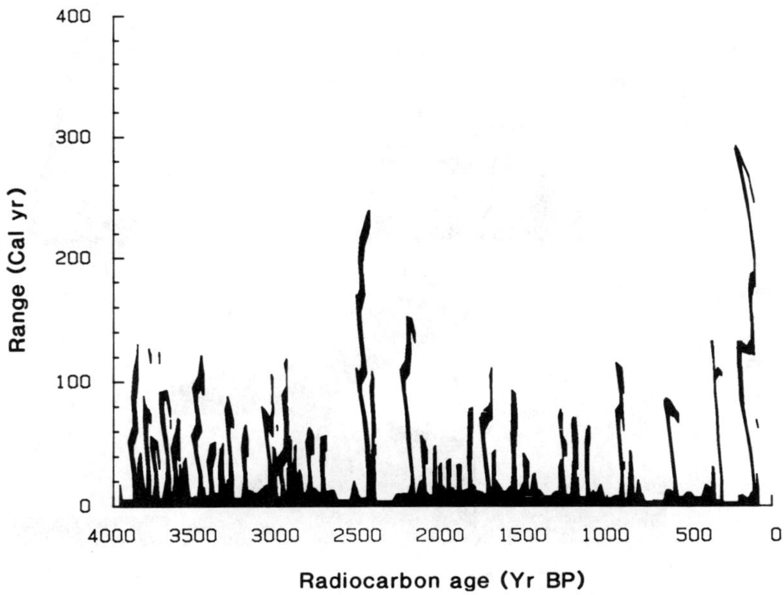

Figure 1. The ranges, in calendar years, obtained from the calibration of radiocarbon ages back to 4000 yr BP; 0 yr BP = AD 1950. The calibration process yields discrete calibrated ages, with a range around these discrete ages that represents the standard deviation in the calibration curve and the sample. The youngest calibrated age of the range(s) obtained for each radiocarbon age was set at zero. The Figure 1 distribution of ranges is for the ideal hypothetical case when the sample standard deviation is zero.

Figure 1 presents graphically the ranges obtained when converting ideal (zero error) sample radiocarbon ages into calibrated ages. Ranges of calibrated ages are produced even for "point source" radiocarbon dates, because the uncertainty in the calibration curve (average standard deviation σ_c of 12 yr, Stuiver and Pearson, 1986; Pearson and Stuiver, 1986) has to be taken into account. The age conversion yields for a specific radiocarbon age either a single range of calibrated ages, or two or more ranges. In Figure 1 these ranges are plotted on the vertical scale relative to the youngest calibrated date, i.e. the youngest calibrated age is set at zero. When the ranges are single and small (Fig. 1), the linear portions of the calibration curve are used where the cal AD/BC ages span an interval comparable in magnitude to the error in the calibration curve. When the wiggly nature of the curve is encountered, the differences between maximum and minimum calibrated age increase substantially. The more detailed fine structure (Fig. 2) of selected Figure 1 "snakes" reveal some surprising configurations.

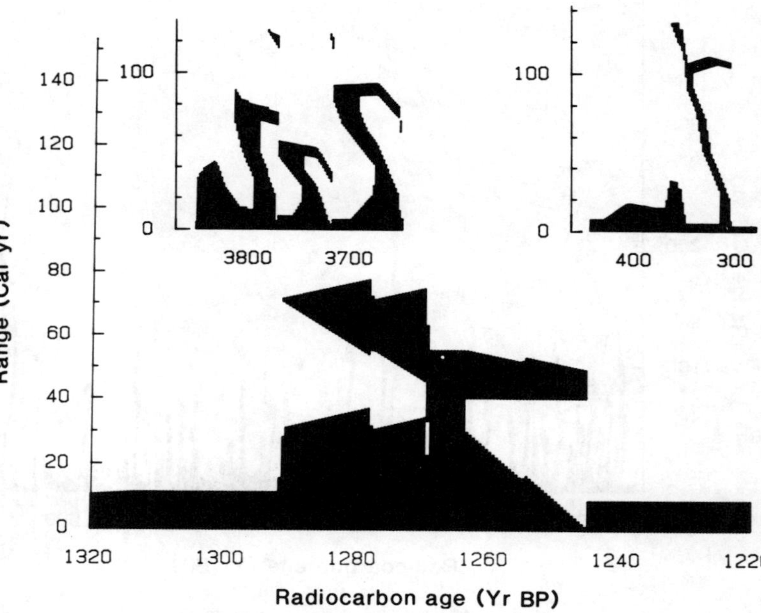

Figure 2. Detailed range distribution of some of the Figure 1 configurations.

The ranges of calibrated ages (defined here as R) given in Figures 1 and 2 are the smallest possible because they are derived for sample radiocarbon ages with presumed zero error in the laboratory determination. Ranges of calibrated ages, similar to those of Figure 1, can also be given for radiocarbon ages with a standard deviation σ_s. Here the radiocarbon age-calibrated age conversion takes into account the standard deviation σ_c in the calibration curve and the σ_s in the sample date. The ranges of calibrated ages are obtained by converting the radiocarbon age with ± boundaries of $\sqrt{\sigma_s^2 + \sigma_c^2}$ (see Stuiver and Pearson, 1986 for further discussion). To compare the multitude of Figure 1 type graphs derived for age errors σ_s of 20, 40, 60 etc years, the parameter $\overline{R}$ is defined as the average range of the maximum and minimum calibrated age encountered for the ± σ_s error interval of individual radiocarbon ages over a 4000 ^{14}C year interval. The $\overline{R}$ value for $\sigma_s = 0$ (Fig. 1) is 48 yr. Of course, $\overline{R}$ values increase with increased standard deviation σ_s in the radiocarbon age. Figure 3 gives this relationship (squares and upper boundary of shaded area). The $\overline{R}$ values are plotted versus 2 σ_s which is the "range" between maximum and minimum radiocarbon age in the ± σ_s error interval.

The Figure 3 $\overline{R}$ values take into account the standard errors in the radiocarbon age and calibration curve, as well as the wiggly nature of the calibration curve. If the relationship between radiocarbon ages and calibrated ages had been ideal, i.e. a one-to-one linear relationship, the results would have resembled the lower curve in Figure 3. Here the hypothetical one-to-one relationship between radiocarbon years and calibrated years yields $\overline{R} = 2\sqrt{\sigma_s^2 + \sigma_c^2}$.

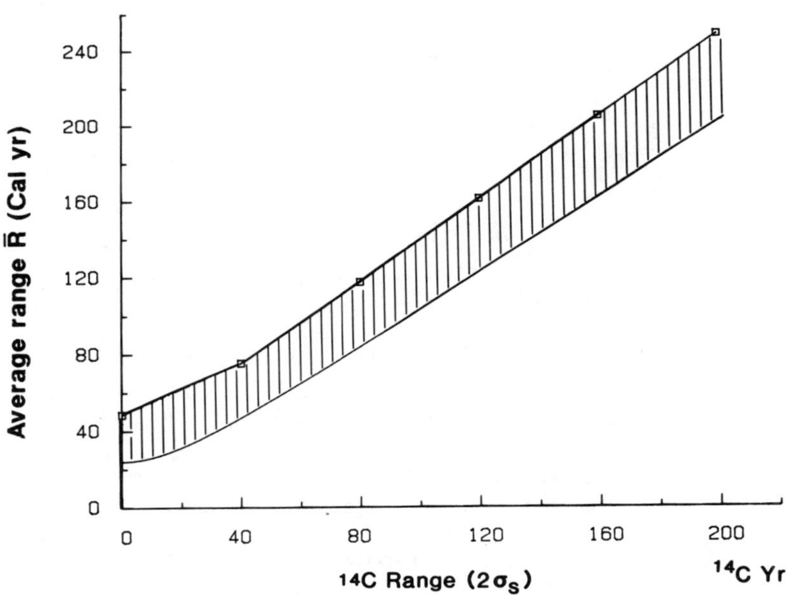

Figure 3. Average difference $\overline{R}$ between maximum and minimum calibrated ages encountered for a 4000 ^{14}C year interval. R is plotted versus the 2 σ_s range in the radiocarbon age. A fluctuating ^{14}C environment increases $\overline{R}$ (shaded area).

The $\bar{R}$ increment caused by the wiggly nature of the calibration curve is given by the shaded area in Figure 2. For a high precision date ($2\sigma_s$ = 40 yr) $\bar{R}$ is about 1.6 times the range expected from the hypothetical linear relationship between radiocarbon years and calibrated years. For larger standard deviations in the radiocarbon age, the influence of the wiggles in the calibration curve becomes relatively less important.

Although average R is substantially increased beyond that expected from the measuring precision alone, there are specific intervals where R values are not influenced by the wiggles in the calibration curve. In fact, even time contraction is possible when converting from radiocarbon yrs to calendar yrs. The histogram of R values derived from converting a 4000 yr stretch of single radiocarbon years into calibrated AD/BC ages is given in Figure 4. For radiocarbon age errors σ_s of 20, 40, and 80 yr R values are below or equal to $2\sigma_s$ in respectively, 36, 28 and 28 percent of all cases.

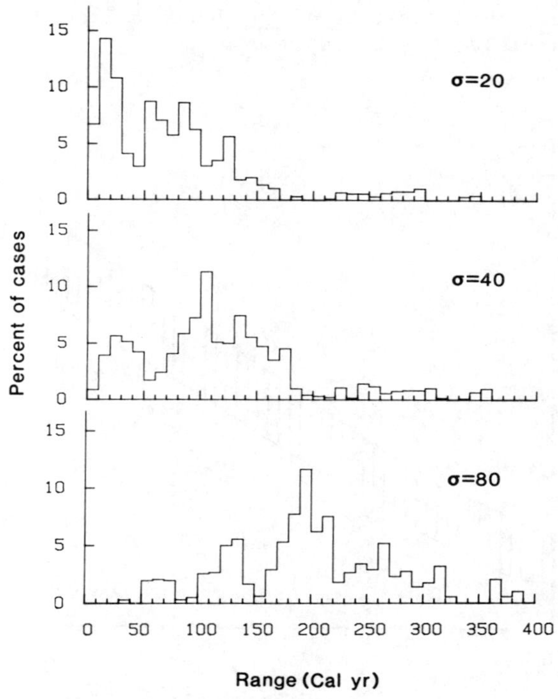

Figure 4. Histograms of age differences R between maximum and minimum calibrated ages for a 4000 yr interval of ^{14}C ages. σ_s is the standard error in the sample radiocarbon age determination.

It should be noted that R values denote differences between the extreme values in the range of calibrated ages. When converting the radiocarbon age of a specific sample, more detailed information will be available from the probability distribution around the intercept ranges (a floppy disk computer program to calculate these distributions can be obtained from the Quaternary Isotope Laboratory) and often only part of the possible ages between the extrema will have a finite probability. The additional information obtained in this manner is not taken into account by the above R analysis.

As seen above, the range R of possible AD/BC ages is enlarged by a fluctuating ^{14}C environment in about two thirds of the cases where high precision dates (σ = 20 yr) are applied. Yet, R values keep declining with improved precision (Figure 3). Thus, although nature does not allow the enjoyment of the fruits of high-precision dating to the fullest extent, the situation is not desperate because additional improvement in measuring precision will still be beneficial.

The ^{14}C research reported here was supported by NSF grants ATM-8318665 and BNS-8701720 to the University of Washington.

REFERENCES

Bradley R.S., 1985. Quaternary Paleoclimatology, Allen and Unwin, Boston, 472pp.
International Study Group, 1982. Nature, **298**, 619-623.
Pearson G.W. and Stuiver M., 1986. Radiocarbon, **28**, 839-862.
Sternberg R.S. and Damon P.E., 1983. Radiocarbon, **25**, 239-248.
Stuiver M., 1978. Nature, **273**, 271-274.
Stuiver M., 1982. Radiocarbon, **24**, 1-26.
Stuiver M. and Pearson G.W., 1986. Radiocarbon, **28**, 805-838.
Stuiver M., Pearson G.W. and Braziunas T.F., 1986. Radiocarbon, **28**.
Stuiver M. and Polach H.A., 1977. Radiocarbon, **19**, 355-363.
Stuiver M. and Quay P.D., 1980. Science, **207**, 11-19.

THE SPECTRAL CHARACTERISTICS OF PRE-QUATERNARY CLIMATIC RECORDS, AN EXAMPLE OF THE RELATIONSHIP BETWEEN THE ASTRONOMICAL THEORY AND GEO-SCIENCES

A. BERGER
Université Catholique de Louvain
Institut d'Astronomie et de Géophysique G. Lemaître
2 Chemin du Cyclotron
B-1348 Louvain-la-Neuve, Belgium

ABSTRACT. Proxy climatic records now available since the mid-Palaeozoic (400 million years ago) show clear rhythmicities which can be related to the main astronomical periods. An explanation of the relationship between these spectra of orbital and geological data is tentatively given in terms of the linear or non-linear characteristic of the response of the climate system to the astronomical forcing. This unification of the astronomical theory for all geological time scales is based on the observation that the main signal in any monthly insolation for any latitude is precession, followed by obliquity which is mainly felt in high latitudes in non-equinoctial seasons and by eccentricity which modulates the yearly amount of energy received by the whole Earth. The eccentricity is also shown to play an important role by itself at some particular times of the year, namely the perihelion and the aphelion.

1. INTRODUCTION

Until 1970 roughly, the Milankovitch theory (Milankovitch, 1941; Imbrie, 1985; Berger, 1988a) was largely disputed, with discussions based on fragmentary geological records supported by incomplete and frequently incorrect radiometric data. The accuracy of the astronomical parameters and of the related insolation fields was not known, and the climate was considered too resilient to react to "such small changes" as observed in the summer half-year caloric insolation of Milankovitch.

In the late 1960's, judicious use of radioactive dating and paleomagnetic techniques gradually clarified the Pleistocene time scale (Broecker et al., 1968). Better instrumental methods came on the scene by for determining oxygen isotopes for ice-age foraminifera relics (Emiliani, 1966), ecological methods of core interpretation were perfected (Imbrie and Kipp, 1971), global climates in the past were reconstructed (CLIMAP, 1976, 1981) and atmospheric general circulation models and climate models became available (Kutzbach, 1985; Crowley, 1988). With these improvements in dating and interpreting geological data in terms of paleoclimates, with the increased use of computers and the de-

A. Berger et al. (eds.), Climate and Geo-Sciences, 47–76.

velopment of astronomical and climate models, a more critical and deeper investigation of the astronomical theory as a whole became necessary. It is such a multidisciplinary systematic study, using powerful modern techniques, of the Earth's orbital elements, of insolation, of climate and of geological data which has brought, mainly since 1975, some major findings confirming progressively the astronomical theory.

These are : 1. the bipartition of the precessional period; 2. the amplitude of the seasonal cycle as described by the daily or monthly insolation values; 3. the diversity of the classical astronomical frequencies in geological data in relation to the diversity of such frequencies in the different insolations available all over the Earth and throughout the year; 4. the phase coherency between the climate proxy data and the orbital/insolation parameters in these frequency bands; 5. the discovery of secondary astronomical periods in some geological records; 6. the presence of the astronomical signal in the Pliocene-Miocene times and in the pre-Cenozoic; 7. the high frequencies as combination tones of astronomical ones; and 8. the success of reproducing climate proxy data in both the frequency and time domains using a hierarchy of climate models, from ice sheets to snapshot general circulation and transient coupled models of the whole climate system.

These findings were reviewed in particular by Imbrie (1985) and Berger (1987, 1988a and 1988b) who has underlined their importance for geo-sciences. The emphasis will be put here on the frequency domain trying to see whether or not it is logical to have some specific frequencies recorded in particular climatic parameters for specific times covering most of the Palaeozoic, Mesozoic and Cenozoic era.

2. ASTRONOMICAL FREQUENCIES IN MIDDLE AND LATE QUATERNARY PALEOCLIMATES

In 1976, Hays, Imbrie and Shackleton demonstrated from spectral analysis of the climate sensitive indicators in selected deep-sea records that the astronomical frequencies (corresponding to the 100, 41, 23 and 19 kyr periods) are significantly superimposed upon a general red noise spectrum. It is the geological observation of the bipartition of the precessional peak (23 and 19 kyr were found instead of the usual 21 kyr), confirmed in the astronomical computations made independently by Berger (1977a), which was one of the first most delicate and impressive of all tests for the Milankovitch theory.

Since 1976, spectral analysis of climatic records of the past 800 kyr or so, has provided substantial evidence that, at least near the frequencies of variation in obliquity and precession, a considerable fraction of the climatic variance is driven in some way by insolation changes accompanying the perturbations of the Earth's orbit (Imbrie and Imbrie, 1980; Berger, 1987). For example, if some long deep-sea cores are used and the uncertainty in the geological time scale and in the spectral analysis is accounted for, Berger and Pestiaux (1987) have shown that the following peaks are significantly present in the astronomical bands: 103 kyr with a standard deviation of 24 kyr, 42 kyr with a standard deviation of 8 kyr, and 23 kyr with a standard deviation of 4 kyr.

As already shown by Hays et al. (1976), the variance components centered near the 100 kyr cycle which dominates most upper Pleistocene climatic records, seem to be in phase with the eccentricity cycle (high eccentricity at low ice volume). It is worth here to remember that Milankovitch requested a high eccentricity for an ice age to occur - which is just the reverse of the correlation claimed here. Considering only the caloric insolations developed to the first degree of eccentricity, he viewed indeed the effect of the eccentricity through the precessional parameter alone. However, if a higher degree of accuracy is used, eccentricity appears independently. For example, the total energy W_E received globally by the Earth over the whole year (or a fraction of it) is inversely proportional to $\sqrt{1-e^2}$:

$$W_E = \frac{S_a}{\sqrt{1-e^2}} \pi R^2 T$$

where R is the radius of the Earth and T the invariable length of the year. S_a is the solar constant, S_0, reduced to the invariable semi-major axis of the ecliptic :

$$S_a = S_0 \left(\frac{r_m}{a}\right)^2 = S_0 \sqrt{1-e^2}$$

As S_0 varies because both the output from the Sun and the mean distance, r_m, from the Earth to the Sun vary :

$$r_m^2 = a^2 \sqrt{1-e^2}$$

a fixed value S_a is generally used for the whole Neogene (1353 Wm^{-2}). Although the absolute effect of this factor $(1-e^2)^{1/2}$ is relatively small (a few per mil at the most), this term increases the annual global insolation at high eccentricity (for example, at e = 0.075, the total solar energy received by the Earth is increased by 0.3 %) and decreases it during low eccentricity, a result coherent with the recent findings. Seemingly $(1-e^2)^{-2}$ appears as a factor in all daily insolations, W, at all latitudinal bands, reflecting the full equation of the elliptical motion and the variation of r in terms of a :

$$W = S_a \frac{(1 + e \cos v)^2}{(1-e^2)^2} \langle \cos z \rangle$$

where v is the true anomaly and $\langle \cos z \rangle$ the mean zenith distance of the Sun over the day. The long-term variation of $\langle \cos z \rangle$ is driven by the obliquity whereas the longitude of the perihelion, $\tilde{\omega}$, appears in $v=\lambda-\tilde{\omega}$, where λ is the longitude of the Earth. Among all possible values of $v (0 \leq v \leq 2\pi)$, the following ones illustrate the total effect of the eccentricity alone :

$$\text{at the perihelion : } W^P = S_a \frac{1}{(1-e)^2} < \cos z >$$

$$\text{at the aphelion : } W^A = S_a \frac{1}{(1+e)^2} < \cos z >$$

$$\text{at orthogonal distances } (v = \frac{\pi}{2} \text{ or } \frac{3\pi}{2}) : W = S_a \frac{1}{(1-e^2)^2} < \cos z >$$

Taking e = 0.075 as a maximum value of e, the factors containing e vary with a maximum amplitude of respectively 17, 14 and 1%. Although it is interesting to see that these insolations are only functions of obliquity and eccentricity and not of precession, one must recognize that the perihelion drifts continuously through all the seasons according to the long term variation of its longitude. The physical mechanisms through which the climate system would response to such a forcing are thus very difficult to conceive. Nevertheless, these examples illustrate how complex the relationship between the insolations and the orbital parameters is. It also underlines the importance that the eccentricity can have individually; terms containing e alone can, therefore, not be ignored any more, especially since they reinforce the impact on climate in the explanation of the strength of 100 kyr cycle.

The exceptional strength of this cycle calls, however, for a sensitivity of the climate system to one (or more) of these insolation parameters having large variations driven by e (this paper), or for a stochastic (Hasselman, 1976; Kominz and Pisias, 1979) amplification of the insolation forcing, or for a non-linear amplification through the deep ocean circulation, the carbon dioxide, the ice sheet related mechanisms and feedbacks, the isostatic rebound of the elastic lithosphere and of the viscous mantle and the ocean-ice interactions. The 100 kyr climatic cycle can indeed be explained both (i) from the eccentricity signal directly, provided an amplification mechanism can be found (as in the double potential theory of Nicolis 1980, 1982 and Benzi et al. 1982); and/or (ii) by a beat between the two main precessional components as shown from non-linear climate theory (Wigley, 1976) and modeling experiments (Ghil and Le Treut, 1981; Le Treut and Ghil, 1983). It must be stressed here that the same arises for the 100 kyr eccentricity cycle in celestial mechanics : the frequency corresponding to the second period of the eccentricity (94945 years in Table 3 of Berger (1977a)) is obtained from precession frequencies number 3 and 1 in Table 2 of the same reference (1/18976 - 1/23716). The other ones (number 1 and 3 to 6) are coming respectively from the following combinations : 2-1, 3-2, 4-1, 4-2 and 3-4 (more details are available in Berger et al., 1987).

However, Imbrie and his collaborators (1984) have made it clear that the coherency of orbital and climatic variables in the 100 kyr band is enhanced significantly when the geologic record is tuned precisely to the obliquity and precession bands. Using the so-called orbitally-tuned SPECMAP time scale (refined by Martinson et al., 1987, for the last 300,000 years), they indeed found a coherency in the astronomical bands significant at more than 99% level of confidence. This supports the second mechanism (beat) rather than the first !

A fairly coherent phase relationship was also reasonably well defined between insolation and ice volume in Kominz and Pisias (1979) where obliquity consistently lead the ^{18}O record by about 10,000 years, whereas precession seemed to be in phase with the 23 kyr geological signal. However, the recent results obtained by SPECMAP show that these leads and lags are more complicated. CLIMAP (1976) and more recently the SPECMAP (Imbrie et al., 1984) teams have indeed shown that the phase lags in the climate response to orbital forcing depend upon the nature of the climatic parameters themselves and upon their geographical location. For example in their data, the sea-surface temperature of the southern oceans seems to lead the response of the northern hemisphere ice sheets by roughly 3,000 years, and the results of the SPECMAP Atlantic study reveal the lead of southern hemisphere responses ahead of the northern hemisphere and highest sensitivity of the high-latitude North Atlantic in all astronomical bands (Imbrie et al., 1989).

Moreover, the shape of the spectrum depends also upon the location of the core and the nature of the climatic parameter analysed (Hays et al., 1976). For example in core V30-97, the 41 kyr cycle is not seen at all whereas the 23 kyr cycle is dominant in Atlantic summer sea-surface temperatures of the last 250,000 years (Ruddiman and McIntyre, 1981; Imbrie, 1982). This is not too surprising as these spectra depend upon the way the climate system reacts to the insolation forcing and upon which type of insolation it is sensitive too. Indeed, contrary to the well-received Milankovitch idea that the high polar latitudes must record the obliquity signal (as shown in the Vostok core, for example, Jouzel et al., 1987) whereas low latitudes record only the precessional one, the latitudinal dependence of the insolation parameters is more complex. Clearly the mid-month high-latitude summer insolation displays a stronger signal in the precession band than in the obliquity one (Berger and Pestiaux, 1984).

Besides the usual 100, 41, 23 and 19 kyr-periods which characterize most of the late Pleistocene data, Ruddiman et al. (1986) succeeded in finding in the geological records one of the secondary astronomical periods that was already predicted in 1976, the 54 kyr one (Berger, 1977a). A similar period of 58 kyr was found in a 400 kyr record of the paleomagnetic field from Summer Lake in South-central Oregon (Negrini et al., 1988). Finally, combination tones have also been found in deep-sea cores with a high sedimentation rate (Pestiaux et al., 1988) and in the Vostok ice core over the last 140,000 years (Yiou et al., 1989).

3. EARLY PLEISTOCENE RECORDS

Unfortunately, the interpretation of the data is not always as clear. The 100 kyr cycle, so dominant a feature of the late Pleistocene record, does not exhibit a constant amplitude over the past 2-3 million years as displayed on a 3-dimensional time evolutionary spectrum (Pestiaux and Berger, 1984a) of deep-sea core V28-238 (Shackleton and Opdyke, 1973 and 1976). Clearly, this periodicity disappears before 10^6 years ago, at a time the ice sheets were much less developed over the Earth, reinforcing the idea that the growth of the major ice sheets may have played a role in the modulation of the 100 kyr cycle.

Although, the most important term of a 412 kyr period in the eccentricity was already predicted by Berger in 1976, it is only two years later that appropriate statistical techniques and geological time series, long enough to discover such a periodicity, became available. Shackleton suggested this periodicity in 1978 and Briskin and Harrell (1980) using 2-million year sediment cores from both the Atlantic and the Pacific found it in the oxygen isotopic record of planktonic foraminifers, in coarse sediment fraction and in the magnetic inclination (which lead them to propose that a relationship may exist between the eccentricity and the Earth's core modulation of the magnetic field). Kominz et al. (1979) reported it also from two spectra covering the 730,000-year long records and the investigation by Moore et al. (1982) of calcium carbonate concentrations in equatorial Pacific sediments (core RC11-209), have shown that the Pacific carbonate spectrum has been dominated for the past 2 million years by variance in the 400 kyr band, with more modest contributions in the 100 kyr and 41 kyr bands (matching the variations of respectively the eccentricity and obliquity).

The sedimentological record of DSDP sites in the Norwegian sea (Jansen et al., 1988) and North Atlantic (Shackleton et al., 1984; Shackleton and Hall, 1985) now permit documentation of the climatic evolution over the last 2.8 Myr. A major expansion of the Scandinavian ice sheet is clearly shown to have occured at about 2.56 Myr and relatively severe glacials appeared until about 2 Myr. Between 2 and 1.2 Myr the glaciations in Scandinavia were small but as indicated by the Matuyama records the orbital forcing by the obliquity factor mainly was able to produce small to medium sized northern hemisphere ice sheets after a change in the climatic boundary conditions had made this possible at around 2.5 Myr. A transition toward larger glacials took place during the period 1.2 to 0.6 Myr. A new change in boundary conditions at about 0.9 Myr increased indeed meridionality of the northern hemisphere climate system, as suggested by Ruddiman et al. (1986) to made even larger ice sheets possible. Thus the appearance of the 100 kyr dominance in climate spectra most likely was a response to such a change which promoted growth of glaciers at lower latitudes than before, probably by a mechanism bearing on increased precipitation from warm oceans and decreased summer ablation on chilled continents. During the Brunhes, larger climatic amplitudes were caused by glacials and interglacials respectively colder (larger ice sheets) and warmer (smaller ice sheets) than in the Matuyama.

4. DATA FROM THE PLIOCENE AND MIOCENE (NEOGENE)

Very interestingly, with the existence of a few deep-sea cores extending through the first appearance of abundant ice rafted debris in North Atlantic, 2.5 million years ago, this most important eccentricity term of a period of 412 kyr, was confirmed in the geological record. Using cross-spectral analysis, Moore et al. (1982) discovered that the 400 kyr eccentricity and sedimentary cycles were in phase over the last 8 million years, but the 100 kyr cycle which dominated climatic variability during the Pleistocene ice ages (29% of the total variability) had only a minor effect 5 to 8.5 Myr ago, where it accounted for 6 times less variability than as today at the DSDP site 158. Prell (1982) also failed to find evidence of a strong 100 kyr cycle in pre-Pleistocene sediments at the DSDP site 502 in the western Caribbean and at the site 503 in the eastern equatorial Pacific. He did apparently find evidence of the 41 kyr cycle in a 7 million-year sediment record, as well as the evidence of a 250 kyr cycle.

Spectral analysis of DSDP Hole 552A reveals also a dominant quasi-periodicity associated with obliquity-induced temperature variations in surface water and weaker peaks at the eccentricity and precession periods (Backman et al., 1986). More precisely, Shackleton et al. (1988) found a 100 kyr-peak for the period 0 to 1 Myr BP, a 100 kyr and a 41 kyr-peaks for 0.5 to 1.2 Myr BP and mainly a 23 kyr-one for 2.5 to 3.5 Myr BP, this section displaying however significantly less variance. In the Mediterranean Pliocene, rhythmic lithological variations in the Trubi and Narbone Formations of Sicily and Calabria show cycles that could be related to precession and eccentricity (Hilgen, 1987). In particular, the precession cycle corresponds well with the mean duration of the deposition of basic rhythmites, which comprise small-scale, bipartite or quadripartite marl rhythms in the Trubi and individual sapropelitic layers in the Narborne Formation. The 100 and 400 kyr cycles of eccentricity would match the average duration of both layer-scale marl rhythms in the Trubi, which approximately comprise 5 and 20 basic rhythmites respepctively, as well as small-scale and large-scale clusters of sapropelitic layers in the Narbone Formation. However, consistent discrepancies in the Rossello composite section (Pliocene of Sicily, Hilgen and Langereis, 1989) could argue for an age re-assignment of the Gilbert and Gauss chrons based on phase relations between the sedimentary and supposedly corresponding orbital cycles. The entire upper Miocene to Holocene sedimentary sequence recovered in a hydraulic piston core at DSDP site 532 on Walvis Ridge, which is under the seaward edge of the Benguela upwelling system, shows distinct cycles in amount of $CaCO_3$ that correlate with dark and light cycles of sediment color, which origin may be fluctuations in the Benguela Current caused by variations in the Earth's obliquity (Meyers et al., 1986). The average periodicities of these carbonate cycles for the Quaternary, upper Pliocene and lower Pliocene are about 35, 46 and 28 kyr respectively, with an overall average of about 36 kyr for the last 5 Myr (Dean and Gardner, 1984).

In fact, the majority of the cycles shown in Neogene pelagic carbonate sediments from the world ocean basins have periodicities in the range of 20 to 60 kyr, but within the limits of dating techniques, these periodicities are not constant through time. Oligocene to Miocene cycles at DSDP site 366 have an average periodicity of about 44 kyr, while the Eocene cycles average 19 kyr. The winnowing cycles from site 586 (western equatorial Pacific) have maximum periodicities of about 60 kyr in the lower Quaternary, minimum periodicities of about 40 kyr in the upper Pliocene and maximum periodicities of 70 kyr in the lower Pliocene and upper Miocene. Moreover the temporal changes in periodicity of these winnowing cycles are almost exactly opposite to those of the carbonate cycles from sites in the South Atlantic, Caribbean and eastern equatorial Pacific (Dean and Gardner, 1986). Such changes in periodicities of sedimentary cycles with time may be entirely real or may, at least in part, be due to uncertainties in the Neogene time scale.

Benthic oxygen isotopic records in the latest Miocene/earliest Pliocene at DSDP site 588 (SW Pacific) and other cores from the N and SW Atlantic (Hodell et al., 1986) are marked by high frequency changes in $\delta^{18}O$ ratios that resemble Quaternary glacial/interglacial cycles, except that the amplitude of these cycles is approximately 1/3 the late Quaternary signal and the wavelength is about 400 kyr. As similar patterns have been observed in six different sites located at different depths and in different ocean basins, this cyclicity may reflect the natural period of glacial-interglacial oscillation in Antarctic ice volume during the late Miocene. Moreover, Hodell et al. (1986) have also suggested a strong correspondance between the trends in the oxygen isotopic data and the deposition of the Messinian evaporites in the Mediterranean. Tiwari (1987) used Walsh spectral analysis to examine the cyclicity in these $\delta^{18}O$ data from Site 588. He showed a dominance of low-frequency power with spectral peaks at approximately 2,000, 1,250, 800, 400 and 93 kyr. However, a number of these periods are very close to the aliased periods predicted for precession sampled at 25,000 year intervals by Pisias and Mix (1988), which means that aliasing can not be excluded.

Time spectral analysis has been performed also on the dates of geomagnetic reversals of the last 20 Myr (Stothers, 1987). Possible evidence is found from the presence of high spectral peaks for two very long periodicities, 400 kyr and 1,300 kyr, that may be associated with variations of the Earth's orbital eccentricity as predicted by Berger (1977a), although a number of tests do not confirm their statistical significance.

This Miocene-Pliocene time scale, thus requires our most urgent attention. The upper Pliocene ($\sim$ 5 Myr BP) is beyond the limit of validity of the astronomical calculations by Berger (1978) as far as the time domain is concerned (Berger, 1984). However, there is still a large confidence in the value of the astronomical frequencies, in such a way that conclusions from a comparison between geological and astronomical data may still be convincingly drawn in the frequency domain. Moreover, a new astronomical solution valid over the last 10 Myr is now underway (Berger et al., 1988). It is thus challenging to see to what extent may the methodology developed for the Mid- and Upper-Pleistocene (Berger, 1987) be extended to Early Pleistocene, Pliocene and Miocene.

5. THE MIDDLE AND EARLY TERTIARY

Late Oligocene evaporite formations in the Rhinegraben deposits, France, are characterized by alternations of thick halitic strata and intervals with thin levels of anhydrite and clay (Foucault and Blanc-Valleron, 1989). Spectral analysis of the thickness of anhydrite laid down during 1000 years shows a period of 20.5 kyr very close to the present average Earth's precession cycle. Another Oligocene shelf sequence, The Boom

Clay Formation in northeastern Belgium has been analysed by van Echelpoel and Weedon (1989). The time series of grain size and thickness of the alternating laterally continuous beds of silt annd clay were dated from planktonic foraminiferal zones. Over a time span extending from 31.6 to 34 Myr BP, a cyclicity of less than 120 kyr was found which may reflect the eccentricity cycle.

In 1929, Bradley used varve counts of the Eocene Green River Formation in the Rocky Mountains to identify a precessional cycle in the alternation of oil shale and dolomitic marlstone characteristic of humid and arid climates, respectively. The rhythms described by Bradley from the upper Parachute Creek Member can be traced from Colorado to Utah. Furthermore, they seem grouped into bundles of 3-6 couplets just as precessional events are grouped by the 100 kyr cycle of eccentricity. These bundles in turns seem grouped into large sets of 3.5 to 5 bundles reflected in the oil-yield potential of the constituent "oil shale" beds. These superbundles would therefore reflect the 400 kyr eccentricity cycle. Crude inspection does not reveal an obliquity cycle but the complexity of the patterns suggest the interference of a number of different waves which can only be properly revealed by further research on the time scale (Fischer, 1986).

6. THE MESOZOIC (Cretaceous, Jurassic, Triassic)

A century ago, Gilbert (1895) suggested that fluctuations of climate and their record in sedimentary successions might be due to the influence of astronomical parameters, mainly precession. On such a basis, he gave an estimate of the timespan of an Upper Cretaceous pelagic sequence in Colorado that fits remarkably well with the results of modern radiometric datings (Fischer, 1980). His concept was based on the work of astronomers, such as Newton (1643-1727) and d'Alembert (1717-1783) who had discovered the precessional cycle and Le Verrier (1811-1877) who was the first to compute the long-term variations of the Earth's orbital elements, works on which Adhémar (1842) and Croll (1875) based their theories on the Quaternary ice ages (Imbrie and Imbrie, 1979; Berger, 1988a).

Despite Gilbert's great challenge, it is only since twenty years roughly that the geologists under the leadersip of W. Schwarzacher and A.G. Fischer (see their early works 1948 and 1964 respectively) start to recognize the effects of orbital cycles on climates and sediments also for pre-Quaternary times (Einsele and Seilacher, 1982; de Boer and Shackleton, 1984; Arthur and Garrison, 1986a; Fischer et al., 1988).

6.1. The Cretaceous

Marine foraminiferal oxygen isotope ratios turned out to be monitors of global ice volume and the enormous sets of data now collected and analyzed by time series methods leave no doubts that the Quaternary glacial regimes were forced by the orbital periodicities. Just how the variations in insolation patterns influence climate remains an active topic of research, but the general consensus is that in glacial times the climatic effects of insolation patterns are strongly amplified and modified by the albedo of the expanding ice sheets and by other feedback mechanisms. But did orbital variations drive climatic oscillations in non-glacial times, like in the Cretaceous without such amplification ?

It seems so, at least from accumulating geological evidences that the orbital variations were linked to climate at periods shorter than 100 kyr during the past few hundred million years at times when major ice masses were probably absent. Rhythmicity is indeed a pronounced characteristic of nearly all Cretaceous and Cenozoic pelagic carbonate sequences whether deposited in shallow or deep-water environments. Intercalations of carbonate-rich and carbonate-poor beds, on the order of tens of centimeters thick, are common in environments where there is also an abundant supply of clay. In anoxic depositional environments, intervals of oxygen-depletion are periodic as well, and enrichment of organic carbon occurs in the relatively clay-rich intervals. In highly compacted pure chalks, the rhythmicity is reflected by decimeter-thick limestone beds separated by thin stylolitic seams (Arthur et al., 1984). The estimated average periodicities of all most obvious types of cycles in Cretaceous- Cenozoic pelagic strata usually are either about 20 kyr or 40 kyr; groups of beds (bundles) often form less obvious cycles with periods of 100 kyr. However, studies of cyclicity in the more ancient record are hampered by several significant problems. Diagenesis has altered the primary chemistry and stratigraphic relationships across cycles; the absolute time scales used for estimating the duration of cycles become more imprecise with increasing age; the periodicities of orbital variations are changing with time; and suitable mechanisms for coupling variations in temporal and latitudinal receipt of solar insolation to sedimentation changes are difficult to find (Arthur and Garrison, 1986b).

In order to by-pass this difficulty, three approaches are used to measurement of rhythm periods in pre-Quaternary geological records : varve counts, direct radiometric determinations and calculation of mean sedimentation rates using the timing assigned to stages or other units by means of radiometric scales. Only varve counts have provided so far sufficient accuracy to answer the question of whether the periodicities of ancient cycles are those of the orbital variations. Radiometric dates and extrapolated periodicities are limited by uncertainties but other factors come also to one's help (Fischer, 1986). The character of the precession index is indeed such that precessional cycles must wax and wane in strength with the eccentricity. Furthermore if the full range of precession cycles is recorded, the number of precessional events in an eccentricity group must lie between 3 and 7 averaging a fraction under 5, according to the length of the precession cycles which extend from 13,900 years to 31,300 years with a mean of 21,740 years, as computed by Berger (1976) for the last 5 Myr. On the other hand, the obliquity signal is more steady and can be expected to yield simple, regular sedimentary oscillations. Thus the distinctive patterns come to be useful criteria for the identification of driving cycles. However, we must also consider the possibilities of other Milankovitch cycles, of their variations in time, of additional cycles induced by the response of the Earth (not only its climate but also its geological environment) to Milankovitch forcing and of cycles unrelated to the orbital parameters.

For testing the astronomical hypothesis from the Late Triassic until the Miocene, the Umbrian pelagic facies exposed in the Appennines of central Italy contains a remarkable sequence of pelagic sediments extending indeed from the mid-Jurassic to the mid-Tertiary (Alvarez and Montanari, 1988). Rhythmicity has been investigated in the Cretaceous portion of this sequence. From a time series analysis of the variations in bedding thickness in the Maiolica and Cenomanian Scaglia Bianca (basal Late Cretaceous) limestones, Fischer and Schwarzacher (1984) arrived at a precession timing for the couplets and at a 100 kyr for the bundles. Studying the upper Fucoid Marls (Upper Albian and Cenomanian part of this Umbria-Marche sedimentary sequence), de Boer (1982, 1983) and de Boer and Wonders (1984) found the precession cycle for the marl-limestone bedding couplets, while bundles of approximately 5 couplets record the 100 kyr eccentricity, and superbundles of 4 bundles record the 400 kyr cycle. The tie of these oscillations to the orbital cycles was confirmed by the detailed analysis of this Fucoid Marls, a typical black shale formation located at Piobbico (Italy). Carbonate production in these pelagic Aptian-Albian (mid-Cretaceous) sediments quantified by calcium carbonate and optical densitometry time series, assuming a constant sedimentation rate of 5 mm/kyr, reflects the orbital eccentricity and precessional cycles (Herbert and Fischer, 1986). Planktonic foram abundance and the nature of color of the rocks allowed recognition of periods close to the obliquity and precessional ones (Tornaghi et al., 1988). Spectral analysis of two other series (percentage carbonate content and photodensitometer record of light-dark variation) from this mid-Cretaceous Piobbico core, after tuning the curves to the bundle peak found from independant time control, show the presence of two major components which match in spacing the 98 and 126 kyr components of the eccentricity curve (Park and Herbert, 1987). The identification of the bundle with the 100 kyr eccentricity cycle has thereby received independent confirmation, which in turn strengthens the identification of other cycles. As in Pleistocene marine sequences, the eccentricity 100-kyr term thus dominates the sedimentary variance in the Umbrian cycles, but anoxic episodes occur in phase with a limestone-marl repetition and reflect minima of the precessional cycle (Herbert et al., 1986).

Somewhat similar patterns appear in the Niobrara chalk of the American Western Interior (Coniacian - basal Campanian stages in the Upper Cretaceous; Bottjer et al. 1986). The bedding rhythms are here the primary depositional result of variations in carbonate productivity, terrigeneous dilution, redox conditions and/or the energy of bottom currents. Here too these bedding couplets reflect the precession and are grouped into 100 kyr cycles. These in turn are grouped into less-well defined superbundles with possible rhythms at 400 to 1,800 kyr. However, the British Chalk shows evidence of both the precessional and obliquity cycles (Hart, 1987). The relative abundance of calcarous nannofossil taxa in a sequence of samples collected from a continuous Upper Maastrichtian section (late Cretaceous) in Alabama has been analysed using statistical techniques (Mayers and Worsley, 1973). Provided a number of reasonable assumptions are made, the dominant frequencies of the periodic component are within the order of magnitude of the high eccentricity frequency (100 kyr). Eccentricity cycles (100, 400 and 2,000 kyr) were also found to match the periodicities characterizing the turbidites from the Maastrichtian Helminthoid Flysh in the Ligurian Alps in Italy (Foucault et al., 1987).

Redox cycles occur abundantly in the Lower Cretaceous sediments of the Atlantic deep-sea floor. Cotillon and Rio (1984) calculated the periods of Valanginian in the Gulf of Mexico, Atlantic Ocean and Vocontian basin (France) at $\sim$ 16 kyr. Their averages for the Hauterivian stage are between 16 and 26 kyr and the couplets are bundled in sets of 4 to 6. A comparative study between $\delta^{18}O$ variations of the bulk carbonate and the thickness of limestone-marl couplets has also been done in the Kimmeridgian-early Valanginian deposit from southern Spain. Foucault and Renard (1987) found that the average thickness of these couplets (27 ± 9 kyr) is compatible with the actual value of precession; they related the increase in the couplets thickness to an increase of sea-water temperature.

In summary, it appears that the precesssion index has been widely recorded in marine sediments and is fingerprinted by the grouping of individual precession events into sets that correspond to the 100 kyr eccentricity cycle. But in contrast to the bundled Cenomanian of Umbria, the Cenomanian of the Maritime Alps shows a fairly regular oscillation in the 40-50 kyr bracket suggesting that the obliquity cycle was the driver for what appears to have been a dilution rhythm (Fischer et al., 1985). A more extensive development of that rhythm is also found in the basin of Lower Saxony, northern Germany, from the Valanginian through the Albian stage at least if the Harland et al. (1982) time scale is used (Schneider, 1964; Fischer, 1986).

What can these spectral characteristics infer in terms of the response of the Cretaceous climate system to orbital forcing ? This Cretaceous climate is in fact characterized by warm poles as evidenced by abundant high latitude floras and faunas and no direct evidence of permanent ice, a globally averaged surface temperature 6-12°C higher than the present day value, warm oceanic deep water and tropical surface ocean temperatures similar or slightly warmer than today's values (Barron, 1988). The question of polar ice through the Cretaceous remains however debated (Kemper, 1987). If this would be the case, growth and decay of even small ice caps and the cooling and warming of the oceans could result in eustatic fluctuations of a few meters enough to cause scour cycles (ROCC Group, 1986) or emergence cycles of carbonate platforms (Hardie et al., 1986).

Nevertheless, the cyclicity in Cretaceous strata deposited during such a period of globally warm climate can not be explained by glacial-interglacial changes in rates of deep-water production in high latitudes and consequent changes in carbonate dissolution. Arthur et al. (1984) proposed, instead, that periodic changes in insolation, evaporation, wind stress and/or rainfall in a wide variety of environments caused changes in input of terrigenous detritus, water mass stratification, surface productivity, deepwater oxygen content and rates of carbonate dissolution. Unfortunately, diagenesis commonly renders the cycles unsuitable for study of primary environmental signals and progress in understanding the cause of such cyclicity remains slow.

In order to achieve such a goal, Barron et al. (1985) used a combination of geochemistry, sedimentology and general circulation model results, to show how cycles in the Cretaceous Greenhorn Formation deposited in the U.S. western interior seaway probably recorded climatic cycles. To test this hypothesis, Glancy et al. (1986) forced their general circulation model with the seasonal maximum and minimum insolation calculated from assumed Cretaceous orbital parameters. Their simulations give northern hemisphere winter precipitation increases of over 6 mm/day, suggesting that cyclical changes in insolation produced cyclical changes in precipitation (climate) which, in turn, produced cyclical changes in terrestrial runoff (a geologic process) that might have created the rhythmic sedimentation observed in the Greenhorn Formation.

Within the limits of plausible mechanisms of maintaining planetary warmth and polar heat transport, the model by Barron (1984) shows also that the suggested reduced obliquity based on Cretaceous-Eocene high latitudes floras (Wolfe, 1980; Douglas and Williams, 1982) would have resulted in a substantial polar cooling, far too cold to be compatible with the paleontologic evidence. Moreover, the notion of such a drastically reduced obliquity received little support from celestial mechanics (Ward, 1982). As demonstrated by Barron (1988), the leading candidates for explaining the Cretaceous warm global temperatures during which the Milankovitch cycles can occur are, instead, the very different Cretaceous continental geometry, the decrease in continental area due to high eustatic sea level, higher atmospheric CO_2 concentrations and amplification of these factors through a greater role of the ocean in poleward heat transport.

6.2. The Jurassic

To demonstrate a possible relation between Jurassic sedimentary microrhythms and orbital patterns, House (1985) considered two sequences chosen for their unambiguous relation to the chronostratigraphical scale. The Lower Lias microrhythms typically comprises a thin laminated, dark, organic-rich shale followed by lighter coloured calcareous shale with bioturbated limestone beds at the top. The Upper Jurassic Kimmeridge Clay comprises alternations of well-laminated oil shale and/or laminated organic-rich shale with more calcareous, lighter coloured shale or mudrock; this sequence is interrupted by dolomitic or coccolith limestone bands which form convenient markers. Although several uncertainties exist in the values used of the length of the Jurassic (60 to 75 Myr) and of the number of Jurassic zones, the agreement of an average of the microrhythm period found in these shale-limestone sequences of the British Jurassic with the value of 41 kyr is very close. The obliquity cycle operating by the latitudinal shift of the climatic zones, these microrhythms might be explained by oscillations in and out of the tropical forest belt, although the model invoking cold-water benthic versus warm-water planktonic regimes is also promising.

The Blue Lias Formation just discussed here has also been the subject of Walsh power-spectral analysis by Weedon (1985/86). Two cycles with duration less than 93 kyr were found, which may record changes in orbital precession and obliquity, the sedimentary cycles being tentatively attributed to changes in the volume of runoff. The same technique was also applied to three Liassic limestone - shale sequences from Breggia Gorge, Switzerland (Weedon, 1989). The regularity of the cycles observed combined with crude direct dating and wavelength ratios implies the detection of the 21 kyr precession cycle in all three cases, of the 100 kyr modulation of this cycle in two cases and of the 41 kyr obliquity cycle in one.

6.3. The Triassic

The occurence of rhythmic bedding in platform limestones of the Alpine Triassic was already recognized in late 1930's and Schwarzacher (1948) first noted the, now well-established, ratio of five couplets to a bundle. Studies by Goldhammer et al. (1987) of the emergence of the Middle Triassic Latemar marine carbonate platforms of the Dolomites (Italy) show also such a cyles ratio of 5:1 suggestive of glacial control forced by precession (20 kyr) and eccentricity (100 kyr). The stacked meter-scale cycles, each made of a subtidal unit with a cap rock, must be the result of relative sea level oscillations which occured with an average frequency in the 10-kyr range (Hardie et al., 1986). Finally, the high variability of this Anisian-Ladinian Latemar carbonate platform sequence has been examined using a prolate spheroidal expansion of the couplet thickness spectrum (Hinnov et al., personal communication). The results indicate the presence of multiple signals with characteristics similar to those predicted from the astronomical theory, confirming how much the orbital cycles are significantly imprinted in pre-Cretaceous sedimentary records.

In eastern North America, Van Houten (1964) first recognized the cyclical nature of Newark Supergroup sediments which he associated with the orbital variations. More recently, Fourier analysis of long sections of the Late Triassic Lockatong and Passaic Formation of the Newark basin show periods in thickness of 25, 44, 100, 133 and 400 kyr corresponding roughly to the astronomical periodicities (Olsen, 1986), as judged by radiometric time scales and varve-calibrated sedimentation rates. If the alternation of the black shale facies deposited in deep water with drab-to-red playa deposits yields a period of 21 kyr by extrapolation of varve thickness, by the same token alternations of chemical and detritus - rich sequences occur on a 100 kyr schedule and red sequences occur at 400 kyr intervals.

7. THE PALAEOZOIC

In the varved Permian Castile evaporites (Anderson, 1984), the initiation and termination of deposition and annual thickness trends are consistent with the 100 kyr eccentricity cycle and the precession cycle is recorded in the rate of calcium sulfate deposition.

The Carboniferous system is distinguished in many places by cyclic sequences which record transgressive cycles called cyclotherm engendered by the rhythmic growth and decay of the Carboniferous ice sheets. These eustatic cyclotherms are best developed in the Late Pennsylvanian of the North American Mid-continent region. They presumably record the 400 kyr eccentricity cycle (Heckel, 1977;1986) and in Kansas, the up to five limestones within these cyclothems would reflect the 100 kyr eccentricity cycle (Fischer, 1986). Already in 1964, Schwarzacher demonstrated such regular cyclicity in limestone bed thickness for the Irish Carboniferous. Spectral analysis of lower Carboniferous limestone sections from NW Ireland shows indeed a stratification cycle of 1 to 3 m in thickness, each cycle consisting of 5 to 7 limestone beds (Schwarzacher, 1987). The scale of the cycles, their regularity and the ratio of 5 beds to one cycle suggest that they are ultimately caused by variations in the precession (21 kyr) and eccentricity (100 kyr) of the Earth's orbit. However, it is difficult to assess the uncertainty of such estimates quantitatively. It is possible that the cycles were in reality shorter and could represent the 40 kyr obliquity period, although there is no supporting evidence for this.

As both fluvial and deltaïc systems are obviously sensitive to climate and eustasy, Van Tassel (1987) has attributed the rhythmicity in sandstone shale alternations in the Devonian pro-delta facies of the Catskills to orbital effects.

Autocyclicity and local tectonic pulsing might have had possible influences on carbonate platform stratigraphy of the Cambrian and Early Ordovician (Hardie and Shinn, 1986) when no continental land masses existed in the polar regions to support waxing and waning ice caps.

8. THE PRECAMBRIAN

As it was the case throughout the Phanerozoic (Fischer, 1964), the development of upward shallowing cycles has been an important part of the evolution of shallow-water platforms. There is much evidence, as in the Rocknest formation exposed in northwest Canada (1.9 Gyr), in the Wallace and Helena formation (1.5 to 1.2 Gyr) and in others (Grotzinger, 1986), that suggests these cycles formed in response to high-frequency eustatic oscillations in sea-level perhaps related to small scale continental or alpine glaciation. Where geochronological constraints permit calculation, the average period of cyclicity is very likely within the Milankovitch band of 20 to 100 kyr.

9. PRE-QUATERNARY FREQUENCIES OF THE EARTH'S ORBITAL PARAMETERS

All these interesting results encourage research on the stability of the solar system in order to determine to which extent the changing Earth-Moon distance (see for instance, Williams, 1989), for example, influenced the length of the main astronomical periods.

Precise knowledge of the Earth's orbital variations are not available prior a few million years as the relative position of the Sun and the planets are not known. Although numerical solutions to the solar system celestial mechanics problem are available that are reliable for the last 5 to 10 Myr (Berger, 1984; Berger et al., 1988), the physical problem is quasi-periodic and has long-period oscillations whose nature is only beginning to be understood (Milani and Nobili, 1984 and 1985). On longer time scales, the solar system may indeed be subject to chaotic disruption (Laskar, 1989). It is likely that its overall structure has been stable during the last hundreds of millions of years (Milani et al., 1986) but unpredictable perturbations to the phase of the astronomical forcing are possible. Simultaneously, the lack of precision of radiometric time scales in geology adds additional uncertainty.

Nevertheless, the largest amplitude Milankovitch frequencies can be computed for the Phanerozoic (last 1 billion years roughly). These frequencies are indeed a function of the precession frequency k which is a function of the slightly asymmetrical lunar and solar torque on the Earth's tidal bulge. At the present the torque exerted by the Moon on the Earth's equatorial bulge accounts for about two thirds of its precession rate, with the rest due to the Sun's torque. During past epochs when the Moon was closer to the Earth, k was larger and hence the periods smaller. Such recent astronomic computations (Berger et al., 1987) show (Table 1) that the precession and obliquity cycles would be reduced prior to 2 billion years ago with the period of obliquity cycles starting to approach the precessional ones if we take into account this varying Earth-Moon distance only. According to this hypothesis, already at 250 Myr ago changes were not negligible : the main precessional periods were 18.4 kyr and 22.1 kyr and the obliquity one was 38.2 kyr.

TABLE I. Changes in the main periods of precession and obliquity due to changes in the Earth-Moon distance only (Loutre and Berger, 1989).

Epoch (billion years BP) (Gyr)	Precession (kyr)		Obliquity (kyr)	k ("/yr)
0	19	23	41	50.4
0.25	18.4	22.1	38.2	52.8
0.5	17.6	21.0	35.1	55.7
1	16.9	20.0	32.3	59.0
2	15.5	18.1	27.6	65.8
2.5	14.7	16.9	25.0	70.8

However, a more accurate model does exist for the last 500 Myr but not for prior times. This model takes into account all the parameters which influence the orbital frequencies, in particular the Earth-Moon distance and the Earth's rotation and moments of inertia (this explains the desagreement between the values at 0.5 Gyr BP in tables 1 and 2). When all these factors are considered, the 19, 23, 41 and 54 kyr orbital periodicities become at the Ordovician-Silurian boundary, 440 Myr ago respectively : 16.4, 19.3, 30.6 and 37.3. Values for the Cretaceous, Permian, Carboniferous and Devonian are given in Table 2.

TABLE II. Estimated values of the astronomical periods over the last 500 Myr taking into account the variation of the Earth-Moon distance, the Earth's rotation and moments of inertia (Berger et al., 1987).

Geological Epoch	Age (Myr BP)	19	23 (kyr)	41	54	k ("/yr)
Holocene	0					50.4
Upper Cretaceous	72	18.6	22.5	39.4	51.2	51.7
Lower Permian	270	17.6	21.0	35.2	44.3	55.7
Upper Carboniferous	298	17.4	20.7	34.3	42.9	56.6
Middle Devonian	380	16.8	19.9	32.1	39.6	59.2
Lower Silurian	440	16.4	19.3	30.6	37.3	61.2

Eccentricity being computed from the (e, π) system (Berger, 1977b) where π is the longitude of the fixed perihelion, its frequencies are not supposed to change with time, at least in the context of the present discussion.

On the other hand, the amplitudes of the main obliquity terms are also changing in time. They are indeed also function of k, being proportional to $\nu_i/(k-\nu_i)$ (ν_i is the frequency corresponding to a given amplitude). During past epochs, the obliquity variations were thus somewhat smaller : 440 Myr BP, the amplitudes of the 41 kyr and 54 kyr periods had decreased respectively by 25 and 30 %. For precession, the amplitude of the most important terms are independent of k; it is only the secondary terms which are also function of the precession frequency. The amplitude of the variations in precession and in eccentricity are thus not expected to change significantly through time.

This effect depends in fact upon the relationship between k and the characteristic frequencies, ν_i, of the planetary perturbations themselves (Berger, 1977b). When $k >> \nu_i$ for all i, the amplitudes of the obliquity variation are relatively small and can be significantly augmented only if k decreases to a value comparable with the ν_i's.

A dramatic event may therefore occur in the future when the Moon moves somewhat further from the Earth. k will decrease and the associated period will increase to a 50 kyr period comparable to one of the ν_i's. When that occurs, the amplitude of the obliquity oscillations will increase by more than an order of magnitude and enormous climatic changes will occur as is the case for Mars (Ward, 1974; Pollack, 1979).

As far as the mean obliquity is concerned, its change could occur because of core-mantle coupling (which should tend to cause a decrease, Pollack, 1982) and because of the lunar torque on the equatorial tidal bulge (which should tend to increase obliquity by $3.5°/10^9$ yr - Mac Donald, 1964). But it is only very gradually through viscous interactions that the mantle can drag the core along. As the time constant for the coupling is very long, it seems likely to be unsignificant for the purpose of modeling climate over the past 0.5 Gyr, and Hunt (1979) concluded for an increase in the mean obliquity. Our computations (Loutre and Berger, 1989) show however that this increase is not significant over the last billion years.

10. IMPRINTS OF ORBITAL FREQUENCIES IN SEDIMENTS AND PHYSICAL PROCESSES

Power spectra for Pleistocene data particularly from ice cores, deep-sea and lacustrine sediments, have been used to identify the role of orbital forcing because of its near regularity and the characteristic duration of the orbital cycles. These principal cycles involved are 100 kyr, 41 kyr, 23 and 19 kyr. Cyclic variations in oxygen-isotope values and percentage carbonate within deep-sea cores have also been ascribed to orbitally forced climatic changes as far back as the Miocene.

Pelagic-hemipelagic sediments (as in the Umbrian pelagic facies exposed in the Apennines of Italy), low-latitudes evaporites and midlatitudes lakes have recorded precessional couplets bundled into sets corresponding to the short cycle of eccentricity as far back as Triassic and Permian time. In addition, there are sequences dominated by the obliquity cycle. For times where no ice was present on the globe, for instance in the Triassic-Cretaceous, it has been speculated that certain cyclic marine and lacustrine rocks owe their cyclicity to a climatic control also related to the astronomical periods.

The Late Carboniferous, a time of large scale glaciation, shows large, rhythmic marine transgressions and regressions. In the resulting well-know cyclothems, the shorter rhythms (precession and 100 kyr-eccentricity) seem to have been suppressed in favor of the longer eccentricity rhythm. This "red shift" seems more pronounced than in the Pleistocene and suggests that Carboniferous climates reacted with greater inertia acting as a low-pass filter (Fischer, 1986). However, it could also be that the Pleistocene records are generally not long enough to record this main eccentricity cycle (with the largest amplitude in the eccentricity expansion; cf Berger, 1977a, Table 3) or that the sampling frequency in the time domain is not large enough in the Carboniferous record to detect the shorter cycles (low sedimentation rate, compaction, hiatuses, bioturbation can prevent such a detailed sampling).

We have thus seen that oscillatory patterns are common in the Phanerozoic and Proterozoic stratigraphic record (in particular varves, pelagic and hemipelagic sediments, emergence of carbonate platforms and transgression patterns in Carboniferous cyclothems). Where geochronological constraints permit calculation, the average cycle period is very likely within the Milankovitch band of 20 to 100 kyr. As these cycles probably formed under climatic control associated or not with eustatic oscillation in sea level, it is possible that Milankovitch forcing has occured for at least the last 2.2 Gyr of Earth history.

However, because of all the uncertainties associated with the ancient records, this relationship in the frequency domain is not conclusive. The cycles could be random events that coincidentally had an average period within the Milankovitch band or the astronomical frequencies can have been totally distorted by geological processes (Algeo and Wilkinson, 1988). Today the Milankovitch paradigm pervades the investigation of the ancient sedimentary record where it is readily recognized in deposits of closed systems such as deep marine and lacustrine sediments. In contrast, in shallow marine sedimentation, the effect of

Milankovitch cycle is easily over printed by local noise or irregularly episodic controls so that orbitally induced rhythms have been obscured or destroyed altogether. As an example, analyses of solitic ironstones, a distinctive kind of shallow-water deposit scattered through space and time (from the Miocene, the Cretaceous, the Liassic, the Devonian and the Silurian), display repetitive stratigraphic successions with a several hundred thousand year cadence which Van Houten (1986) has attributed tentatively to the 400 kyr astronomical cycle, although it is unclear how this would exert a eustatic control without the waxing and waning of continental glaciers. Clearly there is need for a better-constrained temporal framework. There is also need for rigorous criteria for distinguishing between allogenic and autogenic controls. Nevertheless, Milankovitch cycles have become an increasingly reasonable possibility as our understanding of the function has improved and methods of correlation have been made more precise.

The periodicities of all the cycles are highly variable in space and time, even over the past 5 Myr (Dean and Gardner, 1986). This suggests that the primary record of climate forcing, supposed here to be orbitally controlled, at least partly, has been modified by local effects such as tectonic movements, bioturbation, ocean circulation, climate, etc. As an example, according to Anderson (1986), changes in seasonality were probably responsible for recording the precession cycle in the Permian Castille formation. This suggests that seasonal deposition may be a means for transferring long-term climatic change directly into this stratigraphic record.

Consequently the resultant records are complex products of primary climate signals with superimposed local effects. There is ample of possible mechanisms that can serve to amplify the climatic signal generated by orbital variations and to drive sedimentological expression. Productivity cycles, carbonate cycles, dissolution cycles, emergence cycles, redox cycles and other are possible explanations in processes ultimately tuned by orbital variations. These effects have changed through time with changes in the length of orbital cycles as well as with changes in the geographical distribution of oceans and continents, in the chemistry of the atmosphere and ocean, etc.

In order to explain the extremely weak precessional signal but stronger 100 and 400 kyr ones that they found in pelagic Albian Scisti a Fucoidi sequence, Ripepe and Fischer (1989) constructed from the precessional index calculated by Berger (1978) a synthetic stratigraphy in a manner consistent with that observed in the Albian core segment. The model involves the conversion of the precessional index into sedimentation rates, conversion of the time scale into a depth scale and modification of the resultant stratigraphy by bioturbation. It is particularly interesting (as found also by Pestiaux and Berger, 1984b, for Pleistocene records) that each of these 3 steps transfers spectral power from the high-frequency signal to the lower eccentricity ones leading to a final spectrum of the simulated sequence closely comparable to that of the core.

Considering the complex series of linkages between orbital parameters, climate and the sediment record (cf for example Fig. 1 of Dean and Gardner, 1986), it is really surprising, but significant, to find that the sedimentary cycle periodicities are so similar through space and time, or at least coherent, and situated within the range of orbital cycles.

11. UNIFICATION OF THE ASTRONOMICAL CYCLES THROUGH GEOLOGICAL TIMES

Most of the insolation parameters for any latitude display a precessional signal stronger than the obliquity one (Berger and Pestiaux, 1984). Assuming that the atmosphere transfers the astronomical frequencies without changing their relative amplitudes (including those of the combination tones), the precession signal must thus appear as the dominant feature of all climatic proxy data. This is what is found for the Jurassic and Cretaceous, in particular, where no important snow fields nor ice sheets seem to have existed.

However, some climatic parameters could show a secondary obliquity signal, provided they are only not related to equinoctial seasons and/or equatorial regions for which only precession is present. But, if for any reason, latitudes or months where the obliquity plays a role, are given more importance by the climate system itself, the obliquity signal may be reinforced to the detriment of precession. This seems to be the case where and when significant snow fields start to cover high latitudes at the end of the Pliocene and beginning of the Quaternary, the role of these polar latitudes being then magnified by the albedo-temperature feedback.

Moreover, when the ice sheets reach a critical size, the characteristic time of the associated isostatic rebound start to interact with some orbital frequencies leading to combination tones (Wigley, 1976; Ghil, 1980). This is typically the case for the 100 kyr cycle of the late Pleistocene, but also for the higher frequencies found in the range of 1 cycle per 10 kyr and their multiples (Pestiaux et al., 1988; Yiou et al., 1989).

This simple, but fundamental view in the frequency domain which unifies all the results accumulating now from Pre-Quaternary as well as for Quaternary times can only be accepted if the complete seasonal cycle of insolation is taken into account in order to force a non-linear deterministic climate system. The caloric seasons of Milankovitch, linearization of the climate system and purely stochastic forcing, although they have their own advantages, are not allowing to see this basic behaviour of the frequency response of the climate system to the astronomical forcing over the whole time scale of the Earth's history.

12. CONCLUSIONS

In studies of sedimentary rhythmicity in the past, the main problem is in assessing the period, in years, of the cycles. For much of the Phanerozoic, the radiometric scale is certainly not accurate enough for testing which of the orbital forcing patterns might be operating in the past, or for assessing the length of time which individual sedimentary rhythms might have taken to form.

However, we have seen that there is more than a hint emerging from geological data and from astronomical investigations that a quite novel time scale might be possible using orbitally forced characteristics in combination with the radiometric and biostratigraphic time scales. This would give to the time scale both precision and accuracy (House, 1987) allowing further independent tests to be made.

13. ACKNOWLEDGMENT

Thanks are due to Mrs M.F. Loutre (Louvain-la-Neuve, Belgium), Dr. F.J. Hilgen (Utrecht, Netherlands), Dr. P. de Boer (Utrecht, Netherlands), Prof. A.G. Fischer (Los Angeles, USA), Prof. N. Shackleton (Cambridge, UK), Dr. G.P. Weedon (Cambridge, UK) and Mr. M. Ripepe (Firenze, Italy) for stimulating discussions.

I am grateful to Mrs N. Materne-Depoorter who has kindly typed this manuscript.

REFERENCES

Adhémar, J.A. (1842). Révolutions des Mers, Déluges Périodiques. Publication privée, Paris.

Algeo, T.J., Wilkinson, B.H. (1988). Periodicity of mesoscale phanerozoic sedimentary cycles and the role of Milankovitch orbital modulation. J. of Geology, **96**, 313-322.

Alvarez, W., Montanari, A. (1988). The Scaglia Limestones (Late Cretaceous-Oligocene) in the Northeast Apennines carbonate sequence : stratigraphic context and geological significance. In : "The Eocene-Oligocene Boundary in the March-Umbrian basin (Italy)", I. Premoli-Silva, R. Coccioni and A. Montanari (Eds), Int. Subcomm. Paleog. Strat. Ancona, Spec. Publ. I.1.

Anderson, R.Y. (1984). Orbital forcing of evaporite sedimentation. In : "Milankovitch and Climate", A. Berger, J. Imbrie, J. Hays, G. Kukla, B. Saltzman (Eds), 147-162, D. Reidel Publ. Company, Dordrecht, Holland.

Anderson, R.Y. (1986). The varve microcosm : propagator of cyclic bedding. Paleoceanography, 1(4), 373-382.

Arthur, M.A., Dean, W.E., Bottjer, D., and Schole, P.A. (1984). Rhythmic bedding in Mesozoic-Cenozoic pelagic carbonate sequences : the primary and diagenetic origin of Milankovitch like cycles. In : "Milankovitch and Climate", A. Berger, J. Imbrie, J. Hays, G. Kukla, B. Saltzman (Eds), 191-222, D. Reidel Publ. Company, Dordrecht, Holland.

Arthur, M.A., Garrison R.E. (guest eds) (1986). Milankovitch cycles through geologic time. Paleoceanography, 1(4), 369-586.

Arthur, M.A., Garrison R.E. (1986). Cyclicity in the Milankovitch band through geologic time : an introduction. Paleoceanography, 1 n°4, 369-372.

Backman, J., Pestiaux, P., Zimmerman, H., Hermelin, O. (1986). Paleoclimatic and palaeoceanographic development in the Pliocene North

Atlantic : Discoaster accumulation and coarse fraction data. In : C.P. Summerhayes and N.J. Shackleton (Eds), "North Atlantic Palaeoceanography", Geological Society, Special Publ. n°21, 231-242.

Barron, E.J. (1984). Climatic implications of the variable obliquity explanation of Cretaceous,paleogene high-latitude floras. Geology, **12**, 595-598.

Barron, E.J. (1988). Studies of Cretaceous climate. In : "Contributions of Geophysics to Climate Change Studies", A. Berger, R. Dickinson and J. Kidson (Eds), American Geophysical Union, Washington D.C.

Barron, E.J., Arthur, M.A., Kauffmann, E.G. (1985). Cretaceous rhythmic bedding sequences : a plausible link between orbital variations and climate. Earth Planetary Science Letters, 327-340.

Benzi, R., Parisi, G., Sutera, A. and Vulpiani, A. (1982). Stochastic resonance in climatic change. Tellus, **34**, 10-16.

Berger, A. (1976). Obliquity and precession for the last 5,000,000 years. Astronomy and Astrophysics, **51**, 127-135.

Berger, A. (1977a). Support for the astronomical theory of climatic change. Nature, **268**, 44-45.

Berger, A. (1977b). Long term variations of the Earth's orbital elements. Celestial Mechanics, 5, 3-74.

Berger, A. (1978). Long-term variations of daily insolation and Quaternary climatic changes. J. Atmospheric Sciences, **35(12)**, 2362-2367.

Berger, A. (1984). Accuracy and frequency stability of the Earth's orbital elements during the Quaternary. In : "Milankovitch and Climate", A. Berger, J. Imbrie, J. Hays, G. Kukla, B. Saltzman (Eds), 3-40, D. Reidel Publ. Company, Dordrecht, Holland.

Berger, A. (1987). Pleistocene climatic variability at astronomical frequencies. In : "Global Change", H. Faure and N. Rutter (Eds) International Quaternary Association, Ottawa, Ont.

Berger, A. (1988a). Milankovitch Theory and Climate. Review of Geophysics, in press.

Berger, A. (1988b). Milankovitch effects on long-term climatic changes. In : "Climatic Change and Impacts : A General Introduction", R. Fantechi (Ed.), 1st Course of the European School of Climatology and Natural Hazards.

Berger, A., and Pestiaux, P. (1984). Accuracy and stability of the Quaternary terrestrial insolation. In : "Milankovitch and Climate", MA. Berger, J. Imbrie, J. Hays, G. Kukla, B. Saltzman (Eds), 83-112, D. Reidel Publ. Company, Dordrecht, Holland.

Berger, A., Loutre, M.F., Dehant, V. (1987). Influence of the variations of the lunar orbit on the astronomical frequencies of the Pre-Quaternary insolations. Scientific Report, **1987/15**, Institut d'Astronomie et de Géophysique G. Lemaître, Université Catholique de Louvain-la-Neuve.

Berger, A., Loutre, M.F., Laskar, J. (1988). Insolation values for the climate of the last 10 millions years. Scientific Report, **1988/13**, Institute of Astronomy and Geophysics Georges Lemaître, Catholic University of Louvain-la-Neuve.

Bottjer, D.J., Arthur, M.A., Dean, W.E., Hattin, D.E., Savrda, Ch.E. (1986). Rhythmic bedding produced in Cretaceous pelagic carbonate environments : sensitive recorders of climatic cycles. Paleoceanography, 1 **n°4**, 467-482.

Bradley, W.H. (1929). The warves and climate of the Green River. Epoch. US Geol. Survey Prof. Pap. **158**, 87-110.

Briskin, M., Harrell, J. (1980). Time series analysis of the Pleistocene deep-sea paleoclimatic record using periodic regression. Marine Geology, **36**, 1-22.

Broecker, W.S., Thurber, D.L., Goddard, J., Ku, T., Matthews, R.K., Mesolella, K.J. (1968). Milankovitch hypothesis supported by precise dating of coral reefs and deep sea sediments. Science, **159**, 297-300.

CLIMAP Project Members (1976). The surface of the Ice-Age Earth. Science, **191**, 1131-1137.

CLIMAP Project Members (1981). Seasonal reconstruction of the Earth's surface at the Last Glacial Maximum, McIntyre A. and Cline R. (Eds), Geological Society of America. Map and Chart Series MC-36, Boulder, 1-18.

Cottilon, P., Rio, M. (1984). Cyclic sedimentation in the Cretaceous of deep sea drilling sites 535 and 540 (Gulf of Mexico), 534 (Central Atlantic) and in the Vocontian Basin (France). In : "Initial Reports of the Deep Sea Drilling Project", Washington D.C. Govt. Printing Office, **77**, 339-376.

Croll, J. (1875). Climate and Time in their Geological Relations. Appleton, New York.

Crowley, Th.J. (1988). Paleoclimate Modelling. In : M. Schlesinger (Ed.), "Physically-Based Modelling and Simulation of Climate and Climatic Change", 883-949, Kluwer Academic Publishers, Dordrecht, Holland.

Dean, W.E., and Gardner, J.V. (1984). Cyclic variations in calcium carbonate and organic carbon in Miocene to Holocene sediments, Walvis Ridge, South Atlantic ocean. In : "Milankovitch and Climate", A. Berger, J. Imbrie, J. Hays, G. Kukla, B. Saltzman (Eds), 265-266, D. Reidel Publ. Company, Dordrecht, Holland.

Dean, W.E., and Gardner, J.V. (1986). Milankovitch cycles in Neogene deep-sea. Paleoceanoography, 1 **n°4**, 539-554.

de Boer, P.L. (1982). Cyclicity and the storage of organic matter in Middle Cretaceous pelagic sediments. In : "Cyclic and Event Stratification", G. Einsele and A. Seilacher (Eds), 456-475, Springer Verlag, Berlin.

de Boer, P.L. (1983). Aspects of middle Cretaceous pelagic sedimentation in Southern Europe. Geologica Ultraiectina, redelelingen van het Instituut voor Aardwetenschappen der Rijksuniviersiteit te Utrecht, n°**31**.

de Boer, P.L., and Wonders, A.A.H. (1984). Astronomically induced rhythmic bedding in Cretaceous pelagic sediments near Moria (Italy). In : "Milankovitch and Climate", A. Berger, J. Imbrie, J. Hays, G. Kukla, B. Saltzman (Eds), 177-190, D. Reidel Publ. Company, Dordrecht, Holland.

de Boer, P.L., Shackleton, N. (1984). Pre-Pleistocene evidence of orbital forcing. In : "Milankovitch and Climate", A. Berger, J. Imbrie, J. Hays, G. Kukla and B. Saltzman (Eds), 827-830, D. Reidel Publ. Company, Dordrecht, Holland.

Douglas, J.G., and Williams, G.E. (1982). Southern polar forest : the Early Cretaceous floras of Victoria and their paleoclimatic significance. Palaeogeography, -climatology, -ecology, **39**, 171-185.

Einsele, G., and Seilacher, A. (Eds) (1982). Cyclic and Event Stratification. Berlin, Springer Verlag, 536pp.

Emiliani, C. (1966). Isotopic paleotemperatures. Science, **154(3751)**, 851-857.

Fischer, A.G. (1964). The Lofer cyclothems of the Alpine Triassic. Kans. State Geol. Surv. Bull., **169(1)**, 107-149.

Fischer, A.G. (1980). Gilbert-Bedding rhythms and geochronology. In : "The Scientific ideas of G.K. Gilbert", An assessment of the occasion of the Centennial of the United States Geological Survey (1879-1979), E.L. Yochelson (Ed.), The Geological Society of America, Special paper 183, 93-104.

Fischer, A.G. (1986). Climatic rhythms recorded in strata. Ann. Rev. Earth Planet. Sci., **14**, 351-376.

Fischer, A.G., and Schwarzacher, W. (1984). Cretaceous bedding rhythms under orbital control. In : "Milankovitch and Climate", A. Berger, J. Imbrie, J. Hays, G. Kukla, B. Saltzman (Eds), 163-176, D. Reidel Publ. Company, Dordrecht, Holland.

Fischer, A.G., Herbert, T., and Premoli-Silva, I. (1985). Carbonate bedding cycles in Cretaceous pelagic and hemipelagic sediments. In : "Fine-Grained Deposits and Biofacies of the Cretaceous Western Interior Seaway : Evidence of Cyclic Sedimentary Processes", L.M. Pratt, E. Kauffman, F. Zelts (Eds), 1-10, Soc. Econ. Paleontol. Mineral. Guideb. series.

Fischer, A.G. et al. (1988). Cyclostratigraphy. Cretaceous Resources Events Rhythms, a NATO Advanced Research Workshop, Global Sedimentary Geology Program, a Commission of the International Union of Geological Sciences, Perugia, Italy.

Foucault, A., and Renard, M. (1987). Contrôle climatique de la sédimentation marno-calcaire dans le Mésozoïque d'Espagne (Sierre de Fontcalent, Province d'Alicante) : arguments isotopiques. C.R. Acad. Sci. Paris, t. **305**, sér. II, 517-521.

Foucault, A., Powichrowski, L., and Prud'Homme, A. (1987). Le contrôle astronomique de la sédimentation turbiditique : exemple du Flysch à Helminthoïdes des Alpes Ligures (Italie). C.R. Acad. Sci. Paris, t. **305**, Série II, 1007-1011.

Foucault, A., and Blanc-Valleron, M.M. (1989). Orbital and solar control of evaporite sedimentation, an example in Late Oligocene of the Rhinegraben, France. In : "Milankovitch Cyclicity in the Pre-Pleistocene Stratigraphic Record", D.G. Smith, A. Berger, P.L. de Boer (Eds), Terra Abstracts, **1(1)**, 242.

Ghil, M. (1980). Internal climatic mechanisms participating in glaciation cycles. In : A. Berger (Ed.), "Climatic Variations and Variability : Facts and Theories", 539-557, D. Reidel Publ. Company, Dordrecht, Holland.

Ghil, M. and Le Treut, H. (1981). A climate model with Cryodynamics and Geodynamics. J. of Geophysical Research, **86**, 5262-5270.

Gilbert, J.K. (1895). Sedimentary measurement of Cretaceous time. J. of Geology, Feb. Mar., 121-127.

Glancy, T.J., Barron, E.J., and Arthur, M.A. (1986). An initial study of the sensitivity of modeled Cretaceous climate to cyclical insolation forcing. Paleoceanography, 1 **n°4**, 523-538.

Goldhammer R.K., Dunn, P.A., and Hardie, L.A. (1987). High frequency glacio-eustatic sealevel oscillations with Milankovitch characteristics recorded in Middle Triassic platform carbonates in northern Italy. Am. J. Sci., **287**, 853-892.

Grotzinger, J.P. (1986). Upward shallowing platform cycles : a response to 2.2 billion years of low-amplitudes, high-frequency (Milankovitch Band) sea level oscillations. Paleoceanography, 1 **n°4**, 403-416.

Hardie, L.A., and Shinn, E.A. (1986). Carbonate depositional environments, modern and ancient, 3, tidal flats. Colorado School of Mines Quart., **81(1)**, 74p.

Hardie, L.A., Bosellini, A., and Goldhammer, R.K. (1986). Repeated subaerial exposure of subtidal carbonate platforms, Triassic, Northern Italy : evidence for high frequency sea level oscillations on a 10^4 year scale. Paleoceanography, 1 **n°4**, 447-458.

Harlant, W.B., Cox, A.V., Llewellyn, P.G., Pickton, C.A.G., Smith, A.G., and Walters R., 1982. A Geologic Time Scale, Cambridge University Press, Cambridge, 131pp.

Hart, M.B. (1987). Orbitally induced cycles in the chalk facies of the U.K. Cretaceous Res., **8**, 335-348.

Hasselman, K. (1976). Stochastic climate models, part I. Tellus, **28**, 473.

Hays, J.D., Imbrie, J., and Shackleton, N.J. (1976). Variations in the Earth's orbit : pacemaker of the Ice Ages. Science, **194**, 1121-1132.

Heckel, P.H. (1977). Black hole in Pennsylvanian cyclothems. Am. Assoc. Pet. Geol. Bull., **61**, 1045-1068.

Heckel, P.H. (1986). Sea-level curve for Pennsylvanian eustatic marine transgressive-regressive depositional cycles along midcontinent outcrop belt, North America. Geology, **14**, 330-334.

Herbert, T.D., and Fischer, A.G. (1986). Milankovitch climatic origin of mid-Cretaceous black shale rhythms, Central Italy. Nature, **321 (6072)**, 739-743.

Herbert, T.D., Stallard, R.F., and Fischer, A.G. (1986). Anoxic events, productivity rhythms and orbital signature in a mid-Cretaceous deep--sea sequence from central Italy. Paleoceanography, 1 **n°4**, 495-506.

Hilgen, F.J. (1987). Sedimentary rhythms and high resolution chronostratigraphic correlations in the Mediterranean Pliocene. Newsl. Stratigr., **17(2)**, 109-127.

Hilgen, F.J., and Langereis, C.G. (1989). Sedimentary cycles in the Mediterranean Pliocene : discrepancies with the quasi-periods of the Earth's orbital cycles ? In : "Milankovitch Cyclicity in the Pre-Pleistocene Stratigraphic Record", D.G. Smith, A. Berger, P.L. de Boer (Eds), Terra Abstracts, **1(1)**, 241.

Hodell, D.A., Elmstrom, K.M., and Kennett, J.P. (1986). Relationship between latest Miocene Benthic $\delta^{18}O$ changes global ice volume, sea-level, and the Messinian salinity crisis. Nature, **320**, 411-414.

House, M. (1987). Geological rhythms, cycles and other revolutions. Geol. Mag. (Great Britain), **124(3)**, 273-276.

House, M. (1985). A new approach to an absolute time-scale from measurements of orbital cycles and sedimentary micro-rhythms. Nature, **315**, 721-725.

Hunt, B.G. (1979). The effects of past variations of the Earth's rotation rate on climate. Nature, **281**, 188-191.

Imbrie, J. (1982). Astronomical theory of the Pleistocene Ice Ages : a brief historical. Icarus, **50**, 408-422.

Imbrie, J. (1985). A theoretical framework for the Pleistocene ice ages. J. Geol. Soc. London, **142**, 417-432.

Imbrie, J., and Imbrie, K.P. (1979). Ice Ages, Solving the Mystery. Enslow Publishers, Short Hills, New Jersey.

Imbrie, J., and Imbrie, J.Z. (1980). Modelling the climatic response to orbital variations. Science, **207**, 943-953.

Imbrie, J., and Kipp, N.G. (1971). New micropaleontological method for quantitative paleoclimatology : application to a Late Pleistocene Caribbaen Core. In : "Late Canezoic Glacial Ages, Turekian K.K. (Ed.),71-181, Yale University Press, New Haven.

Imbrie, J., Hays, J., Martinson, D.G., McIntyre, A., Mix, A.C., Morley, J.J., Pisias, N.G., Prell, W.L., Shackleton, N.J. (1984). The orbital theory of Pleistocene climate : support from a revised chronology of the marine ^{18}O record. In : "Milankovitch and Climate", A. Berger, J. Imbrie, J. Hays, G. Kukla, B. Saltzman (Eds), 269-305, D. Reidel Publ. Company, Dordrecht, Holland.

Imbrie, J., McIntyre, A., Mix, A. (1989). Oceanic response to orbital forcing in the Late Quaternary : observational and experimental strategies. In : "Climate and Geo-Sciences", A. Berger, S. Schneider and J.Cl. Duplessy (Eds), , Kluwer Academic Publishers, Dordrecht, Holland.

Jansen, E., Bleil, U., Henrich, R., Kringstad, L., and Slettemark, B. (1988). Paleoenvironmental changes in the Norwegian Sea and the Northeast Atlantic during the last 2.8 m.y. : deep-sea drilling project/ocean drilling program sites 610, 642, 643 and 644. Paleoceanography, 3(5), 563-581.

Jouzel, J., Lorius, Cl., Petit, J.R., Genthon, C., Barkov, N.I., Kotlyakov, V.M. (1987). Petrov Vin, Vostok ice core : a continuous isotope temperature record over the last climatic cycle. Nature, **329 (6138)**, 403-408.

Kemper, E. (1987). Das Klima der Kreide-Zeit. Geol. Jb. A96, 5-185.

Kerr, R.A. (1988). A sun-weather connection broken. Science, **242**, 1012-1013.

Kominz, M.A. and Pisias, N.G. (1979). Pleistocene climate : deterministic or stochastic ? Science, **204(4389)**, 171-173.

Kominz, M.A., Heath, G.R., Ku, T.L., Pisias, N.G. (1979). Brunhes time scales and the interpretation of climatic changes. Earth Planet Sci. Letters, **45**, 394-410.

Kutzbach, J.E. (1985). Modeling of paleoclimates. Adv. Geophys., **28A**, 159-196.

Laskar J. (1989). A numerical experiment on the chaotic behaviour of the solar system. Nature, **338**, 237-238.

Loutre, M.F., and Berger, A. (1989). Pre-Quaternary values for the precessional parameter and the amplitudes of obliquity and precession.

Scientific Report, **1989/4**, Institut d'Astronomie et de Géophysique G. Lemaître, Université Catholique de Louvain-la-Neuve.

Le Treut, H., and Ghil, M. (1983). Orbital forcing, climatic interactions and glaciation cycles. J. of Geophysical Research, **88** C9, 5167-5190.

MacDonald, G.J.F. (1964). Earth and Moon : past and Future. Science, **145**, 881-890.

Martinson, D.G., Pisias, N.G., Hays, J.D., Imbrie, J., Moore, T.C., and Shackleton, N.J. (1987). Age dating and the orbital theory of the ice ages : Development of a high-resolution 0 to 300,000-year chronostratigraphy. Quat. Res., **27(1)**, 1-29.

Mayers, I.R., and Worsley, T.R. (1973). Statistical recognition of late Cretaceous cyclic sedimentation by means of calcareous nannofossil population studies. Palaeogeography, -climatology, -ecology, **13**, 81-90.

Meyers, P.A., Dunham, K.W., and Ran, G.H. (1986). Organic geochemical character of opal-rich sediments in light-dark cycles near the Pliocene-Pleistocene boundary in Deep Sea Drilling Project Site 532, Walvis Ridge. Paleoceanography, **1(4)**, 567-575.

Milani, A., and Nobili, A.M. (1984). Resonance packing between Jupiter and Uranus. Nature, **310**, 753-755.

Milani, A., and Nobili, A.M. (1985). Resonant structure of the outer solar system. Celestial Mechanics, **35(3)**, 269-287.

Milani, A., Nobili, A.M., Fox, K., and Carpino, M. (1986). Long-term changes in the semi-major axis of the outer planets. Nature, **319**, 386-388.

Milankovitch, M. (1941). Kanon der Erdbestrahlung. Royal Serbian Academy, Spec. publ. 132, section of Mathematical and Natural Sciences, vol. 33 (published in English by Israel Program for Scientific Translation, for the U.S. Department of Commerce and the National Science Foundation, Washington D.C., 1969).

Moore, T.C., Pisias, N.G., Dunn, D.A. (1982). Carbonate time series of the Quaternary and late Miocene sediments in the Pacific ocean : a spectral comparison. Marine Geol., **46**, 217-233.

Negrini, R.M., Verosub, K.L., Davis, J.O. (1988). The middle to late Pleistocene geomagnetic field recorded in fine-grained sediments from Summer Lake, Oregon and Double Hot Springs, Nevada, USA. Earth and Planetary Science Letters, **87**, 173-192.

Nicolis, C. (1980). Response of the Earth-atmosphere system to a fluctuating solar input. In : "Sun and Climate", 385-396, CNES-CRS-DGRST, Toulouse, October 1980.

Nicolis, C. (1982). Stochastic aspects of climatic transitions-response to a periodic forcing. Tellus, **34**, 1-9.

Olsen, P.E. (1986). A 40-million-year lake record of Early Mesozoic orbital climatic forcing. Science, **234**, 842-848.

Park, J., and Herbert, T.D. (1987). Hunting for paleoclimatic periodicities in a geologic time series with an uncertain time scale. J. Geophys. Res., **92**, B13, 14027-14040.

Pestiaux, P., and Berger, A. (1984a). An optimal approach to the spectral characteristics of deep-sea climatic records. In : "Milankovitch and Climate", A. Berger, J. Imbrie, J. Hays, G. Kukla, B.

Saltzman (Eds), 417-446, D. Reidel Publ. Company, Dordrecht, Holland.

Pestiaux, P., and Berger, A. (1984b). Impacts of deep-sea processes on paleoclimatic spectra. In : "Milankovitch and Climate", A. Berger, J. Imbrie, J. Hays, G. Kukla, B. Saltzman (Eds), 493-510, D. Reidel Publ. Company, Dordrecht, Holland.

Pestiaux, P., van der Mersch, I., Berger, A., and Duplessy, J.Cl. (1988). Paleoclimatic variability at frequencies ranging from 1 cycle per 10,000 years to 1 cycle per 1,000 years : evidence for nonlinear behavior of the climate system. Climatic Change, **12**(1), 9-37.

Pisias, N.G., and Mix, A.C. (1988). Aliasing of the geological record and the search for long period Milankovitch cycles. Paleoceanography, **3**(5), 613-620.

Pollack, J.B. (1979). Climatic change on the terrestrial planets. Icarus, **37**, 479-553.

Pollack, J.B. (1982). Solar, astronomical and atmospheric effects on climate. In : "Climate in Earth History", 68-76, National Res. Council, National Acad. Sci. Washington D.C.

Prell, W.L. (1982). Oxygeen and carbon isotope stratigraphy for the Quaternary of Hole 502B : evidence for two modes of isotopic variability. In : "Initial reports of the Deep Sea Drilling Project", 455-464, W.L. Prell and J.V. Gardner et al. (Eds), vol. LXVIII, Washington D.C.

Ripepe, M., and Fischer, A.G. (1989). Stratospheric rhythms synthesized from orbital variations. In : "Milankovitch Cyclicity in the Pre-Pleistocene Stratigraphic Record", D.G. Smith, A. Berger, P.L. de Boer (convenors), Terra Abstracts, 1(1), p. 241.

R.O.C.C. Group (1986). Rhythmic belding in Upper Cretaceous pelagic carbonate sequences : varying sedimentary response to climatic forcing. Geology, **14**, 153-156.

Ruddiman, W.E., Shackleton, N.J. and McIntyre, A. (1986). North Atlantic sea-surface temperatures for the last 1.1 million years. From Summerhayes C.P. and Shackleton N.J. (Eds), North Atlantic Palaeoceanography Geological Society Special Publication, 21, pp. 155-173.

Ruddiman, W.F., and McIntyre, A. (1981). Oceanic mechanisms for amplification of the 23,000-year ice-volume cycle. Science, **212**, 617-627.

Ruddiman, W.F., Raymo, M., and McIntyre, A. (1986). Matuyama 41,000-year cycles : North Atlantic Ocean and northern hemisphere ice sheets. Earth and Planet. Sc. Lett., **80**, 117-129.

Schneider, F.K. (1964). Erscheinungsbild und Entstehung der rhythmischen Bankung der altkretazisichen Tongesteine Nordwestfalens und der Braunschweiger Bucht. Fortschr. Geol. Rheinl. Westfalen, 7, 353-382.

Schwarzacher, W. (1948). Uber die sedimentäre Rhythmik des Dachsteinkalkes von Lofer. Geol. Bundesamt. Wien Verh., **(10-12)**, 175-188.

Schwarzacher, W. (1964). An application of statistical time series analysis to a limestone-shale sequence. J. Geol., 72, 195-213.

Schwarzacher, W. (1987). Astronomical cycles for measuring geologic time. Modern Geology, **11**, 175-381.

Shackleton, N.J., and Opdyke, N.D. (1973). Oxygen isotope and paleomagnetic stratigraphy of equatorial Pacific core V28-238 : Oxygen isotope temperatures and ice volumes on a 10^5 and 10^6 year scale. Quaternary Research, **3**, 39-55.

Shackleton, N.J., and Opdyke, N.D. (1976). Oxygen isotope and paleomagnetic stratigraphy of Pacific core V28-239 late Pliocene to latest Pleistocene. Geological Society of America, memoir **145**, pp. 449-464.

Shackleton, N.J., and Hall, M.A. (1985). Oxygen and carbon isotope stratigraphy of Deep Sea Drilling Project hole 552A : Plio-Pleistocene glacial history. Initial Rep. Deep Sea Drilling Project, **81**, 599-609.

Shackleton, N.J., Backman, J., Zimmerman, H., Kent, D.V., Roberts, D.G., Schnitker, D., Baldauf, J.G., Desprairies, A., Homrighausen, R., Huddlestun, P., Keene, J.B., Kaltenback, A.J., Krumsiek, K.A.O., Morton, A.C., Murray, J.W., and Westberg-Smith, J. (1984). Oxygen isotope calibration of the onset of ice-rafting and history of glaciation in the North Atlantic region. Nature, **307**, 620-623.

Shackleton, N.J., Imbrie, J., and Pisias, N. (1988). The evolution of oceanic oxygen-isotope variability in the North Atlantic over the past three million years. Phil. Trans. R. Soc. London, **B318**, 679-688.

Stothers, R.B. (1987). Do slow orbital periodicities appear in the record of Earth's magnetic reversals ? Geophys. Res. Letters, **14(11)**, 1087-1090.

Tiwari R.K. (1987). Higher order eccentricity cycles of the middle and late Miocene climatic variations. Nature, **327**, 219-221.

Tornaghi, M.E., Premoli-Silva, I., and Ripepe, M. (1988). Planktonic foraminiferal distribution records productivity cycles : evidence from the Aptian-Albian of central Italy. Rivista Italiana di paleontologia C Stratigrafia Milano.

Van Echelpoel, E., and Weedon, G.P. (1989). Milankovitch cyclicity in the Boom Clay Formation, an Oligocene siliciclastic shelf deposit. Terra Abstracts, 1(1), 242.

Van Houten, F.B. (1964). Cyclic lacustrine sedimentation, Upper Triassic Lockatong formation, central New Jersey and adjacent Pennsylvania. Bull. Kans. State Geol. Surv., **169**(2), 497-531.

Van Houten, F.B. (1986). Search for Milankovitch patterns among Oolitic ironstones. Paleoceanography, **1 n°4**, 459-466.

Van Tassel, J. (1987). Upper Devonian Catskill delta margin cyclic sedimentation : Brallier Scherr and Foreknobs Formation of Virginia and West Virginia. Geol. Soc. Amer. Bull., **99**, 414-426.

Ward, W.R. (1974). Climatic variations on Mars, 1, Astronomical theory of insolation. J. Geophys. Res., **79**, 3375-3387.

Ward, W.R. (1982). Comments on the long-term stability of the Earth's obliquity. Icarus, **50**, 444-448.

Weedon, G.P. (1985/86). Hemipelagic shelf sedimentation and climatic cycles : the basal Jurassic (Blue Lias) of South Britain. Earth and Planetary Science Letters, **76**, 321-335.

Weedon, G.P. (1989). The detection and illustration of regular sedimentary cycles using Walsh power spectra and filtering, with examples from the Lias of Switzerland. J. of Geological Society, London, **146**, 133-144.

Wigley, T.M.L. (1976). Spectral analysis : astronomical theory of climatic change. Nature, **264**, 629-631.

Williams, G.E. (1989). Late Precambrian tidal rhythmites in South Australian and the history of the Earth's rotation. J. Geological Society, **146**, 97-111.

Wolfe, J.A. (1980). Tertiary climates and floristic relationships at high latitudes in the Northern Hemisphere. Palaeogeography, -climatology, -ecology, **30**, 313-323.

Yiou, P., Genthon, C., Jouzel, J., Ghil, M., Le Treut, H., Barnola, J.M., Lorius, Cl., and Korotkevitch, Y.N. (1989). High-frequency paleoariability in climate and in CO_2 levels from Vostok ice-core records. In : "Interaction of the Global Carbon and Climate Systems", R. Keir (Ed.), EPRI report.

POLAR ICE CORES AND CLIMATE

C. J. Lorius
Laboratoire de Glaciologie et Géophysique de l'Environnement
B.P. 96
38402 ST Martin d'Hères Cedex
France

ABSTRACT. Although there are needs for a better understanding of the atmosphere-snow and ice connection and for improving deep ice dating polar ice cores are providing multiple proxy records of climate and related parameters. They show evidences of anthropogenic impact on aerosol concentrations in Greenland snow (i.e. SO_4 and NO_3) and, on a global scale, on greenhouse gases such as CH_4 and CO_2. Over the last climatic cycle (i.e. ~ 150 ky) they indicate temperature changes up to about 10°C with periodicities characteristic of earth orbital parameters. Very cold climatic stages are associated with lower snow accumulation and high concentrations of marine and continental aerosols, likely reflecting strengthened sources and atmospheric transport. Greenland ice has recorded rapid changes of climate during the last ice age and deglaciation. Antarctic cores have shown a close association between CO_2 and climate throughout the last climatic cycle. New results on methane and sulfates further support the picture of large chemical atmospheric modifications, some of them involving the biosphere, which have likely contributed in glacial interglacial climatic changes.

1. INTRODUCTION

The recovery of records of the environmental history is one of the keys to a better understanding of climatic changes and thus may contribute to better assessment of man's influence on climate. In this respect polar ice can provide a wide range of information, as it records atmospheric climatic conditions and sampled aerosols and gases in consecutive datable layers over time scales extending to 150 ky B.P.

It has been less than 20 years since the first papers from the Camp Century (Greenland) ice core were published and recently results from the Vostok (Antarctica) record have become available. A recent workshop* provided a good opportunity to review the major findings of these two decades and to point out further possibilities and needs. This paper

* Dahlem Workshop on the Environmental Record in Glaciers, Berlin, March 13-18, 1988

A. Berger et al. (eds.), Climate and Geo-Sciences, 77–103.

summarizes some characteristics of ice cores as an atmospheric record and presents an overview on the results obtained on various time scales mainly from a very few deep polar ice cores.

2. THE ATMOSPHERIC RECORD

While we cannot expect to find an ideal climatic recorder in nature, glacial ice is proving to be a rather remarkably close approximation to it, in particular when information is viewed on long time scales, and with a time resolution of decades to centuries.

The basis for paleotemperature reconstruction is the existence of current correlations between the proportion of deuterium to hydrogen and oxygen 18 to oxygen 16 atoms measured in deposited snow and temperature conditions at the site, resulting from fractionation processes which take place in the atmospheric water cycle. Although the isotopic composition of polar snow depends on several parameters, there exists a linear relationship in both of the polar ice sheets, between the mean annual surface temperature and the mean $\delta^{18}O$ or δD value of deposited snow (Lorius and Merlivat, 1977 ; Johnsen et al., 1988). The validity of using the observed current relationships for paleotemperature reconstruction has recently been supported by various approaches. For instance in Antarctica the temperature change deduced for the last deglaciation is consistent with values derived from another paleoindicator (Petit et al., 1987) and the observed in situ temperature profile in the Dye 3 bore hole can be best fitted by ice flow model calculations (Dahl-Jensen and Johnsen, 1986) assuming a surface temperature in South Greenland during the glaciation which is essentially in agreement with the one derived from the ice core isotopic record. On a more theoretical basis, the observed slopes (6 °/.. /°C for Antarctic δD and 0.67 °/.. /°C for Greenland $\delta^{18}O$) are in good agreement with values derived from a one-dimensional atmospheric isotope model (Jouzel and Merlivat, 1984); however if General Circulation Models (Jouzel et al., 1987) has been successful in many respects in reproducing the global distribution of δ in precipitation including Antarctica, these models have so far been unable to reproduce important features like the seasonal δ variation in Greenland precipitation. Although more theoretical work remains to be done to fully describe the quantitative isotope-temperature transfer function, the large isotopic changes obtained from ice cores can be interpreted, as a first order approximation, in terms of local atmospheric temperature changes, after correcting for long term isotopic changes of sea water which are driven by global ice volume changes.

Relating gas concentrations obtained from ice core extraction to atmospheric values which are of global significance is much more straightforward ; although precipitation mechanisms may have secondary effects, there is for instance, within the precision of the measurements, no differences between the CO_2 and CH_4 greenhouse gas data monitored by modern observations and those obtained from corresponding polar ice layers (Pearman and Fraser, 1988).

Such precipitation mechanisms are of importance in order to establish a quantitative link between the concentration of aerosols and impurities in surface snow (Davidson, 1988). Although there is no doubt that the concentration of a species in the air is reflected in snow deposits at the site, there is a lack both of direct observations and in understanding aerosol deposition processes. The relative contribution of "wet" (including precipitation scavenging) and "dry" deposition is poorly known and may vary with location and possibly with time. This may make it difficult to make quantitative intercomparisons between various ice cores and, to a lesser extent, between different stages of the same record. When looking at ice core records, we must also be aware that the atmospheric circulation and aerosol transport characteristics introduce a biased view of global patterns although this representativity problem may be addressed by including tracers in atmospheric general circulation models (Joussaume et al., 1988).

There seems to be no post depositional processes of enough importance to cause a disturbance in the atmospheric record. The main concern however is the possible existence of melting which can eventually occur in the warmest part of polar ice sheets as even very slight melting can affect in particular the concentrations of gases. An overview summarizing the climate-related information which can be obtained from ice cores is given in Table 1.

TABLE 1 : Short summary of climatic information and corresponding ice core signal

ATMOSPHERE	ICE CORE
Temperature	D/H, $^{18}O/^{16}O$
Precipitation	D/H, $^{18}O/^{16}O$, ^{10}Be
Humidity	D/H, $^{18}O/^{16}O$
Aerosols	
. natural (continents, sea volcanoes, biosphere)	chemicals (Al, Ca^{++}, Na^{+}, H^{+}, SO_4^{--}, NO_3^{-})
. man made	SO_4^{--}, NO_3^{-}, Pb, radioactive fallout
Circulation	Particles
Gases : natural and man made	O_2, N_2, CO_2, CH_4, N_2O

In ice sheets the deepest ice may originate from remote upstream locations. In such cases the ice core record not only reflects atmospheric changes at the drill site but also integrates significant effects linked with ice flow conditions. There are for instance several drillings reaching back to the last glaciation (or even available sequences exposed at surface) in coastal areas of both ice sheets. Although they are of significant interest for ice sheet dynamics and interaction with climatic changes, they will not be discussed in this paper, in which we restrict ourselves to the records obtained in central parts on Greenland and Antarctica which provide more direct and representative environmental atmospheric data. Even in such cases the deepest part of the records, when the ice is close to the bedrock, can be significantly disturbed by ice flow conditions.

3. ICE CORE SERIES

So far there are only two deep ice cores from Greenland and three from Antarctica (Table 2) which provide long term climatic and environmental records extending over the last ice age and only one provides a complete undisturbed series over the last climatic cycle, back to the ice age which preceeded the previous interglacial. Although specific engineering skills are required to drill in low temperature deep holes filled with fluid (to balance glaciostatic pressure) this technical difficulty is not the main limiting factor. Heavy logistics support is also required to conduct such drilling operations in remote and inhospitable areas. In fact the technical requirements are now rather well established ; the difficulty seems now to get sufficient support to organize new drilling projects.

TABLE 2 : Deep polar ice core records

	Drillings		Current conditions			
	Depth (m)	Year	Ice thickness	Elevation (m)	Mean Temperature (°C)	Mean accumulation ($kg.m^{-2}.a^{-1}$)
Greenland						
Camp Century 77°10'N 61°08'W	1387	1966	1387	1885	- 24	32
Dye 3 65°11'N 43°50'W	2037	1981	2037	2480	- 19.6	50
Antarctica						
Byrd 79°59'S 120°01'W	2163	1968	2163	1530	- 28	16
Dome C 74°40'S 124°10'E	905	1978	3400	3240	- 53	3.4
Vostok 78°28'S 106°48'E	950 2083 2202	1974 1982 1985	3700	3490	- 55.5	2.3

Over the years, the development of new ideas and the availability of more sophisticated measuring equipment, has allowed the determination of an increasing number of environmental parameters. As an illustration, ^{14}C dating of the ice which required several tons of ice for a single measurement few years ago may now be possible with a few kg, using the recently developped accelerator mass spectrometry techniques. In some cases ice from previous drillings is no longer available ; this prevents, for instance, a detailed study correlating rapid changes of temperature and CO_2.

Polar ice drilling projects are required for both ice sheets. They must be performed in various geographical areas if we want to take full advantage of the various potentialities which depend upon glaciological and climatological characteristics. The all problem of climate-ice sheet interaction will necessitate a series of drillings along flow lines ; in turn the results will allow further refinement of the interpretation of the long term environmental records.

4. TIME SCALES

The major objective for ice core dating is to establish high precision continuous chronologies to determine as much as possible the precise timing of major environmental changes and the lead and lag time between various proxy data, to clarify the processes of global change.

For the upper part of the ice sheets the accuracy of the chronology can be very high. Annual layers can be counted from seasonal variations of various parameters such as visual stratigraphy, physical properties, isotopic composition, electrical conductivity, chemicals, etc... The range of applications of such variations varies with factors such as accumulation rate, ice thickness and ice flow but at great depth they are becoming indiscernible.

Prominent features found in ice cores can be used as reference horizons. They can provide ages when causal events are documented (radioactive fallout from nuclear tests over the last decades, ash or aerosol deposits from volcanoes) ; more generally, large scale atmospheric events can be used for relative intercomparison between ice cores (see for instance below the use of ^{10}Be peaks to correlate antarctic ice records), or with other paleodata such as those from sea, land or lake sediments. Many of these other records have been independently dated in particular through the use of radioactive isotopes. When clear features in an ice core record can be identified to correspond to dated events in other records, these dates can be used to help control the ice core chronology. A further possibility exists in using astronomical time scale over the last climatic cycle which has been utilized for instance to tune the deep sea records. However at this stage this approach has not been used for ice core dating ; it seems more appropriate to establish independent chronologies for ice core records if we want to be able to gain information on leads and lags between the different components of the climatic system, and which may throw light on forcing-feedback factors and mechanisms.

Detailed accurate chronologies have been obtained only for the two Greenland ice cores back to about 10 000 ys B.P. Although there are promising possibilities to obtain absolute ages for ice core records from radioactive and other dating techniques long term time scales have been so far based on numerical modelling of the age distribution through ice sheets by ice dynamics (Reeh, 1988). The characteristics of the present ice sheet (surface and bedrock topography, accumulation, temperature, ice velocity and viscosity) can be used to compute such an age distribution using steady state models (Budd and Young, 1983). The accuracy of the method depends on a number of factors. For the upper part of the ice sheets the accuracy can be very high but for regions near the bed complications of the ice flow can make the derived chronology unreliable. In general inland regions of the thick ice sheets have the best prospects for reliable long term chronologies.

Over the Holocene period (~ 10 000 years) where steady state conditions apply, both the Greenland Camp Century and Dye 3 ice cores are accurately dated while much less confidence has to be given to the chronology of deeper ice, covering beyond the Last Glacial Maximum, and which is situated in the bottom two hundred meters above the bedrock.

Ideally, past changes of the features of the ice sheet should be used in time dependent models to recompute a more precise age distribution. For the stable central areas the accumulation rate is the most important time variable to consider. Values derived from the Vostok ice core study have been taken into account in a two-dimensional ice flow model to establish a time scale with a possible accuracy on the order of 10-15 ky, given an age of about 160 ky at a depth of 2080 m, which is about 1400 m above the bedrock (Lorius et al., 1985).

Climatic features have been used to compare the Camp Century and Dye 3 records (Dansgard et al., 1982) ; this has also been done for the three antarctic ice cores (Jouzel et al., 1988). Antarctic and Greenland ice cores chronologies now have to be linked. For this purpose identification of isochronous horizons characterizing large scale atmospheric changes of aerosols (i.e. ^{10}Be) and trace gases (such as methane and CO_2) concentrations could be a very powerful tool.

At this stage it is worth mentioning that the ice chronology cannot be directly applied to the dating of the trace gases record. Owing to the air enclosure process in the ice the air contained in ice bubbles is in fact isolated from the atmosphere after the snow precipitation has been deposited at the surface of the ice sheet (Schwander and Stauffer, 1984). The estimated time difference may reach values up to about 4 000 years for the coldest and lowest snow accumulation conditions (Barnola et al., 1987).

5. CURRENT CHANGES : ANTHROPOGENIC IMPACT

Possibly because of the unfavorable signal to noise ratio, no clear indication of possible current climatic warming trend has been obtained from shallow ice cores. However, besides providing a reconstruction of significant volcanic events (Hammer et al., 1985 ; Legrand and Delmas, 1987) over the last centuries, snow and ice layers have recorded atmospheric changes which may be of importance for climatic changes.

With regard to aerosol, enhancement in nitrates and sulfates in the Arctic during the past 200 years broadly follows the known inventory of anthropogenic emissions. Changes in NO_3^- were not clearly detectable before the 1950s whereas SO_4^{--} has increased progessively during the last century. Qualitatively this agrees with the known shift of fuel usage from coal to oil and to the rapid growth of automobile emissions of NO_3^- since the 1950s. Particularly important are findings from Greenland ice cores of a marked enhancement in SO_4^- and NO_3^- depositions (Mayewski et al., 1986 and Figure 1 ; Neftel et al., 1985a) by factors of 3-4 and almost 2 respectively, that have been detected since the beginning of the industrial era. Such trends do not appear in antarctic records in agreement with the concentration of industrial activities in Northern continental areas and the limitation of interhemispheric atmospheric transport for those relatively short lived species. As previously pointed out the amount of change of airborne concentrations and its geographical representativity cannot however be really quantified utilizing current knowledge.

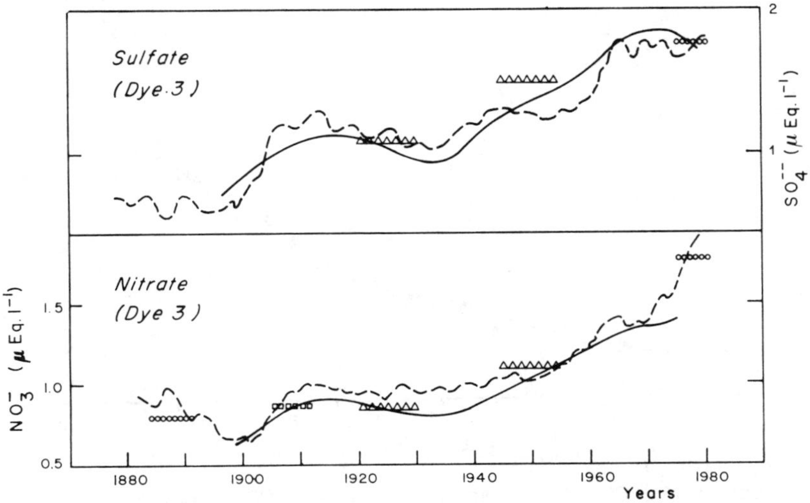

Figure 1. A comparison of the trends of pollutant deposition in Arctic snow for the last 100 years. The SO_4 and NO_3 results at Dye 3 (South Greenland) are taken from Mayewski et al. (1986) (stippled line) and from Neftel et al. (1985a) (solid line). Open circles represent mean values taken from Herron (J. Geophys. Res., 87, 30-52, 1982) , squares from Rasmussen K.L., Clausen H.B. and T. Risbo (Icarus, 58, 100-108, 1984) and triangles from Finkel, Langway and Clausen (J. Geophys. Res., 91, 9849-55, 1986). (R. Delmas, personal communication).

This is not the case for the gases which mix homogeneously in the atmosphere. Recent secular changes of the key greenhouse gases CO_2, CH_4 and N_2O have been precisely documented from polar ice cores. Ice core analysis (Figure 2) shows that the preindustrial concentration of atmospheric CO_2 was about 280 ppmv in the 18th century and increased to 300 (around 1920) and then to about 320 ppmv in early 1960' (Neftel et al., 1985b) mainly as a result of fossil fuel use and deforestation. Ice core data links with current monitored values which are now over 340 ppmv. Ice core information has been of crucial importance in documenting current changes before direct atmospheric measurements, which started only about 30 years ago by providing natural reference base line and variability. Measurements made on ice cores suggest for instance fluctuations of around 10 ppmv in pre-industrial periods (Raynaud and Barnola, 1985 ; Siegenthaler and Oeschger, 1987). A detailed assessment of such variations is potentially of great value for a better understanding of the CO_2 cycle ; for this purpose isotopic carbon data countain information which are helpful to discriminate between different sources which may cause the observed changes in the total atmospheric CO_2 concentration.

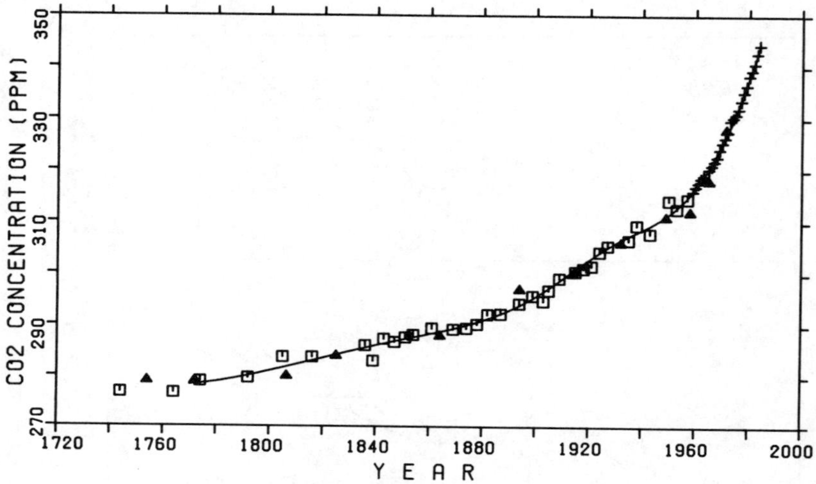

Figure 2. Atmospheric CO_2 increase in the past 200 years as indicated by measurements on air trapped in old ice from Siple Station (Antarctica) (from Siegenthaler and Oeschger, 1987).

Ice core measurements (Figure 3) are at present the exclusive source of information on the concentrations and trends of methane before the middle of this century (Stauffer et al., 1985 ; Pearman et al., 1986 Khalil and Rasmussen, 1987). The ice record is quite complete up to around 1000 y B.P. indicating that methane concentrations were about 700 ppbv 200 years ago while current atmospheric values are over 1600 ppbv. There has been considerable discussion as to the probable cause of this increase which may possibly be explained by more or less equivalent contributions from human and natural origins.

Although to a smaller extent the anthropogenic impact is also affecting N_2O concentrations. Ice core data indicate that the preindustrial concentrations were around 285 ppbv compared to present levels of about 310 ppbv (Khalil and Rasmussen, 1988).

So far it has not been possible to measure chlorofluorocarbon concentrations, in air extracted from ice bubbles, which are of importance for the ozone problem. Such determinations and those of other trace gases (CO, other hydrocarbons) will hopefully be possible in the future ; they would provide useful information on possible changes in average OH concentrations which are playing a role in reactive atmospheric chemistry and therefore for global biochemical cycles.

In summary ice core measurements allow the documentation of significant atmospheric changes over the last centuries which reflect an anthropogenic impact. Besides aerosols, global concentration increases of greenhouse gases over the last 200 years are about 20 % for CO_2, 8 % for N_2O but more than 200 % for methane.

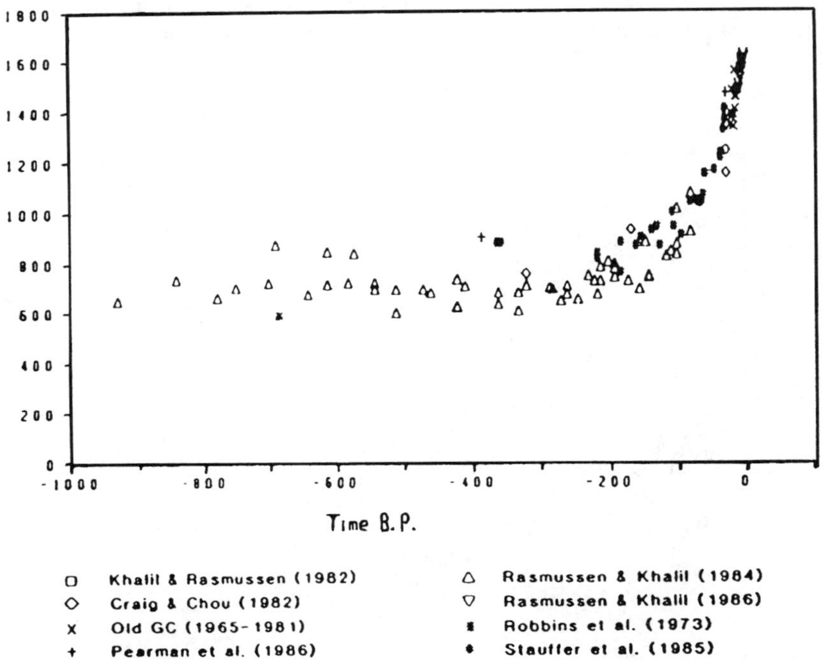

Figure 3. The atmospheric concentrations of CH_4 over the last 1000 years (in ppbv). The figure shows the overall agreement of different studies, the time when CH_4 concentrations began to increase rapidly, and the doubling of CH_4 over the last several hundred years. (From Khalil and Rasmussen, 1987).

6. THE LAST ICE AGE

6.1 Antarctica

Three deep ice cores reaching back at least to the Last Glacial Maximum (LGM, ~ 18 ky B.P.) have been drilled at Byrd (West Antarctica) and at Vostok and Dome C (East Antarctica). The original Dome C ice-flow based chronology (Lorius et al., 1979) has now been revised (Jouzel et al., 1988) to take into account past accumulation changes and the discovery of ^{10}Be peaks (Raisbeck et al., 1987) in both the Dome C and Vostok ice cores.

The isotopic records (Lorius et al. , 1979 ; Johnsen et al., 1972 ; Jouzel et al., 1987) versus depth are interpreted in terms of temperature change versus age (Figure 4). There is a remarkable similarity of the three climatic records over the last ~ 65 ky with a temperature change of about 9°C associated with the deglaciation which seems to take place in two steps. The slightly warmer event observed during the transition may eventually correspond to the Younger Dryas episode recorded in the Greenland ice. Common climatic features appear further back in

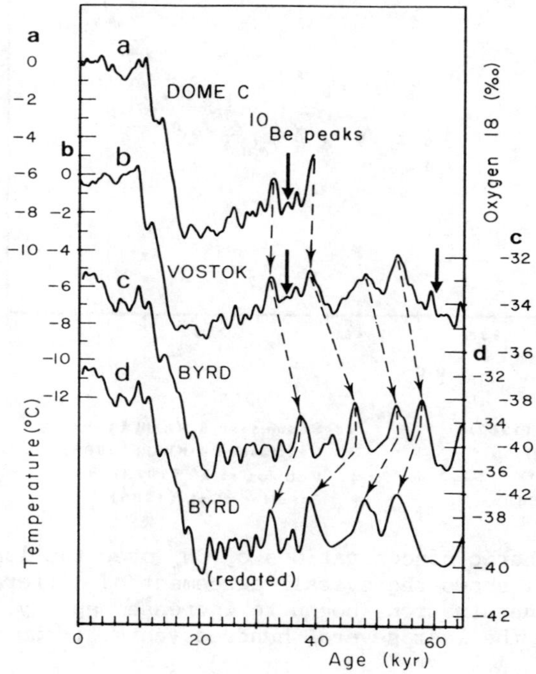

Figure 4. Antarctic ice core climatic records over the last 65 ky. Dome C (a) and Vostok (b) : left scale in °C expressed as a difference with respect to current temperature value. The solid arrows indicate observed ^{10}Be peaks.
Byrd (c) : right scale in $\delta^{18}O$ °/°° (from Johnsen et al., 1972 ; with a time scale from Clausen and Hammer, personal communication). The dotted arrows suggests a possible correspondence of climatic events in the three cores. Curve (d) is the Byrd record using a modified time scale obtained using these events as time markers (From Jouzel et al., 1988).

time allowing the correlation of the Byrd, Dome C and Vostok records. The generally good quantitative agreement between the three climatic records indicates that they are representative at least of all of Antarctica and that climatic events even of relatively small amplitude are in fact documented in the isotopic record. As we will see later paleo accumulation rates derived from both isotope-temperature and ^{10}Be data suggest (Figure 13) a decreased rate of precipitation accumulation during cold climatic conditions, with a value of about 50 % of the current one during the LGM.

Antarctic ice cores show a much larger impurity concentration during the LGM than during the current interglacial (Thompson et al., 1981 ; Palais and Legrand, 1985 ; De Angelis et al., 1987 ; Legrand et al., 1988a and Figure 5). With regard to dust of continental origin the LGM/Holocene ratio is ~ 8 for Byrd, 27 for Dome C and up to 30 for Vostok. Corresponding marine aerosol ratios are of about 2 to 3 for Byrd and 5 both at Dome C and Vostok. Although part of the observed variation may be connected with accumulation rate changes, these high glacial concentration values have been interpreted as resulting from strengthened sources and meridional transport linked to higher wind speeds (likely induced by higher temperature gradients with latitude), more extensive arid areas over the surrounding continents and the greater exposure of continental shelves due to a lower sea level (Petit et al., 1981). The Dome C results (Figure 5) show that continental dust concentrations were close to Holocene characteristics value well before the end of the temperature deglaciation record.

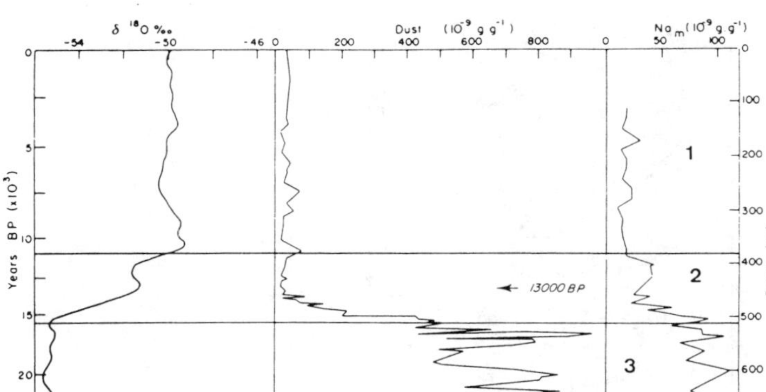

Figure 5. Smoothed stable isotope content, dust and Na marine concentrations measured in the Dome C (Antarctica) ice core and plotted as a function of age and depth (m of ice). Horizontal lines delimit different climatic conditions (From de Angelis et al., 1984).

6.2 Greenland

Both the Camp Century (Johnsen et al., 1972) and Dye 3 (Dansgaard et al., 1982) ice cores cover the last ice age (Figure 6). The total $\delta^{18}O$ shift associated with deglaciation is about 11 and 7°/₀₀ respectively, possibly corresponding to climatic warmings of 16 and 10.5°C, (Dansgaard and Oeschger, 1988). The warming which is apparently stronger in the Camp Century area agrees with data from another Arctic (Ellesmere Island, Fisher et al., 1983 ; Paterson et al., 1977) ice core. However an ice record from Devon Island, Canada, which is almost the same latitude as Camp Century, shows an isotopic shift of only 8 °/₀₀, i.e. the same as at Dye 3. It has been argued that the lower glacial values obtained at Camp Century may be due to a much higher surface elevation connected with the possible existence of an ice bridge between Northwest

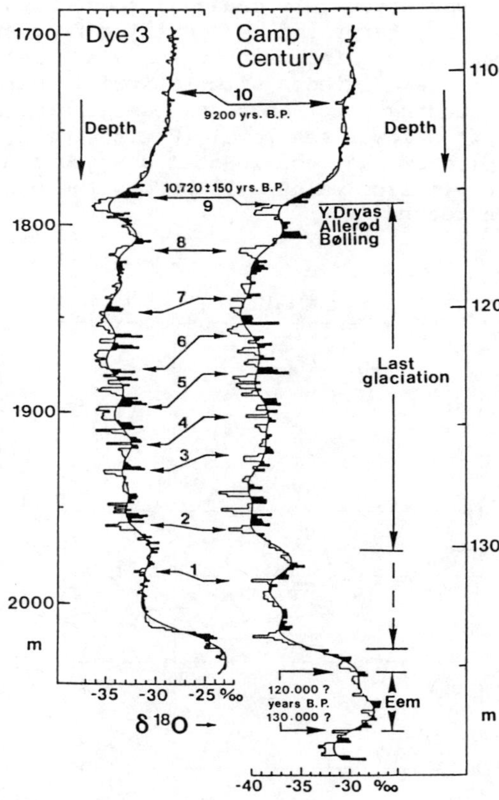

Figure 6. $\delta^{18}O$ profiles along the deepest 300 m of the ice cores from Dye 3 and Camp Century. The numbered arrows indicate layers of simultaneous deposition. (From Dansgaard et al., 1982).

Greenland and Arctic Canada or alternatively that glacial ice at Camp Century is not of local origin but originated from high elevation areas of central North Greenland. Both viewpoints are supported by total gas content measurements but are strongly refuted by ice flow modelling. The current interpretation (Dansgaard and Oeschger, 1988) is that the high δ shift in the Camp Century and Ellesmere Island ice cores reflects a higher temperature change than in South Greenland in agreement with the fact that global climatic changes are generally more pronounced at higher latitudes ; this interpretation leaves the smaller δ shift on Devon Island as an unsolved problem. As in Antarctica, glacial ice is characterized by higher dust levels most likely due to a greater atmospheric turbidity (Hammer et al., 1985).

In Greenland ice records there are clear indications of rapid climatic changes at the end of the Pleistocene. During deglaciation the so-called Bolling-Allerod-Younger Dryas oscillation (Dansgaard and Oeschger, 1988 and Figure 7) which is reasonably well dated in the Dye 3 ice core depicts a mean temperature increase in Greenland of about 7°C over a time period on the order of only 40 years, around 10 720 y B.P. The accumulation rate increased approximately 60 % judging from the rapid increase in annual layer thickness. Concentrations of chemical components in the ice changed drastically, including CO_2 concentrations. The Greenland δ profile matches many of the ^{14}C-dated European records, even in detail (as shown for example in Figure 7). Accurate dating of ice cores in Antarctica could help to answer this question if in fact the feature (of smaller amplitude) observed there during the deglaciation corresponds to the same event despite its origin in the North Atlantic area.

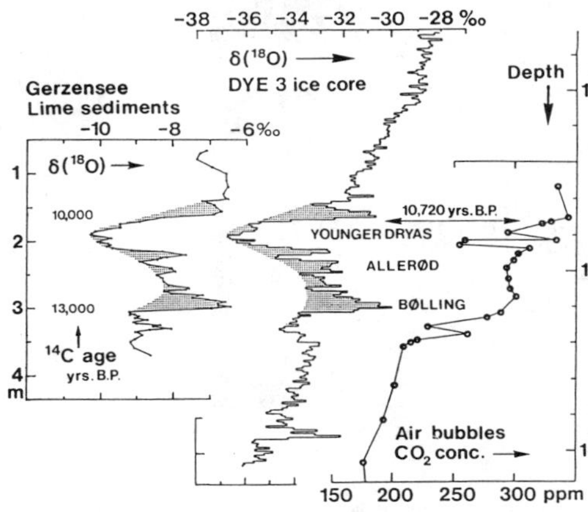

Figure 7. Middle : Detailed $\delta^{18}O$ profile along a 120 m increment of the Dye 3 deep corre containing ice deposited during the entire Pleistocene-et-Holocene transition. Right : CO_2 concentration in air bubbles. Left : $\delta^{18}O$ of lime sediments in the Swiss Lakle Gerzen. (From Dansgaard and Oeschger, 1988 and Dansgaard et al., 1984).

The type of rapid climatic change illustrated by the Bolling-Allerod-Younger-Dryas oscillation is not unique. It appears to be the last of a series of events observed throughout the glaciation. During periods around 30 and 40 ky B.P., $\delta^{18}O$ varied up to about 4 °/.. in the Greenland ice core records, corresponding to temperature changes of about 5-6°C (Dansgaard and Oeschger, 1988) within about a century or less. A detailed plot of the δ peaks from the Dye 3 core reveals a saw tooth shape with more abrupt changes from cold to mild conditions than vice-versa. Over these time intervals the δ temperatures are in phase with dust concentration (Hammer et al., 1985 and Figure 8) with higher values during cold conditions. These changes also correspond to CO_2 variations (Stauffer et al., 1984) with low values in the range of 180-200 ppmv for periods of cold climate and higher values in the range of 240-260 ppmv for mild climatic conditions. Such rapid CO_2 changes, correlated with δ shifts (Figure 9) have also been documented from the Camp Century ice core. This supports the assumption that the atmospheric CO_2 concentration changed dramatically parallel to the climatic variations, since the physical and chemical characteristics of the two ice cores are different and, also because the temperature at Camp Century is lower than at Dye 3 (thus also the contribution of melt layers to the CO_2 excursions seems less probable). This interpretation is however subject to further confirmation ; for instance these rapid CO_2 changes should also be present in Antarctic ice cores.

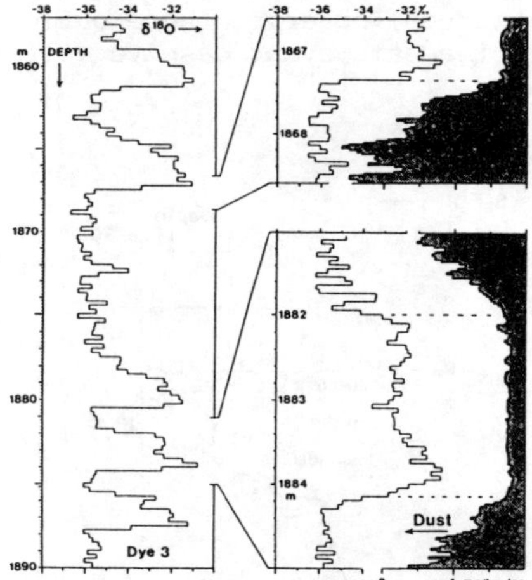

Figure 8. Detailed $\delta^{18}O$ and dust concentration profiles along an increment of the Dye 3 deep core plotted versus depth (From Dansgaard et al., 1984).

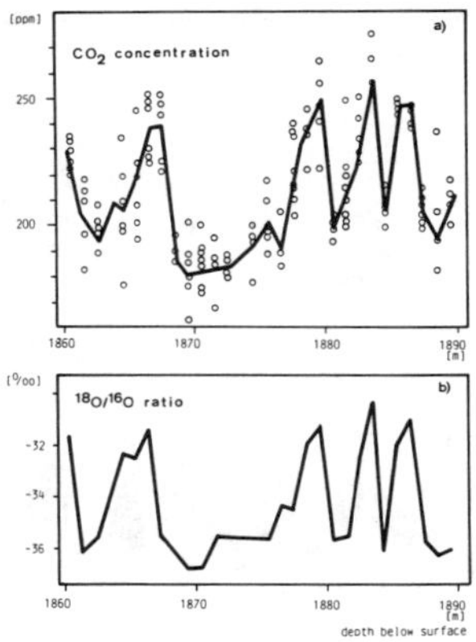

Figure 9. CO_2 and δO^{18} values measured on ice samples from Dye 3. The 30m increment corresponds to about 10 ky (from Stauffer B. et al., 1984)
(a) Circles indicate the results of single measurements of the CO_2 concentration of air extracted from ice samples. The solid line connects the mean values for each depth.
(b) The solid line connects the $\delta^{18}O$ measurements made on one sample every metre.

6.3 CO_2 and CH_4 records

Initial studies of the air trapped in Greenland and Antarctic ice cores revealed that, in close association with climatic changes, the atmospheric CO_2 content could have been around 190-200 ppmv during the Last Glacial Maximum compared with an average close to 270-280 during the Holocene period (Delmas et al., 1980 ; Neftel et al., 1982). More recently a detailed reconstruction of atmospheric CO_2 for the 15 to 50 ky B.P. period has been obtained from the Byrd ice core (Neftel et al., 1988 and Figure 10). The data reveal CO_2 concentrations in the range of 180-220 ppmv in the time interval which corresponds to the one in which rapid changes (with values up to 240-260 ppmv) were observed, in Greenland ice. There is however no indication of CO_2 values as high as the ones observed in Greenland during the last ice age. Several reasons could explain such discrepancies ; in particular, unprecise matching of the Greenland and Byrd time scales, the possible existence of artefacts (melting, complication due to ice flow) in the Greenland records and dampening of CO_2 changes in the Byrd record. None of these explanations appears completely satisfactory and to solve this important question we may have to wait for the recovery of a new ice core from Central Greenland.

CO_2 is not the only greenhouse gas which shows large concentration shifts associated with changing climate. Recent measurements on the Dye 3 and Byrd cores (Stauffer et al., 1988 and Figure 11) show that CH_4 concentrations follow similar patterns with an increase from about 350 ppbv near the end of the last ice age to Holocene values of about 650 ppbv.

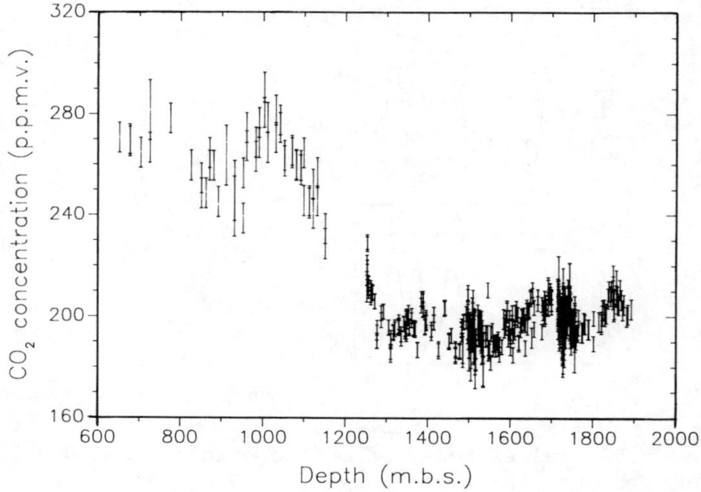

Figure 10. CO_2 concentrations measured in the Byrd ice core plotted versus depth. (From Neftel et al., 1988).

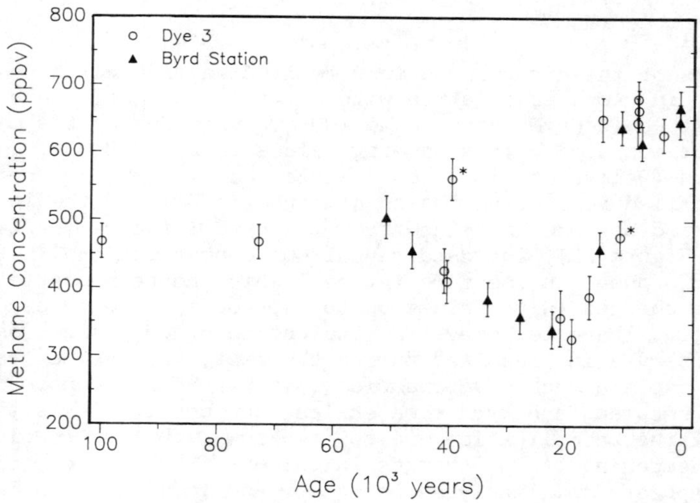

Figure 11. Measured CH_4 concentration plotted against estimated ages of air enclosed in bubbles of ice. Circles represent results from the ice core from Dye 3. Triangles represent results from the ice core from Byrd Station. (From Stauffer et al., 1988).

7. ICE SHEETS AND CLIMATE

The current interpretation of isotopic records relates to surface temperature changes which may reflect both climatic and ice surface elevation changes. Besides ice flow perturbation, discussed above, ice thickness changes may contribute to the isotopic signal. Total gas content (V) measured in ice cores depends, in particular, on changes of surface atmospheric pressure which is elevation dependent (Raynaud and Lebel, 1979). Although ice sheet thickness changes probably do not significantly affect the isotope temperature interpretations, they provide interesting insights to the ice sheet-climate interaction, as previously discussed for Greenland. Scarse Antarctic data (Lorius et al., 1984) support the picture of a present-day Antarctic ice sheet not dramatically different from the one during the Last Glacial Maximum, but somewhat thicker in central parts and thinner in more coastal areas. Such changes in the shape of the ice sheet are in agreement with sensitivity studies stressing the quick response of the central part to changes in the rate of snow accumulation (which was smaller during the ice age) compared to coastal areas, which react primarily to sea level and temperature changes. As a result, temperature changes in central Antarctic areas could be somewhat larger in amplitude than estimated. However, interpretations of V variations in terms of elevation changes have assumed that the atmospheric pressure field over Antarctica remained unchanged. In fact part of the variance observed could be due to changes in the atmospheric pressure pattern and a more definite interpretation of V data requires further efforts regarding a better definition of the transfer function and independent information (possibly from General Circulation Models) on possible changes in the pressure field between an ice age and an interglacial (Lorius et al., 1988b).

8. THE LAST CLIMATIC CYCLE : VOSTOK DATA

Although the record from the Camp Century, Dye 3 and Byrd ice cores extend back to the end of the previous interglacial, the Vostok ice core is the first to provide an environmental record (essentially undisturbed by ice flow conditions) which covers the entire last glacial-interglacial cycle.

The climatic record (Jouzel et al., 1987 and Figures 12 and 15) shows the existence of two interglacials (A, the current Holocene and G) and extends back to the previous ice age (H). The peak of the last interglacial is significantly warmer (about 2°C) than the Holocene period. Conditions equivalent to those prevailing during the LGM (B) were encountered only at the end of the penultimate glacial (H), around 150 ky B.P. The last deglaciation is clearly a two step process with two warming periods interrupted by a 2°C temperature reversal lasting about 1 ky. The last glacial period is characterized by three well marked temperature minima (with the one occurring around 110 ky B.P. about 2°C warmer than full glacial conditions) separated by two interstadials respectively 4 and 6°C above the minima.

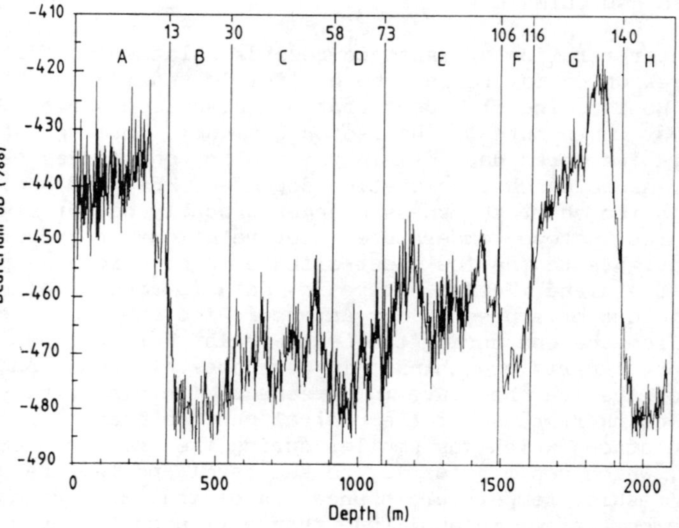

Figure 12. Deuterium content (in °/₀₀ w.r.t. SMOW) against depth in the Vostok ice core with successive stages and the ages corresponding to the limits between these stages. (From Jouzel et al., 1987).

These data offer a unique opportunity to examine, using a continental record, the link between an atmospheric time series and astronomical forcing. Spectral analysis shows that beside the ~ 100 ky oscillation, the isotope record is dominated by a strong ~ 40 ky component (clearly seen by visual inspection of the record). The analysis also shows a weak 20 ky component. The two frequency bands can be associated with the obliquity and precession cycles, respectively.

In addition to this temperature record, the Vostok ice core has allowed an evaluation of past accumulation (precipitation) changes (Figure 13). One approach is based on the fact that current accumulation rates in Antarctica are governed by the amount of water vapour circulating above the inversion layer. This amount is itself controlled by temperature via the saturation vapour pressure (Robin, 1977). Using the linear relationship between the atmosphere (above the inversion) and surface temperatures, the isotope-based temperature record can thus be used to estimate past accumulation. Results give quite similar accumulation values during the two interglacials while precipitation appears to have been reduced to about 50 % of the current value during the coldest periods of the ice age. A second approach (Raisbeck et al., 1987 ; Yiou et al., 1985) is based on ^{10}Be measurements performed along the ice core. Concentrations of this long lived cosmogenic radioisotope are quite similar during the two interglacials ; they increase during colder periods, by up to a factor of 2 during full glacial conditions. Assuming a constant ^{10}Be deposition flux, the measured concentrations are reflecting the amount of snow precipitation. Indeed there is a very good agreement between accumulation rates derived independently from the isotope temperature and ^{10}Be records. The coherence of these results lends some support to this estimate of past precipitation and also indirectly further supports the deuterium-temperature interpretation.

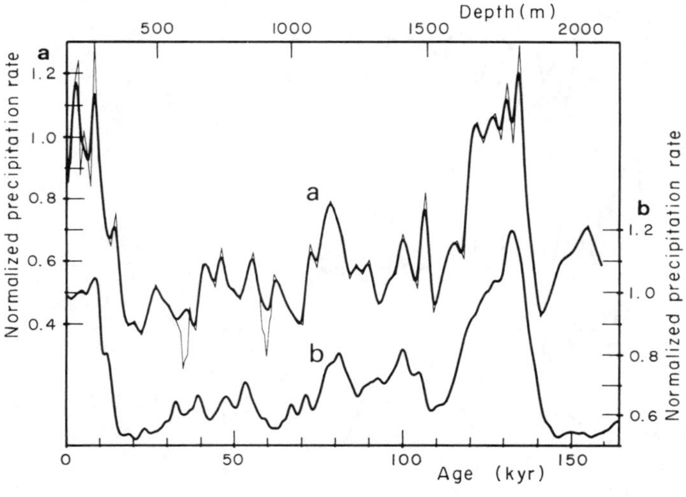

Figure 13. Vostok ice core : normalized precipitation rate with respect to Holocene mean value.
a (upper curve, left scale) from ^{10}Be concentration record (Raisbeck et al., 1987), the thin curve is obtained using all the ^{10}Be data and the thick smoothed curve not taking into account the ^{10}Be peaks at ~ 35 and 60 ky BP.
b (lower curve, right scale) : from the isotopic temperature record (Jouzel et al., 1987).
(From Jouzel et al., 1988).

Marine and continental aerosols as well as all major ions have now been determined (De Angelis et al., 1987 ; Legrand et al., 1988a) in the Vostok ice core. As an illustration, aluminium (an indicator of continental source), marine sodium and acidity (which is potentially linked to volcanic activity) profiles are given Figure 14. The main features of the Al and especially Na records are rather closely related to climate, with high concentrations associated with coldest stages. However this overall agreement is not marked for the relatively cold period around 110 ky B.P. ; the absence of high Al values could possibly suggest that the rather low temperature of stage F reflects a regional climatic feature. In contrast to the marine (Na) and terrestrial (Al) inputs, there is no apparent relationship between H_2SO_4 content and the temperature record ; this suggests no long term correlation between volcanism and climate. Interpretations proposed above for the impurity concentration of ice during the LGM are likely applicable during the entire last climatic cycle.

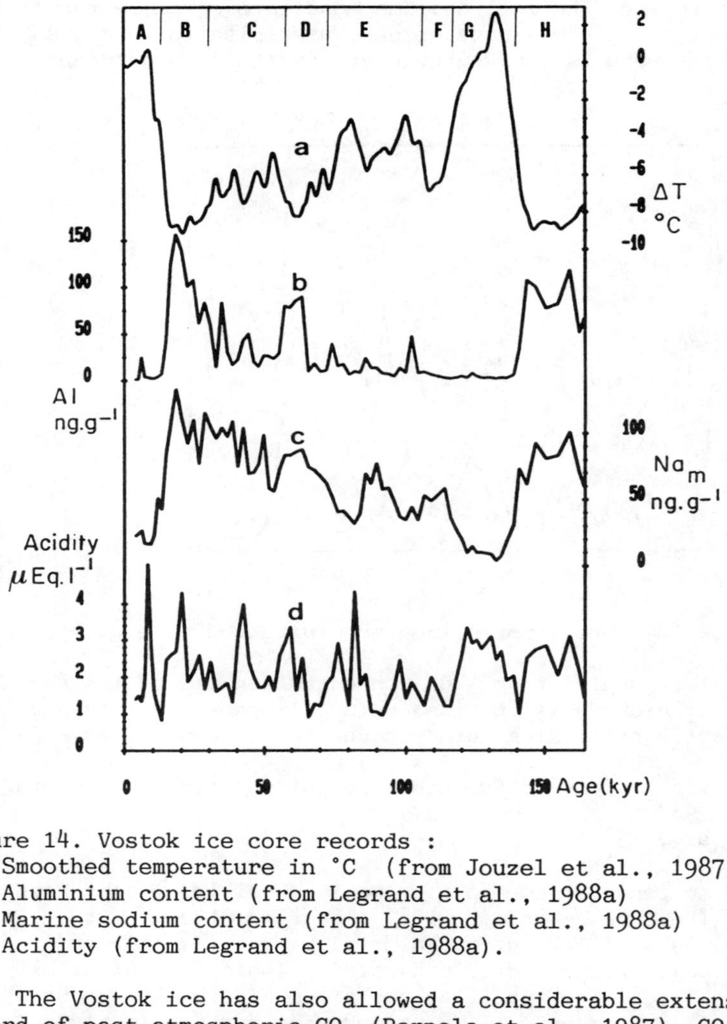

Figure 14. Vostok ice core records :
a : Smoothed temperature in °C (from Jouzel et al., 1987).
b : Aluminium content (from Legrand et al., 1988a)
c : Marine sodium content (from Legrand et al., 1988a)
d : Acidity (from Legrand et al., 1988a).

The Vostok ice has also allowed a considerable extension of the record of past atmospheric CO_2 (Barnola et al., 1987). CO_2 concentrations (Figure 15) exhibit two very large changes between two levels centred around 190-200 and 270-280 ppmv. They correspond to the transition between full-glacial conditions (low CO_2) of the last and penultimate glaciations and the two major warm periods (high CO_2) of the record, the Holocene and the previous interglacial. The high level is comparable with preindustrial CO_2 concentration and the low level is in good agreement with the above mentioned low CO_2 values during the Last Glacial Maximum. The coarse sampling does not allow the search for rapid changes such as those recorded in the Greenland ice which may not be detected anyway due to CO_2 signal averaging associated with the air trapping process under the conditions at the Vostok site.

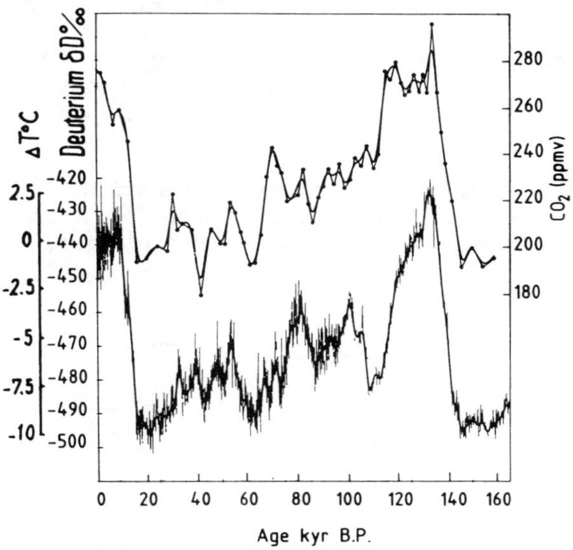

Figure 15. CO_2 concentrations ("best estimates") and smoothed values (spline function) in ppmv plotted against age in the Vostok record (upper curves) and atmospheric temperature change derived from the isotopic profile (lower curve). The deuterium scale corresponds to values after correction for deuterium changes of oceanic water. (From Barnola et al., 1987).

Besides the large glacial-interglacial signal there is a dominant ~20 ky periodic component, close to the orbital precessional cycle. All along the last climatic cycle there is a high correlation between CO_2 and the isotope temperature records. Some differences are nevertheless observed ; there is for instance no very low CO_2 value associated with the rather cold stage depicted around 110 ky B.P. Also both parameters vary simultaneously during transitions from glacial to interglacial while CO_2 clearly lags behind the climatic record when proceeding from warm to cold stages.

There is no conclusive explanation yet for the long term and rapid atmospheric CO_2 changes as described from polar ice core records. Current ideas focus on the role of the oceanic reservoir through a combination of biological productivity and oceanic circulation changes (Broecker and Peng, 1986) and the possible effect of variations of sea ice extent. The Vostok data suggest (Barnola et al., 1987) that the mechanisms involved may have differed depending upon the climatic stages.

Current work (Raynaud et al. 1988) also suggests a close correlation between CH_4 concentration change and climate (Figure 16). A large enhancement (by about 80 %) in CH_4 concentration occured between the end of the penultimate ice age and the following interglacial (roughly in the 160-120 ky B.P. time interval). These and other (Stauffer et al., 1988) results suggest that, like for CO_2, CH_4 and climate have been closely associated throughout the last climatic cycle. Observed CH_4 changes may tentatively be explained (Raynaud et al., 1988) by conside-ring climatic effects on the CH_4 cycle, (i.e. changing source and sink intensities).

Though very different in their biogeochemical behaviour these two greenhouse gases show trends which support the concept of the earth as a complex system with numerous physical, chemical and biological interactions at various time scales.

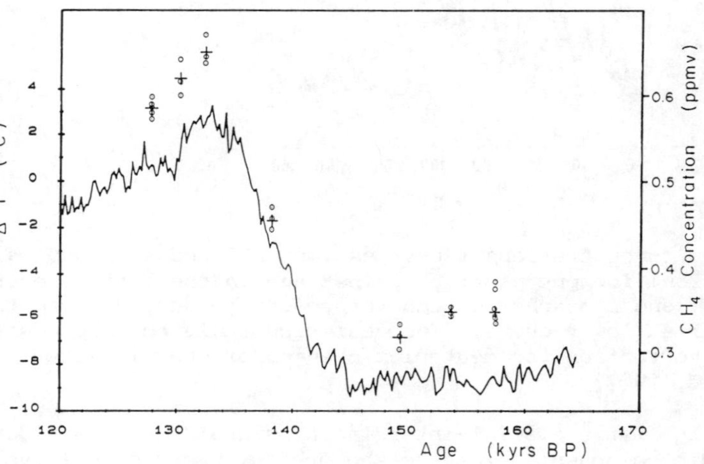

Figure 16. Vostok ice core : CH_4 concentrations (+ indicate mean values) and isotopic temperature profile. (From Raynaud et al., 1988).

9. CONCLUDING REMARKS

Over the last two decades a very small number of polar ice cores have provided a wealth of information about climate and climate related parameters, with different time resolutions, over the last full glacial interglacial cycle.

As outlined in this paper there is need for further work to extract all relevant information which is stored in the ice sheet layers and to quantitatively link them to atmospheric features. Many of the changes recorded are clearly of global significance ; in some cases the spatial validity of the data has to be established. Dating accuracy has to be improved and there is a need to correlate Greenland and Antarctic time series and to compare the ice records with other proxy data.

The hypothesis that orbital and induced insolation changes in the northern hemisphere has controlled paleoclimates (Berger, 1988) has received considerable support from many proxy records including recent data from ice cores which hold probably sufficient information to investigate higher frequency events. There are strong feedbacks in the climatic system and ice cores reveal that besides ice sheet growth and decay, sea ice and snow extents and possibly physical internal oscillations in the atmosphere-ocean system, many other parameters have to be considered in order to understand large past climatic changes. The finding that significant changes in atmospheric chemistry accompagny major climatic changes opens up new questions inasmuch that concentration changes in atmospheric CO_2 and CH_4 during a glacial-interglacial are comparable to the changes now observed in response to human activities and that the reconstructed temperature variations are also of the same order as those predicted by modelled effects of man's impact on climate in the next century.

It has been suggested from ice core data (Genthon et al., 1987) that CO_2 may have played a significant role in past climatic change ; recent results on CH_4 (Stauffer et al., 1988 ; Raynaud et al., 1988) show that other greenhouse gases are also likely involved. Impurity concentrations measured in ice also show significant changes in the past resulting from large variations in the aerosol loading. There are not only changes in the amount of aerosols but also relative changes of species (Delmas and Legrand, 1988). Interglacial snow contains principally a mixture of small amounts of sea salts and mainly two mineral acids, H_2SO_4 and HNO_3 while the continental dust contribution is weak. During the cold ice age stages the input of crustal material becomes more important while the sea salt is also increased, but to a lesser extent. It has been claimed that the resulting increase in the aerosol optical depth could have made a significant contribution to the climatic cooling of the LGM (Harvey, 1988). It is also now suggested from non-sea salt sulphate concentration (which is likely linked with marine biota productivity) profile from the Vostok ice core (Legrand et al., 1988b) that the associated changes in the number of cloud condensation nuclei would have influenced the marine stratus clouds albedo (Charlson et al., 1987) and cause significant cooling at the earth surface during glacial conditions. As all these atmospheric chemistry changes would induce temperature effects in the same direction, there is then growing evidence that they have at least significantly contributed in glacialinterglacial climatic changes (Lorius et al., 1988a).

There is now a need for a "two way" interaction between ice core and modeling communities. Paleo data can help to calibrate and validate models which can in turn provide useful information about gaps in existing past environmental records.

Ice cores appear to have unique potential as they contain various simultaneous proxy records which reflect many features of the past atmosphere on both long and short time scales. Ice cores also show that the variability of chemical parameters is as large as that observed for physical climatic parameters and that both are highly interactive. Such findings outline the need to study the earth as a integrated physical-biological-chemical system. New drilling projects in polar ice sheets should have a high priority. They will hopefully provide further useful information relevant to climatic and environmental changes.

REFERENCES

Barnola, J.M., Raynaud, D., Korotkevich, Y.S., and Lorius, C. 1987: Vostok ice core provides 160,000 year record of atmospheric CO_2, *Nature*, **329**, 408-414.

Berger, A. 1988: Milankovitch theory and climate, *Rev. Geophys.*, in press.

Broecker, W.S. and Peng, T.G. 1986: Glacial to interglacial changes in the operation of the global carbon cycle, *Radiocarbon*, **28**, 309-327.

Budd, W.F., and Young, N.W., 1983: Application of modelling techniques to measured profiles of temperature and isotopes. *The climatic record in polar ice sheets,* ed. G. de Q. Robin, 150-179, Cambridge University Press.

Charlson, R.J., Lovelock, J.E., Andreae, M. O., and Warren, S.G. 1987: Oceanic phytoplankton, atmospheric sulphur, cloud albedo and climate, *Nature*, **326**, 6114, 655-661.

Dahl-Jensen, D. and Johnsen, S.J. 1986: Paleotemperatures still exist in the Greenland ice sheet, *Nature*, **320**, 250-252.

Dansgaard, W., Clausen, H.B., Gundestrup, N., Hammer, C.U., Johnsen,, S.J., Kristinsdottir, P.M. and Reeh, N. 1982: A new Greenland deep ice core, *Science*, **218**, 1273-1277.

——, Johnsen, S.J., Clausen, H.B., Dahl-Jensen, D., Gundestrup, N., Hammer, C.U. and Oeschger, H. 1984: North Atlantic Climatic oscillations revealed by deep Greenland Ice Cores. Climatic Processes and Climate Sensitivity. *Geophysical Monograph*, **29**, Maurice Ewing. 5 : J.E. Hansen, T. Takahashi (eds.), 288-298.

——, and Oeschger, H. 1988: Past environmental long-term records from the Arctic. *The Environmental Record in Glacier and Ice-Sheet,* Eds H. Oeschger and C.C. Langway Jr., John Wiley & Sons Ltd., Dalhem Konferenzen, in press.

Davidson, C.I. 1988: Mechanisms of wet and dry deposition of atmospheric contaminants to snow surfaces, *The Environmental Record in Glacier and Ice-Sheet,* Eds H. Oeschger and C.C. Langway Jr., John Wiley & Sons Ltd., Dalhem Konferenzen, in press.

De Angelis, M., Jouzel, J., Lorius, C., Merlivat, L., Petit, J.R., and Raynaud, D. 1984: Ice age data for climate modelling from an Antarctic (Dome C) ice core, *New Perspectives in Climate Modelling*, Eds. A.L. Berger and C. Nicolis, Elsevier.

——, Barkov, N.I. and Petrov, V.N. 1987: Aerosol concentrations over the last climatic cycle (1660 ky) from an Antarctic ice core, *Nature* **325**, 318-321.

Delmas, R.J., Ascencio, J.M. and Legrand, M. 1980: Polar ice evidence that atmospheric CO_2 20,000 yr B.P. was 50 % of present, *Nature*, **284**, 155-157.

——, and Legrand, M.R.. 1988: Long term changes in the concentrations of major chemical compounds (soluble and insoluble) along deep ice cores, *The Environmental Record in Glacier and Ice-Sheet,* Eds H. Oeschger and C.C. Langway Jr., John Wiley & Sons Ltd., Dalhem Konferenzen, in press.

Fisher, D., Koerner, R., Paterson, W., Dansgaard, W., Gundestrup, N. and Reeh, N. 1983: Effect of wind scouring on climatic records from ice core oxygen-isotope profiles, *Nature*, **301**, 205-209.

Genthon, C., Barnola, J.M., Raynaud, D., Lorius, C., Jouzel, J., Barkov, N.I., Korotkevich, Y.S. and Kotlyakov, V.M. 1987: Vostok ice core : climatic response to CO_2 and orbital forcing changes over the last climatic cycle, *Nature*, **329**, 414-418.

Hammer, C.U., Clausen, H.B., Dansgaard, W., Neftel, A., Kristinsdottir, P. and Johnson, E. 1985: Continuous impurity analysis along the Dye 3 deep core, in "Greenland ice cores : geophysics, geochemistry and the environment", Eds. C.C. Langway, H., Oeschger and W. Dansgaard, *Geophysical Monograph*, **33**, American Geophysical Union, Washington, D.C., 90-94.

Harvey, L.D. 1988: Climatic impact of ice-age aerosols, *Nature*, 334 333-335.

Johnsen, S.J., Dansgaard, W., Clausen, H.B. and Langway, C.C. 1972: Oxygen isotope profiles through the Antarctic and Greenland ice sheets, *Nature*, **235**, 429-434.

——, Dansgaard, W. and White, J. 1988: The origin of Arctic precipitation under glacial and interglacial conditions, *Tellus*, in press.

Joussaume, S., Jouzel, J. and Sadourny, R. 1988: Simulations of the Last Glacial Maximum with an atmospheric General Circulation Model including paleoclimatic tracers. *Proceeding IUGG, Contribution of Geophysical Sciences to Climate Change Studies,* American Geophysical Union, in press.

Jouzel, J. and Merlivat, L. 1984: Deuterium and oxygen 18 in precipitation : modelling of the isotopic effect during snow formation, *J. Geophys. Res.*, **89**, 11749-11757.

——, Lorius, C., Petit, J.R., Genthon, C., Barkov, N.I., Kotlyakov, V.M. and Petrov, V.N. 1987: Vostok ice core : a continuous isotope temperature record over the last climatic cycle (160,000 years), *Nature*, **329**, 403-409.

——, Russel, R., Suozzo, R.J., Koster, R.J., White, J.W.C. and Broecker, W.S. 1987: Simulations of the HDO and $H_2{}^{18}O$ atmospheric cycles using the NASA GISS General Circulation Model : the seasonal cycle for present day conditions, *J. Geophys. Res.*, **92**, 14739-14760.

——, Raisbeck, G., Benoist, J.P., Yiou, F., Lorius, C., Raynaud, D., Petit, J.R., Barkov, N.I., Korotkevitch, Y.S. and Kotlyakov, V.M. 1988: The Antarctic climate over the late glacial period, *Quaternary Research*, in press.

Khalil, M.A.K., and Rasmussen, R.A. 1987: Atmospheric methane : Trends over the last 10,000 years, *Atmospheric Environment*, **21**, 2445-2452.

——, and Rasmussen, R.A. 1988: Nitrous oxide : Trends and global mass balance over the last 3,000 years, *Annals of Glaciology*, **10**, 73-79.

Legrand, M.R. and Delmas, R.J. 1987: Volcanic H_2SO_4 in the antarctic ice sheet : a 220 yr continuous record, *Nature*, **327**, 671-676.

——, Lorius, C., Barkov, N.I. and Petrov, V.N. 1988a: Vostok (Antarctica ice core) : atmospheric chemistry changes over the last climatic cycle (160,000 yr), *Atmospheric Environment*, **22**, 317-331.

——, Delmas, R.J. and Charlson, R.J. 1988b: Climate forcing implications from Vostok ice core sulphate data, *Nature*, in press.

Lorius, C. and Merlivat L. 1977: Distribution of mean surface stable isotope values in East Antarctica : observed changes with depth in a coastal area. *Isotopes and Impurities in Snow and Ice*. Proc. Grenoble Symp. August-September 1975, IAHS, **118**, 127-137.

——, Merlivat, L., Jouzel, J. and Pourchet, M. 1979: A 30 000 y isotope climatic record from Antarctic ice, *Nature*, **280**, 644-48.

——, Raynaud, D., Petit, J.R., Jouzel, J. and Merlivat, L. 1984: Late Glacial Maximum-Holocene atmospheric and ice thickness changes from Antarctic ice core studies, *Annals of Glaciology*, **5**, 88-94.

——, Jouzel, J., Ritz, C., Merlivat, L., Barkov, N.I., Korotkevich, Y.S. and Kotlyakov, V.M. 1985: A 150,000 year climatic record from Antarctic ice, *Nature*, **316**, 591-596.

——, Jouzel, J., Raynaud, D., Barkov, N.I., Korotkevich, Y.S., and Kotlyakov, V.M., 1988a: Antarctic ice core : CO_2 and climatic change over the last climatic cycle, *EOS*, **69**, N° 26, 681, 683-684.

——, Raisbeck, G., Jouzel, J. and Raynaud, D. 1988b: Long-term environmental records from Antarctic ice cores, *The Environmental Record in Glacier and Ice-Sheet*, Eds H. Oeschger and C.C. Langway Jr., John Wiley & Sons Ltd., Dalhem Konferenzen, in press.

Mayewski, P.A., Lyons, W.B., Spencer, M.J., Twickler, M., Dansgaard, W., Koci, B., Davidson, C.I. and Honrath, R.E. 1986: Sulfate and nitrate concentrations from a South Greenland Ice Core, *Science*, **232**, 975-977.

Neftel, A., Oeschger, H., Schwander, J. Stauffer, B. and Zumbrunn, R. 1982: Ice core sample measurements give atmospheric CO_2 content during the past 40,000 yr. *Nature*, **295**, 220-223.

——, Beer, J., Oeschger, H. Zurcher, F. and Finkel, R.C. 1985a: Sulphate and nitrate concentrations in snow from South Greenland 1895-1978, *Nature*, **314**, (6012), 611-613.

——, Moor, E., Oeschger, H. and Stauffer, B. 1985b: Evidence from polar ice cores for the increase in atmospheric CO_2 in the past two centuries, *Nature*, **313** 45-47.

——, Oeschger, H., Staffelbach, T. and Stauffer, B. 1988: CO_2 record in the Byrd ice core 50,000-5,000 years B.P., *Nature*, **331**, 609-611.

Palais, J. and Legrand, M. 1985: Soluble impurities in the Byrd station ice core, Antarctica ; their origin and sources, *J. Geophys.Res.*, **90**, C1, 1143-1154.

Paterson, W.S.B., Koerner, R.M., Fisher, D., Johnsen, S.J., Clausen, H.B., Dansgaard, W., Bucher, P. and Oeschger, H. 1977: An oxygen-isotope climatic record from the Devon Island ice cap, Arctic Canada, *Nature*, **266**, 508-511.

Pearman, G.I., Etridge,F., de Silva, F. and Fraser, P.J. 1986: Evidence of changing concentrations of CO_2, N_2O and CH_4 from air bubbles in antarctic ice, *Nature*, **320**, 248-250.

——, and Fraser, P.J. 1988: Sources of increased methane, *Nature*, 332, 489-490.
Petit, J.R., Briat, M. and Royer, A. 1981: Ice age aerosol content from East Antarctic ice core samples and past wind strength, *Nature*, 293, 391-394.
——, Duval, P. and Lorius, C. 1987: Long-term climatic changes indicated by crystal growth in polar ice, *Nature*, **326**, 62-64.
Raisbeck, G.M., Yiou, F., Bourles, D., Lorius, C., Jouzel, J. and Barkov, N.I. 1987: Evidence for two intervals of enhanced ^{10}Be deposition in Antarctic ice during the last glacial period, *Nature*, **326**, 273-277.
Raynaud, D. and Lebel, B. 1979: Total gas content and surface elevation of polar ice sheets, *Nature*, **281**, 289-291.
——, and Barnola, J.M., 1985: CO_2 and climate : information from Antarctic ice core studies. *Current Issues in Climate Research*, Eds. Ghazi, A. and R. Fantechi, 240-246, D. Reidel.
——, Chappellaz,J., Barnola, J.M., Korotkevich, Y.S. and Lorius, C., 1988: Vostok ice core reveals glacial-interglacial CH_4 change about 140 ky ago : climatic and CH_4 cycle implications, *Nature*, 333, 655-657.
Reeh, N. 1988: Dating by ice-flow modelling : a useful tool or an excercise in applied mathematics. ^font=3The Environmental Record in Glacier and Ice-Sheet;, Eds H. Oeschger and C.C. Langway Jr., John Wiley & Sons Ltd., Dalhem Konferenzen, in press.
Robin, G. de Q. 1977: Ice cores and climatic changes, *Phil. Trans. R. Soc. Land.*, **B 280**, 143-168.
Schwander, J., Stauffer, B. 1984: Age difference between polar ice and the air trapped in its bubbles, *Nature*, **311**, (5981), 45-47.
Siegenthaler, U. and Oeschger, H. 1987: Biospheric CO_2 emissions during the past 200 years reconstructed by deconvolution of ice core data, *Tellus*, **39B**, 140-154.
Stauffer, B., Hofer, H., Oeschger, H., Schwander, J. and Siengenthaler, U., 1984;. Atmospheric CO_2 concentration during the last glaciation, *Annals of Glaciology*, 5, 160-164.
——, Fischer, G., Neftel, A. and Oeschger, H. 1985: Increase of atmospheric methane recorded in antarctic ice, *Science*, **229**, 1386-1388.
——, Lochbronner, E., Oeschger, H. and Schwander, J. 1988: Methane concentration in the glacial atmosphere was only half that of the preindustrial Holocene, *Nature*, **332**, 812-814.
Thompson, L.G. and Mosley-Thompson, E. 1981: Microparticle concentration variations linked with climatic change : evidence from polar ice cores, *Science*, **212**, 812-815.
Yiou, F., Raisbeck, G.M., Bourles, D., Lorius, C. and Barkov, N.I. 1985: ^{10}Be at Vostok Antarctica during the last climatic cycle, *Nature*, **316**, 616-17.

INTERMEDIATE AND DEEP WATER CHARACTERISTICS DURING THE LAST GLACIAL MAXIMUM.

J.C. Duplessy, L. Labeyrie, N. Kallel, and A. Juillet-Leclerc
Centre des Faibles Radioactivités
Laboratoire Mixte CNRS-CEA
F-91198 Gif-sur-Yvette Cedex
FRANCE

INTRODUCTION

Deep water is formed at high latitudes, where the atmosphere extracts heat from the sea surface and where a salinity increase results from sea ice formation during winter conditions. Both processes make the surface water sufficiently dense to sink to great depth and flow toward the equator. Above the deep and bottom water masses of the ocean lie the intermediate water masses, which are associated with the subpolar convergences in both hemispheres.

Gases are exchanged only at the ocean surface. As the diffusion is an extremely slow process, the deep circulation alone is responsible for the continuous supply of dissolved oxygen, permitting life in the deep sea. As the *in situ* oxygen consumption rate decreases exponentially with depth (Richards, 1965), deep and bottom waters are richer in oxygen than intermediate waters, where oxidation of organic matter is important. This results in a well-developed oxygen minimum associated with a CO_2 maximum. These features depend on a balance between the rate of oxygen consumption at depth (which is itself dependent on the flux of organic matter to the abyss and therefore to the surface productivity) and the rate of exchange with well-oxygenated surface water. The concentrations of dissolved oxygen and total dissolved CO_2 are thus depending on the climate which drives the formation processes of both intermediate and deep water masses.

The global deep water circulation has changed dramatically during the last 150,000 years (the last climatic cycle), because conditions prevailing in the high latitude surface water of the northern hemisphere have varied considerably (Duplessy et al., 1980, 1984, 1988a; Curry and Lohmann, 1982; Boyle and Keigwin, 1982, 1985, 1987; Shackleton et al., 1983; Duplessy and Shackleton, 1985; Mix and Fairbanks, 1985; Corliss et al., 1986; Oppo and Fairbanks, 1987; Kallel et al., 1988). For example, the

A. Berger et al. (eds.), Climate and Geo-Sciences, 105–120.

Corliss et al., 1986; Oppo and Fairbanks, 1987; Kallel et al., 1988). For example, the Norwegian Sea was permanently ice-covered during most of the last glaciation (isotope stages 2–4). This sea was thus strongly stratified and did not form any bottom water until the last deglaciation (Duplessy et al., 1988b).

Similarly, the location of intermediate water formation experienced large climate-related changes. Although they are less documented than those of deep water, several modifications of the intermediate water circulation have been reported (Zahn et al., 1987; Oppo and Fairbanks, 1987; Kallel et al., 1988). In this paper, we shall review the major characteristics of the deep and intermediate water masses during the last glacial maximum (isotope stage 2, about 18,000 years ago).

GEOCHEMICAL BASIS

Kroopnick (1985) has demonstrated that the distribution of $^{13}C/^{12}C$ ratio of the total CO_2 (ΣCO_2) dissolved in sea water delineates the general distribution of water masses and that modern $\partial^{13}C$ gradients in the deep ocean follow the net flow of the deep waters. The distribution of the $\partial^{13}C$ value of the ΣCO_2 dissolved in intermediate, deep and bottom waters of the ocean can be used as a tracer of the abyssal circulation under the following conditions: (i) within a deep water mass, the $\partial^{13}C$ value decreases with increasing oxidation of organic matter, such that the $\partial^{13}C$ decrease is related both to the surface productivity and the time elapsed since the water mass was isolated from the atmosphere; (ii) when two water masses are compared, the $\partial^{13}C$ value in each depends not only on their residence time at depth, but also on the ratio between the CO_2 produced from organic matter oxidation and the bicarbonate and carbonate ions derived from carbonate dissolution ; (iii) mixing between two water masses results in $\partial^{13}C$ values intermediate between the two original components (see a detailed discussion in Duplessy and Shackleton, 1985).

Some species of benthic foraminifera (Genus *Cibicides*) closely record the modern $\partial^{13}C$ distribution in the world ocean, providing a good proxy for the reconstruction of the past $\partial^{13}C$ gradients in the ocean (Duplessy et al., 1984; Zahn et al., 1986). The $\partial^{13}C$ value of other species seem much less well related to that of the deep waters in which they are supposed to have lived. For instance, the $\partial^{13}C$ value of *Uvigerina* has been reported to decrease when the flux of organic matter to the sediment is high (Zahn et al., 1986). Gif analyses showed that the $\partial^{13}C$ values of *Bulimina acculeata* is strongly size-dependent whereas that of *Oridorsalis tener* is so erratic, that these two species cannot be used to reconstruct the $\partial^{13}C$ value of ocean water, whereas their $\partial^{18}O$ value is a reliable indicator of ocean water $\partial^{18}O$ (Shackleton and Cita, 1979).

$\partial^{18}O$ variations measured in foraminiferal shells reflect both the global variations of the ocean water $\partial^{18}O$ due to continental ice volume changes and the isotopic fractionation between calcium carbonate and water, which depends upon the temperature at which foraminifera formed their tests (Emiliani, 1955). Recently, Labeyrie et al. (1987) have separated these effects and derived a $\partial^{18}O$ record which is independent of temperature changes and reflects only the mean sea water $\partial^{18}O$

variations during the last 135,000 years due to continental ice volume changes. As this record is common to all benthic foraminiferal records and constitutes 60 to 100% of these records depending on their location and depth, the classical oxygen stratigraphy can be used to recognize unambiguously the sediment deposited during the major events of the last climatic cycle.

The simultaneous determination of the oxygen and carbon isotopic composition of benthic foraminifera in deep sea cores therefore allows one to reconstruct the $\partial^{13}C$ value of abyssal waters during the last glacial maximum.

THE ATLANTIC OCEAN

Data from 30 cores have been used by Duplessy et al. (1988a) to reconstruct a north-south section between 60° N and 45° S for the last glacial maximum, 18 Kyr ago. For each core, the 18 K level has been determined as that of the first maximum value in the oxygen isotope record. The $\partial^{13}C$ value of the total dissolved CO_2 has been estimated as the mean $\partial^{13}C$ value of the foraminifera genus *Cibicides* analyzed in the 18 K level. The majority of the cores come from the eastern basin, but a few cores from the western basin have also been included. In the northern hemisphere, no significant difference is observed between these two sets of cores, probably because their depth was shallower than that of the deepest parts of the mid-Atlantic ridge and water exchange was not inhibited between the eastern and western basins. Results have therefore been plotted in Figure 1 as a depth-latitude section in the Atlantic Ocean, without taking account of the longitude of the analyzed cores. As cores raised from a water depth shallower than one kilometer are rare, the $\partial^{13}C$ of surface water in the high latitudes of the north Atlantic was estimated from the analysis of $\partial^{13}C$ of *N. pachyderma* using the calibration of Labeyrie and Duplessy (1985).

In the very high latitude of the North Atlantic, a Glacial North Atlantic Deep Water mass, with high $\partial^{13}C$ values is found. This water mass has a quite limited extent toward the south. The salient feature in the section (Figure 1) is the presence of low $\partial^{13}C$ values at 2500 m water depth between 20° and 45° N, whereas higher $\partial^{13}C$ values are found in the same depth range both further north and south. These low $\partial^{13}C$ values can be traced to the surface, where they upwelled around 50°–55°N. The infered circulation pattern shown in Figure 1 is in agreement with the divergence already described by Labeyrie and Duplessy (1985), obtained by mapping the $\partial^{13}C$ values of *N. pachyderma* in the north Atlantic during the last glacial maximum. North of this divergence, surface water sank to the bottom and constituted a water mass which extended south to about 40° N at 4000 m depth.

South of the divergence, intermediate waters rich in ^{13}C are found. Carbon isotope records from the eastern Atlantic close to Gibraltar (Zahn et al., 1986; 1987) and from the Mediterranean Sea (Oppo and Fairbanks, 1987) suggest that the influence of Glacial Mediterranean Outflow Water (GMOW) was more important in the intermediate-deep Atlantic than it is today. Oppo and Fairbanks (1987) argue that GMOW extended to the Caribbean during the last glaciation. Therefore GMOW is a

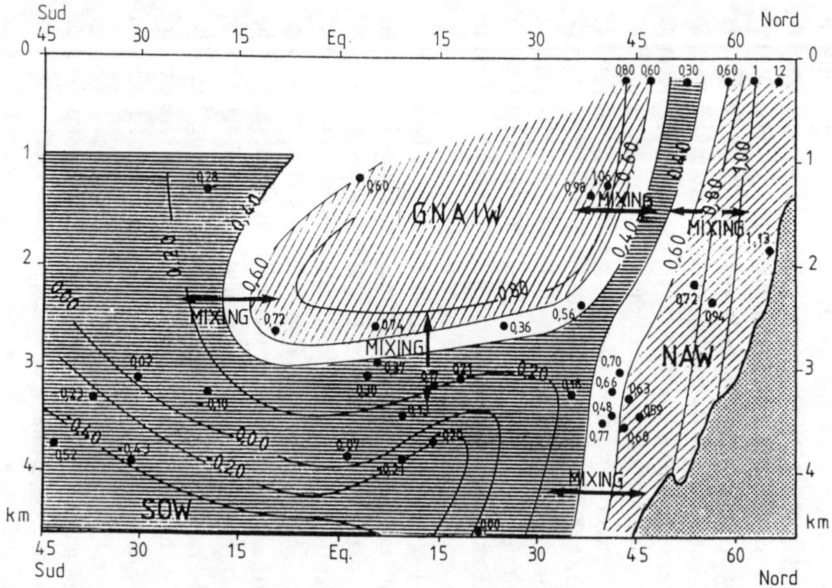

Figure 1: Reconstruction of the $\partial^{13}C$ profile in the eastern Atlantic Ocean during the last glacial maximum, about 18,000 years ago. $\partial^{13}C$ of the total dissolved CO_2 have been estimated from the $\partial^{13}C$ value of benthic foraminifera genus *Cibicides* in 34 sediment cores. Surface $\partial^{13}C$ values have been derived from those of *N. pachyderma* (left coiling) using the correction factor calculated by Labeyrie and Duplessy (1985). SOW indicates a water mass originating from the Southern Ocean, which probably formed by a mechanism similar to that forming the present Antarctic Bottom Water. NAW indicates a water mass originating from the North Atlantic Ocean (and not from the Norwegian Sea). The density increase which caused the surface water to sink was probably due to both cooling and salt content increase during the formation of winter sea ice. GNAIW indicates an intermediate water mass formed by Mediterranean water outflow and possibly sinking at the level of the polar front. The upwelling zone at about 50°N is still hypothetical as the $\partial^{13}C$ surface reconstruction may simply manifest the surface conditions. The flow balance may result from the existence of longitudinal components that our $\partial^{13}C$ reconstruction is not precise enough to detect.

likely component of the Glacial North Atlantic Intermediate Water (GNAIW). However, it is unlikely that GMOW was the only source for GNAIW because strait dynamics implies that the Mediterranean outflow would have been volumetrically smaller than today as the sea level dropped by about 100 m during the last glacial maximum (Assaf and Hecht, 1974).

An additional component for GNAIW may derive from sinking of surface water to intermediate depths south of the polar front. Boyle and Keigwin (1987) have shown that GNAIW was more nutrient depleted, whereas deeper waters were more nutrient enriched. This suggests that the Glacial North Atlantic was an active source of intermediate water production during winter. We do not have enough data to resolve the various components of GNAIW and therefore consider it as a single water mass, which may be traced southward to 15° S.

Below 3000 m, most of the deep eastern Atlantic was filled with bottom water originating from the Southern Ocean so that the relative contribution of water from the southern hemisphere as compared to that from the northern hemisphere in the deep Atlantic was much greater than today. This is in agreement with the Cd data of Boyle and Keigwin (1982;1985), the $\partial^{13}C$ data of Oppo and Fairbanks (1987), the distribution of benthic foraminifera described by Streeter and Shackleton (1979) and Caralp (1987), and the changes of the carbonate compensation depth (Gardner and Hays, 1976).

THE INDIAN OCEAN

In order to determine the impact of the last glaciation on the Indian Ocean's circulation pattern, Kallel et al. (1988) have measured the foraminiferal $\partial^{13}C$ and $\partial^{18}O$ values in seventeen sediment cores from the two major basins of the Northern Indian Ocean, in the depth range from 1250 m to 3420 m (Figure 2).

At 18 K, a very strong and sharp discontinuity separated the high $\partial^{13}C$ values of Intermediate Water to a depth of about 2000 m from negative $\partial^{13}C$ values deeper than 2300 m. This is in sharp contrast with the modern situation where the vertical $\partial^{13}C$ gradient in the North Indian Ocean is lower than 0.1 per mil in this depth range. This change in the vertical profile originates from a decrease of the $\partial^{13}C$ values of Bengal Gulf Deep Water by about 1 ‰ during last glacial maximum. At this time, the mean carbon isotopic composition of the total dissolved CO_2 in the global ocean was 0.32 ‰ smaller than the modern value (Duplessy et al.,1988a) as a result of the reduced mass of the continental biosphere associated with glacial climatic conditions (Shackleton, 1977). The smaller glacial to interglacial $\partial^{13}C$ change for Intermediate Water, compared with this global $\partial^{13}C$ signal, indicates an enhanced glacial ventilation of this water mass, while the larger $\partial^{13}C$ change for Deep Indian Water indicates a very weak ventilation of the deep Indian Ocean during the last glacial maximum.

Cibicides $\partial^{13}C$ values of Glacial Intermediate Water in the western and eastern sectors of the North Indian Ocean are comparable. Yet the deep water in the Arabian

Figure 2: (A) Reconstruction of the isotopic vertical profiles in the Eastern Indian Ocean
$\partial^{13}C$ profiles: (a) for modern conditions (the positive anomaly on this profile is due to the southern position of core RC11-147; (b) for *Cibicides* during the last glacial maximum; (c) $\partial^{13}C$ difference between 18 K and today (the -0.32 of the Northern Arabian Sea vertical line indicates the mean global $\partial^{13}C$ change between 18,000 years B.P. and today due to changes in the mass of the continental biosphere; it represents the expected $\partial^{13}C$ change if there were no changes in the ventilation of water masses.
$\partial^{18}O$ profiles: (d) for modern calcite in isotopic equilibrium with ambient sea water; (e) for LGM *Cibicides;* (f) $\partial^{18}O$ difference between 18 K and today (the 1.1 per mil $\partial^{18}O$ line represents the expected $\partial^{18}O$ difference if there were no deep water temperature changes between 18 K and today. Values > 1.1 indicate lower temperature at 18 K than today).

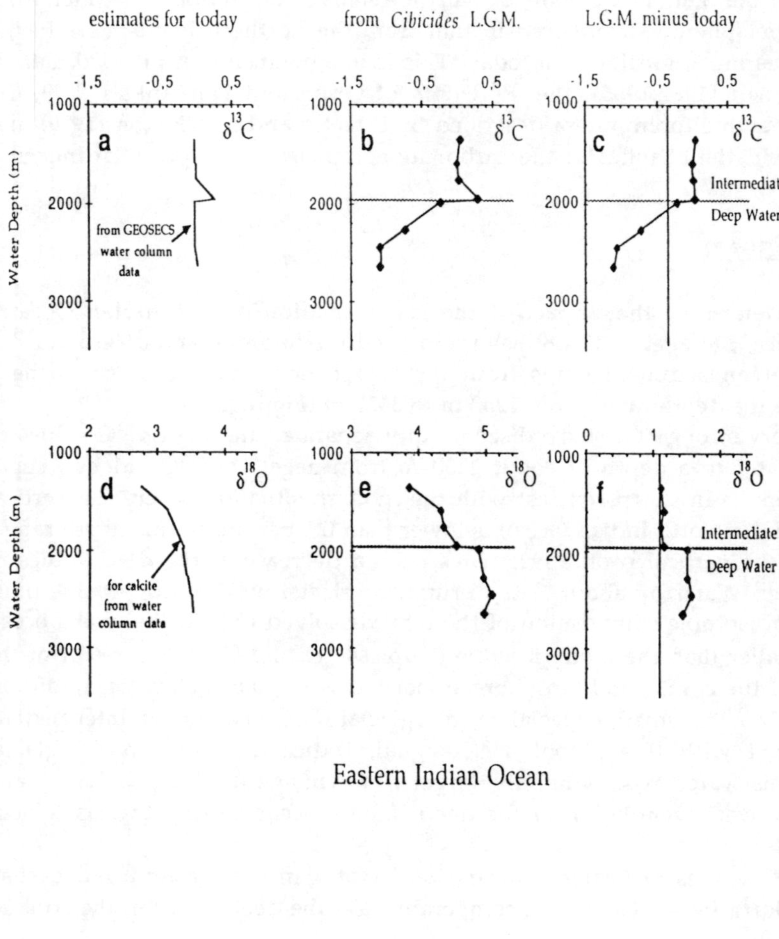

Eastern Indian Ocean

Figure 2: (B) Reconstruction of the isotopic vertical profiles in the Arabian Ocean $\partial^{13}C$ profiles: (a) for modern conditions (the positive anomaly on this profile is due to the southern position of core RC11-147; (b) for *Cibicides* during the last glacial maximum; (c) $\partial^{13}C$ difference between 18 K and today (the -0.32 of the Northern Arabian Sea vertical line indicates the mean global $\partial^{13}C$ change between 18,000 years B.P. and today due to changes in the mass of the continental biosphere; it represents the expected $\partial^{13}C$ change if there were no changes in the ventilation of water masses. $\partial^{18}O$ profiles: (d) for modern calcite in isotopic equilibrium with ambient sea water; (e) for LGM *Cibicides;* (f) $\partial^{18}O$ difference between 18 K and today (the 1.1 per mil $\partial^{18}O$ line represents the expected $\partial^{18}O$ difference if there were no deep water temperature changes between 18 K and today. Values > 1.1 indicate lower temperature at 18 K than today).

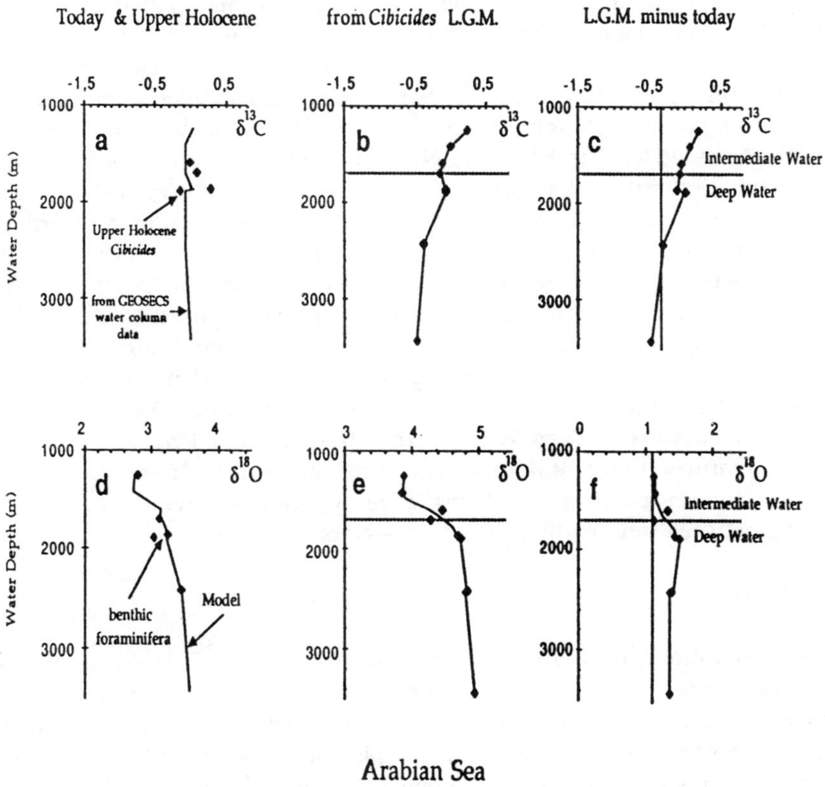

Sea had $\partial^{13}C$ values which were lower than the Intermediate Water, but higher than in the Gulf of Bengal. Such asymmetry, although smaller, exists in the modern northern Indian Ocean. It arises because the passages connecting the basins of the western side are shallower and more narrow than those of the eastern side. As a consequence, the hydrological characteristics of the western deep water are altered by mixing with overlying Intermediate Water during its movement from the Southern Ocean to the Arabian Sea. Deep Water in the eastern basins is more influenced by Antarctic Bottom Water and is thus colder than in the western basins (Fieux et al., 1986).

The profile obtained by plotting the $\partial^{18}O$ values of benthic foraminifera against depth also shows a sharp discontinuity at the same level as the $\partial^{13}C$ profile in both the Bay of Bengal and Arabian Sea (Kallel et al., 1988). This discontinuity is much larger than that observed under modern conditions and indicates that intermediate and deep waters had much more distinct characteristics in glacial times.

Because of a lowering of sea level during glacial conditions, Red Sea and Arabo-Persian Gulf contributions to Intermediate Water were probably very reduced or absent. Three sources are possible for the glacial Intermediate Water in the Northern Indian Ocean: (i) dense water formed in the Northern Arabian Sea by the increased evaporation/precipitation ratio and the possible decreased temperatures of the surface water described by CLIMAP (1976) and Duplessy (1982), (ii) expansion to the North of 10°S of the Antarctic Intermediate Water (AAIW) in the absence of a northern local source and (iii) increased flow of intermediate water from the Pacific Ocean through the Indonesian Archipelago. A more complete coverage is required to determine unambiguously the origin of the glacial North Indian Intermediate Water. Kallel et al. (1988) argued that the vertical salinity/$\partial^{18}O$ gradient in the northern Indian Ocean during 18 K was not significantly different from the modern one. The sharp deep front between Intermediate and Deep Water is thus best explained by a decrease in deep water temperature in the Indian Ocean and the development of a deep thermocline. This would result in a strong density gradient, which prevented noticeable mixing between intermediate and deep waters.

THE PACIFIC OCEAN

Under present conditions, most areas of the Pacific Ocean lie below the carbonate compensation depth, and foraminifera are absent in the sediment. However, it has long been observed that, in contrast with the deep north Atlantic Ocean, dissolution was less intense in most of the Pacific Ocean during isotope stage 2, with a deepening of a few hundred meters of the carbonate compensation depth (Broecker, 1982). This indicates a significant change in the chemistry of the deep waters, with a drop in the water total CO_2 content which may possibly be related to an increase in ventilation at intermediate depth. By contrast with modern conditions, at the last glacial maximum, benthic foraminifera are often present in north Pacific sediments shallower than 4000 m. Duplessy et al. (1988a) have therefore generated a one-dimensional profile reconstructing the mean variations with depth of the $\partial^{13}C$ of the total dissolved CO_2 in the north and equatorial Pacific (Figure 3) by measuring $\partial^{13}C$ values of *Cibicides* in 23 Pacific sediment cores. The large scatter of the $\partial^{13}C$ values for the cores shallower than 2700 m (±0.24 °/oo) may be due to regional differences in water ventilation, as three cores from the eastern Pacific Ocean have a 0.2–0.6 °/oo lower $\partial^{13}C$ value than the cores from the western Pacific Ocean, at equivalent depth. However, the available data are insufficient to reliably interpret these differences. A similar profile was generated for the present conditions (Figure 3) by interpolating from the GEOSECS data (Kroopnick, 1974,1985). The shape of the two profiles is markedly different, as the $\partial^{13}C$ minimum that is present today between 1 and 2.6 km water depth is absent during the L.G.M.. The glacial profile for the Pacific Ocean exhibits $\partial^{13}C$ values which are identical to the modern values for the 1000–2600 m depth range. Deeper than 3000 m, $\partial^{13}C$ values are lower than modern ones by 0.3 °/oo in agreement with the data of Keigwin (1987).

Figure 3 also shows a depth profile of the difference in $\partial^{13}C$ between the Last Glacial Maximum and today's values. A vertical line indicates the global mean shift of -0.32 °/oo. All the $\partial^{13}C$ values from water depths greater than 2600 m show a $\partial^{13}C$ shift indistinguishable from this global mean. By contrast, the mean $\partial^{13}C$ shift for the 1000–2600m depth range is significantly smaller than the global signal. This implies a relatively larger contribution of ventilated water to this water mass during L. G. M. than in the modern ocean. The origin of the ventilated water is difficult to trace with the present data set. Core CH 84-27, located at 700 m depth on the northern slope of East China Sea has significantly lower $\partial^{13}C$ values for 18 K than the other shallow cores. This indicates that there was no winter convection and hence no formation of ventilated deep water in this part of the east China Sea. High $\partial^{13}C$ values are observed in open water of the Pacific down to the equator in a depth range of 712–2600 m. As the sills of the Bering Sea and the Sea of Okhotsk are much deeper than 2000 m, the high $\partial^{13}C$ values cannot be due to deepwater formation by convective mixing in the whole water column during winter in these marginal basins. Duplessy et al. (1988a) therefore believe that the high oxygenation of the water mass in the range 712–2600 m is due to an enhancement of the mechanism which produces intermediate waters in the modern Pacific Ocean, i.e., vertical mixing through the pycnocline in high latitudes and lateral mixing along density surfaces (Reid, 1965).

The contribution indicated by our data of low-nutrient, oxygenated surface water to the Pacific intermediate waters and the great thickness of this water mass is consistent with two features of glacial conditions in the Pacific Ocean: (1) the increase in abundance of *Cibicides* in the north Pacific, since this genus lives preferentially in oxygenated waters (Streeter and Shackleton, 1979) and (2) the decrease in carbonate dissolution and the deepening of the carbonate lysocline (Berger, 1973; Thompson and Saito, 1974).

Kallel (1988) showed that in the equatorial Pacific Ocean, the $\partial^{13}C$ deep front coincides with a sharp variation of the $\partial^{18}O$ value of benthic calcite. This result

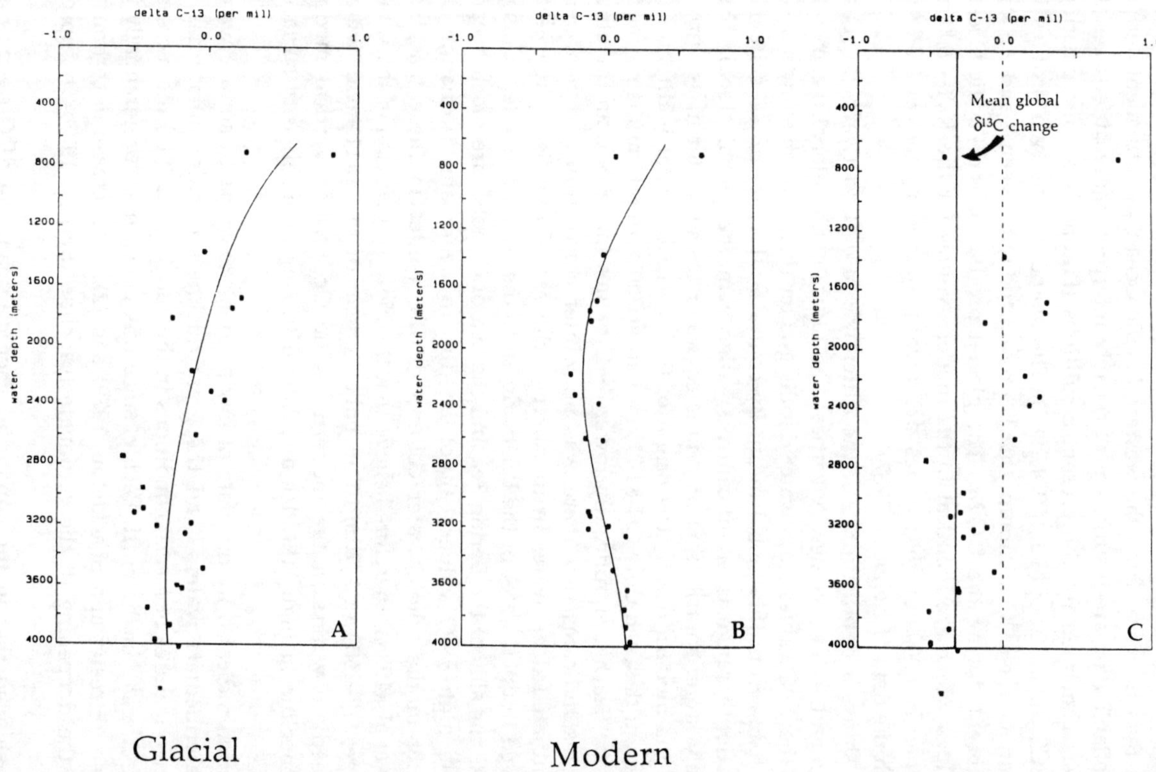

Figure 3: Reconstruction of a one-dimensional $\partial^{13}C$ profile in the Pacific Ocean for the last glacial maximum, about 18,000 years ago (A), for the modern situation (B) and $\partial^{13}C$ difference between the last glacial maximum and today (C). The -0.32 $\partial^{13}C$ vertical line indicates the mean global $\partial^{13}C$ change between 18,000 years B.P. and today due to changes in the mass of the continental biosphere. Dots of the right side of this line correspond to a water mass more oxygenated than today during glacial conditions (as it is the case for intermediate waters).

suggests that, as in the northern Indian Ocean, intermediate and deep waters were strongly separated by a well-developped thermocline and therefore a large density gradient.

THE SOUTHERN OCEAN

Under present conditions, the water which has upwelled from the deep circumpolar current in the vicinity of the Antarctic coast is further cooled by heat exchange with the atmosphere and shelf ice and rendered more saline by sea ice formation, so that it sinks rapidly. Its residence time at the surface is so short that neither phytoplankton utilization nor gas exchange with the atmosphere significantly reduce its CO_2 content (Weiss et al., 1979). As a result, the newly formed Antarctic Bottom Water (AABW) is strongly undersaturated with respect to carbonate, and we observed that calcareous foraminifera are entirely absent from the Weddell Sea. Planktonic and benthic foraminifera are only preserved in the Southern Ocean, far from the Antarctic continent, in sediment shallower than 3.5 km, slightly above the main body of the AABW. However, mixing is strong and the whole Southern Ocean deepwater mass below 1500 m has fairly homogenous $\partial^{13}C$ values south of the polar front (Kroopnick, 1980; 1985).

Core MD84-527 (43°49' S, 51°19' E, 3260 m) has been analyzed for both *N. pachyderma* and *Cibicides*. Figure 4 displays the oxygen isotope record of *Cibicides*, which has been used for dating by stratigraphic correlation, and the carbon isotope records of both *Cibicides* and *N. pachyderma*. The benthic $\partial^{13}C$ record is characterized by very negative values during the glacial stages 2–4. These low $\partial^{13}C$ values in the southern ocean are in agreement with the values reported by Curry and Lohmann (1982) as existing in the Vema Channel during the last glaciation. The *Cibicides* and *N. pachyderma* carbon isotope records from core MD84-527 have the same general structure, and the correlation coefficient between the two records is high (r=0.73), suggesting that strong convection between surface and deep water has occured in the southern ocean during most of the last 150,000 years.

These results agree with the reconstructions of $\partial^{13}C$ gradients in the global ocean deep water during the last interglaciation (Duplessy et al., 1984) and during isotope stages 6 and 5d (Duplessy and Shackleton, 1985): at both times a $\partial^{13}C$ decrease from the southern ocean toward the Indian and Pacific oceans is observed that is analogous to present-day conditions. They also agree with the data plotted in Fig. 1, which show the influx of an important bottom water mass originating from the south into the Atlantic Ocean.

Only three *Cibicides* benthic records from the Southern Ocean are presently available: RC11-120 analyzed by N. J. Shackleton, and MD84-527 and MD73-025 analyzed at Gif (Curry et al., 1988; Kallel, 1988). They all exhibit low $\partial^{13}C$ values during glacial conditions, suggesting that the water which sank at high latitude was much poorly oxygenated than today, probably because the sea-ice cover extended about 1,000 km further north than under modern climatic conditions (CLIMAP, 1976)

Figure 4: $\partial^{18}O$ record of *Cibicides* and $\partial^{13}C$ records of *Cibicides* and *N. pachyderma* (left coiling) in Southern Ocean core MD 84-527. The high degree of correlation of the $\partial^{13}C$ records indicates an important transfer of surface water to the depth (i.e. deep water formation) in the Southern Ocean during the entire last climatic cycle, probably in locations which are active under modern conditions.

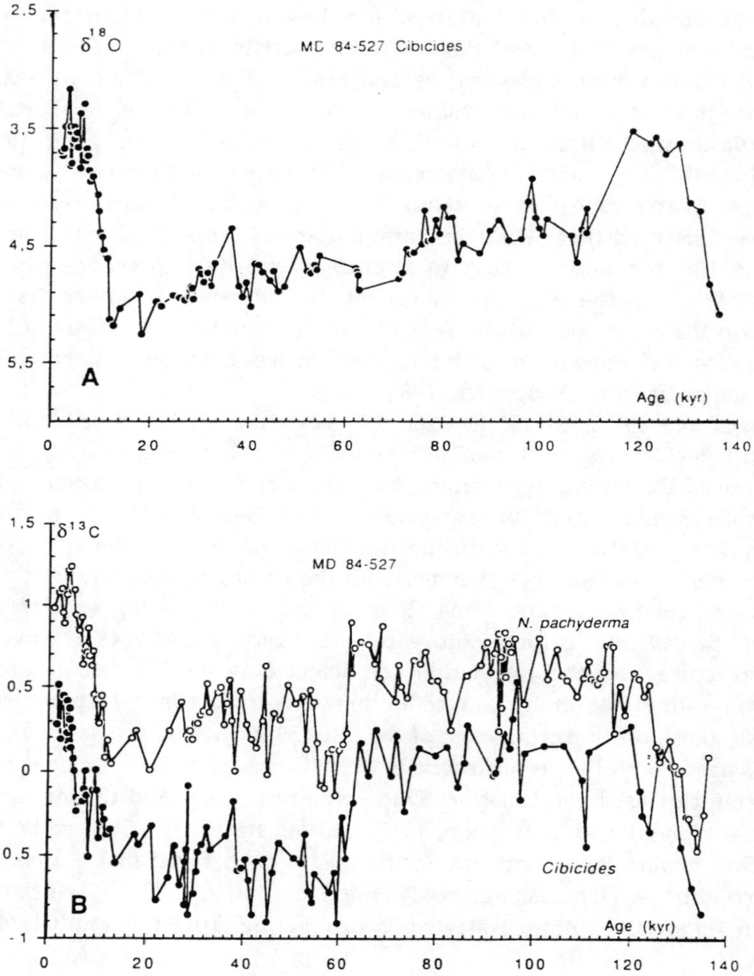

and prevented gas exchange with the atmosphere when deep water upwelled close to the surface. The formation of deep water in the Southern Ocean would thus probably result from the activity of polynya, which developed within the ice-pack.

CONCLUSION

On the time scale of 10^4 years, the circulation of intermediate and deep waters is at least as variable as that of the surface water. It has apparently responded strongly to the climatic changes of the last 150,000 years. This is largely because much of the circulation at 18 K and earlier was driven by processes occurring at high latitudes (as today), i.e. in regions where climatic changes have been largest.

During the last glacial maximum, the northern hemisphere source of deep water was poorly active in the North Atlantic and most of the ocean's bottom and deep water originated from poorly-oxygenated surface water sinking in the high latitudes of the Southern Ocean.

By contrast, Intermediate waters extended somewhat deeper than under modern conditions and were much more oxygenated than today in the Atlantic, Indian and Pacific oceans. We do not have enough data to interpret in detail the hydrology of the Intermediate waters in the Atlantic Ocean. In the Indian and Pacific oceans, intermediate waters were separated from the deep water by a well-developed deep thermocline.

ACKNOWLEDGMENTS

Thanks are due to N.J. Shackleton for numerous stimulating discussions, to J. Antignac and B. Le Coat for help in the isotopic analysis, and to Tom Wigley for a careful review of the manuscript and numerous improvements. The French Commissariat à l'Energie Atomique, Centre National de la Recherche Scientifique (Programme National d'Etude de la Dynamique du Climat and GRECO CO_2), Terres Australes et Antarctiques Françaises (Programme APSARA), and EEC grant EV 4C-0072-F supported the paleoclimatic research program of C.F.R..

This is *C.F.R. Contribution N° 1007*

REFERENCES

Assaf, G., and A. Hecht, Sea straits: A dynamical model, Deep Sea Res., 21, 947–958, 1974.

Berger, W. H., Deep Sea carbonates: Pleistocene dissolution cycles, J. Foram. Res., 3, 187-195, 1973.

Boyle, E. A., and L. D. Keigwin, Deep circulation of the North Atlantic over the last 200,000 years: Geochemical evidence, Science, 218, 784–787, 1982.

Boyle, E. A., and L. D. Keigwin, Comparison of Atlantic and Pacific paleo-chemical records for the last 215,000 years: Changes in deep ocean circulation and chemical inventories, Earth Planet. Sci. Lett., 76, 135–150, 1985.

Boyle, E. A., and L. D. Keigwin, North Atlantic thermohaline circulation during the last 20,000 years: Link to high latitude surface temperature, Nature, 330, 35–40, 1987.

Broecker, W. S., Glacial to interglacial changes in ocean chemistry, Prog. Oceanogr., 11, 151–197, 1982.

Caralp, M., Deep-sea circulation in the northeastern Atlantic over the past 30,000 years: The benthic foraminiferal record, Oceanol. Acta, 10, 27–40, 1987.

CLIMAP Project Members, The surface of the Ice Age Earth, Science, 191, 1131–1137, 1976.

Corliss, B. H., D. G. Martinson, and T. Keffer, Late Quaternary deep-ocean circulation, Geol. Soc. Am. Bull., 97, 1106–1121, 1986.

Curry, W. B., and G. P. Lohmann, Carbon isotopic changes in benthic foraminifera from the western south Atlantic: Reconstruction of glacial abyssal circulation patterns, Quat. Res., 18, 218–235, 1982.

Curry W. B., J. C. Duplessy, L. D. Labeyrie, and N. J. Shackleton, Changes in the distribution of $\partial^{13}C$ of deep water ΣCO_2 between the last glaciation and Holocene, Paleoceanography, 3, 317-341, 1988

Duplessy, J. C., Glacial to interglacial contrasts in the northern Indian Ocean, Nature, 494-498, 1982.

Duplessy, J. C., and N. J. Shackleton, Response of global deep-water circulation to the Earth's climatic change 135,000–107,000 years ago, Nature, 316, 500–507, 1985.

Duplessy, J. C., J. Moyes, and C. Pujol, Deep water formation in the North Atlantic Ocean during the last ice age, Nature, 286, 476–482, 1980.

Duplessy, J. C., N. J. Shackleton, R. K. Matthews, W. Prell, W. F. Ruddiman, M. Caralp, and C. H. Hendy, ^{13}C record of benthic foraminifera in the last interglacial ocean: Implications for the carbon cycle and the global deep water circulation, Quat. Res., 21, 225–243, 1984.

Duplessy, J. C., N. J. Shackleton, R. G. Fairbanks, L. D. Labeyrie, D. Oppo, and N. Kallel, Deep water source variation during the last climatic cycle and their impact on the global deep water circulation, Paleoceanography, 3, 343-360, 1988a.

Duplessy, J. C., L. D. Labeyrie, and P.L. Blanc, Norwegian Sea deep water variations over the last climatic cycle: Paleooceano-graphical implications, in Lectures Notes in

Earth Sciences, 16, 83–116, edited by H. Wanner and U. Siegenthaler, Springer Verlag , New York, 1988b.

Emiliani, C., Pleistocene temperatures, J. Geol., 63, 538-578, 1955.

Fieux, M., F. Schott, and J. C. Swallow, Deep boundary currents in the western Indian Ocean revisited, Deep Sea Res., 33, 415-426, 1986.

Gardner, J.V., and J. D. Hays, Responses of sea-surface temperature and circulation to global climatic change during the past 200,000 years in the Eastern Equatorial Atlantic Ocean, Mem. Geol. Soc. Am., 145, 221–246, 1976.

Kallel, N., Variations de la composition isotopique de l'oxygène et du carbone des foraminifères: traceur de la circulation océanique pendant le dernier maximum glaciaire et lors de la dernière déglaciation, Thesis, Université Paris XI, 1988.

Kallel, N., L. D. Labeyrie, A. Juillet-Leclerc, and J. C. Duplessy, A deep hydrological front between intermediate and deep-water masses in the Glacial Indian Ocean, Nature, 651-655, 1988.

Keigwin, L. D., North Pacific deep water formation during the latest glaciation, Nature, 330, 362–364, 1987.

Kroopnick, P., The dissolved O_2-CO_2-^{13}C system in the eastern equatorial Pacific, Deep Sea Res., 21, 211–222, 1974.

Kroopnick, P., The distribution of ^{13}C in the Atlantic Ocean, Earth Planet. Sci. Lett., 49, 469–484, 1980.

Kroopnick, P., The distribution of ^{13}C of ΣCO_2 in the world oceans, Deep Sea Res., 32, 57–84, 1985.

Labeyrie, L. D., and J. C. Duplessy, Changes in the oceanic C^{13}/C^{12} ratio during the last 140,000 years: High latitude surface water records, Palaeogeogr. Palaeoclimatol. Palaeoecol., 50, 217–240, 1985.

Labeyrie, L. D., J. C. Duplessy, and P. L. Blanc, Variations in mode of formation and temperature of oceanic deep waters over the past 125,000 years, Nature, 327, 477–482, 1987.

Mix, A., and R. G. Fairbanks, North Atlantic surface-ocean control of Pleistocene deep ocean circulation, Earth Planet. Sci. Lett., 73, 231–243, 1985.

Oppo, D. W., and R. G. Fairbanks, Variability in the deep and intermediate water circulation of the Atlantic during the past 25,000 years: Northern hemisphere modulation of the Southern Ocean, Earth Planet. Sci. Lett., 86, 1–15, 1987.

Reid, J. L., Jr., Intermediate waters of the Pacific Ocean, The Johns Hopkins Oceanographic Studies Number 2, Johns Hopkins University Press, Baltimore, Ma. 1965.

Richards, F.A., Dissolved gases other than carbon dioxide, in Chemical Oceanography, edited by J.P. Riley and G. Skirrow, Academic Press, New York, 1965.

Shackleton, N. J., Tropical rainforest history and the equatorial Pacific dissolution cycles. In The Fate of fossil fuel CO_2 in the oceans, edited by N.R. Anderson and NA. Malahoff, , Plenum, New York, pp 401-428, 1977.

Shackleton, N. J., and M. B. Cita, Oxygen and carbon isotope stratigraphy of benthic foraminifers at site 397: Detailed history of climatic change during the Late Neogene, in Initial Reports of the Deep Sea Drilling Project, 47, 433-444, 1979.

Shackleton, N. J., J. Imbrie, and M. Hall, Oxygen and Carbon isotope record of east Pacific core V 19-30: Implications for the formation of deep water in the Late Pleistocene North Atlantic, Earth Planet. Sci. Lett., 65, 233–244, 1983.

Streeter, S. S., and N. J. Shackleton, Paleocirculation of the deep North Atlantic: 150,000 year record of benthic foraminifera and oxygen-18, Science, 203, 168–171, 1979.

Thompson, P. R., and T. Saito, Pacific Pleistocene sediments planktonic foraminifera: Dissolution cycles and geochronology, Geology, 2, 333–335, 1974.

Weiss, R. F., H. G. Östlund, and H. Craig, Geochemical studies of the Weddell Sea, Deep Sea Res., 26, 1093–1120, 1979.

Zahn, R., K. Winn, and M. Sarnthein, Benthic foraminiferal $\partial^{13}C$ and accumulation rates of organic carbon: *Uvigerina peregrina* group and *Cibicidoides wuellerstorfi*, Paleoceanography, 1, 27–42, 1986.

Zahn, R., M. Sarnthein, and H. Erlenkeuser, Benthic isotope evidence for changes of the Mediterranean Outflow during the Late Quaternary, Paleoceanography, 2, 543 559, 1987.

OCEANIC RESPONSE TO ORBITAL FORCING IN THE LATE QUATERNARY:

OBSERVATIONAL AND EXPERIMENTAL STRATEGIES

John Imbrie[1,2], Andrew McIntyre[3], and Alan Mix[4]

[1]Dept. of Geological Sciences, Brown University, Providence, RI 02912

[2]Authors listed alphabetically. No senior authorship implied.

[3]Lamont-Doherty Geological Observatory, Palisades, NY 10964

[4]College of Oceanography, Oregon State University, Corvallis, OR 97331

ABSTRACT

Observations on deep-sea cores demonstrate that late Pleistocene climate is dominated by three broad-band cycles centered near periods of 23 ky, 41 ky, and 100 ky. These cycles permeate the global system, and include changes in the atmosphere, cryosphere, surface ocean, and deep ocean. The periods of these climatic cycles match orbital cycles of precession, obliquity, and eccentricity; and each orbit-climate pair is significantly correlated (coherent). These observations may be explained in different ways. We review various models and conclude that the climatic cycles can be explained as an interaction between orbitally forced and internally driven oscillations of the climate system. Depending on the cycle and model, the external forcing may influence climate either as part of a driving mechanism which determines both the amplitude and phase of the cycle, or as a pacing mechanism which sets the phase of an internal oscillation. Our goal is to search climatic data for clues about the mechanisms which operate within the climate system on orbital time scales. Our strategy is patterned after previous investigations which partition the climatic record into cyclic components, record the phase of system responses in each cycle, and examine these phase sequences for clues about the chain of causal mechanisms.

We review the stratigraphic, chronologic, and statistical methods adopted to achieve these objectives, and then apply the strategy to two sets of data: (1) time series with local implications about conditions at the sea surface at 17 sites ranging from 54°N in the Atlantic to 44°S in the Indian Ocean (faunal estimates of sea-surface temperature, SST); and (2) time series with global implications about ice volume and ocean chemistry ($\delta^{18}O$, $\Delta\delta^{13}C$, and Cd/Ca). The spatial pattern of SST response in each cycle is documented by mapping its amplitude and phase. Each cycle has a characteristic pattern, but all cycles show an early response near 40°S. Information drawn from both data sets is then combined to form a preliminary phase portrait of the entire system. In each cycle, responses

A. Berger et al. (eds.), Climate and Geo-Sciences, 121–164.

occur over a wide range of phases centered on the point of minimum ice, and include indicators which lead and those which lag ice volume. Each cycle is characterized by the same three leading indicators: Cd/Ca, $\Delta\delta^{13}C$, and SST at ~40°S (with the addition of equatorial SST at 23 ky). Lagging indicators include North Atlantic SST at 23 ky.

Data on SST and $\delta^{18}O$ phases are used to test simple models of radiation forcing in the 23 ky and 41 ky bands, and limited to explaining the observations of $\delta^{18}O$, and SST near 50°N and 40°S. The simplest model that fits our data assumes that summer radiation forces the system in the Northern Hemisphere, and that the response at high latitudes is transmitted rapidly to the Southern Hemisphere by modulation of North Atlantic sources of deep water. In the 100 ky band we assume that the phase of an internally generated cycle is set by the amplitude of the 23 ky radiation cycle.

We conclude that the strategy developed here can guide the acquisition of new data, serve as a framework for synthesizing information on different parts of the climate system, and stimulate interactions between data and models. Opportunities now exist for comparing the predicted sequence of system responses with the observed sequence.

INTRODUCTION

The Milankovitch Problem

In the final published form of his theory, Milutin Milankovitch asserted that orbitally driven changes in the annual cycle of incoming radiation -- combined with albedo feedbacks, time delays, and other responses of the system related to the accumulation of ice -- are the cause of the Pleistocene ice ages (Milankovitch, 1941, p. 475-483). The time-dependent patterns of these orbital and radiative changes are quite well known (Berger, 1978a,b). Therefore, given knowledge of the equations which govern the relevant processes in the atmosphere, ocean, cryosphere, and lithosphere, and a coupled model based on this knowledge, it would in principle be possible to make a decisive test of the Milankovitch theory by comparing the model's output in the time domain with the geological record of climate. Unfortunately, our knowledge of the system is so incomplete that it has not yet been possible to conduct such a test. In the view of Saltzman (1987), it is unlikely that our knowledge of key rate constants will ever be sufficiently accurate to achieve a fully quantitative test based on calculations of the boundary fluxes of mass, momentum, and energy.

Measuring and Modeling the Climatic Spectrum

In response to this challenge, some members of the geological and modeling communities have adopted a strategy that puts emphasis on observing and modeling the climatic spectrum, or on explaining the observations in one particular frequency band. Climatic records in late Pleistocene deep-sea cores have received special attention because observations on $\delta^{18}O$ are available to place them in a common stratigraphic and chronologic framework. Proxy records of many different parts of the climate system have been analyzed in this way: the volume of ice on land ($\delta^{18}O$); the circulation pattern and temperature of the surface ocean (planktonic species); the circulation and carbon chemistry of the deep ocean ($\delta^{13}C$, $\Delta\delta^{13}C$, Cd/Ca, $CaCO_3$); and atmospheric transport (continental dust, pollen, diatoms). Examples of spectra derived from these different records can be found in studies by Hays et al. (1976), Ruddiman and McIntyre (1981, 1984), Imbrie et al. (1984), Pisias and Leinen (1984), Boyle (1984); Peterson and Prell (1985), Shackleton and Pisias (1985), Prell and Van Campo (1986), Curry and Crowley (1987), Pokras and Mix (1987), Pisias and Rea (1988), McIntyre et al. (1989).

Although the details of these spectra vary from one climate record to another, and depend in part on the resolution at which the spectrum is calculated, there is general agreement that a substantial fraction of the observed variance is concentrated in bands centered on periods near 23 ky, 41 ky, and 100 ky (ky = 1000 years). The fact that these periods match three prominent peaks in orbital spectra provides a purely empirical argument in favor of the Milankovitch theory. This argument is strengthened by the observation that much of the climatic variance which occurs in each of these bands is coherent with, and therefore linearly related to, the orbital changes which occur in the same band. In the 23 ky (precession) and 41 ky (obliquity) bands, where virtually all the variance in radiation occurs, this statistical argument is supported by physically based, linear models of climatic response; in the 100 ky (eccentricity) band, where the radiation spectrum lacks power, this support is lacking.

Knowledge of the climatic spectrum provides insights about the internal dynamics of the system. The 100 ky cycle, which dominates many but not all climatic spectra over the past 700,000 years, is a case in point. Because the radiation spectrum at any site contains trivial amounts of energy at the 100 ky period, relative to that at other periods, it seems physically unreasonable to account for the 100 ky cycle as the forced response of a linear system (Hays et al., 1976). Thus a number of highly parameterized, nonlinear, time-dependent models have been constructed with the aim of accounting for this cycle in a physically plausible way (Imbrie and Imbrie, 1980; Ghil and Le Treut, 1981; Le Treut and Ghil, 1983; Oerlemans and van der Veen, 1984; Saltzman and Sutera, 1984; Saltzman et al., 1984; Snieder, 1985; Lindzen, 1986; Birchfield, 1987; Peltier, 1987; Saltzman, 1987; Saltzman and Maasch, 1988; Marsiat et al., in press). Each model contains different assumptions about the system, and therefore yields a different explanation of the 100 ky cycle.

The range of these explanations is instructive. We now examine this dynamic panorama, limiting our attention to published models which predict at least one climatic response in addition to ice volume. At one extreme are models which simulate the 100 ky cycle as the _free oscillation of an internally driven, nonlinear system_. In the model of Saltzman (1987), the responses simulated are mean surface ocean temperature, global ocean $\delta^{18}O$, ice volume, atmospheric CO_2, and deep ocean temperature. Such models, which act as nonlinear oscillators driven internally between two quasi-stable states, can generate a 100 ky cycle in the absence of external forcing of any kind (Saltzman and Sutera, 1984). This idea finds empirical support in the abrupt climatic shifts called Terminations that form part of the structure of the 100 ky cycle (Broecker et al., 1985). The unforced solution of the Saltzman model yields responses which occur in an arbitrary time scale but are fixed relative to each other. When the model is subject to a moderately realistic orbital forcing by summer radiation, the forced solution is phase-locked by the external forcing, and has many features which resemble observations (Saltzman, 1986, 1987). Thus the amplitude of the cycle is internally driven whereas the phase is externally forced. Visual inspection of the model solutions suggests that they are coherent with orbital forcing in the three primary bands. Future studies will test this impression and compare the predicted phase sequence with the observed sequence.

Other models simulate the 100 ky cycles of ice volume and global temperature as the response of a nonlinear system driven both externally and internally (e.g., Ghil and Le Treut, 1981; Le Treut and Ghil, 1983). Here the free response of the system, which is dominated by oscillations near 10 ky, is substantially different from the response of the system in the presence of external forcing. A nonlinear resonance mechanism that is based on feedbacks between the model's ice sheet and other parts of the system transfers energy from the precession band (where radiative forcing is strong) into the 100 ky band (where this forcing is very weak). The 100 ky cycle, which dominates the model's ice-volume cycle when external forcing is applied in the precession band, disappears when this forcing is removed. In this sense the model's response at 100 ky can be described as externally forced.

Another class of models simulates the 100 ky ice-volume cycle as the forced response of an externally driven, nonlinear system (Calder, 1974; Imbrie and Imbrie, 1980). In this case, the free response of the system is nil, but nonlinearities in the system's forced response to ice growth at precessional frequencies transfer energy to the 100 ky band. Pisias and Shackleton (1984) improved the simulation of the Imbrie-Imbrie model by adding a term representing CO_2.

These models are important, not because they provide detailed and realistic simulations of the climate record, but because they illustrate the kinds of behavior that might occur under the stated assumptions. Accordingly, we should not ask "which models are correct?" but "which models illustrate a kind of behavior that is consistent with the main features of the observed climate record?"

The Imbrie-Imbrie model offers the narrowest scope for testing against geological data, because it simulates only the ice-volume response. In contrast, the model of Ghil and Le Treut simulates two and the Saltzman model five responses. Both the Ghil and Saltzman models seem to offer the potential for explaining the strong coherencies and the systematic phase relationships observed between orbital and climatic variations. For the model of Ghil and Le Treut, we need only assume that the gain produced by each nonlinear resonance is orbitally forced and constant over time. For Saltzman's model, we assume that the strength of feedbacks operating within the system, and thus the phase of the climatic oscillation, is determined by externally forced modulations in the radiation field. Thus both models seem to provide a reasonable physical framework for explaining the geological record as the output of a nonlinear system responding to a combination of internal and external influences. Depending on the frequency band of interest, and the model adopted, the external influence on a given response can be viewed either as a pacing mechanism which determines the phase of an internally generated oscillation, or as a driving mechanism which combines with the internal oscillation to determine both the amplitude and phase of the response.

Primary and Heterodyne Tones

Our discussion so far has been restricted to the main features of climate that are visible in spectra calculated at rather low resolutions, namely, the broad-band cycles centered at periods near 23 ky, 41 ky, and 100 ky. If spectra of the same climatic records are examined at higher resolution, additional (generally small) peaks are commonly observed at periods which can be predicted as heterodyne tones originating either as combination tones (e.g., 14.7 ky, 30 ky) or as harmonics (e.g., 11.5 ky, 20.5 ky) of the primary orbital oscillations in the precession, obliquity, and eccentricity bands (Pisias and Leinen, 1984; Pokras and Mix, 1987; Pestiaux et al., 1988; Pisias and Rea, 1988). Familiar examples of heterodyne responses occur in ocean tides (Munk and Cartwright, 1966) and musical sounds (Taylor, 1965). In general, combination tones occur in physical systems as nonlinear interactions between primary, forced responses. In the climate system, harmonics of primary responses can originate as artifacts of a Fourier analysis of non-sinusoidal oscillations; as intrinsic features of the climatic response analogous to the semiannual cycles; or as nonlinear interactions between primary responses. Whatever the precise origin of the heterodyne tones observed in Pleistocene spectra might be, their existence is additional, strong evidence that the system is in fact responding to forcing in the primary orbital bands. Hence Ghil and Childress (1987), Le Treut et al. (1988), and Ghil (1989) have argued that combination tones are realistic features of their model's spectra.

The existence of heterodyne tones complicates the task of spectral analysis, especially in those parts of the spectrum where heterodyne tones lie close to the primary cycles. The 19 ky precession cycle is a case in point. Here the direct influence of precession is probably combined with the semi-obliquity harmonic (20.5 ky) and with several combination tones calculated to lie near 19 ky. To avoid this complication, we focus on the 23 ky precession cycle.

An Observational Strategy

We conclude that orbitally driven changes in the radiation balance are an important influence on Pleistocene climate, and that these influences operate through external driving or external pacing mechanisms. This conclusion is based: (1) on empirical evidence from low- and high- resolution spectra cited above; (2) on experiments with nonlinear dynamic models cited above; and (3) on experiments with general circulation models in which the influence of orbital boundary conditions is evaluated (Kutzbach and Otto-Bliesner, 1982; Kutzbach and Guetter, 1984; Kutzbach and Street-Perrott, 1985; Prell and Kutzbach, 1987; COHMAP Members, 1988; Kutzbach and Gallimore, 1988). In our view, therefore, observational and modeling strategies should now focus on understanding the mechanisms of the climate system's response to orbital forcing, and how this response interacts with internally driven oscillations.

Toward this end, one of the strategies adopted by the SPECMAP project is to partition the climatic record into components which correspond to the main frequencies of orbital forcing, and to observe the amplitude and phase of each component at different sites and different parts of the climate system. This strategy is basically a geographic extension of work already accomplished along these lines (e.g., Hays et al., 1976; Ruddiman and McIntyre, 1981, 1984; Prell and Van Campo, 1986; Lyle, 1988; McIntyre et al., 1989), with an emphasis on spectral as opposed to filtering techniques. We believe that this strategy will provide useful insights about the geographic and temporal structure of mechanisms operating in each frequency band, and identify specific links between these responses and changes in the radiation field. Our procedures are patterned after classical studies of tides (Munk and Cartwright, 1966) and climatological studies of the annual and semiannual cycles (White and Wallace, 1978; Horel, 1982), where useful insights into complex physical processes have been obtained by examining response fields in terms of their coherency, amplitude, and phase.

Although our strategy is based on evidence that orbital influences exert a strong control on late Pleistocene climate, we acknowledge that other influences are at work over the frequency band of interest. The coherent variance we observe at orbital periods is probably superimposed on a stochastic background (Mitchell, 1976; Kominz and Pisias, 1979; Ghil, 1989). Moreover, the system itself is not stationary, as demonstrated by changes in the climate spectrum over time (Pisias and Moore, 1981), and may well be responding to tectonic as well as orbital influences (Ruddiman et al., 1986).

Objectives

This paper concentrates on changes in the surface Atlantic and southern Indian Oceans during the last 300,000 years, and examines these changes in the context of variations in global ice volume. Specific objectives are: (1) to present time series of sea-surface temperature (SST) estimates at 17 sites in the Atlantic and southern Indian Oceans, and make them widely available in digital form; (2) to measure the SST spectrum at these sites and determine to what extent the variance is concentrated at the primary frequencies predicted by orbital theory; (3) to partition each of these climatic time series, as well as time series of $\delta^{18}O$ and $\Delta\delta^{13}C$, into frequency components whose central frequency and bandwidth correspond to the primary Milankovitch cycles in the 23-ky, 41-ky, and 100-ky bands; (4) to determine for each component what fraction of the total observed variance in that band is coherent with the orbital forcing, and estimate the amplitude of this component; (5) to estimate the mean phase of the coherent variance in each band with respect to the phase of the orbital forcing; and (6) to display these phases in such a way that the relationships to changes in the global radiation field can be examined for clues about the mechanisms of climatic response.

GAIN AND PHASE MODEL

To develop this strategy, and provide a basis for explaining our statistical methods and interpreting the results, we first examine the Milankovitch theory in the context of a simple systems model (Fig. 1). The model is designed to focus attention, first, on certain <u>physical</u> relationships between orbital geometry and the geographic and seasonal pattern of radiation, and second, on certain <u>empirical</u> relationships between orbital geometry and the observed climate. These relationships, summarized below, are the basis for a statistical procedure designed to associate any given change in the earth's orbit with a particular sequence of climatic responses. The orbital changes occur in discrete bands, and drive changes in radiation which exhibit distinct geographic and temporal patterns that differ from one band to another. For this reason it is important to monitor the climatic responses separately in each band.

The chain of cause and effect begins with the orbital system which, as described by Berger (1978a), transforms perturbations in the gravitational field into output signals that record changes in the earth's orbital geometry: eccentricity, $e(t)$; obliquity or tilt, $\varepsilon(t)$; and the longitude of perihelion measured from the moving vernal point, $\omega(t)$. Conveniently, the influence these geometric changes have on the pattern of incoming radiation can be calculated from two parameters. One of these is simply the obliquity signal, $\varepsilon(t)$. The other is the precession index, $P(t) = e(t) \sin \omega(t)$, which has a nearly linear relationship to the earth-sun distance on June 21.

The radiation system transforms these two orbital signals into a set of top-of-the-atmosphere irradiation curves, $Q_{l,s}(t)$, one for each latitude l and season s. To a close approximation, each of these curves can be modeled as the weighted sum of P and ε, if the phase of P is shifted according to the season (Imbrie and Imbrie, 1980).

The climate system responds to these changes in the radiation field and yields a set of j climatic time series, $C_j(t)$. Given assumptions about the mechanism of the response to particular radiation curves, C_j can be modeled directly. But our objective here is not to test assumptions about the mechanisms of the response to <u>radiation</u>, but to

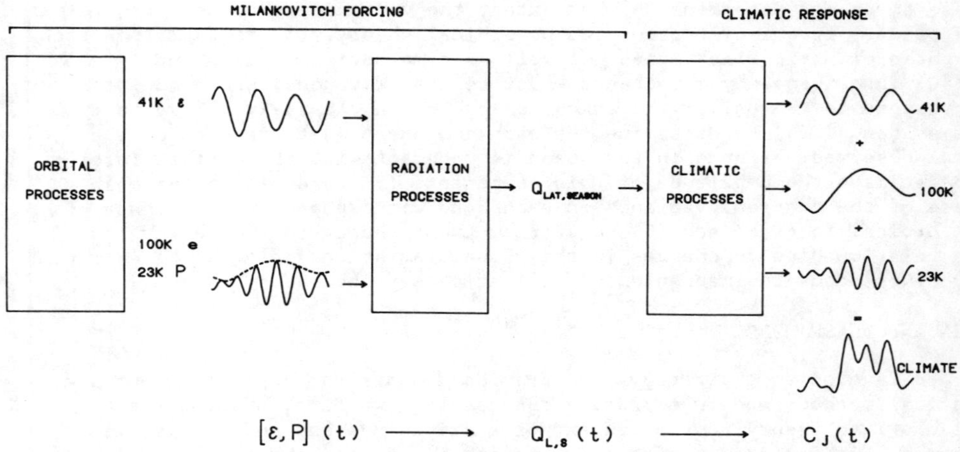

Fig. 1. Systems model of the Milankovitch theory of climate. The radiation system receives two signals from the orbital system: obliquity (ε) and precession (P), the latter with an envelope e (eccentricity). This system transforms ε and P into a radiation curve ($Q_{l,s}$) for each latitude l and season s. The climate system transforms these perturbations of the radiation field into a climate response C_j. Because the observed responses are coherent with the orbital signals, each component of C_j can be modeled as an orbital curve e, ε, or P that has been multiplied by a certain gain and shifted by a certain phase. The sum of these components yields a simulation of the coherent part of the observed record. The example illustrated is Stage 7 of the $\delta^{18}O$ record.

obtain information about these mechanisms by an empirical study of the phase relationships between orbital forcing and the climatic response. Once these linkages are known, the phase relationships to radiation changes can easily be worked out and studied for clues about the mechanism.

For any given climatic record, our procedure begins with an inspection of a cross-spectrum calculated against a set of orbital signals (e, ε, and P). In any frequency band where the coherency is significant, some component of the climatic record must be linearly related to the orbital signal. The phase of that component can be read from the phase spectrum. Proceeding in this way, C_j is analyzed linearly into components, each of which is an orbital signal (e, ε, or P) multiplied by a certain gain and shifted by a certain phase. Addition of these components yields a simulation of the coherent fraction of the observed time series (Imbrie, 1985). The quality of the simulation depends upon the fraction of the total variance that is coherent with the forcing.

An application of this gain-and-phase model is given on Figure 1, where the $\delta^{18}O$ record of Stages 6-8 (250 - 350 ka) is resolved into frequency components (ka = 1000 years before present). Each component is an orbital curve (e, ε, or P) whose amplitude and phase have been determined by cross-spectral analysis of orbital signals vs. $\delta^{18}O$ data. As discussed by Imbrie (1987), this simulated climate curve reproduces reasonably well the timing (and to a lesser extent, the amplitude) of step-like changes towards a glacial or interglacial state which are a familiar part of the $\delta^{18}O$ record, including those steps called Terminations (Broecker and Van Donk, 1970). The implication here is that these steps represent instabilities in the system that are triggered when the coherent part of the orbitally forced response crosses certain thresholds; and that a portion of each step becomes part of the incoherent variance. The mechanism we envision resembles that which produces a tidal bore, or a sudden onset of monsoon rains.

STATISTICAL METHODS AND ASSUMPTIONS

For each climatic record C_j we: (1) select an appropriate bandwidth for analysis; (2) estimate the variance spectrum; (3) calculate the coherency spectrum against e, ε, and P and estimate the coherent variance in each Milankovitch band; (4) estimate the mean phase of the coherent variance in each band and adopt convenient definitions of zero phase; and (5) consider the stationarity of the system.

Bandwidth

The resolution of a spectrum is described by its bandwidth, BW. In this study, BW = 1.33/MΔt, where M is the number of lags, Δt is the time step, and 1.33 is a factor corresponding to our lag window (Jenkins and Watts, 1968). BW is therefore under our control through the choice of M and Δt. The first consideration in choosing BW is to match its width to the process under investigation. The maximum BW that will permit us to distinguish these processes (acting at frequencies of 23 ky^{-1}, 41 ky^{-1}, and 100 ky^{-1}) is 0.015 ky^{-1} (= 0.015 c/ky). With this resolution, bands centered on the three main Milankovitch frequencies include periods of 20-28 ky, 31-59 ky, and 57-400 ky, respectively, with greater weight given to periods near the center of each band. The second consideration is to maximize the reliability of our estimates. Since the confidence interval of these estimates varies as the ratio of lags to record length (M/N), where N is the number of equally spaced sample points, we desire M as small as possible for any given N. But M can be no smaller than a value which yields a BW of 0.015. We therefore analyze all of our records with a standard bandwidth of 0.015. The resulting confidence intervals for estimates of variance density (s^2/f) and coherency (k) will be a function of M/N, and that for phase a function of k and M/N. It should be noted that some heterodyne responses (e.g., 20.5 ky, 50 ky, 52 ky, 69 ky) can be expected to occur within these bands.

Variance Spectrum

After removing any linear trend from the data, a variance density spectrum, $[s^2/f](f)$, is calculated using standard time series procedures (Jenkins and Watts, 1968), and setting BW = 0.015 c/ky. Here s^2 = the observed variance, f = frequency, and s^2/f = variance density. Values of s^2/f are estimated for 101 equally spaced frequencies extending from 0 to the Nyquist frequency, each estimate being averaged over BW. When this spectrum is displayed graphically, the scaling of the ordinate is convenient for many purposes because the area under the curve is proportional to variance. But variance is a statistical concept based on the square of mean deviations: SST, for example, has a variance scaled in $°C^2$. For our purposes it is useful to scale the ordinate so that values relate to the amplitude, a, of the signal in °C. This can be done by making use of the relationship $a^2 = 2s^2$. The result is an amplitude-density spectrum in which a/f, the <u>total amplitude density</u> (in units of °C/f = °C × ky), is a function of f. To obtain an estimate of the total amplitude of the variance contained in a particular Milankovitch band, the variance density spectrum is integrated over a band of width BW centered on the frequency of interest (23 ky^{-1}, 41 ky^{-1}, or 100 ky^{-1}), and then scaled in terms of amplitude. These values are referred to as the <u>total observed amplitudes</u> (in units of °C).

Coherency spectrum

Standard procedures of cross-spectral analysis are employed to define patterns of covariation between two signals (Jenkins and Watts, 1968). In applying these procedures to a deep sea record, important information about system behavior can be obtained by calculating cross-spectra either (1) between two system responses, e.g. $\delta^{18}O$ and SST; or (2) between a system response and an orbital signal. In this paper we limit our attention to cross-spectra between geological time series and orbital signals (e, ε, or P). Because the orbital signals occupy different frequency bands, we simplify the procedure by using a composite orbital signal (ETP) created by adding (normalized) signals of eccentricity, tilt, and precession. A single cross-spectrum calculated between ETP and any climatic response of interest will then describe the responses at all of the primary Milankovitch frequencies (Imbrie et al., 1984). Results of the cross-spectral calculations are examined in the form of a coherency spectrum and a phase spectrum.

Coherency, k, is a measure of the linear correlation between two signals over a given frequency band when the phase difference between them is set to zero. Values of k run from 0 to 1, k^2 being that fraction of the variance in one signal which is linearly related to the variance in the other. A coherency of 1 corresponds to an exact proportionality between the two signals over each oscillation; a coherency of 0 indicates no correlation. The coherency spectrum, k(f), is examined and evaluated for values significantly different from zero against an 80% fiducial level (Julian, 1975).

At each of 101 frequencies between 0 and the Nyquist frequency, we now have a value of k and s^2/f. By calculating $k^2(s^2/f)$, and scaling in terms of amplitude, we obtain an estimate of the amplitude density of the variation that is coherent with the orbital forcing. These values are referred to as <u>coherent amplitude densities</u> (°C × ky). To obtain an estimate of the amplitude of the coherent variance contained in a particular Milankovitch band, $k^2(s^2/f)$ is integrated over a band of width BW centered on the frequencies of interest, and then scaled in terms of amplitude. These values of the <u>coherent amplitude</u> (°C) determine the length of vectors plotted on our maps. The unexplained residuals (i.e., the difference between the total amplitude and the coherent amplitude) reflect a combination of influences, including responses to orbital forcing that are not linear in e, ε, and P, stochastic climate processes, and observational noise.

Phase spectrum

Our objectives here are (1) to estimate the phase of each system response relative to other system responses; and (2) to estimate the phase of system responses relative to the orbital cycles. These estimates are obtained by standard time series procedures which yield a spectrum of phase differences, $\phi(f)$, with 80% confidence intervals (Jenkins and Watts, 1968). The first objective is achieved by calculating a spectrum of phase differences between each local signal (e.g. SST) and $\delta^{18}O$. The results of these calculations can be portrayed as a fan of vectors whose angular positions with respect to $\delta^{18}O$ are set by the observed phase differences.

The second objective amounts to orienting the fan of response vectors with respect to the orbital cycles. This is done by calculating phase spectra between each climatic response (e.g. $\delta^{18}O$, SST) and a set of orbital signals. To eliminate redundancy in presentation, we agree to express the results of all phase calculations as phase differences between climatic and orbital time series. Here we must adopt useful definitions of zero phase and agree on plotting conventions. Variations in e and ε have natural zeros and natural definitions of zero phase. For our purpose, they also have a convenient definition of sign in that higher values of both parameters are associated with higher temperatures. In the case of e, the association is mainly empirical; but higher values of ε

are theoretically and empirically associated with higher temperatures in high latitudes. In contrast, the precession index P has an arbitrary definition of zero phase (June 21), and a definition of sign that yields an empirical association of higher values (greater distance) with lower temperatures. To avoid confusion in the interpretation of phase, we therefore multiply the precession index by -1 before calculating a cross-spectrum. Thus an SST phase difference of zero means that the warmest temperatures occur in phase with maximum e, maximum ε, or minimum earth-sun distance on June 21 (June perihelion). To give $\delta^{18}O$ the same sense, we also reverse its sign before calculating a cross-spectrum. Thus a $-\delta^{18}O$ phase difference of zero means that minimum ice occurs at the orbital configurations just defined.

It is convenient to plot our estimates of SST and $\delta^{18}O$ phase in vector format on diagrams and maps. Since each value of ϕ in the phase spectrum estimates the phase difference between a climatic record and an orbital forcing function, we need only choose the position of zero phase-difference on a phase dial and define the sign. Our plotting conventions are shown on Figure 2. The zero position is placed at the top of each phase diagram (corresponding to the orbital configurations just defined: maximum e, maximum ε, and June perihelion). Phase differences representing climatic responses which lag orbital forcing are defined as positive angles and depicted as vectors measured clockwise from the defined zeros. These conventions are the same as those of Lyle (1988). They differ from those of McIntyre et al. (1989), who plot minimum SST and maximum ice on a phase diagram with a zero set at December perihelion.

Although the zero phase positions as defined on Figure 2 correspond broadly with orbital geometries associated with interglacial conditions, it is important to note that with one exception, we have no firm a priori knowledge of which orbital geometry represents the maximum value of the forcing towards an interglacial response. The exception is in the obliquity cycle, where it seems certain that the maximum forcing towards an interglacial response would coincide with maximum obliquity and the associated summer radiation maxima in both hemispheres. For the eccentricity and precession cycles, however, the plotting conventions adopted here should be regarded as an arbitrary bookkeeping procedure -- but one that is useful for displaying observations in a condensed format and for stimulating ideas about underlying mechanisms.

Stationarity

Over the last 400 ky, the climate system appears to be approximately stationary (Pestiaux and Berger, 1984). But there is evidence that around 400 ka the system's $\delta^{18}O$ response to orbital forcing underwent subtle changes (Imbrie, 1985). Further back in time, around 700 ka, there is clear evidence that the system changed radically. Pleistocene records older that about 700 ka are dominated by the 41-ky cycle, younger records by the 100-ky cycle (Pisias and Moore, 1981; Start and Prell, 1984; Ruddiman et al., 1986; Ruddiman and Raymo, 1988; Shackleton et al., 1988). Thus for our purposes in studying the late Quaternary, the optimum record length would be about 400 ky. Ideally, all our records should have the

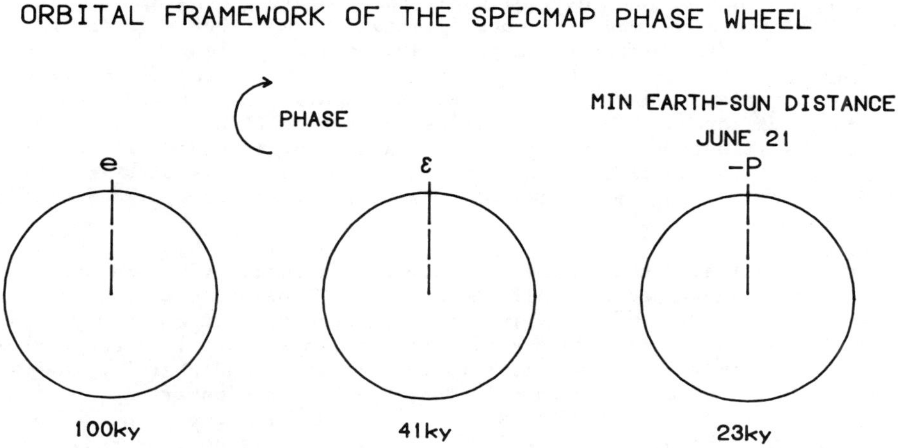

Fig. 2. Conventions of the SPECMAP phase wheel. Cyclical changes in orbital geometry for the 100-ky eccentricity cycle (e), the 41-ky obliquity cycle (ε), and the 23-ky precession-index cycle (P) are represented by vectors rotating clockwise (+) on phase dials. Zero phase on each dial is defined as the position of a vector pointing to the top of the page, and is arbitrarily defined as maximum e, maximum ε, and June 21st perihelion (-P). These definitions correspond to orbital geometries that are empirically associated with forcing toward interglacial conditions. Except for ε, however, there is no _a priori_ physical basis for assuming that these geometries actually represent the maximum interglacial forcing.

same length. Although we have aimed at this ideal, the nature of the geological record has made it impossible to achieve. Our records of sea-surface temperature, for example, have an average length of 300 ky, and range from 150 ky to 400 ky. There are two strategies for dealing with this problem. We can either present calculations derived from records of different length, and assume the 80% confidence limits will include the effects of modest non-stationarity, or truncate all records to the length of the shortest. We adopt the former strategy in this paper, basing our choice on the results of sensitivity tests conducted with individual cores.

Statistical evidence on the question of stationarity can be obtained from the coherency spectrum. If the system is linear and a response has achieved a steady state, peaks in forcing, response, and coherency spectra will exactly coincide at the forcing frequency. But if steady state has not been achieved, slow shifts in the amplitude and phase of the response will be expressed as an off-center peak in the coherency spectrum. On this criterion, our experience with $\delta^{18}O$ and SST signals in the 23 ky and 41 ky bands are characteristic of a linear system at steady state. But in the 100 ky band, coherency peaks are typically off-center, both in $\delta^{18}O$ records (Imbrie et al., 1984, Fig. 10) and in SST records. We take this as evidence that for the 100 ky cycle, the system does not achieve a steady state over a 400 ky interval.

CHRONOLOGIC METHODS AND ASSUMPTIONS

Our primary observations are made as a function of depth in core. Procedures for transforming these observations into time series begin with developing a stratigraphy. Isotopic events in each core are correlated with a standard sequence of globally recognized events whose relative position has already been fixed in the depth domain. This makes it possible to identify local sedimentary hiatuses and local anomalies in accumulation rate before developing a chronology.

Stratigraphy

Pisias et al. (1984) and Prell et al. (1986) used a global set of $\delta^{18}O$ records to develop a standard stratigraphy of major and minor isotopic events that can be widely recognized in deep-sea cores. The assumption here is that a large component of the oxygen isotope signal is global (Shackleton and Opdyke, 1973; Labeyrie et al., 1987). If variations in foraminiferal $\delta^{18}O$ are dominated by changes in the mean isotopic composition of the oceans related to ice volume, as has generally been accepted over the last two decades, then this is a valid assumption, and the chronological precision of the correlations is limited mainly by ocean mixing (~1 ky). If local influences of temperature and salinity are strong, and out of phase with respect to the glaciation cycle, then this assumption is less reliable.

Chronology

Imbrie et al. (1984) and Martinson et al. (1987) placed the standard $\delta^{18}O$ stratigraphy into a chronology referred to as the SPECMAP time scale. For planktonic $\delta^{18}O$ records, this time scale was developed by a three-step procedure. In Step 1, an initial chronology was fixed by linear interpolation between control points having reliable radiometric ages. In selecting these points, attention was limited to dates which could be tied unambiguously to the marine record. Portions of the deep-sea record were chosen which could be dated directly by radiometric measurements on cores; or which could be dated indirectly (1) by $\delta^{18}O$ correlation with corals in radiometrically dated high-stand reefs, or (2) by paleomagnetic correlation with dated lavas. Three control levels were selected: 0-24 ka (Mix and Ruddiman, 1985; Bard et al., 1987); 127 ka (Broecker and van Donk, 1970; Fairbanks and Matthews, 1978); and 730 ka (Mankinen and Dalrymple, 1979). In Step 2, dates for these levels were held fixed while the initial, interpolated dates for other levels were tuned, i.e., iteratively shifted toward a chronology in which oscillations in the filtered $\delta^{18}O$ record -- in both the tilt and precession bands -- were phase-locked as closely as possible with the corresponding orbital oscillations. Prior to tuning, the $\delta^{18}O$ phases relative to tilt and precession were determined from the relatively well-dated record of the last 150 ky. Stationarity was assumed, and these phases were then used to set the timing of the remainder of the $\delta^{18}O$ record. At a late stage in the tuning process (Step 3), the constraints at 127 ka and 730 ka were removed. Based on the standard deviation of the phase differences observed in the 23 ky band, and other considerations, the accuracy of points in the time scale is estimated as ~±3 ky for the past 400 ky.

The resulting oxygen-isotope time scale is reproduced in Figure 3. Six lines of evidence support this chronology and narrowly constrain errors that might stem from circularity. First, when one frequency band is tuned, the other is phase locked with the appropriate orbital signal. Second, the tuned record is coherent with ε and P. Since we do not alter the observed amplitudes of individual oscillations, it would not be possible to force-fit responses that are strongly coherent with the orbital signals, each of which has a distinctive temporal pattern of oscillation amplitudes (Pisias, 1983). Third, in all primary frequency bands, signals other than oxygen isotopes become coherent with the orbital variations when their chronology is fixed by tuning the $\delta^{18}O$ record in the same core (e.g., $\Delta\delta^{13}C$ of Shackleton and Pisias, 1985; and the SST records described below). Fourth, at the end of the tuning process the ages estimated for the two older levels of initial control (128 ka and 734 ka) are within the radiometric errors. Fifth, radiometric ages for sealevel events obtain since the SPECMAP chronology was developed support the essential features of this chronology near 127 ka (coral reef dates by Kaufman, 1986, and Edwards et al., 1987) and between 127 ka and 180 ka (the glacial geological record from Antarctica by Denton et al., 1986, and C. Hendy, personal communication). Sixth, the results of a different tuning strategy applied by Martinson et al. (1987) to a higher-resolution set of benthic $\delta^{18}O$ records match the planktic chronology within the estimated errors of both chronologies.

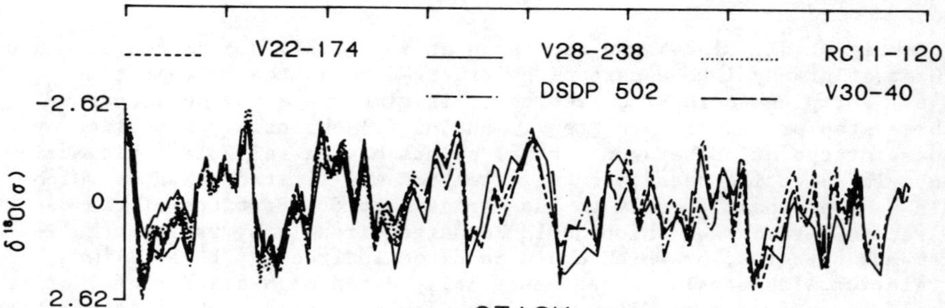

Fig. 3. Foraminiferal $\delta^{18}O$ record of the last ~780,000 years, generated by tuning each of five records to tilt and precession and stacking into a single composite. Values are normalized to zero mean and unit variance. Top: Original normalized data. Middle: Normalized data interpolated at 1-ky intervals and averaged. Bottom: Averaged data smoothed with a 11-point Gaussian filter. From Imbrie et al. (1984).

Winograd et al. (1988) obtained an $\delta^{18}O$ record of groundwater from a calcite vein along a fault zone in Nevada, and dated the record by U-Th methods over the interval 50 ka - 310 ka. Finding that the configuration of their time series matched the marine $\delta^{18}O$ record between Stages 4 and 8, they assumed that the two records are synchronous and concluded that the marine $\delta^{18}O$ chronology is in need of revision. In particular, they concluded that Termination II (128 ka ±3 ka in our chronology) should be 147 ka or older. But unlike the dated sealevel record, the groundwater record dated in Nevada is not physically tied directly to the marine $\delta^{18}O$ record. In our view, the Nevada record is likely to have a regional climatic overprint or a local chemical overprint. To evaluate the possibility of climatic overprinting, one would have to give careful consideration to the results of atmospheric modeling experiments which show that regional changes in effective moisture and winds are significantly out of phase with Laurentide ice (COHMAP, 1988). To evaluate the possibility of chemical overprinting, one would have to examine groundwater records from different fault zones. We therefore give greater weight to the radiometric dates for sealevel events from the coral reef and glaciomarine records referenced above, because these are unambiguously tied to the $\delta^{18}O$ record (Fairbanks and Matthews, 1978). These dates not only confirm an important control point used in developing our chronology (the 127 ka date of Broecker and van Donk (1970), but are wholly consistent with the tuned age of 128 ka ±3 ka.

By correlating each of the cores studied in this paper against the standard $\delta^{18}O$ stratigraphy, the SPECMAP time scale is transferred to these cores. The accuracy of this transfer is checked by calculating a phase spectrum between the standard SPECMAP $\delta^{18}O$ chronology and the $\delta^{18}O$ record in the target core. The transfer is judged adequate when the phase differences in the main Milankovitch bands are near zero (McIntyre et al., 1989). It is important to note that when regional temperature responses (based on non-isotopic data) are considered, the time scale is not modified to improve the fit of local temperature to orbital change. Only the oxygen isotope record is used for establishing the chronology. For this study, we have chosen to transfer the chronology of Imbrie et al. (1984). But the same procedures can be used to transfer any other isotope chronology that has been or might be developed.

DATA

Time series analyzed in this paper are derived from 17 piston cores ranging from 54°N in the Atlantic to 44°S in the Indian Ocean (Table 1). Basic data are of two types: measurements of $\delta^{18}O$ on benthic or planktonic foraminifera, and species counts of planktonic foraminifera or radiolaria. $\delta^{18}O$ data establish a stratigraphy and chronology, using methods described above. Species data provide information about the surface layers of the ocean. This information is extracted using the transfer-function methods of Kipp (1976). The method provides estimates of down-core changes in sea-surface temperature (SST), based on regression equations calibrated against data on the distribution of foraminifera and radiolaria in core-tops. Tests of the method by Molfino et al. (1982), using different biotic groups with different ecological preferences, give essentially the same results on South Atlantic cores. In addition, several studies have demonstrated that SST estimates derived in this way, and the faunal assemblages on which they are based, are linked with the geometry of ocean gyres and patterns of advection (Kipp, 1976; McIntyre et al., 1976; Howard and Prell, 1984; Ruddiman and McIntyre, 1984; Prell et al., 1979; McIntyre et al., 1989).

TABLE 1. Core locations, record lengths (T), number of samples (NS), and interpolated sampling interval (Δt).

No.	Core	Latitude (°)	Longitude (°)	T (ky)	NS	Δt(ky)
1	K708-7	54 N	24 W	242	125	2
2	V27-116	53 N	30 W	245	129	2
3	K708-1	50 N	24 W	242	184	1
4	V30-97	41 N	33 W	241	193	1
5	V25-21	26 N	42 W	172	46	3
6	V12-122	17 N	74 W	400	80	5
7	V30-49	18 N	21 W	325	205	1
8	V25-59	1 N	33 W	146	160	1
9	V25-56	3 S	35 W	256	146	2
10	V30-40	0 N	23 W	257	252	1
11	RC24-07	1 S	12 W	193	148	1
12	RC24-16	5 S	10 W	236	251	1
13	RC24-30	1 S	3 E	400	163	2
14	V22-174	10 S	13 W	400	109	4
15	RC13-228	22 S	11 E	305	225	1
16	RC12-294	37 S	10 W	362	192	2
17	RC11-120	44 S	80 E	290	182	1

The type, source and location of the isotopic and faunal data used in this study are identified in Table 2. Isotopic data on 13 of the 17 cores, and SST estimates on 10 cores, have already been published. Among these records are several which have been extensively analyzed and widely discussed, including four records from the North Atlantic (Ruddiman and McIntyre, 1981, 1984), three from the equatorial Atlantic (McIntyre et al., 1989), and one from the Indian Ocean (Hays et al., 1976).

TABLE 2. Types and sources of stratigraphic data and faunal data for estimating sea surface temperatures. Letter codes: M - multiple stratigraphic indicators, including discontinuous isotope data, volcanic ash layers, and calcium carbonate curves. B - continuous benthic $\delta^{18}O$ record. P - continuous planktonic $\delta^{18}O$ record. F - foraminiferal data. R - radiolarian data. S1 - numerical data available from National Center for Atmospheric Research, Boulder, Colorado as SPECMAP Archive #1. This archive includes N.G. Kipp's core-top foraminiferal data and equation AOF20 used in estimating SST for cores 5 and 7-16.

		Stratigraphy			Faunal Data		
No.	Core	Type	References	Archive	Type	References	Archive
1	K708-7	M	15		F	7,13,15	S1
2	V27-116	M	15		F	7,13,15	S1
3	K708-1	M	15		F	7,13,15	S1
4	V30-97	B	14,15	S1	F	7,13,15	S1
5	V25-21	P	1,2	S1	F	1	S1
6	V12-122	P	4	S1	F	5	S1
7	V30-49	B	11	S1	F	16	S1
8	V25-59	B	10	S1	F	10	S1
9	V25-56	P	16	S1	F	18	S1
10	V30-40	P	10	S1	F	10	S1
11	RC24-07	P	19		F	19	
12	RC24-16	P	10	S1	F	10	S1
13	RC24-30	P	19		F	19	
14	V22-174	P	6	S1	F	18	S1
15	RC13-228	B	12	S1	F	18	S1
16	RC12-294	B	17	S1	F	18	S1
17	RC11-120	P	3,9	S1	R	3,8,20	S1

(1) Crowley and Matthews, 1983
(2) Curry and Crowley, 1987
(3) Hays et al., 1976
(4) Imbrie and Kipp, 1971
(5) Imbrie et al., 1973
(6) Imbrie et al., 1984
(7) Kipp, 1976
(8) Lozano and Hays, 1976
(9) Martinson et al., 1987
(10) McIntyre et al., 1989
(11) Mix and Fairbanks, 1985
(12) Morley and Shackleton, 1984
(13) Ruddiman and Glover, 1975
(14) Ruddiman and McIntyre, 1981
(15) Ruddiman and McIntyre, 1984
(16) A.C. Mix
(17) N.J. Shackleton
(18) N.G. Kipp
(19) A. McIntyre
(20) J.D. Hays

Anticipating that many investigators will wish to perform their own analysis on these data, we are archiving them in digital form at the National Center for Atmospheric Research, Boulder, Colorado as SPECMAP Archive #1. The archive includes values of $\delta^{18}O$, species counts, and SST estimates as a function of core depth; a description of the chronology we have developed for each core; and counts of core-top foraminifera. The archive is also available from the national geophysical data center, Boulder, Colorado.

RESULTS

SST Records

The complexity of the surface ocean record is demonstrated in Figure 4, where 17 time series of cold-season temperature estimates are displayed as a function of latitude. All areas do not respond in the same way. The high-latitude North Atlantic has high-amplitude sea-surface temperature signals (cores 1-4). Moving south, the subtropical North Atlantic and western equatorial Atlantic are relatively stable (5,6,8,9). Temperature variability is again large near the equator (10-13). The southern-hemisphere subtropics display low variability (14), but the signal is relatively large in the mid-southern latitudes (15,16), and small at 44°S (17). The character of the record also changes between areas. The high latitudes are dominated by longer period thermal variability than the tropics.

As discussed in the next section, many features of the SST patterns shown on Figure 4 were discovered by earlier workers who examined time series long enough to permit spectral analysis. In addition, CLIMAP (1984) studied a global array of short records spanning the last interglacial maximum (isotope Stage 5e). They found that the oceanic response in the Southern Hemisphere leads global ice volume, whereas the oceanic response in the North Atlantic and Gulf of Mexico lags ice volume. And Mix et al. (1986a,b) found that two spatial modes of glacial- interglacial climate change have existed in the tropical Atlantic during the past 20,000 years. The first mode, defined by cold glacial and warm interglacial conditions, fluctuates symmetrically with respect to the equator. The second mode, defined by temperature extremes centered on the deglaciation, is hemispherically asymmetrical, with the Atlantic relatively warm and the southern area cold during deglaciation. Although these modes do not separate effects into component frequencies, when considered together they reflect the spatial patterns of phase observed here, with early responses in the Southern Hemisphere and late responses in the Northern Hemisphere.

SST Spectra

When expressed in the frequency domain, most of the climate records we have analyzed show distinct concentrations of total variance in the primary Milankovitch bands, centered at periods near 23 ky, 41 ky, and 100 ky (Table 3; Fig. 5). In nearly every record, coherent variance occurs at the forcing frequencies, even where a distinct peak in total amplitude is lacking. The reason the time series in Figure 4 look so different from each other, in spite of the dominance of orbital frequencies in most of

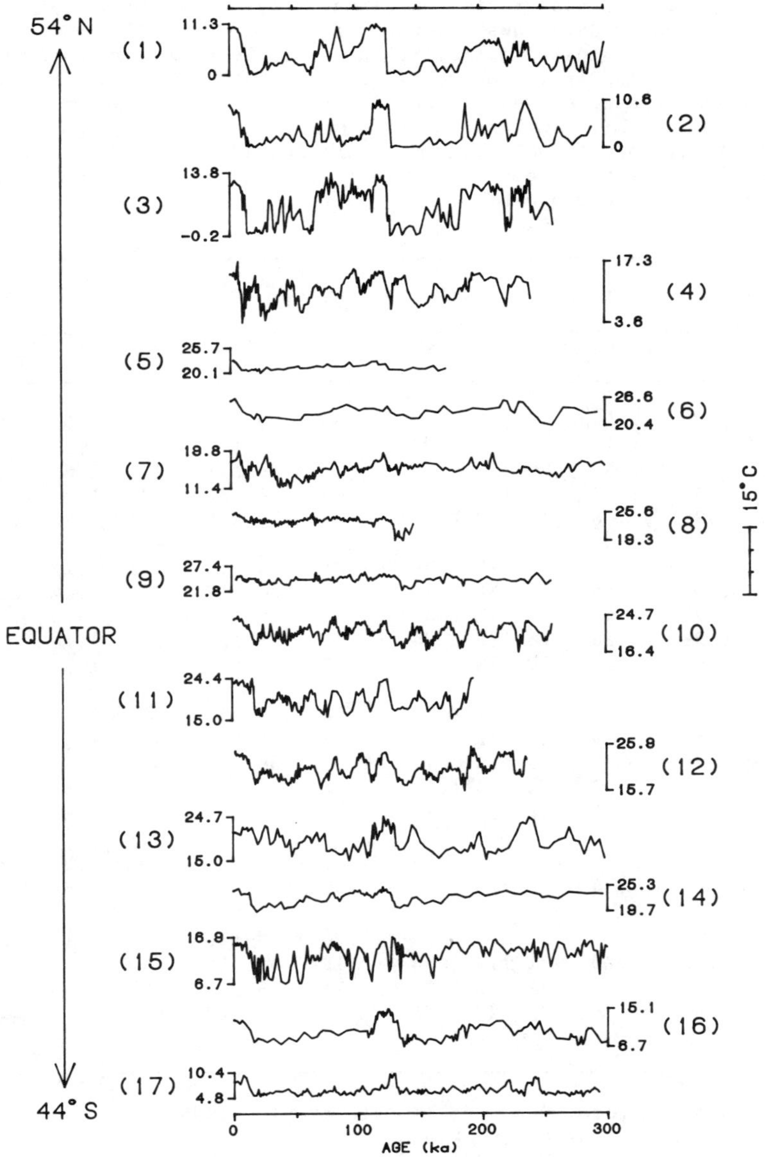

Fig. 4. Variation of foraminiferal estimates of sea-surface temperature (cold season) versus age ranging from the high-latitude North Atlantic (core 1, top) to the southern Indian Ocean (core 17, bottom). The scale factor is the same for all records. This transect illustrates the spatial variability of the SST records. Core names and locations are in Table 1.

TABLE 3. Statistical description of climatic responses in the primary Milankovitch bands. Responses include: $\delta^{18}O$, $\Delta\delta^{13}C$, Cd/Ca, and faunally estimated cold-season SST responses at 16 Atlantic and 1 Indian Ocean sites. Values tabled are: coherency (k); coherent amplitude (a); phase angle (ϕ) with 80% confidence interval; and 80% test statistic for non-zero coherency, k_o. ( ) indicates values of k that are not significant at the 80% level. Amplitudes of SST variations are scaled in °C; amplitudes of $\delta^{18}O$ and $\Delta\delta^{13}C$ variations are scaled in o/oo. All values calculated with a bandwidth of 0.015 c/ky.

No.	Core	100 ky			41 ky			23 ky			k_o
		k	a(°C)	Ø	k	a(°C)	Ø	k	a(°C)	Ø	
1	K708-7	.79	2.4	0 ±17	.80	1.8	93 ±20	.66	.64	133 ±28	.67
2	V27-116	(.62)	1.1	-34 ±31	.77	1.5	92 ±21	.89	1.1	99 ±13	.67
3	K708-1	.79	2.8	20 ±17	.92	2.4	89 ±12	.91	1.6	112 ±12	.68
4	V30-97	.88	1.6	- 1 ±12	(.61)	.69	58 ±32	.87	1.3	170 ±15	.68
5	V25-21	.96	.46	26 ±10	(.70)	.26	110 ±29	.87	.22	155 ±18	.79
6	V12-122	.67	.68	14 ±19	(.28)	.21	115 ±52	(.31)	.12	63 ±49	.52
7	V30-49	(.51)	.55	6 ±34	.80	.49	120 ±17	.72	.44	122 ±22	.58
8	V25-59	.95	.46	48 ±11	(.47)	.18	56 ±48	(.78)	.25	69 ±26	.86
9	V25-56	(.56)	.26	23 ±27	.74	.21	18 ±23	.86	.16	-13 ±15	.66
10	V30-40	.65	.50	20 ±28	(.17)	.13	143 ±70	.95	1.3	44 ± 9	.66
11	RC24-07	.84	.77	-32 ±19	(.57)	.72	98 ±38	.76	1.3	49 ±24	.76
12	RC24-16	.88	1.09	-12 ±15	(.25)	.29	15 ±61	.96	1.5	40 ± 8	.68
13	RC24-30	.78	1.1	-95 ±16	.73	1.2	188 ±19	.55	.58	41 ±29	.52
14	V22-174	.56	.50	26 ±28	(.40)	.22	82 ±40	.67	.32	68 ±22	.52
15	RC13-228	.62	.91	18 ±28	(.38)	.32	39 ±46	.85	.64	70 ±15	.60
16	RC12-294	.87	1.6	-22 ±12	.73	.77	68 ±20	.70	.50	77 ±21	.55
17	RC11-120	.84	.54	-36 ±15	.74	.42	62 ±22	.94	.58	41 ± 9	.62
	$-\delta^{18}O$ c	.81	.38	12 ±15	.84	.20	78 ±13	.91	.18	87 ± 9	.52
	$-\Delta\delta^{13}C$ d	.86	.20	-21 ±13	.84	.12	57 ±15	.72	.06	51 ±21	.65
	$-\Delta\delta^{13}C$ e	.90	.21	-24 ±12	.86	.14	28 ±15	(.55)	.03	79 ±34	.63
	-Cd/Ca f	.80		-72 ±27	.84		-14 ±22	.67		43 ±41	.64

c SPECMAP stack record, 0-400 ka (Imbrie et al., 1984)
d E. Eq. Pac. core V19-30 (Shackleton and Pisias, 1985)
e Tropical Atlantic stack, 0-281 ka (Curry and Crowley, 1987)
f S. Atl. core AII 107-131, 0-302 ka (Boyle, 1984) - we have slightly revised the author's original time scale to conform to that of Imbrie et al. (1984).

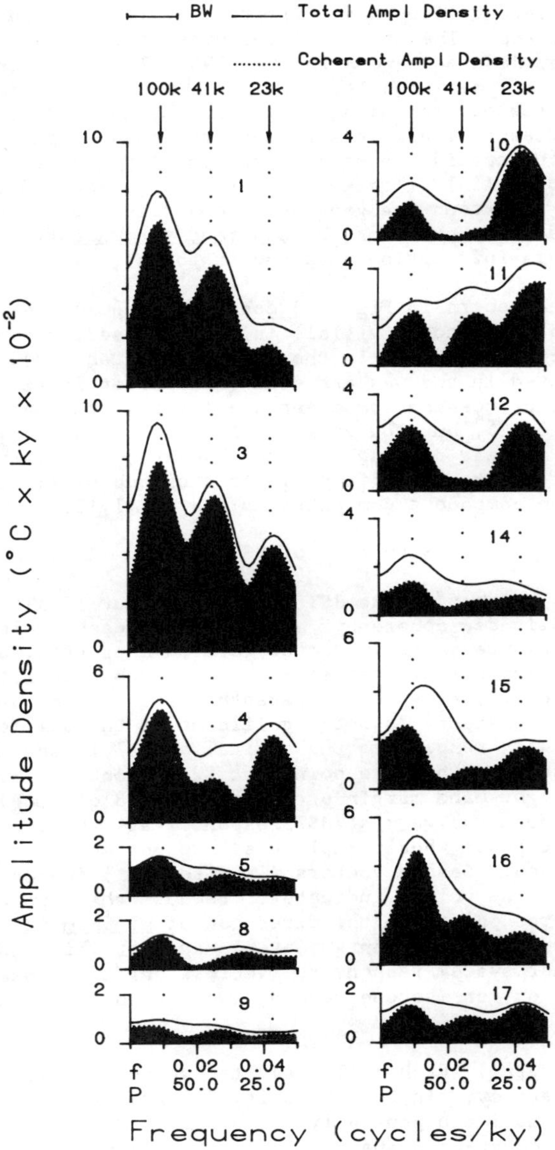

Fig. 5. Amplitude-density spectra of SST records in the Atlantic and southern Indian Oceans. Lines indicate the smoothed spectrum of total amplitude, while solid fields represent the estimates of coherent amplitude density. Estimates of the amplitudes of the coherent variance (°C) in the three Milankovitch bands centered on periods of 23 ky, 41 ky, and 100 ky are derived by integration and plotted on Figure 6. Each of these bands is of width BW.

them, is that different sites have different amplitudes and phases in the various orbital bands. The essential features of this geographic pattern were discovered by Ruddiman and McIntyre (1981, 1984) working in the mid- to high-latitudes of the North Atlantic, by Mix et al. (1986) and McIntyre et al. (1989) in the equatorial Atlantic, and by Hays et al. (1976) in the southern Indian Ocean. These workers used the peak heights in variance spectra to document spatial changes in the total variance observed in different bands, and digital filters to portray the phase. Our contribution in this paper is to extend the geographic coverage, to extract information about the amplitude and phase of the coherent variance from cross-spectra, and to display this information in a vector format.

The amplitude spectra in Figure 5 confirm and quantify the clear geographic pattern of amplitudes visible in the time series of Figure 4. Highest values occur near 50°N in the Atlantic (sites 1,3). Low amplitudes are documented in the western equatorial Atlantic (sites 8,9). Amplitudes of the precession component are large in the eastern equatorial Atlantic (sites 10-12), but low at sites further south (14-17). Many of the spectra show a clear dominance of the 100-ky cycle measured either in terms of total or coherent amplitudes. In the equatorial Atlantic, however, the precession component dominates (sites 10,11,12).

Structure of SST Cycles

A statistical summary of the SST response at our 17 sites includes values of the amplitude, coherency, and phase for each cycle (Table 3). This information is presented in vector form on maps (Fig. 6). The vector representing the response at each site has a length proportional to the coherent amplitude of the cycle. A parenthesis is placed around site numbers where the coherency is not significant at the 80% level. The phase difference with respect to the orbital signal is shown by an angle measured clockwise from the zero point (at 12 o'clock). Thus, a vector pointing to the right-hand margin of the page (at 3 o'clock) would represent a cycle in which the warmest SST response lags the orbital signal by 90°. In the inset phase wheel, amplitudes are not shown, but all phases observed are plotted. Dashed vectors (labeled "IV") indicate the phase of the oxygen-isotope ice volume indicator. Because the sign of $\delta^{18}O$ is reversed, the vector points in the direction of minimum ice. The $-\delta^{18}O$ phase provides a convenient reference against which to examine the phases of other system responses. Vectors shifted clockwise from the IV vector lag the oxygen-isotope record; vectors shifted counterclockwise lead $-\delta^{18}O$.

Amplitude patterns. Each cycle has its own geographic pattern of amplitude and coherency. In the obliquity (41 ky) band, the response in low latitudes is small and generally incoherent. Only one tropical site exhibits a large, coherent signal (site 13). This might be expected from the geometry of the radiation cycle: changes in obliquity have little effect on tropical insolation. In the precession band (23 ky), however, the amplitudes are as high in the tropical Atlantic as in the high latitudes, and all responses but two are coherent. In the 100 ky band, the response is coherent at most sites, with the largest amplitudes occurring

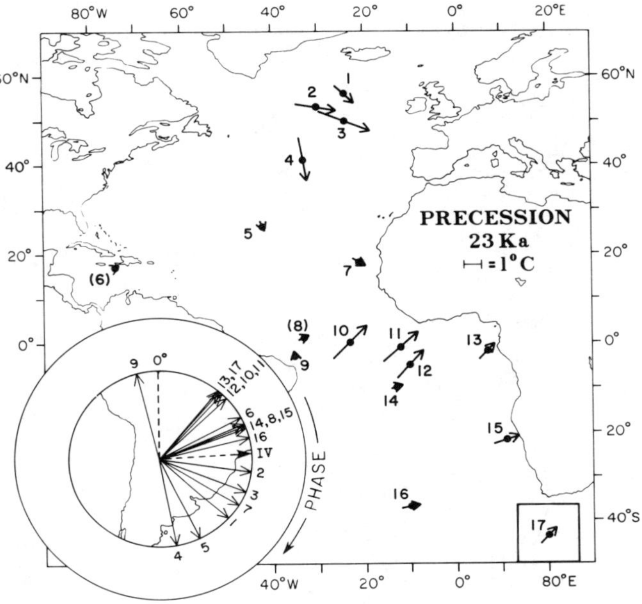

Fig. 6. Milankovitch cycles of cold-season sea-surface temperature at 16 sites in the Atlantic Ocean and site 17 in the southern Indian Ocean. On the map, the coherent amplitude of each cycle in °C is indicated by the length of the arrow, according to the scale shown. The phase of the maximum temperature is indicated by the direction of the arrow, using the SPECMAP phase convention (Fig. 2). A parenthesis around the site number indicates a coherency with the orbital signal that is not significant at the 80% level. Solid vectors on the phase wheel show the observed phase at all sites; the dashed vector shows the phase of minimum ice volume (IV).

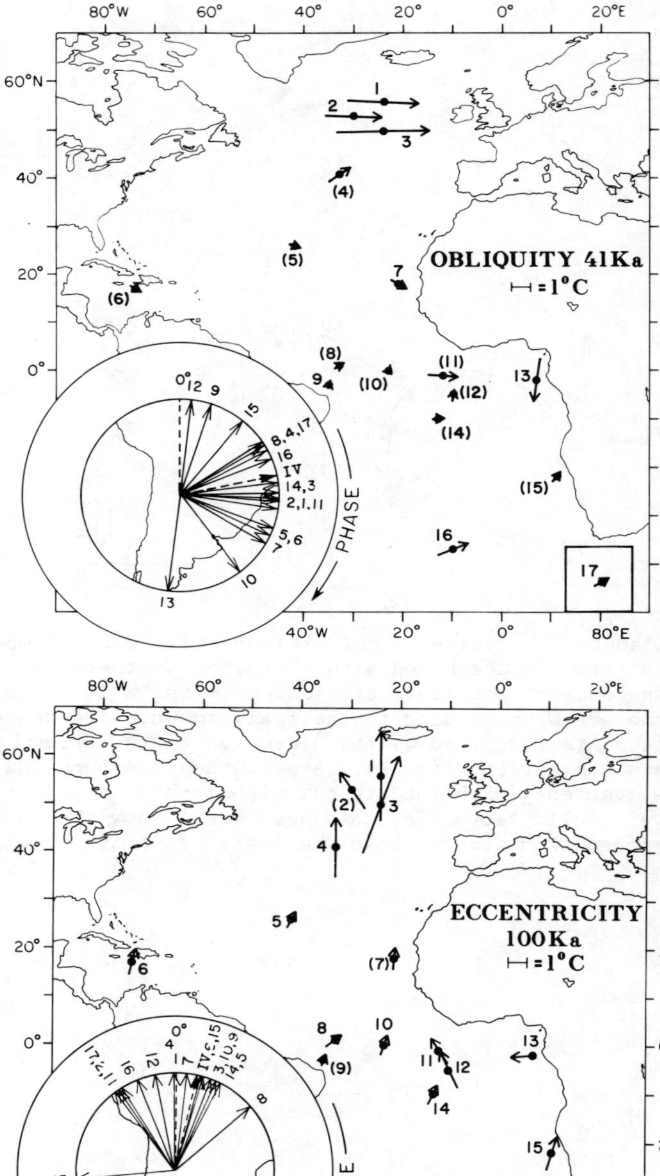

80°W
60°
40°
20°
0°
20°E
60°N
40°
20°
0°
20°
40°S
OBLIQUITY 41Ka
=1°C
PHASE
0°
IV
13
80°E
40°W
ECCENTRICITY
100Ka
=1°C
PHASE

in the Atlantic in mid-ocean regions away from the centers of subtropical gyres.

Phase patterns. Each cycle has its own characteristic pattern of phases (Fig. 6). Our data in the precession (23 ky) band suggests a significant phase shift across the thermal equator. Without exception, all coherent temperature responses south of the equator lead changes in ice volume, and all coherent North Atlantic responses lag ice volume. The greatest lags occur within the subtropical gyre and near its northern margin (sites 4,5). Ignoring the low-amplitude response at site 9, the greatest leads occur as an essentially synchronous cluster of responses: at three equatorial Atlantic sites (10,11,13), one nearby site in the Atlantic (12), and one site from the Southern Ocean (17). In short, along the transect of cores studied in this paper, the Southern Hemisphere ocean leads and the Northern Hemisphere ocean lags changes in ice volume over the precession band; and the observed range of high-amplitude SST phases is large: ~130° between sites 4 and 17. Mechanisms to explain this pattern of "lagging warmth" in the high-latitude North Atlantic have been proposed by Ruddiman and McIntyre (1984). Mechanisms to explain the early SST response in the equatorial Atlantic have been proposed by Mix et al. (1986) and McIntyre et al. (1989).

Coherent records of phase in the 41 ky band are available at seven sites, mainly in the high northern and southern latitudes. Our data for these sites show a phase shift between the hemispheres, with the SST response at high southern latitudes (sites 16,17) leading the response in the North Atlantic (sites 1-3,7). The only exception is the eastern boundary site (13). But unlike the pattern at 23 ky, the total range of SST phases is small (58° between sites 7 and 17); and the 16° SST lead with respect to ice volume at sites 16 and 17 is not statistically significant (Table 3). Responses in the North Atlantic (sites 1-3,7) show lags that are statistically significant with respect to ice volume, but small (~25°), in essential agreement with the findings of Ruddiman and McIntyre (1984).

Coherent records of SST phase in the 100 ky band are available at 14 widely distributed sites. Our observations demonstrate a clear difference in the oceanic response of the two hemispheres, and a latitudinal dependence of responses in the Southern Hemisphere. North of 1°N, all coherent SST responses are essentially on phase with changes in ice volume. This is consistent with the results of Ruddiman and McIntyre (1984), who conclude that the phases of SST at 41 ky and 100 ky are regulated by the size of Northern Hemisphere ice sheets. The total range of responses (27° between sites 4 and 5) is tightly clustered around the ice volume minimum at 12°, and none of the individual observations differ significantly from this minimum (Table 3). In contrast, coherent responses between 1°N and 30°S exhibit a very wide range. Even if we eliminate the two extreme responses near the ocean margin (sites 8,13), the range is twice that of the Northern Hemisphere (62° between sites 14 and 17). South of 30°S, the SST response at sites 16 and 17 leads ice volume significantly, and leads most of the rest of the Atlantic Ocean. The exceptions are two sites in the equatorial Atlantic (11,12) which respond on phase with the southern sites; and the eastern boundary site 13, which responds earlier. We infer that this latitudinal pattern reflects a combination of influences: (1) the geographically broad influence exerted by the Northern Hemisphere ice sheets on the atmosphere and surface ocean at one end of our transect (Ruddiman and McIntyre, 1984; Keffer et al., 1988); (2) the somewhat narrower influence exerted by Southern Hemisphere sea-ice at the other end (Manabe and Broccoli, 1985); (3) the relatively small influence exerted by the cryosphere in between these areas, which would be freer to respond to radiative forcing (Prell and Kutzbach, 1987; Kutzbach and Gallimore, 1988; McIntyre et al., 1989); and (4) the global influence that changes in CO_2 would exert (Broccoli and Manabe, 1987). From the work of Barnola et al. (1987), it seems clear that this influence would be strong in the 100 ky band with a phase near that of ice volume. Our phase map suggests that along the margin of Africa (sites 13,14) local influences of the wind field on the oceanic heat budget overwhelm the global pattern.

Summary. Each cycle has a characteristic pattern of amplitude and phase. The main features of the phase pattern are summarized in Figure 7 (top):

SST responses in high southern latitudes lead SST responses at high northern latitudes in all three cycles. Southern responses also lead with respect to changes in ice volume, but the observed lead at 41 ky is probably not statistically significant.

The 23 ky cycle, which is coherent over the entire array of cores investigated, exhibits a significant phase shift between the hemispheres. Southern Hemisphere sites lead and Northern Hemisphere sites lag changes in ice volume.

The 41 ky cycle is expressed mainly at high latitudes in both hemispheres, with southern responses leading.

The 100 ky cycle is coherent at most of the sites investigated. In the North Atlantic, the response is everywhere on phase with ice volume. Between the equator and 30°S, responses vary widely from site to site over a range of phases centered on the point of minimum ice. South of 30°S, the SST response leads ice volume at the two sites available for study here.

DISCUSSION

Sequence of System Responses

Strategy. One goal of this study is to improve our understanding of the processes of climate change which operate within the climate system on orbital time scales. Our strategy is to record the sequence of responses in each of the Milankovitch cycles, and then to examine each sequence for clues about the chain of causal mechanisms. We assume that some of the links in the chain will involve relatively simple, one-way forcings between elements of the system. Others will involve a web of feedbacks. In either case, our chances for understanding the underlying physics should be improved by knowledge of the actual response sequence. The same observational strategy has proved useful in unraveling the physics of such complex phenomena as the Indian monsoon, ENSO, and oceanic tides.

The main challenges we face in applying this strategy to the ice-age problem are the enormous physical dimensions of the climate system and the diversity of processes operating within it. Most investigators would agree that a satisfactory level of understanding should be based on knowledge of physical and chemical changes occurring in the atmosphere, cryosphere, surface ocean, deep ocean, and lithosphere. Among the changes that can be monitored in deep-sea cores, in addition to SST and ice volume, we single out two for special attention here: a measure of the vertical gradient of ^{13}C in the ocean ($\Delta\delta^{13}C$), and a measure of the phosphate content of deep water (Cd/Ca). Both measures have been linked to the efficiency with which the ocean drives nutrients into the deep water, and they presumably monitor oceanic mechanisms for changing atmospheric CO_2 (Broecker, 1982; Shackleton et al., 1983; Boyle, 1988). The $\Delta\delta^{13}C$ data are taken from two studies, the first from an eastern equatorial Pacific core (Shackleton and Pisias, 1985), and the second from a stacked record in the tropical Atlantic (Curry and Crowley, 1987). For our present purpose, we calculate the vector sum of the responses given in Table 3. The Cd/Ca record, obtained by Boyle (1984), monitors changes in the phosphate content of the South Atlantic at a depth of 2925 m. To simplify the study of phase relationships between chemical and other records, and give all indicators a positive interglacial sense, we reverse the sign of the chemical indicators. Thus a positive direction in $-\Delta\delta^{13}C$ records a minimum in the vertical ^{13}C gradient of tropical oceans, which may be associated with a minimum in the transfer of carbon into deeper waters. A positive direction in -Cd/Ca records a minimum in phosphate content which may be closely associated with a maximum contribution of northern sources to South Atlantic deep waters, and roughly associated with high atmospheric CO_2 (barring significant contributions from mechanisms unrelated to the nutrient-carbon pump).

Phase portrait. To display essential phase information we construct a phase portrait, i.e., a diagram which includes a selection of the phase observations available, in a format that emphasizes the location of each response within the system (Fig. 7, top). Each circle is a schematic map of the world. Vectors on these maps display information about the phase of the SST response along a mid-ocean transect (Fig. 6, Table 3). Each phase is calculated from the vector sum of the coherent responses at sites near 50°N (sites 1-3), near 20°N (5,7), 5°S (10-12,14), and near 40°S (16,17).

Above each map is a vector fan which displays phase information derived from records which we judge to represent global, as opposed to purely local, responses. These are the records of Cd/Ca, $\Delta\delta^{13}C$, and $\delta^{18}O$. As noted previously, the sign conventions of these vectors are such that they point in the direction of minimum ice ($-\delta^{18}O$), maximum contribution of North Atlantic source waters to deep water in the South

Atlantic (-Cd/Ca), and lowest efficiency in the pumping of carbon from surface to deeper waters ($-\Delta\delta^{13}C$). The vector fan can be considered as a portion of a phase dial, oriented as described in Figure 2. The curved vector placed above each fan is a 90° arc of a circle which designates a sector of the phase dial. The sector starts at the point of maximum ice melting and terminates at the point of minimum ice. SST and other responses observed in this sector can be considered leading indicators with respect to ice volume. Responses which occur over the succeeding 90° sector, between the points of minimum ice and maximum ice growth, are lagging indicators. (This sector is not designated on Figure 7.) Together, the two 90° sectors represent the interglacial hemicycle. Interpretation of these phase portraits should be accompanied by consideration of confidence intervals (Table 3).

With respect to $\delta^{18}O$ and $\Delta\delta^{13}C$, the phase patterns displayed in the vector fans on Figure 7 are essentially the same as those discovered by Shackleton et al. (1983) and Shackleton and Pisias (1985). Additional important information can be gained by examining the entire phase portrait:

(1) Each cycle is characterized by the same three leading indicators: Cd/Ca, $\Delta\delta^{13}C$, and SST (~40°S). In the 23 ky cycle, equatorial Atlantic SST is also a leading indicator.

(2) In the 100 ky and 41 ky cycles, the three leading indicators display exactly the same sequence: Cd/Ca -- $\Delta\delta^{13}C$ -- SST (~40°S) -- $\delta^{18}O$. In the 23 ky band, the sequence differs only in that SST and Cd/Ca occur at the same point in the cycle.

(3) In the 41 ky and 100 ky cycles, changes in Cd/Ca occur essentially on phase with the start of the interglacial hemicycle, i.e., at the point of maximum melting. In the 23 ky band, no responses occur at this position.

Although a full interpretation of these phase portraits is beyond the scope of this paper, we conclude that changes in the circulation and carbon chemistry of the deep ocean, coupled in some way with changes in the surface ocean at high southern latitudes, are an early and essential part of the chain of responses which leads towards an interglacial world in all three cycles. Recent advances in time-dependent modeling create an important opportunity to compare the phase sequence predicted by models with the sequence observed in proxy data. An example of such a comparison is given below in the section on the 100 ky cycle.

We emphasize that the marine database used in our study needs to be expanded to include new observations on: (1) indicators of intermediate and deep water such as ^{13}C, Cd/Ca, and $CaCO_3$; (2) indicators of other parts of the surface ocean, including high latitude sites where the density of surface waters changes on phase with deep water indicators; and (3) indicators of changes in the wind field and conditions on land. In addition, it is important to include information about the phase of atmospheric properties derived from long ice cores (Barnola et al., 1987).

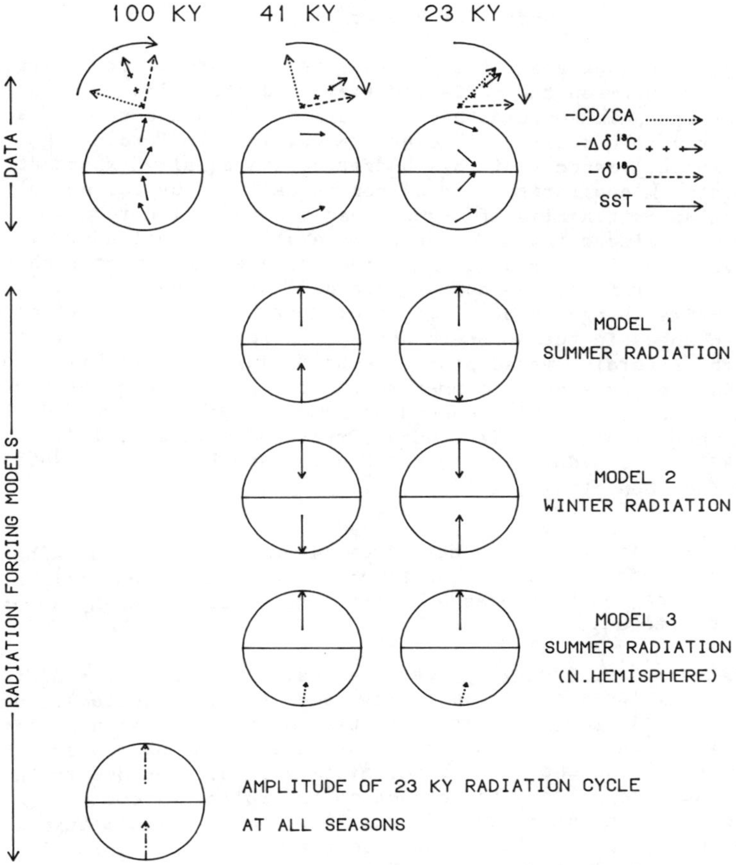

Fig. 7. Phase portraits of the observed climatic response (top) and models of radiation forcing (bottom). Each complete circle is a schematic map of the world on which geographic patterns of phase are shown by vectors, using the conventions of Fig. 2. TOP: Vectors on the maps represent observed SST responses along a mid-ocean Atlantic transect at 50°N, 20°N, 5°S, and 40°S (Fig. 6, Table 3). Vector fans above these maps record the phase of isotopic and chemical time series with global implications, using sign conventions which designate interglacial responses: minimum Cd/Ca (maximum NADW flux), minimum vertical ^{13}C gradient (minimum transfer of carbon to deeper waters), and lightest $\delta^{18}O$ (minimum ice). The 90° arc marks the sector where responses lead the ice volume minimum. BOTTOM: Vectors on maps represent models of radiation forcing for the three climate cycles. Solid vectors in Models 1-3 represent June or December radiation, depending on the hemisphere and season; dotted vectors represent transmission of a Northern Hemisphere response with a 1 ky delay. Dash-dot vectors represent the phase of the upper envelope (amplitude) of the 23 ky radiation cycle.

Models of Radiation Forcing

Strategy. Another goal of this study is to improve our understanding of causal links between the radiation field and the climate system. Since the birth of the Milankovitch theory, there has been uncertainty about which aspects of the radiation budget are critical in forcing the climatic response. Not all aspects of this budget can be equally important, because any gain in one latitude or season is balanced by losses in another. Milankovitch himself argued that the radiation received during summer is the dominant influence on climate at every latitude, because these changes are of an order of magnitude greater than winter changes. In his model, therefore, icecaps at both ends of the earth respond to local summertime changes. Many recent workers have focused attention on explaining changes in the Northern Hemisphere ice sheets. Following Milankovitch, several have adopted June 65°N, July 65°N (Imbrie and Imbrie, 1980), or some other summer month at high northern latitudes, as the forcing function for time-dependent models. Others have focused on explaining phenomena in low latitudes (Prell and Kutzbach, 1987; McIntyre et al., 1989); in the Southern Ocean (Crowley and Parkinson, 1988a); or in Antarctica (Genthon et al., 1987).

Our strategy is to use the data summarized in Figure 7 as a basis for testing simple models of radiation forcing. Our attention will be limited to explaining the observed phases of $\delta^{18}O$ and of SST at high southern and northern latitudes.

Precession and obliquity cycles. We examine three models of radiation forcing in the precession and obliquity bands (Fig. 7, bottom). Model 1 assumes that local summer radiation forces the response at high latitudes in both hemispheres. We define summer as June in one hemisphere and December in the other. In the 41 ky band, maximum insolation occurs in both hemispheres at zero phase with respect to the obliquity cycle, i.e., at the point of maximum obliquity. In the 23 ky band, maximum insolation occurs 180° out of phase with respect to the precession cycle, i.e., at zero phase (June perihelion) in the Northern Hemisphere, and at 180° phase (December perihelion) in the Southern Hemisphere.

Are the observed phases consistent with this model? For the $\delta^{18}O$ phases, the answer is yes in both cycles, because the $\delta^{18}O$ record is dominated by the Northern Hemisphere ice sheets which are thought to have a time constant on the order of 17 ky (Weertman, 1964; K. Herterich, pers. comm.). The observed phases of $\delta^{18}O$ (78° at 41 ky, and 87° at 23 ky) are consistent with the response of a linear system having a time constant on the order of 17 ky (Imbrie and Imbrie, 1980).

Before examining the pattern of SST phases, we must make assumptions about the mechanisms of the SST change. We assume that the SST responses we observe near 50°N and 40°S are driven in both hemispheres by nearfield influences of the Laurentide, Scandinavian, and Antarctic ice sheets, together with their flanking fields of sea ice. Any lag in nearfield temperature behind the local ice field would reflect an atmosphere-ocean response to the ice. The time constant of this response can be expected to vary from one hemisphere to another and from one frequency band to another. We assume that the maximum lag of SST behind the ice would be 90° (e.g. the local meltwater chilling of Ruddiman and McIntyre). The minimum lag would be 0°. Thus the possible lags of SST behind the primary forcing range from 0° to 180°, depending first on the phase of the local cryospheric response, which does not exceed 90°, and second on the time constant of the SST response to the cryosphere.

Using these criteria, we can ask if the observed SST phases (Fig. 7, top) are consistent with the summer forcing of Model 1. For the 41 ky band, the answer is yes. For the 23 ky band, the answer is yes for the Northern Hemisphere, but no for the Southern Hemisphere, where the observed SST phases range from 41° to 71° (Table 3), and cannot be considered a local response to Southern Hemisphere radiation with a phase of 180°. We therefore reject Model 1 because it fails to explain the 23 ky Southern Hemisphere SST pattern (Table 4).

We are thus led to examine <u>Model 2</u>, which assumes that local <u>winter</u> radiation forces the response at high latitudes of both hemispheres (winter being defined as December and June). As shown in Figure 7, changes in winter and summer radiation are out of phase by 180°. We reject this model as inconsistent with the observed responses except for Southern Hemisphere SST at 23 ky (Table 4).

TABLE 4. Tests of three radiation models. Agreement between predicted and observed phases of $\delta^{18}O$ and high-latitude SST is coded as yes/no for the Northern (N) and Southern Hemispheres (S). Agreement is defined as a lag of less than 180° with respect to the forcing. See text and Figure 7 for details.

		41 ky		23 ky	
		$\delta^{18}O$	SST	$\delta^{18}O$	SST
MODEL 1	N	YES	YES	YES	YES
	S	--	YES	--	NO
MODEL 2	N	NO	NO	NO	NO
	S	--	NO	--	YES
MODEL 3	N	YES	YES	YES	YES
	S	--	YES	--	YES

One possible solution is to assume that summer radiation is the dominant forcing -- except for the Southern Hemisphere in the precession band, where winter forcing dominates. Rejecting this solution as undesirably complex, we are led to Model 3. This model assumes that summer radiation forces the climatic response at high latitudes and both frequencies in the Northern Hemisphere, and that the response there is transmitted to high latitudes of the Southern Hemisphere through modulations in the flux of North Atlantic deep water (Weyl, 1968; Crowley and Parkinson, 1988b), with a delay of ~1 ky. This influence is assumed to dominate local radiation effects at high southern latitudes, as the heat budget modeling of Crowley and Parkinson (1988a) suggests. Observed $\delta^{18}O$ and SST phases are consistent with this model in both cycles and both hemispheres (Table 4).

If we accept Model 3, the implication is that changes in the density of surface waters in the Norwegian Sea, or some other high-latitude area that lies beyond the range of our data, occur very early in the interglacial hemicycle (near 0° in both obliquity and precession bands). If suitable long records can be discovered, Model 3 can be tested empirically.

The 100 ky eccentricity cycle. What role does eccentricity play in forcing or setting the phase of the 100 ky climate cycle? Since Croll (1875) first considered the problem, two general mechanisms have been considered. The first is based on the direct control eccentricity exerts on the globally and annually averaged amount of insolation. Although this effect is small (~0.1%), modeling experiments have been conducted to explore its climatic influence (Le Treut and Ghil, 1983).

The second mechanism, proposed by Croll, is based on the indirect control that eccentricity exerts on the radiation budget through its modulation of the amplitude of the precession cycle ($e \sin \omega$). Thus the upper and lower envelopes of the 23 ky radiation cycle for any latitude and season are determined by eccentricity (Berger, 1978a). Croll himself focused on the lower envelope, and the effect that colder winters would have on interhemispheric heat transport in the Atlantic Ocean. Many workers since Croll have focused on the upper envelope (Fig. 1) and the effect that warmer summers due to precession (and tilt) would have on Northern Hemisphere ice sheets (Mesolella et al., 1969; Broecker and van Donk, 1970). Hays et al. (1976) suggested that a nonlinear response of the ice sheets would bring out the 100,000-year cycle in the geologic record by forcing the time-dependent mean of the 23 ky climate cycle to approach values directly proportional to the upper envelope of the 23 ky precession cycle (i.e. eccentricity). Imbrie and Imbrie (1980) tested this idea numerically, and found that a significant amount of 100 ky power can be generated in this way, provided that the mean time constant of the system is of the same order as the precession period. In this model, the mean time constant is 17 ky. However, the phase of the model ice volume curve (43°) is significantly different from observation (12°), as pointed out by Pisias and Shackleton (1984) who improved the model's performance by adding a non-feedback CO_2 term. This suggests the need for a term representing feedback between the ocean carbon system and ice volume.

More recently, Saltzman et al. (1984) have forced their three-component feedback system with a representation of the radiation effects of orbital variations, and found that the phase of the model ice volume curve closely matches the observed phase in the 100 ky band. The largest time constant in their model is ~10 ky. The free oscillatory solution of this model is phase-locked by the external forcing. Experimenting with another model of the same type, Saltzman and Maasch (1988) have shown that the orbitally forced solution yields a response interpreted in terms of NADW flux that leads ice volume and eccentricity by about 30° in the 100 ky band. This prediction is fully consistent with the phase information on Cd/Ca and $\Delta\delta^{13}C$ on Figure 7.

The data and modeling experiments now available seem consistent with the idea first suggested by Mesolella et al. (1969) that eccentricity influences climate through its control on the upper envelope of the 23 ky radiation curve (Fig. 7, bottom). More work is needed to understand the nature of this influence. Investigations by Saltzman and his colleagues suggest that this influence takes the form of pacing an internally driven 100 ky cycle. Investigations by Ghil and his colleagues suggest that an externally forced response combines with internal oscillations to yield a 100 ky cycle that would otherwise be absent.

CONCLUSIONS

Observations on deep-sea cores demonstrate that late Pleistocene climate is dominated by three broad-band cycles centered near periods of 23 ky, 41 ky, and 100 ky. The periods of these climatic cycles match orbital cycles of precession, obliquity, and eccentricity. Over the past 400 ky, climatic oscillations in each band are coherent with orbital variations, i.e. they exhibit a significant linear correlation with the orbital signal when the phase difference is set to zero.

We conclude that data on all three cycles are best explained by a version of the Milankovitch theory in which the climatic record is viewed as the response of a nonlinear system forced both internally and externally. The external forcing may influence climate either as part of a driving mechanism which determines both the amplitude and phase of the cycle, or as a pacing mechanism which sets the phase of an internal oscillation.

In the Milankovitch theory, the insolation changes that influence the climatic system can be accurately predicted. This provides a rare opportunity to link climatic response to an external forcing, and thereby discern mechanisms of global climate change on scales longer than the annual cycle. Our strategy for investigating these mechanisms is to partition the climatic record into cyclic components; to record the phase sequences of system responses in each cycle; and to examine the sequences for clues about the chain of causal mechanisms.

Making these links between forcing and response requires several steps. The first step, establishing a global stratigraphy and chronology, has been accomplished. Although future work will certainly refine our abilities to tell time in the distant past, at present we have a reasonable system that allows us to compare the responses of different proxy indicators and to place local climate records into the global chronologic framework via oxygen isotopes.

The second step, analyzing regional climate changes, is under way. Large-scale patterns of climate change in the Milankovitch bands, as illustrated here by the SST records of the Atlantic Ocean, and by the chemical and isotopic indicators of ocean chemistry and circulation, are now being assembled. Global patterns are beginning to emerge. Future work will refine these patterns and add data from the rest of the global oceans.

The third step, developing hypotheses and testing them with appropriate quantitative models, is just beginning. Ultimately this effort will require linking atmospheric, cryospheric, and oceanic models on a variety of scales, ranging from highly parameterized dynamic models to high-resolution general circulation models. The latter will be essential to understand how changes in the global wind field force changes in the ocean heat budget, and thereby influence the SST phases we observe. These models will guide further data acquisition, and the data will in turn test the models. It is this interplay between data and modeling that will advance our understanding most rapidly. For example, a careful documentation of the sequence of responses on the globe, as outlined here for Atlantic SST, will narrow the choice of models.

ACKNOWLEDGEMENTS

This paper is a contribution of the SPECMAP project, and we are grateful to our colleagues J. E. Kutzbach, D. G. Martinson, J. J. Morley, D. Murray, N. G. Pisias, W. L. Prell, N. J. Shackleton, and J. R. Toggweiler for their many contributions to this paper. It is a pleasure to acknowledge the assistance of other colleagues who read our first draft and made valuable suggestions for its improvement. These include E. A. Boyle, G. E. Birchfield, T. J. Crowley, M. Ghil, W. R. Howard, R. K. Matthews, B. Molfino, D. Murray, W. F. Ruddiman, and B. Saltzman. Mrs. Angeline Duffy developed algorithms for estimating amplitude spectra and coherent amplitudes, carried out all the calculations, and made valuable suggestions for interpreting the results. We acknowledge the financial support of the following NSF grants: ATM-8319372 (Brown University), ATM-8318850 (Columbia University), and ATM-8319371-03 (Oregon State University).

REFERENCES CITED

Bard, E., Arnold, M., Duprat, J., Moyes, J., and Duplessy, J.-C., 1987, Reconstruction of the last deglaciation: deconvolved records of $\delta^{18}O$ profiles, micropaleontological variations and accelerator mass spectrometric ^{14}C dating. Climate Dynamics, v. 1, p. 101-112.

Barnola, J.M., Raynaud, D., Korotkevich, Y.S., and Lorius, C., 1987, Vostok ice core provides 160,000-year record of atmospheric CO_2: Nature, v. 329, p. 408-414.

Berger, A.L., 1978a, Long-term variations of daily insolation and Quaternary climatic changes. Jour. Atmos. Sci., v. 35, p. 2362-2367.

Berger, A.L., 1978b, A simple algorithm to compute long term variations of daily or monthly insolation. Contr. No. 18, Institut d' Astronomie et de Geophysique G. Lemaitre, Univ. Cath. de Louvain, Belgium, 39 pp.

Birchfield, G.E., 1987, Ice sheet dynamics and the Pleistocene ice ages. In C. Nicolis and G. Nicolis, eds., Irreversible Phenomena and Dynamical Systems Analysis in Geosciences, NATO ASI Ser. C, D. Reidel Publ. Co., v. 192, p. 381-398.

Boyle, E.A., 1988, The role of vertical chemical fractionation in controlling late Quaternary atmospheric carbon dioxide. Jour. Geophys. Res., v. 93, p. 15,701-15,714.

Boyle, E.A., 1984, Cadmium in benthic foraminifera and abyssal hydrography : evidence for a 41 kyr obliquity cycle. In Hansen P.E. and T. Takahashi, eds., Climate Processing and Climate Sensitivity, Am. Geophys. Union, Washington, D.C., p. 360-368.

Broccoli, A.J., and Manabe, S., 1987, The influence of continental ice, atmospheric CO_2, and land albedo on the climate of the last glacial maximum. Climate Dynamics, v. 1, p. 87-100.

Broecker, W.S., 1982, Glacial to interglacial changes in ocean chemistry. Prog. Oceanogr., v. 11, p. 151-197.

Broecker, W.S., Peteet, D.M., and Rind, D., 1985, Does the ocean-atmosphere system have more than one stable mode of operation? Nature, v. 315, p. 21-26.

Broecker, W.S., and van Donk, J., 1970, Insolation changes, ice-volumes, and the ^{18}O record in deep-sea cores. Rev. Geophys. & Space Phys., v. 8, p. 169-198.

Calder, N., 1974, Arithmetic of ice ages. Nature, v. 252, p. 216-218. See also Imbrie and Imbrie, 1980, op. cit.

CLIMAP Project Members, 1984, The last interglacial ocean. Quat. Res., v. 21, p. 123-224.

COHMAP Project Members, 1988, Climatic changes of the last 18,000 years: observations and model simulations. Science, v. 241, p. 1043-1052.

Croll, J., 1875, Climate and Time, Appleton & Co., New York.

Crowley, T.J., and Matthews, R.K., 1983, Isotope-plankton comparisons in a late Quaternary core with a stable temperature history. Geol., v. 11, p. 275-278.

Crowley, T.J., and Parkinson, C.L., 1988a, Late Pleistocene variations in Antarctic sea ice I: effect of orbital insolation changes. Climate Dynamics, v. 3, p. 85-91.

Crowley, T.J., and Parkinson, C.L., 1988b, Late Pleistocene variations in Antarctic sea ice II: effect of interhemispheric deep-ocean heat exchange. Climate Dynamics, v. 3, p. 93-103.

Curry, W.B. and Crowley, T.J., 1987, the $\delta^{13}C$ of equatorial Atlantic surface waters: implications for ice age pCO_2 levels. Paleoceanography, v. 2, p. 489-517.

Denton, G.H., Hughes, T.J., and Karlen, W., 1986, Global ice-sheet system interlocked by sea level. Quat. Res., v. 26, p. 3-26.

Edwards, R.L., Chen, J.H., Ku, T.-L., and Wasserburg, G.J., 1987, Precise timing of the last interglacial period from mass spectrometric determination of Thorium-230 in corals. Science, v. 236, p. 1547-1553.

Fairbanks, R.G., and Matthews, R.K., 1978, The marine oxygen isotope record in Pleistocene coral, Barbados, West Indies. Quat. Res., v. 16, p. 181-196.

Genthon, C., Barnola, J.M., Raynaud, D., Lorius, C., Jouzel, J., Barkov, N.I., Korotkevich, Y.S., and Kotlyakov, V.M., 1987, Vostok ice core: climatic response to CO_2 and orbital forcing changes over the last climatic cycle. Nature, v. 329, p.414-418.

Ghil, M., 1989, Deceptively-simple models of climatic change. In Berger et al., eds., Climate and Geosciences, Kluwer Academic Publishers. In press.

Ghil, M., and Childress, S., 1987, Topics in Geophysical Fluid Dynamics: Atmospheric Dynamics, Dynamo Theory and Climate Dynamics. Springer-Verlag, New York, 485 pp.

Ghil, M., and Le Treut, H., 1981, A climate model with cryodynamics and geodynamics. J. Geophys. Res., v. 86, p. 5262-5270.

Hays, J.D., Imbrie, J., and Shackleton, N.J., 1976, Variations in the earth's orbit: pacemaker of the ice ages. Science, v. 194, p. 1121-1132.

Horel, J.D., 1982, On the annual cycle of the tropical Pacific atmosphere and ocean. Monthly Weather Rev., v. 110, p. 1863-1878.

Howard, W.R., and Prell, W.L., 1984, A comparison of radiolarian and foraminiferal paleoecology in the southern Indian Ocean: new evidence for the interhemispheric timing of climatic change. Quat. Res., v. 21, p. 244-263.

Imbrie, J., 1985, A theoretical framework for the Pleistocene ice ages. Jour. Geol. Soc., v. 142, p. 417-432.

Imbrie, J., 1987, Abrupt terminations of late Pleistocene ice ages: a simple Milankovitch explanation. In W.H. Berger and L.D. Labeyrie, eds., Abrupt Climatic Change, D. Reidel Publ. Co., p. 365-367.

Imbrie, J., Hays, J.D., Martinson, D.G., McIntyre, A., Mix, A.C., Morley, J.J., Pisias, N.G., Prell, W.L., and Shackleton, N.J., 1984, The orbital theory of Pleistocene climate: support from a revised chronology of the marine $\delta^{18}O$ record. In Milankovitch and Climate, Part 1, A.L. Berger et al., eds.: D. Reidel Pub. Co., p. 269-305.

Imbrie, J. and Imbrie, J.Z., 1980, Modeling the climatic response to orbital variations. Science, v. 207, p. 943- 953.

Imbrie, J. and Kipp, N.G., 1971, A new micropaleontological method for quantitative paleoclimatology: Application to a late Pleistocene Caribbean core. In Late Cenozoic Glacial Ages, K.K. Turekian, ed., p. 77-181, Yale University Press, New Haven, Conn.

Imbrie, J., van Donk, J., and Kipp, N.G., 1973, Paleoclimatic investigation of a late Pleistocene Caribbean deep-sea core: comparison of isotopic and faunal methods. Jour. Quat. Res., v. 3, p. 10-38.

Jenkins, G.M. and Watts, D.G., 1968, Spectral analysis and its applications. Holden Day, San Francisco, 525 pp.

Julian, P.R., 1975, Comments on the determination of significance levels of the coherence statistic. Jour. Atmos. Sci., v. 32, p. 836-837.

Kaufman, A., 1986, The distribution of $^{230}Th/^{234}U$ ages in corals and the number of last interglacial high-sea stands. Quat. Res., v. 25, p. 55-62.

Keffer, T., Martinson, D.G., and Corliss, B.H., 1988, The position of the Gulf Stream during Quaternary glaciations. Science, v. 241, p. 440-442.

Kipp, N.G., 1976, New transfer function for estimating past sea-surface conditions from sea bed distribution of planktonic foraminiferal assemblages in the North Atlantic. Mem. Geol. Soc. Am. 145, p. 3-42.

Kominz, M.A. and Pisias, N.G., 1979, Pleistocene climate: deterministic or stochastic? Science, v. 204, p. 171-173.

Kutzbach, J.E. and Gallimore, R.G., 1988, Sensitivity of a coupled atmosphere/mixed-layer ocean model to changes in orbital forcing at 9000 years BP. J. Geophys. Res., v. 93, p. 803-821.

Kutzbach, J.E. and Guetter, P.J., 1984, Sensitivity of late-glacial and Holocene climates to the combined effects of orbital parameter changes and lower boundary condition changes: "snapshot" simulations with a general circulation model for 18, 9, and 6 ka BP. Ann. of Glaciology, v. 5, p. 85-87.

Kutzbach, J.E., and Otto-Bliesner, B.L., 1982, The sensitivity of the African-Asian monsoonal climate to orbital parameter changes for 9000 years B.P. in a low-resolution general circulation model. Jour. Atmos. Sci., v. 39, p. 1177-1188.

Kutzbach, J.E. and Street-Perrott, F.A., 1985, Milankovitch forcing of fluctuations in the level of tropical lakes from 18 to 0 kyr BP. Nature, v. 317, p. 1-5.

Labeyrie, L.D., Duplessy, J.C., and Blanc, P.L., 1987, Variations in mode of formation and temperature of oceanic deep waters over the past 125,000 years. Nature, v. 327, p. 477-482.

Le Treut, H. and Ghil, M., 1983, Orbital forcing, climatic interactions, and glaciation cycles. Jour. Geophys. Res., v. 88, p. 5167-5190.

Le Treut, H., Portes, J., Jouzel, J. and Ghil, M., 1988, Isotopic modeling of climatic oscillations: implications for a comparative study of marine and ice core records. Jour. Geophys. Res., v. 93, p. 9365-9383.

Lindzen, R.S., 1986, A simple model for 100K-year oscillations in glaciation. Jour. Atmos. Sci., v. 43, p. 986-996.

Lorius, C., Jouzel, J., Ritz, D., Merlivat, L., Barkov, N.I., Korotkevich, Y.S., and Kotlyakov, V.M., 1985, A 150,000-year climatic record from Antarctic ice. Nature, v. 316, p. 591-596.

Lozano, J.A., and Hays, J.D., 1976, Relationship of radiolarian assemblages to sediment types and physical oceanography in the Atlantic and western Indian Ocean sectors of the Antarctic Ocean. In Investigation of Late Quaternary Paleoceanography and Paleoclimatology, R.M. Cline and J.D. Hays, eds., Geol. Soc. Amer. Mem., p. 303-336.

Lyle, M., 1988, Climatically forced organic carbon burial in equatorial Atlantic and Pacific Oceans. Nature, v. 335, p. 529-532.

Manabe, S., Broccoli, A.J., 1985, The influence of continental ice sheets on the climate of an ice age. J. Geophys. Res., v. 90, p. 2167-2190.

Mankinen, E.A., and Dalrymple, G.B., 1979, Revised geomagnetic polarity time scale for the interval 0-5 m.y. B.P. Jour. Geophys. Res., v. 84, p. 615-626.

Marsiat, I., Berger, A., Fichefet, T., Gallee, H., and Tricot, C., in press, Modelling the transient response of a coupled model to the astronomical forcing over the last glacial to interglacial cycle. In: The Beginning of an Inland Glaciation, Facts and Problems of Climate Dynamics, B. Frenzel, ed., Mainz.

Martinson, D.G., Pisias, N.G., Hays, J.D., Imbrie, J., Moore, T.C., and Shackleton, N., 1987, Age dating and the orbital theory of the ice ages: development of a high-resolution 0 to 300,000-year chronostratigraphy. Quaternary Res., v. 27, p. 1-30.

McIntyre, A., Kipp, N.G., et al., 1976, Glacial North Atlantic 18,000 years ago: a CLIMAP reconstruction. In Investigation of Late Quaternary Paleoceanography and Paleoclimatology, R.M. Cline and J.D. Hays, eds., Geol. Soc. Amer. Mem. 145, p. 43-76.

McIntyre, A., Ruddiman, W.F., Karlin, K., and Mix, A.C., 1989, Surface water response of the equatorial Atlantic Ocean to orbital forcing. Paleoceanography.

Mesolella, K.J., Matthews, R.K., Broecker, W.S., and Thurber, D.L., 1969, The astronomical theory of climatic change: Barbados data. Jour. Geol., v. 77, p. 250-274.

Milankovitch, M., 1941, Kanon der Erdbestrahlung und seine Andwendung auf das Eiszeitenproblem, Royal Serb. Acad., Spec. Publ., 133, Belgrade, 633 p. English translation published in 1969 by Israel Program for Scientific Translations available from U.S. Dept. Comm. Page reference in text is to this edition.

Mitchell, J.M.,Jr., 1976, An overview of climatic variability and its causal mechanisms. Quat.Res., v. 6, p. 481-493.

Mix, A.C., and Fairbanks, R.G., 1985, North Atlantic surface control of Pleistocene deep-ocean circulation. Earth and Planet. Sci. Lett., v. 73, p. 231-243.

Mix, A.C., and Ruddiman, W.F., 1985, Structure and timing of the last deglaciation: oxygen-isotope evidence. Quat. Sci. Rev., v. 4, p. 59-108.

Mix, A.C., Ruddiman, W.F., and McIntyre, A., 1986a, Late Quaternary paleoceanography of the tropical Atlantic, 1: spatial variability of annual mean sea-surface temperatures, 0-20,000 years B.P. Paleoceanography, v. 1, p. 43-66.

Mix, A.C., Ruddiman, W.F., and McIntyre, A., 1986b, Late Quaternary paleoceanography of the tropical Atlantic, 2: the seasonal cycle of sea surface temperatures, 0-20,000 years B.P. Paleoceanography, v. 1, p. 339-353.

Molfino, B., Kipp, N.G., and Morley, J.J., 1982, Comparison of foraminiferal, coccolithophorid, and radiolarian paleotemperature equations: Assemblage coherency and estimate concordancy. Quat. Res., v. 17, p. 279-313.

Morley, J.J., and Shackleton, N.J., 1984, The effect of accumulation rate on the spectrum of geologic time series: evidence from two South Atlantic sediment cores. In A. L. Berger et al., eds., Milankovitch and Climate, Part 1, D. Reidel Pub. Co., p. 467-480.

Munk, W.H., and Cartwright, D.E., 1966, Tidal spectroscopy and prediction. Phil.Trans.Roy.Soc. London, v. 259, p. 533-581.

Oerlemans, J. and van der Veen, C.J., 1984, Ice Sheets and Climate. D. Reidel Pub. Co., Dordrecht, 217 pp.

Peltier, W.R., 1987, A relaxation oscillator model of the ice age cycle. In C. Nicolis and G. Nicolis, eds., Irreversible phenomena and dynamical systems analysis in geosciences. NATO ASI Series C, v. 192, p. 399-416.

Pestiaux, P., and Berger, A.L., 1984, An optimal approach to the spectral characteristics of deep-sea climatic records. In Milankovitch and Climate, v. 1, A. Berger et al., eds., Reidel, Dordrecht, p. 417-445.

Pestiaux, P., Van der Mersch, I., and Berger, A., 1988, Paleoclimatic variability at frequencies ranging from 1 cycle per 10000 years to 1 cycle per 1000 years: evidence for nonlinear behaviour of the climate system. Climatic Change, v. 12, p. 9-37.

Peterson, L.C. and Prell, W.L., 1985, Carbonate preservation and rates of climatic change: an 800 kyr record from the Indian Ocean. Geophys. Monog. 32, p. 251-269.

Pisias, N.G., 1983, Geologic time series from deep-sea sediments: time scales and distortion by bioturbation. Mar. Geol., v. 51, p. 99-113.

Pisias, N.G. and Leinen, M., 1984, Milankovitch forcing of the oceanic system: Evidence from the Northwest Pacific. In Milankovitch and Climate, v. 1, A. Berger et al., eds., Reidel, Dordrecht, p. 307-330.

Pisias, N.G., Martinson, D., Moore, T.C. Jr., Shackleton, N.J., Prell, W.L., Hays, J. and Boden, G., 1984. High resolution stratigraphic correlations of benthic oxygen isotopic records spanning the last 300,000 years. Mar. Geol., v. 56, p. 119-136.

Pisias, N.G., and Moore, T.C. Jr., 1981, The evolution of Pleistocene climate: a time series approach. Earth and Planet. Sci. Lett., v. 52, p. 450-458.

Pisias, N.G. and Rea, D.K., 1988, Late Pleistocene paleoclimatology of the central equatorial Pacific: sea surface response to the southeast trade winds, Paleoceanography, v. 3, p. 21-37.

Pisias, N.G. and Shackleton, N.J., 1984, Modeling the global climate response to orbital forcing and atmospheric carbon dioxide changes. Nature, v. 310, p. 757-759.

Pokras, E.M., and Mix, A.C., 1987, Earth's precession cycle and Quaternary climatic change in tropical Africa. Nature, v. 326, p. 486-487.

Prell, W.L., Hutson, W.H. and Williams, D.F., 1979, The subtropical convergence and late Quaternary circulation in the southern Indian Ocean. Marine Micropaleo., v. 4, p. 225-234.

Prell, W.L., Imbrie, J., Martinson, D., Morley, J., Pisias, N., Shackleton, N., and Streeter, H., 1986, Graphic correlation of oxygen isotope stratigraphy application to the late Quaternary. Paleoceanography, v. 1, p. 137-162.

Prell, W.L., and Kutzbach, J.E., 1987, Monsoon variability over the past 150,000 years. Jour. Geophys. Res., v. 92, p. 8411-8425.

Prell, W.L., and Van Campo, E., 1986, Coherent response of Arabian Sea upwelling and pollen transport of late Quaternary monsoonal winds. Nature, v. 323, p. 526-628.

Ruddiman, W.F., and Glover, L.K., 1975, Subpolar North Atlantic circulation at 9300 yr BP: faunal evidence. Quat. Res. v. 5, p. 361-389.

Ruddiman, W.F., and McIntyre, A., 1981, Oceanic mechanisms for amplification of the 23,000-year ice-volume cycle. Science, v. 212, p. 617-627.

Ruddiman, W.F., and McIntyre, A., 1984, Ice age thermal response and climatic role of the surface Atlantic Ocean, 40°N - 63°N. Geol. Soc. Amer. Bull., v. 95, p. 381-396.

Ruddiman, W.F. and Raymo, M.E., 1988, Northern Hemisphere climate regimes during the past 3 Ma: possible tectonic connections. In N.J. Shackleton et al., eds., The Past Three Million Years: Evolution of Climatic Variability in the North Atlantic Region, Phil. Trans. R. Soc. Lond. B 318, p. 411-430.

Ruddiman, W.F., Raymo, M., and McIntyre, A., 1986, Matuyama 41,000-year cycles: North Atlantic Ocean and northern hemisphere ice sheets. Earth and Planet. Sci. Lett., v. 80, p. 117-129.

Saltzman, B., 1987, Carbon dioxide and the $\delta^{18}O$ record of late-Quaternary climatic change: a global model. Climate Dynamics, v. 1, p. 77-85.

Saltzman, B., Hansen, A.R., Maasch, K.A., 1984, The late Quaternary glaciations as the response of a three-component feedback system to earth-orbital forcing. J. Atmos. Sci., v. 41, p. 3380-3389.

Saltzman, B., and Maasch, K.A., 1988, Carbon cycle instability as a cause of the late Pleistocene ice age oscillation: modeling the asymmetric response. Global Biogeochem. Cycles, v. 2, p. 177-185.

Saltzman, B. and Sutera, A., 1984, A model of the internal feedback system involved in late Quaternary climatic variations. J. Atmos. Sci., v. 41, p. 736-745.

Shackleton, N.J., Hall, M.A., Line, J., and Shuxi, C., 1983, Carbon isotope data in core V19-30 confirm reduced carbon dioxide concentration of the ice age atmosphere. Nature, v. 306, p. 319-322.

Shackleton, N.J., Imbrie, J., and Pisias, N.G., 1988, The evolution of oceanic oxygen-isotope variability in the North Atlantic over the past three million years. Phil. Trans. R. Soc. Lond., B 318, p. 679-688.

Shackleton, N.J., and Opdyke, N.D., 1973, Oxygen isotope and paleomagnetic stratigraphy of equatorial Pacific core V28-238: oxygen isotope temperatures and ice volumes on a 10^5 and 10^6 year scale. Quat. Res., v. 3, p. 39-55.

Shackleton, N.J., and Pisias, N.G., 1985, Atmospheric carbon dioxide, orbital forcing, and climate. In The Carbon Cycle and Atmospheric CO_2: Natural Variations Archean to Present, E. Sundquist and W. Broecker, eds., Geophysical Monograph 32, American Geophysical Union, Washington, D.C., p. 303-317.

Snieder, R.K., 1985, The origin of the 100,000 year cycle in a simple ice age model. Jour. Geophys. Res., v. 90, p. 5661-5664.

Start, G.G. and Prell, W.L., 1984, Evidence for two Pleistocene climatic modes: data from DSDP site 502. In A. Berger and C. Nicolis, eds., New Perspectives in Climate Modeling, Elsevier Publ. Co., Holland, p. 3-22.

Taylor, C.A., 1965, The Physics of Musical Sounds, The English University Press, London, 196 p.

Weertman, J., 1964, Rate of growth or shrinkage of nonequilibrium ice sheets. Jour. Glaciology, v. 38, p. 145-158.

Weyl, P.K., 1968, The role of the oceans in climatic change: A theory of the ice ages. Meteorological Monographs v 8., p. 37-62.

White, G.H. and Wallace, J.M., 1978, The global distribution of the annual and semiannual cycles in surface temperature. Mon. Wea. Rev., v. 106, p. 901-906.

Winograd, I.J., Szabo, E.J., Copley, T.B., Riggs, A.C., 1988, A 250,000-year climatic record from Great Basin vein calcite: implications for Milankovitch theory. Science, v. 242, p. 1275-1280.

RECENT DEVELOPMENTS IN QUATERNARY PALEOCLIMATOLOGY

Raymond S. Bradley
University of Massachusetts
Amherst, MA 01003, U.S.A.

ABSTRACT. Sources of paleoclimatic data for the Quaternary have greatly increased in the last twenty years. Long terrestrial records, in ice cores, lake sediments, peat and loess-paleosol sequences have greatly expanded knowledge of climatic changes over the last interglacial-glacial cycle; the reconstruction of past atmospheric composition has been of particular significance. Developments in dating methods (accelerator-based ^{14}C, thermoluminescence [TL] and electron spin resonance [ESR] dating) have been important in understanding the paleoclimatic record in sediments. New techniques are needed for recovering deep ice cores from ice sheets and long cores of lacustrine sediments in order to expand the limited number of terrestrial records which cover the last interglacial-glacial (and earlier) cycles.

INTRODUCTION

Twenty years ago, the record of past climate was primarily based on continental deposits, but these were rarely continuous sedimentary records and so the picture of climatic variations which developed was consequently incomplete. In the next decade, studies of marine sediments revolutionised our understanding of climatic variations and enabled models of the causes of climatic change to be tested. Extensive sampling programs for marine sediments also provided a global perspective on the observed changes (CLIMAP, 1976; 1984). Undoubtedly, studies of marine sediments provided an explosion of information which has continued to grow in quantity and quality (Ruddiman, 1985). However, the last ten years have seen a renewed focus on continental records of climate which complement the perspective provided by marine sediments (COHMAP, 1988). Continental deposits often provide more detailed information about short-term (high frequency) changes of climate than most marine records, raising new questions about the causes of climatic variations and feedbacks within the climate system (Broecker et al., 1986).

During the last decade much research has focused on the role of carbon dioxide (and other 'greenhouse gases') in the energy balance of

A. Berger et al. (eds.), Climate and Geo-Sciences, 165–171.

the earth-atmosphere system. Of critical importance has been the need to provide a long-term perspective on natural CO_2 variations prior to any anthropogenic effects. Studies of ^{13}C in planktonic and benthic foraminifera provided the first view of how CO_2 concentrations might have varied in the distant past, over an interglacial-glacial cycle (Shackleton et al., 1983; Pisias and Shackleton, 1984; Shackleton and Pisias, 1985). CO_2 levels were significantly lower in glacial times, and as high as, or higher than, Holocene levels in the previous interglacial. In broad outline, this view has been confirmed by studies of air trapped in ice from Antarctica (Barnola et al., 1987; Neftel et al., 1988). CO_2 concentrations of only 180-200 p.p.m.v. characterized the last glacial maximum as well as the preceding glacial maximum (marine isotope stage 6). However, there are important differences between the ice core and marine sediment-based reconstructions of former CO_2 levels which point to new and potentially significant lines of future research.

In addition to the CO_2 studies in ice from Antarctica, equally important parallel studies of $\delta^{18}O$ and δD have enabled paleotemperature estimates to be made (Lorius et al., 1985; Jouzel et al.,1987). Comparison with the CO_2 record suggests that orbitally-induced radiation changes are amplified significantly by interglacial to glacial CO_2 changes (Shackleton and Pisias, 1985; Genthon et al., 1987). This is an important development in the quest for understanding the causes and mechanisms of glaciation and deglaciation.

Ice cores have also provided fertile ground for studies of shorter-term climatic variations and of detailed measurements of past atmospheric composition. For example, studies of Antarctic ice indicate CO_2 levels in the mid-18th century averaged ~280 p.p.m.v. (Neftel et al., 1985) compared to 348 p.p.m.v. in 1987. A long ice core from Peru has provided a very important ~1500 year record from tropical South America where detailed paleoclimatic records are very sparse (Thompson et al., 1985; 1986). In particular, the Little Ice Age, from 1490 to ~1880 is clearly recognizable. More recently, the recovery of several cores from the Dunde Ice Cap on the Qinghai-Tibetan plateau (38°N, 96.4°E, 5300m) promises to provide very significant information about past climate and loess deposition in this area (Thompson et al., 1988). Other extra-polar ice caps in South America and Central Asia have the potential of providing detailed paleoclimatic records from areas about which very little is known. Efforts to recover such cores are likely to be extremely rewarding. In addition, new deep ice cores from Greenland will provide an important northern hemisphere data set for comparison with the Antarctic records.

Continuous continental records spanning more than the last interglacial-glacial cycle are extremely rare compared to those from the marine realm. Relatively few well-dated records exist. One of the most detailed records to emerge in the past decade is that of Hoochiemstra (1984) who studied in great detail pollen variations in a 357m core from a former lake basin in the High Plain of Bogota, Colombia (~5°N). This monumental study utilized radiocarbon, fission track, K/Ar and paleomagnetic reversals to provide a chronology of vegetation changes over the last 3.5 million years. These changes were then compared with the marine sedimentary record of paleoclimate (Hoochiemstra, 1984, 1988; van der Hammen, 1985). Hoochiemstra's study is unique; no other continental paleoclimate record exists which is so well-dated and which has been examined in such detail. Other long cores spanning the last interglacial-glacial cycle have been studied (e.g. Grande Pile: Woillard, 1978; Woillard and Mook, 1982; Clear Lake: Adam et al., 1981; Adam and West, 1983) but only a few lacustrine records exceed this time period (e.g. Lake Biwa: Fuji, 1979; Horie, 1987).

Regions of loess deposition, beyond the former margins of continental ice sheets, provide some of the longest and best paleoclimatic records from the continents. Alternating periods of loess deposition and soil formation are well-recorded in many areas (Pye, 1987) but the longest continuous records are from central China, spanning over 2.4 million years (Liu et al., 1985a, 1985b; Heller and Liu, 1982; Kukla, 1987a). Studies of soil, pollen and faunal remains have enabled paleoclimatic conditions to be estimated (Liu et al, 1985b). Until recently, dating was based mainly on a paleomagnetic reversal stratigraphy, but recent studies have supplemented this chronology with TL and amino acid age estimates. Variations in magnetic susceptibility also hold great promise for further defining the loess/paleosol chronostratigraphy (Kukla et al., 1988). Paleoclimatic interpretation and correlation of loess deposits from eastern Asia to western Europe (e.g. Kukla, 1987b) and from North to South America remains a goal for future research.

Dating of continental deposits remains problematical, particularly when examining short-term, 'abrupt' changes in climate (e.g. Woillard, 1979; Seret, 1983; Frenzel and Bludau, 1987). Tandem accelerator mass spectrometry techniques for ^{14}C have proven to be very useful in constructing detailed chronologies of paleoclimatic variations (e.g. Bard et al., 1987). Thermoluminescence (TL) and electron spin resonance (ESR) dating methods have received much theoretical and experimental attention over the last decade and their application to dating sediments such as loess (at least over the last interglacial-glacial cycle) has been an important development (Wintle and Huntley, 1982; Wintle, 1987; Hennig and Grun, 1983; Radtke and Grun, 1988). Recent developments in optical dating of sediments suggest that this will also become a useful technique (Huntley et al., 1985; Godfrey-Smith et al., 1988).

^{10}Be measurements in ice cores (Yiou et al, 1985) and in loess (Shen, 1987) have provided a very important new method of checking the chronologies of continental deposits. Further work on the long-term flux of ^{10}Be is an important area of future chronological research with direct significance for paleoclimatology.

Computer models of the general circulation (GCMs) have been employed more frequently in the last decade to examine changes related to orbital forcing and differing boundary conditions at various times in the past (Kutzbach, 1985; Crowley, 1988). Models have also been used to simulate $\delta^{18}O$ and desert dust dispersal during glacial time (Joussame and Jouzel, 1987). Comparison of model output with paleoclimatic records provides a basis for placing sparse proxy records into a global perspective and for checking both the accuracy of the model reconstructions and of the interpretation of proxy data (e.g. Kutzbach and Street-Perrott, 1985; Kutzbach and Wright, 1985).

The last decade has witnessed important developments in paleoclimatology, particularly with respect to records from the continents. However, since most climate proxies are sedimentary deposits, in which material from one moment in time is buried by ever-younger sediments, reconstruction of past climate must begin by recovering a well-preserved sedimentary record. This seemingly trivial problem is, in fact, a significant barrier to future paleoclimatic research. For example, the technological problems of recovering a high quality ice core from depths exceeding 3500m in East Antarctica are quite profound. Similarly, many potentially valuable lake sedimentary deposits await the development of coring devices which can be deployed with minimal logistical problems in what are often relatively isolated locations. Analytical techniques and laboratory facilities are already available to examine the materials recovered. With improvements in the technology of core recovery, the prospects are excellent for obtaining many new continuous continental climatic records spanning the entire Quaternary.

REFERENCES

Adam, D.P. and West, G.J., 1983. Temperature and precipitation estimates through the last glacial cycle from Clear Lake, California pollen data. Science, **219**, 168-170.

Adam, D.P., Sims, J.D. and Throckmorton, C.K., 1981. 130,000 year continuous pollen record from Clear Lake County, California. Geology, **9**, 373-377.

Bard, E., Arnold, M., Duprat, J., Moyes, J. and Duplessy, J-C., 1987. Bioturbation effects on abrupt climatic changes recorded in deep sea sediments: correlation between $\delta^{18}O$ profiles and accelerator ^{14}C dating. In: Abrupt Climatic Change, W.H. Berger and L.D. Labeyrie (eds.), D. Reidel, Dordrecht, 263-278.

Barnola, J.M., Raynaud, D., Korotkevich, Y.S. and Lorius, C., 1987. Vostok ice core provides 160,000 year record of atmospheric CO_2. Nature, **329**, 408-414.

Broecker, W.S., Peteet, D.M. and Rind, D., 1986. Does the ocean-atmosphere system have more than one stable mode of operation? Nature, **315**, 21-25.

CLIMAP Project Members, 1976. The surface of the ice age earth. Science, **191**, 1131-1137.

CLIMAP Project Members, 1984. The last interglacial ocean. Quaternary Research, **21**, 123-224.

COHMAP Projects Members, 1988. Climatic changes of the last 18,000 years: observations and model simulations. Science, **241**, 1043-1052.

Crowley, T.J., 1988. Paleoclimate modelling. In: M.E. Schlesinger (ed.) Physically-Based Modelling and Simulation of Climate and Climatic Change, Part II, 883-949. Kluwer Publishers.

Frenzel, B. and Bludau, W., 1987. On the duration of the interglacial to glacial transition at the end of the Eemian interglacial (deep sea stage 5e): botanical and sedimentological evidence. In: Abrupt Climatic Change, W.H. Berger and L.D. Labeyrie (eds.), D. Reidel, Dordrecht, 151-162.

Fuji, N., 1979. A preliminary report of the pollen analyses on the 1000m core samples from the east coast of Lake Biwa, Japan. Paleolimnology of Lake Biwa and the Japanese Pleistocene, **Vol. 9**, S. Horie (ed.), Kyoto, Japan, 303-319.

Genthon, C., Barnola, J.M., Raynaud, D., Lorius, C., Jouzel, J., Barkov, N.I., Korotkevich, Y.S. and Kotlyakov, V.M., 1987. Vostok ice core: climatic response to CO_2 and orbital forcing changes over the last climatic cycle. Nature, **329**, 414-418.

Godfrey-Smith, D.I., Huntley, D. J. and Chen, W-H., 1988. Optical dating studies of quartz and feldspar sediment extracts. Quaternary Science Reviews, **7**, 373-380.

Heller, F. and Liu, T-S., 1982. Magnetostratigraphical dating of loess deposits in China. Nature, **300**, 431-433.

Hennig, G.J. and Grun, R., 1983. ESR dating in Quaternary geology. Quaternary Science Reviews, **2**, 157-238.

Hoochiemstra, H., 1984. Vegetational and climatic history of the High Plain of Bogota, Colombia: a continuous record of the last 3.5 million years. Dissertationes Botanicae, **Vol. 79**. J. Cramer, Vaduz.

Hoochiemstra, H., 1988. The orbital-tuned marine oxygen isotope record applied to the Middle and Late Pleistocene pollen record of Funza (Colombian Andes) Palaeogeography, Palaeoclimatology, Palaeoecology, **66**, 1/2, 9-18.

Horie, S. (ed.), 1987. History of Lake Biwa Contrib. No. 553, Institute of Paleolimnology and Paleoenvironment of Lake Biwa, Kyoto, 242 pp.

Huntley, D.J., Godfrey-Smith, D.I. and Thewalt, M.L.W., 1985. Optical dating of sediments. Nature, **313**, 105-107.

Joussame, S. and Jouzel, J., 1987. Simulation of climatic tracers using atmospheric general circulation models. In: Abrupt Climatic Change, W. H. Berger and L. D. Labeyrie (eds.), D. Reidel, Dordrecht, 369-381.

Jouzel, J., Lorius, C., Petit, J.R., Genthon, C., Barkov, N.I., Kotlyakov, V.M. and Petrov, V.M., 1987. Vostok ice core: a continuous isotope temperature record over the last climatic cycle (160,000 years). Nature, **329**, 403-408.

Kukla, G., 1987a. Loess stratigraphy in central China. Quaternary Science Reviews, **6**, 191-219.

Kukla, G., 1987b. Pleistocene climates in central China and Europe compared to oxygen isotope record. Paleoecology of Africa, **18**, 37-45.

Kukla, G., Heller, F., Liu, X-M., Xu, T-C., Liu, T-S. and An, Z-S., 1988. Pleistocene climates in China dated by magnetic susceptibility. Geology, **16**, 811-814.

Kutzbach, J.E., 1985. Modeling of paleoclimates. Advances in Geophysics, **28A**, 159-195.

Kutzbach, J.E. and Street-Perrott, A., 1985. Milankovitch forcing of fluctuations in the level of tropical lakes from 18 to 0 kyr B.P. Nature, **317**, 130-134.

Kutzbach, J.E. and Wright, H.E., 1985. Simulation of the climate of 18,000 years B.P.: results for the North American/North Atlantic/ European sector and comparison with the geologic record of North America. Quaternary Scienc Reviews, **4**, 147-187.

Liu, T-S., et al., 1985a. Loess and the Environment China Ocean Press, Beijing.

Liu, T-S., An, Z-S., Yuan, B-Y. and Han, J-M., 1985b. The loess-paleosol sequence in China and climatic history. Episodes, **8**, 21-28.

Lorius,C., Jouzel, J., Ritz, C., Merlivat, L., Barkov, N.I., Korotkevich, Y.S. and Kotlyakov, V.M., 1985. A 150,000 year climatic record from Antarctic ice. Nature, **316**, 591-596.

Neftel, A., Moor, E., Oeschger, H. and Stauffer, B. 1985. Evidence from polar ice cores for the increase in atmospheric CO_2 in the past two centuries. Nature, **315**, 45-47.

Neftel, A., Oeschger, H., Staffelbach, T. and Stauffer, B., 1988. CO_2 record in the Byrd ice core 50,000 - 5,000 years B.P. Nature, **331**, 609-611.

Pisias, N.G. and Shackleton, N.J., 1984. Modelling the global climate response to orbital forcing and atmospheric carbon dioxide changes. Nature, **310**, 757-759.

Pye, K., 1984. Loess. Progress in Physical Geography, **8**, 176-217.

Radtke, U. and Grun, R., 1988. ESR dating of corals. Quaternary Science Reviews, **7**, 465-470.

Ruddiman, W.F., 1985. Climate studies in ocean cores. In: Paleoclimate Analysis and Modeling, A. Hecht (ed.), J. Wiley, New York, 197-257.

Seret, G., 1983. Rather long duration of the transient climatic events in the Grande Pile (Vosges, France). In: Paleoclimatic Research and Models, A. Ghazi (ed.), D. Reidel, Dordrecht, 139-143.

Shackleton, N.J. and Pisias, N.G., 1985. Atmospheric carbon dioxide, orbital forcing and climate. In: The Carbon Cycle and Atmospheric Carbon Dioxide: Natural Variations, Archean to Present, E.T. Sundquist and W.S. Broecker (eds.), American Geophysical Union, Washington D.C., 303-317.

Shackleton, N.J., Hall, M.A., Line, J. and Cang, S., 1983. Carbon isotope data in core V19-30 confirm reduced carbon dioxide concentration in the ice age atmosphere. Nature, **306**, 319-322.

Shen, C-D. 1986. Beryllium-10 in Chinese Loess. Unpublished doctoral thesis, University of Bern, Switzerland.

Thompson, L.G., Mosley-Thompson, E., Bolzan, J.F. and Koci, B.R., 1985. A 1500 year record of tropical precipitation recorded in ice cores from the Quelccaya ice cap, Peru. Science, **229**, 971-973.

Thompson, L.G., Mosley-Thompson, E., Dansgaard, W. and Grootes, P.M., 1986. The "Little Ice Age" as recorded in the stratigraphy of the tropical Quelccaya ice cap. Science, **234**, 361-364.

Thompson, L.G., Wu. X-L, Mosley-Thompson, E. and Xie, Z-C., 1988. Climatic records from the Dunde Ice Cap, China. Annals of Glaciology, **10**, 178-182.

van der Hammen, T., 1985. The Plio-Pleistocene climatic record of the tropical Andes. J. Geological Society of London, **42**, 483-489.

Wintle, A.G., 1987. Thermoluminescence dating of loess. Catena, Supplement **9**, 103-115.,

Wintle, A.G. and Huntley, D.J., 1982. Thermoluminescence dating of Quaternary sediments. Quaternary Science Reviews, **1**, 31-53.

Woillard, G., 1978. Grande Pile peat bog: a continuous pollen record of the last 140,000 years. Quaternary Research, **9**, 1-21.

Woillard, G., 1979. Abrupt end of the last interglacial s.s. in North East France. Nature, 281, 558-562.

Woillard, G. and Mook, W.G., 1982. Carbon-14 dares at Grande Pile: correlation of land and sea chronologies. Science, **215**, 159-161.

Yiou, F., Raisbeck, G.M., Bourles, D., Lorius, C. and Barkov, N.I., 1985. ^{10}Be in ice at Vostok, Antarctica during the last climatic cycle. Nature, **316**, 616-617.

PALEOCLIMATE DATA FOR STUDIES OF GLOBAL CLIMATE CHANGE AND EARTH SYSTEM SCIENCE

T.L. HOLCOMBE
National Oceanic and Atmospheric Adminstration
National Geophysical Data Center E/GC3
325 Broadway
Boulder, Colorado 80303
U.S.A.

ABSTRACT. Paleoclimate data are scattered by region and discipline and paleoclimate-related studies are being conducted in many subject areas. Sources of paleoclimate data include tree-rings, lake and bog sediments, ice cores, marine sediments, and historical records. Diverse paleoclimate data of various time scales need to be assembled into global data bases to validate climate models and support studies of processes governing climate change.

A program in paleoclimatology has been implemented at the U.S. National Geophysical Data Center (NGDC), the goals of which are (1) to acquire paleoclimate data bases from all disciplines and time scales and integrate these into a global data base, and (2) to cooperate with national and international science programs in the use of these data to better understand and model climate change.

Cooperation to build paleoclimate data bases is needed between national and international science organizations, funding agencies, and data centers.

Background

Paleoclimatology has been the subject of continually expanding research and exciting progress during the last 25-30 years. Principal results of paleoclimate studies have been summarized in a comprehensive review paper by Thomas Crowley (1983). Types of data and analytical methods employed in paleoclimatic reconstruction are systematically reviewed by R.S. Bradley (1985). Except for historical and instrumental records, mostly from the last few hundred years, paleoclimatology data are "proxy" data -- measurements of natural phenomena which contain a record of past climate. Taken together, these data are obtained from diverse sources and vary as to time span and time resolution. Although data which record climate extend throughout the geological record, the great body of paleoclimate-proxy information records climate during the Quaternary epoch, with the greatest diversity and volume of available information coming from the last 20,000 years. Primary data sources are ocean sediments (to 1,000,000 years BP and beyond, with 100 to several thousand years resolution); tree rings (to 8,000 years BP, with 1 year resolution); lake and bog sediments (to 100,000 years BP, with 1-10 year resolution for varves, 100-500 year resolution for radiocarbon-dated chronologies); and ice cores (to 100,000 years BP, with 1-10 year resolution to 15,000 years BP, 100-1000 year resolution to 100,000 years BP). Measurements which are climate-related include oxygen-isotope ratios, tree ring width and wood density, abundance of foraminiferal species, characteristics of other fossil assemblages,

A. Berger et al. (eds.), Climate and Geo-Sciences, 173–177.

percentages of pollen types, atmospheric composition (from air bubbles in ice), and many other parameters. Most of these data are converted to time series, calibrated by occurrence of seasonal banding (to 10,000 years BP), by radiometric dating (100-100,000 years BP), or by direct or indirect correlation with events such as orbital cycles and magnetic field reversals (>10,000 years BP).

Paleoclimatology at the U.S. National Geophysical Data Center

Because data which record paleoclimate are so diverse, and paleoclimate-related studies are being conducted in so many different disciplines, available paleoclimate data are widely dispersed. For each discipline there is an associated research community and program management structure which is distinct from the others. Furthermore, data from the various sources are typically scattered in regional data bases. While moves are underway to coordinate research efforts between disciplines, and combine various data types into global data bases, most of these efforts are in the initial stage.

In response to the recognized need to assemble global paleoclimate data bases in order to make significant advances in the study of global climate change and earth systems science, a program in paleoclimatology has been implemented at the U.S. National Geophysical Data Center (NGDC). After much consideration, primary goals for the NGDC program have been formulated (1) to establish a series of data bases containing proxy data from tree rings, pollen, lake and bog sediments, ice cores, marine sediments, and other sources, (2) to integrate paleoclimate data bases which span all time scales and disciplines, (3) to participate in national, and international programs in climate and global change, and (4) to assist climate modelers in defining and understanding climate change.

We plan to work with each of the research communities separately in the design and structure of data bases, format and content of data, and assembling of data from diverse regional sources. There seems to be agreement among most investigators that we should be archiving "level 1 proxy data", that is, actual field and laboratory measurements of climate-related natural phenomena (Bradley, 1985), with somewhat less emphasis on time-series reconstructions and climate interpretations, which are often controversial and may be superseded as interpretations and models of past climate evolve. Proxy data from marine sediments are one step removed from climate interpretation because they record conditions within the ocean rather than the atmosphere. Therefore, for the oceans, it may also be desirable to archive paleo-oceanographic parameters such as sea-surface temperatures.

Our challenge is first to establish one or more viable data bases for each discipline area. Other paleoclimate-proxy data bases from miscellaneous sources (loess deposits, corals, histories of agricultural harvests, lake levels, movement of animal habitats, for example) will not be emphasized at this time but will be added when possible. Once established, each data base can be updated and expanded by taking advantage of every opportunity to add data. Finally, these data will be integrated into a single global data base. Each data base would be supported with capability to search by various criteria -- geographic area, data type, time interval, for example -- and each data base would be supported by image processing and graphics capability.

Data volume is not a limiting factor for paleoclimate data; an intuitive estimate is that the sum total of level 1 proxy data would not exceed 500 megabytes. The principal limiting factor for building paleoclimate data bases is the large amount of time, labor, and resources which will be required to prepare, assemble, code, and organize such a diverse body of basic information.

Progress to Date

The International Tree-Ring Data Bank (IT-RDB) has been established at the University of Arizona. NGDC has agreed to assume data management responsibilities for IT-RDB under the direction of its advisory committee. Additional U.S. data bases at the University of Arizona, University of Arkansas, Columbia University and the University of Washington would be integrated as soon as reasonably possible. We would also hope to include other international data bases which reside in the U.K., Switzerland, USSR, and Australia. This expanded international data base would consist of ring widths, densities, and isotopic concentrations as a function of year, and a normalized proxy climate data set per site. The present IT-RDB size of 40 megabytes of digital data from 72,000 cores from 36,000 trees from 1,200 sites is anticipated to expand to over 150 megabytes of data from 2,500 sites. Climatic reconstructions, are considered to be of great interest but still in the research domain.

As part of an overall effort to develop paleoclimate data bases, an initial start has been made toward a worldwide data base for palynological studies. Three levels of data must be considered: (1) raw data as a function of depth, (2) raw data as a function of time, and (3) climate syntheses. Current attention is centered on assessing information about data holdings and regional expertise. The next stages of development include assembling a directory of data sources and expertise, establishing a computer network of project participants, and combining the expertise of regional palynology experts to design a common approach to data management. This last effort, perhaps involving a mini-workshop, will concentrate on designing the framework for the multi-level data bases and recommending methods for their development.

In a related development, the coordinating committee for the "Analysis of Rapid and Recent Climatic Change" (ARRCC), a U.S. project associated with the U.S. National Science Foundation through the Climate Dynamics Program, has recently invited the participation of NGDC for the purpose of establishing a long-term data management effort for the reconstruction and modeling of interannual to century-scale climate change. An initial task of the ARRCC committee will be a pilot project reconstructing paleo-climate for the "Little Ice Age." Longer-term objectives will then be evaluated and a more comprehensive program will be designed. A workshop is planned for early 1989, possibly hosted by NGDC, which will help define the data management effort for ARRCC.

As part of the NGDC's program to establish data bases on global ice parameters, an inventory of ice cores was prepared and published in 1980 (Glaciological Data Report GD-8). This document compiled data from most of the ice cores collected up to 1980 and described literature sources and research activities. This inventory will be brought up to date as a first step towards the development of a comprehensive digital data base of ice core parameters. Final decisions on the optimal data format will evolve after we have had the opportunity to confer with key researchers in this field. With respect to current research, NGDC has been identified as the organization best able to meet the data management and archiving needs of the Greenland Ice Sheet Program (GISP-II). GISP-II is the largest and most comprehensive ice core program currently being planned by U.S. and European scientists, with field work to begin in 1990.

A large body of ocean paleoclimate data, and some new types of information from marine sediments, are being generated at institutions such as Columbia University and Brown University. Compilation and integration of these existing data will be a major portion of the effort to construct a coherent data base of paleoclimate data from marine sediments. NGDC already holds the CLIMAP (Climate: Long-Range Mapping and Prediction) data base. The CLIMAP data file consist of micropaleontology, percentages of opal, quartz, carbon and carbonate, age dates, and sea-surface temperature estimates for over 600 sediment cores. NGDC also maintains a broad base of marine sediment data in digital and analog form which can be tapped for a paleoclimate data base, including a large body of data collected by the Deep Sea Drilling Program (DSDP) -- micropaleontology, geochemistry, and paleomagnetism -- as well as a comprehensive index to core material available for further analysis at major U.S. repositories.

We have addressed planning for data management and the establishment of data bases by scheduling meetings with representatives of each community; alternatively, we are participating in their own workshops or meetings, and in several instances we have invited experts to visit NGDC for seminars and talks. Later, possibly in 1989, we may convene a workshop at NGDC to take up issues related to establishment of the multidisciplinary global paleoclimate data base. By the end of 1989 we expect to have established "skeleton" data bases in each of the aforementioned discipline areas. We anticipated that we will also have accumulated substantial holdings in some of these data bases.

The type of computer on which the integrated, multidisciplinary data base will ultimately reside at NGDC is as yet undecided. Paleoclimate files at NGDC presently reside on a PC-AT microcomputer with an 80 megabyte hard drive which is connected to the NGDC VAX 11/750 via Ethernet, and which has access to a WORM drive mass storage system. A study of relational Data Base Management Systems (DBMS's) that function on, and can communicate between, main frames/mini-computers/micro-computers is underway to determine if such a DBMS would be appropriate for paleoclimate data. Considerations being taken into account include future efficient interface with Geographic Information Systems, speed of operation, and ease of access/data sharing with users and contributors at remote sites. Remote access to data systems at NGDC is expected to be greatly facilitated later this year by connection of University of Colorado high-speed network lines into NGDC. This will allow remote users to access data resident on a variety of machines within NGDC through Ethernet from a wide range of outside systems including Telenet and BITNET. However, since local PCs are single-user and the VAX cannot manage an increased number of users, NGDC will have to explore possible equipment configurations necessary to allow remote use. One possibility is to channel remote access to data bases through 32-bit multiuser microcomputer or workstations.

Recommendations

Because of the magnitude of the task of building a global paleoclimate data base, the diversity of data types, and the range of subject matter disciplines, cooperative efforts to build such data bases are needed between the various paleoclimate research projects, national and international science organizations, funding agencies, and the national and international data centers. Specifically, the following future efforts are recommended:

1. Research projects studying paleoclimate and collecting new information should include in their research plans provision for data management and for transmittal of new data to established national and international data repositories such as NGDC and those of the ICSU World Data Center System.

2. Agencies sponsoring paleoclimate research should review their policies regarding data management and transmittal of data to national and international repositories to ensure that these items are not neglected. Funding should be earmarked for these purposes to ensure implementation.

3. Projects are needed in various discipline areas, for example, pollen data bases from lake and bog sediments, to assemble already existing but scattered data into regional or continental compilations, and make sure that these data are assimilated into the main global data bases.

4. Informal links need to be established or strengthened between those who collect and assemble paleoclimate data bases, those who are building and archiving global paleoclimate data bases, and those who study and model global climate. One of NGDC's important needs is advice regarding the form and content of information needed from global multidisciplinary paleoclimate data bases which support climate modeling.

Bradley, R.S., 1985, Quaternary Paleoclimatology, Winchester, Mass. 01890, USA, Allen and Unwin, Inc. 472 pages.

Crowley, T.J., 1983, The geologic record of climate change, Reviews of Geophysics and Space Physics, v. 21, p. 828-877.

PALEOCLIMATE PERSPECTIVES ON A GREENHOUSE WARMING

Thomas J. Crowley
Applied Research Corporation
305 Arguello Drive
College Station, Texas 77840
U.S.A.

ABSTRACT. Results from paleoclimatology are used as a frame-of-reference for interpreting possible consequences of a future greenhouse warming. The warming will probably represent a first-order change in the earth's climate. Planetary temperatures could rise to levels normally associated with an ice-free state, yet retain cold poles due to the inertial presence of ice sheets. The magnitude and rate of warming could be comparable to or exceed any well-documented geological event, with the resultant climate representing perhaps a unique occurrence in earth history. Although there is a reasonably good agreement between atmospheric climate models and inferred climate change over the last 20,000 years, it is suggested that this agreement may have limited usefulness in evaluating climate model projections of a greenhouse warming. The best agreements between models and observations involve situations where the system is either strongly forced locally (e.g., ice sheets in the northern hemisphere) or where there is an approximately linear relation between forcing and response (orbital forcing and seasonal surface temperature response in the monsoon belt). Neither of these categories is applicable to a greenhouse warming. Paleoclimate results also suggest that ocean heat transport may have varied in the past. However, we are at too early a stage in paleo-ocean modeling to evaluate whether the inferred changes can be satisfactorily simulated.

1. Introduction

Some time in the next century the earth's climate is expected to undergo a major warming as a result of increasing levels of carbon dioxide and other trace gases. A doubling of CO_2 is estimated to cause an increase in global average temperatures of 2.8-5.2°C (e.g., Schlesinger, 1989), and a comparable increase may come from the contribution of other radiatively important trace gases (Dickinson and Cicerone, 1986). Since warming of several degrees exceeds the ± 0.5°C range of "natural variability" (Fig. 1) during the past one hundred years (e.g., Jones *et al.*, 1986), we do not have any direct experience with a climate change of this magnitude. Furthermore, climate model simulations of the regional responses to a greenhouse warming, although instructive, are not so reliable that we can place great credence in their results. It is therefore appropriate to inquire how the projected large greenhouse warming compares with major changes of climate in the geologic record, and whether validation of climate models against past boundary conditions has helped us formulate some level of confidence in climate models. The purpose of this paper is to synthesize results from paleoclimate studies (cf. Crowley and North, 1989) in order to establish a frame-of-reference for interpreting projections about the consequences of a future greenhouse warming.

A. Berger et al. (eds.), Climate and Geo-Sciences, 179–207.

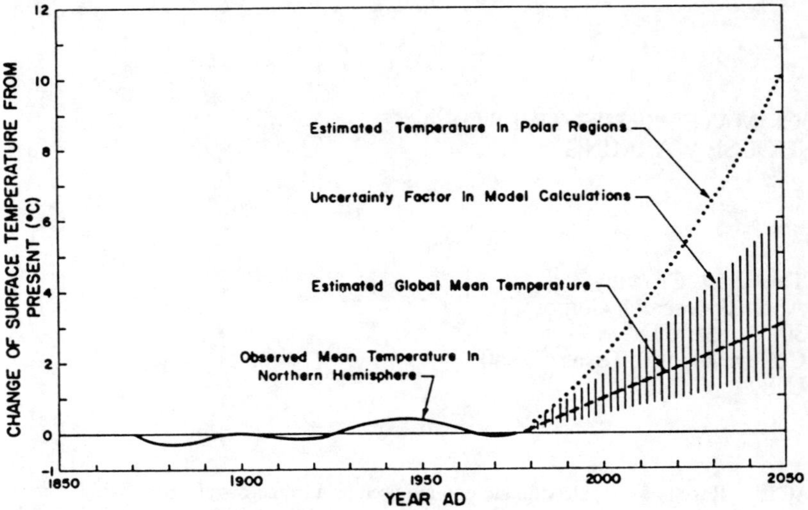

Figure 1. The mean surface temperature for the northern hemisphere since 1860. Dotted line shows future global mean surface temperature change. Crosshatching shows uncertainties in model calculations. [Figure from Stuiver *et al.* (1981), adapted from Mitchell (1977) and Kellogg (1977).]

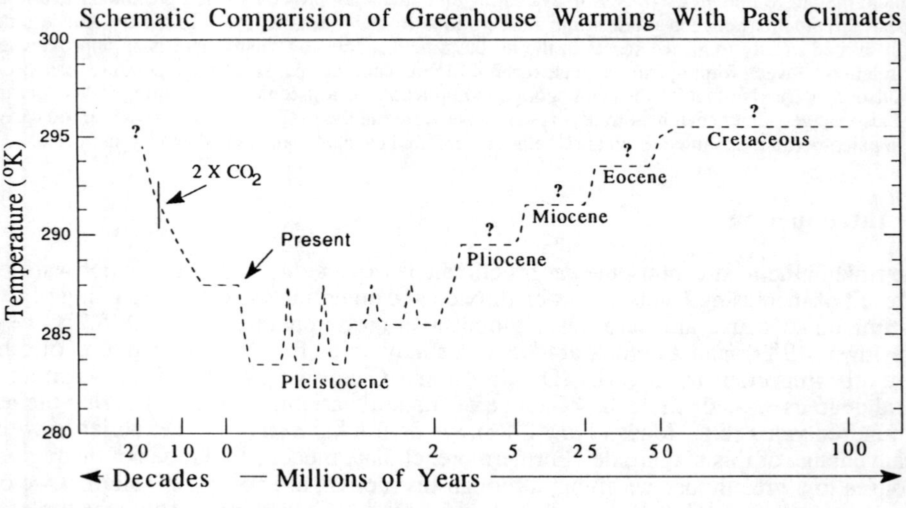

Figure 2. Schematic comparison of future greenhouse warming with past changes in temperature. Note that pre-Pleistocene changes are not well fixed in magnitude, but their relative warmth is approximately correct. Maximum warming in the Cretaceous based on estimates by Barron and colleagues (Thompson and Barron, 1981; Barron and Washington, 1985). Time intervals in between have been scaled accordingly.

2. Lessons From the Geologic Record

2.1 MAGNITUDE OF GREENHOUSE CLIMATE PERTURBATION

One of the first questions to ask about a potential greenhouse warming of several degrees is: "How significant is the projected warming?" The answer is "Quite significant" (Fig. 2). A CO_2 doubling of 2.8-5.2° C (mean = 3.9°C) is comparable in magnitude to the global-average temperature decrease of about 3.5-4.0°C that occurred during the last glacial maximum (Hansen *et al.*, 1984; Kutzbach and Guetter, 1986). Even the highly unlikely minimum increase of 1°C (Hansen *et al.*, 1988) would raise temperatures to levels not reached since the last interglacial (120,000 years before present, BP).

The mean greenhouse increase of 3.9°C represents one-third to one-half the estimated global temperature increase during the maximum warmth of the mid-Cretaceous (Thompson and Barron, 1981; Barron and Washington, 1985). Since CO_2 and other trace gases could exceed the 2X level by a significant amount, a future greenhouse warming represents a first-order change in climate. Unless there is a major shift in energy sources, atmospheric temperatures could eventually approach levels assumed for an ice-free earth (+6-10°C). Furthermore, the rate of change of global temperatures (~4°C/century) exceeds by perhaps an order of magnitude the greatest rates that occurred during the catastrophic deglaciations of the Pleistocene - the most well-documented example of an extreme abrupt climate change in the geologic record. [Although there is good evidence for an asteroid impact at the end of the Cretaceous, the environmental consequences still require clarification, and there is a need for more data to test environmental scenarios; cf. Crowley and North, 1989].

Even though future temperatures may increase greatly, perhaps to levels normally found for ice-free climates, it is uncertain whether an extreme greenhouse warming would completely destroy land-based ice sheets. The atmospheric CO_2 perturbation would eventually be neutralized by the deep-ocean carbonate reservoir after 5,000-10,000 years (Sundquist, 1985). About ten thousand years were required to melt the great Pleistocene ice sheets. Melting time for the East Antarctic Ice Sheet would probably be longer, because part of it is situated on a high plateau and would be less susceptible to some of the feedbacks invoked to explain Pleistocene ice sheet disintegration.

Since the time constants for temperature response of the East Antarctic Ice Sheet are so long, perhaps tens of thousands of years (e.g., Whillans, 1981), the ice sheet might not reach an equilibrium response to the CO_2-induced perturbation until after the CO_2 has disappeared from the system. The ice sheets perturbed response will then further modify global climates. Thus, the fully coupled climate system may never reach full equilibrium with the greenhouse perturbation, and the system will remain out of equilibrium for thousands, perhaps even tens of thousands of years, after the initial perturbation. (Because the ice sheet response is so much slower than other components of the climate system, these other components may approach equilibrium even though they technically will not reach equilibrium.)

If CO_2 values reach several times the present level, the earth may therefore experience nonglacial-level air temperatures and yet retain cold poles due to the inertial presence of the ice sheets. There is only one moderately well-documented example of this conjunction in the geologic record - the Late Ordovician glaciation 450 million years ago (Ma; see discussion in Crowley *et al.*, 1987). Since this glaciation occurred after the atmosphere warmed, the sequence of events is opposite to that which will occur in the future. One may have to look to the Late Precambrian (about 650 Ma) to find an example of a possible CO_2 increase following a major period of ice growth (e.g., Crowley and

North, 1988). The conjectured CO_2 increase, driven by changes in sea floor spreading, would occur at a rate orders of magnitude slower than the anthropogenic CO_2 pulse.

The CO_2 perturbation therefore represents a major climate change, which will occur at a rate comparable to or significantly exceeding any which occurred in earth history. The magnitude and duration of the event is significant even on geological scales. Since there are no completely satisfactory geological analogues for the warm atmosphere/cold pole combination, the future greenhouse climate may represent a unique realization in earth history.

2.2 IMPLICATIONS OF HISTORICAL CLIMATE FLUCTUATIONS FOR DETECTION AND MODULATION OF A GREENHOUSE WARMING

Some climate models predict that we should already be experiencing a greenhouse warming. Despite uncertainties in the predictions, in the absence of other forcing one might expect continuously rising temperatures after the greenhouse effect has manifested itself. Geological data suggest that such a simple picture might not be valid. There have been significant centennial- and millenial- scale climate fluctuations in historical times (e.g., Crowley, 1983). Temperatures during the Little Ice Age (~1450-1890) were on the order of 1.0-1.5°C cooler and were broadly synchronous over much of the earth (Mosley-Thompson and Thompson, 1982; Thompson *et al.*, 1986; Zhang and Crowley, 1989). Although the origin of these fluctuations is not well understood, they may well occur in the future and modify the CO_2 warming, especially during its early phases.

Dansgaard *et al.* (1971) utilized time-series analysis of Greenland ice core $\delta^{18}O$ fluctuations to suggest that a "natural" minimum in atmospheric temperatures would occur in the latter part of the 20'th century (Fig. 3), i.e., at about the same time that the climate signal was supposed to rise above the noise level of natural variability. The coincidence of a "natural" minimum with a CO_2 warming may affect our interpretation of reasons for the suspected delayed response in warming. For example, sequestering of heat in the deep ocean has been advanced as an explanation for the delayed warming (Hansen *et al.*, 1984; Schlesinger, 1986). Instead, the delay may reflect a temporary coincidence of a natural minimum with a CO_2 rise, and that this cancellation will shortly be negated. Since more severe "Little Ice Age" climate fluctuations have a longer (1,000-2,000 years) recurrence interval (cf. Crowley, 1983), and the most recent severe cooling ended less than 300 years ago, there may be less reason to expect a future reversal to more severe cooling in the next few centuries.

2.3 REGIONAL RESPONSES TO A GREENHOUSE WARMING

Paleoclimate studies also provide some information concerning possible regional consequences of a greenhouse warming. For example, a greenhouse warming might cause a considerable increase in summer aridity in central North America and Eurasia (e.g., Manabe and Wetherald, 1986; cf. Washington and Meehl, 1984). Similar drying occurred during the early Holocene warm period (6,000 - 9,000 BP; COHMAP, 1988), but since the forcing in the latter case was primarily due to seasonal changes in orbital insolation forcing (Kutzbach and Guetter, 1986), it is probably not wise to press the analogy too closely.

Another greenhouse effect may involve an increase of about 5-m in world sea level, due to postulated melting of the marine-based West Antarctic Ice Sheet (Mercer, 1978; Stuiver *et al.*, 1981). A similarly high sea level stand occurred during the last interglacial at 125,000 BP (Mesolella *et al.*, 1969; Dodge *et al.*, 1983). Such a response might therefore be attainable with only a modest change in forcing. [However, Robin (1985) has

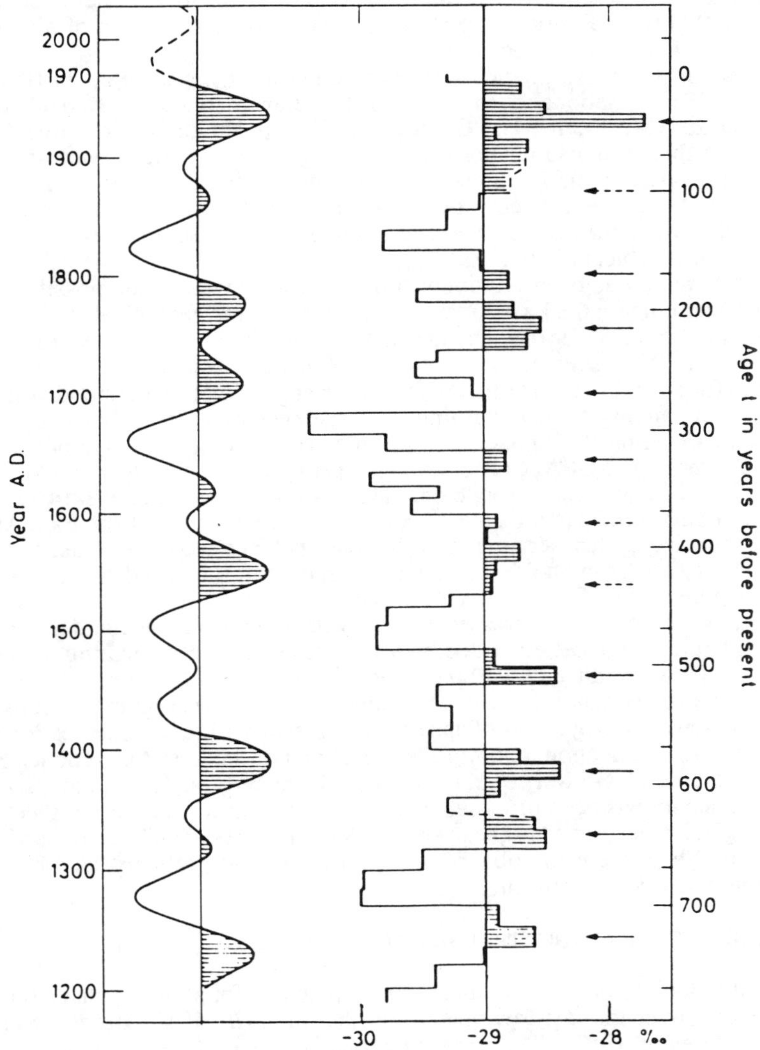

Figure 3. Right: $\delta^{18}O$ values of increments of an ice core from Camp Century (Greenland) plotted against time **t** since the deposition of the ice. The hatched areas correspond to relatively warm periods (Johnsen *et al.* 1970). Left: A synthesis of the two harmonics that dominate the stepped curve. The dashed (1970) extrapolation suggests the probable future climatic development as a continued cooling through the 1970's and perhaps 1980's, followed by a warming trend toward a new climatic optimum around the year 2015 A.D. [From Dansgaard *et al.*, 1971].

suggested that the 5-m sea level increase may have been due to a thinner East Antarctic Ice Sheet at 120,000 BP]. Bentley (1984) has estimated that a minimum of 500 years would be required for this melting to occur.

Climate models also predict changes in high-latitude sea ice cover (Parkinson and Kellogg, 1979; Manabe and Stouffer, 1980; Parkinson and Bindschadler, 1984; Washington and Meehl, 1984). Geological data appear to set an upper limit on the magnitude of this response during the early stage of greenhouse warming. Although northern hemisphere midlatitude glaciation was not initiated until about 2.4-3.2 Ma., the Arctic Ocean has apparently been covered by seasonal pack ice for at least the last 5 Ma. (Clark, 1982). Since the history of Arctic Ocean ice cover requires better documentation, this conclusion is subject to revision.

To date there has been relatively little attention given to the possible effects of a greenhouse warming on the East Antarctic Ice Sheet. Some geological studies suggest that the East Antarctic Ice Sheet may have been much smaller at 4-5 Ma (Denton, 1985). This suggestion must be better documented before it can be accepted. The mechanisms responsible for such a postulated climate change are not well understood. Given that the changes accompanying a CO_2 warming are large even on a geological scale, it might be necessary to examine the impact on the East Antarctic Ice Sheet more closely. For example, a summary of GCM CO_2-doubling experiments (Schlesinger and Mitchell, 1987) indicates that winter warming around coastal Antarctica may increase 8-14°C, presumably as a result of decreased sea ice extent. Since the mass balance of the East Antarctic Ice Sheet is controlled by temperature, and the mass balance would increase with increased temperatures (warmer air holds more water vapor), an increased greenhouse warming might change the size of East Antarctic Ice Sheet.

The large temperature increases coupled with a sea level rise could affect portions of the East Antarctic Ice Sheet grounded below sea level, possibly resulting in glacial surges (cf. Wilson, 1969; Budd and McInnes, 1978; Hollin, 1980; Thomas, 1984). Hughes (1987a) has suggested that ice streams that drain the large ice sheet may be unstable to sea level fluctuations and trigger a collapse of ice presently draining sub-sea level basins in East Antarctica. A reduction in the sub sea-level part of the East Antarctic Ice Sheet to the postulated volume of 4-5 Ma could result (Fig. 4), raising sea level perhaps 25-30 m. A similar mechanism has been proposed as an important component of ice sheet collapse at the last deglaciation (e.g., Hughes, *et al.*, 1977; Hughes, 1987b). If part of the East Antarctic Ice Sheet were to collapse, the process would probably take a minimum of several thousand years to transpire.

2.4 POSSIBILITY OF ABRUPT TRANSITIONS

Interest in the stability of climate states, and the potential for abrupt climate transitions, has grown steadily over the last few years (e.g., North *et al.*, 1981; Broecker *et al.*, 1985; Berger and Labeyrie, 1987). There are numerous examples of abrupt climate change in the geologic record (e. g., Crowley and North, 1988, 1989), with the long-term trend in climate during the last 100 million years involving the evolution from an ice-free state through several stages to one with polar ice caps. Transitions often involved abrupt steps, with data suggesting that the northern hemisphere has most recently undergone the transition.

If similar abrupt steps were to occur in the future, physical considerations and experience with the record suggest that the most likely candidates for such a transition are the Arctic Ice Cap, the West Antarctic Ice Sheet, and possibly production rates of North Atlantic Deep Water (NADW). The latter could change due to decreasing salinities in the subpolar North Atlantic as a result of melting sea ice, partial melting of the Greenland Ice

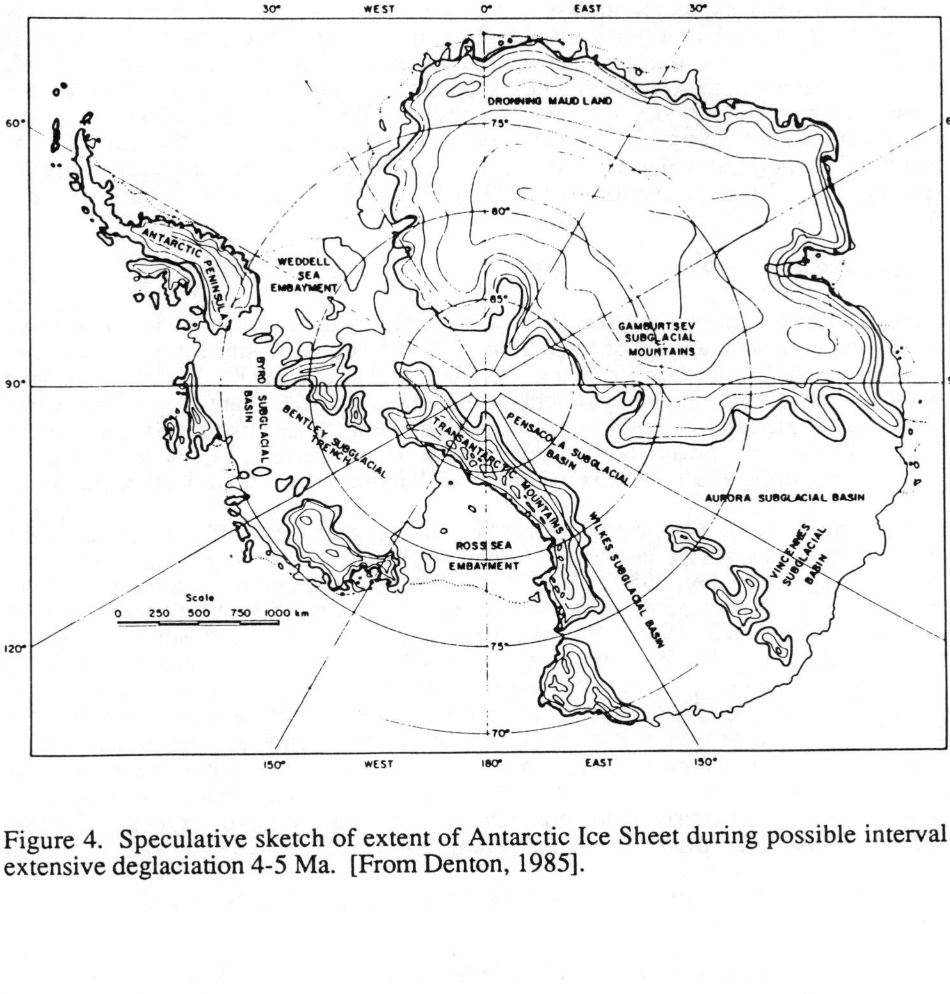

Figure 4. Speculative sketch of extent of Antarctic Ice Sheet during possible interval of extensive deglaciation 4-5 Ma. [From Denton, 1985].

Sheet, and changes in precipitation-minus-evaporation. If NADW production rates decreased there would be a significant reduction in poleward heat transport by the Gulf Stream/North Atlantic Current system. This peculiar feedback could short-circuit greenhouse warming in the European sector and possibly cause local cooling. Studies of a Greenland ice core (Fig. 5) and northwest European pollen sequences indicate that there have been numerous abrupt flips in climate in this sector over the last few tens of thousands of years (mainly during cold intervals), and that such transitions are a distinct possibility (e.g., Woillard and Mook, 1982; Dansgaard *et al.*, 1984; Broecker *et al.*, 1985).

2.5 CLIMATE-BIOSPHERE INTERACTIONS

A comparison of past abrupt transitions with the evolutionary record indicates that some transitions occur near intervals of extinction, with the largest biotic response occurring during the early stages of a climate change (Crowley and North, 1988). We therefore have the potential for a climate-induced extinction event in the next 200 years. At this stage, it is not possible to estimate the magnitude of such an event. In all likelihood it would be difficult to separate the climate-induced extinctions from those due to habitat destruction by man-induced changes, which will likely have a much more devastating effect on the earth's ecosystem.

There may also be a positive feedback between the biosphere and climate. Changes in total biomass can result in a mass transfer of carbon from the biosphere to the atmosphere (e.g., Broecker, 1982). Such a change may have occurred at the end of the Cretaceous and resulted in an estimated increase of atmospheric CO_2 levels by a factor of 2-3X (Hsü and McKenzie, 1985). Partial collapse of the ocean's thermohaline circulation (due to warming of high-latitude surface waters) would also affect the global carbon cycle, because it would decrease the rate of nutrient cycling from deep waters to the surface (Bryan and Spelman, 1985). Carbon transfer from the ocean biosphere to the atmosphere could produce a climate feedback as strong as that caused by an enhanced uptake of heat from the atmosphere (Bryan and Spelman, 1985). This scenario has some geological support. During the Cretaceous warm period, the thermohaline circulation was apparently less intense, and there is fairly good geological evidence indicating lower levels of ocean productivity at that time (Bralower and Thierstein, 1984).

2.6 POTENTIAL FOR ORBITAL INSOLATION INDUCED ICE GROWTH DURING THE NEXT 10,000 YEARS

Time series analyses of Pleistocene records indicate that Pleistocene ice volume changes were strongly influenced by orbital insolation changes - the Milankovitch effect (Hays *et al.*, 1976). As the characteristic time between ice growth events is about 20,000 years, and the last glacial maximum was 18,000 years ago, could the later stages of a future greenhouse warming be interrupted by a return to glacial growth? Several factors mitigate against such a possibility. Insolation changes over the next 10,000 years will not be nearly as large as they were for some post-interglacial intervals of the past (Fig. 6), because eccentricity levels, which modulate the 20,000 year precession signal, are very low (however, a small minimum in summer forcing is expected at 4,000 after present, AP). Utilization of a simple climate model (Imbrie and Imbrie, 1980), tuned to past orbital-ice volume relationships, predicts a more significant ice growth at 23,000 AP, significantly after the greenhouse purturbation has vanished. This model may not be applicable to the future, however, because a greenhouse warming may significantly alter

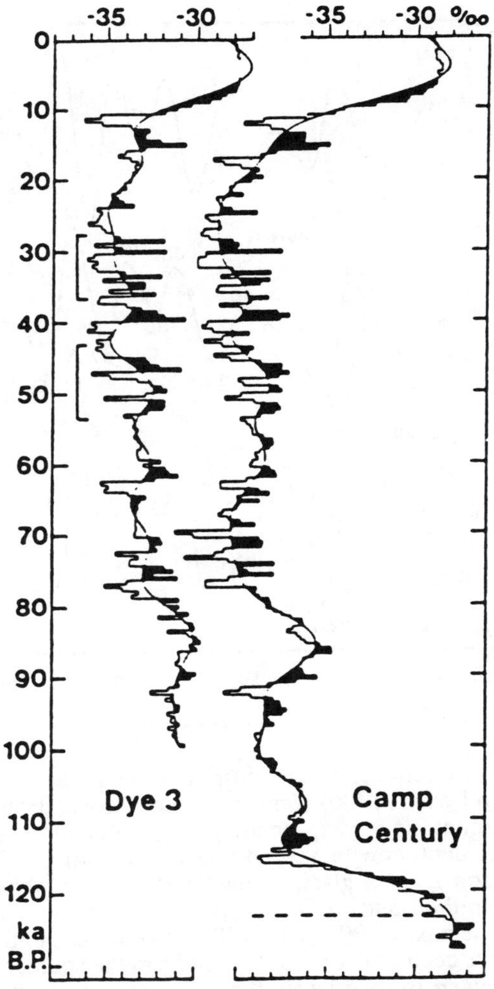

Figure 5. $\delta^{18}O$ profiles along the Dye 3 and Camp Century (Greenland) ice cores, plotted on a common linear time scale, illustrating a tendency for the system to have bistable modes, especially during the Pleistocene. [From Dansgaard *et al.*, 1984]

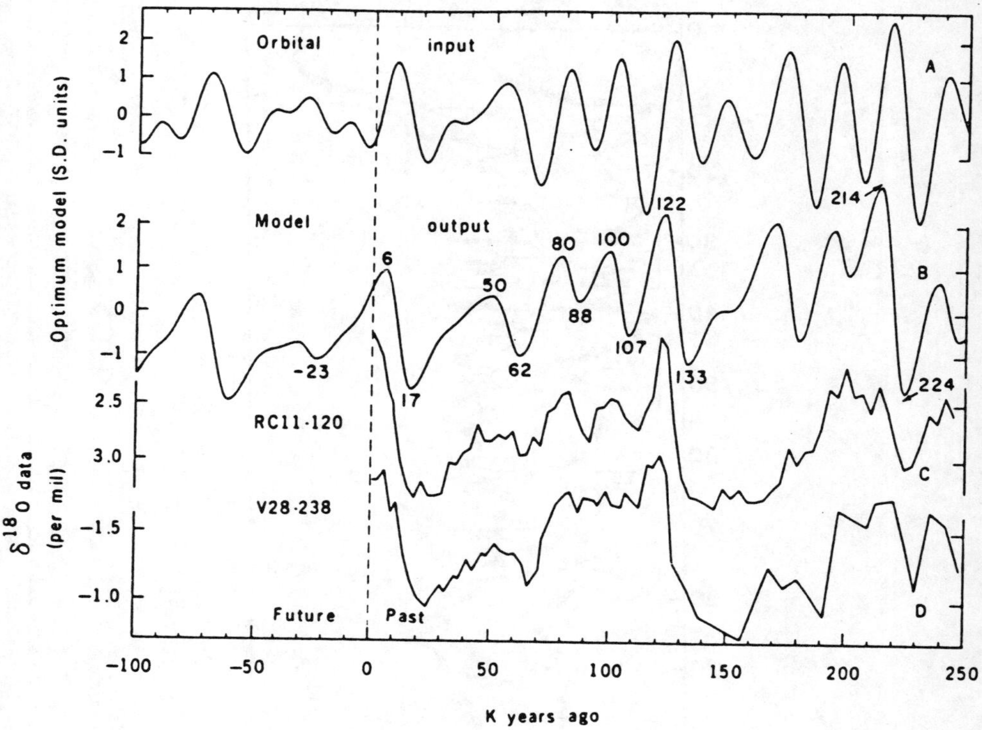

Figure 6. Orbital insolation input and output of response model for ice volume fluctuations (Imbrie and Imbrie, 1980) compared with isotopic data on climate of the past 250,000 years. (A) Orbital input corresponding to an irradiation curve for July at 65°N. (B) Output of a system function with a mean time constant of 17,000 years and a ratio of 4 : 1 between the time constants of glacial growth and melting. Ages of selected maxima and minima are given in thousands of years. According to this model, the influence of orbital variations over the next 23,000 years will be to enlarge continental ice sheets. (C) Oxygen isotope curve for deep-sea core RC11-120 from the southern Indian Ocean (Hays *et al.*, 1976). (D) Oxygen isotope curve for deep-sea core V28-238 from the Pacific Ocean. (Shackleton and Opdyke, 1973). [From Imbrie and Imbrie, 1980].

boundary conditions used to tune the Pleistocene model (i.e., the assumption of stationarity will be violated).

3. Validation of Climate Models with Paleoclimate Data

Different GCMs make different predictions about the consequences of a CO_2 warming (e.g., Schlesinger and Mitchell, 1987). It is desirable to test these models against radically different geological boundary conditions and validate them against an independent set of geological data. This section will focus on comparisons with the climate of the last 18,000 years, i.e., the last glacial maximum (10,000-20,000 BP), and the present Holocene interglacial (0-10,000 BP) and utilize data primarily from the CLIMAP (Climate: Long-Range Investigation, Mapping, and Prediction) and COHMAP (Cooperative Holocene Mapping Project) groups.

3.1 COMPARISON WITH THE LAST 20,000 YEARS

During the last glacial maximum, large ice sheets developed in North America and Europe (Fig. 7), sea ice cover expanded in both hemispheres, the tropics may have been drier (especially Africa and South America), SSTs decreased several degrees in high latitudes but apparently little in the tropics, and atmospheric CO_2 levels decreased about 80 ppm (CLIMAP, 1976, 1981; Barnola *et al.*, 1987; Crowley and North, 1989). The massive ice sheets had a significant effect on the atmospheric circulation in northern high- and mid-latitudes. In both the NCAR (National Center for Atmospheric Research) and GFDL (Geophysical Fluid Dynamics Laboratory) general circulation models (GCMs), the great dome of the North American ice sheet split the jet stream, with one branch displaced north of the ice divide, and another, stronger branch located along the southern ice margin (Fig. 8). Convergence of these two airstreams along the oceanic polar front in the central North Atlantic resulted in maximum jet velocities 10 m/s greater than present (Manabe and Broccoli, 1985b; Kutzbach and Guetter, 1986).

Changes in the upper air circulation were associated with significant simulated changes at the surface in northern midlatitudes, almost all of which are supported by geological data from different regions (Fig. 8). Southwesterly flow in Alaska resulted in generally mild ice-age climate (Kutzbach and Wright, 1985). Some north Pacific storms tracked along the southern branch of the jet (Fig. 8), increasing rainfall in the southwestern U.S. This result is supported by high ice-age lake levels in this region. In the ice divide between the Laurentide and Greenland ice sheets, strong northwesterly winds advected very cold air into the northwest Atlantic. Cooling and advection probably account for the significant equatorward displacement of sea ice and the oceanic polar fronts in this region (Manabe and Broccoli, 1985b; cf. McIntyre *et al.*, 1976). The extensive area of sea ice upwind of Europe during the winter resulted in a marked change of conditions from the present. Advection of this cold air, combined with local cooling due to the Fennoscandian Ice Sheet, resulted in winter temperature anomalies in Europe larger (20-25°C) than for any other land area (Kutzbach and Wright, 1985).

In the high latitudes of the southern hemisphere the most significant changes in climate involved expansion of Antarctic sea ice. Manabe and Broccoli (1985a; Broccoli and Manabe, 1987) calculated that lower atmospheric CO_2 levels may have been responsible for the expansion (Fig. 9). Changes in interhemispheric deep-ocean heat transport, due to changes in the production rates of warm NADW, have also been proposed as a mechanism affecting Antarctic ice cover (Weyl, 1968; Broecker, 1984). However, such changes may account for only 20-30% of the glacial-interglacial changes in

Figure 7. Northern hemisphere ice distribution at 18,000 years BP. [After Denton and Hughes, 1981]

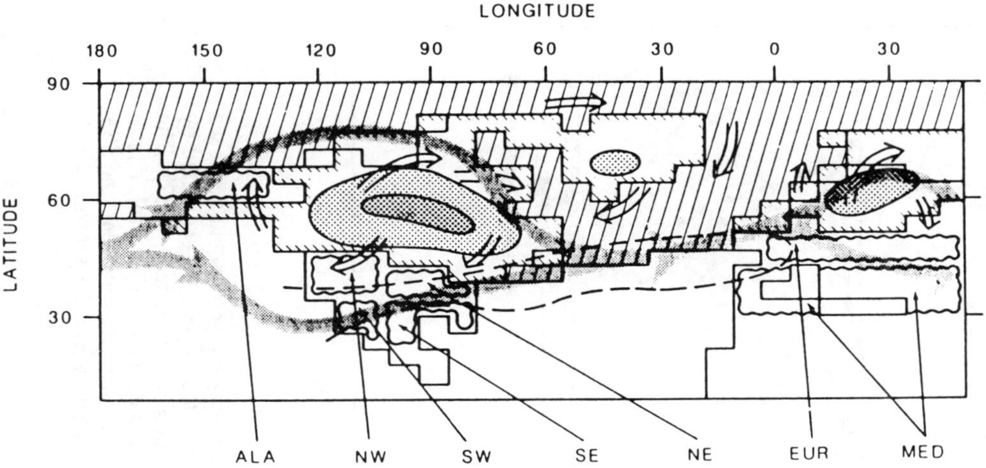

	ALA	NW	SW	SE	NE	EUR	MED
ΔT (K) JAN	5/10	-5/-10	0/-5	0/-5	-5/-10	-20/-25	-10/-15
ΔP (%) JAN	-25/-30	~0	30/35	~0	~0	-10/-15	-25/-30
Δ(P-E) (%) ANN	-45/-50	-15/-20	80/85	-25/-30	-10/-15	-70/-75	0/-5
ΔT (K) JUL	0/-5	-5/-10	0/-5	0/-5	-5/-10	0/-5	0/-5
ΔP (%) JUL	-40/-45	-25/-30	-25/-30	-25/-30	-10/-15	-5/-10	5/10

Figure 8. NCAR climate model simulations in northern midlatitudes at 18,000 BP and comparison with geological data: winds aloft (stippled arrows), surface winds (double-shafted arrows). Location and highest elevation of the North American, Greenland, and European ice sheets are shown with hatched lines. The dashed lines along the southern jet stream enclose the bad of increased precipitation. Area-average values of departures (18,000 BP-minus-present) of temperate (°K), precipitation (%), and precipitation-minus-evaporation (%) are shown for seven areas: Alaska (ALA), northwest USA (NE) Europe (EUR), and the Mediterranean (MED). The values are expressed by the nearest 5°K or 5% increments that bracket the area average. For example, in area NW in January 18,000 BP the area-average temperature was 6°K below present; the table entry is -5/-10. Note that this is not necessarily the range of temperatures over the area. Underlining indicates that the departure from the model control was significant above the 90% level compared to the model's natural variability. [From Kutzbach and Wright, 1985].

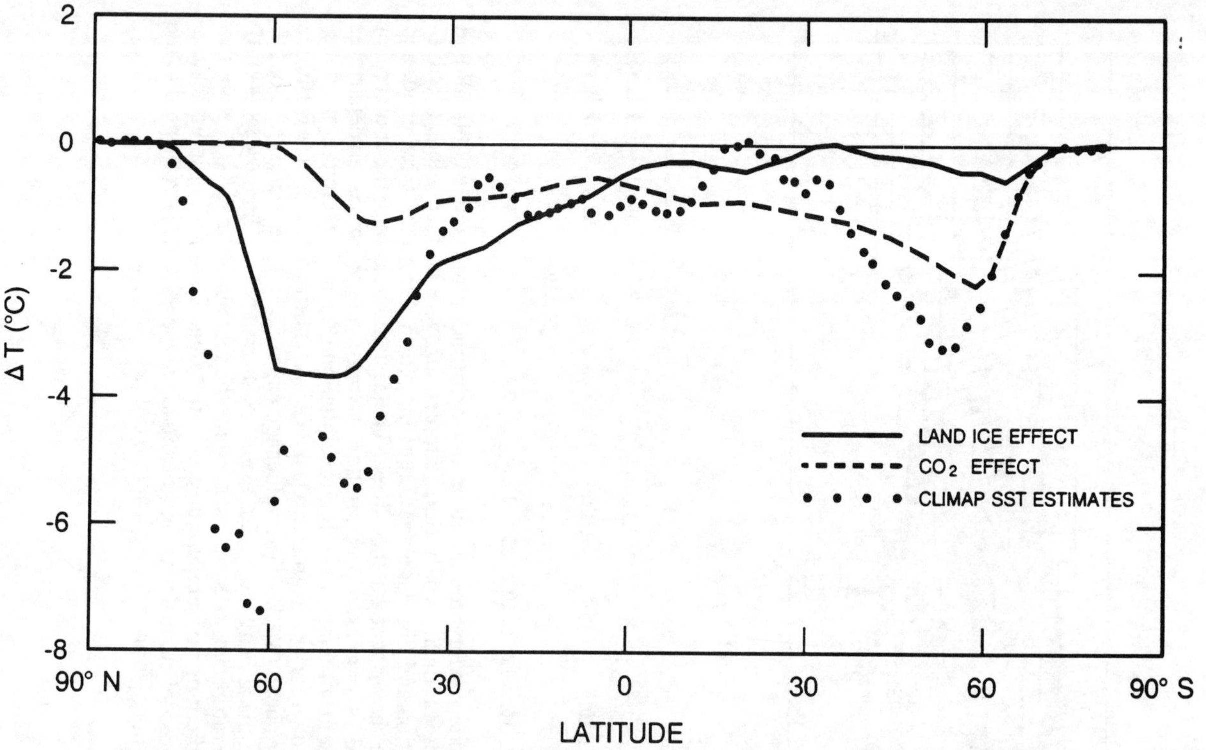

Figure. 9. Latitudinal distribution of the difference in zonally-averaged annual mean sea surface temperature between the late glacial maximum and the present, as estimated by CLIMAP (dots) and as simulated by the GFDL/GCM (Broccoli and Manabe, 1987). The solid line represents the zonally-averaged SST difference (ice age minus control) due to changes in land ice as prescribed by CLIMAP (1981). The dashed line represents the SST difference due to lower ice-age CO_2 levels. [From Crowley and Parkinson, 1988; redrawn & relabelled from Broccoli and Manabe, 1987].

SECTION 6 - CONFERENCE OVERVIEW AND RECOMMENDATIONS

sea ice (Crowley and Parkinson, 1988). The latter calculation suggests that another mechanism is needed to explain Antarctic sea ice variations - a conclusion compatible with Manabe and Broccoli's results suggesting that CO_2 was more important.

There are still some problems in the high southern latitudes. GCMs simulate central Antarctic mean annual temperature decreases of about 4.0-5.0°C (Hansen *et al.*, 1984; Broccoli and Manabe, 1987). This result is only about one-half that estimated from ice core measurements (Jouzel *et al.*, 1987). Part of the GCM changes are due to the prescribed greater ice sheet elevation of ~500 m, a prescription no longer considered justified (Denton, 1985). The elevation change should cause about a 4°C decrease in a region near the dry-adiabatic lapse rate (cf. Broccoli and Manabe, 1987). Thus, a "corrected" GCM estimated cooling might be only about 1.0-2.0°C, one-fourth the observed range. The origin of this discrepancy is not well understood.

In the tropics, models and observations do not agree. Calculations with the NCAR and NASA/GISS (Goddard Institute for Space Studies) GCMs suggest that temperature (Fig. 10) and precipitation decreases are too small to explain the extent of tropical aridity and snowline depressions (Hansen *et al.*, 1984; Kutzbach and Street-Perrott, 1985; Kutzbach and Guetter, 1986; evidence summarized in Crowley and North, 1989). [Note that lowland tropical aridity is much better documented for Africa and South America than it is for southeast Asia and Indonesia (Crowley and North, 1989). Although there is good evidence for snowline depressions in southeast Asia and Indonesia, (limited) evidence from lower elevations does not suggest significant changes in aridity (Caratini and Tissot, 1988; Stuijts *et al.*, 1988). However, the data in this region are still not in agreement with GCMs, for the most extensive area of model-generated increased aridity at 18,000 BP occurs in Indonesia and southeast Asia - cf. Rind and Peteet, 1985.]

The generally limited GCM response at 18,000 BP is well-understood, because temperatures and moisture in the tropical atmosphere are very closely tied to SSTs. If the SSTs did not change much, then there is no strong reason to suspect a strong model response. However, there are several model deficiencies which might explain part of the discrepancy-precipitation or cloud parameterizations may be in error, atmospheric aerosols may be important, biosphere feedback due to decreased evapotranspiration might be significant (cf. Sellers *et al.*, 1986), or the stomatal response of plants to lower CO_2 levels may result in decreased growth (Hansen *et al.*, 1984; Kutzbach and Street-Perrott, 1985; Wigley and Jones, 1985). As an example of the magnitude of these feedbacks, Coakley and Cess (1985) have calculated that increased tropospheric aerosols could reduce precipitation over central Africa by 20% - enough to account for lower ice-age lake levels in this region.

Rind and Peteet (1985) suggested that the model-data discrepancies in the tropics might be resolved if the CLIMAP SSTs were lowered by 2°C. A much better model-data agreement results. However, the CLIMAP SSTs have been tested with a number of different empirical approaches, all of which support the main result of the project - that SSTs in the tropics did not change much (Prell, 1985; Brassell *et al.*, 1986; Broecker, 1986). Furthermore, a recent modeling study suggests that CLIMAP SSTs were in energy balance with the ocean-atmosphere system at 18,000 BP (Hyde *et al.*, 1989). The model-data discrepancy in the tropics at 18,000 BP represents one of the most important unsolved problems in paleoclimatology and is a significant impediment to enhanced confidence about model performance.

Subsequent changes in climate involving the evolution of the African-Asian monsoon also indicate good agreement between models and data (Kutzbach and Street-Perrott, 1985). During the early part of the present Holocene Interglacial (about 6,000-11,000 BP), lake level evidence indicates high levels throughout Africa and extending eastward to India. The primary forcing mechanism apparently involves significant

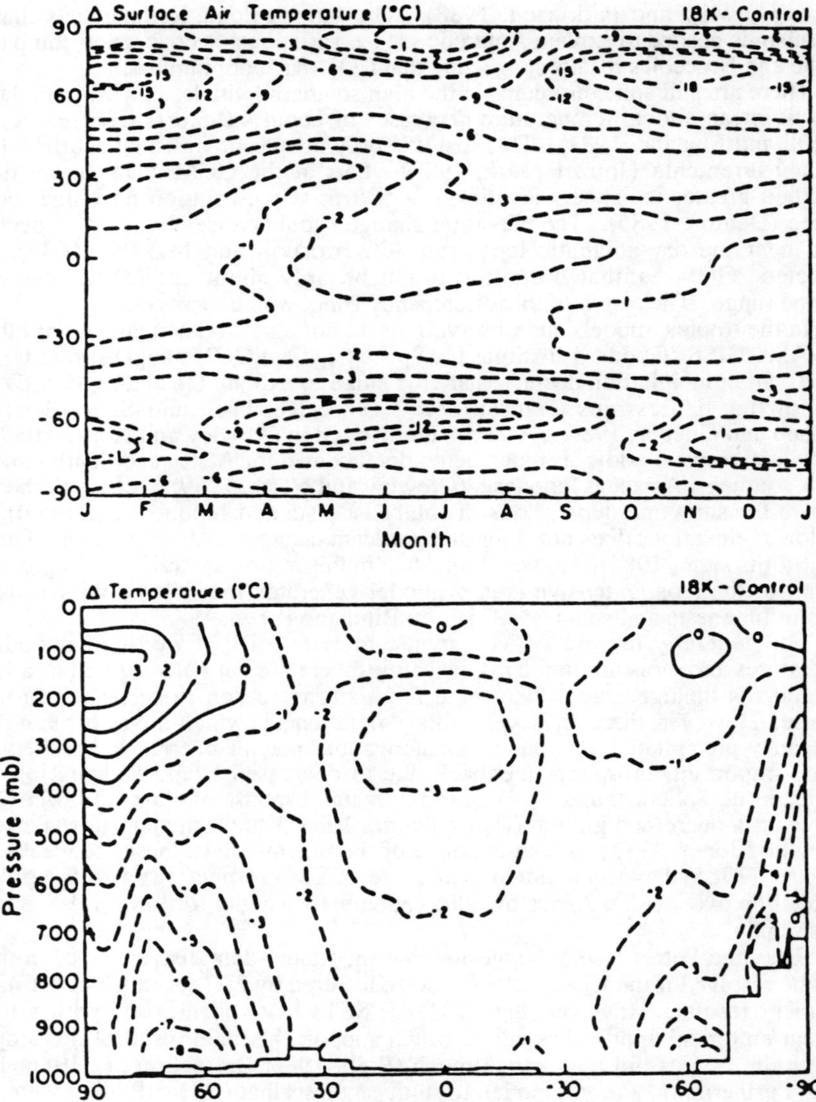

Figure 10. NASA/GISS GCM estimated temperature differences (ice age minus modern control) for month, latitude, and altitude. This figure illustrates that at all latitudes and seasons tropical temperature departures are small at 18,000 BP. [After Rind and Peteet, 1985].

orbitally-induced insolation changes in the seasonal cycle (Kutzbach and Otto-Bliesner, 1982; Kutzbach and Guetter, 1986; Kutzbach and Street-Perrott, 1985; COHMAP, 1988). At its maximum 11,000 BP, July insolation values were more than 40 W/m^2 above present at some latitudes (an 8% increase).

Experiments with GCMs of varying levels of complexity indicate that increased July insolation caused large changes in summer temperatures over Northern Hemisphere land masses and small changes over the ocean. The largest change occurred on the largest land mass and was about 2-4°C (Kutzbach and Guetter, 1986). Increased heating caused enhanced low-level inflow and decreased sea level pressure over the center of Eurasia (Fig. 11). Stronger southwest winds resulted in enhanced moisture convergence in the northern summer Intertropical Convergence Zone. A 10-20% increase in precipitation is simulated by the NCAR CCM (Kutzbach and Guetter, 1986). Comparisons of net precipitation-minus-evaporation with a hydrologic model for lakes (Fig. 12) shows good agreement between model and data (Kutzbach and Street-Perrott, 1985; COHMAP, 1988).

Intercomparison of different climate models at 9,000 BP (Fig. 13) indicates that many of the models have the same sensitivity to seasonal forcing. Since the energy balance model is linear, the good agreement suggests that, on the seasonal time scale, the temperature response to solar forcing in continental interiors is linear (Crowley, 1988; however, the actual evolution of monsoon rain systems clearly involves nonlinear processes). This conclusion is supported by direct comparisons of the harmonics of solar insolation forcing and surface temperature response for the present (K.-Y. Kim and T.J. Crowley, in preparation). If there were a significant nonlinear temperature response, one would expect substantial variance in the higher modes of the surface temperature field. Such is not the case. This result cannot be extrapolated to mean-annual (e.g., CO_2) forcing, because it is known that system sensitivity varies by time scale (North *et al.*, 1984), a result in part attributable to the fact that the seasonal system response on local scales is not in thermodynamic equilibrium (Hansen *et al.*, 1981; Cess and Potter, 1984).

3.2 OCEAN CIRCULATION FEEDBACKS

One of the largest uncertainties concerning greenhouse predictions involves the role of ocean circulation feedbacks in modifying climate change. Some general statements about the ocean circulation can be made about past climates. Because paleo-ocean modeling has lagged behind paleo-atmosphere modeling, the statements will be less robust than those for atmospheric model results, and conclusions in this section should therefore be considered preliminary.

Some important oceanographic changes during the last glacial maximum involved a redirection of the Gulf Stream flow from northeasterly at present to easterly (i.e., parallel to the oceanic polar fronts, cf. Fig. 8), probable increase in the equatorial ocean circulation due to stronger trades, a reduction in NADW formation rates, and probable increase in intermediate water formation (e.g., CLIMAP, 1981; Boyle and Keigwin, 1987; Duplessy *et al.*, 1988). Geologists have also suggested that there were significant ocean circulation changes during times when the northern hemisphere is presumed to have been ice-free (Savin *et al.*, 1975; Shackleton and Boersma, 1981). Such a conclusion is based on geological data (e.g., Fig. 14) that suggests changes in gradient of past SSTs (Keigwin and Corliss, 1986). The changes have been interpreted in terms of increased poleward ocean heat transport.

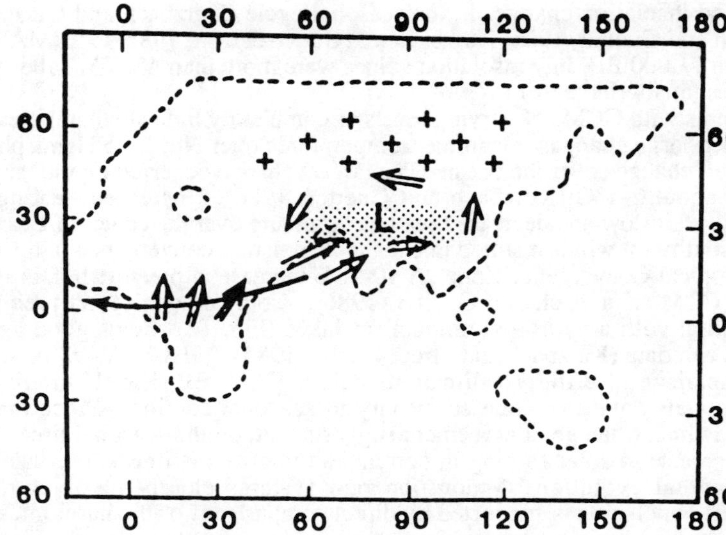

Figure 11. Schematic diagram of temperature and circulation changes that resulted from greater seasonal forcing of the African-Asian monsoon during the early part of the present interglacial. Plus signs mean warmer surface temperatures; **L**, intensified monsoon low; broad arrows, stronger low-level winds; narrow arrows, stronger upper level wind; shaded area, increased precipitation. [From Kutzbach and Otto-Bliesner, 1982].

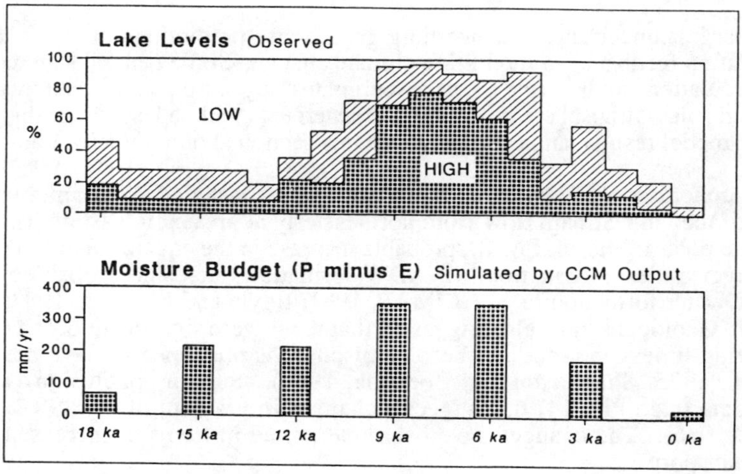

Figure 12. Comparison of GCM generated net precipitation-minus-evaporation in the African monsoon belt with estimated changes based on lake level variations. [From COHMAP, 1988].

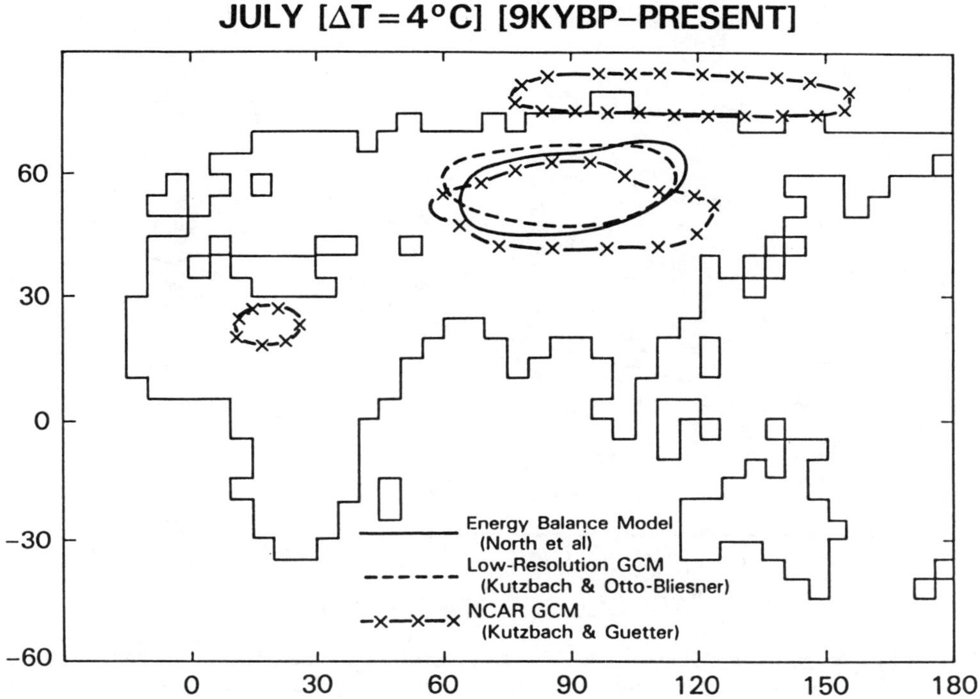

Figure 13. A comparison of linear energy balance (EBM) and general circulation model (GCM) responses to increased summer insolation. Time interval is about 10,000 years ago. The 4°C contours indicate the difference between the model response for 10,000 years ago and the present (earlier time period warmer). The good agreement between a linear EBM and the GCMs suggests an approximately linear response of the temperature field to seasonal forcing over land. The energy balance model is from North *et al.* (1983); the low-resolution model and the NCAR models are from Kutzbach and Otto-Bliesner (1982) and Kutzbach and Guetter (1984), respectively. [From Crowley *et al.*, 1986].

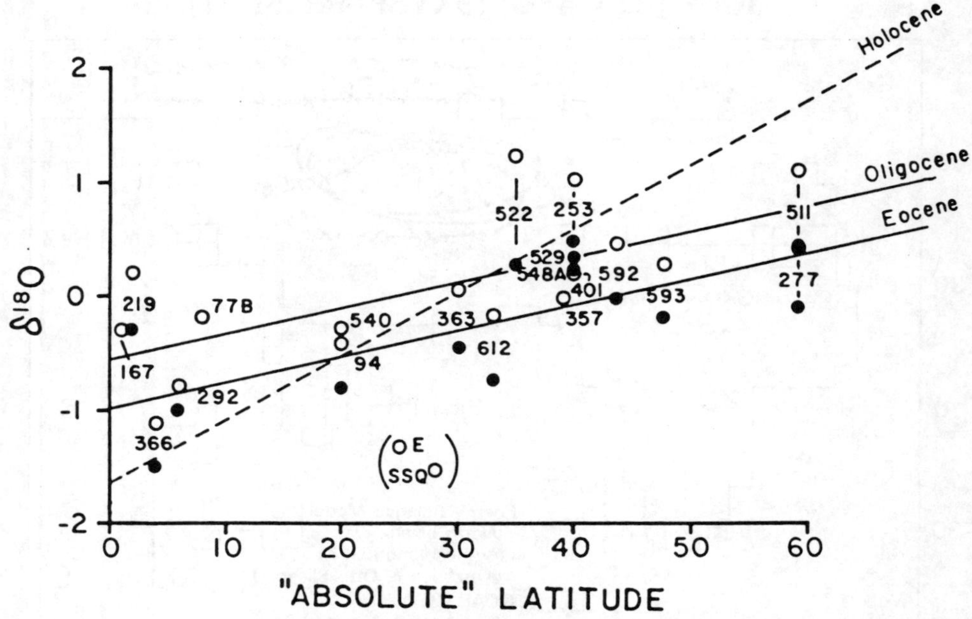

Figure 14. Average late Eocene (40 Ma) and early Oligocene (35 Ma) oxygen-isotope results in surface-dwelling planktonic foraminifera plotted against paleolatitude. The Eocene and Oligocene $\delta^{18}O$ latitudinal gradients are considerably less than the Holocene gradient and have often been interpreted in terms of changes in ocean heat transport for older intervals. [From Keigwin and Corliss, 1986].

We are still at too early a stage in paleo-ocean modeling to determine whether we satisfactorily understand the inferred changes in both Pleistocene and pre-Pleistocene oceans. There are a few statements that can be made, however. Keffer *et al.* (1988) suggest that the changes in Gulf Stream flow at 18,000 BP are due to downstream modification of the surface flowfield over the subpolar North Atlantic by the huge Laurentide Ice Sheet (cf. Figs. 7 & 8). Manabe and Bryan (1985) found that expanded sea ice tends to suppress vertical convection in a coupled ocean-atmosphere model. This result may explain decreased NADW formation at 18,000 BP. Finally, Horrell (1989) has suggested that postulated changes in Cenozoic ocean heat transport (Fig. 14) are too large and too difficult to reconcile with theory. Observations may be contaminated by recrystallization of samples, which considerably alters $\delta^{18}O$ values (e.g., Killingley, 1983).

3.3 MODEL-DATA COMPARISONS FOR WARMER CLIMATES

There have been an increasing number of paleoclimate simulations for the Cenozoic and Cretaceous, when climates were clearly warmer than the present in high latitudes (e.g., Barron and Washington, 1982, 1984, 1985; Barron, 1985; Crowley *et al.*, 1986). The frequent conclusion seems to be that increased levels of CO_2 are required to explain the postulated very warm climates of the mid-Cretaceous. The higher levels of CO_2 would result from increased volcanism due to higher rates of plate motion (Berner *et al.*, 1983; Lasaga *et al.*, 1985). Climate and geochemical models suggest that past CO_2 levels could have been anywhere from 4X-10X present levels.

Although higher levels of atmospheric CO_2 may sometimes be required for raising high latitude temperatures, there are some problems with considering these scenarios for validation of climate models. For example, it has not been adequately demonstrated from geological data that global temperatures increased everywhere, a response expected from greenhouse scenarios. Coupled ocean-atmosphere models predict that such high CO_2 should increase tropical SSTs perhaps 4-5°C (Manabe and Bryan, 1985). Yet it is difficult to find unambiguous evidence that tropical SSTs (in the open ocean) were that much warmer (Crowley and North, 1989). It is also unclear whether higher CO_2 will eliminate winter freezing in high latitudes, a result suggested by the data. Additionally, it is the author's opinion (Crowley *et al.*, 1986) that some of the floral and faunal evidence from high latitudes may indicate increases more in seasonal than in mean-annual temperatures, with migration or dormancy occurring in the winter season. Thus, any attempts to validate greenhouse-predicted winter temperatures may result in misleading conclusions. Finally, it is very difficult to find geochemical data substantiating higher CO_2 levels. Some of the evidence suggests that CO_2 levels did not increase as much as estimated (Shackleton, 1985). Thus, although modeling pre-Pleistocene climates remains an interesting exercise, I am not convinced we can use the results presently available as examples of model validation for warmer climates.

3.4 SUMMARY OF MODEL-DATA COMPARISONS

To summarize model-data comparison at 18,000 BP, there is rather good agreement between models and climate in northern midlatitudes. Matters are being clarified in the southern hemisphere, but there are continuing significant problems in the tropics. There is good agreement between models and observations in the tropics at 9,000 BP, and the jury is still out on paleo-ocean model-data agreements. Based on these comparisons, with one major exception there appears to be reasonably good agreement between models and observations over the last 20,000 years (Fig. 15).

"Paleo-Modeling Scorecard"	
Region and Time	**Model-Data Agreement**
N. High-Lat (18k)	Very Good
Low-Lat. (18k)	Poor
Low-Lat. (9k)	Very Good
S. High-Lat. (18k)	Good
Ocean Circ. (18k)	Insufficient Work

Figure 15. Model-data agreement for various times during the last 20,000 years. Note that evaluation of model performance represents the author's opinions and not some consensus tabulation. However, there is considerable support for the different ratings given (see text).

Such agreement might tempt the reader to feel more confident about GCM predictions for the future. This temptation should be resisted. The two cases where models agree best with the data (glacial maximum at northern latitudes and monsoon changes) may in some respects involve simple levels of forcing. In one case the North American ice sheet is so large that it may dominate any model response, with the type of response constrained by the enormity of the forcing. Likewise, the monsoon system is directly forced by seasonal solar heating of a large land mass, and seasonal changes in temperature appear to be linearly related to changes in forcing. (This interpretation assumes that the linearly-driven temperature changes are primarily responsible for precipitation changes, which are clearly nonlinear. The assumption appears justified because of the very good model-data agreements discussed in the text.)

For mean annual forcing in the tropics at 18,000 BP the models and observations diverge quite severely. This is the most significant time/space slice in Fig. 15 where the system is not radically forced or linearly connected. In a sense, this disagreement in the tropics may be a better measure of model performance than the two successful cases, for the greenhouse climate will not involve locally-forced or strong linear connections. Satisfaction with paleoclimatic simulations cannot be achieved until the model-data discrepancy in the tropics at 18,000 BP is resolved.

4. Summary and Conclusions

Geological data provide a surprisingly rich set of results that enable us to peer into the future with some enhanced level of insight into consequences of a greenhouse warming. The warming will probably be a very major climate change. Planetary temperatures could rise to levels normally associated with an ice-free state, yet retain cold poles due to the inertial presence of ice sheets. The magnitude of warming could produce atmospheric temperatures comparable to that of an ice-free state. The rate of warming could be comparable to or exceed any that has occurred in earth's history (except perhaps for an asteroid impact). The resultant climate would represent perhaps a unique realization in earth history. I personally consider the potential magnitude of the greenhouse warming as so large that it is difficult to comprehend, even after considerable reflection on the matter and comparison with geological data.

The greenhouse warming could significantly reduce Arctic and Antarctic sea ice cover, trigger melting of the marine-based West Antarctic Ice Sheet, and perhaps even affect the much larger East Antarctic Ice Sheet. Decreased ocean heat transported in the subpolar North Atlantic could result from salinity variations. The abrupt warming may also affect biotic extinction rates, but such extinctions will likely be dwarfed by anthropogenic-induced extinctions due to habitat destruction of ecosystems. Geological data provide considerable support for some of the above scenarios.

Although there is reasonably good agreement between climate models and inferred climate change over the last 20,000 years, the agreement may have limited usefulness in evaluating climate model projections of a greenhouse warming. Paleoclimatological results also suggest that the role of ocean heat transport may have varied for nonglacial and glacial climates, but it is premature to apply this conclusion to projections of the ocean's role in future climate change.

ACKNOWLEDGMENTS

This work was supported by NSF grant ATM-8722145 and U.S. DOE grant B-R7113-A-E. I thank M. R. Riches, M. E. Schlesinger and H. Virji for comments on an earlier draft.

REFERENCES

Barnola, J. M., D. Raynaud, Y. S. Korotkevich and C. Lorius, 1987: 'Vostok ice core provides 160,000-year record of atmospheric CO_2.' *Nature* **329**, 408-414.

Barron, E. J., 1985: 'Explanations of the Tertiary global cooling trend,' *Palaeogeog., Palaeoclim., Palaeoecol.*, **50**, 45-61.

Barron, E. J., and W. M. Washington, 1982: 'Cretaceous climate: A comparison of atmospheric simulations with the geologic record.' *Palaeogeog., Palaeoclim., Palaeoecol.,* **40**, 103-133.

Barron, E. J., and W. M. Washington, 1984: 'The role of geographic variables in explaining paleoclimates: Results from Cretaceous climate model sensitivity studies.' *J. Geophys. Res.*, **89**, 1267-1279.

Barron, E. J., and W. M. Washington, 1985: 'Warm Cretaceous climates: High atmospheric CO_2 as a plausible mechanism.' In *The Carbon Cycle and Atmospheric CO_2: Natural Variations Archean to Present,* eds. E. T. Sundquist and W. S. Broecker, Amer. Geophys. Union, Washington, DC, 546-553.

Bentley, C. R., 1984: 'Some aspects of the cryosphere and its role in climatic change.' In *Climate Processes and Climate Sensitivity,* eds. T. Takahashi and J. E. Hansen, Geophys. Mono. 29, Amer. Geophys. Union, Washington, DC, 207-220.

Berger, A. L. and L. D. Labeyrie (Eds.), 1987: *Abrupt Climatic Change: Evidence and Implications,* D. Reidel Publishing Co., Dordrecht, Holland, 417 pp.

Berner, R. A., A. C. Lasaga and R. M. Garrels, 1983. 'The carbonate-silicate geochemical cycle and its effect on atmospheric carbon dioxide over the last 100 million years.' *Amer. Jour. Sci.*, **283**, 641-683.

Boyle, E. A., L. D. Keigwin, 1987: 'North Atlantic thermohaline circulation during the past 20,000 years linked to high-latitude surface temperatures.' *Nature*, **330**, 35-40.

Bralower, T. J., and H. R. Thierstein, 1984: 'Low productivity and slow deep-water circulation in mid-Cretaceous oceans.' *Geology*, **12**, 614-618.

Brassell, S. C., E. Eglinton, I. T. Marlowe, U. Pfaumann, and M. Sarnthein, 1986: 'Molecular stratigraphy: A new tool for climatic assessment.' *Nature* , **320**, 129-133.

Broccoli, A. J., Manabe, S. 1987: 'The influence of continental ice, atmospheric CO_2, and land albedo on the climate of the last glacial maximum.' *Clim. Dynam.*, **1**, 87-100.

Broecker, W. S., 1982: 'Glacial to interglacial changes in ocean chemistry.' *Prog. Oceanogr.*, **11**, 151-197.

Broecker, W. S., 1984: 'Terminations.' In *Milankovitch and Climate, Part II*, eds. A. Berger, J. Imbrie, J. D. Hays, G. J. Kukla and B. Saltzman, D. Reidel, Dordrecht, The Netherlands, 687-698.

Broecker, W. S., 1986: 'Oxygen isotope constraints on surface ocean temperatures.' *Quat. Res.* **26**, 121-134.

Broecker, W. S., D. M. Peteet, and D. Rind, 1985: 'Does the ocean-atmosphere system have more than one stable mode of operation?' *Nature*, **315**, 21-26.

Bryan, K. and M. J. Spelman, 1985: 'The ocean's response to a CO_2-induced warming,' *J. Geophys. Res.*, **90**, 11679-11688.

Budd, W., and B. McInnes, 1978: 'Modelling surging glaciers and periodic surging of the Antarctic ice sheet.' In *Climate Change and Variability - A Southern Perspective*, eds. A. B. Pittock L. A. Frakes, D. Jenssen, J. A. Peterson and J. W. Zillman, Cambridge Univ. Press, 228-234.

Caratini, C. and C. Tissot, 1988: Paleogeographical evolution of the Mahakam Delta in Kalimantan, Indonesia during the Quaternary and late Pliocene,' *Rev. Palaeobot. Palynol.*, **55**, 217-228.

Cess, R. D., and G. L. Potter, 1984. 'A commentary on the recent CO_2-climate controversy,' *Clim. Change*, **6**, 365-376.

Clark, D. L., 1982: 'Origin, nature, and world climate effect of Arctic ocean ice cover,' *Nature*, **300**, 321-325.

CLIMAP Project Members, 1976: 'The surface of the ice-age earth.' *Science*, **191**, 1131-1137.

CLIMAP Project Members, 1981: 'Seasonal reconstruction of the earth's surface at the last glacial maximum.' *Geol. Soc. Amer.*, Map and Chart Series, No. 36.

Coakley, J. A., and R. D. Cess, 1985: 'Response of the NCAR Community Climate Model to the radiative forcing by the naturally occurring tropospheric aerosol,' *J. Atmos. Sci.*, **42**, 1677-1692.

COHMAP Members, 1988: 'Climatic changes of the last 18,000 years: Observations and model simulations.' *Science*, **241**, 1043-1052.

Crowley, T. J., 1983: 'The geologic record of climatic change,' *Rev. Geophys. Space Phys.*, **21**, 828-877.

Crowley, T. J., 1988: 'Paleoclimate modelling.' In *Physically-Based Modelling and Simulation of Climate and Climatic Change*, ed M. E. Schlesinger, Kluwer Academic Publishers, Dordrecht, The Netherlands, 883-949.

Crowley, T. J., and G. R. North, 1988: 'Abrupt climate change and extinction events in earth history,' *Science*, **240**, 996-1002.

Crowley, T. J. and C. L. Parkinson, 1988: 'Late Pleistocene variations in Antarctic sea ice II: Effect of interhemispheric deep ocean heat exchange,' *Clim. Dynam.*, **3**, 93-105.

Crowley, T. J., and G. R. North, 1989: *Paleoclimatology*, Oxford University Press (in preparation).

Crowley, T. J., D. A. Short, J. G. Mengel, and G. R. North, 1986: 'Role of seasonality in the evolution of climate during the last 100 million years,' *Science*, **231**, 579-584.

Crowley, T. J., J. G. Mengel, D. A. Short, 1987: 'Gondwanaland's seasonal cycle,' *Nature*, **329**, 803-807.

Dansgaard, W., S. J. Johnsen, H. G. Clausen, and C. C. Langway, Jr., 1971: 'Climatic record revealed by the Camp Century ice core.' In *The Late Cenozoic Glacial Ages*, ed. Karl K. Turekian, Yale University Press, New Haven and London, 37-56.

Dansgaard, W., Dansgaard, W., S. J. Johnsen, H. B. Clausen, D. Dahl-Jensen, N. Gundestrup, C. U. Hammer, and H. Oeschger, 1984: 'North Atlantic climatic oscillations revealed by deep Greenland ice cores.' In *Climate Processes and Climate Sensitivity*, eds. J. E. Hansen and T. Takahashi, Geophys. Mono **29**, Amer. Geophys. Union, Washington, DC, 288-298.

Denton, G. H., 1985: 'Did the Antarctic Ice Sheet influence Late Cainozoic climate and evolution in the Southern Hemisphere?' *S. African J. Sci.*, **81**, 224-229.

Denton, G. H., and T. J. Hughes, Eds., 1981: *The Last Great Ice Sheets,* John Wiley, New York,.483 pp.

Dickinson, R. E., and R. J. Cicerone, 1986: 'Future global warming from atmospheric trace gases,' *Nature*, **319**, 109-115.

Dodge, R. E., R. G. Fairbanks, L. K. Benninger, F. Maurrasse, 1983: 'Pleistocene sea levels from raised coral reefs of Haiti,' *Science,* **219**, 1423-1425.

Duplessy, J.-C., N. J. Shackleton, R. G. Fairbanks, L. Labeyrie, D. Oppo, and N. Kallel, , 1988: 'Deepwater source variations during the last climatic cycle and their impact on the global deepwater circulation,' *Paleoceanog.*, **3**, 343-360.

Hansen, J., D. Johnson, A. Lacis, S. Lebedeff, P. Lee, D. Rind, and G. Russel, 1981: 'Climatic impact of increasing atmospheric carbon dioxide,' *Science*, **213**, 957-966.

Hansen, J., A. Lacis, D. Rind, G. Russell, P. Stone, I. Fung, R. Ruedy, and J. Lerner, 1984: 'Climate sensitivity: Analysis of feedback mechanisms.' In *Climate Processes and Climate Sensitivity,* eds. J. E. Hansen and T. Takahashi, Amer. Geophys. Union, Washington, DC, 130-163.

Hansen, J., I. Fung, A. Lacis, D. Rind, S. Lebedeff, R. Ruedy, and G. Russell, 1988: 'Global climate changes as forecast by Goddard Institute for Space Studies three-dimensional model.' *J. Geophys. Res.* **93**:9341-9364.

Hays, J. D., J. Imbrie, and N. J. Shackleton, 1976: 'Variations in the earth's orbit: Pacemaker of the ice ages,' *Science,* **194**, 1121-1132.

Hollin, J. T., 1980: 'Climate and sea level in isotope stage 5: An East Antarctic ice surge at 95,000 BP?' *Nature,* **283**, 629-633.

Horrell, M. A., 1989: '0-18 based sea surface paleotemperature estimates: How reliable are they?' *Palaeogeog., Palaeoclim., Palaeoecol.* (submitted).

Hsü, K. J., and J. A. McKenzie, 1985: 'A "Strangelove" ocean in the earliest Tertiary. In *The Carbon Cycle and Atmospheric CO_2: Natural Variations Archean to Present,* eds. E. T. Sundquist and W. S. Broecker, Amer. Geophys. Union, Washington, DC, 487-492.

Hughes, T. J., 1987a: 'Deluge II and the continent of doom: Rising sea level and collapsing Antarctic ice.' *Boreas,* **16**, 89-100.

Hughes, T. J., 1987b: 'Ice dynamics and deglaciation models when ice sheets collapsed.' In *North America and Adjacent Oceans During the Last Deglaciation,* eds. W. F. Ruddiman and H. E. Wright, Jr., Geol. Soc. Amer., Boulder, CO., 183-220.

Hughes, T. J., G. H. Denton, and M. G. Grosswald, 1977: 'Was there a late-Würm Arctic ice sheet?', *Nature,* **266**, 596-602.

Hyde, W. T., T. J. Crowley, K.-Y. Kim, and G. R. North, 1989: 'Comparison of GCM and energy balance model simulations of seasonal temperature changes over the past 18,000 years.' *J. Clim.* (in press).

Imbrie, J., and J. Z. Imbrie, 1980: 'Modeling the climatic response to orbital variations,' *Science,* **207**, 943-953.

Johnsen, S. J., W. Dansgaard, H. B. Clausen, and C. C. Langway, Jr., 1970: 'Climatic oscillations 1200-2000 AD,' *Nature*, **227**, 482

Jones, P. D., T. M. L. Wigley, and P. B. Wright, 1986: 'Global temperature variations between 1861 and 1984,' *Nature*, **322**, 430-434.

Jouzel, J., C. Lorius, J. R. Petit, C. Genthon, N. I. Barkov, V. M. Kotlyakov, and V. M. Petrov, 1987: 'Vostok ice core: A continuous isotope temperature record over the last climatic cycle (160,000 years),' *Nature,* **329**, 403-418.

Keffer, T., D. G. Martinson, B. H. Corliss, 1988: 'The position of the Gulf Stream during Quaternary glaciations,' *Science* **241**, 440-442.

Keigwin, L. D., and B. H. Corliss, 1986: 'Stable isotopes in late middle Eocene to Oligocene foraminifers,' *Geol. Soc. Amer. Bull.*, **97**, 335-345.

Kellogg, W. W., 1977: 'Effects of human activities on global climate,' *World Meteorological Organization.*

Killingley, J. S., 1983: 'Effects of diagenetic recrystallization on $^{18}O/^{16}O$ values of deep-sea sediments,'*Nature,* **301**, 594-597.

Kutzbach, J. E., and B. Otto-Bliesner, 1982: 'The sensitivity of the African-Asian monsoonal climate to orbital parameter changes for 9000 years BP in a low-resolution general circulation model,' *J. Atmos. Sci.*, **39**, 1177-1188.

Kutzbach, J. E., and P. J. Guetter, 1984: 'The sensitivity of monsoon climates to orbital parameter changes for 9000 years BP: Experiments with the NCAR general circulation model.' In *Milankovitch and Climate, Part II,* eds. A. Berger, J. Imbrie, J. D. Hays, G. J. Kukla and B. Saltzman, Reidel, Dordrecht, The Netherlands, 801-820.

Kutzbach, J. E., and F. A. Street-Perrott, 1985: 'Milankovitch forcing of fluctuations in the level of tropical lakes from 18 to 0 kyr BP,' *Nature*, **317**, 130-134.

Kutzbach, J. E., and H. E. Wright, 1985: 'Simulation of the climate of 18,000 yr BP: Results for the North American/North Atlantic/European sector and comparison with the geologic record,' *Quat. Sci. Rev.*, **4**, 147-187.

Kutzbach, J. E., and P. J. Guetter, 1986: 'The influence of changing orbital parameters and surface boundary conditions on climate simulations for the past 18,000 years.' *J. Atmos. Sci.*, **43**, 1726-1759.

Lasaga, A. C., R. A. Berner, and R. M. Garrels, 1985: 'An improved geochemical model of atmospheric CO_2 fluctuations over the past 100 million years.' In *The Carbon Cycle and Atmospheric CO_2: Natural Variations Archean to Present*, eds. E. T. Sundquist and W. S. Broecker, Geophys. Mono. **32**, Amer. Geophys. Union, Washington, DC, 397-411.

Manabe, S., and R. J. Stouffer, 1980: 'Sensitivity of a global climate model to an increase of CO_2 concentration in the atmosphere,' *J. Geophys. Res.*, **85**, 5529-5554.

Manabe, S., and R. T. Wetherald, 1980: 'On the distribution of climate change resulting from an increase of CO_2 content in the atmosphere,' *J. Atmos. Sci.*, **37**, 99-118.

Manabe, S., and A. J. Broccoli, 1985a: 'A comparison of climate model sensitivity with data from the last glacial maximum,' *J. Atmos. Sci.*, **42**, 2643-2651.

Manabe, S., and A. J. Broccoli, 1985b: 'The influence of continental ice sheets on the climate of an ice age,' *J. Geophys. Res.*, **90**, 2167-2190.

Manabe, S., and K. Bryan, 1985: 'CO_2-induced change in a coupled ocean-atmosphere model and its paleoclimatic implications,' *J. Geophys. Res.*, **90**, 11689-11708.

Manabe, S., and R. T. Wetherald, 1986: 'Reduction in summer soil wetness induced by an increase in atmospheric carbon dioxide,' *Science*, **232**, 626-628.

McIntyre, A., N. G. Kipp, A. W. H. Bé, T. Crowley, T. Kellogg, J. V. Gardner, W. Prell, and W. F. Ruddiman, 1976: 'The glacial North Atlantic 18,000 years ago: A CLIMAP reconstruction,' *Geol. Soc. Amer. Mem.,* **145**, 43-76.

Mercer, J. H., 1978: 'West Antarctic ice sheet and CO_2 greenhouse effect: A threat of disaster,' *Nature,* **271**, 321-325.

Mesolella, K. J., R. K. Matthews, W. S. Broecker, and D. L. Thurber, 1969: 'The astronomical theory of climatic change: Barbados data,' *J. Geol.*, **77**, 250-274.

Mitchell, J. M., Jr., 1977: 'Carbon dioxide and future climate,' *Environmental Data Service,* 3-9.

Mosley-Thompson, E., and L. G. Thompson, 1982: 'Nine centuries of microparticle deposition at the South Pole,' *Quat. Res.*, **17**, 1-13.

North, G. R., R. F. Cahalan, J. A. Coakley, 1981: 'Energy balance climate models,' *Rev. Geophys. Space Phys.*, **19**, 91-121.

North, G. R., J. G. Mengel and D. A. Short, 1983: 'Simple energy balance model resolving the seasons and the continents: Application to the astronomical theory of the ice ages.' *J. Geophys. Res.,* **88**, 6576-6586.

North, G. R., J. G. Mengel, and D. A. Short, 1984: 'On the transient response patterns of climate to time dependent concentrations of atmospheric CO_2.' In *Climate Processes and Climate Sensitivity*, eds., J. E. Hansen and T. Takahashi, Geophys. Mono **29**, Amer. Geophys. Union, Washington, DC, 164-170.

Parkinson, C. L., and W. W. Kellogg, 1979: 'Arctic sea ice decay simulated for a CO_2 induced temperature rise,' *Clim. Change*, **2**, 149-162.

Parkinson, C. L., and R. A. Bindschadler, 1984: 'Response of Antarctic sea ice to uniform atmospheric temperature increases.' In *Climate Processes and Climate Sensitivity,* eds. J. E. Hansen and T. Takahashi, Amer. Geophys. Union, Washington, DC, 254-264.

Prell, W. L., 1985: 'The stability of low-latitude sea-surface temperatures: An evaluation of the CLIMAP reconstruction with emphasis on the positive SST anomalies.' *U. S. Dept. of Energy Contract Report*, Contract No. DE-ACO2-83ER60167, 60 pp.

Rind, D., and D. Peteet, 1985: 'Terrestrial conditions at the last glacial maximum and CLIMAP sea-surface temperature estimates: Are they consistent?' *Quat. Res.*, **24**, 1-22.

Robin, G. de Q., 1985: 'Contrasts in Vostok core--changes in climate or ice volume?' *Nature,* **316**, 578-579.

Savin, S. M., R. G. Douglas, and F. G. Stehli, 1975: 'Tertiary marine paleotemperatures,' *Geol. Soc. Amer. Bull.,* **86**, 1499-1510.

Schlesinger, M. E., 1986: 'Equilibrium and transient climatic warming induced by increased atmospheric CO_2,' *Clim. Dynam.,* **1**, 35-51.

Schlesinger, M. E., 1989: 'Model projections of the climatic changes induced by increased atmospheric CO_2.' In *Climate and Geo Sciences,* ed. A. Berger, Kluwer Academic Publishers, Dordrecht, The Netherlands, (this volume).

Schlesinger, M. E., and J. F. B. Mitchell, 1987: 'Climate model simulations of the equilibrium climatic response to increased carbon dioxide,' *Reviews Geophys.,* **25**, 760-798.

Sellers, P. J., Y. Mintz, Y. C. Sud, and A. Dalcher, 1986: 'A simple biosphere model (SiB) for use within general circulation models,' *J. Atmos. Sci.*, **43**, 505-531.

Shackleton, N. J., 1985: Oceanic carbon isotope constraints in oxygen and carbon dioxide in the Cenozoic atmosphere. In *The Carbon Cycle and atmospheric CO_2: Natural Variations Archean to Present*, eds. E. T. Sundquist and W. S. Broecker, Geophys. Mono. 32, Amer. Geophys. Union, Washington, DC, 412-418.

Shackleton, N. J., and N. D. Opdyke, 1973: 'Oxygen isotope and paleomagnetic stratigraphy of equatorial Pacific core V28-238: Oxygen isotope temperatures and ice volume on a 100,000 and 1,000,000 year scale'. *Quat. Res.* **3**, 39-55.

Shackleton, N. J., and A. Boersma, 1981: 'The climate of the Eocene ocean,' *J. Geol. Soc. (London),* **138**, 153-157.

Stone, P. H., 1978: 'Constraints of dynamical transports of energy on a spherical planet,' *Dyn. Atmos. Oc.*, **2**, 123-139.

Stuijts, I., J. C. Newsome, and J. R. Flenley, 1988: 'Evidence for late Quaternary vegetational change in the Sumatran and Javan highlands.' *Rev. Palaeobot. Palynol.,* **55**, 207-216.

Stuiver, M., G. H. Denton, T. J. Hughes, and J. L. Fastook, 1981: 'History of marine ice sheet in West Antarctica during the last glaciation: A working hypothesis.' In *The Last Great Ice Sheets,* eds. G. H. Denton and T. J. Hughes, John Wiley, New York, 319-436.

Sundquist, E. T., 1985: 'Geological perspectives on carbon dioxide and the carbon cycle.' In *The Carbon Cycle and Atmospheric CO_2: Natural Variations Archean to Present*, eds. E. T. Sundquist and W. S. Broecker, Amer. Geophys. Union, Washington, DC, 5-60.

Thomas, R. H., 1984: 'Ice sheet margins and ice shelves.' In *Climate Processes and Climate Sensitivity,* eds. J. E. Hansen and T. Takashashi, Amer. Geophys. Union, Washington, DC, 265-274.

Thompson, L. G., E. Mosley-Thompson, W. Dansgaard, and P. M. Grootes, 1986: 'The Little Ice Age as recorded in the stratigraphy of the tropical Quelcaya ice cap.' *Science*, **234**, 361-364.

Thompson, S. L., and E. J. Barron, 1981: 'Comparison of Cretaceous and present earth albedos: Implications for the causes of paleoclimates,' *J. Geol.*, **89**, 143-167.

Washington, W., and G. Meehl, 1984: 'Seasonal cycle experiment on the climate sensitivity due to a doubling of CO_2 with an atmospheric general circulation model coupled to a simple mixed layer ocean,' *J. Geophys. Res.*, **89**, 9475-9503.

Weyl, P. K. 1968: 'The role of the oceans in climatic changes: A theory of ice ages.' *Met. Monogr.*, **8**, 37-62.

Whillans, I. M., 1981: 'Reaction of the accumulation zone portions of glaciers to climatic change,' *J. Geophys. Res.*, **86**, 4274-4282.

Wigley, T. M. L., and P. D. Jones, 1985: 'Influence of precipitation changes and direct CO_2 effects on streamflow,' *Nature*, **314**, 149-152.

Wilson, A. T., 1969: 'The climatic effects of large-scale surges of ice sheets,' *Can. J. Earth Sci.*, **6**, 911-918.

Woillard, G. M., and W. G. Mook, 1982: 'Carbon-14 dates at Grande Pile: Correlation of land and sea chronologies,' *Science*, **215**, 159-161.

Zhang, J. and T. J. Crowley, 1989: 'Historical climate records in China and reconstruction of past climates,' *J. Clim.*, (in press).

SECTION 2 – UNDERSTANDING CLIMATE AND CLIMATIC CHANGES

DECEPTIVELY-SIMPLE MODELS OF CLIMATIC CHANGE

M. Ghil
Climate Dynamics Laboratory, Department of Atmospheric Sciences, and Institute of Geophysics and Planetary Physics
University of California
Los Angeles, CA 90024-1565
U.S.A.

ABSTRACT. Climate dynamics provides a meeting and testing ground for two areas of recent inquiry: (a) the theory and modeling of complex biogeophysical systems, and (2) the theory and applications of nonlinear dynamical systems. The purpose of this brief review is to illustrate, by example more than by generalities, recent progress in theoretical climate dynamics. The simplicity of the models used as examples allows a rather complete analysis, physical and mathematical, of the underlying concepts: (i) internal variability of the climatic system, manifested by self-sustained, free oscillations; (ii) quasi-equilibrium response to external forcing; and (iii) interaction between the free and forced variations of the system. The illustration proceeds through these three steps to show how simple, forced climatic oscillators may explain most of the features of Quaternary isotopic records. Along the way, the tools of dynamical systems and bifurcation theory are explained and applied.

1. INTRODUCTION AND MOTIVATION

1.1. Linear and Nonlinear Description of Climatic Change

Climatic changes exhibit considerable complexity, in space as well as in time. Variations of climate in space, from point to point, region to region, and continent to continent or to adjacent ocean basin, are the object of numerous studies, in this volume and elsewhere. By restricting ourselves to sufficiently-long time scales, physical intuition suggests that the spatial details might become less important (e.g., Mitchell, 1976). As we shall see, this physical intuition is supported by certain rigorous results from the mathematical theory of nonlinear dynamical systems (Témam, 1988).

Temporal complexity of climatic records prevails on scales from decades, through centuries and millenia, to millions of years and beyond (U.S. Committee for GARP, 1975, Appendix A). This complexity is apparent in the time domain, as well as in the spectral domain: climatic records for all these time scales show a tantalizing mix of regularity and

A. Berger et al. (eds.), Climate and Geo-Sciences, 211–240.

irregularity. The regularity manifests itself in the spectral domain as *lines* or at least as narrow *peaks*, indicative of significant power at or near a given frequency or a few such frequencies. The irregularity appears spectrally as broad-band "noise," indicative of power distributed over a large range of frequencies.

Until recently, a tendency existed in climate studies to explain the appearance of such mixed spectra by the superposition of two different types of effects: the lines as due to a simple kind of agency, acting outside and independently of the climatic system, the simply- or multiply-periodic *forcing;* the broad-band, continuous part of the spectrum as due to, internal or external, *stochastic noise*, with simple statistics at least. An example of simply-periodic forcing, important on decadal time scales, is the annual cycle of insolation. Multiply-periodic forcing, important on the time scale of tens and hundreds of thousands of years, is exemplified by the orbital perturbations of insolation due to gravitational interactions between the Earth and other planets. Both kinds of forcing result from mechanical effects, which act outside the climatic system, and are simpler in nature than the physical, chemical and biological effects within the system.

An example of internal stochastic forcing is given by synoptic-scale weather phenomena, when considering climatic time scales of decades and beyond (Hasselmann, 1976; Mitchell, 1976). Weather is far from simple in itself, but it suffices to consider its gross statistics, such as variance, on the time scales defined above. External stochastic forcing is exemplified by meteoritic bombardment or volcanic erruptions.

Underlying this approach to the explanation of mixed spectra is the basic assumption that the climatic system is *linear*, i.e., that it can be usefully-described by linear equations. The superposition principle and linearity are essentially synonimous in mathematics and physics. Recently, however, awareness has grown throughout the physical and biological sciences that complex behavior and mixed spectra can be explained in a more unified, less reductionist way by the tools of *nonlinear* dynamics (Arnold, 1983; Ghil and Childress, 1987; Gleick, 1987). I shall try to illustrate in the present review how these tools can be brought to bear on the problems of climate dynamics.

1.2. Climate Dynamics, GCMs and EBMs

Climate dynamics, as an analytic and quantitative, rather than descriptive and phenomenological discipline, starts in this half century with two related developments, one in the mid-1950s, the other in the early 1970s. The first one was the advent of general circulation models (GCMs). GCMs are fairly detailed models of global atmospheric flow (Phillips, 1956; Chang, 1977; Washington and Parkinson, 1986), with millions of variables. Their strength lies in simulating the present climate, and allowing for limited comparisons with actual past and hypothetical future changes. Other reviews in this volume deal with atmospheric, oceanic and coupled GCMs.

The second development was the advent of much simpler, energy-balance models (EBMs) in the late 1960s (Budyko, 1969; Sellers, 1969), and their in-depth understanding in the mid-1970s (Held and Suarez,

1974; North, 1975; Ghil, 1976). Both EBMs and GCMs, in spite of their widely-differing degree of detail, are in a certain sense equilibrium models, good at describing a stationary or very slowly varying climate.

Although this is changing with today's and tomorrow's computing and communications technology, GCMs have been by-and-large "closed-shop", ponderous learning devices. EBMs, by contrast, put understanding of climate at everybody's fingertips. They showed how basic physical mechanisms, such as radiation balance, the ice-albedo feedback, the greenhouse effect, and eddy-diffusive heat transfer, can be modeled semi-empirically, and that the resulting models are capable of providing a good approximation to the present climate. The success of these models in approximating the present climate depends, of course, in part, on information taken from observations, as well as from GCM simulations.

1.3. Multiple Equilibria, EBMs and Bifurcations

In addition to their ability to reproduce the present climate, EBMs showed the first unmistakable signature of strong nonlinearity in climate dynamics: the appearance of multiple equilibria, or multiple stable stationary solutions. Let T be global-average surface temperature, nondimensionalized by an appropriate physical scaling, and μ be the fractional change in solar input Q from a reference value Q_*, so that $\mu = Q/Q_* - 1$. It turns out that a highly-simplified, bare-bones version of a globally-averaged, i.e., spatially zero-dimensional (0-D) EBM can be written as

$$\dot{T} = \mu - T^2, \tag{1.1}$$

where $\dot{T} \equiv dT/dt$ is the rate of change of the nondimensional temperature T. Eq. (1.1) has two steady-state solutions

$$T_{\pm} = \pm \sqrt{\mu} \tag{1.2}$$

for $\mu > 0$, i.e. for $Q > Q_*$, and none for $\mu < 0$.

The linear stability of these two solutions is obtained by considering infinitesimal perturbations about them, say $T(t) = T_+ + \theta(t)$. The equation for the perturbations θ follows by substituting the perturbed $T(t)$ into (1.1), taking (1.2) into account, and neglecting terms quadratic in θ:

$$\dot{\theta} = - 2\sqrt{\mu}\, \theta \, . \tag{1.3}$$

The solution $\theta(t) = \theta_o \exp(-2\sqrt{\mu}\, t)$ decays with time t for arbitrary initial perturbations θ_o. Hence the upper stationary solution branch $T_+(\mu) = \sqrt{\mu}$ is linearly stable.

The stability of the lower branch $T_- = -\sqrt{\mu}$ is analyzed in the same way. But the sign in the equation for infinitesimal perturbations is reversed with respect to (1.3). Hence initial perturbations grow, and the branch is unstable. The upper branch has been associated with the present climate, and the lower branch with an ice age (Ghil, 1976, 1981). At $\mu = 0$, or $Q = Q_*$, the two branches coalesce, and the corresponding, unique solution, $T_+(0) = T_-(0) = 0$, is neutrally stable, i.e., infinitesimal perturbations neither grow nor decay.

Eq. (1.1) provides the simplest example of multiplicity of solutions. This multiplicity is associated with the exchange of stability between the two branches of solutions at the point ($\mu = 0$, $T = 0$). The appearance at this point of the two branches is called *bifurcation*, from the Latin word for the forking (of *two* branches). More precisely, (1.1) is the so-called normal form of a saddle-node bifurcation, or turning point (Ghil and Childress, 1987, pp. 161, 177, 319; Guckenheimer and Holmes, 1983, Section 3.4).

In fact, EBMs exhibit typically *three* branches of solutions: the upper- and lower-most branch are stable; they correspond to the present climate and to a completely ice-covered earth, respectively. These two stable branches are separated by an unstable one, to which no particular physical significance can be attached. The three branches form an S-shaped curve, as shown in Fig. 1. The point (T_c, μ_c)

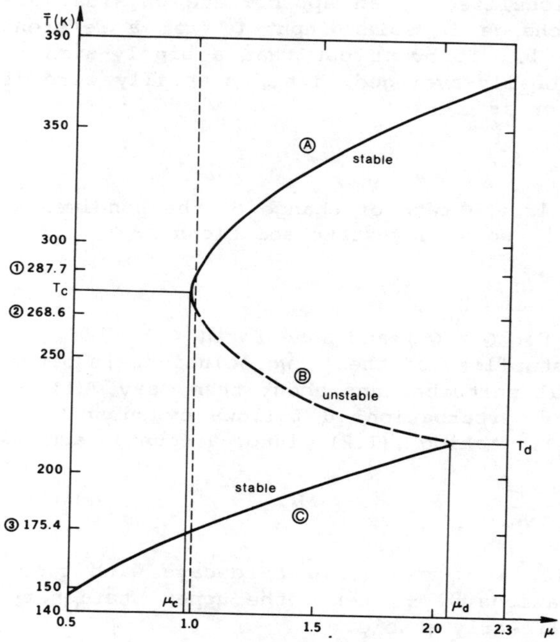

Fig. 1. A typical S-shaped bifurcation curve for an EBM. $\bar{T}$ is global average surface temperature (T in the text) and μ the insolation parameter Q/Q_0, where Q_0 is present-day insolation (this differs slightly from μ as defined in the text). The values of $\bar{T}$ for $Q = Q_0$ are labeled 1, 2 and 3. $\bar{T}_1 = 287.7\text{K}$ is an excellent approximation to the present-day climate (after Ghil, 1976).

corresponds exactly to the normal form (1.1), while the point (T_d, μ_d) is also a saddle-node bifurcation point, with normal form

$$\dot{T} = \mu + T^2. \tag{1.4}$$

Eq. (1.4) is equivalent to (1.1) via the change of coordinates $\mu \to -\mu$ and $T \to -T$ (Ghil and Childress, 1987).

Now this is simple, neat and not too intimidating. But what good is it? Well, if it applied only to 0-D EBMs, not much good. But in fact this bifurcation phenomenon was discovered first in latitudinally-dependent, zonally-averaged EBMs, and it was realized only later that its essence could be explained, physically and mathematically, in a 0-D model (Ghil, 1981; North *et al.*, 1981). In the language of climate modelers, an EBM with latitudinal resolution is called (spatially) one-dimensional (1-D), but from the viewpoint of nonlinear dynamics this is an infinite-dimensional system, since it takes an infinite number of Legendre polynomial coefficients to follow the evolution of the temperature profile in time.

The parabolic dependence (1.2) of equilibrium mean temperature on insolation for the present-climate branch, as well as the multiplicity of possible climates, have also been confirmed by a spatially three-dimensional (3-D) GCM (Wetherald and Manabe, 1975). Bimodality has now been studied for low-frequency atmospheric variability with models of varying complexity (Charney and DeVore, 1979; Benzi *et al.*, 1986), for 0-D and 1-D ice-sheet models (Weertman, 1976; Källén *et al.*, 1979), for ocean GCMs (Bryan, 1986), and even for coupled atmosphere-ocean GCMs (Manabe and Stouffer, 1988). The bimodality concept has also been considered, at least qualitatively, in the context of geochemical processes in the atmosphere-ocean system (Broecker *et al.*, 1985).

The success in applying simple bifurcation concepts to models with such a wide range of physical content and spatial detail, and to nature itself, is due to a general mathematical result called the *center manifold theorem* (e.g., Guckenheimer and Holmes, 1983, Section 3.2). It states, roughly speaking, that only the unstable modes, of which in practice there are few, determine the interesting behavior in a bifurcation. All the stable modes, even infinitely many, will only change the quantitative, but not the qualitative part of the behavior.

Thus GCMs can help EBMs by providing correct numerical values for empirical parameters, and EBMs can help GCMs by providing interesting ideas to be tested. So much for equilibria, saddle-node bifurcation, and the synergistic relation between EBMs and GCMs.

1.4. Climatic Variability, External and Internal

If climate were really stationary, understanding and predicting it would be quite easy. The point in studying climate is, of course, the fact that it changes. Climatic changes in response to external forcing only are still relatively easy to understand and to predict, provided the external agency which does the forcing is itself highly predictable, e.g. purely-periodic or slowly-varying. Thus if μ were to change sufficiently slowly in Eq. (1.1), $\mu = \mu(\varepsilon t) \geq \mu_o > 0$, for $0 < \varepsilon << \sqrt{\mu_o} < 1$, the stable solution would be, to a good approximation, $T(t) \cong \sqrt{\mu(\varepsilon t)}$.

As discussed in Section 1.1, the complexity of climatic changes is not exhausted by such a description. It warrants an attempt at a more complete, but also consistent description, using the tools of nonlinear dynamics and bifurcation theory. In this approach, both peaks and continuous background in complex climatic spectra can be explained to a large extent by the internal variability of the system, comprised of periodic and aperiodic, self-sustained oscillations. The internal variability is further enriched and modulated by the external forcing, periodic, quasi-periodic or stochastic. It is this variability which I shall try to describe, using simple models. Due to limitations of time and space, the present review is restricted to the study of paleoclimatic variability.

I shall start with a quick answer to the question: what do we really know about Quaternary glaciation cycles? While this question is addressed in greater detail in Part I of this volume, in Section 2 of this review I take the dynamicist's point of view, i.e., what do we know that we can model? The short answer is "preciously little", while the longer answer is "a lot of interesting and hard stuff".

In Section 3, I proceed to show how the variance of the recorded changes can be modeled by purely-periodic, self-sustained oscillations. The modeling of the more detailed spectral characteristics of the Quaternary proxy records requires external, orbital forcing of climatic oscillators, sketched in Section 4. Conclusions and suggestions for future research follow in Section 5.

Some of the things said here are described at greater length in Ghil (1985), Ghil and Childress (1987, Chapters 5, 6, 10, 11 and 12), and Saltzman (1985). Many books on nonlinear dynamics exist by now: Arnold (1983), Guckenheimer and Holmes (1983), and Jordan and Smith (1987) complement each other in level and material covered; Gleick (1987) provides a colorful introduction for the general reader.

2. INFORMATION CONTENT OF QUATERNARY RECORDS

2.1. Proxy Records, EBMs and Oscillatory Models

Since no thermometers can be taken into the past, all paleoclimatic records are based per force on rather indirect data. The most reliable proxy data providing continuous paleoclimatic time series are microfaunal relative abundance counts and stable isotope ratios. Among the latter, the mainstay of Quaternary studies over the last dozen years has been $\delta^{18}O$ records in marine sediment cores and in ice cores. In marine cores $\delta^{18}O$ represents largely, but not solely, global ice volume (Shackleton and Opdyke, 1973), while in ice cores it represents largely local near-surface air temperature (Lorius *et al.*, 1985).

The $\delta^{18}O$ record of four marine cores is shown in Fig. 2. Two of these, RC11-120 (Hays *et al.*, 1976) and V19-240 (Shackleton, 1977) cover much of the late Pleistocene and have moderate sedimentation rates on the average. The other two, V28-238 and V28-239 (Shackleton and Opdyke, 1973 and 1976, respectively) cover most or all of the Quaternary, having much smaller sedimentation rates. The corresponding resolution is between a few thousand to tens of thousands of years. The trade-off between resolution and total length of such a time series has been imposed by the limit in physical length of piston cores (Crowley, 1983). This trade-off implies that each core provides a band-limited view of climatic variability.

All four records show irregular changes occurring over the shortest, as well as the longest time scales resolved. The range of $\delta^{18}O$ variations is normalized in the figure by the variance of the record in each panel. This $\delta^{18}O$ range translates into a change of global ice volume by a factor of 2-to-3 over the Quaternary. Microfaunal counts in marine sediments and $\delta^{18}O$ in ice cores indicate a corresponding change of global temperature by a few degrees Kelvin (Crowley, 1983; Lorius *et al.*, 1985).

Occam's razor requires us to look for the simplest explanation of

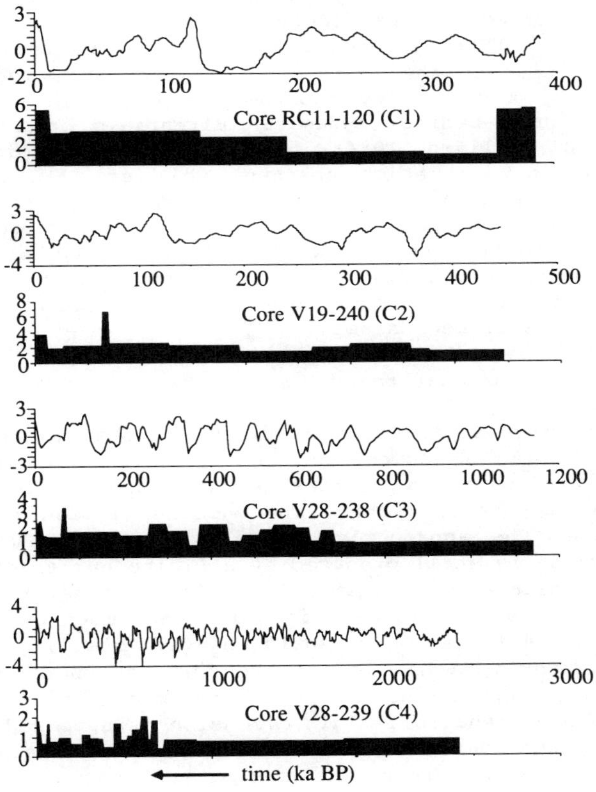

Fig. 2. Four marine records of $\delta^{18}O$, a proxy indicator of global ice volume. The ordinate is centered on the mean and normalized by the standard deviation of each record. Estimated sedimentation rates are given as solid bars underneath each core, in cm/ka (after Pestiaux, 1984).

this variability, that is, we look for external forcing. External forcing is present on these time scales in the form of insolation variations due to changes in the Earth's orbital parameters (Milankovitch, 1941; Berger, 1978). But this forcing provides only a change of net insolation ΔQ, averaged over the Earth's surface and the seasons, of about one part in a thousand. The sensitivity of EBMs, and GCMs for that matter (Ghil and Childress, 1987, p. 320) is of $\Delta\bar{T} \cong 1^{\circ}$K for $\Delta Q/Q_o = 0.01$, where $\bar{T}$ is the global mean temperature; Q_o is today's insolation, which is slightly larger (thanks goodness) than the bifurcation value Q_* corresponding to $\mu = 0$ in Eq. (1.1).

Hence direct orbital forcing of a quasi-equilibrium model only results in roughly one-tenth of the inferred temperature change between last glacial maximum (LGM), say, 18ka (1ka = 1000 years, the basic time unit for this review) B.P. (before present) and the present (Schneider and Thompson, 1979; Ghil, 1985). This leads to a search for internal variability mechanisms, which may either amplify the response to this external forcing, or act independently of it.

Since the late 1970s, a large number of such mechanisms have been proposed, in addition to the ice-albedo feedback and greenhouse effect already modeled by EBMs. They include hydrologic effects (Källén *et al.*, 1979), ice sheet-bedrock dynamics (Birchfield *et al.*, 1981; Ghil and Le Treut, 1981; Pollard, 1984), ice-sheet thermodynamics (Oerlemans, 1982), thermohaline circulation changes (Ghil *et al.*, 1987; Saltzman *et al.*, 1984), and changes in the carbon cycle (e.g. Broecker *et al.*, 1985). We shall see how nonlinearly-interactive mechanisms are modeled, and how they produce the required variability, as regular and irregular oscillations. Before that, however, it is incumbent upon us to examine whether the existing records permit one to distinguish between different mechanisms or groups of mechanisms.

2.2. Dimensionality of Quaternary Time Series

One very popular application of dynamical systems theory concerns the determination of the number of active degrees of freedom *(d-o-f)* manifest in a particular time series (Eckmann and Ruelle, 1985; Farmer *et al.*, 1983; Mayer-Kress, 1986). The underlying idea is that if this number is finite, the apparently irregular, chaotic character of the series (see Fig. 2) is still produced by a finite number of interactive, deterministic, nonlinear mechanisms; if it is infinite, the irregularity results from a stochastic process, in which case much less information, such as means and variances only, would be available.

This intriguing idea has led to the formulation of a large number of algorithms to compute the so-called *dynamical dimension D* of the system's phase-space behavior. This dimension is supposed to be, within a factor of two, the number of *d-o-f* in the preceding paragraph. Among these algorithms, the one due to Grassberger and Procaccia (1983) is easy to comprehend and implement. It was applied to cores V28-238 and V28-239 by Nicolis and Nicolis (1984) and by Grassberger (1986). The first application resulted in an estimate of $D \cong 3.1$, the second in the statement that the data did not allow a finite estimate.

The difficulty in applying this algorithm, or any other one of the same general type, to real time series is due to the fact that these algorithms do not include a systematic way to discriminate between measurement noise and the internal dynamics. A different approach to determining the *significant* number of *d-o-f* was proposed by Broomhead and King (1986), based on statistical concepts from signal processing (Pike *et al.,* 1984). Similar concepts have been applied to the nonlinear dynamics of short-term climatic variability by Ghil (1987).

Vautard and Ghil (1988) have developed and refined this approach, and applied it to the four cores in Fig. 2. Their Singular Spectral Analysis (SSA) estimates the *statistical dimension S*, rather than attempting to determine *D*. SSA deals explicitly with the problems of measurement noise, and with the strong nonstationarity due, at least in part, to the large changes in sedimentation rate. The values of *S* given by SSA range between 5, for core RC11-120, and 10, for V28-239, with confidence intervals of ±1. The difference in *S* between the two agrees well with visual inspection of the time series, which indicates a greater range of time scales in the variability of the latter than the former core.

It follows that a deterministic nonlinear model with a few lumped, or global, variables, interacting through a comparable number of physical mechanisms, should be able in principle to reproduce the dynamic information available in Quaternary proxy records, to within the accuracy of these records. Much greater spatial detail is available for selected times in the past, such as LGM, the preceding glacial maximum at about 125ka B.P., and a few times during the Holocene. But the additional *d-o-f* associated with this spatial detail do not seem to be crucial for the long-term dynamics.

2.3. Spectral Information: the 100ka Dilemma

So far we have seen that the range of Quaternary temperature changes requires internal variability mechanisms for its explanation, and that five-to-ten such mechanisms should suffice to explain the recorded variability. What specific features need to be explained? The shortest answer to this question can be provided by spectral analysis.

The four cores in Fig. 2, as well as most other such cores, reveal by visual inspection a dominant near-periodicity of 100ka (e.g., Broecker and van Donk, 1970). The visual result is confirmed by spectral analysis, which in addition yields smaller peaks near 40ka and 20ka (Hays *et al.,* 1976), and still smaller ones between 10ka and 15ka (Pestiaux, 1984; Pestiaux *et al.,* 1988). These peaks are superimposed on a continuous background, which contains most of the variance. The spectral continuum has increasing power with decreasing frequency, as do stochastic processes of the type called *red noise* (Mitchell, 1976).

The two types of forcing considered important in explaining this spectrum are the orbital changes, discussed already in Section 2.1, and white-noise forcing by shorter-lived climatic phenomena (Hasselmann, 1976). White noise has a flat spectrum; it produces a red-noise spectrum when passing through a linear, stable system, which acts as a low-pass filter. Orbital changes are quasi-periodic (Ghil and Childress, 1987, Section 12.3), with an obliquity peak at 41ka, precessional peaks at 23ka and 19ka, and eccentricity peaks at 413, 95, 124, 100 and 131ka, in decreasing order of amplitude.

The idealized spectra of Quaternary records and of the relevant forcing are shown in Figs. 3a and 3b, respectively. The climatic peaks at 40ka and 20ka are most likely related to the obliquity and precessional peaks, respectively. Part of the red-noise climatic background, but far from all, can be explained by the white-noise forcing. This leaves the so-called "100ka dilemma" of explaining the largest peak, near 100ka, which cannot be produced by direct eccentricity forcing of a quasi-equilibrium model, neither in its amplitude (see Section 2.1), nor in its location, due to the widely-distributed nature of the eccentricity variations, with only the fourth-largest peak at 100ka.

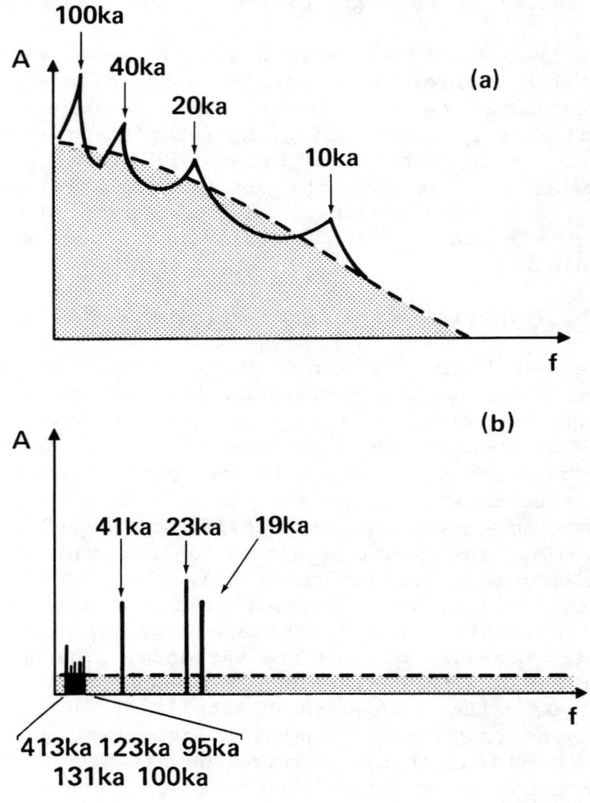

Fig. 3. The 100ka dilemma: a composite power spectrum of marine $\delta^{18}O$ (a) *vs.* the composite spectrum of orbitally-caused insolation changes (b). The units on the ordinate are arbitrary (A); the continuous part of each spectrum is stippled.

Many modeling efforts have focussed on this 100ka dilemma. A complete explanation of the Quaternary record, however, requires modeling most of the variance in the "red-noise" continuum, as well as the high-frequency peaks to the right of the precessional ones in Fig. 3a. The latter peaks are the ones related to the relative suddenness of "terminations" (Broecker and van Donk, 1970), as well as to large-amplitude, rapid "spikes" in the $\delta^{18}O$ records, such as the Allerød-Younger Dryas event between 9ka and 13ka B.P. (Duplessy *et al.*, 1981; Jouzel *et al.*, 1987). The ingredients of such a complete explanation will be sketched in the following two sections, with references to complementary approaches. The last section also contains a comparison of results from various models.

3. FREE OSCILLATIONS OF THE CLIMATIC SYSTEM

3.1. Schematic Models of Oscillations

The simplest type of internal variability can be modeled as purely-periodic, self-sustainend oscillations. I shall give two examples, and develop one of them in greater detail. The first one goes like this:

Ice-albedo feedback, which plays a key role in EBMs, states that the radiation balance becomes more negative, and hence temperature T goes down, when the planetary reflectivity, or albedo, α goes up,

$$\dot{T} \cong -\alpha \quad . \tag{3.1a}$$

Albedo in turn increases with the ice volume V,

$$\alpha \cong V \quad . \tag{3.1b}$$

Eliminating α between (3.1a) and (3.1b) yields

$$\dot{T} \cong -V \quad . \tag{3.1c}$$

The schematic symbol "$\cong$" above should be read as "positively-correlated with". It only indicates variations in the same sense, not direct, linear proportionality. This apparently vague concept of "co-variation" has been given an exact meaning within the mathematical framework of Boolean delay equations (BDEs), developed by Ghil and Mullhaupt (1985) and Ghil *et al.* (1987) for the rigorous analysis of so-called conceptual models of climatic change.

In addition to Eq. (3.1c) for T, we need an equation for the evolution of V in order to close the system. Ice-sheet models (ISMs) are governed by an ice-mass balance equation, which can be schematized in this short-hand notation as

$$\dot{V} \cong p \quad , \tag{3.2a}$$

where p is net solid precipitation over the entire ice sheet. This in turn is given by the difference

$$p \cong p_{ac} - p_{ab} , \tag{3.2b}$$

where the two terms are net accumulation and net ablation. Net accumulation rates on ice sheets are larger during warm climatic episodes (Lorius *et al.*, 1985; Jouzel *et al.*, 1987),

$$p_{ac} \cong T \ ; \tag{3.2c}$$

this has been called the *precipitation-temperature feedback* by Le Treut and Ghil (1983). Let us assume for the moment that the effect of T on p_{ac}, Eq. (3.2c), is stronger than its obviously opposite effect on p_{ab}, so that

$$p \cong T \ . \tag{3.2c'}$$

Eliminating p between (3.2a) and (3.2c') leads to

$$\dot{V} \cong T \ . \tag{3.2d}$$

Eqs. (3.1c) and (3.2d) together form an oscillatory system

$$\dot{T} \cong - V , \tag{3.3a}$$

$$\dot{V} \cong T \ . \tag{3.3b}$$

The two mechanisms in the symbolic system (3.3) are at the basis of the model of two nonlinear ordinary differential equations (ODEs) formulated by Källén *et al.* (1979), which will be presented in the next subsection. The results of this model show indeed a self-sustained oscillation, as suggested by (3.3), and provide supporting evidence for the assumption made here in replacing Eqs. (3.2b,c) by (3.2c').

The second example of coupling between two physical mechanisms which may yield an oscillation concerns ISMs (Birchfield *et al.*, 1981; Oerlemans, 1982; Pollard, 1984). The first equation is that for ice-mass balance (3.2a). The second has been called *load-accumulation feedback* by Ghil (1985). It states that the net solid precipitation rate goes down as the ice load increases, since the ice sheet sinks into the bedrock, causing the snow line to move poleward,

$$\dot{p} \cong - V \ . \tag{3.4a}$$

Rewriting (3.2a) for convenience here as

$$\dot{V} \cong p \ , \tag{3.4b}$$

we see that system (3.4) also represents an elementary oscillation.

More precisely, the physical mechanisms schematized in system (3.3) or (3.4) give rise, through their coupling, to the *possibility* of an oscillation. Whether an oscillation does indeed occur and is stable depends on the numerical value of parameters in the respective system of nonlinear ODEs. The oscillation schematized by (3.4) has been modeled in a 1-D ISM with bedrock dynamics by Birchfield and Grumbine (1985) and by Hyde and Peltier (1985).

Substantial differences between the two models exist, which are not reflected by the simple conceptual equations (3.4). The parameter values used, especially for the visco-elastic properties of the bedrock, are also different. Consequently, the Birchfield and Grumbine model exhibits an oscillation with a period of 60-80ka, for a reasonable range of parameter values. The preliminary simple-forcing experiment of Peltier and Hyde (1984) also indicates the presence of an oscillation with a period of 100 ka, although no complete analysis is provided.

Various other oscillatory mechanisms for paleoclimatic variability have been considered by Sergin (1979), Saltzman, Sutera and associates (see Saltzman, 1985, for a detailed review) and by Ghil *et al.* (1987). A similar suite of coupled mechanisms and oscillatory 0-D, 1-D and multi-dimensional oscillatory models have been developed to explain the El Niño/Southern Oscillation (ENSO) (McWilliams and Gent, 1978; Cane and Zebiak, 1985; Suarez and Schopf, 1987; Neelin, 1988).

ENSO is an irregular oscillation in the coupled atmosphere-ocean system, with a mean period of 42 months; the best-known, "warm" phase of the cycle occurs about every 3-to-7 years. Aside from the phenomenon being more tropical than polar, and the much shorter time scale, ENSO has many qualitative features in common with Quaternary glaciations. In this case the external forcing is the seasonal cycle, much stronger and more clear-cut than the insolation changes of the Quaternary. But internal mechanisms still produce the broad interannual peak and the intermittent character of the oscillations.

3.2. A Nonlinear Climatic Oscillator

Here we examine how the potential for an oscillation contained in the interactive mechanisms (3.3) is realized in a system of two coupled ODEs, formulated by Källén *et al.* (1979) and further analyzed by Ghil and Tavantzis (1983), Nicolis (1984) and Harvey and Schneider (1984). The model couples the radiation balance of EBMs, with their ice-albedo feedback (3.3a), to the ice-mass balance of ISMs, with the temperature-precipitation feedback (3.3b). It is globally-averaged, but distinguishes between open ocean, marine ice, bare land and continental ice sheet (Fig. 4).

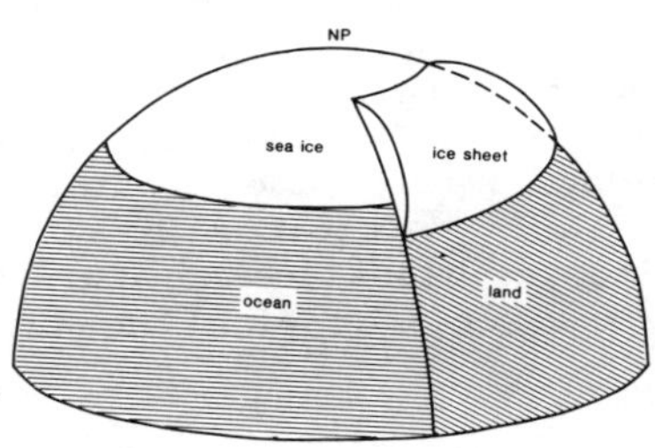

Fig. 4. Surface features of a nonlinear oscillatory climate model, Eq. (3.6a) (from Ghil and Childress, 1987).

The ice sheet, following Weertman (1976), is axially symmetric, and its meridional cross-section is symmetric about a vertical axis half-way between the Arctic coastline and its southern tip (see Fig. 5). Due to plastic flow, the profile of the ice sheet is parabolic, which is an excellent approximation for present-day ice sheets, caps and domes of widely-varying sizes (Paterson, 1972; Ghil and Childress, Sec. 11.2). The mass balance of the ice sheet is given by net accumulation with rate a north of the snowline, and net ablation with rate a' south of it,

$$\dot{V} = aA - a'A' \quad , \tag{3.5}$$

where A and A' are the accumulation and ablation areas, respectively. Eq. (3.5) corresponds, in physical variables, to the conceptual Eqs. (3.2a,b).

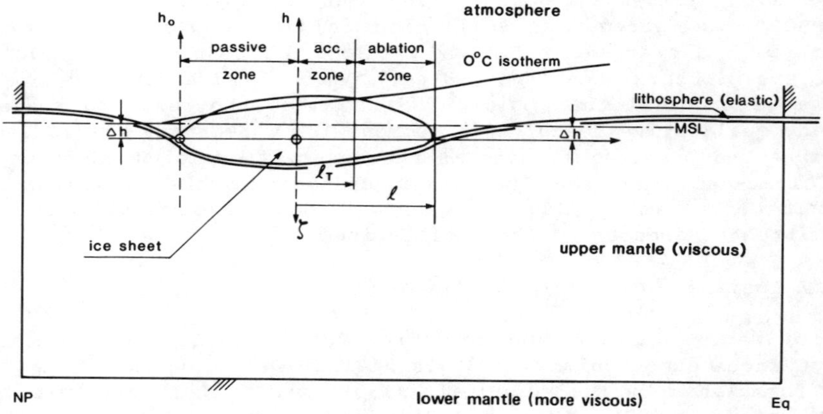

Fig. 5. Meridional cross section through the Earth's upper strata, Eq. (3.6b) (after Ghil and Childress, 1987).

Empirically, the ratio $\varepsilon = a/a'$ is close to 1/3. The precipitation-temperature feedback of Eq. (3.2c) is expressed by making $\varepsilon = \varepsilon(T)$ a ramp function with a portion where it increases with T, as shown in Fig. 6. The snowline is the intersection of the 0°C isotherm with the surface of the ice sheet (see Fig. 5), at a meridional distance ℓ_T from the vertical symmetry axis. The position ℓ_T of the snowline, separating area A from A', is determined within the model by assuming that the 0°C isotherm rises as global temperature T rises.

The radiation balance associated with the surface features in Fig. 4 is given by

$$c_T \dot{T} = Q \left\{1 - \gamma(\alpha_0 + \alpha_1 \ell) - (1-\gamma)\alpha_{oc}(T)\right\} - \kappa(T-T_\kappa) \quad . \tag{3.6a}$$

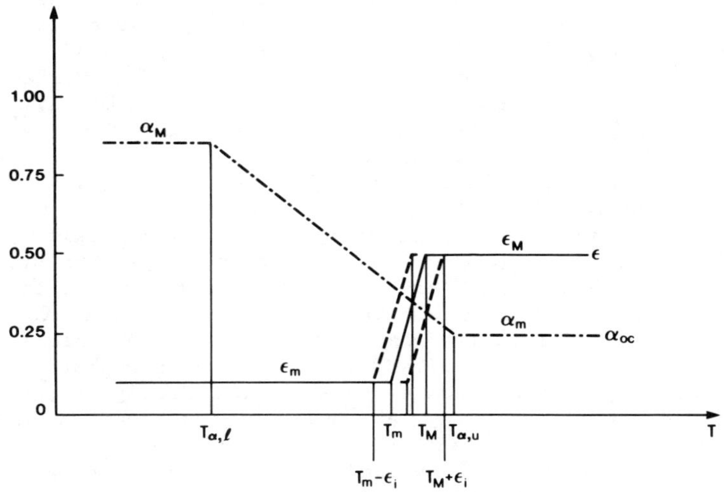

Fig. 6. The accumulation-to-ablation ratio $\varepsilon(T)$ (solid line) and the oceanic albedo $\alpha_{oc}(T)$ (dash-dotted line) as a function of globally-and-annually averaged temperature T (from Le Treut and Ghil, 1983). The two parallel dashed lines are explained in Section 4.2, Eqs. (4.3a-c).

The first term on the right-hand side (RHS) of this equation is the solar radiation absorbed by the system, the second is outgoing infrared (IR) radiation. The heat capacity is c_T, γ is the fraction of the Earth's surface occupied by land, 2ℓ is the meridional extent of the ice sheet, and T_κ a reference temperature. The linear approximation of IR radiation is well justified by both the data (Budyko, 1969) and model results (Källén *et al.*, 1979). The value of c_T is based on ocean overturning times of 1ka and larger (see Ghil *et al.*, 1987).

Of the insolation Q at the top of the atmosphere only an amount $Q(1-\alpha)$ is absorbed. The albedo contains contributions from land and sea, $\alpha = \gamma\alpha_{land} + (1-\gamma)\alpha_{oc}$. The land contribution is simply linear in the ice extent ℓ. The ocean contribution is a ramp function of temperature, also shown in Fig. 6 (cf. Sellers, 1969). For fixed ℓ, Eq. (3.6a) is essentially an EBM; it exhibits multiple equilibria, arising from back-to-back saddle-node bifurcations (1.1, 1.4), as discussed in Section 1.3 (see Fig. 1).

Taking into account the geometry of the ice sheet, as shown in Fig. 5, the mass-balance equation (3.5) can be written as:

$$c_L\dot{\ell} = \ell^{-1/2}\left\{\left[1 + \varepsilon(T)\right]\ \ell_T(T,\ell) - \ell\right\}, \tag{3.6b}$$

where c_L is a relaxation constant for the ice sheet. At fixed T, Eq.

(3.6b) yields two stable equilibria: a vanishingly-small ice cap, and a sizable ice sheet. These are separated by an unstable equilibrium, and result from two saddle-node bifurcations (1.1, 1.4), just like in EBMs (Ghil, 1984; Ghil and Childress, 1987). The coupled system (3.6a,b) also has multiple equilibria. Of these, an equilibrium (T_S, ℓ_S) close to the present climate is the only one to be stable for any parameter values at all (Ghil and Tavantzis, 1983).

System (3.6a,b) is the detailed quantitative expression of the coupled conceptual mechanisms (3.3a,b), and of additional, secondary mechanisms, not subsumed by the minimal conceptual system (3.3). The model (3.6) does in fact exhibit stable oscillatory solutions, as suggested by (3.3) and shown in Fig. 7. When T is near its maximum (point A' in the figure), the hydrologic cycle is most active, with large evaporation over the low-to-midlatitude oceans, and larger snow accumulation rates, making ℓ grow (point A''). As ℓ grows, albedo increases and temperature starts to decrease, gradually cutting back evaporation and precipitation ($B' = B$). The ice sheet is still expanding, as the snow mass accumulated along its surface flows through the interior of the sheet to its periphery ($B'' = B$). Temperature keeps sinking (C'), while the ice reaches its maximum extent (C''), further lowering temperature (D').

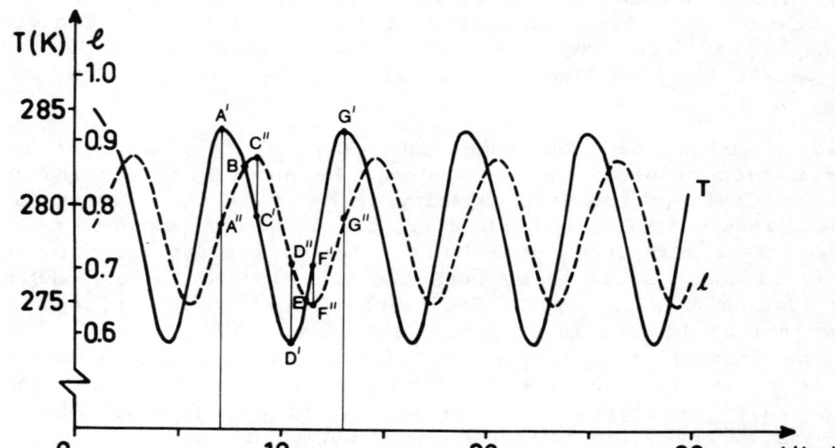

Fig. 7. Self-sustained oscillation of a coupled EBM-ISM (from Ghil and Childress, 1987).

Due to small snow accumulation, continued ablation causes the ice sheet to start shrinking (D''). This makes temperature rise again ($E' = E$), leading to a renewal of snow accumulation. The increased accumulation competes with the continued shrinking, due to the delay in the plastic flow of the ice ($E'' = E$). As the ice sheet starts to expand again (F''), the temperature continues to rise due to the thermal delay in the system (F'). This in turn accelerates the hydrologic activity and the continued expansion of the ice sheets, completing the cycle (G', G'').

The physics of this oscillation, as discussed in the preceding two paragraphs, contains the basic mechanisms (3.1c) and (3.2a,b,c). Aside from these, it also contains the physics of the change of 0°C isotherm position, affecting the ratio A/A'. Together, these mechanisms yield an oscillation with T leading V by about one quarter phase. Thus $\dot{T}$ and V are negatively correlated, while $\dot{V}$ and T are positively correlated, just as indicated by Eqs. (3.3a,b). This justifies *a posteriori* our replacing (3.2c) by (3.2c').

The presence or absence of the oscillation in Fig. 7 depend most strongly on the parameter $\mu = c_T/\kappa c_L$. Large μ corresponds physically to a very small, possibly pre-Quaternary ice cap (see Ghil, 1984; also Saltzman and Sutera, 1987). At these values, no oscillation exists and the stationary solution (T_s, ℓ_s) is stable. For Quaternary-size ice sheets, c_L is comparable in value to c_T. As μ decreases to $O(1)$, i.e., as we proceed from very small to more substantial ice sheets, the equilibrium (T_s, ℓ_s) transfers its stability to a branch of periodic solutions, by a bifurcation mechanism. In contradistinction to EBMs (Section 1.3), which involve only (multiple) branches of stationary solutions, this transfer of stability is associated with a *Hopf bifurcation*.

The normal form of such a bifurcation from stationary to periodic solutions is slightly more complicated than (1.1). It is written out and stability analysis performed in Ghil and Childress (1987, Sec. 12.2, *i.a.*). Here we only show the simplest bifurcation diagram for it (Fig. 8). Hopf

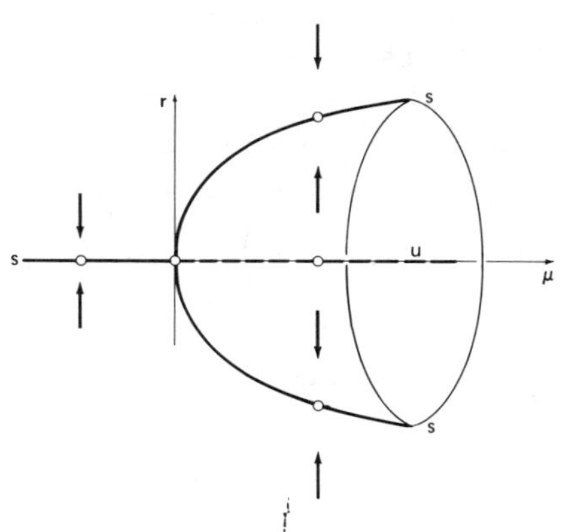

Fig. 8. Hopf bifurcation of a stable periodic solution from a stationary one. Solid branches are stable (s), dashed branches are unstable (u). The mean radius r of the periodic orbit (shown in false perspective as light solid) behaves parabolically, $r = \sqrt{\mu}$, near the bifurcation point (compare Eq. (1.2) for saddle-node bifurcation). The open circles indicate: the bifurcation point ($r = 0$, $\mu = 0$), a unique stable equilibrium ($r = 0$, $\mu = \mu_- < 0$), an unstable equilibrium ($r = 0$, $\mu = \mu_+ > 0$), and a stable periodic solution ($r = \sqrt{\mu_+} > 0$). The arrows visualize attraction by the stable solutions, stationary or periodic, and repulsion by the unstable equilibrium (after Ghil and Childress, 1987).

bifurcation occurs in most physical situations where a periodic solution arises from a stationary one. The center manifold theorem applies to this type of bifurcation, as well as to saddle-node bifurcations (Section 1.3). Hence Hopf bifurcations are not restricted to systems of two nonlinear ODEs, but apply to systems of partial differential equations (PDEs), as long as a single oscillatory mode becomes unstable. Such a bifurcation represents essentially the growth to finite amplitude, and nonlinear equilibration, of an oscillatory instability in a system of arbitrarily many degrees of freedom.

The oscillation in Fig. 7 has about the right amplitude in temperature T and ice volume V, compared to paleoclimatic records. But its period is about 10ka, while the largest fluctuations during the Quaternary are associated with the 100ka peak. Furthermore, Figs. 2 and 3 show the records to be irregular, rather than purely-periodic, with additional peaks and with a red-noise background. These additional features will obtain by adding one mechanism, bedrock dynamics, and orbital forcing to the model (3.6).

4. ORBITALLY-FORCED CLIMATIC OSCILLATIONS

4.1. A Forced Oscillator: Nonlinear Resonance

The augmented system of three coupled ODEs now includes all three mechanisms of (3.3) and (3.4): radiation balance, with ice-albedo feedback; ice-mass balance, with precipitation-temperature feedback; and isostatic rebound, with load-accumulation feedback. The third independent variable in the nonlinear ODE system is bedrock deflection ζ; its rate of change is due to the elasticity of the lithosphere and the viscous flow of the upper mantle (see Fig. 5). The complete system is coupled through the ice-mass balance equation, which is modified from the earlier form (3.6b), while the energy-balance stays the same:

$$\dot{T} = Q(t)\left\{1-\gamma(\alpha_o + \alpha_1\ell) - (1-\gamma)\alpha_{oc}(T)\right\} - k(T-T_k) \,, \tag{4.1a}$$

$$\dot{\ell} = \left\{1+(2/3)C\,\zeta\,\ell^{-1/2}\right\}^{-7}\left\{(3/2)\,\mu\ell^{-1/2}\left[1+\varepsilon(T,t)\right]\,\ell_T(T,\ell,\zeta)-\ell\right] - (2/3)\ell^{-1/2}(A\zeta\ell^{-2} + B\ell^{-3/2})\right\} , \tag{4.1b}$$

$$\dot{\zeta} = A\zeta\ell^{-2} + B\ell^{-3/2} \;; \tag{4.1c}$$

A, B and C are constants, and the relaxation constants c_T and c_L of (3.6a,b) have been absorbed into the parameter μ and time t by a suitable scaling.

For fixed insolation Q and accumulation-to-ablation ratio ε, the equilibria of system (4.1) are in one-to-one correspondence with those of (3.6), and the self-sustained oscillation of (4.1) also differs only slightly from that shown in Fig. 7 (Ghil and LeTreut, 1981). Time-dependent external forcing, however, combines with this simple internal variability to yield more complex and more realistic model behavior.

The orbitally-caused insolation forcing is of two types: eccentricity forcing, which has a globally-and-annually averaged net effect, and obliquity and precessional forcing, which produces only larger or smaller differences in heating rates between hemispheres and seasons. The former is included by allowing Q in Eq. (4.1a) to change with the two representative periods τ_4 = 100ka and τ_5 = 400ka,

$$Q(t) = Q_o \left\{1 + \delta_4 \sin (2\pi t/\tau_4) + \delta_5 \sin\left[(2\pi t/\tau_5) + \phi_5\right]\right\} , \qquad (4.2)$$

where ϕ_5 is an arbitrary phase. The amplitudes δ_4 and δ_5 are proportional to $(\Delta e)^2/2$, where Δe is the total change in eccentricity e. Since $\Delta e \leq 0.04$ over the Quaternary, δ_4 and δ_5 should not exceed 0.001.

Using δ_4 = 0.001, δ_5 = 0, and a value of μ large enough for the system not to oscillate internally, the solution asymptotes to a small-amplitude forced oscillation with period 100ka and amplitude of roughly 0.1^oK. This is consistent with the sensitivity results for quasi-equilibrium models cited in Section 2.1. For the same forcing, however, and a smaller μ value, at which internal oscillations would be present for δ_4 = 0, the forced solution shows oscillations with an amplitude comparable to that in Fig. 7 and a period near 10ka, but modulated now at 100ka (LeTreut and Ghil, 1983).

This represents a *nonlinear resonance* phenomenon, in which part of the internal variability is transferred to the forcing frequency. Linear resonance only occurs for the forcing frequency being very-nearly equal to the natural, internal frequency. Both linear and nonlinear resonance involve an increase in the amplitude of the internal oscillation. But nonlinear resonance also occurs for large differences between internal and external frequencies. The amplitude of the modulation at 100ka, when the internal, 10ka-oscillation is present, is much larger than the amplitude of the purely-forced oscillation when the internal one is absent (compare panels *a* and *b* of Fig. 13 in Ghil, 1985). This is what I mean by *transfer of variability* from the internal to the external frequency.

4.2. Entrainment, Combination Tones and Chaotic Solutions

Forcing associated with obliquity and precessional changes has no net effect on the globally-and-annually averaged $Q(t)$. The usual approach, dating back to Milankovitch (1941), has been to calculate the insolation at a particular latitude and season as representative for their effect on the waxing and waning of the ice sheets. It seems simpler and less arbitrary to incorporate these orbital-forcing components through their net effect on the hydrologic cycle. In model (4.1), this effect can be expressed by allowing the effective hydrologic temperatures T_m and T_M, which determine the ramp portion of $\varepsilon(T)$ in Fig. 6, to vary with the respective orbital changes. Thus in Eq. (4.1b)

$$\varepsilon(T,t) = \varepsilon(\,T(t);\; T_m(t),\; T_M(t)) \quad , \qquad (4.3a)$$

while

$$(\,T_m(t),\; T_M(t)) = (T_m^o,\; T_M^o) \left\{1 + \varepsilon_j \sin\left[(2\pi t/\tau_j) + \phi_j\right]\right. . \qquad (4.3b,c)$$

The periods $\tau_1 = 19$ka and $\tau_2 = 23$ka are associated with precession, while $\tau_3 = 41$ka is associated with obliquity (see also Fig. 3). This modeling of orbital effects does not change the maximum or minimum accumulation or ablation rate. It only changes the range over which $\varepsilon = a/a'$ is sensitive to temperature variations. Values of ε_j used in the model are typically between 0.001 and 0.005.

We start by applying hydrologic forcing (4.3) to the model at a single frequency $f_j = 1/\tau_j$, $j = 1,2$ or 3. When this is done in the model's oscillating regime, increasing ε_j gradually, one observes partial and then total transfer of variability from the internal frequency $f_o \cong 0.1\text{ka}^{-1}$ to the forcing frequency f_j or to an integer multiple thereof, kf_j, $k = 1,2,3,\ldots$. This phenomenon is called *entrainment*, or frequency locking, and it was first observed by C. Huygens to occur between two clocks with different pendulum periods when brought close enough together. It is also found in the biologic realm, where many animals and insects have internal clocks with periods close to, but not equal to, 24 hours, so-called circadian rhythms, which are entrained by the diurnal light cycle (Gleick, 1987; Winfree, 1980).

Further increase of ε_j leads to loss of entrainment, or detrainment: the solutions no longer have period f_o or kf_j, but become irregular and their power spectrum becomes continuous. Such irregularity, produced by a few *d-o-f* interacting deterministically, is called *chaos*, rather than randomness. As indicated already in Section 2.2, this possibility has attracted considerable attention recently in the physical and biological sciences. The estimate given there of $5 \leq S \leq 10$ is compatible with deterministic chaos explaining much of the spectral continuum in Fig. 3a.

When precessional forcing with both ε_1 and ε_2 nonzero is applied to the model's internal oscillations, the transfer of variability from f_o no longer occurs just to one frequency, nor to f_1 and f_2 only. *Combination tones* $k_1f_1 + k_2f_2$, with arbitrary integer k_1 and k_2, are excited. If both k_1 and k_2 are positive, we have a sum tone; if $k_1k_2 < 0$, we have a difference tone.

In music, difference tones were discovered by the German organist Sorge and the Italian violinist Tartini in the middle of the 18th century. Sum tones were discovered by Helmholtz (1885), who developed a rigorous theory for their auditive perception. Difference tones produced in the inner ear by sound propagation from musical instruments are deeper and more easily perceived than sum tones, which explains the earlier discovery of the former.

Subjecting the model to both hydrologic and insolation forcing, with appropriate values of ε_1, ε_2, ε_3, δ_4 and δ_5 (Le Treut *et al.*, 1988), one obtains the spectrum shown in Fig. 9a for a simulated marine core, and in Fig. 9b for a simulated ice core. The first panel uses log-log coordinates, showing clearly the continuous background with a negative slope, and the dominant peak near 100ka. The second panel uses log-linear coordinates to emphasize the range of frequencies for which most Quaternary records provide information.

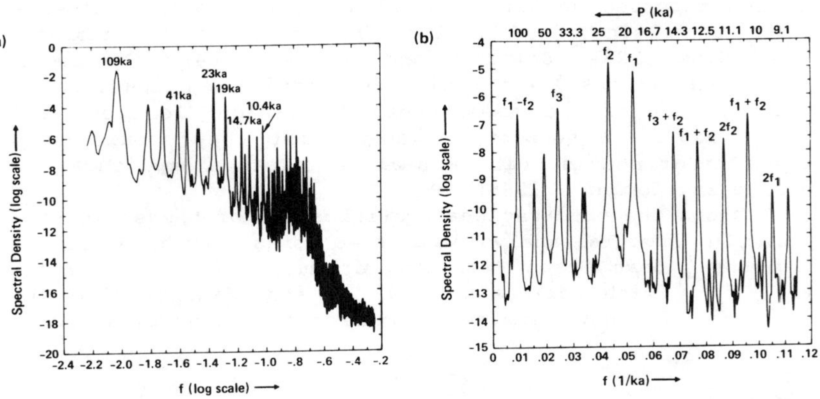

Fig. 9. Power spectrum of simulated marine $\delta^{18}O$ (a) and ice-core $\delta^{18}O$ (b) records. The periods associated with the labeled peaks in panel (b) are, from left to right, 109ka, 41ka, 23ka, 19ka, 14.7ka, 13ka, 11.5ka, 10.4ka and 9.5ka (from Le Treut *et al.*, 1988).

The climatic variables T, V and ζ (not shown here) are given by the same forced solution of model (4.1), with the same parameter values in both panels. The isotopic variables in either panel are derived from the climatic ones by simple diagnostic relationships, similar to those used in interpreting actual isotopic data (Le Treut *et al.*, 1988). The peaks in both panels are exactly at the same frequencies, and both panels use a logarithmic scale for the ordinate. The differences between the two panels in relative amplitude of the peaks are thus due entirely to the fact that marine $\delta^{18}O$ is dominated by ice volume V, while ice-core $\delta^{18}O$ is dominated by temperature T. Given the model's nonlinearity these differences are not surprising, and are consistent with such differences between spectra of distinct paleoclimatic indicators which are known to occur even in the same marine core (Hays *et al.*, 1976; Ruddiman and McIntyre, 1981).

The dominant peak in the marine record (Fig. 9a) corresponds to a difference tone, 109ka = $1/(f_1 - f_2)$. The directly-forced orbital peaks f_1, f_2 and f_3 are somewhat smaller. Finally, sum tones 14.7ka = $1/(f_2+f_3)$, 13ka = $1/(f_1+f_3)$ and 10.4ka = $1/(f_1+f_2)$, as well as harmonics 11.5ka = $1/2f_2$ and 9.5ka = $1/2f_1$, are in evidence in the high-frequency band to the right of the orbital peaks. The dominance of the difference and sum tones $f_1 \pm f_2$ is probably due to the proximity of the two precessional frequencies, relative to the obliquity frequency f_3 (Le Treut and Ghil, 1983). The magnitude of the peak near 100ka is affected but little by changing the amplitude of direct eccentricity forcing.

The sum tones and harmonics are harder to perceive in the records, because they require higher sedimentation rates in order to be resolved, and because they are more affected by bioturbation and other post-depositional processes (Dalfes *et al.*, 1984; Pestiaux, 1984). They have not been reported so far in oscillatory models with internal frequencies to the left of obliquity (Birchfield and Grumbine, 1985; Peltier and Hyde, 1984; Saltzman and Sutera, 1984). The presence of the high-frequency tones $\sum_{j=1}^{3} k_j f_j$, with substantial amplitudes, is thus a specific prediction of the model under discussion (Le Treut and Ghil, 1983; Ghil, 1985). I suspect, however, that they should be present in other nonlinear models, with a smaller relative amplitude (cf. G.E. Birchfield, pers. commun., 1988).

Table I shows the observational verification of these high-frequency peaks by careful spectral analysis of three cores with high sedimentation rates from the Indian Ocean (Pestiaux *et al.*, 1988). Six out of the eight peaks $f_i^{(d)}$ with periods between 5ka and 13ka, resolved in these cores, are very well approximated by combination tones with low order K:

$$K = \sum_{j=1}^{3} \left|k_j\right| \le 3 , \quad \left|f_i^{(d)} - \sum_{j=1}^{3} k_j^{(i)} f_j\right| \le 10^{-2} \quad . \tag{4.4a,b}$$

The constraint (4.4a) of low order is necessary because of Kronecker's lemma, which states that any given $f^{(d)}$ can be approximated arbitrarily well by a sum tone $kf' + \ell f''$, provided f'/f'' is irrational and $|k|+|\ell|$ is large enough.

TABLE I. High-frequency peaks $f^{(d)}$ in marine cores (Pestiaux, 1984; Pestiaux *et al.*, 1988) and in theory $\Sigma k_j f_j$ (Le Treut and Ghil, 1983; Ghil, 1985; Le Treut *et al.*, 1988). Agreement according to Eq. (4.4) is indicated by an asterisk in the last column.

$1/f^{(d)}$ (Data)	k_1	k_2	k_3	K	$1/\Sigma k_j f_j$ (Theory)	Good Agreement
13.0	1	0	1	2	12.98	*
12.3	2	0	-1	3	12.37	*
10.2	1	1	0	2	10.41	*
9.5	2	0	0	2	9.51	*
8.8	0	2	1	3	8.99	*
7.4	1	2	0	3	7.17	*
5.6	0	4	0	4	5.76	
5.5	1	3	0	4	5.47	

The spectral information in Fig. 9 and Table I is also reflected in the time domain. The model-simulated time series of $\delta^{18}O$ in a marine core, whose power spectrum appears in Fig. 9a, is shown in Fig. 10. In contradistinction from Pollard (1984) or Saltzman (1987), no effort whatsoever was made to "tune" the solution in Fig. 10 to any particular record. In my view, such tuning should not be done, for two reasons. First, it is true that the four records in Fig. 2 (and other Quaternary records, not shown here) exhibit a number of common, agreed-upon features; still, considerable discrepancies in timing and magnitude of these features persist (Duplessy, 1978; Jouzel *et al.*, 1987; Winograd *et al.*, 1988). Furthermore, any efforts at forcibly reconciling these differences by orbital tuning (Imbrie *et al.*, 1984) only reduce the precious information content of the records (Grassberger, 1986; Vautard and Ghil, 1988).

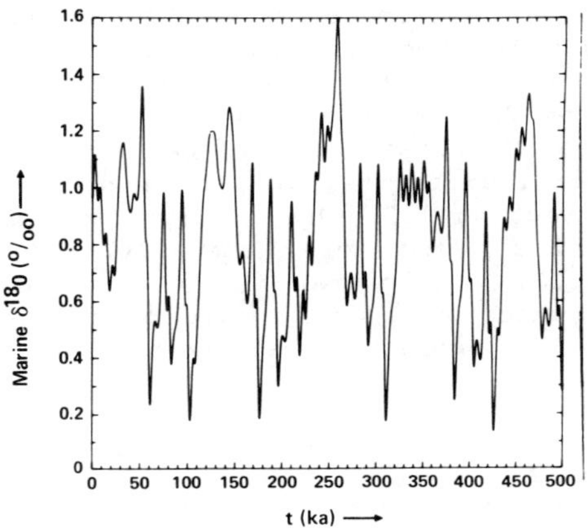

Fig. 10. Model-simulated marine $\delta^{18}O$ record; its power spectrum appears in Fig. 9a (from Le Treut *et al.*, 1988).

Second, deterministically-chaotic time series, while more predictable than stochastically-random ones, still have severely limited predictability, and hence reproducibility (Ghil and Childress, 1987, Sec. 12.6; Le Treut and Ghil, 1983). This is illustrated, for instance, by Pollard's (1984) failure, in spite of skillful multi-parameter tuning, to achieve a good fit between his model solution and the $\delta^{18}O$ record he used, over more than a few 100ka. *A fortiori*, it is remarkable that the qualitative aspect of Fig. 10 matches most of the important features in Fig. 2 for the late Quaternary (recall that time advances to the left in Fig. 2, and to the right in Fig. 10).

The rapidity of terminations and the sharp "spikes" are associated with the high-frequency peaks in the spectrum. The mean distance between terminations is near 100ka: there are five over 500ka in the model (Fig. 10), as compared to seven since the well-dated Brunhes-Matuyama magnetic reversal near 700ka B.P. in the data (Fig. 2). The irregularity in the distance between terminations, and the varying number of spikes between them, are associated with the continuous background and the limited predictability that it entails.

Among the short events, or spikes, in Fig. 10, some have almost full glacial-to-interglacial amplitude, and follow the terminations, especially at 490ka, near the "present". The latter event has remarkable similarity with the Younger Dryas, during the last deglaciation. A few authors (e.g., Duplessy *et al.*, 1981) have studied only the features of this event in and around the North Atlantic, while others have claimed that the event was in fact restricted to this region of the Northern Hemisphere (e.g., Broecker *et al.*, 1985). But recent evidence indicates that the event affected the entire Northern Hemisphere (Kallel *et al.*, 1988), and even the whole globe (Jouzel *et al.*, 1987). This makes an explanation of the Younger Dryas along the lines of global mechanisms, as in Fig. 10, more plausible.

Table I and the comparison of Figs. 9 and 10 with Figs. 3 and 2, respectively, leave us with both surprise and satisfaction that such a simple model as (4.1) can explain so many features of Quaternary records. Obviously the model is semi-empirical, and many parameters are but best guesses. To obviate the need for this empiricism, much more detailed models would be necessary. But we have seen in Sections 1.3, 2.2 and 3.2 that the same mechanisms, if embedded into a much more complex, spatially-detailed and realistic model, can still produce the same results. This is supported by the similar behavior of 0-D and 1-D EBMs, and even of GCMs (Section 1.3), by the limited information content of the data (Section 2.2), and more generally, by the center manifold theorem (Section 3.2). Therefore, it is only the computer power and human perseverance which have to be larger in order to arrive at the same, or very similar, results when using less empirical, but more cumbersome models.

5. CONCLUDING REMARKS

5.1. Summary

Simple nonlinear models are amazingly versatile tools in exploring the mechanisms of irregular climatic variability. The nonlinearity manifests itself in multiple stationary solutions, stable periodic, and chaotic ones. Deterministic chaos is not restricted to the simplest, 0-D models: it is pervasive in models with varying spatial detail and hierarchic complexity (Témam, 1988), as well as in the climatic system itself. Dynamical systems and bifurcation theory provide a common thread for the study of more and more complex behavior in models with increasing detail. This thread can be followed in moving back and forth between simple and detailed, to gain ideas and develop intuition from the simple ones, and confidence and realism from the detailed ones. It is the thread which can lead not only through the labyrinthine couloirs of modeling, but through the bewildering labyrinth of nature herself.

A systematic investigation of the information content of Quaternary records indicates that 5-to-10 active mechanisms should suffice to model the records' main, statistically-significant features. This number is comparable to that of mechanisms proposed to explain Quaternary glaciation cycles, not in any individual model, but in all of them combined. A quick comparison of some of these models against the records is given in Table II.

TABLE II. Comparison of spectral features in the four cores, C1 through C4 of Fig. 2, with model results. The periods in the table heading correspond to predictions of the descriptive astronomical theory (A), of the ISM of Birchfield and Grumbine (1985:B), of the model presented here (G), and of the marine-ice-sheet oscillatory models of Saltzman, Sutera and associates (S). The entries refer to principal components (PCs) of SSA: the higher the order of the PC, the lower its variance (from Vautard and Ghil, 1988).

<table>
<tr><td rowspan="2">Period</td><td colspan="3">Low frequencies</td><td colspan="2">Orbital frequencies (A,B,G,S)</td><td colspan="2">High frequencies</td></tr>
<tr><td>109ka*</td><td>100ka**</td><td>60-80ka**</td><td>41ka</td><td>19-23ka</td><td colspan="2">14.7ka, 13ka, 10.4ka</td></tr>
<tr><td>Prediction</td><td>G</td><td>S</td><td>B</td><td>Obliquity</td><td>Precession</td><td colspan="2">G</td></tr>
<tr><td>C1</td><td colspan="3">130ka (PC#1, irregular)</td><td>40ka (PC#2, variable)</td><td>20-25ka (PC#3)</td><td>15-20ka (#4)</td><td>10ka (PC#5)</td></tr>
<tr><td>C2</td><td colspan="3">100ka (PC#1)</td><td>30-40ka (PC#2)</td><td></td><td colspan="2">10-15ka (PC#4)</td></tr>
<tr><td>C3</td><td colspan="3">103ka (PC#1, #2)</td><td>40-50ka (PC#3)</td><td>20-30ka (PC#4)</td><td colspan="2"></td></tr>
<tr><td>C4</td><td colspan="2">100ka (PC#1,#2)</td><td>80ka (PC#3) also 300-500 ka</td><td>40-45ka(#9-10)</td><td></td><td colspan="2"></td></tr>
</table>

* Difference tone of the precessional frequencies.
** Self-sustained oscillation.

It is clear that all models reproduce the relatively-small obliquity and precession peaks (PCs #2, 3, 4 or 9). Each of *B*, *G* and *S* reproduces some lower-frequency peak (PCs #1, 2 or 3), and only *G* yields the high-frequency peaks. An interesting feature emphasized by Vautard and Ghil's *SSA* is the peak at 80ka, given only by the Birchfield and Grumbine (1985) model. Thus neither model by itself reproduces all the features of the data. This is not too surprising, when considering the fact that each model only uses two or three mechanisms, rather than five or ten, as suggested by the data.

5.2. *Suggestions for Future Research*

i) Continue the retrieval and study of climatic data, with objectivity and without theoretical bias.
ii) Use nonlinear concepts systematically in model formulation and analysis, at all levels of model complexity.
iii) Strengthen the dialog between data gatherers and analyzers, theoreticians using simple models, and "modelers" using GCMs.

Acknowledgements. It is a pleasure to thank colleagues too numerous to mention, except in the references, for many fruitful discussions, pre-prints and reprints. Thoughtful and penetrating comments by G. E. Birchfield and by one of two anonymous referees helped improve the presentation of the material. The work reported in a few sections of this review, and the writing, were supported by NSF under grant ATM86-

15424. B. Gola turned her usual magic trick of finishing to type almost before I finished to write. W. Soll drafted and prepared the figures for publication.

REFERENCES

Arnold, V. I., 1983. Geometrical Methods in the Theory of Ordinary Differential Equations. Springer-Verlag, New York/Heidelberg/ Berlin, 334 pp.

Benzi, R., P. Malguzzi, A. Speranza, and A. Sutera, 1986. The statistical properties of general atmospheric circulation: Observational evidence and a minimal theory of bimodality. Quart. J. Met. Soc., 112, 661-674.

Berger, A. L., 1978. Long-term variations of daily insolation and Quaternary climatic changes. J. Atmos. Sci., 35, 2362-2367.

Berger, A., J. Imbrie, J. Hays, G. Kukla and B. Saltzman (eds.), 1984. Milankovitch and Climate: Understanding the Response to Astronomical Forcing, vols. I & II. D. Reidel, Dordrecht/Boston/Lancaster, 895 pp.

Birchfield, G. E. and R. W. Grumbine, 1985. "Slow" physics of large continental ice sheets, the underlying bedrock and the Pleistocene Ice Ages. J. Geophys. Res., 90, 11294-11302.

Birchfield, G. E., J. Weertman and A. T. Lunde, 1981. A paleoclimate model of northern hemisphere ice sheets. Quartern. Res., 15, 126-142.

Broecker, W. S., and J. van Donk, 1970. Insolation changes, ice volumes, and the 0^{18} record in deep-sea cores. Rev. Geophys. Space Phys., 8, 169-197.

Broecker, W. S., D. M. Peteet, and D. Rind, 1985. Does the ocean-atmosphere system have more than one stable mode of operation? Nature, 315, 21-25.

Broomhead, D. S., and G. P. King, 1986. Extracting qualitative dynamics from experimental data. Physica 20D, 217-236.

Bryan, F., 1986. High-latitude salinity effects and interhemispheric thermohaline circulations. Nature, 323, 301-304.

Budyko, M. I., 1969. The effect of solar radiation variations on the climate of the Earth. Tellus, 21, 611-619.

Cane, M. A., and S. E. Zebiak, 1985. A theory for El Niño and the Southern Oscillation. Science, 228, 1085-1087.

Chang, J. (ed.), 1977. General Circulation Models of the Atmosphere. Methods Comput. Phys., 17, Academic Press, New York, 337 pp.

Charney, J. G., and J. G. DeVore, 1979. Multiple flow equilibria in the atmosphere and blocking. J. Atmos. Sci., 36, 1205-1216.

Crowley, T. J., 1983. The geologic record of climatic change. Rev. Geophys. Space Phys., 21, 828-877.

Dalfes, H. N., S. H. Schneider and S. L. Thompson, 1984. Effects of biotrubation on climatic spectra inferred from deep sea cores, in Berger et al., pp. 481-492..

Duplessy, J.-C., 1978. Isotope studies. In Gribbin, pp. 46-67.

Duplessy, J.-C., G. Delibrias, J. L. Turon, C. Pujol, and J. Duprat, 1981. Deglacial warming of the Northeastern Atlantic Ocean: correlation with the paleoclimatic evolution of the European continent. Paleogeogr. Paleoclimatol. Paleoecol. 35, 121-144.

Eckmann, J.-P., and D. Ruelle, 1985. Ergodic theory of chaos and strange attractors, Rev. Mod. Phys., 57, 617-656.

Farmer, J. D., E. Ott and J. A. Yorke, 1983. Physica 7D, 153.

Ghil, M., 1976. Climate stability for a Sellers-type model. J. Atmos. Sci., 33, 3-20.

Ghil, M., 1981. Energy-balance models: an introduction. In A. Berger (ed.), D. Reidel, Dordrecht/Boston, pp. 461-480.

Ghil, M., 1984. Climate sensitivity, energy balance models, and oscillatory climate models. J. Geophys. Res., 89D, 1280-1284.

Ghil, M., 1985. Theoretical climate dynamics: an introduction. In Turbulence and Predictability in Geophysical Fluid Dynamics, edited by M. Ghil, R. Benzi and G. Parisi, pp. 347-402, North-Holland, Amsterdam.

Ghil, M., 1987. Dynamics, statistics and predictability of planetary flow regimes. In Irreversible Phenomena and Dynamical Systems Analysis in the Geosciences, C. Nicolis and G. Nicolis (eds.), D. Reidel, Dordrecht/Boston/Lancaster, pp. 241-283.

Ghil, M., and S. Childress, 1987. Topics in Geophysical Fluid Dynamics: Atmospheric Dynamics, Dynamo Theory and Climate Dynamics. Springer-Verlag, New York, 485 pp.

Ghil, M., and H. LeTreut, 1981. A climate model with cryodynamics and geodynamics. J. Geophys. Res., 86, 5262-5270.

Ghil, M., and A. Mullhaupt, 1985. Boolean delay equations, II: periodic and aperiodic solutions. J. Stat. Phys., 41, 125-173.

Ghil, M., and J. Tavantzis, 1983. Global Hopf bifurcation in a simple climate model. SIAM J. Appl. Math., 43, 1019-1041.

Ghil, M., A. Mullhaupt and P. Pestiaux, 1987. Deep water formation and Quaternary glaciations. Climate Dyn., 2, 1-10.

Gleick, J., 1987. Chaos: Making a New Science. Viking, 354 pp.

Grassberger, P., 1986, Do climatic attractors exist? Nature, 323, 609-612.

Grassberger, P., and I. Procaccia, 1983. Phys. Rev. Lett., 50, 346.

Gribbin, J. (ed.), 1978. Climatic Change. Cambridge Univ. Press, Cambridge/New York/Melbourne, 280 pp.

Guckenheimer, J., and P. Holmes, 1983. Nonlinear Oscillations, Dynamical Systems, and Bifurcations of Vector Fields. Springer-Verlag, New York/Berlin/Heidelberg/Tokyo, 453 pp.

Harvey, L. D. D., and S. H. Schneider, 1984. Sensitivity of internally-generated climate oscillations to ocean model formulation. In Berger et al. (1984), pp. 653-667.

Hasselmann, K., 1976. Stochastic climate models, Part I. Theory. Tellus, 28, 473-485.

Hays, J. D., J. Imbrie and N. J. Shackleton, 1976. Variations in the earth's orbit: pacemaker of the ice ages. Science, 194, 1121-1132.

Held, I. M., and M. J. Suarez, 1974. Simple albedo feedback models of the icecaps. Tellus, 36, 613-628.

Helmholtz, H. L. F., 1885. On the Sensations of Tone as a Physiological Basis for the Theory of Music, 2nd English ed. Reprinted by Dover, New York (1954), 576 pp.

Hyde, W. T., and W. R. Peltier, 1985. Sensitivity experiments with a model of the ice age cycle: The response to harmonic forcing. J. Atmos. Sci., 42, 2170-2188.

Imbrie, J., J. D. Hays, D. G. Martinson, A. McIntyre, A. C. Mix, J. J. Morley, N. G. Pisias, W. L. Prell and N. J. Shackleton, 1984. The orbital theory of Pleistocene climate: support from a revised chronology of the marine $\delta^{18}O$ record. In Berger et al. (1984), pp. 269-305.

Jordan, D. W., and P. Smith, 1987. Nonliner Ordinary Differential Equations, 2nd ed. Clarendon Press, Oxford, 381 pp.

Jouzel, J., C. Lorius, J. R. Petit, C. Genthon, N. I. Barkov, V. M. Kotlyakov and V. M. Petrov, Vostok ice core: a continuous isotopic temperature record over the last climatic cycle (160,000 years), Nature, 329, 403-408.

Kallel, N., L. D. Labeyrie, M. Arnold, H. Okada, W. C. Dudley and J.-C. Duplessy, 1988. Evidence of cooling during the Younger Dryas in the western North Pacific. Oceanol. Acta, in press.

Källén, E., C. Crafoord and M. Ghil, 1979. Free oscillations in a climate model with ice-sheet dynamics. J. Atmos. Sci., 36, 2292-2303.

Le Treut, H., and M. Ghil, 1983. Orbital forcing, climatic interactions, and glaciation cycles. J. Geophys. Res., 88C, 5167-5190

Le Treut, H., J. Portes, J. Jouzel, and M. Ghil, 1988. Isotopic modeling of climatic oscillations: Implications for a comparative study of marine and ice-core records. J. Geophys. Res., 93D, 9365-9383.

Lorius, C., J. Jouzel, C. Ritz, L. Merlivat, N. I. Barkov, Y. S. Korotkevitch, V. M. Kotlyakov, 1985. A 150,000-year climatic record from Antarctic ice. Nature, 316,

Manabe, S., and R. J. Stouffer, 1988. Two stable equilibria of a coupled ocean-atmosphere model. J. Climate, in press.

Mayer-Kress, G., 1986. Dimensions and Entropies in Chaotic Systems, Springer-Verlag, Berlin.

McWilliams, J. C., and P. R. Gent, 1978. A coupled air and sea model for the tropical Pacific. J. Atmos. Sci., 35, 982-989, and 36(1979), 181.

Milankovitch, M., 1941. Canon of Insolation and the Ice Age Problem (in German). Roy. Serb. Acad. Spec. Publ., 133, 1-633; Engl. transl.: Israel Program for Scientific Translations, Jerusalem, 1969, 482 pp. (available from Natl. Tech. Info. Svce., US Dept. of Commerce).

Mitchell, J. M.,Jr., 1976. An overview of climatic variability and its causal mechanisms. Quartern. Res., 6, 481-493.

Neelin, J. D., 1988. Interannual oscillations in an ocean-GCM/simple-atmosphere coupled model. Phil. Trans. Roy. Soc. Lond., A309, to appear.

Nicolis, C., 1984a. Self-oscillations, external forcings, and climate predictability. In Berger et al. (1984), pp. 637-652.

Nicolis, C. and G. Nicolis, 1984, Is there a climatic attractor? Nature, 311, 529-532.

North, G. R., 1975a. Analytical solution to a simple climate model with diffusive heat transport. J. Atmos. Sci., 32, 1301-1307.

North, G. R., R. F. Cahalan and J. A. Coakley, Jr., 1981. Energy balance climate models. Rev. Geophys. Space Phys., 19, 91-121.

Oerlemans, J., 1982. Glacial cycles and ice sheet modelling. Climatic Change, 4, 353-374.

Paterson, W. S. B., 1972. Laurentide ice sheet: estimated volumes during Lake Wiusconsin. Rev. Geophys. Space Phys., 10, 885-917.

Peltier, R., and W. Hyde, 1984. A model of the ice age cycle. In Berger et al. (1984), pp. 565-580.

Pestiaux, P., 1984. Approache Spectrale en Modélisation Paléoclimatique. Ph.D. Thesis, Univ. Catholique de Louvain, Louvain-la-Neuve, Belgium.

Pestiaux, P., I. van der Mersch, A. Berger and J. C. Duplessy, 1988. Paleo-climatic variability at frequencies ranging from 1 cycle per 10000 years to 1 cyle per 1000 years: Evidence for nonlinear behaviour of the climate system, Climatic Change, 12, 9-37.

Phillips, N. A., 1956. The general circulation of the atmosphere: a numerical experiment. Quart. J. Roy. Meteorol. Soc., 82, 123-164.

Pike, E. R., J. G. McWhirter, M. Bertero and C. de Mol, 1984. Generalized information theory for inverse problems in signal processing, IEEE Proc., 131, 660-667.

Pollard, D., 1984. Some ice-age aspects of a calving ice-sheet model. In Perger et al. (1984), pp. 541-564.

Ruddiman, W. F., and A. McIntyre, 1981. Oceanic mechanisms for amplification of the 23,000-year ice-volume cycle. Science, 212, 617-627.

Saltzman, B., 1985. Paleoclimatic modeling, in Paleoclimate Analysis and Modeling, A. D. Hecht (ed.), J. Wiley, New York, pp. 341-396.

Saltzman, B., and A. Sutera, 1984. A model of the internal feedback system involved in late Quaternary climatic variations. J. Atmos. Sci., 41, 736-745.

Saltzman, B., A. R. Hansen and K. A. Maasch, 1984. The late Quaternary glaciations as the response of a three-component feedback system to earth-orbital forcing, J. Atmos. Sci., 41, 3380-3389.

Saltzman, B., and A. Sutera, 1987. The mid-Quaternary climatic transition as the free response of a three-variable dynamical model. J. Atmos. Sci., 44, 236-241.

Schneider, S. H. and S. L. Thompson, 1979. Ice ages and orbital variations: Some simple theory and modeling. Quat. Res., 12, 1888-203.

Sellers, W. D., 1969. A climate model based on the energy balance of the earth-atmosphere system. J. Appl. Meteorol., 8, 392-400.

Sergin, V. Ya., 1979. Numerical modeling of the glacier-ocean-atmosphere global system. J. Geophys. Res., 84, 3197-3204.

Shackleton, N. J., 1977. The oxygen isotope stratigraphic record of the late Pleistocene. Philos. Trans. R. Soc. Lond. B280, 169.

Shackleton, N. J., and N. D. Opdyke, 1973. Oxygen isotope and paleomagnetic stratigraphy of equatorial Pacific core V28-238: oxygen isotope temperatures and ice volumes on a 10^5 and 10^6 year scale. Quatern. Res., 3, 39-55.

Shackleton, N. J., and N. D. Opdyke, 1976. Oxygen isotope and paleomagnetic stratigraphy of Pacific core V28-239 Late Pliocene to Latest Pleistocene. Geol. Soc. Amer. Memoir, 145, 449-463.

Suarez, M. J., and P. S. Schopf, 1988. A delayed action oscillator for ENSO. J. Atmos. Sci., sub judice.

Témam, R., 1988. Infinite-Dimensional Dynamical Systems in Mechanics and Physics. Springer-Verlag, New York, 500 pp.

U.S. Committee for GARP, 1975. Understanding Climatic Change, A Program for Action. National Academy of Sciences, Washington, D.C., 239pp.

Vautard, R. and M. Ghil, 1988. Singular spectrum analysis in nonlinear dynamics, with applications to paleoclimatic time series. Physica D., in press.

Washington, W. M., and C. L. Parkinson, 1986. An Introduction to Three-Dimensional Climate Modeling. Oxford Univ. Press, Oxford/New York, 422p.

Weertman, J., 1976. Milankovitch solar radiation variations and ice age ice sheet sizes. Nature, 261, 17-20.

Wetherald, R. T., and S. Manabe, 1975. The effect of changing the solar constant on the climate of a general circulation model. J. Atmos. Sci., 32, 2044-2059.

Winfree, A. T., 1980. The Geometry of Biological Time. Springer-Verlag, New York, 530 pp.

Winograd, I. J., T. B. Coplen, B. J. Szabo and A. C. Riggs, 1988. A 250,000-year climatic record from Great Basin vein calcite - major implications for the Milankovitch theory. Science, to appear.

CLIMATE PREDICTABILITY : A DYNAMICAL VIEW

C. Nicolis[1] and C.L. Keppenne[2]

[1]Institut d'Aéronomie Spatiale de Belgique, Avenue Circulaire 3
B-1180 Bruxelles, Belgium
[2]Institut des Sciences Naturelles Appliquées, Place Croix du Sud 3,
Boîte 7, B-1348 Louvain-la-Neuve, Belgium

ABSTRACT. Evidence is reported that the dynamics involved in the 500 mb geopotential record is associated to deterministic chaos. The fractal dimension of the attractor and the average time beyond which predictions on individual trajectories break down are estimated. The possibility of carrying out predictions beyond this limit on the basis of a statistical approach is discussed.

1. INTRODUCTION

One of the basic reasons behind the well-known difficulty of long term predictions in the evolution of the atmosphere and climate, is in the fact that the principal variables undergo complex dynamics, typical features of which are aperiodicity in time and irregular patterning in space. According to the common wisdom, this is however only a temporary drawback : thanks to the availability of computational facilities, it is believed that a detailed description will eventually be achieved, leading to a theoretically unlimited predictive capability.

Our purpose in this work is to present a rather different view. First, we shall report on some data analysis suggesting that the dynamics associated with low frequency atmospheric variability corresponds to deterministic chaos. As a consequence, the underlying system will turn out to be extremely sensitive to initial conditions, in the sense that initially nearby states diverge exponentially in the course of time (Guckenheimer and Holmes, 1983). Because of the inherent uncertainty related to the finite precision of measurements, this will entail that individual trajectories cannot be predicted beyond a certain interval of time (Lorenz, 1984). We shall show that this time is an intrinsic property of the system, and give an estimate for the particular set of atmospheric data we shall analyze. Finally, we shall suggest a possibility for carrying out predictions beyond this intrinsic limit, based on the idea that a description in terms of individual trajectories must be abandoned in favor of a statistical approach.

A. Berger et al. (eds.), Climate and Geo-Sciences, 241–251.

2. THE DATA SET

In previous work we have analyzed the oxygen isotope record of a deep sea core of the last million years using the above described point of view, and produced evidence of chaotic dynamics on this long time scale (Nicolis and Nicolis, 1984; 1985). Here we shall deal with variability on a shorter time scale, associated with the 500 mb geopotential record (Keppenne and Nicolis, 1988; see also Fraedrich, 1986; Essex et al., 1987). The particular data set we shall consider is the daily record over a period of 24 years starting from January 1st, 1961 of the following nine western European stations : Bordeaux, De Bilt, Lisbon, London, Marseille, Paris, Reykjavik, Rome and Stockholm. Fig. 1 depicts a representative time series drawn from the data at the Marseille station.

We want to identify the salient features of the dynamical system associated with these data, independent of any modelling. To this end, we embed the evolution in phase space, the space spanned by a set of linearly independent variables taking the dominant part in the dynamics. We do not know offhand the number of these variables, but a result established in 1981 by Takens stipulates that they can be picked up from the initial time series and the hierarchy of variables obtained from it by applying successively higher shifts τ, 2τ, ... etc.

$$
\begin{aligned}
X_1(t_i) &: X_o(t_1), \ldots\ldots\ldots\ldots, X_0(t_N) \\
X_2(t_i) &: X_o(t_1+\tau), \ldots\ldots\ldots\ldots, X_o(t_N+\tau) \\
X_n(t_i) &: X_o(t_1+(n-1)\tau), \ldots\ldots, X_o(t_N+(n-1)\tau)
\end{aligned}
\tag{1}
$$

An alternative procedure is to combine Takens' method and the familiar empirical orthogonal function analysis. EOF's are just the eigenvectors of the covariance matrix i.e. the matrix of quadratic averages. They therefore describe variables that are statistically independent up to third order correlations. Using Takens' reconstruction, the covariance matrix takes the form :

$$
\Phi_{ij} = \frac{1}{N_T}\sum_{k=1}^{N_T} X_o(t_k + i\tau)\, X_o(t_k + j\tau) \; ; \; i,j = 0,1,\ldots,n-1 \tag{2}
$$

where N_T is the number of data points.

Let $\lambda_1 > \lambda_2 > \ldots \lambda_n$ be the eigenvalues of the covariance matrix and $c_1, \ldots, c_n$ the corresponding eigenvectors. The space spanned by these latter is referred to as singular space (Broomhead and King, 1986). Embedding our data into this space amounts therefore to switching from the state vector $\underset{\sim}{X} = \{X_1, \ldots, X_n\}$ to a state vector $\underset{\sim}{Y} = \underset{\sim}{X}.\underset{\approx}{c}$, where $\underset{\approx}{c}$ is an nxn matrix whose columns are the eigenvectors c_i.

3. THE WEATHER ATTRACTOR

It is now clear that in the space spanned by the time lagged variables or by the EOFs the time evolution given by our data set will

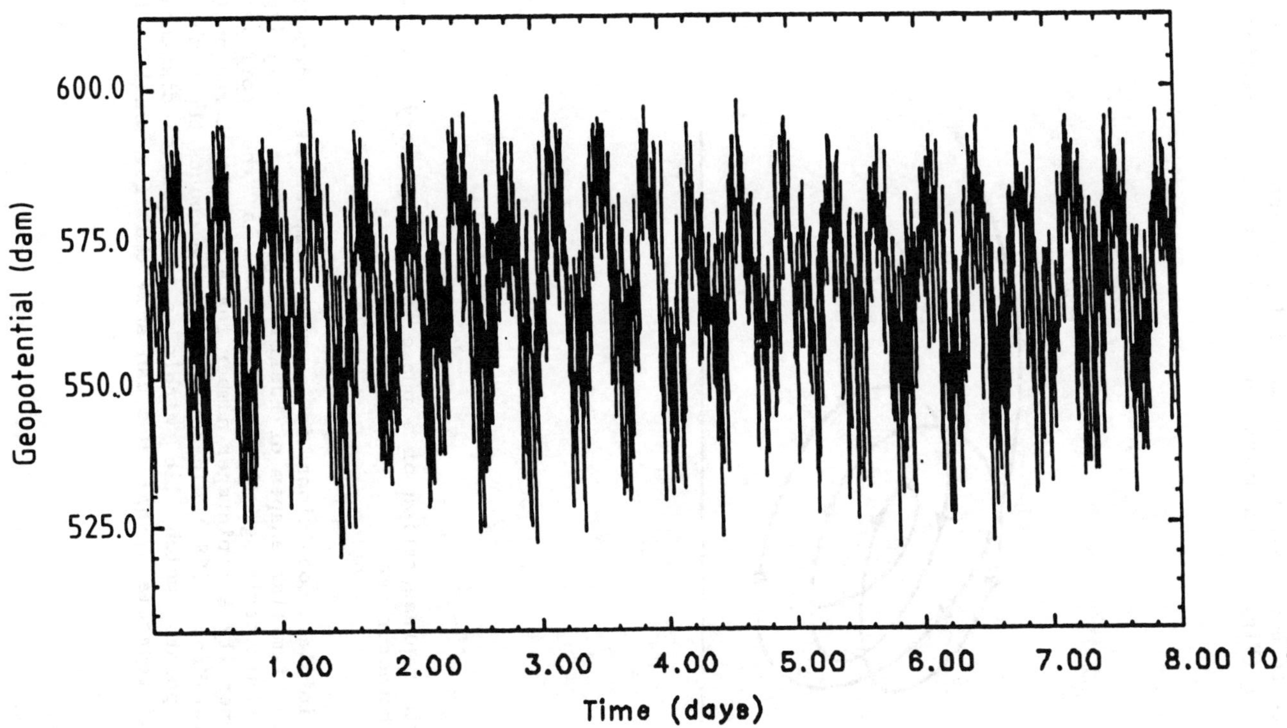

Fig. 1. Time evolution of the 500 mb geopotential height at Marseille.

lie on a curve, the phase space trajectory. Sooner or later this trajectory will converge to a certain set (or manifold), which will be referred to as the attractor (Fig. 2). An attractor describing chaotic motion with

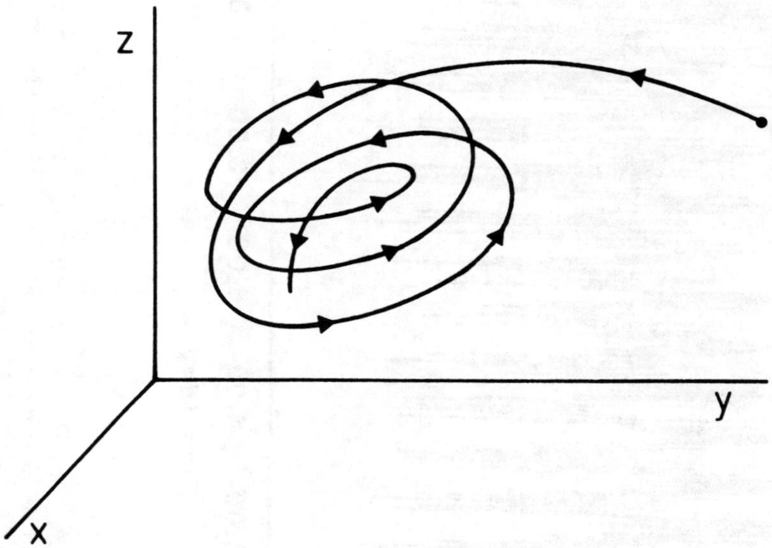

Fig. 2. Schematic representation of a phase space trajectory converging to an attractor set.

sensitivity to initial conditions is a very complex set, generally different from the familiar shapes of Euclidean geometry. A very important quantity characterizing this difference is the correlation dimension, ν, generally a noninteger number for a chaotic attractor (see for instance Mayer-Kress, 1986). To determine ν one counts the number of pairs of data points which lie within a prescribed distance r (Grassberger and Procaccia, 1983). One can show that for small r, this number scales as

$$C(r) \sim r^{\nu} \tag{3}$$

Obviously for ν = 1, 2, ... one obtains the familiar Euclidean manifolds (lines, surfaces, etc), but for ν non-integer and greater than 2 one deals with a fractal object.

These considerations suggest the following algorithm :

(i) Choose increasingly large values of embedding dimension n, and for each n plot the values of $\underset{\sim}{X} = (X_1, \ldots, X_n)$ for the N_T data points (cf. eq. (1)).

(ii) For each n, determine the correlation dimension, ν of the above plotted data set.

(iii) Study the dependence of ν for increasing n (Fig. 3). If this dependence saturates to some value ν_s beyond a certain reasonably small n_s, we will conclude that our system is a deterministic dynamical system. As for n_s, it will represent the minimum number of variables needed to describe the dynamics.

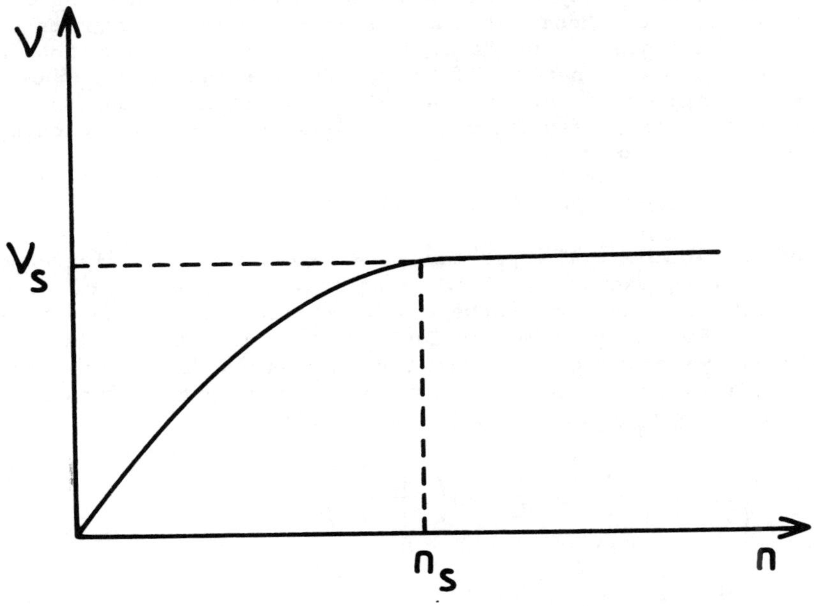

Fig. 3. Dependence of dimensionality, ν on the number of phase space variables n for a deterministic dynamical system.

Table 1 gives the numbers obtained from five of our data sets. We see that the two methods of phase space reconstruction give comparable results for each time series as well as for computations involving the space average of all signals. All attractors analyzed have a correlation

TABLE I : Correlation dimension.

Signal	Method of delays	Singular phase space
Lisbon	7.6	7.4
Marseille	6.8	6.7
Reykjavik	7.2	7.4
Roma	8.3	8.2
Stockholm	7.8	8.1
Space averaged	7.7	8.0

dimension in a narrow range with a mean value of ~ 7.5 and a dispersion of 10%. Consequently it is reasonable to ascertain that the individual time series refer to a well defined dynamical system describing the short term variability of the western European climate. Since the error bar associated to these calculations is not easy to estimate despite the relative abundance of the data, let us, however, be cautious and forget about the fractal nature of these numbers until we show that our dynamical system displays the most important property of fractal attractors, namely, sensitivity to initial conditions. We carry out this analysis in the following section.

4. LYAPOUNOV EXPONENTS AND PREDICTABILITY

Let us formulate the problem of sensitivity to initial conditions in a quantitative manner. We imagine at time t = 0 a set of data included in a small n-dimensional sphere, whose center is on the attractor (see Fig. 4). The long time evolution of this sphere is subsequently monitored. We order the principal axes of this object from most rapidly to least rapidly growing and compute the mean growth rate σ_i of the ith principal axis p_i over a long period of time :

$$\begin{aligned} \sigma_i &= \lim_{t\to\infty} \frac{1}{t} \int_o^t d\tau \frac{d}{d\tau} \ell n\left(\frac{p_i(\tau)}{p_i(0)}\right) \\ &= \lim_{t\to\infty} \frac{1}{t} \ell n\left(\frac{p_{i(t)}}{p_i(0)}\right) \end{aligned} \tag{4}$$

$p_i(0)$ being the radius of the initial sphere. The set of σ_i is referred to as <u>Lyapounov exponents</u> of the underlying dynamical system. It can be shown that there exist as many Lyapounov exponents as phase space dimensions (Guckenheimer and Holmes, 1983). One of them is necessary equal to zero, expressing the fact that the relative distance of

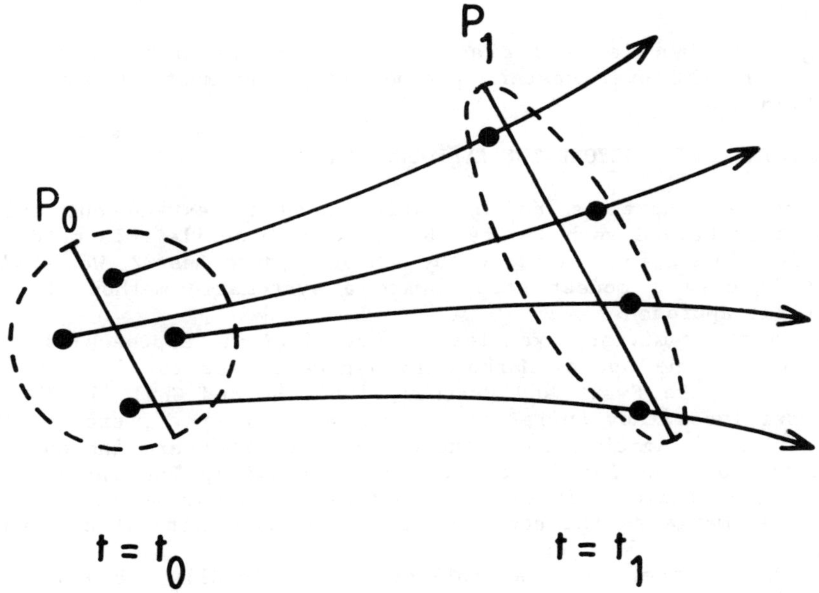

Fig. 4. Schematic representation of the exponential divergence of initially nearby states on a chaotic attractor.

initially close states on a given trajectory varies slower than exponentially. Others are negative, expressing the exponential approach of initial states to the attractor. If the dynamical system at hand is chaotic the sphere will evolve to a complex ellipsoidlike form reflecting the exponential divergence of nearby initial conditions along at least one direction on the attractor. We will conclude in this case that there is at least one positive Lyapounov exponent. Note that in a well-behaved dissipative system the sum of all exponents must be strictly negative (Guckenheimer and Holmes, 1983).

Our analysis is mainly based on the work of Eckmann *et al.* (1986). The algorithm developed by these authors allows the estimation of large magnitude exponents with a relatively good accuracy. Since we are interested in the predictability problem of chaotic attractors we will focus only on the largest positive σ_i's. We found two such positive Lyapounov exponents of comparable magnitude. Therefore one deals here with a *hyperchaotic attractor*. The sum of the positive Lyapounov exponents gives an estimate of the Kolmogorov entropy K, whose inverse is the mean predictability time of our variable (Schuster, 1984). As it turns out

$$K^{-1} \sim 3 \text{ weeks} \tag{5}$$

meaning that this is as far as we can go to predict the 500 mb geopotential height over western Europe using the most elaborate weather prediction model.

5. PREDICTABILITY BEYOND THE LYAPOUNOV TIME

Obviously, there is an imperative need to extend our prediction capabilities beyond such limits. But how to reconcile this need with the intrinsic limitations imposed by chaotic dynamics ? We shall here suggest briefly a modest step toward a systematic method, based on a statistical approach.

The most familiar examples of statistical approach are linear regression or the use of Markov chain models (see for instance, Spekat *et al*., 1983; De Swart and Grasman, 1987; Mo and Ghil, 1987). Despite their usefulness they suffer from a fundamental flaw : the inability to select in a systematic manner those variables that are indeed likely to give rise to a well-defined stochastic dynamics, for instance of the Markov type (Feller, 1968). In contrast, in our method statistical properties emerge as the consequence of the complexity of the underlying dynamics.

The basic idea is as follows. Operationally, because of the inherently limited precision of measurements, atmospheric or climatic states are defined in a non-local fashion (Lorenz, 1984). This "coarse-graining" or "lumping" maps the evolution based on the full set of the primitive equations into a complex sequence of transitions between cells C_1, C_2, ... in the phase space (see Fig. 5). Among the large number of ways to monitor the variables is there, then, one that leads to a clearcut stochastic process like for instance a first order Markov chain ? We have given a partial answer to this question by finding out conditions on the dynamics and on the partitioning of the state space, for which a Markov evolution is an exact image of the full dynamics described by the primitive equations (Nicolis and Nicolis, 1988).

Our method is based on the concept of *Markov partitions* (C_i) for which the boundaries between cells are preserved by the dynamics. We have identified some classes of dynamical systems, which include low order atmospheric circulation models, for which the probabilities $\{P_n(j)\}$, for being in cell C_j of such a partition at time n evolve according to the Markovian master equation (Nicolis, 1988)

$$P_n(j) = \sum_i W_{ij} P_{n-1}(i) \tag{6}$$

The *conditional probability matrix* W is a positive matrix, whose row sums are equal to unity. Its structure can be explicitly determined from the dynamical system, once a Markov partition has been determined.

Eq. (6) allows us to estimate the statistical state of the system at time n on the sole basis of the time series (assumed stationary) and of the initial data. Now, eq. (6) is a linear equation possessing a unique

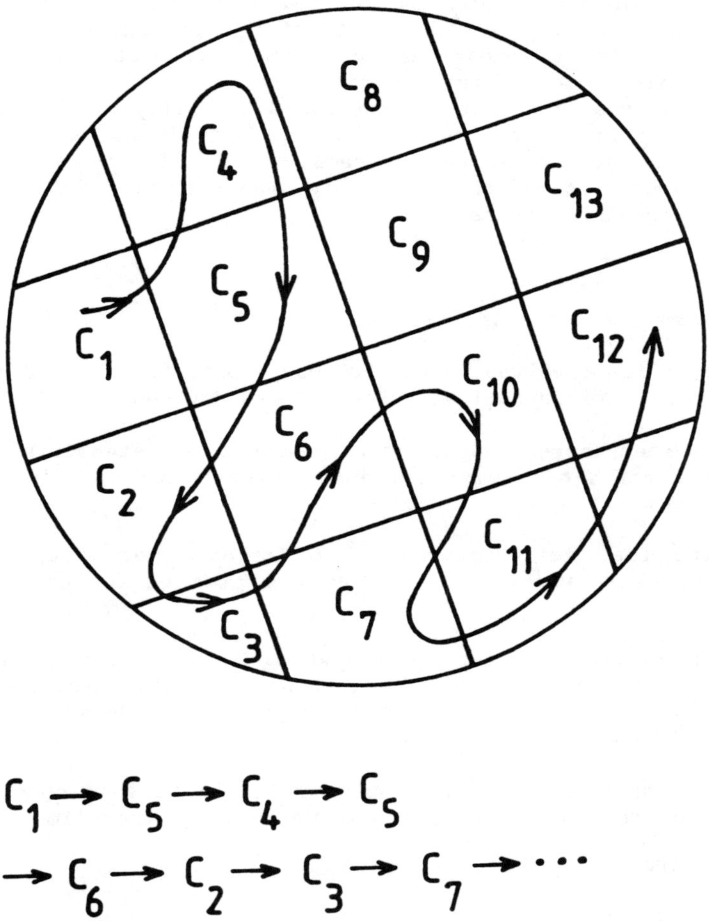

Fig. 5. Time evolution of a phase space trajectory viewed as a sequence of transitions between the cells of a "coarse-graining" partition.

stationary solution. It can be shown that starting from any initial condition, the system converges to this unique stationary solution as $n \to \infty$. Put differently, small changes in the (probabilistic) initial conditions will give rise to only slight differences in the time evolution of the variable of interest. This fact, to be contrasted with the sensitivity to initial conditions of the underlying dynamical system, guarantees the fiability of the statistical forecasting. Notice that sensitivity to initial conditions has been instrumental in allowing us to cast the dynamics in a Markovian form.

It should be realized that the increased fiability of the statistical description is obtained at the cost of a partial loss of information, arising from the coarseness of the cells C_i. In this view an "optimal" forecasting would therefore amount to reconciling, in a judicious manner, these opposing trends.

6. CONCLUSIONS

Let us summarize our main results.

(i) The dynamics involved in the 500 mb geopotential record stems from a deterministic dynamical system of few degrees of freedom.

(ii) This system shares two essential features of deterministic chaos : fractal attractor dimension and at least two positive Lyapounov exponents.

(iii) Despite the limited number of variables involved the system is intrinsically unpredictable beyond a time of the order of a few weeks.

(iv) It is possible to set up a systematic statistical description, which is an exact image of the dynamics. This description allows one to extend our predictive capabilities beyond the limits imposed by chaotic dynamics.

We believe that these approaches are likely to shed some new light into the basic problem of prediction in meteorology and climatology.

ACKNOWLEDGEMENTS

We are grateful to J.F. Royer of the Centre National des Recherches Météorologiques, Toulouse for providing us with the data. We thank M. Salazar for his help in the preparation of the manuscript. C. Keppenne is Research Assistant with the FNRS (Belgium). This work is supported, in part by the European Economic Community under contract number ST2J-0079-1-B(EDB).

REFERENCES

Broomhead, D.S. and King, G.P., 1986 : Extracting qualitative dynamics from experimental data. Physica, **20D**, 217-236.

De Swart, H.E. and Grasman, J., 1987 : Effect of stochastic perturbations on a low-order spectral model of the atmospheric circulation. Tellus, **39A**, 10-24.

Eckmann, J.P., Ollifson Kamphorst, S., Ruelle, D. and Ciliberto, S., 1986 : Lyapounov exponents from a time series. Phys. Rev., **34A**, 4971-4979.

Essex, C., Lookman, T. and Nerenberg, M.A.H., 1987 : The climate attractor over short time scales. Nature, **326**, 64-66.

Feller, W., 1968 : An introduction to probability theory and its applications. Wiley, New York.

Fraedrich, K., 1986 : Estimating the dimensions of weather and climate attractors, J. Atmos. Sci., **43**, 419-432.

Grassberger, P. and Procaccia, I., 1983 : Measuring the strangeness of strange attractors. Physica, **9D**, 189-208.

Guckenheimer, J. and Holmes, P., 1983 : Nonlinear oscillations, dynamical systems and bifurcations of vector fields. Springer-Verlag.

Keppenne, C.L. and Nicolis, C., 1988 : Global properties and local structure of the weather attractor over western Europe. J. Atmos. Sci., submitted.

Lorenz, E.N., 1984 : Some aspects of atmospheric predictability, in Problems and prospects of long and medium range weather forecasting. Barridge, D.M. and Källen, E., eds., Springer, 1-20.

Mayer-Kress, G., Ed., 1986 : Dimensions and entropies in chaotic systems. Springer-Verlag, Berlin.

Mo, K.C. and Ghil, M., 1987 : Statistics and Dynamics of persistent anomalies, J. Atmos. Sci., **44**, 877-901.

Nicolis, C. 1988 : Chaotic dynamics, Markov processes, and climate predictability. Tellus, submitted.

Nicolis, C. and Nicolis, G., 1984 : Is there a climatic attractor ? Nature, **311**, 529-532.

Nicolis, C. and Nicolis, G., 1986 : Reconstruction of the dynamics of the climate system from time-series data. Proc. Natl. Acad. Sci. USA, **83**, 536-540.

Nicolis, G. and Nicolis, C., 1988 : A master equation approach to deterministic chaos. Phys. Rev., **38A**, 427-433.

Schuster, H.G., 1984 : Deterministic chaos. Physik Verlag, Weinheim.

Spekat, A., Heller-Schulze, B. and Lutz, M., 1983 : Ueber Grosswetter und Markov-Ketten. Meteor. Rundsch., **36**, 243-248.

Takens, F., 1981 : Detecting strange attractors in turbulence, in Dynamical Systems and Turbulence. Springer-Verlag, 366-381.

Coupled Ocean and Sea-Ice Models: Review and Perspectives

J.-P. van Ypersele
Institut d'Astronomie et de Géophysique G. Lemaître,
Université Catholique de Louvain,
Chemin du Cyclotron 2,
B-1348 Louvain-la-Neuve,
Belgium

ABSTRACT. This article reviews the existing coupled models of ocean and sea ice used in climate studies, and describes the methods used to realize their coupling. The role of sea ice in the climate system is first described. Then a brief history of uncoupled sea-ice modelling is presented, following a hierarchical approach. The coupling of ocean and sea ice is divided into three areas: heat, momentum, and salt. Each area of coupling is reviewed in detail. Typical examples of coupled sea-ice/ocean models are discussed. An example application of a coupled model to the simulation of the Weddell Polynya is presented. It is shown that above-freezing sea-suface temperature in the polynya area is associated with intense convection, and that ice divergence helps to precondition the area for overturning. Perspectives on possible and needed progress in coupled ocean/sea-ice modelling are outlined.

1. Introduction: Sea Ice, Ocean, and Climate

Sea ice, produced by the freezing of a thin layer of polar sea-water (of the order of 3 m in the Arctic, 2 m in the Antarctic), affects the climate system in several ways. Through its high albedo (0.6 to 0.7 for bare ice, and up to 0.9 for snow-covered ice), sea ice reduces the proportion of solar energy absorbed at the surface; this effect participates in the famous ice-albedo positive feedback (Budyko, 1969). It also replaces a relatively warm ocean surface (above -2 °C) by a snow or ice surface well isolated from the ocean heat reservoir, and whose surface temperature can drop to -30 °C or less in winter. This means a reduction of outgoing surface infrared radiation from 300 to 200 W m^{-2}.

Sea ice represents an obstacle to air-sea exchanges of sensible heat, momentum, water vapour, carbon dioxide and other gases. For example, Maykut (1978) calculated that the total heat loss through 3 m of arctic ice in winter is only $\approx$ 1.5% of that from the open ocean. The air-ice drag coefficient is similar to the air-water drag over close, smooth sea ice (Banke *et al.*, 1980) but is up to 3 times larger than the latter in the marginal ice zone, where the ice is broken and dispersed by wave activity (Andreas *et al.*, 1984). Sea ice also replaces the wind stirring and wave breaking activity driving the ocean mixed layer by the largely inferior stirring by a slowly moving solid (Semtner, 1984a).

The various phenomena and effects mentioned above are complicated by the fact that sea ice is not an homogeneous medium: it can be deformed, or fractured in individual floes, which tend to act as rafts under the ocean and wind stresses. These movements can produce areas of ridging, with sea-ice thickness locally larger than 10 m, and can also create open water areas intertwined in the sea-ice cover. If these open water areas are longitudinal,

A. Berger et al. (eds.), Climate and Geo-Sciences, 253–277.

they are called "leads", otherwise they are called "polynyas". Polynyas can be as large as the Weddell Sea Polynya, which had an area of 300 000 square kilometres between 1974 and 1976 (Zwally *et al.*, 1983).

When it forms or melts, sea ice affects the upper ocean salt budget, because most of the salt contained in sea-water is expelled from the ice by the building of the ice crystals. Young sea ice has a typical salinity lower than 0.015, while older sea ice, in which excess salt has had more time to be drained or to migrate downward, has a typical salinity of 0.005 after one year, 0.002 after a few years (Weeks and Ackley, 1986). Sea ice formation thus increases the density of the under-ice water, which can induce convection and overturning, while sea ice melting has the effect of stabilizing the water column. It is interesting to remark that, on the seasonal time scale, the thermodynamic processes of sea ice act to replace the seasonal cycle of sea surface temperature with one of sea surface salinity (Semtner, 1984a). Brine rejection is thought to contribute to the maintenance of the cold upper halocline in the Arctic Ocean (Aagaard *et al.*, 1981), and to the formation of dense shelfwater in the Antarctic (Foster and Carmack, 1976), which can influence the deep ocean circulation (Rooth, 1982). These important effects of brine rejection and meltwater flux make the inclusion of the salt conservation equation in sea-ice/ocean coupled models essential. For a more comprehensive review of the multitude of physical processes associated with sea ice (see Figure 1), the reader is referred to the book edited by Untersteiner (1986a).

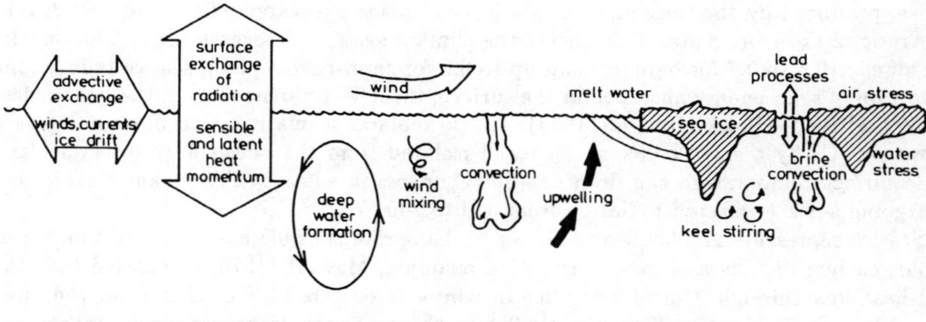

Figure 1: Processes near the ice margin. (From Lemke, 1986).

Sea ice is one of the climate system elements most sensitive to small changes in the surface energy balance. Untersteiner (1986b) mentions that the latent heat released when a one-meter thick ice layer is formed is an order of magnitude smaller than the mean annual total of either shortwave or longwave radiation at the surface. Since in a single year, these annual totals may easily deviate by ten percent from their average value, it is not

Table 1 Sea-ice area [10^6 km^2] (from Untersteiner, 1975).

Southern Ocean		Arctic Ocean	
Minimum	Maximum	Minimum	Maximum
2.5	20.0	8.4	15

surprising that small changes in the climatic forcing are accompanied by large changes in sea-ice extent. Table 1 gives estimates of the range of sea-ice extent for both hemispheres.

The fact that modelling studies of the CO_2-effect have shown the maximum impact to occur in the polar regions as a result of seasonal reductions in sea-ice coverage (see for example, Manabe and Stouffer, 1980) emphasizes the importance of properly representing sea ice in climate models.

The introduction of sea ice in climate models is progressively taking into account the various processes affecting it. Logically, the first approach is to model the thermodynamic aspects, which govern the creation and melting of sea ice. The second step is to incorporate the dynamic processes, which can redistribute the sea ice produced by the thermodynamic model. The two steps above are usually done in uncoupled models, where the sea-ice forcing, in particular the ocean heat and salt fluxes, is imposed without allowing feedbacks. The ultimate sea-ice model now appears to be one in which sea ice is fully coupled to oceanic and atmospheric models. In the following sections, a hierarchy of existing models is reviewed, some results of a coupled model application to the Weddell Polynya area are presented, and perspectives for future work are outlined.

2. Uncoupled modelling

Before discussing the coupling of models of different subsystems, it is important to understand as much as possible the behaviour of individual components. In particular, it is important to know the sensitivity of the main variables of each subsystem to changes in forcing by other components. A short review of sea-ice models follows.

Sea ice is mostly characterized by its thickness, extent, concentration (fraction of ice-covered water), and velocity. Existing sea-ice models can be broadly categorized in two ways: according to their spatial dimension (one: thickness or latitude; two: thickness + latitude, or latitude + longitude; three: thickness + latitude + longitude), or to the physical processes (thermodynamic or dynamic) they include. Without reviewing the complete history of sea-ice modelling, I would like to mention a few important steps.

2.1 Thermodynamic models

The simplest thermodynamic sea-ice "model" is probably that made by the empirical freezing days rule (Anderson, 1961), which correlates sea-ice thickness and the cumulative number of freezing-degree days, without attempting to use fundamental physical principles. Maykut (1986) shows that this correlation has a physical basis, but contains implicit assumptions about the oceanic heat flux and the snow thickness, which makes it inapplicable out of the area for which it was derived.

An important step was made with the Maykut and Untersteiner (1971) model, which is a very high resolution (40 layers) one-dimensional thermodynamic model which predicts the time evolution of thickness for an ice slab subjected to climatological atmospheric energy fluxes, constant ocean heat flux, and snowfall. Internal heating due to penetrating solar radiation is taken into account. This model essentially solves the heat diffusion equation for a medium whose heat conductivity and specific heat depend on temperature and salinity, and balances the upper and lower surface heat budget with the latent heat of freezing or melting.

Semtner (1976) simplified the Maykut and Untersteiner model, reduced the number of layers to a maximum of three, changed the algorithm, and produced a model which was more suitable to climate studies. In this model, Semtner used a simple formulation for the absorption of penetrating solar energy and the retarding effect it has on sea-ice melting and cooling. A fixed-depth, purely thermodynamic mixed-layer model replaces the ice model when no ice is present, so that the onset of freezing can be computed. To a certain extent, this mixed-layer model could be viewed as coupled to the ice model, but the two models do not co-exist in time, so this is not a real coupling. Variants of Semtner's thermodynamics are included in most of today's sea-ice models.

Figure 2 compares the seasonal cycle of ice thickness simulated for standard arctic conditions with the models of Maykut and Untersteiner (1971) and Semtner (1976). One can see that the three-layer version of Semtner's model gives a better reproduction of the reference seasonal cycle computed by Maykut and Untersteiner than the "zero-layer" version, which neglects internal storage of sensible and latent heat.

It is possible to make a brute force extension of a thermodynamic model to two or even three dimensions by assuming that no lateral interactions take place, and by simply applying the model to a series of gridpoints. Pease (1975) followed this approach for the Antarctic with a one-dimensional thermodynamic model applied to a series of points along the 155°E meridian, and was able to compute the advance and retreat of ice. The maximum and minimum ice extent were not accurately predicted. This failure is most likely related to the lack of representation of leads and ice dynamics. Indeed, assuming that no lateral interactions take place is not satisfactory because sea ice is not a continuum: it breaks with waves and floe-to-floe interactions, forming leads, and as soon as it has sides in contact with water, lateral heat exchanges tend to increase or decrease the ice concentration. An important positive feedback is present here in summer: solar energy heats the mixed-layer water in the ice-free area, and this accelerates the sea-ice lateral melting.

To improve the Semtner model extension to the horizontal, it was thus necessary to include leads parameterization, which was done by Washington *et al.* (1976) for the zero-layer version. Its principle, originally suggested by Semtner (1976) is as follows. An additional variable, representing the amount of open water in leads, is carried at each gridpoint, and the heat balance equation is applied to this fraction of open water. On account of variability in mechanical forcing by the wind, or long-term divergence, a minimum value of 2%

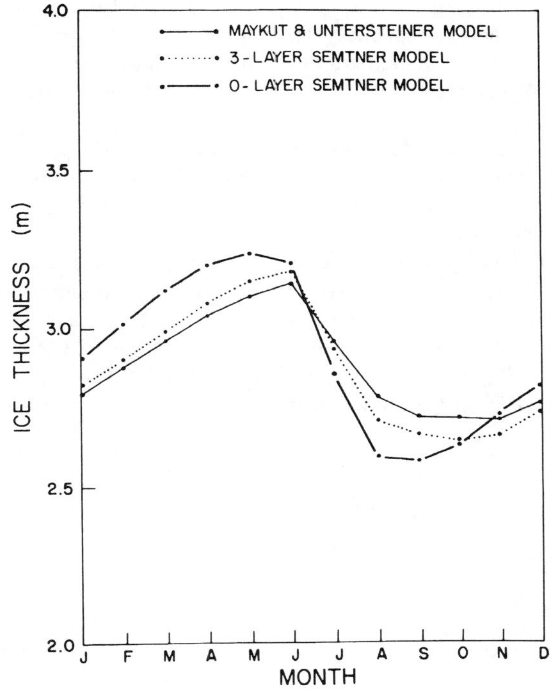

Figure 2: Seasonal cycle of ice thickness simulated for standard arctic conditions with the models of Maykut and Untersteiner (1971) and Semtner (1976). (From Washington and Parkinson, 1986).

is assumed to exist for the fraction of open water. Any surplus of heat absorbed in leads during spring or summer is used for lateral melting of ice at its existing thickness, and tends to increase the fraction of open water. On the other hand, new thin ice is formed in the lead when the heat balance is negative. So as not to require the recording of two species of ice—thin and thick—a new average thickness is computed and the ice concentration is increased to the maximum allowable value. With this parameterization, open areas enlarge rather slowly but freeze over rapidly. The study by Washington *et al.* showed the improvement in the antarctic seasonal sea-ice cycle produced by the introduction of leads.

Parkinson and Washington (1979) built up on the Washington *et al.* (1976) model, and used a more sophisticated lead parameterization in which for a positive lead heat balance, the larger the lead area, the larger the energy used directly to warm the lead rather than in melting the ice laterally. For a negative heat balance, the ice fraction is increased, but the ice thickness remains unchanged. Should this result in an ice concentration tendency to exceed an assumed maximum value (99.5% for any grid square in the Arctic and 98% in the Antarctic), the remaining energy deficit is offset by a cooling of the water under the ice, possibly leading to bottom ice accretion. This parameterization might be more realistic than the Semtner scheme, especially for low ice concentrations. More elaborate leads parameterizations are used in dynamic models discussed in the next section. A revised version of Parkinson and Washington (1979) model is described in Crowley and Parkinson (1989).

It should be stressed that those thermodynamic models need appropriate forcing at their upper and lower surfaces. The atmospheric forcing must be derived from observed or modelled variables such as radiation, air temperature, humidity, wind speed, and snowfall (see, e.g., Parkinson and Washington, 1979).

The ocean heat flux must be specified, as long as no coupling takes place with the ocean. This is very unfortunate because 1) there are virtually no measurements of this flux, and 2) ice thickness is very sensitive to it. Experiments made for arctic conditions by Maykut and Untersteiner (1971) and by Semtner (1976) show that changing the oceanic heat flux from 2 to 0 or 4 W m^{-2} multiplies the average ice thickness by 2 or 0.5, respectively (see Figure 3). A similar sensitivity exists for the Antarctic (Demuth and van Ypersele, 1987).

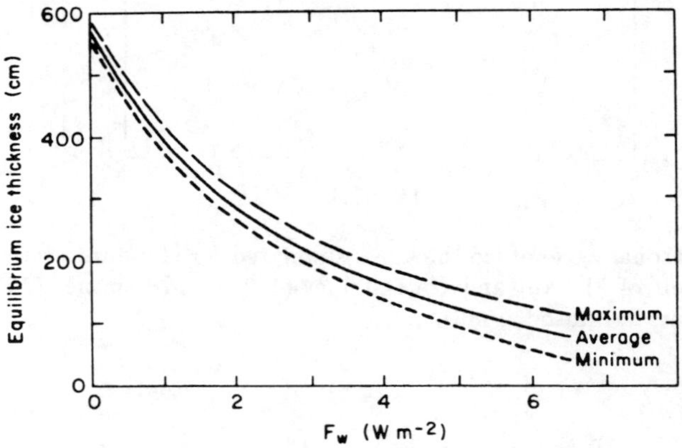

Figure 3: Sea-ice thickness sensitivity to vertical ocean-to-ice heat flux at equilibrium, for arctic conditions. (From Maykut and Untersteiner (1971) redrawn by Maykut (1986)).

This high sensitivity, combined with the fact that this flux is poorly known, gives one of the reasons why coupled sea-ice/ocean models are needed: one of their advantages is that they can compute this heat flux.

2.2 Dynamic models

Sea ice is not only subject to heat fluxes: it receives momentum from both the atmosphere and the ocean, and moves in response. To complicate the matter, ice is neither rigid nor a nice compressible material, and ice floes also hit each other, deform, and break. This is the domain of ice dynamics. Ice dynamic processes have an important role in the climate system. They govern the ice motion, which transmits mechanical energy into the ice-covered ocean. Ice motion towards the equator also represents a non-negligible poleward heat transport (Semtner, 1984). Dynamic processes control the ice concentration, which strongly influences the heat exchange between ocean and atmosphere. They can also export newly-formed ice continuously from a certain area, which causes a net annual average salt influx in that area. This can strongly affect the upper ocean dynamics, and even the deep ocean circulation. An example of the effect of interaction between ice dynamics, ocean dynamics, and the ice thermodynamics is discussed in section 4.3.

When Ekman (1905) extracted from Nansen's observations of ice movement in the Arctic a law relating wind and ice velocity, he actually provided the foundation of a first sea-ice dynamic model. Nansen observed that ice drifted at about 2 percent of wind speed, 30 degrees to the right of the surface wind direction in the northern hemisphere. This is also known as the Zubov Law, which can equivalently express the ice velocity as being parallel to the geostrophic wind, with a speed proportional to the pressure gradient (see, e.g., Wadhams, 1986). This simple relation can actually be justified by neglecting the internal ice stress in the full ice dynamics equations (Thorndike and Colony, 1982).

If we consider ice as a continuum, we can write an ice momentum equation, where ice acceleration (measured in a reference frame attached to the earth) is due to the combined effect of wind stress at the top surface, water stress at the bottom, a force due to the tilt of the water surface, the Coriolis force, and divergence of the internal ice stress (see, e.g., Washington and Parkinson (1986, p. 152–159)).

The complete momentum equation cannot be solved easily, mostly because the relation between internal ice stress and deformation, called the *constitutive law,* is not well known. Since large scale behaviour of sea ice cannot be observed in a laboratory, field studies are necessary.

The Arctic Ice Dynamics Joint Experiment (AIDJEX), which took place between 1970 and 1977 (Pritchard, 1980), provided measurements of the different terms entering the momentum equation, and led to the development of the AIDJEX model, which is to ice dynamics modelling what Maykut and Untersteiner (1971) is to ice thermodynamics modelling.

The AIDJEX model (Coon, 1980) consists mainly of a complete momentum balance equation, an *elastic-plastic* constitutive law which relates the internal ice stress to deformation, and an ice thickness distribution law, which relates the dynamic and the thermodynamic terms that change the ice thickness. In essence, this model assumes that ice resists compression as an elastic material until stress causes ridging, from which point the ice compresses as a plastic material, characterized by a stress independent of deformation rate. The goal of this model was to simulate the thermodynamics and dynamics of sea ice on space scales of 100 km and time scales of 1 day in interior parts of the pack. The AIDJEX model was not appropriate for low ice concentration areas.

Hibler (1979) implemented a model based on the AIDJEX work, in which he proposed a *viscous-plastic* constitutive law, considering the ice as a linear viscous fluid for small deformation rates and as a rigid plastic for high deformation rates. In general, a *viscous* rheology is defined as one in which stress can only be sustained through non-recoverable dissipation of energy by deformation, while a special case is the ***linear viscous*** rheology, where the stress depends linearly on the strain rates (Hibler, 1986). An important advantage of this viscous-plastic formulation is that it is easier to compute numerically (implicit method, Eulerian formulation) than the elastic-plastic one. Hibler (1986, 1988) documents the need for such a nonlinear rheology, and reviews the studies in which this approach has been used.

The Hibler (1979) model is still too heavy to be used routinely in climate studies, in particular because it requires an iterative solution, and Hibler proposed around 1985 a simplified rheology, called ***bulk viscous*** below, appropriate for climate time scales, when the forcing varies slowly (Washington and Parkinson, 1986, p. 157–158; Hibler, 1988). It implies the use of a nonlinear bulk viscosity having a null value when ice diverges, whereby convergence is resisted in a plastic fashion but divergence is allowed to occur without resistance. No shear viscosity exists in this rheology. Hibler (1988) reports, on the basis of a test in the Arctic, that this rheology treats at least the first-order effects of ice dynamics.

Semtner (1987) used such formulation in a coupled ocean/sea-ice model of the Arctic, and found results similar to those obtained with the full dynamics of Hibler. Semtner used an explicit timestepping method instead of the relaxation method used by Hibler, and this led to a threefold computer time reduction over the full viscous-plastic approach.

Another, very rough, approach to the modelling of sea-ice dynamics was suggested by Semtner (1984b). It consists in replacing the full momentum equation by a relationship giving the ice velocity as a function of ocean current and geostrophic wind. Thorndike and Colony (1982) analysed data from automatic data buoys deployed in 1979 in the Arctic Ocean, and found that a simple vector relationship could be used to compute the long-term (several months) average ice motion away from the coasts. The resulting ice velocity can then be used to compute the ice advection, convergence, and divergence, which affects the leads area. This parameterization was used by van Ypersele (1986) in a coupled three-dimensional ocean and sea-ice model of the Weddell Sea area, and by Demuth and van Ypersele (1989) in a coupled mixed-layer and sea-ice model of the same region.

3. Coupling ocean and sea ice

3.1 Introduction

Ideally, sea ice should just be seen as a particular state of sea-water, and sea-ice physics should be included in ocean models. In an ideal world, there should thus be no need to couple different models, but in practice, one often tries to build a coupled model from existing parts, which were not meant originally to work together. This may prove difficult, and the following sections attempt to explain some of the intricacies involved.

Coupling ocean and sea ice involves ocean model changes that might be categorised in two types:

- type I changes correspond to the *passive* role of ice in the system, seen as an obstacle to the atmospheric forcing of the ocean and affecting the vertical heat, salt, and momentum exchanges;

- type II changes are related to the *active* role of ice; it is limited to the salinity forcing due to freezing and melting.

Three important principles underlying an ideal coupling are that 1) priority should be given to interactive coupling of *sensitive* variables (determined by separate sensitivity analyses of ice and ocean models); 2) physically-based parameterizations of the interactions should be used whenever possible; and 3) conservation laws for mass, energy, and momentum should be respected by the physical and numerical set-up.

Adherence to those principles does not guarantee that the results obtained from coupled models will immediately be better than those of uncoupled models. Indeed, the coupling introduces new degrees of freedom, which may lead to new feedbacks and instabilities. Replacing certain variables that were *specified* in the uncoupled models by *computed* variables can also show that these are influenced by other processes than those included in the coupling scheme.

Sea ice and ocean must be coupled in three areas: heat, salt, and momentum. It is important to keep the coupled model flexible, so that the coupling in each of these three areas can be switched on and off independently, and that the original submodels can even be run in a completely uncoupled mode.

Most modifications for coupling are typically implemented with logical tests: if (ice present) then effect=X else effect=Y. If the ice-covered fraction A of each grid cell is computed in the ice model, it is necessary to weight by $(1 - A)$ and A, respectively, the effects of ice-free and ice-covered areas.

Each of the three domains of coupling is discussed below.

3.2 HEAT

Ocean atmospheric forcing by heat fluxes must be modified when sea ice is present. Below sea ice, the only part remaining from the atmospheric forcing is a small fraction of the solar energy (Grenfell and Maykut, 1977). If we neglect the solar term (for a discussion of the importance of this term in certain cases, see Fichefet and Gaspar, 1988), the only way the ice-covered ocean can gain or lose energy at its upper surface is by latent heat of freezing or melting.

First attempts at computing the oceanic heat flux were based on the temperature difference between ice bottom and ocean at a certain depth (Pease, 1975). This approach requires a somewhat arbitrary proportionality constant. Hibler (1984) calculated the heat flux by setting it equal to 25% of any cooling of the mixed layer and/or of the ice growth.

The hypothesis of thermodynamic equilibrium leads to a more physical computation of this heat flux. If we assume that the ice is in thermodynamic equilibrium with the underlying water, the bottom surface of sea ice must have a temperature close to the freezing temperature of sea-water (≈ -1.9 °C). The sea-water density dependence on temperature, with the density maximum observed at the freezing temperature (when the salinity > 0.0247), is such that the whole mixed-layer temperature must be close to the water temperature below the ice. If advection or convection occurring in the ocean model tends to increase the heat content of this mixed layer, the only way to satisfy the fixed temperature constraint is to pass this heat to the ice, melting some of it. This method allows to compute the oceanic heat flux that is specified in uncoupled sea-ice studies; it was used by, e.g., Toole (1981), Hibler and Bryan (1987), and Semtner (1987).

Actually, the sea-water freezing temperature depends on salinity. Fichefet and Gaspar (1988) showed that this dependence produces another term in the oceanic heat flux: if salinity increases, the freezing temperature decreases and the mixed layer must give some of its excess heat to the sea ice. The importance of this term under real sea ice has not yet been assessed.

An additional problem is posed by the sea-ice movement relative to the ocean. When ice is advected in an area where no ice was present, it must melt proportionally to the difference between the sea-water temperature and the freezing temperature. This is also equivalent to an oceanic heat flux.

If the ice cover is not uniform, two additional remarks are in order. First, the above computations must be weighted by the fraction of ice-covered water, keeping energy conservation as a guide. Second, any energy absorbed in areas of open water entrapped in ice (leads, polynyas) may be used to melt ice at the bottom, or for lateral melting. The way this energy must be distributed between bottom and sides is not well known, and different approaches have been used (cfr the discussion in section 2.1).

3.3 Momentum

Coupling sea ice and ocean models means that wind stress over the ocean must be replaced by an ice-water stress when ice is present. The drag of a gas on the wavy ocean surface is replaced by the drag of a solid (the ice floe keel) on the water.

Air-ice drag coefficients are reviewed in Brown (1981), Overland (1985), and McBean (1986), while McPhee (1986) discusses water-ice drag coefficients. Ideally, ice-water drag should depend on the amount of ice ridging, but this kind of refinement is not yet included in coupled ice-ocean models.

As seen above, the ice movement is not simply in the direction of the wind, and the velocity of ice relative to water depends not only on the wind, but also on internal ice factors. When internal ice stress tends to oppose the ice motion, it reduces the turning angle between the ice motion and wind.

When a thermodynamic - dynamic model is used, one possible method to compute the surface forcing under the ice cover is to compute explicitly the force due to ice interaction and subtract it from the wind stress. The remainder is used as the stress term transferred into the ocean. The water stress term in the ice momentum equation is a quadratic function of the difference between ice velocity and oceanic surface current. For simplicity, the oceanic surface current can be approximated by the horizontal ocean velocity in the second level of the ocean model. This method is used by Hibler and Bryan (1987).

If the ice cover is not continuous, the full wind stress must be allowed in open water areas, and a weighted stress due to air and ice must be used to drive the ocean.

Another method is that used by van Ypersele (1986), who computed the ice velocity directly from the Thorndike and Colony (1982) parameterization (cfr section 2.2), and then use that velocity in an ice-water stress term. This approach gives good results in the Antarctic, where ice resistance is low, but would probably not work well for the Arctic.

For coupling with mixed-layer models, the kinetic energy input into the mixed layer due to wind stress and ice keel stirring is typically proportional to the third power of the ice velocity relative to the ocean below the frictional layer (Lemke and Manley, 1984; Lemke, 1987).

3.4 Salt

As for heat and momentum, ice represents an obstacle to freshwater exchange with the atmosphere, but it is also a source of salt or freshwater in its own right.

3.4.1. *Ice as an Obstacle:* Evaporation from the ocean to the atmosphere must be set to zero in the ice-covered fraction; this is equivalent to a local increase of freshwater influx. On the other hand, snowfall is intercepted by the ice, and the freshwater contained in it is not available to the ocean before it melts, months or years later. This temporary decrease in freshwater flux may introduce a phase shift in freshwater time distribution, or even a net decrease of local freshwater flux if snow-covered ice moves away from the area and is replaced by snow-free ice.

3.4.2. *Ice as an Agent:* Since most of the salts dissolved in sea-water are rejected at the time of sea-ice formation, there is a salt flux proportional to the rate of freezing, which is maximal for thin ice. In addition to that initial process, sea-ice slowly continues to lose salt with time (Untersteiner, 1968). Sea-ice melting produces the inverse effect, or a freshwater input. This process tends to destabilize the water column when ice forms, and to stabilize it at the time of melting.

The ice contribution to the surface salt flux can be computed by putting it equal to the growth rate of the ice multiplied by the difference between the water and ice average salinities (weighted by the ratio of ice over sea-water density). An ice salinity between 0.0025 and 0.010 was used by van Ypersele (1986), Lemke (1987), and Fichefet and Gaspar (1988). Hibler and Bryan (1987) assumed that no salt at all is retained when sea ice forms, which has the advantage of being globally conservative for salt.

To illustrate the complexity of ice-ocean interactions, let us examine some consequences of those effects. The salt flux feeds back into the ice thermodynamics, since the convection induced by the brine rejection can upwell warmer water susceptible to oppose the freezing process (negative feedback when freezing). On the other hand, the stabilization associated with melting tends to decrease the part of the oceanic heat flux due to deep water convection (negative feedback) but, if leads are present, the stratification helps the water to be heated quickly by the solar energy (positive feedback).

Integrated on an annual cycle, the net salt flux through the upper surface due to the freezing/melting process is equal to zero when ice is stationary. However, the horizontal ice movement can produce positive or negative net salt fluxes. For example, coastal areas subject to off-shore katabatic winds around the Antarctic can be regions of strongly positive net salt influx, since sea ice is continuously formed and evacuated before it can melt.

4. Examples of existing coupled models

Given the different kinds of ice models (thermodynamic, thermodynamic + dynamic) and of ocean models (mixed layer, general circulation), several variants are possible. This section will be limited to models coupled by heat, salt, and momentum applied to climatic studies. It must be mentioned that steps on the way of coupling were made by Pease (1975), Pollard *et al.* (1983), and Toole (1981). This brief review of coupled ocean/sea-ice models is by no means exhaustive, and the reader is also referred to the studies by Demuth and van Ypersele (1989), Foster (1972), Häkkinen (1986, 1987), Ikeda (1985, 1988), Killworth and Paldor (1985), Motoi *et al.* (1987), Overland and Pease (1988), Røed (1983), Røed and O'Brien (1983), Stigebrandt (1981), and Wilmott and Mysak (1989), some of which address mesoscale or local processes not yet included in climate-oriented models.

A first category of coupled models is made by the one-dimensional (vertical) models, sometimes applied repeatedly on a series of gridpoints along a meridian, and a second category is made by three-dimensional models including effects of horizontal movements of ice and water.

4.1 One dimensional coupled models

In the *first category* we find, for example, the models of Lemke (1987) and Fichefet and Gaspar (1988). Lemke (1987) couples a variable-depth mixed-layer/pycnocline model (Lemke and Manley, 1984) with a simplified version of Semtner's (1976) model without leads. The modelled mixed layer overlies a pycnocline where temperature and salinity vary exponentially to their fixed bottom boundary value (set at 3000 m). The oceanic heat flux to the ice is computed as an entrainment heat flux, whereby warm pycnocline water is entrained into the upper layer, using the hypothesis of thermodynamic equilibrium described above. Heat and salt budgets interact since salt fluxes are associated with freezing and melting, influenced themselves by the oceanic heat flux.

This model is applied to antarctic conditions, and the results show, as expected, that the winter mixed-layer depth is larger under the sea-ice cover than in the open ocean, because of the strong brine convection. In spring, however, the reverse is observed due to the ice melt freshwater flux. As shown in Figure 4, near the ice edge (64°S curve), the entrained oceanic heat flux is maximum at the time of initial freezing (around day 190), when the mixed layer deepens, and minimum at the time of melting, when the water column stabilizes. Closer to the Antarctic continent (70°S curve), there is a second maximum at the end of winter (day 270), at a time when the thermocline is rather sharp and upward heat diffusion (present in the model through an inferior limit to the thermocline thickness) is more important than entrainment.

Lemke experiments also showed that a fixed mixed layer and a constant oceanic heat flux cannot give a good description of the seasonal sea-ice and upper ocean temperature cycle, since the crucial variable for freeze-up is the heat content of the mixed layer, a function of both temperature and depth. Lemke compared simulations made with fixed and interactive mixed layer, and showed that because of the space and time dependence of the entrained oceanic heat flux and the variability in the upper ocean thermal inertia, both amplitude and phase of the seasonal cycle of sea ice were modified when an interactive

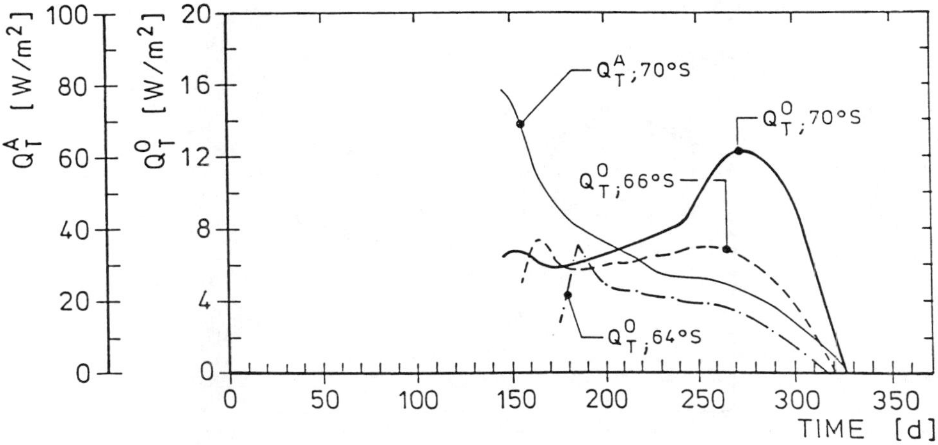

Figure 4: Entrainment oceanic heat flux (Q_T^O) at 64, 66, and 70°S, as modelled with a coupled sea-ice/mixed-layer model (Lemke, 1987).

mixed layer was used. In the fixed mixed-layer case, at 64°S, the freezing occurs about 1 month earlier and the melting occurs about 10 days earlier than in the prognostic mixed-layer case. In agreement with this longer freezing period, the fixed mixed-layer simulation shows a significantly larger sea-ice extent than the interactive simulation shows.

Fichefet and Gaspar (1988) coupled the mixed-layer model of Gaspar (1988) to the three-layer version of Semtner (1976) ice model. Leads are parameterized by using the mass conservation law to redistribute the new ice formed when the lead heat budget is negative, and lateral melting when the heat balance is positive. The oceanic heat flux receives particular attention, and the study shows that for the Arctic, terms usually neglected (solar energy input through the ice, fictive flux associated with freezing temperature changes) can be important. To compare these results with those of Lemke (1987), these differences in oceanic heat flux definition should be kept in mind.

Figure 5 shows the seasonal cycle of the different components of the oceanic heat flux at 84.5°N. The modelled entrainment and diffusion heat fluxes at that latitude closely follow each other, being positive in winter and spring and maximum in April, when the temperature gradient at the bottom of the mixed layer is greatest. The total oceanic heat flux vanishes in summer, mostly due to the negative heat flux (indicated by black dots in Figure 5) associated to the higher freezing temperature of less saline meltwater. At lower latitudes, below thinner summer ice, the solar irradiance penetrating into the mixed layer through the ice is found to produce a maximum oceanic heat flux in summer.

4.2 Three-dimensional coupled models

The first example of coupled models in the *second category* is that of Hibler and Bryan

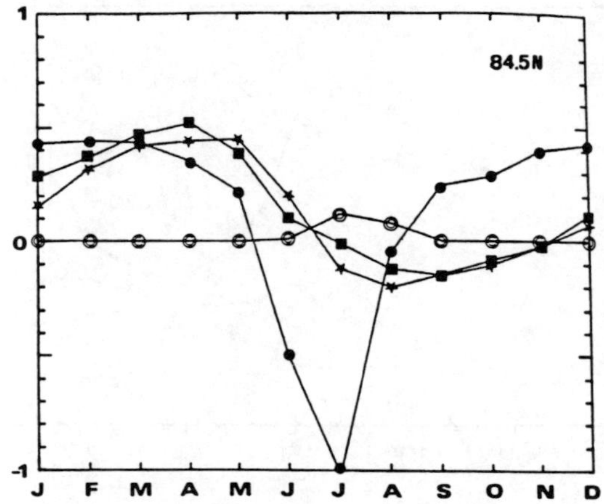

Figure 5: Seasonal evolution of the vertical ocean-to-ice heat flux components (W m^{-2}) at 84.5°N, as simulated by Fichefet and Gaspar (1988) with a coupled sea-ice/mixed-layer model. The black dots represent the heat flux associated with salinity-induced freezing-point changes, stars stand for the diffusive flux at the base of the mixed layer, squares indicate the entrainment heat flux, and circles represent the absorbed fraction of the solar flux transmitted by the ice. (From Fichefet and Gaspar, 1988).

(1987). They coupled the GFDL oceanic general circulation model (Bryan, 1969) to the thermodynamic-dynamic sea-ice model of Hibler (1979), and applied it to the Arctic, with a grid spacing of 160 km and up to 14 levels to resolve the bottom topography.

The ice model supplies the temperature, salinity and ocean-forcing stress (equal to the wind stress plus the internal ice stress) at the upper surface of the ocean, while the ocean model provides the horizontal water velocity needed in the water stress term, the horizontal gradient of pressure, and the vertical oceanic heat flux.

Observed temperature and salinity values were used to weakly force the climatological aspects of the ocean model so that its equilibrium time scale is similar to that of the ice model, following the "robust-diagnostic" method of Sarmiento and Bryan (1982). No diagnostic relaxation is used in the uppermost level of the ocean, which is included in the sea-ice model, and a 3-year time constant is used in the bulk of the ocean. The model is forced with monthly climatological temperatures and daily winds from the FGGE year (December 1978 to November 1979). No snow is allowed over the ice. The coupled model is integrated for 5 years.

Hibler and Bryan compared the coupled model results with those obtained from an ice-only model, which included only a motionless fixed-depth, 30 m mixed layer. They found that the most major difference between these two simulations is that the coupling improves significantly the simulation of winter ice margin in the Greenland and Norwegian

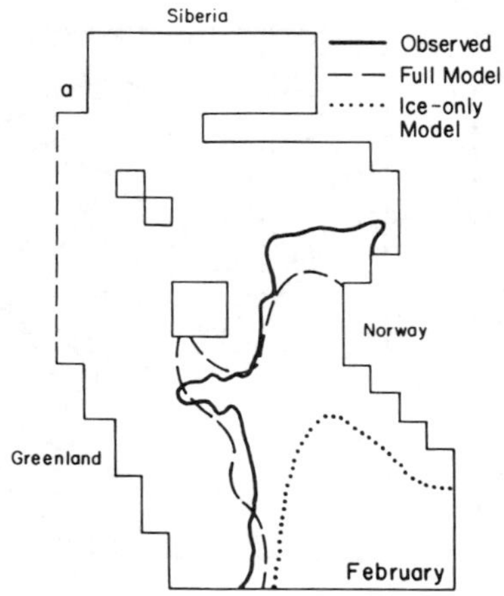

Figure 6: February Greenland and Norwegian Seas ice edge simulated by Hibler and Bryan (1987) with a coupled diagnostic ocean/sea-ice (dashed line), with the ice model and a fixed mixed layer (dotted line), and compared to observations (continuous line). (From Hibler and Bryan, 1987).

Seas (see Figure 6). Analysis of the ice edge advance and retreat of this model shows there is a sensitive balance between convection and a sealing of the upper ocean by large amounts of fresh water. For example, no ice forms in the ice-free portion of the Norwegian Sea due to overturning of the upper 600 m, a fact which effectively means that the whole upper ocean must be cooled before ice can form. Further in from the ice edge, oscillations tend to occur in the system, with large melting rates followed by less heat flux, which allows further freezing etc. A sensitivity experiment without lateral transport showed that these heat fluxes are strongly dependent on the lateral transport of heat in the upper ocean. In summary, the picture emerging from these experiments is that the arctic ice-edge position is largely determined by a competition between ice advection from the north and heat advection from the south, which ultimately reaches the ice mainly through vertical convection.

Semtner (1987) obtained similar results with a fully prognostic general circulation ocean model (Semtner, 1974, 1986), coupled to a three-layer snow-covered ice model including thermodynamics (Semtner, 1976) and the "bulk-viscous" dynamics described in section 2.2. The ocean model uses a grid spacing of 110 km and up to 13 vertical levels. The coupled model is integrated for 20 years. His results show that oceanic heat flux not only governs the winter sea-ice extent in the Greenland and Barents Seas, but also governs the minimum summertime extent in the Arctic Basin. Since no diagnostic forcing is used in this ocean model, Semtner's conclusions on the ocean heat flux role are even stronger than Hibler and Bryan's.

By replacing the ocean model by a motionless fixed-depth mixed layer, Semtner showed

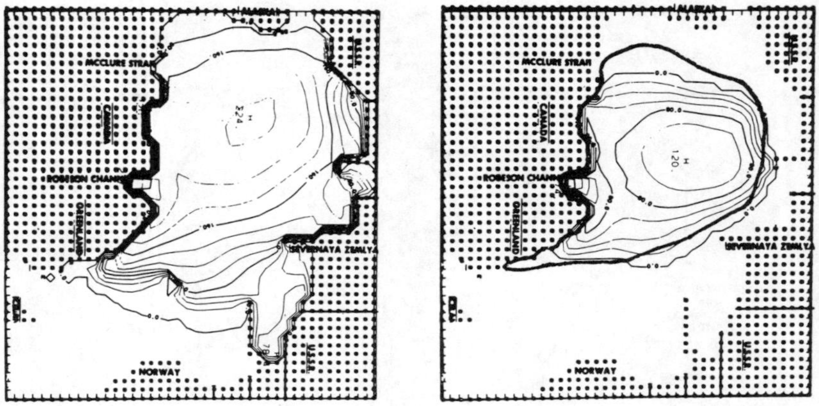

Figure 7: Upper: September arctic ice thickness simulated with a prognostic ocean/sea-ice model (Semtner, 1987). The thick line represents observations. Lower: September ice thickness when only a fixed-depth ocean mixed layer is used. The contour interval is 20 cm. (From Semtner, 1987).

that both Central Arctic ice thickness and compactness are reduced by the full ocean coupling (Figure 7). He also showed that surface convergence in the Central Arctic might contribute, through a compensating upwelling of above-freezing water around the margins, to a thinning of the existing winter ice cover that predisposes the ice to have an enlarged summer areal retreat relative to the case of the mixed layer only.

4.3 A coupled model study of the Weddell Polynya

In the antarctic domain, the Semtner (1974, 1986) oceanic general circulation model was

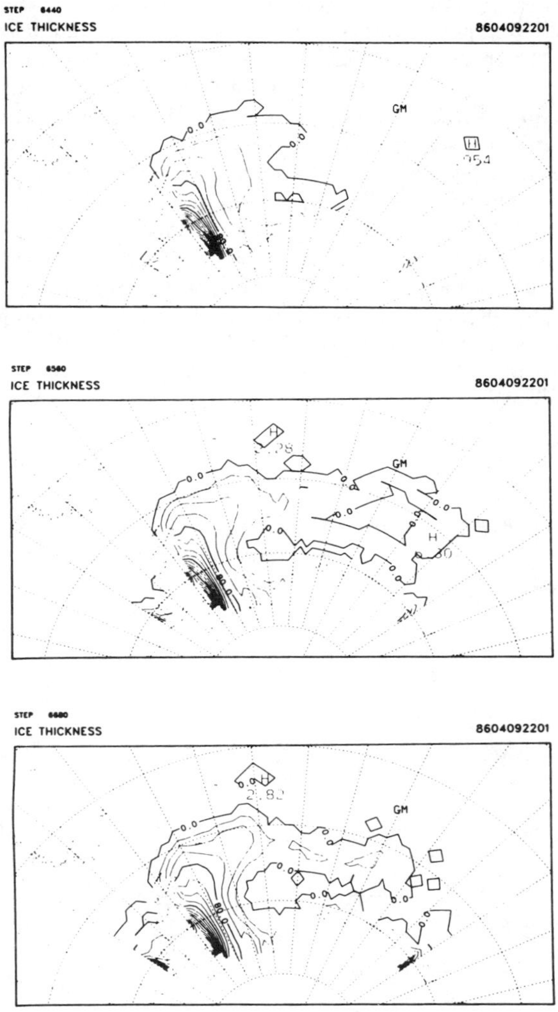

Figure 8: Formation of a Weddell Sea Polynya. Ice thickness contours simulated with a coupled ocean/sea-ice model (van Ypersele, 1986) for June, July, and August (top to bottom). The contour interval is 20 cm. The Greenwich meridian is marked by GM. (From van Ypersele, 1986).

coupled by van Ypersele (1986) with a three-layer thermodynamic snow-covered ice model (Semtner, 1976) including parameterized dynamics. The ocean model used has a resolution of 1° in latitude by 2° in longitude, and up to 12 levels in the vertical. A weak robust-diagnostic forcing (3–year time constant) is used as in Hibler and Bryan (1987), but only below the 500 m depth, while they applied it everywhere below the uppermost level. In this study, the upper levels stay fully prognostic.

The dynamic part of the ice model is even simpler than in Semtner (1987): it consists in computing the ice velocity as a simple function of wind and current velocity derived by Thorndike and Colony (1982), and in using this ice velocity to compute the ice-water stress and changes in ice concentration and thickness due to ice convergence or divergence.

This coupled model was applied to a sector of the Antarctic Ocean including the Drake Passage and the Weddell Sea, and forced by wind and fluxes computed from monthly climatological averages of atmospheric variables (Taljaard *et al.*, 1969). The coupled model is integrated for 5 years.

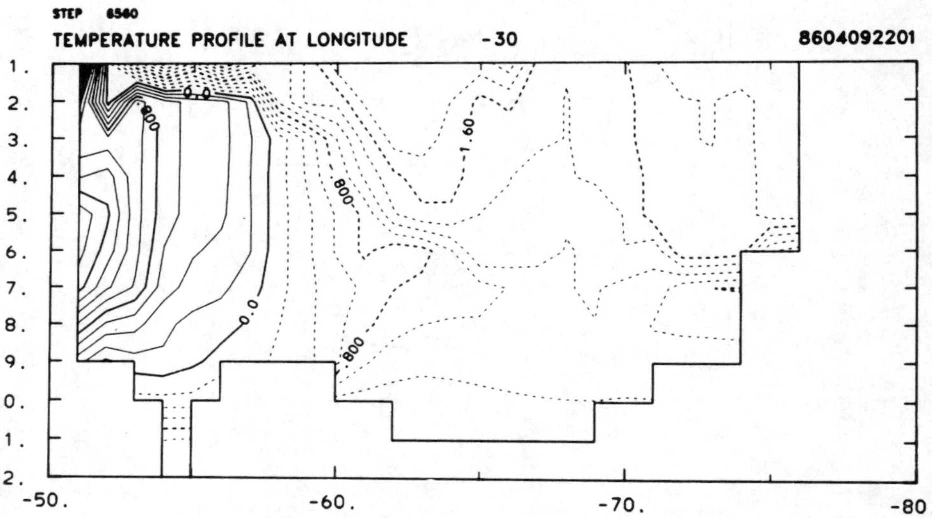

Figure 9: Simulated July potential temperature profile along the 30°W meridian. The X-axis indicates latitude, the Y-axis the model level number. The contour interval is 0.2 °C. (From the same experiment as in Figure 8.)

The coupling provided a cycle of sea-ice extent with correct minimum and maximum values but insufficient ice in part of winter. This is due to a large simulated polynya (Figure 8) in the area where the Weddell Sea Polynya has occasionally been observed. Since this polynya has been attributed by previous authors to several possible causes, involving both

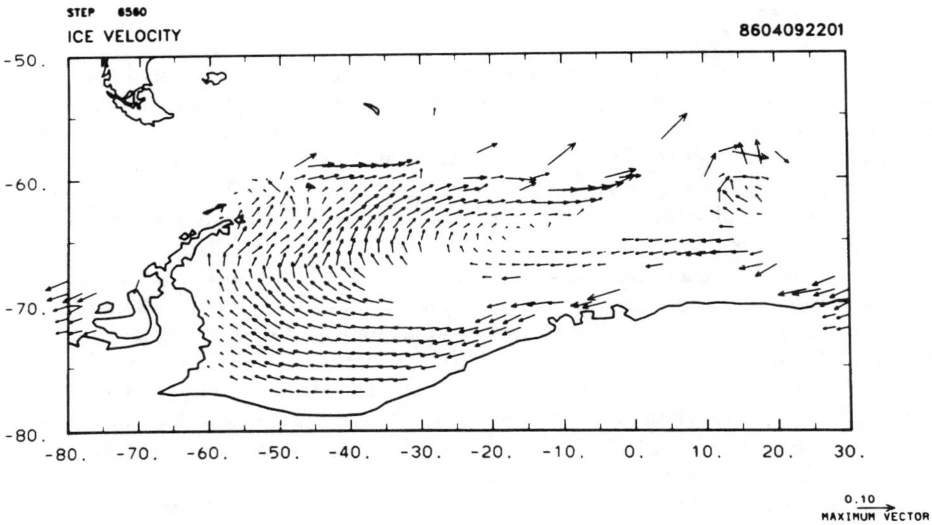

Figure 10: Simulated July ice velocity field. The scaling vector represents 0.1 m s^{-1}. (From van Ypersele, 1986).

atmospheric forcing (Parkinson, 1983) and oceanic convection (Martinson *et al.*, 1981), it is interesting to know that in this coupled experiment, above-freezing temperature is maintained through the winter in the polynya area. In July, the maximum sea surface temperature is −1.31 °C close the (30°W, 68° S) point.

Vertical profiles of temperature (Figure 9) and salinity along the 30°W meridian in July show that this high temperature is associated with intense convection. Another insight into the mechanism of this polynya is given by the ice velocity field, shown in Figure 10 for the same month. Ice divergence is observed away from the polynya area, and it certainly contributes to the reduction in ice concentration. It is likely that before the convective event has taken place, the ice divergence and the high freezing rates associated with ice formation in the central area of the divergence help to precondition the area for overturning. It must be noted that the fact that such a large polynya occurs in this coupled simulation with climatological forcing may indicate that the convective adjustment scheme classically used in three-dimensional models of the ocean could be inappropriate for this polar region. The possible polynya mechanism involving a combination of both ice divergence and high oceanic heat flux associated with overturning is also discussed by Lemke (1987) and Motoi *et al.* (1987).

5. Perspectives

This brief review has shown the multiple ways by which sea ice and ocean interact and the need for interactive coupling. Progress in coupled sea-ice/ocean modelling for climate studies can be made in several directions, some of which are outlined below. The reader is also referred to WCP (1984, 1987, 1989) for additional recommendations.

Very generally, better ocean models must be coupled with better sea-ice models (including thermodynamics and dynamics), and be forced with better air data or coupled with atmospheric models.

Due to the high sensitivity of sea-ice thickness to the oceanic heat flux, if ocean general circulation models are too slow to consider, at least mixed-layer models should be used to compute this flux and allow it to vary in space and time. Detailed mixed-layer studies will also be useful to derive practical heat flux parameterization to be used in other coupled studies, or may be in atmospheric general circulation modelling studies which tend to use oversimplified thermodynamic sea-ice models with a constant oceanic heat flux.

There is an urgent need to compare the model-generated oceanic heat flux to observations, both in different areas and seasons. The heat flux cannot be measured directly in the ocean, but can be derived in the presence of ice from the difference between heat conduction in the ice and the amount of heat of fusion associated with a change of ice thickness (McPhee and Untersteiner, 1982). This method has not yet been widely applied.

It should be noted that the heat and salt budgets interact very strongly, and that climate models which couple only ocean and sea ice by the heat flux, without salt coupling, are missing the very important feedbacks associated with brine rejection or meltwater flux (see section 3.4). Particular attention should also be given in future studies to the treatment of gravitational instabilities, as the convective adjustment scheme used in most ocean models might not be appropriate for certain polar areas.

In general, to compute better energy and mechanical forcing, the representation of both atmospheric and oceanic boundary layers should be improved, over both ice and leads (where strong energy exchanges take place). Specification of atmospheric variables such as wind, cloudiness, and precipitation also need improvement.

To assess the real performance of coupled models, they should be applied to interannual variability studies, which are more stringent tests than equilibrium studies (see Walsh *et al.*, 1985). The simulation of occasional features such as the Weddell Polynya at the proper times also constitutes an interesting test of coupled models. Interannual variability studies imply the need for better forcing data (atmospheric energy fluxes, wind) and accurate validation data in both hemispheres (ice extent, velocity, concentration, thickness distribution, and albedo) for periods of at least 10 years (Lemke *et al.*, 1980). This calls at least for routine, satellite-based, passive microwave measurements, combined with ice buoys observations in both hemispheres.

The ice dynamics, which should be included in every sea-ice simulation made for climatic purpose, needs to be better known, especially in the southern hemisphere and in the marginal ice zones. In that respect, drifting ice buoys programs, with measurements of (at least) atmospheric pressure, air temperature, and ice motion, should be expanded.

It should be stressed that improving the models also means looking for computationally efficient methods, adequate for climate studies. Additional knowledge should also be used, whenever possible, to streamline the description of physical processes involved. An example is given by ice dynamics: the bulk-viscous plastic approach in Semtner (1987), or the derivation of ice motion from wind and current in van Ypersele (1986), combined with leads parameterization, are more usable in long term climate integrations with coupled models than the full dynamics approach of Hibler (1979). However, their validity for different circumstances needs further assessment.

A final area in which coupled sea-ice/ocean models might prove useful is geochemistry. Indeed, for example, an overlooked effect of sea ice is that of a barrier to CO_2 exchanges at the air-sea interface. The inclusion of carbon cycle modelling in coupled ocean and sea-ice models could improve our understanding of global geochemical cycles.

6. Acknowledgements

Comments on an early version of this paper by Drs. Claude Demuth and Thierry Fichefet are gratefully acknowledged.

REFERENCES

Aagard, K., L.K. Coachman, and E.C. Carmack, 1981: On the halocline of the Arctic Ocean. *Deep-Sea Res.*, **28**, 529–545.

Anderson, D.L., 1961: Growth rate of sea ice. *J. Glaciol.*, **3**, 1170–1172.

Andreas, E.L., W.B. Tucker III, and S.F. Ackley, 1984: Atmospheric boundary-layer modification, drag coefficient, and surface heat flux in the antarctic marginal ice zone. *J. Geophys. Res.*, **89**, 649–661.

Banke, E.G., S.D. Smith, and R.J. Anderson, 1980: Drag coefficients at AIDJEX from sonic anemometer measurements. In *Sea Ice Processes and Models,* ed. R.S. Pritchard, University of Washington Press, Seattle, 430–442.

Brown, R.A., 1981: Modeling the geostrophic drag coefficient for AIDJEX. *J. Geophys. Res.*, **86**, 1989–1994.

Bryan, K., 1969: A numerical method for the study of the circulation of the World Ocean. *J. Comp. Phys.*, **4**, 347–376.

Budyko, M.I., 1969: The effect of solar radiation on the climate of the Earth. *Tellus,* **21**, 611-619.

Coon, M.D., 1980: A review of AIDJEX modeling. In *Sea Ice Processes and Models,* ed. R.S. Pritchard, University of Washington Press, Seattle, 12–27.

Crowley, T.J., and C.L. Parkinson, 1989: Late Pleistocene variations in antarctic sea ice I: effect of orbital isolation changes. *Climate Dynamics*, **3**, 85–91.

Demuth, C., and J.-P. van Ypersele de Strihou, 1987: Sea-ice simulations in the Weddell Sea. *Proceedings of the Belgian National Colloquium on Antarctic Research*, Prime

Minister's Office, Science Policy Office, Brussels, 195–212. (Available from the Science Policy Office, 8 Rue de la Science, B–1040 Brussels, Belgium).

Demuth, C., and J.-P. van Ypersele de Strihou, J.-P., 1989: Simulations of the annual sea-ice cover in the Weddell Sea. *Final report, Contract ANTAR/CVT/01, Belgian National Research Programme on the Antarctic*, Prime Minister's Office, Science Policy Office, Brussels, 47 pp.

Ekman, V.W., 1905: On the influence of the earth's rotation on ocean currents. *Ark. Mat. Astron. Fys.* **2**, 1–52.

Fichefet, T., and P. Gaspar, 1988: A model study of upper ocean–sea ice interaction. *J. Phys. Oceanogr.*, **18**, 181-195.

Foster, T.D., 1972: Haline convection in polynyas and leads. *J. Phys. Oceanogr.*, **2**, 462–469.

Foster, T.D., and E.C. Carmack, 1976: Frontal zone mixing and Antarctic Bottom Water formation in the southern Weddell Sea. *Deep-Sea Res.*, **23**, 301–317.

Gaspar, P., 1988: Modeling the seasonal cycle of the upper ocean. *J. Phys. Oceanogr.*, **18**, 161–180.

Grenfell, T.C., and G.A. Maykut, 1977: The optical properties of ice and snow in the Arctic Basin. *J. Glaciol.*, **18**, 445–463.

Häkkinen, S., 1986: Coupled ice-ocean dynamics in the Marginal Ice Zone: upwelling/downwelling and eddy generation. *J. Geophys. Res.*, **91**, 819–832.

Häkkinen, S., 1987: A coupled dynamic-thermodynamic model of an ice-ocean system in the Marginal Ice Zone. *J. Geophys. Res.*, **92**, 9469–9478.

Hibler, W.D., III, 1979: A dynamic thermodynamic sea ice model. *J. Phys. Oceanogr.*, **9**, 815–846.

Hibler, W.D., III, 1984: The role of sea ice dynamics in modeling CO_2 increases. In *Climate Processes and Climate Sensitivity*, ed. J.E. Hansen and T. Takahashi, Geophysical Monograph 29, American Geophysical Union, Washington, D.C., 238–253.

Hibler, W.D., III, 1986: Ice dynamics. In *The Geophysics of Sea Ice,* Proc. NATO Advanced Study Institute on Air–Sea–Ice Interactions, ed. N. Untersteiner, Plenum Press, New York, 577–640.

Hibler, W.D., III, 1988: Modelling sea ice thermodynamics and dynamics in climate studies. In *Physically-Based Modelling and Simulation of Climate and Climatic Change, Part 1,* ed. M.E. Schlesinger, Kluwer Academic Publ., Dordrecht, 509–563.

Hibler, W.D., III, and K. Bryan, 1987: A diagnostic ice-ocean model. *J. Phys. Oceanogr.*, **17**, 987–1015.

Ikeda, M. 1985: A coupled ice-ocean model of a wind-driven coastal flow. *J. Geophys. Res.*, **90**, 9119–9128.

Ikeda, M. 1988: A three-dimensional coupled ice-ocean model of coastal circulation. *J. Geophys. Res.*, **93**, 10731–10748.

Killworth, P.D. and Paldor N., 1985: A model of sea-ice front instabilities. *J. Geophys. Res.*, **90**, 883–888.

Lemke, P., 1986: Stochastic description of atmosphere – sea ice – ocean interaction. In *The Geophysics of Sea Ice,* Proc. NATO Advanced Study Institute on Air–Sea–Ice Interactions, ed. N. Untersteiner, Plenum Press, New York, 785–823.

Lemke, P., 1987: A coupled one-dimensional sea ice–ocean model. *J. Geophys. Res.*, **92**, 13164–13172.

Lemke, P., and T.O. Manley, 1984: The seasonal variation of the mixed layer and the pycnocline under polar ice. *J. Geophys. Res.*, **89**, 6494–6504.

Lemke, P., E.W. Trinkl, and K. Hasselmann, 1980: Stochastic dynamic analysis of polar sea ice variability. *J. Phys. Oceanogr.*, **10**, 2100–2120.

Manabe, S., and R.J. Stouffer, 1980: Sensitivity of a global climate model to an increase of CO_2 concentration in the atmosphere. *J. Geophys. Res.*, **85**, 5529–5554.

Martinson, D.G., P.D. Killworth, and A.L. Gordon, 1981: A convective model for the Weddell Polynya. *J. Phys. Oceanogr.*, **11**, 466–488.

Maykut, G.A., 1978: Energy exchange over young sea ice in the central Arctic. *J. Geophys. Res.*, **83**, 3646–3658.

Maykut, G.A., 1986: The surface heat and mass balance. In *The Geophysics of Sea Ice,* Proc. NATO Advanced Study Institute on Air–Sea–Ice Interactions, ed. N. Untersteiner, Plenum Press, New York, 395–463.

Maykut, G.A., and N. Untersteiner, 1971: Some results from a time dependent, thermodynamic model of sea ice. *J. Geophys. Res.*, **76**, 1550–1575.

McBean, G., 1986: The atmospheric boundary layer. In *The Geophysics of Sea Ice,* Proc. NATO Advanced Study Institute on Air–Sea–Ice Interactions, ed. N. Untersteiner, Plenum Press, New York, 283–337.

McPhee, M.G., 1986: The upper ocean. In *The Geophysics of Sea Ice,* Proc. NATO Advanced Study Institute on Air–Sea–Ice Interactions, ed. N. Untersteiner, Plenum Press, New York, 339–394.

McPhee, M.G., and N. Untersteiner, 1982: Using sea ice to measure vertical heat flux in the ocean. *J. Geophys. Res.*, **87**, 2071–2074.

Motoi T., N. Ono, and M. Wakatsuchi, 1987: A mechanism for the formation of the Weddell Polynya in 1974. *J. Phys. Oceanogr.*, **17**, 2241–2247.

Overland, J.E., 1985: Atmospheric boundary layer structure and drag coefficients over sea ice. *J. Geophys. Res.*, **90**, 9029–9049.

Overland, J.E., and C.H. Pease, 1988: Modeling ice dynamics of coastal seas. *J. Geophys. Res.*, **93**, 15619–15637.

Parkinson, C.L., 1983: On the development and cause of the Weddell Polynya in a sea ice simulation. *J. Phys. Oceanogr.*, **13**, 501–511.

Parkinson, C.L., and W.M. Washington, 1979: A large-scale numerical model of sea ice. *J. Geophys. Res.*, **84**, 311–337.

Pease, C.H., 1975: A model for the seasonal ablation and accretion of antarctic sea ice. *AIDJEX Bull.*, **29**, 151–172.

Pollard, D., M.L. Batteen, and Y.-J. Han, 1983: Development of a simple upper-ocean and sea-ice model. *J. Phys. Oceanogr.*, **13**, 754–768.

Pritchard, R.S. (ed), 1980: *Sea Ice Processes and Models,* University of Washington Press, Seattle, 474 pp.

Røed, L.P., 1983: Sensitivity studies with a coupled ice–ocean model of the Marginal Ice Zone. *J. Geophys. Res.*, **88**, 6039–6042.

Røed, L.P., and J.J. O'Brien, 1983: A coupled ice–ocean model of upwelling in the Marginal Ice Zone. *J. Geophys. Res.*, **88**, 2863–2872.

Rooth, C., 1982: Hydrology and ocean circulation. *Prog. Oceanog.,* **11**, 131–149.

Sarmiento, J.L., and K. Bryan, 1982: An ocean transport model for the North Atlantic. *J. Geophys. Res.*, **87**, 394–408.

Semtner, A.J., Jr., 1974: An oceanic general circulation model with bottom topography. *Numerical Simulation of Weather and Climate,* Tech. Rept. No. 9, Dept of Meteorology, Univ. of California, Los Angeles, 99 pp.

Semtner, A.J., Jr., 1976: A model for the thermodynamic growth of sea ice in numerical investigations of climate. *J. Phys. Oceanogr.*, **6**, 379–389.

Semtner, A.J., Jr., 1984a: Sensitivity of ocean circulation to sea-ice processes. In *Report of the meeting of experts on sea ice and climate modelling*, WCP–77, World Climate Research Programme, World Meteorological Organization, Geneva, Appendix D, 1–8.

Semtner, A.J., Jr., 1984b: On modelling the seasonal thermodynamic cycle of sea ice in studies of climatic change. *Climatic Change*, **6**, 27–37.

Semtner, A.J., Jr., 1986: Finite-difference formulation of a World Ocean model. In *Advanced Physical Oceanographic Numerical Modelling,* ed. J.J. O'Brien D. Reidel Publ. Co., Dordrecht, 187–202.

Semtner, A.J., Jr., 1987: A numerical study of sea ice and ocean circulation in the Arctic. *J. Phys. Oceanogr.*, **17**, 1077-1099.

Stigebrandt, A., 1981: A model for the thickness and salinity of the upper layer in the Arctic Ocean and the relationship between the ice thickness and some external parameters. *J. Phys. Oceanogr.*, **11**, 1407–1422.

Taljaard, J.J., H. van Loon, H.L. Crutcher, and R.L. Jenne, 1969: *Climate of the Upper Air, I, Southern Hemisphere. Vol. 1, Temperatures, Dew Points and Heights at Selected Pressure Levels,* NAVAIR 50-1C-55, U.S. Naval Weather Service, Washington, D.C., 135 pp.

Thorndike, A.S., and R. Colony, 1982: Sea ice motion in response to geostrophic winds. *J. Geophys. Res.*, **87**, 5845–5852.

Toole, J.M., 1981: Sea ice, winter convection and the temperature minimum layer in the Southern Ocean. *J. Geophys. Res.*, **86**, 8037–8047.

Untersteiner, N., 1968: Natural desalination and equilibrium salinity profile of perennial sea ice. *J. Geophys. Res.*, **73**, 1251–1257.

Untersteiner, N., 1975: Dynamics of sea ice and glaciers and their role in climatic variations. In *The Physical Basis of Climate and Climate Modelling,* GARP Publ. Series **16**, World Meteorological Organization, Geneva, 206–224.

Untersteiner, N., (ed.) 1986a: *The Geophysics of Sea Ice,* Proc. NATO Advanced Study Institute on Air–Sea–Ice Interactions, Plenum Press, New York, 1196 pp.

Untersteiner, N., 1986b: The geophysics of sea ice: overview. In *The Geophysics of Sea Ice,* Proc. NATO Advanced Study Institute on Air–Sea–Ice Interactions, ed. N. Untersteiner, Plenum Press, New York, 1–8.

van Ypersele, J.-P., 1986: *A numerical study of the response of the Southern Ocean and its sea ice to a CO_2-induced atmospheric warming.* Cooperative thesis NCAR CT-99, Université Catholique de Louvain and National Center for Atmospheric Research, Louvain-la-Neuve and Boulder, 135 pp. (NTIS Order No: PB87–163218/GAR).

Wadhams, P., 1986: The seasonal ice zone. In *The Geophysics of Sea Ice,* Proc. NATO Advanced Study Institute on Air–Sea–Ice Interactions, ed. N. Untersteiner, Plenum Press, New York, 825–991.

Walsh, J.E., W.D. Hibler III, and B. Ross, 1985: Numerical simulation of northern hemisphere sea ice variability, 1951–1980. *J. Geophys. Res.*, **90**, 4847–4865.

Washington, W.M., and C.L. Parkinson, 1986: *An Introduction to Three-Dimensional Climate Modeling.* University Science Books and Oxford University Press, Mill Valley and Oxford, 422 pp.

Washington, W.M., A.J. Semtner, C. Parkinson, and L. Morrison, 1976: On the development of a seasonal change sea-ice model. *J. Phys. Oceanogr.*, **6**, 679–685.

WCP, 1984: *Report of the meeting of experts on sea ice and climate modelling,* WCP–77, World Climate Research Programme, World Meteorological Organization, Geneva, 34 pp + 10 appendices.

WCP, 1987: *Report of the second session of the WCRP Working Group on Sea Ice and Climate,* WCP–128, World Climate Research Programme, World Meteorological Organization, Geneva.

WCP, 1989: *Report of the third session of the JSC Working Group on Sea Ice and Climate,* WMO/TD–272, World Climate Research Programme, World Meteorological Organization, Geneva, 23 pp + 11 annexes.

Weeks, W.F., and S.F. Ackley, 1986: The growth, structure, and properties of sea ice. In *The Geophysics of Sea Ice,* Proc. NATO Advanced Study Institute on Air–Sea–Ice Interactions, ed. N. Untersteiner, Plenum Press, New York, 9–164.

Wilmott, A.J., and L.A. Mysak, 1989: A single steady-state coupled ice–ocean model, with application to the Greenland-Norwegian Sea. *J. Phys. Oceanogr.*, **19**, 501–518.

Zwally, H.J., J.C. Comiso, C.L. Parkinson, W.J. Campbell, F.D. Carsey, and P. Gloersen, 1983: *Antarctic Sea-Ice Cover 1973–1976: Satellite Passive-Microwave Observations.* Special Publication SP-459, NASA, Washington, D.C., 206 pp.

ATMOSPHERIC CHEMISTRY-CLIMATE INTERACTIONS

Guy Brasseur
Atmospheric Chemistry Division
National Center for Atmospheric Research
P. O. Box 3000
Boulder, CO 80307-3000
USA

and Michel M. Verstraete
Office for Interdisciplinary Earth Studies
University Corporation for Atmospheric Research
P. O. Box 3000
Boulder, CO 80307-3000
USA

ABSTRACT. Atmospheric chemistry plays an important role in driving climate and its variability at all space and time scales. The chemical composition of the atmosphere has evolved since the formation of the planet 4.5 billion years ago, and is currently changing at an accelerated rate, primarily as a result of human activities. This paper presents a description of the most important chemical processes occurring in the troposphere and in the stratosphere, and discusses potential climatic impacts of changes in the chemical composition of the atmosphere. Interdisciplinary research will be needed to address the most crucial issues, because of the strong coupling between the atmosphere, the biosphere, the hydrosphere, the chemosphere and the technosphere.

1. Introduction

The chemical composition of the atmosphere has been—and will continue to be—a major controlling factor of the climate system. This results from the fact that atmospheric gases absorb radiation in specific spectral bands, and thereby affect the amount of solar radiation reaching the surface of the Earth or the amount of thermal infrared radiation exiting the atmosphere.

Repetitive measurements of atmospheric composition over the last few decades have documented rather large increases in many trace constituents, including carbon dioxide, methane, chlorofluorocarbons, as well as variations in other constituents, such as ozone in the stratosphere. Severe large scale and possibly long term consequences are expected from these chemical changes, including a general warming of the troposphere, cooling of

A. Berger et al. (eds.), Climate and Geo-Sciences, 279–302.

the stratosphere, increase in solar ultraviolet at the surface, etc. Since the bulk of these chemical changes is associated, directly or indirectly, with human activities, these causal relations are cause for concern because human population is increasing at a fast rate, and industrialization and deforestation are not expected to level off soon. Since societies are largely responsible for these changes, they should take the actions required to revert these processes or reduce potential impacts.

Human kind is conducting a giant chemical experiment in the atmosphere which should lead to substantial environmental and climatic perturbations on a global scale. This paper reviews some of the most important aspects of these questions, and provides some background on the intimate relation between climate and atmospheric chemistry.

2. Evolution of the chemical composition of the atmosphere

Planet Earth was formed about 4.5 billion years ago, from accretion of materials in what would become the Solar System. It is generally believed that this original planet lost its primary atmosphere and regained a secondary atmosphere through the progressive release of volatile materials included within the planet (Walker, 1977).

Multiple factors probably have influenced the chemical composition of the atmosphere as well as the climate of the Earth. Firstly, the mass of the planet was sufficiently small to allow the escape to space of light gases such as hydrogen, thereby increasing the oxidizing capacity of the atmosphere. Hence the photodissociation of water molecules resulted in the appearance of oxygen and ozone molecules, even before the biospheric production from photosynthesis started. It is also widely believed that the intensity of solar irradiance has gradually increased since the formation of the solar system. This must have produced substantial modifications of the atmospheric circulation, as well as of the chemical composition of the atmosphere and oceans. Long term decreases in volcanic activity also has had a significant influence on the chemical composition of the atmosphere and on the climate.

This secondary atmosphere must have been somewhat similar to the current atmospheres of Venus and Mars, with less than 1% of oxygen present, but large amounts of carbon dioxide and, to a lesser extent, nitrogen. By 3.5 billion years ago, large quantities of water must have been released into the original ocean basins, and the hydrological cycle must have been active. Since water contributes to the formation of sediments in the oceans, it has played an important role in the evolution of the carbon cycle and therefore on the amount of CO_2 present in the atmosphere. Life probably appeared at or even before that time, in a non-oxidizing atmosphere. The "invention" of photosynthesis by early bacteria resulted in the massive production of molecular oxygen and ozone, is responsible for the current relatively large (20%) concentration of O_2 in the Earth's atmosphere, and made aerobic life forms possible. In this way, life itself had an early and very significant role in determining atmospheric composition (Margulis, 1982). Furthermore, oceanic biota, by maintaining an imbalance in carbonated concentrations between the deep ocean and the surface layers, regulates the uptake of atmospheric carbon dioxide by ocean, which has a significant influence on the radiative balance at the Earth surface. Other greenhouse gases such as methane or nitrous oxide are also largely produced by the biosphere.

There is evidence that the total mass of the atmosphere did not change much once this secondary atmosphere was established, but also that its composition evolved in time. Space is missing here for even a cursory treatment of this chemical history, but it is worth

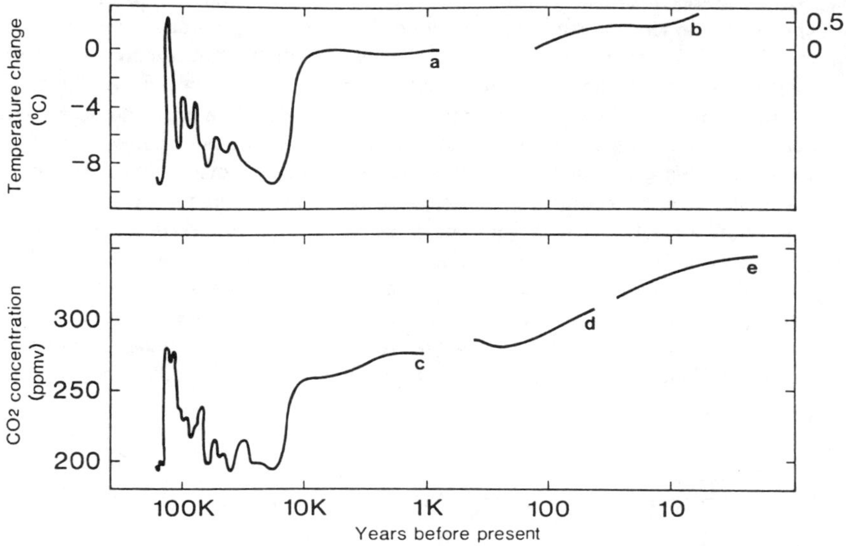

Figure 1: Temperature and CO_2 variations for the past 160,000 years. (a) Surface ice temperature variations in Antarctica (Barnola *et al.*, 1987); (b) Global annual mean surface temperature (Jones *et al.*, 1986); (c) CO_2 variations from ice-core analysis (Barnola *et al.*, 1987); (d) CO_2 variations from ice-core analysis (Pearman *et al.*, 1986); and (e) Modern Antarctic CO_2 observations (Bacastow and Keeling, 1981; Komhyr *et al.*, 1985). Reproduced from Pearman, 1988.

mentioning that the carbon dioxide concentration of the atmosphere probably evolved significantly even on short geological periods such as the glacial-interglacial time scale of the last 100,000 years (Fig. 1).

In the remainder of this paper, we will be concerned mostly with the current accelerated chemical changes taking place mainly as a result of human activities. In particular, we will review the emission of trace gases at the surface and their impact on the oxidizing capacity of the atmosphere, and discuss tropospheric and stratospheric chemical perturbations.

3. Current trace gas emissions

Because the most abundant constituents of the atmosphere (N_2 and O_2) are transparent in the infrared spectral region, other less abundant gases, such as the so-called greenhouse gases (CO_2, H_2O, O_3, CH_4, N_2O, the chlorofluorocarbons or CFCs) play an important role in the energy budget of the climate system. The distribution of these gases is also affected by solar radiation through photodecomposition or oxidation processes, as well as by atmospheric transport. In the presence of sunlight, these gases initiate a variety of chemical processes which illustrate the complex interactions between solar radiation, the atmosphere, the biosphere, the ocean, etc.

Because of the development in many parts of the world of industrial and agricultural activities, the release in the atmosphere of various chemical compounds is increasing rapidly. It is now well known for example that, primarily as a result of the burning of fossil fuels, the concentration of carbon dioxide, which was of the order of 285 ppmv in the beginning of the 19th century, is now (1989) approaching 350 ppmv. Beside a photodestruction mechanism, which takes place only above 80 km altitude, no chemical process destroys CO_2 in the atmosphere: this gas dissolves into the ocean, where it forms carbonate ions, or is incorporated in the biosphere through photosynthesis in the green plants.

In this section, we will focus on the long-lived gases (also referred to as source gases) which are produced at the surface and released in the atmosphere (essentially CO_2, CH_4, N_2O and the CFCs). The rates of emission of these gases on the global scale, or even for individual environments, is not well established. Several studies have provided information on local sources but, because of their great variability in time and space, a generalization to the entire world of these observations is not possible.

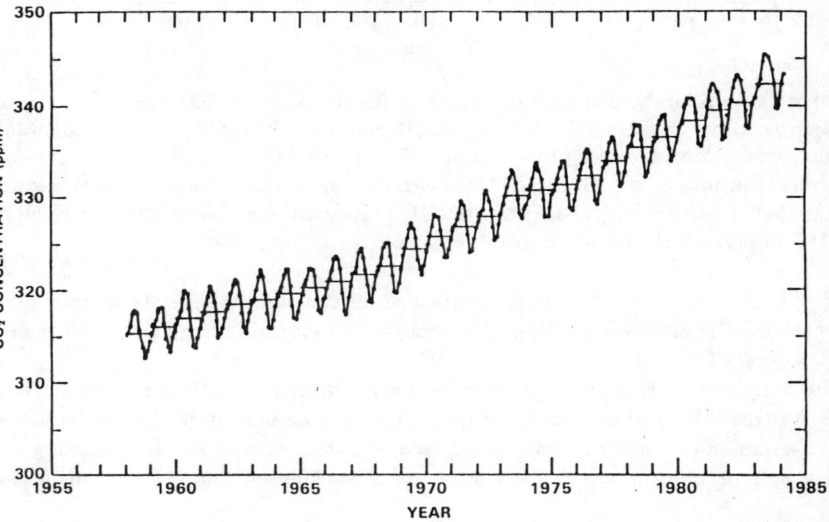

Figure 2: Mixing ratio of atmospheric CO_2 (ppmv) measured at Mauna Loa Observatory, Hawaii. The continuous trend associated with fossil fuel consumtion, as well as the seasonal variation related to biological activity are visible.

In the past, interest in the greenhouse effect has focused mainly on carbon dioxide (CO_2). Each year, nearly 5 Gt of carbon is released into the atmosphere primarily as a result of fossil fuel consumption. The rise in the atmospheric density of this gas (Fig. 2), could be the cause of the observed upward trend in the global mean temperature over the last 100 years (Fig. 3). Measurements made at different locations in the world, indicate that the CO_2 mixing ratio is currently increasing by about 0.4% per year. Analyses of air trapped in ice show that the CO_2 mixing ratio was of the order of 285 ppmv during the 17th and 18th centuries, and that temperature variations over the 160,000 years were accompanied by significant changes in carbon dioxide. CO_2 is expected to produce a significant climatic change in the next century since its atmospheric abundance is predicted to grow further in the future.

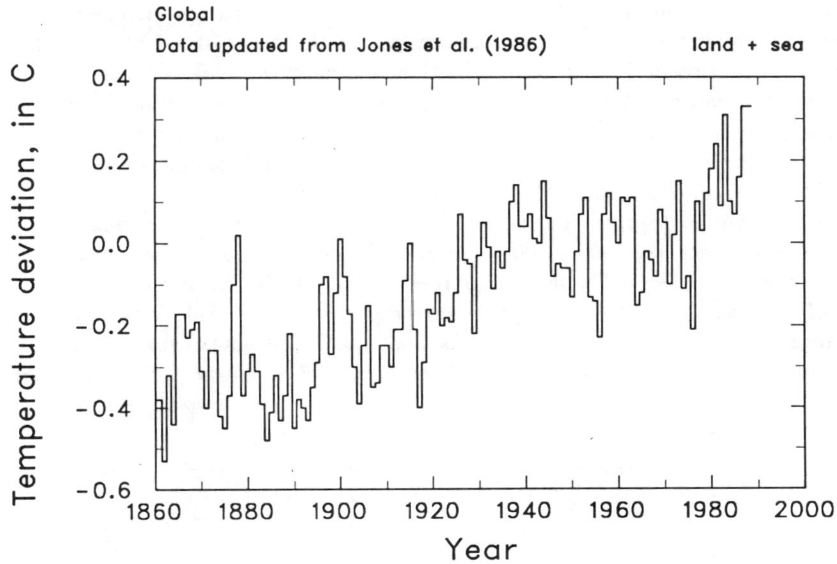

Figure 3: Global annual mean temperature evolution since 1860, adapted from Jones *et al.*, 1986), with updated data from Wigley (Personal communication, 1988).

The study of the global carbon cycle, however, requires an accurate determination of the storage and exchanges of carbon compounds between the different parts of the geosphere. For example, the yearly transfers between the atmosphere and the ocean and between the atmosphere and the biosphere are roughly 13 percent and 10 percent of the atmospheric CO_2 content, respectively. The CO_2 input due to fossil fuel consumption and deforestation represents only 0.8 percent and 0.15 percent of the CO_2 content. The capacity of the ocean to take up atmospheric CO_2 is limited by the slow mixing processes in the deep ocean and the buffering factor of the carbonate system. Consequently, the rate of increase in the concentration of total carbon in water at equilibrium, is an order of magnitude lower than the rate of increase of the CO_2 concentration in the atmosphere. Only the well-mixed upper layer (50–100 m) of the ocean effectively contributes to the uptake of CO_2 on time scales of decades. Therefore, the size of the oceanic reservoir available for sequestering carbon dioxide on these time scales is comparable to the effective size of the atmosphere, and about half of the carbon dioxide released by fossil fuel burning remains airborne. The anthropogenic perturbation, while small compared to the natural exchange fluxes, enhances by about 1.5 ppmv per year the global amount of carbon dioxide in the atmosphere.

It is important to note that carbon dioxide, a linear molecule with a relatively simple absorption spectrum, has a strong absorption band at 15 μm. Since this band is located in a spectral region in which the emission of the terrestrial environment is very intense, it significantly affects the energy budget of the atmosphere. Another intense band at 4.3 μm has a less important influence because both solar and terrestrial emissions are weak around this wavelength.

Methane (CH_4), the most abundant organic compound in the atmosphere, presently has an average mixing ratio at the Earth's surface of the order of 1.7 ppmv. This gas is produced essentially by biological sources (wetlands, swamps and lakes, rice paddies, tundra and the boreal marshes, enteric fermentation in cattle, termites and other insects, biomass and solid waste burning, natural gas losses and coal mining). About 85 percent of methane emitted at the surface is destroyed in the troposphere through oxidation by the hydroxyl radical (OH). Recent measurements have shown that methane concentration is globally increasing by about 0.018 ppmv per year (Fig. 4). Analyses of ice core samples shows that this concentration increase began 200 years ago and accelerated during the 20th century. This trend probably reflects changes in sources but could also be due to variations in the loss rate of this gas due to decreased OH density, perhaps as a result of increasing releases in the atmosphere of pollutants such as carbon monoxide. Methane concentrations were of the order of 1.15 ppmv in the middle of this century and as low as 0.35 ppmv during the ice age.

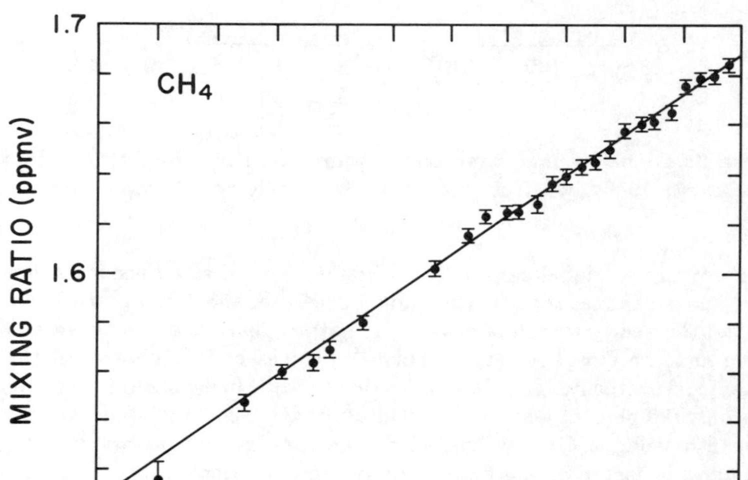

Figure 4: Trend in globally averaged atmospheric methane mixing ratio (ppmv) measured between January 1978 and September 1987 (From Blake and Rowland, 1988).

Methane has both direct and indirect effects on climate. It interacts directly with terrestrial infrared radiation, essentially through an absorption band at 7.66 μm and indirectly by forming carbon dioxide, water vapor and ozone through chemical oxidation. These constituents further contribute to the greenhouse effect of the atmosphere.

Although the dominant hydrocarbon in the atmosphere is methane, other organic species are released at the Earth's surface and are known to play an important role in the lower atmosphere, especially in polluted areas. Terpenes (such as isoprene) are naturally emitted by vegetation but other non-methane hydrocarbons, such as alkanes, alkenes, etc. are released in the atmosphere as a result of biomass burning, gasoline combustion, incineration, solvent and petroleum evaporation, natural gas leaks, refinery wastes, etc. Natural productions of these hydrocarbons include emissions by oceans, microbial activity in soils, as well as emission by foliage.

The importance of non-methane hydrocarbons has long been recognized for the chemistry of the urban atmosphere. However, even in some remote areas of the Northern Hemisphere, more carbon can be processed in atmospheric reactions from the oxidation of non-methane hydrocarbons than from the oxidation of methane. The non-methane hydrocarbons also generate additional products such as acetaldehyde, acetic acid, and peroxy acetylnitrate (PAN). All these organic species have strong absorption bands in the infrared.

Nitrous oxide (N_2O) is also produced at the Earth's surface by bacterial activity associated with complex nitrification and denitrification processes in soils. The present concentration is nearly 310 ppbv, with an increase of 0.8 ppbv per year. Natural sources of N_2O include microbiological processes on the continent and in the ocean; anthropogenic productions are associated with burning of biomass and fossil fuels, as well as extensive use of nitrogen fertilizers. The major sink of N_2O is its photodissociation in the stratosphere. Its growth is of the order of 0.25% per year. Figure 5 presents measurements of N_2O in Tasmania between 1978 and 1987. Analyses of ice cores show that the average concentration of this compound in the atmosphere was about 285 ppbv before the beginning of industrialization. This gas also contributes to the greenhouse effect of the atmosphere.

Sulfuric compounds produced at the Earth's surface play an important role in the climate system. Biogenic processes generally emit sulfur species in reduced form. For example, dimethylsulfide ($(CH_3)_2S$) results from biological activity in the oceans and carbonyl sulfide (COS) is probably produced both in soils and in the oceans. Hydrogen sulfide (H_2S) results from the decomposition of organic matter and volcanic eruptions. Sulfur dioxide (SO_2) is an intermediate in the oxidation of these sulfur species. Large amounts of SO_2 are also continually produced by coal combustion and volcanically active regions. These two compounds are rapidly destroyed in the troposphere, while carbonyl sulfide is very stable and reaches the stratosphere. These sulfur-based compounds, when oxidized or photolyzed, result in the formation of sulfuric acid aerosols. These particles play a major role in the climate system by modifying the radiative characterisitics of the atmosphere. They may also trigger a number of chemical reactions on their surface. Consequently, changes in temperature as well as in the concentration of atmospheric trace gases (such as ozone) should be expected after massive volcanic eruptions.

Finally, the chlorofluorocarbons (CFCs), which are now widely used in a variety of industrial and domestic applications, are the dominant sources of inorganic chlorine in the stratosphere. Among these, CFC–11 ($CFCl_3$) and CFC–12 (CF_2Cl_2) play the major role but the contributions of carbon tetrachloride (CCl_4), methylchloroform (CH_3CCl_3), and

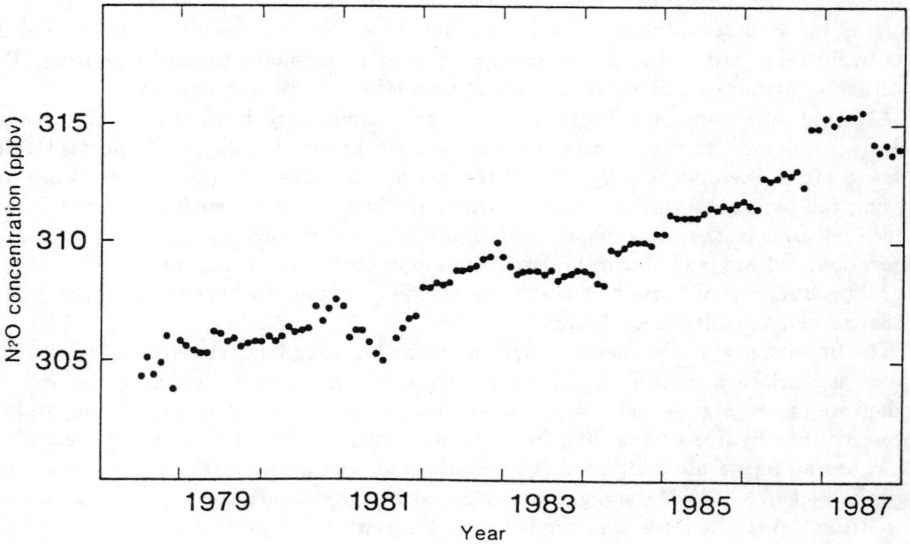

Figure 5: Atmospheric N_2O measured in surface air at the Cape Grim Observatory, Tasmania. Updated from Fraser and Derek (1987) and in a preliminary calibration scale. Reproduced from Pearman, 1988.

CFC–22 ($CHClF_2$) are also substantial. The destruction of these species, by photodissociation or eventually by oxidation, leads to the formation of free chlorine, a potential ozone depletion agent. The emission of these gases and consequently their mixing ratio in the atmosphere have increased substantially over the last three decades. Figure 6 shows the evolution in the production of CFC-11 and CFC-12 since 1961.

With the exception of methyl chloride (CH_3Cl) and, to a certain extent, of methyl bromide (CH_3Br), these compounds are almost exclusively of industrial origin. They are used for example as aerosol propellants, refrigerants, agents for blowing plastic foams, solvents, etc. Methyl chloride, which also contributes to the formation of inorganic chlorine, appears to be essentially of natural origin.

A number of bromine-containing gases have been measured in the atmosphere. The most abundant are methyl bromide (CH_3Br) and bromoform ($CHBr_3$). Concentrations of Halon 1301 (CF_3Br) and Halon 1211 (CF_2BrCl) are rising rapidly, and because of their high ozone depletion potential, could become a threat for the ozone layer.

Fully halogenated halocarbons such as CCl_4 (CFC-10), $CFCl_3$ (CFC-11), CF_2Cl_2 (CFC-12), $C_2F_3Cl_3$ (CFC-113), $C_2F_4Cl_2$ (CFC-114), C_2F_5Cl (CFC-115) as well as the brominated compounds such as CF_2BrCl (halon-1211), CF_3Br (halon-1301), $C_2F_4Br_2$ (halon-2402) are photodecomposed by shortwave solar ultraviolet in the stratosphere. The lifetime of these molecules is relatively long (about 70 years for CFC-11, 140 years for CFC-12, 100 years for CFC-113). Since the rate of release of these chemical compounds at the surface exceeds substantially their rate of destruction in the stratosphere, these gases accumulate in the atmosphere. Partially halogenated halocarbons such as CH_3CCl_3 (methylchlo-

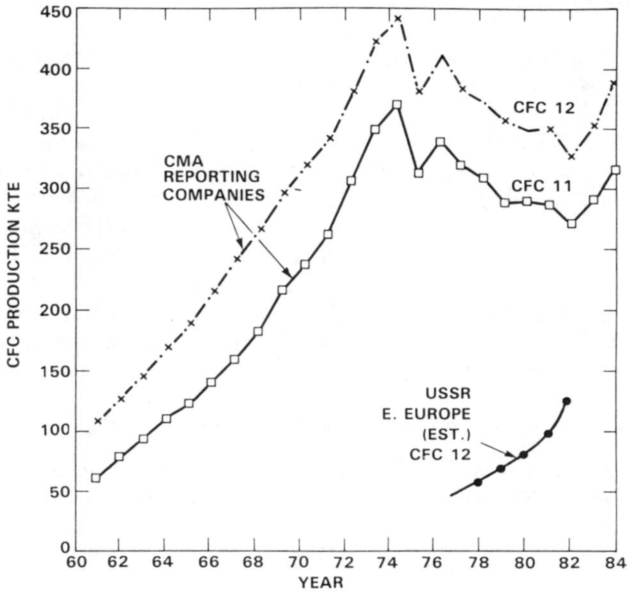

Figure 6: Chlorofluorocarbon production history for CFC-11 and CFC-12 between 1961 and 1984. Upper curves: production by the manufacturing companies, as reported to the Chemical Manufacturers Association (CMA). Lower curve: estimated production by the Soviet Union and Eastern Europe for CFC-12 only.

roform), or other industrial products such as $CHClF_2$ (HCFC-22), CF_3CHCl_2 (HCFC-123), CF_3CHClF (HCFC-124), CCl_2FCH_3(HCFC-141b) and $CClF_2CH_3$(HCFC-142b) are destroyed in large part in the troposphere by OH radicals. The lifetime of these compounds is generally of the order of 10 years.

The photochemical or chemical decomposition of these halocarbons leads to the production of chlorine atoms which contribute efficiently to the destruction of ozone in the stratosphere. The potential for ozone depletion is substantially larger in the case of the fully halogenated than in the case of partially halogenated halocarbons. Several molecules such as CF_3CHF_2 (HFC-125), CF_3CH_2F (HFC-134a), CF_3CH_3 (HFC-143a) or CHF_2CH_3 (HFC-152a) include fluorine atoms but do not contain any chlorine or bromine atoms. Since most of the inorganic fluorine in the atmosphere is sequestred as HF, these chemical compounds do not result in substantial destruction of ozone. All the fully or partially halogenated halocarbons contribute however to the greenhouse effect.

In summary, the rapid rise in the density of several source gases and the related change in ozone and temperature are a direct consequence of the modification of natural ecosystems, the consumption of fossil fuel and the release in the atmosphere of gases produced by industry. These perturbations on a global scale, which are a major environmental issue to be faced by humanity, can be attributed ultimately to rapid growth of industrialization and agriculture. The solutions to these problems require a better understanding of the strong couplings between the atmosphere, the biosphere, the soils and the ocean.

4. The oxidizing capacity of the atmosphere

Perhaps the most important role of the troposphere from the chemical point of view is its ability to oxidize gases which are released at the surface. The most important oxidant is the OH radical but other oxidants are the peroxy radicals (HO_2)—or higher forms RO_2, where R is an alkyl such as CH_3—nitrogen trioxide, ozone, chlorine or oxygen atoms. Oxidation in the aqueous phase is also of significance for species which hydrolyze in droplets, the most likely compound to play a key role in this process being hydrogen peroxide (H_2O_2).

Hydroxyl radicals (OH) play an important role not only in the stratosphere and mesosphere, where they participate in the destruction of ozone, but also in the troposphere where they regulate the removal of species as diverse as CH_4, CO, CH_3Cl, $CHCl_2F$, CH_3CCl_3, CH_3Br, NO_2, NH_3, SO_2, H_2S and $(CH_3)_2S$. The formation of hydroxyl radicals results from the presence of ozone in the atmosphere. Solar radiation at wavelength shorter than 310 nm, when it dissociates ozone, produces an oxygen atom in its electronically excited ^1D state:

$$O_3 + h\nu\ (\lambda < 310\ \mathrm{nm}) \longrightarrow O(^1D) + O_2$$

Although the excited atoms are rapidly quenched to the ground state ^3P by O_2 and N_2, a fraction of them react with water vapor to form the OH radical:

$$H_2O + O(^1D) \longrightarrow 2OH$$

Hydroxyl radicals are destroyed by a reaction with carbon monoxide

$$OH + CO \longrightarrow CO_2 + H$$

followed by a recombination of atomic hydrogen with molecular oxygen, which produces the peroxy radical HO_2:

$$H + O_2 + M \longrightarrow HO_2 + M$$

where M(N_2 or O_2) is a third body that allows conservation of momentum. In NO-poor environments (mixing ratio less than a few pptv), HO_2 is converted back to OH by

$$HO_2 + O_3 \longrightarrow OH + 2O_2.$$

The net result of this sequence of reactions is a loss of ozone and a conversion of carbon monoxide into carbon dioxide. In this case, the budget of odd hydrogen is not affected.

The oxidation of methane and other hydrocarbons leads to an additional loss of ozone and also affects the concentration of hydroxyl radicals (Crutzen, 1974, 1988). Non-methane hydrocarbons initiate an extremely complex organic chemistry which interacts with other chemical species, especially in the boundary layer.

In the presence of a sufficiently large amount of nitric oxide (NO), HO_2 is converted back to OH by

$$HO_2 + NO \longrightarrow OH + NO_2$$

Subsequently, in the presence of sunlight, nitrogen dioxide (NO_2) is photodissociated

$$NO_2 + h\nu \longrightarrow NO + O$$

and an ozone molecule is produced by

$$O + O_2 + M \longrightarrow O_3 + M.$$

The oxidation of methane by the OH radical

$$CH_4 + OH \longrightarrow CH_3 + H_2O,$$

followed by the production of methyl peroxide

$$CH_3 + O_2 + M \longrightarrow CH_3O_2 + M,$$

also leads, in NO_x-rich environments, to the production of ozone since CH_3O_2 (as well as other organic peroxy radicals produced by the oxidation of several non-methane hydrocarbons) converts NO into NO_2

$$NO + CH_3O_2 \longrightarrow NO_2 + CH_3O,$$

similarly to the reaction of NO with HO_2. Hence in polluted regions, where the abundance of NO is higher than a few pptv, the simultaneous presence of carbon monoxide, methane (or other hydrocarbons) and nitrogen oxides leads, in the presence of sunlight, to the formation of ozone and odd hydrogen. Thus, ozone is photochemically produced in urbanized and industrial areas, and these rather localized sources may have an impact on the regional climate (*e.g.*, ozone episodes over the Eastern United States or Europe) and probably on the global (hemispheric) climate. Despite the relatively short lifetime of species such as CO and NO, local pollution sources may affect a wide area through the transport of peroxyacetylnitrate (PAN), a reservoir of odd nitrogen which is stable in cold air but decomposes in warmer environments.

Photochemical conditions in the troposphere have been changing with time, especially since the beginning of the industrial era. The increases in carbon dioxide and methane have already been mentioned. Larger emissions of carbon monoxide and nitric oxide (*e.g.*, as a result of combustion) are taking place. Ozone, in the temperate zone of the Northern hemisphere, is believed to be increasing on the average by about 0.5–2% per year. The highest densities in these regions appear to be observed in summer (suggesting a strong photochemical component in the formation of O_3) rather than in spring (as observed in unpolluted areas where ozone transport from the stratosphere probably influences the conditions in the troposphere).

The largest production of carbon monoxide results from the oxidation of methane and other natural and anthropogenic hydrocarbons. Since CO is also produced by incomplete combustion, fossil fuel burning, forest clearing, savanna and wood burning represent substantial additional sources. CO is also emitted by vegetation and in some oceanic processes. This gas, which is responsible for the conversion of OH into HO_2, is destroyed by OH. Other tropospheric losses are due to uptake by soils and transport into the stratosphere. Since as much as 60 percent of the CO production is probably related to anthropogenic processes, and since the lifetime of these molecules is only of the order of 2 months, the observed mixing ratios are significantly different in both hemispheres (typically 150–200 ppbv in the Northern hemisphere and 50 ppbv in the Southern hemisphere). Seasonal cycles in surface concentrations are observed and probably associated—at least in part—with variations in the vegetation and biomass burning.

An accurate budget of nitrogen oxides in the troposphere is still difficult to establish because of uncertainties in the different sources which have been identified: fossil fuel burning at the surface, emission by aircraft in the troposphere, lightning activity primarily in the tropics, deforestation, intrusion of odd nitrogen from the stratosphere. Because the average lifetime of NO_x is of the order of a day, the NO-rich environments should be found mainly in the temperate regions of the Northern hemisphere and in the boundary layer in the tropics. Because pollution and thunderstorms take place primarily over continental regions, it is estimated that more than 90% of NO is produced over the continents, even though these areas represent less than 30% of the Earth surface. Figure 7 shows the increase in time of the emission of NO_x over the United States.

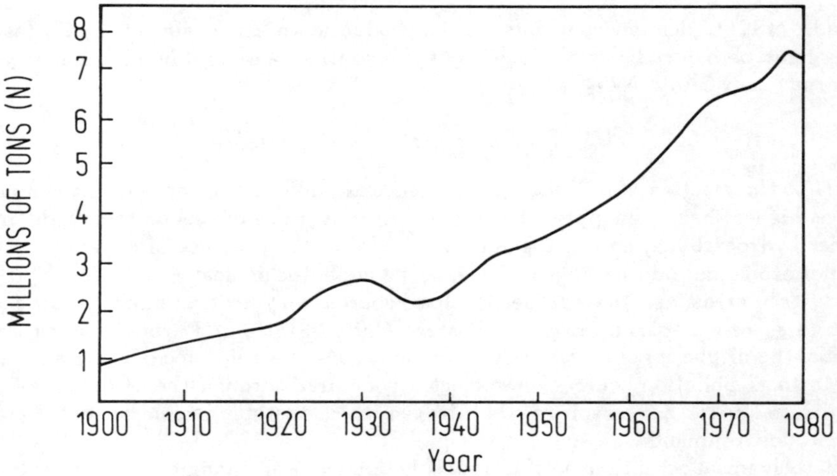

Figure 7: Estimates of NO_x emissions in the US (10^{12} g N per year). Reproduced from Crutzen (1988), adapted from NAPAP (1984).

To summarize, the concentration of OH is controlled, to a large extent, by the presence of ozone, carbon monoxide and nitrogen oxides. This radical, which is a major oxidant in the troposphere, determines the lifetime of several gases emitted at the surface and, for example, prevents the build up in the atmosphere of many hydrocarbons. Ozone is produced photochemically in the troposphere in the presence of methane and other hydrocarbons, carbon monoxide and nitrogen oxides. The sources of these gases are unevenly distributed over the globe and strongly influenced by human activity. The emission of these species is increasing with time, essentially in industrialized regions. A direct consequence of these human activities is a probably significant increase in the density of tropospheric ozone in the Northern hemisphere as well as an enhanced oxidation capacity of the troposphere. The following Table gives a summary of parameters related to selected trace gases in the troposphere (Ehhalt, 1986).

Despite the importance of these changes in chemical composition and the associated effects on climate and health, tropospheric studies suffer from a lack of global observations and therefore require extensive use of aircraft and satellite-borne instruments, in addition to an extension of ground-based measurement stations. The interpretation of these observations will require the use of global and regional three-dimensional models whose development becomes possible due to the rapid growth in computer power.

Current tropospheric mixing ratio, lifetime, global burden, increase rate and human contribution to global sources for selected trace gases.

Gas	Mixing Ratio (ppbv)	Lifetime (years)	Burden (10^6 tonnes)	Increase Rate (% per year)	Anthropogenic Sources (%)
CH_4	1700	10	3900	1	60
N_2O	305	100	1900	0.2	30
CO	100	0.2	400	1 ?	60
O_3	40	0.1	270	1 ?	50
NO_2	variable (0.5)	0.002	(0.3)	1	60
CF_2Cl_2	0.35	90	6	5	100
$CFCl_3$	0.20	70	3.9	5	100

5. Chemistry of stratospheric ozone

Perhaps the easiest way to summarize the chemical and photochemical processes occurring in the stratosphere is to discuss the formation and destruction mechanisms of ozone in this region of the atmosphere.

Ozone is produced by a three-body reaction

$$O + O_2 + M \longrightarrow O_3 + M$$

following the photolysis of molecular oxygen (O_2), by which 2 oxygen atoms (O) are formed:

$$O_2 + h\nu(\lambda < 242 \text{ nm}) \longrightarrow O + O.$$

Ozone is also photolyzed by UV and visible light

$$O_3 + h\nu \longrightarrow O_2 + O$$

but the oxygen atom recombines almost immediately with O_2 to reform ozone. The net loss of odd oxygen (O_x is either O or O_3) occurs through a recombination of ozone and atomic oxygen

$$O + O_3 \longrightarrow 2O_2.$$

This recombination process can be catalyzed by different radicals present in the atmosphere. For example, above 55 km or so, the most efficient destruction mechanisms of O and O_3 involves the action of hydrogen compounds through the following cycles (Bates and Nicolet, 1950)

$$H + O_3 \longrightarrow OH + O_2$$

$$OH + O \longrightarrow H + O_2$$

$$\text{Net: } O + O_3 \longrightarrow 2O_2$$

and

$$
\begin{aligned}
OH + O &\longrightarrow H + O_2 \\
H + O_2 + M &\longrightarrow HO_2 + M \\
HO_2 + O &\longrightarrow OH + O_2 \\
\hline
\text{Net: } O + O &\longrightarrow O_2
\end{aligned}
$$

Hydroxyl radicals and hydrogen atoms are produced by dissociation or oxidation of water vapor, methane and molecular hydrogen. These gases are originating at the Earth's surface and transported into the middle atmosphere. In the stratosphere, the destruction of odd oxygen is catalyzed by the presence of nitrogen oxides (Crutzen, 1970)

$$
\begin{aligned}
NO + O_3 &\longrightarrow NO_2 + O_2 \\
NO_2 + O &\longrightarrow NO + O_2 \\
\hline
\text{Net: } O_3 + O &\longrightarrow 2O_2
\end{aligned}
$$

Nitric oxide is produced ***in situ*** in the stratosphere by oxidation of nitrous oxide, a gas produced at the surface and transported into the stratosphere. Another catalytic destruction of ozone (Stolarski and Cicerone, 1974)

$$
\begin{aligned}
Cl + O_3 &\longrightarrow ClO + O_2 \\
ClO + O &\longrightarrow Cl + O_2 \\
\hline
\text{Net: } O_3 + O &\longrightarrow 2O_2
\end{aligned}
$$

essentially by UV photodecomposition of industrially produced chlorofluorocarbons (Molina and Rowland, 1974). In the lower stratosphere and in the troposphere, the most efficient ozone destruction mechanism is due to a cycle involving the presence of hydrogen radicals

$$
\begin{aligned}
OH + O_3 &\longrightarrow HO_2 + O_2 \\
HO_2 + O_3 &\longrightarrow OH + 2O_2 \\
\hline
\text{Net: } 2O_3 &\longrightarrow 3O_2
\end{aligned}
$$

Figure 8 presents a schematic diagram of the chemical transformations of trace gases occurring in the stratosphere and indicates how these mechanisms affect ozone. This diagram shows that the source-gases produced at the Earth's surface and penetrating in the stratosphere, when photolyzed or oxidized, lead to the formation of active radicals (OH, NO, Cl, etc.), which contribute to the destruction of ozone. In the presence of sunlight, these radicals are nearly in equilibrium with more stable constituents (named reservoirs) such as HNO_3, $ClONO_2$, $HOCl$, N_2O_5, HCl, H_2O_2. These chemical reservoirs are inactive towards ozone and, since their chemical lifetimes are relatively long in the lower stratosphere, their concentrations are significantly higher than those of the fast-reacting radicals and their distributions are affected by transport in this region of the atmosphere.

The density of ozone in the stratosphere and consequently the ozone column amount as well as the depth of penetration of solar ultraviolet light and the heating balance in the middle atmosphere are highly dependent on the chemical reactions affecting all these minor constituents.

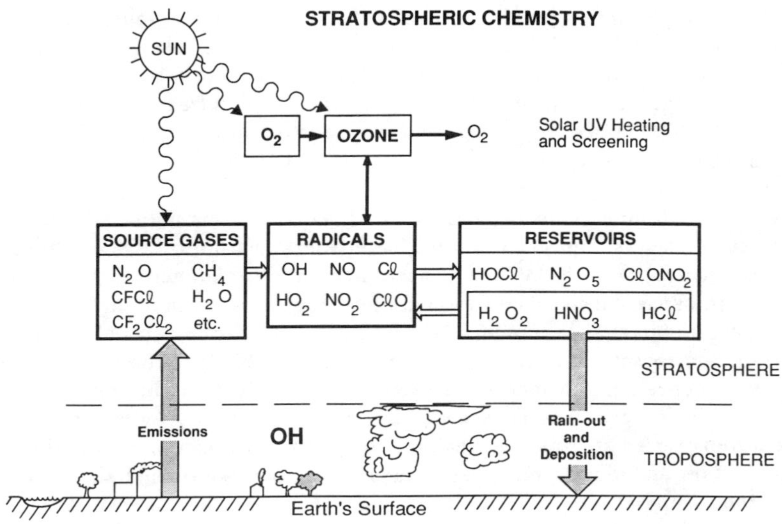

Figure 8: Schematic diagram of chemical and photochemical conversions in the stratosphere in relation with ozone chemistry.

6. Perturbations in the atmospheric composition

Trace gases released in the atmosphere do not only affect the radiation balance in the atmosphere; they also alter the global composition of the atmosphere, in particular the distribution of ozone. For example, the production of chemically active radicals belonging to the nitrogen, hydrogen or chlorine families, which are a direct product of the decomposition of the source gases produced at the Earth's surface, contribute significantly to the destruction of ozone in the stratosphere and in the mesosphere. A thinning of the ozone layer would lead to increased penetration in the atmosphere of harmful solar ultraviolet radiation with substantial environmental consequences at the Earth's surface. Furthermore, enhanced levels of ultraviolet radiation will increase the photochemical activity in the troposphere and particularly the processes affecting the production and destruction rates of species such as ozone and O H. More generally, a modification of the oxidizing capacity of the troposphere will affect the destruction rate and consequently the lifetime and the density of a variety of species, including the greenhouse gases and chemical pollutants. Chemical perturbations are expected to have consequences not only on the entire climatic system but also on the global environment.

Estimation of possible changes in the chemical composition of the atmosphere can be made by numerical simulations using sophisticated coupled chemical-dynamical-radiative models of the atmosphere. Different scenarios of the growth in the emission of source gases, changes in solar activity, and plausible increase in the aerosol content of the atmosphere should be considered to assess the climatic impact of these natural and anthropogenic forcings. In this overview, only particular cases will be briefly discussed.

Figure 9a shows, for example, the change in the meridional distribution of temperature for a doubling of the carbon dioxide amount calculated by the two-dimensional model of Brasseur *et al.*, (1989). Besides the warming of the troposphere resulting from the greenhouse effect, a cooling of the stratosphere due to increased infrared emission to space is clearly observable. The largest temperature decrease is found near the stratopause, particularly at high latitude (−8 K at the equator, −10 K at the summer pole and −17 K at the winter pole).

Because of the temperature dependencies of the rate constants involved in the destruction of ozone, the cooling of the stratosphere resulting from a CO_2 doubling tends to increase the concentration of ozone and hence its column abundance. Figure 9b shows the relative change in ozone density as a function of altitude and latitude in June, and Fig. 9c presents the percentage change in the vertically integrated ozone concentration as a function of latitude and season. According to the model, a doubling of the CO_2 concentration should lead to a concentration increase as large as 25 to 30% in the upper stratosphere. Total ozone should increase on the average by about 1 percent; increases larger than 2 percent are however predicted at high latitude in spring. This seasonal modulation results from several factors including atmospheric transport of ozone-enriched air from the upper to the lower stratosphere.

Another model simulation attempts to describe the effect of the chlorofluorocarbons on ozone. Figures 10a and 10b show the meridional distribution of the percentage change in the temperature and ozone density for an increase of odd chlorine from 2.8 to 7.5 ppbv. Such an increase in Cl_x corresponds to a change from present-day conditions to a situation reached in the middle of the 21st century if the emission of the chlorofluorocarbons do not increase more than authorized by the Montreal Protocol on the protection of the ozone layer. Temperature changes of the order of 10 K and ozone reductions larger than 50 percent are predicted in the upper stratosphere at high latitudes. Because the largest ozone depletion takes place above the layer where the density of this gas is highest, the calculated reduction in the ozone column for this particular case is limited to about 2 percent at the equator but reaches more than 6 percent at the poles in Spring (Figure 10c). It should be indicated however that the effect of heterogeneous chemistry (on the surface of liquid or solid particles present in the atmosphere) is not taken into account in this particular model simulation. These processes however are known to be important over Antarctica, where, in the presence of polar stratospheric clouds, they initiate a complex chemistry leading to a springtime "ozone hole", and perhaps also at other latitudes, especially after large volcanic eruptions.

Monitoring of total column ozone abundances has been performed for more than 30 years. The determination from these observations of a possible long-term trend is difficult to achieve due to existing dynamical oscillations and variations in solar activity (which affect the photochemical production of O_3), and because of the poor geographical coverage of the ground-based instruments measuring ozone and possible instrument drift. A careful analysis of the data however shows that, at mid- and high latitudes in the Northern hemisphere, the total ozone amount has decreased by about 2 to 3 percent over the 1970–1986 period. Satellite data, corrected for instrument drift through a normalization with the well calibrated ground-based instruments, suggests a trend of about −2.5 percent in the region between 65°N and 65°S for the period 1979 to 1986.

In 1985, scientists of the British Antarctic Survey (Farman *et al.*, 1985) reported that the ozone column measured in October over the station of Halley Bay in Antarctica had

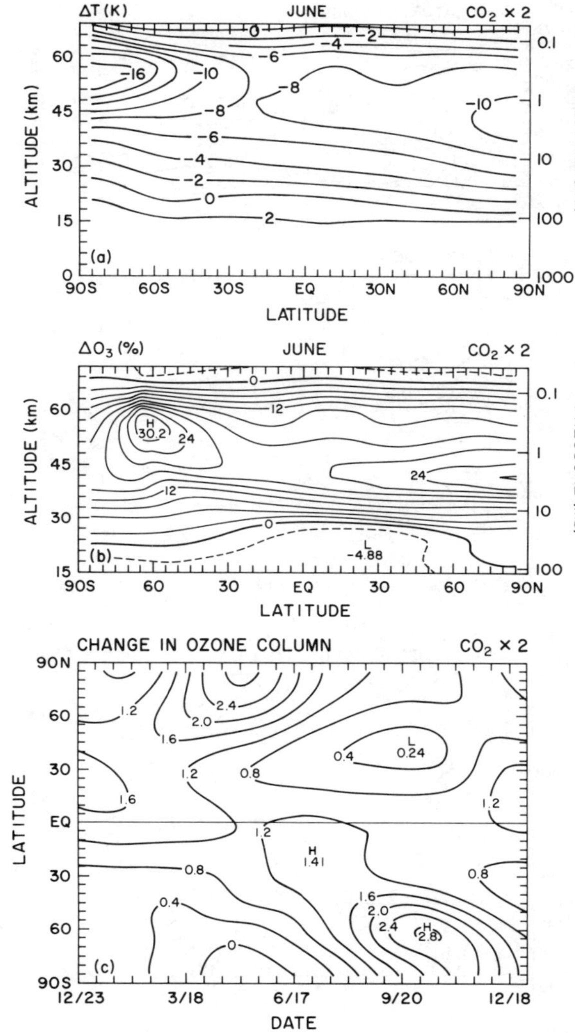

Figure 9: Response of the middle atmosphere for a doubling in CO_2: (a) Temperature change (K) for June conditions; (b) percentage ozone change for June conditions; (c) percentage change in the ozone column abundance as a function of latitude and season.

gradually decreased by as much as 40 percent between 1979 and 1984. Satellite measurements (Stolarski *et al.*, 1986) showed that this dramatic ozone depletion takes place from September to November in a region extending over the entire Antarctic continent. It was

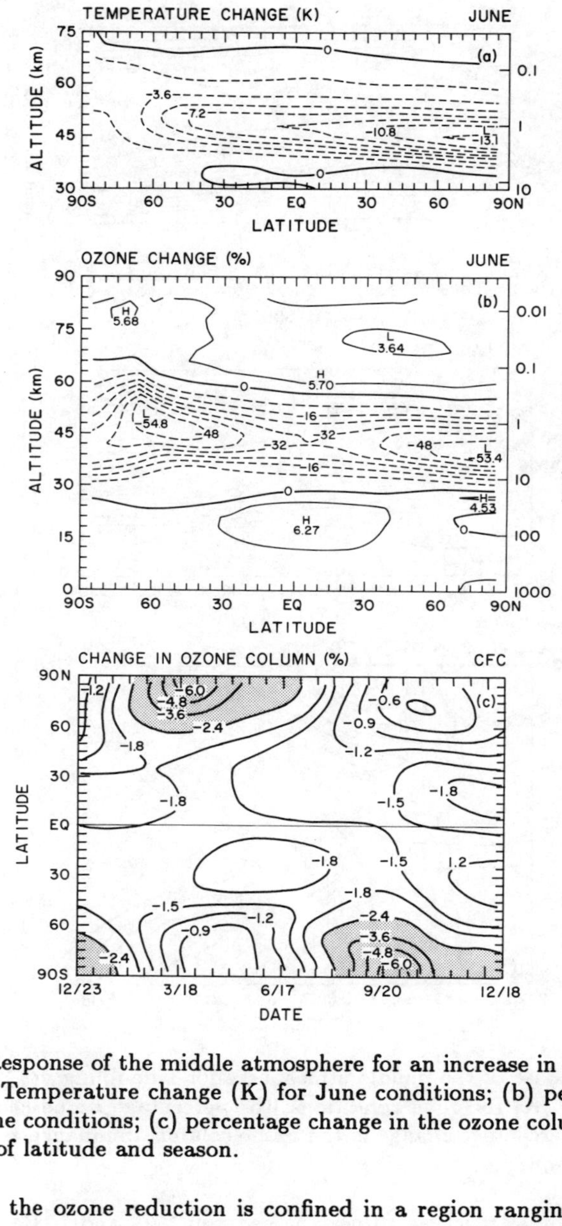

Figure 10: Response of the middle atmosphere for an increase in Cl_x from 2.8 to 7.5 ppbv: (a) Temperature change (K) for June conditions; (b) percentage ozone change for June conditions; (c) percentage change in the ozone column abundance as a function of latitude and season.

also observed that the ozone reduction is confined in a region ranging from 12 to about 22 km altitude.

The ozone reduction observed from satellites during the 1979–1986 period, is of the

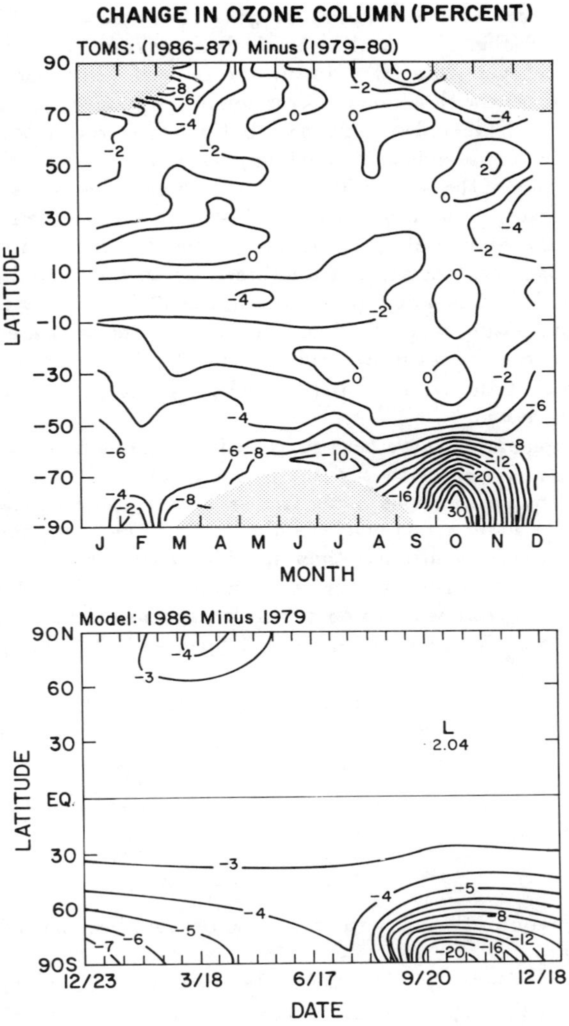

Figure 11: Upper frame: Relative change in the ozone column abundance measured by the TOMS instrument on board Nimbus 7 satellite (after correction based on information provided by ground-based Dobson instruments) between 1979–1980, and 1986–1987. The observed values have been averaged over two years to remove the effect of the quasi-biennial oscillation. Lower frame: Relative change in the ozone column abundance derived by the chemical radiative dynamical two-dimensional model of Brasseur *et al.*, 1989, in which a crude parameterization of heterogeneous processes in the polar region has been introduced.

order of 1–2 percent at the equator but generally increases with latitude, especially in winter and spring (Fig. 11). A clear hemispheric asymmetry is visible. The appearance of the ozone hole over Antarctica during the 80's is clearly shown. These observations are broadly consistent with a model estimation of the ozone change during the same period (Fig. 11), if this model includes a parameterization of heterogeneous processes in the regions where the temperature is sufficiently cold for polar stratospheric clouds to be produced. The observed magnitude of the ozone reduction in the hole, however, is larger than predicted by the model. However, because of interannual variations in the temperature over the poles, the "depth" of the ozone hole fluctuates from year to year, while the model provides an average value that does not account for these interannual variations of dynamical origin. The observed depletion in global ozone during the 1979–1986 period is believed to be due, in part, to the increasing concentrations of chlorine in the atmosphere (of industrial origin) and, in part, to the decrease over this period of solar activity (Brasseur *et al.*, 1988), except for high latitude winter and spring when the ozone decrease is attributed only to anthropogenic activity.

Different explanations have been suggested to account for the recently observed change in the ozone climatology of the Southern Hemisphere. It appears today, as shown for example by recent observations made from a high-altitude aircraft, that active chlorine (*e.g.*, ClO) in the lower stratosphere is about 100 times more abundant inside the polar vortex of the Southern hemisphere than at mid-latitudes, and that, in late September, O_3 and ClO densities are clearly anticorrelated on different space scales (Fig. 12). At the same time, the polar vortex appears to be denitrified and dehydrated. Under these conditions, the concentration of ClO may reach about 1 ppbv and the following cycle (Molina *et al.*, 1987)

$$\begin{array}{rcl} 2\,(\mathrm{Cl} + \mathrm{O_3} & \longrightarrow & \mathrm{ClO} + \mathrm{O_2}) \\ \mathrm{ClO} + \mathrm{ClO} + M & \longrightarrow & \mathrm{Cl_2O_2} \\ \mathrm{Cl_2O_2} + h\nu & \longrightarrow & \mathrm{ClOO} + \mathrm{Cl} \\ \mathrm{ClOO} + M & \longrightarrow & \mathrm{Cl} + \mathrm{O_2} + \mathrm{M} \\ \hline \mathrm{Net}: 2\,\mathrm{O_3} & \longrightarrow & 3\,\mathrm{O_2}, \end{array}$$

which does not require the presence of atomic oxygen, becomes efficient to destroy ozone at low altitudes.

The existence of high ClO densities in the lower stratosphere over Antarctica is believed to be associated with the presence of polar stratospheric clouds (PSCs), which are produced in these regions because of the extremely low temperatures (less than 195 K) occurring in the polar vortex of the Southern hemisphere during winter. Molina *et al.* (1987) as well as Wofsy *et al.* (1988) have suggested that, if HCl molecules dissolve in the ice crystals of the polar stratospheric clouds, a reaction of gas phase $ClONO_2$ with dissolved HCl

$$\mathrm{ClONO_2}(g) + \mathrm{HCl}(s) \longrightarrow \mathrm{Cl_2}(g) + \mathrm{HNO_3}(s)$$

would produce nitric acid in the condensed phase and Cl_2 in the gas phase. As the Sun returns over Antarctica, Cl_2 would be photolyzed and active chlorine (Cl, ClO) produced in sufficiently large quantity to destroy ozone. The details of the clouds formation and the composition of the particles as a function of temperature are not yet fully understood. Two

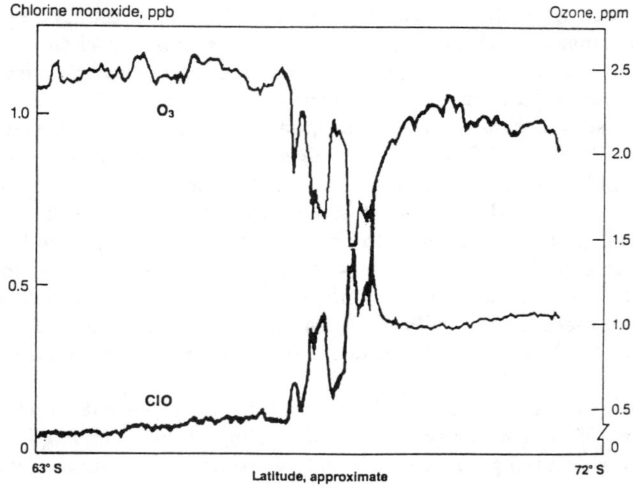

Figure 12: Simultaneous measurement of chlorine monoxide and ozone as observed from NASA's ER-2 research airplane at 18.5 km altitude between 63°S and 72°S. The data were collected on September 16, 1987, as the plane entered the ozone hole. Concentration of chlorine monoxide increased to about 500 times the levels predicted at mid-latitudes (Data from J. Anderson, Harvard University).

types of PSCs are usually distinguished: Clouds of type I, which are formed below about 195 K, are composed of nitric acid ice in the predominant form of nitric trihydrate. The average diameter of these particles is of the order of 1 μm. These clouds may be dominant in catalyzing heterogeneous reactions but the sedimentation velocity of its particles is too low to account for the removal of substantial quantities of nitrogen oxide. Type II clouds, which are produced when the temperature decreases below approximately 187 K, are composed of water ice and have a substantially larger particle mass than type I PSCs. The particles involved have a diameter of the order of 10–100 μm, and are therefore large enough to carry condensed material downward by several kilometers on time scales of hours to days. The conditions for the temperature to become sufficiently low for these clouds to be produced results from the particular dynamical situation over Antarctica. The air masses inside the polar vortex are cold and nearly isolated from the rest of the stratosphere. The possible intrusion of air in the vortex on small scales should however be investigated.

7. Concluding remarks

By releasing in the atmosphere large amounts of gases of industrial origin, by burning increasing quantities of fossil fuel, by destroying the rainforest in the tropics, humankind is progressively changing on a global scale the conditions prevailing in the atmosphere as well as the Earth's climate. Atmospheric chemistry plays a major role in the determination of the present and future climatic conditions because of the interactions between trace gases and radiation. Furthermore, important couplings between the different components of the geosphere occur through mass transfer of chemical compounds.

Further research is needed to study the various feedbacks influencing the climate system. It is important, for example, (1) to assess how climatic changes, associated with increasing emissions of atmospheric gases, will in turn influence the emission level of these gases, (2) to determine to what extent warmer temperatures will accelerate the chemical transformation of source gases in the troposphere, (3) to examine how the absorption of CO_2 by oceans will be modified if the ocean surface layers warm up, and (4) to assess how climatically-induced changes in both the oceanic and terrestrial biosphere will affect not only the emission and deposition of trace gases, but also the albedo of the surface, which itself affects the photochemistry in the atmosphere. Finally, clouds, which introduce major feedbacks in the climate system, also play an important role in tropospheric chemistry and produce convective transports of key chemical species in the free atmosphere. Increasing levels of methane, after its oxidation into carbon dioxide and water vapor, could lead to additional cloudiness near the tropopause and thereby contribute further to climatic changes.

The International Geosphere-Biosphere Program will represent an important milestone in the investigation of the Earth's system as a whole and of the complex processes—including chemical processes occurring in the atmosphere—that drive climatic changes. It is for example important to systematically measure on the global scale the distributions, the sources and sinks of trace gases in the atmosphere, and simultaneously to develop a new generation of high resolution climate models which include chemical and biological processes in addition to atmospheric dynamics. Such task requires interdisciplinary approaches which need to be coordinated at the international level.

References

It is impossible in this brief article to mention all important publications dealing with atmospheric chemistry and its importance for climate research. Numerous references can be found in the WMO Report No. 16 entitled *Atmospheric Ozone 1985*, Geneva, Switzerland. Other references will be found in the books by G. Brasseur and S. Solomon, entitled *Aeronomy of the Middle Atmosphere* (Reidel Publishing Co., Second Edition, 1986), by R. P. Wayne entitled *Chemistry of Atmospheres* (Claredon Press, 1985), by J. S. Levine (Ed.) entitled *The Photochemistry of Atmospheres* (Academic Press, 1985), and by P. Warneck entitled *Chemistry of the Natural Atmosphere* (Academic Press, 1988).

Bacastow, R. B. and C. D. Keeling (1981) 'Atmospheric Carbon Dioxide Concentration and the Observed Airborne Fraction', in *Carbon Cycle Modelling*, SCOPE **16**, Ed. by B. Bolin, John Wiley, Chichester, 390 pp.

Barnola, J. M., D. Raynaud, Y. S. Korotkevich, and C. Lorius (1987) 'Vostok Ice Core Provides 160,000-year record of Atmospheric CO_2', *Nature*, **329**, 408–414.

Bates, D. and M. Nicolet (1950) 'The Photochemistry of Water Vapor', *Journal of Geophysical Research*, **55**, 301–327.

Blake, D. R. and F. S. Rowland (1988) 'Continuing Worldwide Increase in Tropospheric Methane, 1978 to 1987', *Science*, **239**, 1129–1131.

Brasseur, G., M. H. Hitchman, P. C. Simon and A. De Rudder (1988) 'Ozone Reduction in the 1980's: A Model Simulation of Anthropogenic and Solar Perturbation', *Geophysical Research Letter*, **12**, 1361–1364.

Brasseur, G., M. H. Hitchman, M. Dymek, M. Pirre and E. Falise (1989) 'An Interactive Chemical, Dynamical, Radiative Two-dimensional Model of the Middle Atmosphere', in preparation.

Crutzen, P. J. (1970) 'The Influence of Nitrogen Oxides on the Atmospheric Ozone Content', *Quarterly Journal of the Royal Meteorological Society*, **96**, 320–325.

Crutzen, P. J. (1974) 'Photochemical Reactions Initiated by and Influencing Ozone in Unpolluted Tropospheric Air', *Tellus*, **26**, 47–57.

Crutzen, P. J. (1988) 'Tropospheric Ozone: An Overview', in *Tropospheric Ozone*, Ed. by I. S. A. Isaksen, Reidel Publishing Co., 3–32.

Ehhalt, D. H. (1986) 'Atmospheric Trace Gases: Interactions with the Biosphere', in *Climate-Vegetation Interactions*, Ed. by C. Rosenzweig and R. Dickinson, UCAR, Office for Interdisciplinary Earth Studies, Report OIES–2.

Farman, J. C., B. G. Gardiner and J. D. Shanklin (1985) 'Large Losses of Total Ozone in Antarctica Reveal Seasonal ClO_x/NO_xInteraction', *Nature*, **315**, 207–210.

Jones, P. D., T. M. L. Wigley, and P. B. Wright (1986) 'Global Temperature variations Between 1861 and 1984', *Nature*, **322**, 430–434.

Komhyr, W. D., R. H. Gammon, T. B. Harris, L. S. Waterman, T. J. Conway, W. R. Taylor, and K. W. Thoning (1985) 'Global Atmospheric CO_2 Distribution and Variations from 1968–1982 NOAA/GMCC CO_2 Flask Sampling Data', *Journal of Geophysical Research*, **90**, 5567–5596.

Margulis, L. (1982) *Early Life*, Van Nostrand Reinhold Co., New York, 160 pp.

Molina, M. J. and F. S. Rowland (1974) 'Stratospheric Sink for Chlorofluoromethanes: Chlorine Atom Catalyzed Destruction of Ozone', *Nature*, **249**, 810–814.

Molina, M. J., T. L. Tso, L. T. Molina and F. C.-Y. Wang (1987) 'Antarctic Stratospheric Chemistry and Chlorine Nitrate, Hydrogen Chloride: Release of Active Chlorine in the Antarctic Stratosphere', *Science*, **238**, 1253–1257.

NAPAP (National Acid Precipitation Assessment Program) (1984) *Annual Report to the President and Congress 1984*, Director of Research, 726 Jackson Place, N.W., Washington, D.C. 20503, USA.

Pearman, G. I. (1988) 'Greenhouse gases: Evidence for atmospheric changes and anthropogenic causes', in *Greenhouse, Planning for Climate Change*, Ed. by G. I. Pearman, E. J. Brill, Leiden, The Netherlands, 752 pp.

Pearman, G. I., D. M. Atheridge, F. de Silva, and P. J. Fraser (1986) 'Evidence of Changing Concentrations of Atmospheric CO_2, N_2O, and $CH_4$from Air Bubbles in Antarctic Ice', *Nature*, **320**, 248–250.

Stolarski, R. S. and R. J. Cicerone (1974) 'Stratospheric Chlorine: A Possible Sink for Ozone', *Canadian Journal of Chemistry*, **52**, 1610–1615.

Stolarski, R. S., A. J. Krueger, M. R. Schoeberl, R. D. McPeeters, P. A. Newman and J. C. Alpert (1986) 'Nimbus 7 Satellite Measurements of the Springtime Antarctic Ozone Decrease', *Nature*, **322**, 808–811.

Walker, J. C. (1977) *Evolution of the Atmosphere*, Macmillan Publishing Co., New York, 318 pp.

Wofsy, S. C., M. J. Molina, R. J. Salawitch, L. E. Fox and M. B. McElroy (1988) 'Interactions Between HCl, NO_x and H_2O Ice in the Antarctic Stratosphere: Implications for Ozone', *Journal of Geophysical Research*, **93**, 2442–2450.

ICE AND CLIMATE STUDIES

J. OERLEMANS
Institute of Meteorology and Oceanography
University of Utrecht, The Netherlands

and

Alfred-Wegener-Institut für Polar- und Meeresforschung
Bremerhaven, FRG

1. Introduction

During the last 10 or 20 years, interest in the continental cryosphere (I do not consider sea ice) as a component of the climate system has increased tremendously. There seem to be two main reasons for this. In the first place, deep drilling on the major ice sheets and subsequent analysis of the ice cores has established a unique record of climate variability and history of the atmospheric composition over the last 160000 years. Secondly, physical modelling of large ice sheets and related geodynamics has shown that ice sheets are not passive elements just documenting ice ages (this view has been taken in many classical studies, and is still taken in some recent ones !), but instead form an active component.

Still, it is probably fair to state that our knowledge of glaciers and land ice is underdeveloped as compared to that of the ocean, and particularly the atmosphere. On the other hand, many initiatives are taken in climatic research groups to employ physics-based models of glaciers and ice sheets, and chemists and physicists involved in ice-core analysis show an increasing interest in placing their results in a framework of world-wide climatic change.

Since the ice-core aspect is treated in another contribution, I will concentrate on the dynamic role land ice plays in the climate system.

2. Land ice in climate studies, a brief overview

I think it is useful to distinguish between the following aspects/subjects: (1) the environmental record contained in ice sheets; (2) the extension of land ice as a climate indicator; (3) the dynamic interplay between ice sheets, ocean and atmosphere; (4) greenhouse warming and land ice. Although the latter is actually contained in (3), the time scale of interest

A. Berger et al. (eds.), Climate and Geo-Sciences, 303–307.

and the detail of information requested justifies to mention this as a distinct issue.

(1) *The environmental record in ice sheets.* Pioneering work was done by Langway (1967) and Dansgaard and Johnson (1969). In recent years there has been a tremendous development, culminating in records of past variations in atmospheric carbon dioxide over the last full glacial cycle (Jouzel et al. 1987, Barnola et al. 1987). In this discipline, there is an increasing need for sophisticated ice-flow models to check and improve time scales.

(2) *Land ice as climate indicator.* The discovery of traces of widespread glaciation has lead to the final acceptation of the glacial theory, stating that there have been periods on earth with a substantially colder climate. As such the extension of the major ice sheets is a very clear climate indicator for long time scales. However, a proper interpretation of former ice-sheet limits certainly requires knowledge of ice dynamics, and the notion that the timing of advances and retreats need not to be uniform over the globe. Recent detailed mapping, based on many reliable C^{14} dates [e.g. Berglund and Mörner, 1984] certainly offers new possibilities.

On a much smaller scale valley glaciers are found to be sensitive indicators of climatic change. They have the advantage that they do not influence the global climate. From several glaciers in Europe the length variations are known since about 1600 AD, and from many other places in the world maximum stands are known from dated geomorphological features like moraines. However, the response of a glacier will strongly be effected by its geometry and the local micro-climatological conditions [e.g. Oerlemans, 1986a]. This implies that numerical models of glacier flow and the micro-climatology of the glacial environment are needed to obtain an optimal interpretation. In recent years several studies in this direction are being undertaken: see e.g. Bindschadler (1977), Kruss (1984), Oerlemans (1986b), Huybrechts et al. (1988).

(3) *Ice sheets, oceans and atmosphere.* Here the slow physics of the climate system are the central focus. Concerning the glaciological aspect, the height-mass balance feedback is the dominating element. Any increase in land ice extent normally implies increasing surface elevation and hence colder conditions at the ice sheet surface. Without this strong positive feedback, no *realistic* climate model would ever be able to simulate large northern hemisphere ice sheets. The importance of the height-mass balance feedback was explicitly mentioned by Bodvarsson (1955), and analysed in a very transparent way by Weertman (1961a). He also mentioned the potential importance of bedrock sinking as a destabilizing factor in the ice-age rhythm (Weertman, 1961b), and was the first to impose the Milankovitch radiation variations as forcing to an ice-sheet model (Weertman, 1976). His model was very simple (perfectly plastic, i.e. fixed profile), and did not generate 100 kyr power. This was however achieved in a model of the present author (Oerlemans, 1980), and appeared as a consequence of explicitly modelling latitude-dependent ice flow and bedrock sinking. Subsequently, more refined latitude-dependent

ice sheet modelling by various workers has shown that the coupling of ice-sheet dynamics with geodynamics may generate a significant response to the Milankovitch insolation variations (Birchfield et al., 1981; Pollard, 1982; Oerlemans, 1982; Hyde and Peltier, 1986). Broadly speaking, in these studies the 20 and 40 kyr power is an essentially linear response, while the generated 100 kyr power is a nonlinear response inherent to the characteristics of ice-sheet dynamics.

Although latitude-dependent modelling represents a gross simplification of reality, studies have been done with even more schematic pictures of the climate system. The few-component models of Källen et al. (1979), Ghil and LeTreut (1981), Saltzman et al. (1984), etc. have shown that free cycles of any desired period can be obtained. The stream of papers based on few-component models still continues, and here the climate system serves as a playground for experiments in dynamical systems theory [recipe: select three or four variables (say from: deep ocean temperature, ice shelf extent, sea ice cover, carbon dioxide, land ice, bedrock sinking, mixed-layer temperature, tropical forest extent, snow fall), postulate some relations, introduce some time scales, add periodic forcing (or, if desired, random) and generate the cycle you like].

It is quite obvious that progress should now come from more detailed modelling, because this allows a much *closer comparison with proxy data records*. Asynchronic coupling of the ocean, atmosphere and cryosphere in space resolving models is important, but it should be done in a careful way. Although the resulting models will be computer-time consuming, it is very important to extend integrations over more than one glacial cycle. *A model that is tuned to produce one 100 kyr cycle does not generate 100 kyr power !!*

Another point I want to make is that even sophisticated statistical modelling is useful, but never forms a *real alternative to physics-based models*.

(4) *Greenhouse warming and land ice.* Sea level is closely tied to land ice volume, and the response of glacier and ice sheets to a greenhouse warming thus deserves ample study. Although several projections of the response of glaciers and ice sheets have been made (e.g. DOE, 1985), many uncertainties exist. These stem from the very complicated dynamics of West Antarctica (e.g. Van der Veen, 1987), as well as our scarce knowledge concerning the relation between mass balance and meteorological conditions. A particular problem is the poor job GCM's do in simulating the hydrology and micrometeorology of very high latitudes. It is thus of crucial importance that physical parameterization of the cold and stable boundary layer in GCM's is further developed. Also, classical mass-balance measurements, though less appealing from a scientific point of view, will remain essential for a long time to come. In particular, long series of mass balance and meteorological parameters (from the same spot) are needed from various types of climates.

3. Summary of recommendations

1. Ice-core analysis should be developed further, implying the retrieval of *new* deep cores. Detailed models of ice flow should back such studies and help in the interpretation.
2. Numerical models of ice flow and mass balance should be used to interprete historic glacier variations (to replace purely statistical considerations).
3a. Modelling of pleistocene/holocene climatic fluctuations should be done with models that have *spatial resolution*.. This will enable a more precise comparison with proxy data (a crucial step).
3b. Integrations with Milankovitch forcing should be done over at least *500 kyr* (to see whether 100 kyr power really shows up).
4a. To arrive at more reliable estimates concerning land ice volume in the future, *mass balance measurements* on glaciers and ice sheets are very important.
4b. The holocene histories of the Antarctic and Greenland ice sheets must be known. Studies in the *marginal zones* of those ice sheets (shallow cores, temperature profiles in the ice, geomorphological work) should be enforced to achieve this.

References

Barnola J M, D Raynaud, Y S Korotkevitch and C Lorius (1987): Vostok ice core provides 160,000-year record of atmospheric CO_2. *Nature* **329**, 408-414.

Berglund B E and N-A Mörner (1984): Late Weichselian deglaciation and chronostratigraphy of Southern Scandinavia: Problems and present 'state of the art'. In: *Climatic Changes on a Yearly to Millennial Basis* (eds.: N-A Mörner and W Karlén), Reidel, 17-24.

Birchfield G E, J Weertman and A T Lunde (1981): A paleoclimate model of northern hemisphere ice sheets. *Quaternary Res.* **15**, 126-142.

Dansgaard W and S J Johnson (1969): A flow model and a time scale for the ice core from Camp Century, Greenland. *J. Glaciology* **8**, 215-223.

DOE (1985): *Glaciers, Ice Sheets and Sea Level: Effect of a CO_2-induced Climatic Change.* Department of Energy Report DOE/ER/60235-1, Washington DC, 330 pp.

Ghil M and H Le Treut (1981): A climate model with cryodynamics and geodynamics. *J. Geophys. Res.* **86**, 5262-5270.

Huybrechts P, J de Nooze and H Decleir (1988): Numerical modelling of Glacier d' Argentière and its historic front variations. In press.

Hyde W T and W R Peltier (1985): Sensitivity experiments with a model of the ice-age cycle: the response to harmonic forcing. *J. Atmos. Sci.* **42**, 2170-2188.

Jouzel J, C Lorius, J R Petit, C Genthon, N I Barkov, V M Kotlyakov and V M Petrov (1987): Vostok ice core: a continuous isotope temperature record over the last climatic cycle (160,000 years). *Nature* **329**, 403-408.

Källen E, C Grafoord and M Ghil (1979): Free oscillations in a climate model with ice-sheet dynamics. *J. Atmos. Sci.* **36**, 2292-2303.

Kruss P D (1984): Climatic change in East Africa: numerical modeling from the 100 years of terminus record of Lewis Glacier, Mount Kenya. *Z. Gletscherk. Glazialgeol.* **19**, 43-60.

Langway C C (1967): Stratigraphic analysis of a deep ice core from Greenland. CRREL Report 77.

Oerlemans J (1980): Model experiments on the 100 000 yr glacial cycle. *Nature* **287**, 430-432.

Oerlemans J (1982): Glacial cycles and ice-sheet modelling. *Climatic Change* **4**, 353-374.

Oerlemans J (1986): Glaciers as indicators of the carbon dioxide warming. *Nature* **320**, 607-609.

Oerlemans J (1986): An attempt to simulate historic front variations of Nigardsbreen, Norway. *Theor. Appl. Climatol.* **37**, 126-135.

Pollard, D (1982): A simple ice-sheet model yields realistic 100 kyr glacial cycles. *Nature* **296**, 334-338.

Saltzman B, A R Hansen and K A Maasch (1984): The late quaternary glaciations as the response of a three-component feedback system to earth-orbital forcing. *J. Atmos. Sci.* **41**, 3380-3389.

Veen C J van der (1987): The West Antarctic Ice Sheet: the need to understand its dynamics. In: *Dynamics of the West Antarctic Ice Sheet* (eds: C J van der Veen and J Oerlemans), Reidel, 1-16.

Weertman J (1961a): Equilibrium profiles of ice caps. *J. Glaciology* **3**, 953-964.

Weertman J (1961b): Stability of ice-age ice sheets. *J. Geophys. Res.* **66**, 3783-3792.

Weertman J (1976): Milankovitch solar radiation variations and ice age ice sheet sizes. *Nature* **261**, 17-20.

VOLCANOES AND CLIMATE

ALAN ROBOCK
Department of Meteorology
University of Maryland
College Park, Maryland 20742
USA

ABSTRACT. Volcanic eruptions which inject large amounts of sulfur-rich gas into the stratosphere produce dust veils which last several years and cool the earth's surface. At the same time, these dust veils absorb enough solar radiation to warm the stratosphere. Since these temperature changes at the earth's surface and in the stratosphere are both in the opposite direction of hypothesized effects from greenhouse gases, they act to delay and mask the detection of greenhouse effects on the climate system. Tantalizing recent research results have also suggested regional effects of volcanic eruptions, including effects on ENSO.

Complete understanding of the above effects of volcanoes is hampered by inadequacies of data sets on volcanic dust veils and on climate change. The effects of volcanoes have not adequately been separated from ENSO events, and climate modeling of the effects of volcanoes is in its infancy. Specific suggestions are made for future work to improve our knowledge of this important component of our climate system.

1. Introduction

Since 1784, when Benjamin Franklin suggested that the Hecla eruption in Iceland in 1783 might have been responsible for the abnormally cold winter of 1783-4 (Franklin, 1784), emissions from volcanoes have been implicated as a possible cause of weather and climate variations. Although conventional wisdom (Ramanathan, 1988; Self and Rampino, 1988) holds that the effects of volcanic eruptions on climate have not yet been demonstrated, much work has already shown that volcanoes can be important causes of hemispheric temperature changes for several years following large eruptions and even on a 100-year time scale, when their cumulative effects are taken into account. It has also been suggested that they can influence atmospheric circulation, and interact with the El Niño/Southern Oscillation (ENSO) in complex ways, perhaps even causing ENSOs. Because the climatic signal of volcanic eruptions is of approximately the same amplitude as that of ENSO, and because there have been so few large eruptions in the past century, it has been difficult to separate the volcanic signal from that of other simultaneous climatic variations.

In order to understand the timing and amplitude of future global warming, and in order to understand ENSO, more research is necessary on the effects of volcanoes on climate.

A. Berger et al. (eds.), Climate and Geo-Sciences, 309–314.

2. Data

2.1. VOLCANIC ERUPTIONS

It has become clear in the last decade that the effect of a volcano on climate is most directly related to the sulfur content of emissions that reach into the stratosphere, and not to the explosivity of the eruption. These sulfur gases convert to small sulfate particles, which persist for several years in the stratosphere and efficiently scatter the incoming sunlight, reducing the direct and total solar radiation reaching the ground. The most comprehensive survey of past volcanic eruptions (Simkin *et al.*, 1981), however, produced the Volcanic Explosivity Index (VEI) (Newhall and Self, 1982) which gives a geologically-based measure of the power of the volcanic explosion. Eruptions with a high VEI may also have a large stratospheric impact, such as Tambora (1815) or Krakatau (1883), but 2 recent examples demonstrate the danger in using the VEI for climate studies. Mt. St. Helens (1980) had a high VEI, and while it had a large local temperature impact (Robock and Mass, 1982), it had a negligible stratospheric impact (Robock, 1981). El Chichón (1982), on the other hand, had a very large stratospheric impact (Robock, 1983) but relatively small VEI. Several studies have been done using the VEI as an index for the climatic effect of volcanoes, and then treated Mt. St. Helens and El Chichón as special cases. What about all the other special cases in the past that we do not know about?

By combining the VEI with information about the typical type of eruption that each volcano produces, it *may* be possible to produce a "climatic VEI," but it will probably be necessary to include other information to produce a good index of the climatic impact of past eruptions. The Dust Veil Index (DVI) of Lamb (1970) and Mitchell (1970) was designed to be such an index. It has an undeserved reputation for including temperature data in the derivation of the index, but in fact, temperature data were only used for some of the eruptions, and it is possible to derive a temperature-independent index (Robock, 1981a).

Ice core analysis (*e.g.*, Lyons *et al.*, 1988, which has an extensive bibliography) can give the chemistry and particle content of well-dated layers, which can give a measure of the important volcanic parameters. Unfortunately, small local eruptions can give as large a signal as distant large eruptions. By comparing the acidity and particle records from Greenland, Antarctic, and tropical ice cores, it may be possible to produce a global or hemispheric record from signals that appear simultaneously in all 3 regions or in the tropics and one of the high latitude cores.

2.2. CLIMATE

The recent work that has combined land-based and ocean-based information to produce global data sets of surface air temperature and other climatic elements will be very useful to compare to the volcanic record. These data sets still have problems, and must be used with caution.

3. Relationship to surface air temperature

Theoretical studies of the radiative effects (Pollack *et al.*, 1976; Harshvardhan, 1979), superposed epoch analyses using surface air temperature data (Mitchell, 1961; Oliver,

1976; Mass and Schneider, 1977; Taylor *et al.*, 1980, Self *et al.*, 1981, Rampino and Self, 1982), and numerical modeling experiments (Schneider and Mass, 1975; Hunt, 1977; Hansen *et al.*, 1978; Robock, 1978, 1979, 1981b, 1984; Gilliland, 1982; Gilliland and Schneider, 1984) have all demonstrated that volcanic eruptions are an important cause of climatic change. The effects can start as soon as 2 months after the eruption (Kelly and Sear, 1974; Sear *et al.*, 1987) and last for several years, with hemispheric average surface cooling of as much as 0.5°C. At the same time the stratosphere is warmed by more than 10 times this amount by absorption of solar and infrared radiation by the volcanic aerosols. The important absorbing species is silicates (dust), not sulfates (Ogren *et al.*, 1981; Clarke *et al.*, 1983).

Since these temperature changes at the earth's surface and in the stratosphere are both in the opposite direction of hypothesized effects from greenhouse gases, they can act to delay and mask the detection of greenhouse effects on the climate system. Thus it is crucial to understand the climatic and weather effects of large volcanic eruptions in order to make any realistic assessment of future greenhouse warming. It is important to realize that the first half of this century was anomalous in that it had very few volcanic eruptions as compared to the past 500 years (Lamb, 1970; Mitchell, 1970). Volcanic aerosols in the stratosphere are the norm, not the exception.

4. Relationship to ENSO

On 4 April 1982, the El Chichón volcano in Mexico erupted, putting more aerosols into the stratosphere than any other volcano in this century, with the possible exception of Agung in 1963 (Robock, 1983). Although climate model calculations (Robock, 1984; Hansen *et al.*, 1988) suggested a large drop in hemispheric average surface air temperatures for several years following the eruption, 1983 was in fact one of the warmest years on record (Hansen and Lebedeff, 1988). It is believed that this was due to the unprecedented ENSO of 1982.

This is only the most dramatic example of the combination of ENSO and volcanic signals in the climate system. Even if the effects are completely independent physically, they complicate the interpretation of either. Therefore, both modeling and data analysis studies (such as Angell, 1988) are necessary to separate and understand the climate responses to this combination of externally and internally-caused variations.

It has even been suggested that the El Chichón eruption affected the strength of the 1982-83 ENSO (*e.g.*, Hirono, 1988). This idea merits further investigation.

5. Recommendations: Volcanoes and climate

The following projects should receive high priority in future research:

- **An index of volcanic aerosols should be produced that is related to the climatic effect of volcanoes, not their explosivity.** It would start with the Dust Veil Index and Volcanic Explosivity Index, and then modify them with information obtained from ice cores, radiation measurements, surveys of the sulfur content of volcanic emissions, descriptive reports of eruptions and of optical effects, and modern information from lidars and satellite measurements.

- **Comprehensive analysis of past temperature, pressure and precipitation**

data should be performed, searching for the volcanic signal while accounting for the internal climate variations related to ENSO. Of course, much of interannual variability is unrelated to volcanic eruptions, greenhouse warming or ENSO, and is of a stochastic nature (Lorenz, 1968; Hasselmann, 1976; Robock, 1978). It is crucial to understand the volcanic signal along with the other deterministic variations in order to calculate the unpredictable part of the signal.

- **General circulation model simulations of volcanic eruptions should be carried out.** These would produce the signal of a volcanic eruption that would be compared to that found in the analysis above. Simultaneous forcings by tropical sea surface temperature anomalies and volcanoes should also be used to attempt to understand whether the climatic response is related or independent. The 1982-83 El Chichón-ENSO would be a good case to study.

- **The global aerosol monitoring effort should be continued, with special consideration to allow comprehensive observations of future volcanic eruptions.**

6. References

Angell, J. K., 1988: Impact of El Niño on the delineation of tropospheric cooling due to volcanic eruptions. *J. Geophys. Res.*, **93**, 3697-3704.

Clarke, A. D., R. J. Charlson and J. A. Ogren, 1983: Stratospheric aerosol light absorption before and after El Chichón. *Geophys. Res. Lett.*, **10**, 1017-1020.

Franklin, B., 1784: Meteorological imaginations and conjectures. *Manchester Literary and Philosophical Society Memoirs and Proceedings,* **2**, 122 or 375.

Gilliland, R. L., 1982: Solar, volcanic and CO_2 forcing of recent climatic changes. *Clim. Change,* **4**, 111-131.

Gilliland, R. L. and S. H. Schneider, 1984: Volcanic, CO_2 and solar forcing of Northern and Southern Hemisphere surface temperatures. *Nature,* **310**, 38-41.

Hansen, J., I. Fung, A. Lacis, S. Lebedeff, D. Rind, R. Ruedy, and G. Russell, 1988: Global climate changes as forecast by the GISS 3-D model. *J. Geophys. Res.*, **93**, 9341-9364.

Hansen, James and Sergej Lebedeff, 1988: Global surface air temperatures: update through 1987. *Geophys. Res. Let.*, **15**, 323-326.

Hansen, J. E., Wei-Chung Wang and A. A. Lacis, 1978: Mount Agung provides a test of a global climatic perturbation. *Science,* **199**, 1065-1068.

Hasselmann, K., 1976: Stochastic climate models, I. Theory, *Tellus,* **28**, 473-485.

Hirono, M., 1988: On the trigger of El Niño Southern Oscillation by the forcing of early El Chichón volcanic aerosols. *J. Geophys. Res.*, **93**, 5365-5384.

Hunt, B. G., 1977: A simulation of the possible consequences of a volcanic eruption on the general circulation of the atmosphere. *Monthly Weather Review*, **105**, 247-260.

Kelly, P. M. and C. B. Sear, 1984: Climatic impact of explosive volcanic eruptions. *Nature*, **311**, 740-743.

Lamb, H. H., 1970: Volcanic dust in the atmosphere; with a chronology and assessment of its meteorological significance. *Phil. Trans. Roy. Soc. London*, **A266**, 424-533.

Lorenz, E. N., 1968: Climatic determinism, *Meteorol. Monograph.*, **30**, 1-30.

Lough, J. M. and H. C. Fritts, 1987: An assessment of the possible effects of volcanic eruptions on North American climate using tree-ring data, 1602 to 1900 A.D. *Climatic Change*, **10**, 219-239.

Lyons, W. B., P. A. Mayewski, M. J. Spencer and M. S. Twickler, 1988: A Northern Hemisphere volcanic chemistry (1869-1984) and climatic implications using a South Greenland ice core. Submitted to *J. Geophys. Res.*

Mass, C. and S. H. Schneider, 1977: Statistical evidence on the influence of sunspots and volcanic dust on long term temperature trends. *J. Atmos. Sci.*, **34**, 1995-2004.

Mitchell, J. M., Jr., 1961: Recent secular changes of the global temperature. *Ann. N. Y. Acad. Sci.*, **95**, 235-250.

Mitchell, J. M., Jr., 1970: A preliminary evaluation of atmospheric pollution as a cause of the global temperature fluctuation of the past century. In *Global Effects of Environmental Pollution*, S. F. Singer (ed.), (Reidel, Dordrecht), 139-155.

Newhall, C. G. and S. Self, 1982: The volcanic explosivity index (VEI) – an estimate of explosive magnitude for historical volcanism. *J. Geophys. Res.*, **87**, 1231-1238.

Ogren, J. A., R. J. Charlson, L. F. Radke and S. K. Domonkas, 1981: Absorption of visible radiation by aerosols in the volcanic plume of Mt. St. Helens, *Science*, **211**, 834-836.

Oliver, R. C., 1976: On the response of hemispheric mean temperature to stratospheric dust: an empirical approach. *J. Appl. Meteor.*, **15**, 933-950.

Pollack, J. B., O. B. Toon, C. Sagan, A. Summers, B. Baldwin and W. Van Camp, 1976: Volcanic explosions and climatic change: a theoretical assessment. *J. Geophys. Res.*, **81**, 1071-1083.

Ramanathan, V., 1988: The greenhouse theory of climate change: a test by an inadvertant global experiment. *Science*, **240**, 293-299.

Rampino, M. R. and S. Self, 1982: Historic eruptions of Tambora (1815), Krakatoa (1883) and Agung (1963), their stratospheric aerosols and climatic impact. *Quat. Res.*, **18**, 127-163.

Robock, Alan, 1978: Internally and externally caused climate change. *J. Atmos. Sci.*, **35**, 1111-1122.

Robock, Alan, 1979: The "Little Ice Age": Northern Hemisphere average observations and model calculations. *Science,* **206**, 1402-1404.

Robock, Alan, 1981a: A latitudinally dependent volcanic dust veil index, and its effect on climate simulations. *J. Volcanol. Geotherm. Res.*, **11**, 67-80.

Robock, Alan, 1981b: The Mount St. Helens volcanic eruption of 18 May 1980: minimal climatic effect. *Science,* **212**, 1383-1384.

Robock, Alan, and Clifford Mass, 1982: The Mount St. Helens volcanic eruption of 18 May 1980: large short-term surface temperature effects. *Science,* **216**, 628-630.

Robock, Alan, 1983: The dust cloud of the century. *Nature,* **301**, 373-374.

Robock, Alan, 1984: Climate model simulations of the effects of the El Chichón eruption. *Geofísica Internacional,* **23**, 403-414.

Schneider, S. H. and C. Mass, 1975: Volcanic dust, sunspots and temperature trends. *Science,* **190**, 741-746.

Sear, C. B., P. M. Kelly, P. D. Jones and C. M. Goodess, 1987: Global surface-temperature responses to major volcanic eruptions. *Nature,* **330**, 365-367.

Self, Stephen and Michael R. Rampino, 1988: The relationship between volcanic eruptions and climate change: still a conundrum? *Eos,* **69**, 74-75, 85-86.

Self, S., M. R. Rampino and J. J. Barbera, 1981: The possible effects of large 19th and 20th century volcanic eruptions on zonal and hemispheric surface temperatures. *J. Volcanol. Geotherm. Res.*, **11**, 41-60.

Simkin, T., L. Siebert, L. McClelland, D. Bridge, C. Newhall and J. H. Latter, 1981: *Volcanoes of the World.* (Hutchinson Ross, Stroudsburg, PA), 240 pp.

Taylor, B. L., Tsvi Gal-Chen and S. H. Schneider, 1980: Volcanic eruptions and long-term temperature records: an empirical search for cause and effect. *Q. J. Roy. Meteor. Soc.*, **106**, 175-199.

GLOBAL CLIMATIC EFFECTS OF A NUCLEAR WAR: AN INTERDISCIPLINARY PROBLEM

STEVEN J. GHAN
Lawrence Livermore National Laboratory
P.O. Box 808, L-262
Livermore, CA, 94550, USA

ABSTRACT. Evaluation of the global climatic and ecological effects of a nuclear war requires consideration of a wide range of issues involving disciplines such as urban geography, fire mechanics, plume dynamics, cloud microphysics, aerosol physics, radiative transfer theory, atmospheric chemistry, mesoscale dynamics, global climate simulation, plant physiology, agriculture and ecology. Thus, in addition to addressing the global-scale implications of a nuclear war, the global-effects effort also serves as an excellent example of an interdisciplinary research program.

1. INTRODUCTION

The direct local effects of a nuclear war due to blast, fire and radiation are well recognized. The prospect that a major nuclear war might also have significant indirect global-scale effects on the climate of both combatant and non-combatant nations has spurred an international research program which is necessarily interdisciplinary because of its wide scope. Time and space scales of potential effects range from minutes to years and from micrometers to global. Disciplines involved include urban geography, fire mechanics, plume dynamics, cloud microphysics, aerosol physics, radiative transfer theory, atmospheric chemistry, mesoscale dynamics, global atmospheric modeling, ocean and sea-ice modeling, sub-surface thermodynamics and hydrology, plant physiology, agriculture and ecology. Here I shall review the global-effects problem from the perspective of an interdisciplinary framework.

2. DISCUSSION

The possibility of a global-scale climatic response to a major nuclear war was first recognized by Crutzen and Birks (1982) who found that sufficient smoke could be generated by fire following such a war to cover the Northern Hemisphere with an absorption optical depth exceeding one. Subsequent research by others has focused on the details of the climatic response to such a large atmospheric loading of smoke, and on refinements to Crutzen and Birks' estimate of the smoke optical loading.

To illustrate the interdisciplinary nature of the problem, consider how the hemispheric mean absorption optical depth $\tau_a = F\epsilon(1 - f_r)\sigma_a/A$ is determined. Here F is the mass of fuel consumed by fire, ϵ is the smoke emission factor, f_r is the fraction of smoke removed from the atmosphere by prompt scavenging (before the smoke has spread to continental scales), σ_a is the specific absorption coefficient of the smoke, and A is the hemispheric area. This notation follows that of Penner (1986).

A. Berger et al. (eds.), Climate and Geo-Sciences, 315–319.

Estimation of the fuel consumption F requires estimates of the inventories of fuel in the targeted regions for each fuel type. Crutzen and Birks (1982) originally considered wildfires near missile silos to be the principal source of smoke. Subsequent research by Small and Bush (1985) demonstrated that wildfires are less important than urban firestorms as a smoke source for most war scenarios. Crutzen et al. (1984) and Bing (1985) estimated the total fuel loading for wood, petroleum and petroleum derivatives in the NATO and Warsaw Pact countries using different techniques, but arrived at similar numbers. However, not all of the fuel loading would be consumed by fires. An estimate of the fraction consumed requires an understanding of fire mechanics. Reitter et al. (1985) applied a fire-spread model to a burning city, but must still assume a fuel-burn fraction for the areas over which the fire passes. Because the fraction consumed depends on the specific war scenario, the total fuel loading of the combatant nations is frequently used as an upper bound to the amount of fuel that would actually be consumed. For climate simulations a plausible fraction is typically assumed for the fuel consumed.

Smoke emission factors depend primarily on the type of fuel consumed, with much higher emission factors for petroleum and petroleum derivatives than for wood. However, combustion experiments by Williamson et al. (1987) demonstrate that emission factors can also depend quite strongly on the ventilation. There is also some experimental evidence that emission factors may depend on the scale of the fire. Given the large difference between the scale of the combustion experiments (meters) and that of projected urban fires (kilometers), uncertainty due to scaling cannot practically be refined much further.

Estimates of the fraction of smoke removed by prompt scavenging have relied on controlled experiments and on simulations of the smoke plume by dynamical models. (These simulations also provide predictions of the vertical distribution of the smoke injections for larger-scale models.) Cotton (1985) first addressed this problem in a simulation of a firestorm over Denver. However, Cotton neglected nucleation scavenging and hence found that only some 2% of the smoke would be removed from the atmosphere. Subsequent calculations by Bradley (1987) and others which account for nucleation scavenging found that a significant fraction of the smoke could be removed promptly in warm rain systems. Other models including ice physics suggest removal rates may be low. However, these plume-scale models are forced to highly parameterize the cloud microphysics. Controlled laboratory experiments (Hansen and Novakov, 1988) suggest that, with the exception of smoke from petroleum fuels and derivatives, most smoke particles can serve as cloud condensation nuclei and hence are readily scavenged under conditions of supersaturation. Theoretical models of cloud microphysics (Edwards and Penner, 1987) yield similar predictions, but the conclusions depend upon the values of uncertain parameters.

The specific absorption coefficient of the smoke depends on the refractive index of the smoke material and on the size and shape of the smoke particle. The size and shape of the smoke particle can change through processing by clouds and through coagulation with other particles. The dependence of the radiative properties on these characteristics of the smoke is often determined on the basis of the classical Mie theory for effectively-spherical aerosols. However, the Mie theory does not work well for the chained agglomerates that characterize smoke. More sophisticated methods using fractal theory (Berry and Percival, 1986) and general numerical techniques (Iskander et al., 1987) are being developed to address these problems. Meanwhile, the initial optical properties of smoke are based on measurements of small fire emissions.

Estimates of the hemispheric-mean absorption optical depth following a major nuclear war range from about 0.2 to 6 (Penner, 1986). Given such a wide range of optical loading, the climatic effects of the smoke can range from local effects lasting a few weeks to global effects lasting years. For the lower case, the primary climatically-significant response would be a brief surface cooling in or adjacent to the combatant nations. For the larger smoke loadings the local effects are more severe and reductions in precipitation are also predicted (Ghan et al., 1988). Perhaps more importantly, general circulation model simulations by Malone et al. (1986) and others predict strong summertime lofting of the smoke into the stratosphere where aerosol residence times are typically one year. Thus, for moderate and large smoke injections, global-scale interactions with the stratosphere, the ocean and sea ice must also be considered.

If perturbations extend into the stratosphere, interactions with ozone are potentially important. Stevens and Birks (1987) have conducted laboratory experiments that suggest the interaction between ozone and soot would not deplete soot more rapidly from the stratosphere than would gravitational settling or advection. On the other hand, recent simulations with a general circulation model (Thompson and Crutzen, 1988) indicate that ozone concentrations in the stratosphere could be reduced by 50% due to the combination of transport of tropospheric air into the stratosphere and heating of the stratosphere associated with solar absorption by smoke.

Interactions with the ocean and sea ice were first considered by Robock (1984) using an energy balance climate model. Only recently have general circulation models been applied to the chronic effects problem. Preliminary simulations by Ghan et al. (1987) and Thompson and Covey (1988) confirm many features of Robock's simulations.

For the larger smoke loadings the biosphere can be adversely affected. The reduction in agricultural crop yield from extreme conditions of low temperature and reduced sunlight has been studied experimentally for selected crops by Palta (1987). The sensitivity of natural ecosystems to such conditions is not well known, the models of such systems having been calibrated to much more equable conditions. The feedback effect on the atmosphere of the possible failure of cultivated and uncultivated ecosystems has been studied using general circulation models in other contexts.

Much of the present uncertainty lies in the smoke source term τ_a. Accordingly, further work is required to reduce the uncertainty associated with: (a) the fraction of combustible material consumed; (b) the smoke emission factor for those materials with the greatest potential for optical absorption; (c) the fraction removed from the atmosphere by prompt scavenging; and (d) the optical properties of the smoke as it ages in the atmosphere. In addition, some processes affecting the response to the smoke source have not yet been adequately treated. These include interactions with stratospheric ozone and NO_x, and with agricultural systems and natural ecosystems. To date the only simulations of the chronic climatic effects with interactive oceans have used prescribed smoke distributions. Future simulations should represent smoke transport and removal explicitly.

3. CONCLUSIONS

In summary, an elucidation of the global-scale response to a nuclear war is a problem of great breadth, involving many of the sub-disciplines of aerosol physics, meteorology, oceanography, atmospheric chemistry and ecology. As diverse as these fields are, communication between the sub-disciplines has been remarkably effective, with two major interdisciplinary reports published in the last few years (NRC, 1985; SCOPE, 1986). It is my belief that, in addition to addressing the global-scale implications of a nuclear war, the global-effects effort also serves as an excellent example of an interdisciplinary research program.

This work was performed under the auspices of the U.S. Department of Energy by the Lawrence Livermore National Laboratory under Contract W-7405-Eng-48.

REFERENCES

Berry, M. V., and I. C. Percival, 'Optics of fractal clusters of smoke,' *Optica Acta*, **33**, 577–591, 1986.

Bing, G., 'Estimates of total combustible material in NATO and Warsaw Pact Countries,' Lawrence Livermore National Laboratory Report UCRL-93192, 1985.

Bradley, M. M., 'Nucleation scavenging of smoke aerosol above intense fires: Three dimensional simulations.' In *Proceedings of the IUGG Symposium on Aerosols and Climate*, Aug. 9–22, 1987, Vancouver, British Columbia.

Cotton, W. R., 'Atmospheric convection and nuclear winter,' *American Scientist*, **73**, 275–280, 1985.

Crutzen, P. J., and J. W. Birks, 'The atmosphere after a nuclear war: Twilight at noon,' *Ambio*, **11**, 114, 1982.

Crutzen, P. J., I. E. Galbally, and C. Bruhl, 'Atmospheric effects from post-nuclear fires,' *Climatic Change*, **6**, 323–364, 1984.

Edwards, L. L., and J. E. Penner, 'Potential nucleation scavenging of smoke particles over large fires: A parametric study.' In *Proceedings of the IUGG Symposium on Aerosols and Climate*, Aug. 9–22, 1987, Vancouver, British Columbia.

Ghan, S. J., M. C. MacCracken, and J. J. Walton, 'Chronic effects of large atmospheric smoke injections: Interactions with the ocean mixed layer, sea ice and ground hydrology.' In *Proceedings of the IUGG Symposium on Aerosols and Climate*, Aug. 9–22, 1987, Vancouver, British Columbia.

Ghan, S. J., M. C. MacCracken, and J. J. Walton, 'The climatic response, to large atmospheric smoke injections: Sensitivity studies with a tropospheric general circulation model.' *J. Geophys. Res.*, **93**, 8315-8337, 1988.

Hansen, A. D. A., and T. Novakov, 'Water nucleation properties of JP4, diesel, and crude oil smoke particles,' Presented at DNA Global Effects Program Technical Meeting, April 19–21, 1988, Santa Barbara, CA.

Iskander, M. F., H. Y. Chen, T. V. Doung, and J. E. Penner, 'Scattering and absorption by elongated aersol particles in the resonance frequency range.' In *Proceedings of the Third International Conference on Carbonaceous Particles in the Atmosphere*. American Association for Aerosol Research. October 5–8, 1987, Berkeley, CA.

Malone, R. C., L. H. Auer, G. A. Glatzmeier, M. C. Woods, and O. B. Toon, 'Nuclear winter: Three-dimensional simulations including interactive transport, scavenging, and solar heating of smoke,' *J. Geophys. Res.*, **91**, 1039–1054, 1986.

NRC (National Research Council), *The Effects on the Atmosphere of a Major Nuclear Exchange*, National Academy Press, Washington, D.C., 1985.

Palta, J., 'Experimental results,' Presented at DNA Ecological Effects Workshop, October 26–28, 1987, Albuquerque, NM.

Penner, J. E., 'Uncertainties in the smoke source term for "nuclear winter" studies,' *Nature,* **324**, 222–226, 1986.

Reitter, T. A., S.-W. Kang, and A. N. Takata, 'Toward quantifying the source term for predicting global climatic effects of nuclear war: Application of urban fire codes,' Lawrence Livermore National Laboratory Report 53647, 1985.

Robock, A., 'Snow and ice feedbacks for prolonged effects of nuclear winter.' *Nature,* **312**, 667–670, 1984.

SCOPE (Special Committee on Problems of the Environment), *Environmental Consequences of Nuclear War*, 2 Vols, SCOPE Ser., vol. 28, John Wiley, New York, 1986.

Small, R. D., and R. W. Bush, 'Smoke production from multiple nuclear explosions in non-urban areas,' *Science,* **229**, 465,1985.

Stevens, S. L., and J. W. Birks, 'Ozone as a sink for atmospheric carbon aerosols: Today and following a nuclear war.' In *Proceedings of the Third International Conference on Carbonaceous Particles in the Atmosphere.* American Association for Aerosol Research. October 5–8, 1987, Berkeley, CA.

Thompson, S. L., and C. C. Covey, 'Simulations of long-term climate effects of nuclear aerosol using a passive ocean mixed layer GCM,' Presented at DNA Global Effects Program Technical Meeting, April 19–21, 1988, Santa Barbara, CA.

Thompson, S. L., and P. J. Crutzen, 'GCM simulations of nuclear war effects on stratospheric ozone,' Presented at DNA Global Effects Program Technical Meeting, April 19–21, 1988, Santa Barbara, CA.

Williamson, R. B., R. L. Dod, F. W. Mowrer, N. J. Brown, and T. Novakov, 'Smoke emission factors from medium scale experiments.' In *Proceedings of the Third International Conference on Carbonaceous Particles in the Atmosphere.* American Association for Aerosol Research. October 5–8, 1987, Berkeley, CA.

LAND SURFACE PROCESSES IN CLIMATE MODELS: STATUS AND PROSPECTS

Michel M. Verstraete
National Center for Atmospheric Research
P. O. Box 3000
Boulder, CO 80307–3000
USA

ABSTRACT. Surface processes determine to a large extent the nature of the Earth's climate and must therefore be incorporated to some degree in any model used to study the climate system. These processes are briefly reviewed in the context of their incorporation in atmospheric general circulation models, the most sophisticated climate models to date. New developments in surface processes modeling are driven by scientific objectives and constraints in computer power availability. Since the design and implementation of large software codes (such as those needed to include sophisticated surface processes in GCMs) may take years of work, trends in atmospheric sciences (in particular climate modeling) and computer technology are evaluated to assess future needs and opportunities for surface processes parameterizations in global models. A better appreciation of the constraints may be useful in the selection of priorities and objectives, and the assignment of resources.

1. Introduction

Climate is defined differently for different purposes. In this paper, it will designate the state and evolution of the climatic system, which itself comprises the atmosphere, hydrosphere, cryosphere, lithosphere, and biosphere sub-systems. Climatology, the study of the climate system, includes the description of these sub-systems as well as their interactions, both between themselves and with outer space.

Each one of these climatic sub-systems evolves in time as a response to its own internal dynamics and to forcing from the other sub-systems through common interfaces. The interface between the gaseous (atmosphere) and the condensed phases (oceans and continents) is often referred to as the Earth's surface, or the surface. While each sub-system can be studied separately, it is the nature of climate research to focus on interactions between the components of the climate system on a large scale and long term basis and in particular on the exchanges of mass, energy and momentum at these interfaces.

On very long (geological) time scales, convection in the Earth's mantle results in the observed plate tectonics and represents a major factor in the evolution of the climate system,

A. Berger et al. (eds.), Climate and Geo-Sciences, 321–340.

including the biosphere. But on time scales of a thousand years or less, which are of interest here, solar energy is the only important external source of energy in the climate system. This energy is redistributed among the sub-systems before being emitted back to space as thermal radiation. The uneven distribution of heating in the fluid components of the climate system generates large scale circulations, and the differential motion of these fluids results in the transfer of momentum at the interface between these sub-systems. Water vapor, carbon dioxide, and many other chemicals are also exchanged between the climatic sub-systems, and the cycling of these compounds is intimately coupled to the energy cycle through biophysical processes such as evapotranspiration and photosynthesis.

We are now faced with serious issues that are threatening to alter the evolution of our environment. Changes in atmospheric composition including air and water pollution, deforestation, desertification, and accelerated soil erosion are all largely attributable to human activities, and may result in significant climatic changes over the next decades to centuries. Moreover, these shifts in climatic parameters have the potential to further impact the biosphere. As a result, it is vitally important that we understand better the many facets of our complex environment, so that we can take both corrective and preventive action.

Among many pressing issues, the modification of land surface characteristics (such as its reflectance and roughness) by land use and degradation, the role of the biosphere in the climate system and in particular in controlling the fast components of the biogeochemical cycles, and the impact of these changes on the environment and our own societies need to be studied urgently. These goals require the development of tools to observe and study the Earth as a global system, such as satellite observing platforms and global models with adequate representations of the relevant processes.

Of all the sub-systems, the atmosphere is the most responsive. It also interfaces with all other sub-systems, and provides a natural medium to transport mass, momentum and energy between them. It is therefore natural that atmospheric models (in particular general circulation models, GCMs) would be at the core of climate models. At the same time, the interactions between the climatic sub-systems as well as their evolution are controlled to a large extent by biospheric processes. This paper briefly reviews the processes that control the exchanges between the atmosphere and continental surfaces, and describes one of the most important types of models used to investigate climatic issues. Computing power is identified as one of the major constraints on model development and use, and future opportunities for surface processes research are assessed in the light of likely modeling and computing trends.

2. Land surface processes

Over 70% of the Earth surface is covered by oceans, but the continental surfaces provide much of the spatial and temporal variability that makes the weather and climate. They also host the bulk of the living biosphere (Table 1), which is located at or near the interface between the atmosphere and the lithosphere or hydrosphere, and is instrumental in controlling the coupling between these sub-systems (Verstraete and Dickinson, 1986).

Table 1: Biomass distribution, in 10^{12} kg (dry matter)

	Plant biomass	Animal biomass	Total	Percentage
Continents	1837.0	1.005	1838.0	99.73
Oceans	3.9	0.997	4.9	0.26
Total	1840.9	2.002	1842.9	100.00
Percentage	99.9	0.1	100.0	

Source: Ajtay (1979), p. 133.

Research on the Earth's paleo-atmosphere suggests that over geological time scales, the composition of the atmosphere is determined by surface processes, including biological ones. At the other end of the spectrum of time scales, the concentrations of variable constituents in the atmosphere are largely controlled by surface processes. Since radiation is the ultimate external forcing mechanism of the climate system, all factors that affect its transfer through the atmosphere (for example by modifying its composition) or its absorption by the climate system (for instance by changing the reflectance of the system) are bound to have some impact on the climate.

The oceans are the major source of atmospheric water, which plays a particularly important role in this respect, both as a radiatively active gas and as the main constituent of clouds. The marine biosphere is also responsible for the emission of sulfur compounds such as dimethyl sulfide (DMS), an important source of cloud condensation nuclei. However, the majority of other gases such as carbon dioxide, methane, nitrogen oxides, freons, and many other chemicals which play an increasingly important role because their concentrations in the atmosphere are rapidly increasing, originate from continental areas.

Dust and, more generally, suspended particulate matter are also important for climate studies. Dust is picked up from the soil surface by winds when a threshold speed is exceeded. This threshold wind speed is dependent on the mechanical state of the soil, soil moisture and vegetation cover, and the amount of material raised is proportional to a high power of the wind speed above that threshold (*e.g.*, Bagnold, 1943; Fuchs, 1964; Gillette, 1980; Skidmore, 1986). Dust storms have been observed in arid zones for as long as history can recall, and the large scale nature of the transport has been known for some time and is well documented (*e.g.*, Carlson and Prospero, 1972). The advent of satellite imagery has confirmed this large scale transport of dust across oceans and allowed further systematic study of such processes (*e.g.*, McKee, 1979; Morales, 1979; Breed and McCauley, 1986).

Exchanges and transfers of mass, energy and momentum at the interface between the atmosphere and the land surface are the result of specific processes, and these are, in turn, controlled by particular surface parameters. The albedo and emissivity of the surface control the surface radiation balance, which provides the energy needed for the sensible and latent heat released to the atmosphere. The roughness of the surface similarly controls the transfer of momentum between the two sub-systems, while surface temperature affects the infrared cooling, as well as the sensible and latent heat fluxes from the surface to the atmosphere.

These land surface characteristics are typically much more spatially variable than any atmospheric field, except perhaps cloudiness. While synoptic scale temperature or pressure may vary appreciably over hundreds of kilometres, soil properties and vegetation may differ significantly over hundreds of metres. Another aspect of the differences in scale between the atmosphere and the land surface is that the former, being a fluid, homogenizes the variations it undergoes much more rapidly than either the soil or the vegetation. Typically, vegetation will evolve on the seasonal time scale, and soils will change on much longer periods, except during accelerated erosion events.

In this context, the role of the biosphere cannot be overlooked. Continental vegetation, which composes the bulk of its mass, is in close contact with the atmosphere thanks to its large leaf surface (conservatively estimated at 6.12 10^8 km^2), which exceeds the total area of the planet (Verstraete and Dickinson, 1986, p. 359). The biosphere controls to a large extent the many gaseous exchanges with the atmosphere, including water, carbon dioxide, nitrogen, and other elements or compounds. It also is in close contact with the soil through the extensive plant root systems. It is therefore clear that we need to develop global climate models which include many of these processes if we are to understand and predict our changing environment.

3. Global climate models

The climate system is too huge to perform meaningful controlled experiments with it. It is also the only one we have. We can only observe its evolution as a response to both external and internal perturbations, such as changes in solar irradiation or in the atmospheric CO_2 concentration level. As a result, we must investigate its likely evolution with models which mathematically describe our understanding of the physics involved.

The complexity of the climate system has prompted the development of many different models, each with its own range of applications. Sophisticated 3-dimensional models provide the basic tools to understand the multiple interactions between the components of the climate system, while simpler models can be used to perform preliminary investigations, or determine the most important processes or parameters through sensitivity analyses (Schneider and Dickinson, 1974). Models can also help conceptualize our understanding of the system, and suggest better or different observations. Their use in simulation studies allows us to perform impossible or undesirable experiments, to determine the likely impact of planned or possible changes, to reconstruct the most probable evolution of the climatic system in the distant past, or to help interpret remote sensing data.

3.1. The nature of general circulation models (GCM)

Current atmospheric GCMs integrate the primitive equations of motion of the atmosphere either in grid space or spectral space, or a combination of both. They typically include from 8 to 20 layers in the vertical, and have a horizontal resolution of a few degrees in latitude and longitude. The time integration proceeds in steps of a fraction of an hour (Henderson-Sellers and McGuffie, 1987, p. 54, 137; Washington and Parkinson, 1986, p. 167). A number of GCMs are now in operation at large research or operational institutions. (See Washington and Parkinson, 1986, p. 254–258, for an extensive list of these models, including bibliographical references to their description in the literature.) The discussion below refers specifically to the National Center for Atmospheric Research (NCAR) Community Climate Model (CCM1) model, unless otherwise indicated.

GCMs are rather complex models and therefore expensive to use: they place considerable demands both on the central processing unit and on peripheral resources, and their use requires substantial amounts of time even on the most powerful super-computers available today. For instance, NCAR's CCM1 needs from 57 to over 1000 seconds of central processing unit (CPU) time per day of integration to execute on one of the four processors of the Cray X-MP/48, depending on the horizontal resolution and number of levels. Table 2 shows the proportion of time spent on some selected tasks in this model, for an R–15 and a T–42 configuration† with 12 levels.

Table 2: Relative time spent in selected CCM1 tasks

Selected processes	R–15		T–42	
	(s)	(%)	(s)	(%)
Radiation	206.3	36.3	974.3	28.1
Dynamics	152.7	27.1	1184.8	34.1
Input/Output	79.6	14.0	536.8	15.4
Convective adjustment	58.6	10.3	329.2	9.3
Diffusion	22.3	3.8	140.9	4.0
Surface processes	20.4	3.6	135.0	3.8
Cloud parameterization	5.1	0.9	37.2	1.0
Total (all processes)	567.6	100.0	3477.7	100.0

All statistics relative to 10 days of integration.

The processes identified in the table account for about 96% of the total time, the rest being spent in miscellaneous tasks and overhead. The treatment of radiative and dynamical processes account for over 60% of the total CPU time spent integrating the model. Surface processes are very crude in this model: the surface temperature is computed by solving an energy balance equation, assuming instantaneous equilibrium. Surface hydrology is represented by a simple bucket parameterization. Dickinson, Henderson-Sellers, and collaborators have designed a much improved surface package, known as Biosphere-Atmosphere Transfer Scheme (BATS), as explained below. When this package is incorporated into the CCM, the total time spent by the model increases to some 10% of the total time (Dickinson, *et al.*, 1981, 1986; Dickinson and Henderson-Sellers, 1988).

3.2. The evolution of GCMs

Atmospheric general circulation models were initially developed to study the dynamics of the atmosphere, with the specific objective of improving short-term weather prediction (Henderson-Sellers and McGuffie, 1987, p. 42–45; Washington and Parkinson, 1986, p. 1–7). On the space and time scales of relevance for this problem, the numerical solution of the equations describing the motion and the large scale radiation forcing of the atmosphere are sufficient, and the complexities of the surface can be ignored, at least in first approximation.

While great progress was achieved along these lines, these models became progressively used for studying atmospheric problems on much longer time scales. They are now widely recognized as powerful research tools for climate studies, and have been used for a number of investigations ranging from the description of paleoclimates during ice ages (*e.g.*, Kutzbach, 1981; Manabe and Broccoli, 1985; Kutzbach and Guetter, 1984a, 1984b) to the prediction of the impact of an increase of CO_2 in the atmosphere as a result of agricultural practices

† Spectral GCMs are described by codes where the letter indicates the type of spectral truncation in use (R for rhomboidal, T for triangular) and the number indicates the highest wavenumber allowed by the representation. See Henderson-Sellers and McGuffie, 1987, p. 140–146.

(slash and burn, deforestation) and industrial activity (*e.g.*, Washington and Meehl, 1983, 1984, 1986).

Although the existence of the surface has always been acknowledged in these models, its proper parameterization is a recent development, which is by no means complete. The orography and thermal properties of the land surfaces (by contrast with the oceans) were the first characteristics to be recognized in theoretical investigations. These continental features were identified in connection with cyclogenesis and upper atmospheric waves (*e.g.*, Lorenz, 1967; Palmén and Newton, 1969; Dickinson, 1980), or with the Monsoon-type of circulation (Ramage, 1971). It is not until the early seventies that attempts were made to incorporate a crude hydrological cycle into GCMs (*e.g.*, Manabe, 1969; Manabe and Holloway 1975), and until the eighties that vegetation was considered as a distinct surface type, different than soil (Dickinson, 1983, 1984; Sellers *et al.*, 1986; Dickinson and Henderson-Sellers, 1988). Eagleson (1982) and Mintz (1984) reviewed the models then available and discussed the most significant results. Even today, the most sophisticated surface parameterization packages for use in GCMs are very simple compared to reality: For example, all vegetation in the entire grid square may be conceptually replaced by an ensemble of leaves, uniformly distributed over a set fraction of the area, and all soil variations within this area are simply ignored. In other words, even though some of the basic processes may be incorporated to a certain extent, their scale is basically off by two or more orders of magnitude. As mentioned before, work is actively continuing to improve this state of affairs.

3.3. The limitations of GCMs

Although numerical weather prediction has proven rather successful at relatively short time scales (a few days or less), their forecasting skill decreases rapidly on the time scale of 5 to 15 days. Beyond that period, deterministic predictive skill is negligible compared to the long term climatology, except for specific locations or times, although some skill remains for longer-lived features such as the El-Niño/Southern Oscillation (ENSO) phenomenon, or persistence patterns such as blocking events, or for statistical forecasting.

The difficulties of deterministic forecasting are generally attributed to one or more of the following factors:

- An incomplete understanding of the physical factors involved prevents us from being able to describe exactly and completely the observed phenomena and processes. For example, turbulence on a wide range of space and time scales directly affects the weather and climate, but there is no general theory of turbulence and no completely satisfactory way to parameterize it in climate models.
- Even if our understanding of the physics of the climate system was complete, and if all relevant processes could be adequately described mathematically, model predictions would progressively diverge after about a week from actual realization of the climate because of incomplete or inaccurate initialization. Errors of measurement, observations limited to a small subset of all relevant variables, as well as a highly inhomogeneous geographical distribution of the reporting stations, make it impossible to know the exact state of the atmosphere even at a single moment, and therefore also to initialize models accurately.
- The finite spatial and temporal resolution of numerical models implies that all processes too small or too fast to be represented explicitly must be parameterized, and that those processes which can be represented explicitly are so described at only a relatively small number of locations and times. This is a result of the finite nature of such

models, as well as the finite amount of computing power available to solve the problems. Nevertheless, parameterizations of small scale processes in terms of mean values of larger scale processes always lead to approximations.

- Last but not least, there are intrinsic limits to the predictability of the atmosphere, as a result of the stochastic behavior of many of the relevant processes, as well as the nature of the mathematical equations used to describe the atmosphere (Lorenz, 1982).

3.4. Surface processes in GCMs

In part as a result of the recent trend towards interdisciplinary studies of the global climate system, general circulation models are being expanded to include many more processes and interactions between climatic subsystems. The role of surface processes is increasingly important in this perspective, and the challenges associated with this are reviewed next.

The incorporation of detailed surface processes in atmospheric general circulation models presents three types of difficulties. First and foremost, there is a fundamental discrepancy between, on the one hand, the space and time scales at which these processes occur and, on the other hand, the spatial and temporal resolution of these models. Soil and vegetation characteristics can vary appreciably over distances orders of magnitude smaller than the typical distance between consecutive grid points in a GCM. As a result, surface processes must be parameterized instead of being described explicitly. The spatial resolution of current general circulation models is grossly inadequate to resolve these processes, even with high resolution configurations.

The second difficulty lies in the number and complexity of the physical, chemical, and biological processes occurring at or near the surface. Many of the climatically important characteristics of land surfaces result from the action of numerous small scale processes. For example, evapotranspiration is controlled by plant stomatal openings as well as the diffusion of liquid water and water vapor in soil pores and root systems; major biogeochemical cycles are controlled by the status of the highly variable plant canopies; energy fluxes at the surface are much more dependent on the turbulent state of local air circulation than on the large scale average wind field, etc.

Thirdly, the incorporation of more detailed surface processes in large scale models such as GCMs generate significant additional demands in terms of initial and boundary condition specifications. A multi-layer soil submodel is necessary to account for the amplitude and phase of the surface temperature, or for the computation of water storage and runoff. But heat and water transfer in soils depend, in turn, on the type of soil and its moisture content, which are usually quite variable, both horizontally and vertically. Similarly, as long as these models do not include a dynamic vegetation cover, it will be desirable to be able to impose the vegetation type and cover as a function of time, to simulate the seasonal changes.

Recent GCMs attempt to incorporate more detailed parameterizations of the hydrology at the surface, at least on an experimental basis. The BATS scheme developed at NCAR (Dickinson *et al.*, 1981, 1986; Wilson *et al.*, 1987; Dickinson and Henderson-Sellers, 1988) is an example of the current state of the art in this respect: This parameterization includes two soil layers, computes infiltration and soil moisture, and recognizes 12 different soil texture classes for which parameters such as soil porosity, soil water potential, and hydraulic conductivity can be specified as functions of soil moisture. The actual evaporation is computed as the minimum of the potential evaporation and the maximum moisture flux that the soil can sustain. Gravitational drainage is allowed between the two layers, and surface runoff is computed as a function of soil moisture. When vegetation is present at the specified grid point, transpiration, interception and re-evaporation of precipitation are computed explicitly.

Many other processes will be required, however, to study transient climate problems. Dust and other aerosols (of volcanic, maritime, or industrial origin), as well as smoke produced by forest and grass fires, or even emission of volatile organics by vegetation (terpenes and isoprenes), have a definite impact on the radiation transfer in the atmosphere (*e.g.*, Carlson and Benjamin, 1980). Dust has also been identified as a possible cause for climatic change (Bryson and Baerreis, 1967; and Bryson, 1974). The incorporation of dust in GCMs is complicated by its relatively short residence time in the atmosphere, as well as by the complex physical processes associated with its mobilization at the surface and removal by wet and/or dry deposition. The problem is compounded by the mathematical difficulties of designing a satisfactory transport scheme for trace substances in GCMs. This issue has however received a lot of attention lately in the context of 'Nuclear Winter' studies (*e.g.*, Pittock *et al.*, 1986).

4. Modeling and computing trends

We now take a close look at the likely trends in atmospheric modeling and computing, and attempt to outline the prospects for surface processes parameterization schemes in the next 5 to 10 years.

4.1. The trend towards higher resolution models

As explained above, the first atmospheric general circulation models resolved approximate equations under a set of simplifying initial and boundary conditions, on a low resolution grid. However, fundamental research in atmospheric sciences, a better understanding of the numerical aspects of discrete models, and advances in computer technology allowed researchers to design more sophisticated and realistic climate models.

Perhaps the most powerful drive towards higher resolution models is the hope that more detailed global models may be more representative of the actual climate system and therefore more reliable in their predictions. There are two aspects to this premise: First, assuming the relevant physical processes have been included in the equations, a higher resolution model should provide better diagnostic and prognostic capabilities, everything else being constant, because numerical integration is more accurate with smaller space and time steps.

No less importantly, however, is the widespread belief that complex physical processes can be better incorporated explicitly in models with higher spatial and temporal resolution, because less reliance is placed on approximate parameterizations of sub-grid scale processes. Furthermore, even if many important processes are still too small in scale to be incorporated explicitly, the discrepancy between the physical scale of these processes and the smallest scale that can be resolved by the model may be lessened and, presumably, the parameterizations might be more realistic or easier to implement. This belief will become an accepted fact only if and when these parameterizations will be thoroughly validated against independent observations.

There are a number of atmospheric processes that are only poorly included in current models. For example, there is growing evidence that gravity wave drag or cumulus convection are important dynamical processes for the energetics of the atmosphere. The proper treatment of these processes will require higher resolution models, although spatial and temporal resolution does not appear to be the only limiting factor: new parameterizations, based on a better understanding of the physics involved, will have to be developed and validated.

The drive towards higher resolution models is therefore not coming only or even mostly from the need to include better surface processes, but finds its roots in the requirements of numerical weather prediction, atmospheric energetics, and other fields of atmospheric research.

The GCMs used for numerical weather prediction are usually of rather high resolution (T–63 and above are not uncommon), because of the need for the highest possible accuracy. This is feasible only for short time integrations because of the cost involved. When used for climate studies, GCMs are configured at a lower resolution, so that they can be integrated for longer time periods, typically a few months to a few years, and even decades if the objective is to study transient climatic change.

The current standard resolution of the NCAR CCM1 is R–15, which has 40 latitudes and 48 longitudes, although the configuration T–42, with 64 latitudes and 128 longitudes, may soon be adopted as the new standard of reference. Higher resolutions are already available, such as T–63, with 96 latitudes and 192 longitudes. While these are still rather expensive to use for long time integrations, it is likely that such resolutions (approaching 1° in both latitude and longitude) will become more common in the foreseeable future. Table 3 shows various resolutions available for the NCAR CCM1, and the corresponding number of gridpoints:

Table 3: Impact of the model resolution on the number of grid points and time steps.

Model resolution	Latitudes	Longitudes	# of grid points	Time step (s)	# of time steps per day	Factor increase
R–15	40	48	1,920	1800	48	1.00
T–42	64	128	8,192	900	96	8.53
T–63	96	192	18,432	600	144	28.80
T–106	160	320	51,200	300	288	160.00

4.2. The factors limiting GCM resolution

The spatial resolution of GCMs cannot be increased indefinitely, however, because the time step of integration cannot be selected arbitrarily. The failure of early numerical forecasts, as attempted by Richardson, has been attributed in part to the fact that his set of equations accepted gravity waves and other fast propagating waves as solutions: These perturbations generated rather high instantaneous time rates of changes, which, combined with long time steps, resulted in instabilities.

It is now well established that the time step of integration of a GCM using conventional numerical schemes is constrained by the Courant-Friedrichs-Lewy stability criterion:

$$\Delta t \leq \frac{\Delta x}{c}$$

where Δt is the time step of integration, Δx is the spatial resolution, *i.e.*, the distance between two grid points, and c is the speed of the fastest signal that is being propagated in the model (Henderson-Sellers and McGuffie, 1987, p. 137–140; Washington and Parkinson, 1986, Chapter 4). The fastest waves in the atmosphere are the sound waves, which travel at approximately 0.3 km s^{-1} in the lower atmosphere, but these are filtered out of the model by suitable manipulations of the primitive equations (*i.e.*, by imposing hydrostatic equilibrium). Other waves such as those associated with the jet stream can propagate at up to a hundred metres per second or more, and must be used to set the time step.

Because of this, higher resolution models become very expensive. The last column in the above table indicates how many more grid points and time steps there are in the higher resolution models (with the same vertical resolution) than in the current standard reference configuration R–15: it is an indication of the increase in CPU time required to carry out the integration. (This is an upper bound on the CPU time needed, because not all radiation computations are made at each time step; however, the actual monetary cost may increase even faster than this, if the charging algorithm is not linear with CPU and other resources usage.) Between 8 and 160 times more computing power will therefore be needed for doing just the same things as today in higher horizontal resolution, assuming that the vertical resolution remains at 12 levels. If the latter is also increased, even more computing power will be required, although new developments in computational methods may ease somewhat this constraint, as will be seen below.

It is also important to keep in mind that higher resolution versions of the CCM and other spectral models have their own problems: while they may better represent synoptic scale perturbations and follow the evolution of weather over a few days, their climate after longer integrations may not be as good as the one obtained from lower resolution versions. The reasons for this are not clear: It may be due to a fortuitous compensation of errors at a given resolution, and the fact that some errors increase while others decrease when the resolution increases. Indeed, parameterizations that are developed specifically for one resolution cannot be expected to work necessarily at other resolutions. Of course, lower resolution models would allow much longer integrations, but since the essence of surface processes resides in their subgrid scale, it is likely that higher resolution models will be more desirable to investigate them.

A second important consideration is the problem of the initialization of models. The higher the model resolution, the more data is required to set up the initial and boundary conditions. The acquisition, storage, and retrieval of huge data sets poses its own constraints and can itself be a computer intensive task. The addition of detailed physical, chemical and biological processes at the surface will only enhance the problem.

Last but not least, there always remains the problem of validating such models. While detailed and intensive field experiments such as FIFE (First ISLSCP Field Experiment) or HAPEX (Hydrologic-Atmospheric Pilot Experiment) can provide the basic data to design or improve parameterizations for subgrid scale processes, large scale observational programs (ISCCP, ISLSCP, TOGA, WOCE, WCP, etc.), whether surface-based or by satellite remote sensing, will have to provide the global coverage and uniform quality of data needed to validate the models. There is quite a lot of interest currently in the use of satellite remote sensing data to initialize and constrain the evolution of atmospheric models, but the merging of these two technologies is still in its infancy (Thomas and Henderson-Sellers, 1987).

4.3. Computing trends

The feasibility of carrying out the scientific objectives and modeling trends described above must be evaluated in the context of the probable evolution of the computing environment. Even though the power and capability of computers has increased dramatically over the past few decades, the lack or limitations of computing power remains the single most important constraint on the development and implementation of comprehensive global dynamic coupled climate models.

The following table shows the expected evolution of the computing power of Cray computers over the next 5 years or so. Similar trends can be expected from competitors.

Table 4: Projection of Cray super-computer power

Model	Avail. (year)	CPUs (#)	Cycle time (ns)	Memory (Mwords)	SSD (Mwords)	Pk. Perf. (MFLOPS)	Factor increase
X-MP	1985	4	8.5	16	512	941	1.00
Y-MP	1988	8	6.0	128	1024	2666	2.83
Cray 3	1989	16	2.0	512	0	16000	17.00
MP-64	1992	64	2.0	1024	4096	64000	68.00

SSD: Solid State Disk

Pk. Perf.: Peak performance rating

MFLOPS: Million floating point operations per second

It must be kept in mind that the actual sustained performance of a super-computer is only a fraction of the peak performance (Hack, 1986), but the last column of the table remains unchanged. Super-computers in the mid nineteen nineties are expected to be some 68 times more powerful (in terms of MFLOPS) than the X-MP currently in use at such places as the European Center for Medium Range Weather Forecasting or NCAR. This increase is obtained not only by manufacturing faster components and increasing the clock speed of the computer, but also by using multiple processors running in parallel. Throughput increases can also be expected from improvements in the area of input/output and in storage capacities.

It is obvious that the software will have to be adapted or written specifically to take advantage of such features. At NCAR, much of the radiation and dynamics of the CCM1 are currently vectorized, but only preliminary attempts have been made so far to exploit parallel computing facilities. Furthermore, the extensive use of parallel computing poses its own problems and cannot be viewed yet as a cure-all solution (Hack, 1988). More details on the future of computing will be given in Buzbee (1988), later in this volume. Equally evident is the fact that the expected increase in computing power could be entirely spent on increasing the model resolution alone, without any additional physics included.

5. Discussion

With this information, it is possible to make projections of the availability of computing power to handle surface processes over the next 5 years or more. This is achieved by considering various scenarios, in which the projected computing power is allocated to improved surface parameterizations, longer time integrations, higher resolution models, or some combination of these opportunities.

5.1. Low resolution model

To quantify this allocation process, let f_s be the fraction of time currently spent by the CCM1 on surface processes (4%), and f'_s the fraction allocated in 1992, under each scenario. If P designates the factor increase in computing power for a supercomputer at that time (68 as per Table 4), then

$$i_s = \frac{f'_s}{f_s} P$$

$$i_o = \frac{1 - f'_s}{1 - f_s} P$$

where i_s is the factor increase in computing power that should become available for surface processes, and i_o the factor increase in computing power that should be available for all other processes, or for carrying out longer time integrations than now.

Four scenarios will be considered, to give a range of options. The simplest possible scenario (R–15/1) would be to leave the CCM1 as it currently exists at NCAR, in R–15 configuration. The parameterizations could be improved, but only to the extent that the total requirements in computing resources are unchanged. In the second case (R–15/2), the allocation of surface processes will be decreased from the current 4% to some 3% of the total time required by the model. The rationale could be to free up as much time as possible to accomodate other computer intensive tasks, such as a mixed-layer ocean. The third case (R–15/3) could be to increase the current allocation from 4% to 6%, to allow better parameterizations of surface processes while leaving a sizeable increase for other processes, and the final option (R–15/4) could be to increase the allocation of surface processes to 12%, in recognition of their role and importance in many geochemical and biological processes. One application of such a model could be to study the transient response of the climate system to changes in the carbon cycle with an interactive model of the biosphere. The net increases in computing power available for surface processes and for other uses under these four scenarios are indicated in the following table:

Table 5: Low resolution scenarios

Scenario	f'_s/f_s (%/%)	P (×)	i_s (×)	i_o (×)
R–15/1	4/4	68.00	68.00	68.00
R–15/2	3/4	68.00	51.00	68.71
R–15/3	6/4	68.00	102.00	66.58
R–15/4	12/4	68.00	204.00	62.33

Clearly, even a large relative increase of allocation to such a small model component as surface processes still allows sizeable increases in other areas, longer integrations, or both. For example, five years from now, R–15 simulations that currently cover a couple of model years could extend over a century, irrespective of the increase in allocation for surface processes, and still leave some room for improvements in other areas of the model.

5.2. Higher resolution models

It is quite possible, however, that the standard spatial and temporal resolution of the models will increase during this period, for the reasons explained above. The same scenarios can be considered with higher resolution models, if the availability of increased computing power is corrected for the impact of the different space and time resolution. Table 6 shows the results for the T–42 and T–63 configurations:

Table 6: Higher resolution scenarios

Scenario	f_s'/f_s (%/%)	P (×)	i_s (×)	i_o (×)
T–42/1	4/4	68.00/8.53	7.97	7.97
T–42/2	3/4	68.00/8.53	5.98	8.05
T–42/3	6/4	68.00/8.53	11.96	7.81
T–42/4	12/4	68.00/8.53	23.92	7.31
T–63/1	4/4	68.00/28.8	2.36	2.36
T–63/2	3/4	68.00/28.8	1.77	2.39
T–63/3	6/4	68.00/28.8	3.54	2.31
T–63/4	12/4	68.00/28.8	7.08	2.16

From these tables, it can be seen that model resolution has a much stronger impact on the availability of computing power for 'other objectives' (i_o) than changes in allocation for surface processes. It will therefore clearly be possible to increase the allocation of computing resources for surface processes parameterization in future models, and other questions must be raised, such as "What spatial and temporal resolution would be required to model surface processes in a useful way for climatic purposes?"

5.3. New surface packages

It therefore appears that surface parameterizations can be improved considerably without affecting much the rest of the model. However, as noted above, soil, vegetation and land use vary on much smaller scales than those explicitly resolved by high resolution GCMs. Indeed, even a space resolution of 1° latitude by 1° longitude, which is quite remarkable for the atmospheric system, still covers thousands of km^2 of land, depending on latitude.

Since important processes occur at all space scales, it will never be possible to design a surface package with high enough a resolution to describe all processes explicitly. Furthermore, the higher the resolution, the higher the variability and complexity of the surface fields. Since there will always be important smaller scale processes (transpiration through plant stomates, for example), parameterizations will always be necessary.

One way to specify a meaningful spatial resolution is to consider some of the spatial gradients of interest, and choose the resolution necessary to adequately describe them. For example, the Sahel region of Africa is of great scientific interest because, within a latitudinal band of a few degrees, the vegetation changes from tropical forest to desert. Biomass, and in particular primary productivity in this region is extremely dependent on the timing and amount of precipitations, and climate variability in general. To study these problems, a minimum latitudinal resolution of 50 km is typically required. Similarly, if the ecological variations along a topographical gradient must be resolved, an even higher resolution may be required.

Another objective approach to the same question consists in looking at the mixing and advection scales of the processes involved. If some characteristic time scale can be defined for a process of interest, and if the horizontal wind speed characterizes the advection of a relevant quantity, then a horizontal scale is automatically implied. The following table shows the space scale implied by various time scales and wind speeds.

Table 7: Mixing and advection scales

	1 min	1 hr	1 day
1 m s^{-1}	60 m	3.6 km	86.4 km
5 m s^{-1}	300 m	18.0 km	432.0 km
10 m s^{-1}	600 m	36.0 km	864.0 km

The idea behind this approach is therefore that the elementary horizontal areas should be large enough to allow mixing and some 'smoothing' of the natural variability, and small enough to describe the larger scale variance. Clearly, if one is to resolve the daily cycle, a spatial scale of one to a few tens of km would be most appropriate. The next question is then to see whether this would be realistic at all.

5.4. Feasibility

Such a high spatial resolution, however, would be impractical and probably not necessary for the atmospheric component of the model, because of the very short time steps that would be required with current numerical methods. Since the atmospheric component of GCMs requires a moderate spatial but a rather high temporal resolution, and since the detailed treatment of surface processes requires a very high spatial and a moderate temporal resolution, it follows that the best modeling strategy would be to separate the two climatic sub-systems inside the GCM, and couple them as and when needed, possibly asynchronously.

This suggestion is intriguing but not without precedent: There is considerable interest currently in the nesting of a limited-extent mesoscale model inside a global GCM. Such a procedure results in much higher horizontal resolution over a region of interest. The modeling of surface processes as suggested above would in effect be equivalent, except that it would cover only a limited vertical extent at all continental grid points. Another example is provided by the treatment of oceanic processes in atmospheric GCMs: the space and time constants of the oceans are so different from those of the atmosphere that long integrations must be conducted with asynchronous coupling.

The challenge, then, would be to develop a high resolution surface package that would describe the state and evolution of the surface (albedo, temperature, fluxes of water, energy, carbon, etc.) on the basis of the large scale atmospheric forcing, and feed information back to the atmospheric component at the space and time scales it needs. Such a package should be designed with enough flexibility to adapt to various resolutions of the atmospheric component, and could be validated at a scale more compatible with the data from field campaigns and remote sensing platforms.

Let's assume, then, that the surface could be described for 'grid squares' of 40 km on the side. According to Table 7, this scale would be be appropriate to study surface processes that needed to be resolved at least every hour, with horizontal winds of up to 10 m s^{-1}. Such a resolution might allow convenient comparisons with data from satellite remote sensing platforms and/or applications at a regional or watershed level. Since the continental area is 1.48 10^8 km^2, surface properties would have to be specified for approximately 9.25 10^4 elementary areas. Current detailed surface packages need of the order of 40 to 80 parameters to specify the state of the surface: Pinty and Mascart (1988) use 39, Sellers (1986) needs 44, and Dickinson *et al.* (1986) specify over 80. I will assume that 100 parameters or variables may be needed for adequately describing the surface at each grid point.

Two extreme approaches can be considered to estimate the strain on the computing resources, and therefore the feasibility of the project: Either all surface parameters are to be computed by the CPU, or they all are imposed at each time step, therefore loading the I/O channels. For the first case, the number of floating point operations involved becomes the critical factor. From the statistics provided in Tables 2 and 3, it is found that the R–15 and T–42 configurations of the CCM use from 6 10^{-5} to 7.5 10^{-5} second of computing time per continental point and per time step. Assuming further that the sustained performance of a super-computer is about half its rated peak performance (Table 4), this corresponds to 26,000 to 34,000 floating point operations per grid point and per time step on a Cray X-MP.

Since the improved surface package suggested above may require 3 to 10 times more parameters per grid point than currently, I will further assume that this new package will need an order of magnitude more floating point operations, or 3 10^5 FLOPs per grid point. The total amount of computation for the entire continental surface would therefore represent 2.78 10^4 MFLOPs, and this could be achieved in less than a second by 1992, if the projections of Table 4 are realized.

In the second case, the values of all surface parameters are reset at each grid point and at each time step by specifying their value, presumably from observations. The 9.25 10^6 numbers needed to describe the global continental surface represent some 3.0 10^8 bits. If the super-computer was able to integrate the entire model in 6 s, a data transfer rate of the order of 50 Mbps would be required. This is well within current or expected technological capabilities. (NASA' EOS platforms are expected to yield operating data rates of up to 290 Mbps (Butler *et al.*, 1986, p. 17)).

Both approaches would be technically feasible, even though these would be extreme situations: In the case of climate simulations, some of the parameters would be initialized once and for all, some would be reset from observational data bases periodically, and the rest would be recomputed as and when needed.

5.5. Other issues

The incorporation of a high resolution surface package into a lower resolution atmospheric model will pose new problems. For example, if the latter generates synoptic scale precipitations over a GCM grid square, the surface package will have to distribute it among the smaller surface grid squares, probably using a statistical approach, accounting for orography. Transfering information back to the GCM might be somewhat easier, and would involve some intelligent aggregation of data.

Another question will be: Where and when should the coupling take place? It will make sense to separate the sub-systems at those locations and times when the interactions are least variable or sensitive to the internal behavior of either sub-system. In this sense, the surface—as defined in the introduction—is probably not the best interface. It might be much better to interface a free atmosphere model with a shallow boundary layer model that would include the surface. Further investigations on the nature of coupling sub-systems with very different response times will be required to help choose the optimal sub-system boundaries, and their respective space and time resolutions (Schneider and Harvey, 1986; Harvey, 1986).

The development of high resolution surface packages to be included in GCMs would offer a host of new opportunities, such as the dynamical modeling of primary biogeochemical cycles, or at least their fast-response components. Water is the only chemical whose cycle is explicitly included in general circulation models today. It would be of great interest to develop a dynamic carbon cycle in the atmosphere, biosphere and upper ocean, and this will require sources and sinks at the surface to represent the biosphere as well as human activities. It is also likely that once parameterizations have been developed for one chemical, it will become easier to incorporate others, or to address similar issues, such as the mobilization, transport, and deposition of dust on a global basis. This problem has already received some attention in the context of the nuclear winter issue, but many more interesting questions, such as the interpretation of paleo-climatic data, could be investigated with a climate model that could pick-up and transport volcanic dust or pollen.

The availability of an interactive biosphere at the scale described above would allow one to generate quantitative climate impact assessments on a global basis. Issues such as agricultural productivity, ecological succession, or desertification in the face of climatic change or as factors of climatic change could be addressed in a very different light with such a tool. Processes of soil formation and erosion may have to be accounted for in very long integrations, since plant primary productivity may decrease faster due to losses of soil than to climatic change, if the current rates of accelerated soil erosion are not abated over the next decade (WRI/IIED, 1986, p. 54).

It therefore seems that a promising approach, as far as climate research is concerned, would be to maintain a low to moderate resolution atmospheric GCM (such as R–15 or T–42), incorporate an oceanic component to generate interactive sea surface temperatures and a high resolution land surface package such as described above, and integrate for long periods of model time. This would be a very powerful tool of climate research.

6. Conclusion

In this paper, I have tried to assess the likely evolution of atmospheric GCMs in general, and the representation of surface processes in these models in particular. There is a trend towards higher resolution models for various reasons, including better numerical accuracy and the capability to incorporate more detailed processes, although preliminary computations at higher resolution do not always and necessarily improve the predictive skill or the model climate.

Significant increases in computing power are expected over the next few years, but these will not be sufficient to accomodate atmospheric GCMs with a resolution high enough for the study of Planet Earth as an integrated system. Environmental degradation, paleoclimatology, understanding the global water, energy and biogeochemical cycles, climate impact studies and many other theoretical or applied research objectives would greatly benefit from a global model describing all components of the climate system, and in particular the role of the biosphere in these interacting sub-systems.

It was suggested that the atmospheric and surface components of the climate system could be modeled separately, and possibly coupled asynchronously as and when needed.

This would allow the design of surface packages of very high resolution, which could be nested in lower resolution atmospheric models. The technical feasibility of describing continental surfaces at the linear spatial resolution of 40 km was quantitatively investigated in terms of its requirements for computing resources. It was found that the computational load could be as low as 1 second of CPU time before the mid-1990's if the projections of computing power increase for the next few years are realized. Some of the issues inherent in such an approach were raised, and potential uses of this powerful research tool were outlined.

Acknowledgments

J. Hack, L. Bath, B. Briegleb, R. Chervin and P. Kennedy provided the raw data displayed in Tables 2 to 4 or used in the text. In the course of writing this paper, I have benefited from discussions with J. Kiehl, P. Rasch and J. Hack, and their input is greatly appreciated. A. Henderson-Sellers, P. Martin, I. Muszynski and B. Pinty reviewed the manuscript and made a number of suggestions for improving both the content and the language. Y. Sud, S. Schneider and R. Dickinson also brought up additional issues to my attention. I am grateful to all of them for their support and encouragements. This research was supported by NASA under grant NASA-S-56469.

References

Ajtay, G. L., P. Ketner, and P. Duvigneaud (1979) 'Terrestrial Primary Production and Phytomass', in *The Global Carbon Cycle*, Ed. by B. Bolin, E. Degens, S. Kempe, and P. Ketner, SCOPE 13, Wiley & Sons, Chichester, 129–181.

Bagnold, R. A. (1943) *The Physics of Blown Sand and Desert Dunes*, William Morrow & Co., New York.

Breed, C. and J. F. McCauley (1986) 'Use of Dust Storm Observations on Satellite Images to Identify Areas Vulnerable to Severe Wind Erosion', *Climatic Change*, **9**, 243–258.

Butler, D. M. *et al.*. (1986) *From Pattern to Process: The Strategy of the Earth Observing System*, EOS Science Steering Committee Report, Volume II, NASA, Washington D.C.

Bryson, R. A. (1974) 'A Perspective on Climatic Change', *Science*, **184**, 753–760.

Bryson, R. A. and D. A. Baerreis (1967) 'Possibilities of Major Climatic Modification and Their Implications: Northwest India, A Case for Study', *Bulletin of the American Meteorological Society*, **48**, 136–142.

Buzbee, W. (1988) 'Super Computing Versus Parallel Computing', Paper presented at the Conference on "Climate and Geo-Sciences", Louvain-la-Neuve, Belgium, May 1988.

Carlson, T. N. and J. M. Prospero (1972) 'The Large Scale Movement of Saharan Air Outbreaks over the Northern Equatorial Atlantic', *Journal of Applied Meteorology*, **11**, 283–297.

Carlson, T. N. and S. G. Benjamin (1980) 'Radiative Heating Rates for Saharan Dust', *Journal of the Atmospheric Sciences*, **37**, 193–213.

Dickinson, R. E. (1980) 'Planetary Waves (Theory and Observations)', in World Meteorological Organization GARP Publications Series No. **23**, 53–87.

Dickinson, R. E., J. Jäger, W. Washington, and R. Wolski (1981) *Boundary Subroutine for the NCAR Global Climate Model*, NCAR Technical Note TN-173+IA, 75 pp.

Dickinson, R. E. (1983) 'Land Surface Processes and Climate—Surface Albedos and Energy Balance', in *Advances in Geophysics*, **25**, Academic Press, 305–353.

Dickinson, R. E. (1984) 'Modeling Evapotranspiration for Three Dimensional Global Climate Models', in *Climate Processes and Climate Sensitivity*, Ed. by J. Hansen and T. Takahashi, American Geophysical Union, 58–72.

Dickinson, R. E., A. Henderson-Sellers, P. J. Kennedy, and M. F. Wilson (1986) *Biosphere-Atmosphere Transfer Scheme (BATS) for the NCAR Community Climate Model*, NCAR Technical Note TN-275+STR, 69 pp.

Dickinson, R. E. and A. Henderson-Sellers (1988) 'Modelling Tropical Deforestation: A Study of GCM Land Surface Parameterizations', *Quarterly Journal of the Royal Meteorological Society*, in press.

Eagleson, P. S., Editor (1982) *Land Surface Processes in Atmospheric General Circulation Models*, Cambridge University Press, Cambridge, United Kingdom.

Fuchs, N. A. (1964) *The Mechanics of Aerosols*, Translated by R. E. Daisley, M. Fuchs, and C. N. Davies, Revised and Enlarged Edition, Pergamon Press, New York, 408 pp.

Gillette, D. (1980) 'Major Contributions of Natural Primary Continental Aerosols: Source Mechanisms', *Annals of the New York Academy of Sciences*, **338**, 348–358.

Hack, J. J. (1986) 'Peak versus Sustained Performance in Highly Concurrent Vector Machines', *Computer Magazine*, September 1986, IEEE, 11–19.

Hack, J. J. (1988) 'On the Promise of General-Purpose Parallel Computing', *Parallel Computing*, in press.

Harvey, L. (1986) 'Computational Efficiency and Accuracy of Methods for Asynchronously Coupling Atmosphere-Ocean Climate Models. Part 2: Testing with Seasonal Cycle', *Journal of Physical Oceanography*, **16**, 11–24.

Henderson-Sellers, A. and K. McGuffie (1987) *A Climate Modelling Primer*, J. Wiley & Sons, Chichester, United Kingdom, 217 pp.

Kutzbach, J. E. (1981) 'Monsoon Climate of the Early Holocene: Climatic Experiment Using the Earth's Orbital Parameters for 9000 Years Ago', *Science*, **214**, 59–61.

Kutzbach, J. E. and P. J. Guetter (1984a) 'Sensitivity of Late-Glacial and Holocene Climates to the Combined Effects of Orbital Parameter Changes and Lower Boiundary Changes: Snapshot Simulations with a General Circulation Model for 18, 9, and 6 ka BP', *Annals of Glaciology*, **5**, 85–87.

Kutzbach, J. E. and P. J. Guetter (1984b) 'The Sensitivity of Monsoon Climates to Orbital Parameter Changes for 9000 Years BP: Experiments with the NCAR General Circulation Model', *Milankovitch and Climate*, Part 2, A. Berger *et al.*, editors, D. Reidel Publishing Co., Dordrecht, 801–820.

Lorenz, E. N. (1967) *The Nature and Theory of the General Circulation of the Atmosphere*, World Meteorological Organization, WMO No. **218**, TP 115, 161 pp.

Lorenz, E. N. (1982) 'Atmospheric Predictability Experiments With a Large Numerical Model', *Tellus*, **34**, 505–513.

Manabe, S. (1969) 'Climate and Ocean Circulation. I. The atmospheric Circulation and the Hydrology of the Earth's Surface', *Monthly Weather Review*, **97**, 739–774.

Manabe, S. and J. L. Holloway (1975) 'The Seasonal Variation of the Hydrologic Cycle as Simulated by a Global Model of the Atmosphere', *Journal of Geophysical Research*, **90**, 1617–1649.

Manabe, S. and A. J. Broccoli (1985) 'The Influence of Continental Ice Sheets on the Climate of an Ice Age', *Journal of Geophysical Review*, **90**, 2167–2190.

McKee, E. D. Editor (1979) *A Study of Global Sand Seas*, Geological Survey Professional Paper 1052, 429 pp.

Mintz, Y. (1984) 'The Sensitivity of Numerically Simulated Climates to Land-Surface Boundary Conditions', in *The Global Climate*, Ed. by D. D. Houghton, Cambridge University Press, 79–105.

Morales, H. C. Editor (1979) *Saharan Dust: Mobilization, Transport, Deposition*, SCOPE Report 14, John Wiley.

Palmén, E. and C. W. Newton (1969) *Atmospheric Circulation Systems*, Academic Press, 603 pp.

Pinty, J. P. and P. Mascart (1988) 'Evaluation d'un modèle d'interface sol/végétation/atmosphère à l'aide de données SAMER et HYDRA de la campagne HAPEX-MOBILHY 1986', Note OPGC No. 96, Clermond-Ferrand, France.

Pittock, A. B., T. Ackerman, P. Crutzen, M. MacCracken, C. Shapiro, and T. Turco (1986) *Environmental Consequences of Nuclear War, Vol. 1, Physical and Atmospheric Effects*, SCOPE 28, John Wiley & Sons, Chichester, 359 pp.

Ramage, C. S. (1971) *Monsoon Meteorology*, Academic Press, 296 pp.

Schneider, S. and R. Dickinson (1974) 'Climate Modeling', *Journal of Geophysical Research*, **12**, 447–493.

Schneider, S. and L. Harvey (1986) 'Computational Efficiency and Accuracy of Methods for Asynchronously Coupling Atmosphere-Ocean Climate Models. Part 1: Testing with a Mean Annual Model', *Journal of Physical Oceanography*, **16**, 3–10.

Sellers, P. J., Y. Mintz, Y. Sud, and A. Dalcher (1986) 'A Simple Biosphere Model (SiB) for Use within General Circulation Models', *Journal of the Atmospheric Sciences*, **43**, 505–531.

Skidmore, E. L. (1986) 'Wind Erosion Climatic Erosivity', *Climatic Change*, **9**, 195–208.

Thomas, G. and A. Henderson-Sellers (1987) 'Evaluation of Satellite Derived Land Cover Characteristics for Global Climate Modelling', *Climatic Change*, **11**, 313–347.

Verstraete, M. M. and R. E. Dickinson (1986) 'Modeling Surface Processes in Atmospheric General Circulation Models', *Annales Geophysicae*, **4**, 357–364.

Washington, W. M. and C. L. Parkinson (1986) *An Introduction to Three-Dimensional Modeling*, University Science Books and Oxford University Press, 422 pp.

Washington, W. M. and G. A. Meehl (1983) 'General Circulation Model Experiments on the Climatic Effects Due to a Doubling and Quadrupling of Carbon Dioxide Concentration', *Journal of Geophysical Research*, **88**, 6600–6610.

Washington, W. M. and G. A. Meehl (1984) 'Seasonal Cycle Experiment on the Climate Sensitivity Due to a Doubling of CO_2 with an Atmospheric General Circulation Model Coupled to a Simple Mixed-Layer Ocean Model', *Journal of Geophysical Research*, **89**, 9475–9503.

Washington, W. M. and G. A. Meehl (1986) 'General Circulation Model CO_2 Sensitivity Experiments: Snow-Sea Ice Albedo Parameterizations and Globally Averaged Surface Air Temperature', *Climatic Change*, **8**, 231–241.

Wilson, M. F., A. Henderson-Sellers, R. E. Dickinson, and P. J. Kennedy (1987) 'Sensitivity of the Biosphere-Atmosphere Transfer Scheme (BATS) to the Inclusion of Variable Soil Characteristics', *Journal of Climate and Applied Meteorology*, **26**, 341–362.

TROPICAL DEFORESTATION AND CLIMATIC CHANGE

Norman Myers
Consultant in Environment and Development,
Upper Meadow, Old Road,
Headington, Oxford, United Kingdom

Introduction: the Putative Linkages

So far as we can discern, tropical forests play a significant role in atmospheric patterns, whether at local, regional or global levels. In turn, this means they contribute to the workings of climate, notably via their impact on rainfall regimes among other repercussions (for some overview treatments, see Dickinson, 1987; Houghton, 1984; Lawson, 1986; Prance, 1986; Rosenzweig and Dickinson, 1986; Reynolds and Thompson, 1987). This raises important implications for development in the humid tropics with their two billion people--especially for agriculture and hence for the capacity of human communities to feed themselves. A number of tropical crops are more vulnerable to climate than is usually the case with temperate-zone crops; and in many parts of the humid tropics, a marginal shortfall in precipitation can cause a substantial shortfall in the outputs of several staple crops (Oram, 1986; Parry and Carter, 1985). In addition, tropical deforestation may eventually prove disruptive for climatic patterns in regions far outside the humid tropics.

The key question is this: "How far--that is, how much, and to what extent--can tropical deforestation change climate?" The question is increasingly asked. Yet climatologists are far from being able to supply substantive answers, let alone definitive answers. The best they can offer is some insights into some linkages. According to a number of observers (Dickinson and Henderson-Sellers, 1988; Hansen and Takahashi, 1984; Hare, 1985; Henderson-Sellers and Robinson, 1986; Hutchison and Hicks, 1985; Moore, 1986; and Shukla and Mintz, 1982), the putative interactions amount to this. Forests of all kinds exchange moisture and energy with the atmosphere more intensively that do other types of land-surface cover. The main mechanism lies with evapotranspiration, which in turn depends on three requirements: moisture in the soil; vegetation that transfers moisture from the soil to the atmosphere; and energy that converts the moisture into water vapour. Most of the energy in question comes from radiational heating of the land surface, and thus it depends on surface albedo (or reflectivity of the surface). In turn, the albedo depends on the vegetation, which in turn again depends on the soil moisture. As vegetation is eliminated, so there is less scope for evapotranspiration, and less moisture is dispatched into the atmosphere for recycling as rain. In addition, forests, by virtue of the water vapour they evapotranspire, tend to increase the cloudiness of the atmosphere above them, and clouds alter the climate of forest regions in terms of both temperature and rainfall.

A. Berger et al. (eds.), Climate and Geo-Sciences, 341–353.

Within this analytic framework, tropical forests are exceptionally important. Because of their geographic location, they receive a maximum of visible and especially of photochemically active, ultraviolet solar radiation. They comprise a large fraction of the Earth s biomass, and an even larger proportion of global productivity. As human disruption of these biomass stocks increases, so tropical forest ecosystems will become still more significant for climate change. Thus tropical forests, and human modification of their ecosystem workings, bear significantly on climate--all the more so if the tracts in question are large. Of course the exact response will be subject to a good deal of variation in accord with local conditions, depending too on how the largescale atmospheric circulation is altered.

In this paper, we shall explore some of the purported linkages between tropical deforestation and climatic change. But first, a couple of qualifications are in order. This is an unusually complex, under-researched and controversial issue. Second, and more important, we shall not gain a conclusive understanding of the linkages for quite a time to come. So policy makers will have to respond while in the semi-dark--a prospect from which many policy makers tend to shrink. But let us remind ourselves that decisions on the future of tropical forests will be made willy-nilly: if not by policy makers with the backing of systematized science, then by commercial loggers, cattle ranchers, slash-and-burn farmers and other agents of forest depletion who will take their decisions in light of such limited information as they have to guide their "planning processes."

Review of Tropical Deforestation Rates

Before we look at some empirical studies, let us briefly remind ourselves how fast tropical forests are being depleted. The amount of forest remaining in the humid tropics is now about 8 million sq. kms., or roughly half of what once existed (according to bioclimatic data). Almost 100,000 sq. kms. are being eliminated each year in complete and permanent fashion, mainly through conversion to agriculture for largescale ranching and smallscale cultivation; and at least another 100,000 sq. kms. are being grossly disrupted each year by heavy logging and intensive slash-and-burn farming (FAO and UNEP, 1982; Melillo *et al.*, 1985; Molofsky *et al.*, 1986; and Myers, 1980 and 1984). This means that 2.5 percent of the biome is being depleted each year in terms of viable forest cover with full biomass and ecological complexity. Some regions are being affected more rapidly than others.

If present exploitation patterns and land-use trends persist (for the most part they are accelerating), we shall find that by shortly into the next century there will be little forest left outside of two large blocs in the Zaire basin and western Brazilian Amazonia, plus some outlier portions in New Guinea and the Guyana highlands. Even these remnant tracts appear little likely to survive beyond a further few decades (Myers, 1985).

Impacts at Local Level

At local level there are some alleged linkages between decline in forest cover and decline in rainfall regimes. This has long been a contentious issue, and certain observers (notably Pereira, 1973; Lee, 1978; Lockwood, 1980; and Thompson, 1980) assert there is not, and cannot be, any connection between the two. But some recently emergent evidence suggests otherwise, based on findings from localities in widely different parts of the humid tropics.

1. PANAMA CANAL AREA

Deforestation in the Panama Canal area from the start of this century has resulted in the loss of most of the forest cover in the Canal area. At the same time there has been a steady decline in rainfall as measured at Barro Colorado Island in the middle of the Canal. The decline has occured at an average rate of 6.1 mm. per year (and 9.6 mm. at a site 8 kms. further north), until it has fallen to the mid-1980s level of 2650 mm. per year (Windsor et al., 1986). Yet this trend observed from stations in mid-Isthmus is not observed at stations on or near the Atlantic and Pacific coasts. Thus the process of convection--the daily heating and subsequent rise of moist air, followed by turbulence and thunderstorms (which is the principal source of mid-Canal rainfall)--has weakened over the middle of the area, but there has been no change in the supply of moisture advected from adjacent areas by prevailing winds along the coasts. Theoretically convection should be weakened by a reduction in local stocks of moisture; and the change in vegetation from forest cover to grassland (as has occurred in much of the Panama Canal area) diminishes the amount of moisture stored in the soil-vegetation layer.

2. IVORY COAST

In the southern half of Ivory Coast there has been progressive replacement of forest with croplands during the past 30 years. There has also been a steady dropoff in evapotranspiration rates, from 60 to 35 percent, i.e. from a level typical of forests to a level typical of savannah (while runoff has increased eight-fold). True, climate in the Ivory Coast is more affected by nearby oceanic influences than is the case with Amazonia. Nonetheless there seems to have been an association between deforestation and drying-out of formerly forested lands during a period of 30 years. Areas that were formerly excellent for cocoa growing are being abandoned because of less rainfall, less humidity, and longer and harsher dry seasons, plus lowering of water tables and drying up of wells (Spears, 1986).

3. SEVERAL PARTS OF INDIA

There are parallel phenomena reported in several parts of India. In the Western Karnataka area of the Western Ghats, for instance, Mener-

Homji (1985) has examined deforestation and rainfall records at 28 stations during the 1886-1982 period, and has evaluated them in terms of 12 rainfall criteria (6 related to rainfall overall, 6 to the number of rainy days). He has found that when deforestation has exceeded 15 percent within areas up to 16 kms. of the stations, there has been some degree of decline in rainfall regimes. At the Ooty station, for instance, the number of rainy days outside the monsoon season (monsoonal rainfall derives from outside the area) totalled 416 days for the period 1886-1890, but steadily fell away throughout the following eight decades until it totalled only 271 days (a 35-percent decline) during the 1978-1982 period. At Chotanagpur there used to be sufficient convectional rainfall to support tea plantations, but in the wake of deforestation there is now so little such rainfall that the tea plantations have been abandoned.

Similar accounts and analyses are reported, albeit with less detail, from Assam (Sarmah, 1976), the Ranchi Plateau (Warren, 1974), the Andaman and Nicobar Islands (Biswas, 1980), and the Western Ghats of southern India (Green and Minkowski, 1977).

4. PENINSULAR MALAYSIA

In the northwestern States of Penang and Kedah in Peninsular Malaysia, there has been not only a decline in rainfall associated with decline of forest cover, but a less predictable distribution of rainfall in both time and space. This has been documented throughout the past 75 years, and particularly since 1960 (Chan, 1986; Goh, 1980). As a result, some 20,000 hectares of paddy ricefields have been abandoned in this "rice bowl" of the Peninsula, and another 72,000 hectares have registered a marked dropoff in production, leading to an overall shortfall in the Peninsula's rice output of 27.5 percent in 1978.

5. PARTS OF THE PHILIPPINES

Several areas in the Philippines appear to feature a drop-off in rainfall associated with deforestation. On Mount Apo there used to be such dependable rainfall that the city of Davao installed no waterworks to ensure water supplies year-round. But during the 1960s' coffee boom there was much forest felling to make way for coffee plantations; and today there is an extended drought each year, to the extent that coconut plantations are drying out (Alfonso, 1987). Similar phenomena have been noted in Bicol, also in Mindanao; and in the Cagayan Valley of northeastern Luzon.

In addition to these five cases, there is a number of further instances reported, albeit with less documented detail. In southern China, the 20,000-sq. km. Xishuangbanna area has lost more than half its tropical forest cover in the last 30 years. Data from three stations during the 1954-1980 period show that annual rainfall has dropped off by an average of 4 percent, and by 15 percent during the dry season. Overall the climate has become drier and hotter during the warm season, and drier and colder during the cool season (Zhang, 1986). Similarly in Guanacaste Province of northwestern Costa Rica there has been a decline in lowland rainfall extending over 40-50 years, and apparently associated with broadscale deforestation (Fleming, 1986).

In none of these cases is it possible to postulate a direct and/or causative linkage between changes in vegetation cover and shifts in rainfall patterns. All manner of other variables could well be at work. Very few problem-specific experiments have been carried out, largely by virtue of the scale and complexity of the processes involved. Nonetheless, in light of the severe repercussions that may arise from the putative linkages, there is urgent need for research of systematic scale.

Let us hypothesize, meantime, that an additional key factor could lie with land surface roughness (Garrat, 1978; Hare, 1985; Rowntree et al., 1985; Sud and Smith, 1985). Through its influence on horizontal water-vapour transport, land surface roughness can significantly affect the distribution of convective rainfall. The main determinant of surface roughness is generally vegetation cover, in the form of grass, bushes and trees (also buildings). This roughness exerts a "drag" on winds moving across them, dissipating some of the winds' energy. Since deforestation reduces the roughness of the surface, this change could well diminish rainfall in some areas. Indeed it is sometimes thought (Sud et al., 1985) that changes in local surface roughness can be as important an influence as changes in surface albedo.

Impacts at Regional Level

1. AMAZONIA'S HYDROLOGICAL CYCLE

According to research with water isotopes (Lettau et al., 1979; Salati and Vose, 1984; Salati et al,. 1988; Shuttleworth et al., 1984), much of the moisture in Amazonia--between half and four-fifths in central and western sectors--remains within the ecosystem. That is, it is constantly evapotranspired by plants into the atmosphere, where it gathers in storm clouds before being precipitated back onto the forest (mean recycling time, 5.5 days). Thus much of the Amazonian forest represents a significant source of its own moisture: it does not have to depend on moisture imported from the ocean surrounds.

Were Amazonia to be widely deforested, there could be a pronounced decrease in the amount of moisture being evapotranspired into the atmosphere, leading to a significant decline in rainfall (Henderson-Sellers and Gornitz, 1984; Shuttleworth, 1988). A decline of just this scale alone would entrain profound and irreversible ecological changes in many parts of the basin. But more important still, it could trigger a self-reinforcing process of growing desiccation for remaining forest cover, with declining moisture stocks followed by yet more desiccation, and so forth. Eventually the repercussions could extend outside Amazonia, even to southern Brazil with its major agricultural lands.

How far a similar drying-out phenomenon could arise in other large tracts of tropical forest, notably that of the Zaire basin, is scarcely considered to date, let alone researched.

2. THE ALBEDO EFFECT

Much of the energy that converts surface moisture into water vapour comes from the sun's radiational heating of the land surface. The energy thus depends on surface albedo, or reflectant "shininess" of the land surface. In turn, the albedo depends on vegetation, which absorbs more heat than bare soil. Over thick vegetation there are vigorous thermal upcurrents taking moisture (provided by the same plant cover) up into the atmosphere, where it condenses as rain. Because of its influence on convection patterns, wind currents and hence rainfall regimes, the albedo effect constitutes a basic factor in controlling climate. In tropical-forest zones, albedo appears to be, together with surface moisture, the most important factor by which the land surface can affect climate.

When vegetation is removed from the Earth's surface in large amounts, the result is often a self-promoting cycle of albedo enhancement, leading to a new stable state of less warm soil, lower rainfall and sparser vegetation. There ensues a significant decrease in rainfall and evapotranspiration, as in cloud cover. Were albedo-derived processes to become disrupted through tropical deforestation, the repercussions could well extend, via altered patterns of air circulation, throughout an entire region. If, say, the whole of Amazonia's almost 5 million sq. kms. were to be deforested during a period of 35-50 years, giving way to grasslands and crops, the surface albedo is estimated to increase from 0.11 to 0.19, and rainfall to decrease by 0.5-0.7 mm. per day, with both evapotranspiration and cloud cover being reduced (Henderson-Sellers and Gornitz, 1984).

Impacts at Global Level: The Greenhouse Effect

By the first quarter of the next century at latest, and perhaps even by the turn of this century, we shall surely be experiencing the climatic dislocations of a planetary warming, stemming from buildup of carbon dioxide and other "greenhouse gases" in the global atmosphere (Bolin _et al._, 1986; U.S. Department of Energy, 1985). Associated with this general warming will be some redistribution of precipitation patterns. The link-up with tropical forests is significant, insofar as they appear to be a significant source of greenhouse gases.

First of all, consider carbon dioxide as a greenhouse gas. The buildup of carbon dioxide in the atmosphere stems primarily from combustion of fossil fuels, to an extent of some 5.2 gigatons a year. But there is also a sizeable contribution from burning of biomass, notably in tropical forests (Houghton *et al.*, 1987; Woodwell *et al.*, 1983). An acceptable figure for this biotic contribution to date is 1.9 gigatons a year, within a range of 0.9 to 2.7 gigatons.

In addition to carbon dioxide, there are a number of trace gases that molecule for molecule are far more efficient than carbon dioxide in absorbing infrared radiation from the planetary surface. These gases include nitrous oxide and methane, some of the main sources of which could well be tropical forests (Bolle *et al.*, 1986; Goreau and de Mello, 1988; Harriss *et al.*, 1988; McElroy and Wofsy, 1986; Keller *et al.*, 1986; Mooney *et al.*, 1987). Other sources include hydrocarbon fuels, farms and oceans, whose contributions are moderately well documented--by contrast with the case for tropical forests, which have been all too little investigated.

Potential Synergistic Interactions

While the individual sector linkages as set out above appear to be significant in themselves, their most potent repercussions could eventually lie with synergistic interactions. These have scarcely been identified to date, let alone defined and analysed. But we should surely consider what might happen, for instance, when certain tropical forests desiccate in the wake of hydrological disruptions (as per the Salati analysis for Amazonia above): this could lead to an increase in wild fires, thus compounding the emissions of trace gases. In a greenhouse-affected world, warmer temperatures may well upset albedo workings; and many plant species could suffer acutely if not terminally through respiratory losses of carbohydrates (though, conversely, increased concentrations of carbon dioxide could promote their survival). Greater warmth may also cause convective rainfall to become more intense and concentrated, possibly leading to greater flooding and erosion, while leaving less moisture in the soil and leading to more stressful dry seasons. Methano-genic production appears to expand with warmer temperature, so methane emissions could increase, thus strengthening the greenhouse-gas feedback. For further discussion, see Dickinson and Virgi (1987).

Many other synergistic responses could surely be at issue. They remain almost entirely uninvestigated.

Policy Repercussions

How shall policymakers respond to a rapidly-changing situation? Clearly the deforestation issue carries profound implications for development questions of several sorts. Conversely of course, conservation of tropical forests, with their high metabolic and biogeochemical recycling rates, could increase the ability of the biosphere overall to buffer all manner of changes in atmospheric chemistry arising from human activities--most notably the greenhouse effect.

Despite these key considerations, many long-range policy and planning decisions are being taken with respect to food production in the humid tropics, especially irrigation systems (plus many other broad-scope decisions as concerns, for example, hydropower facilities) on the assumption that climatic patterns of the future will be a simple extension of the past. Yet as we have seen, there is fast-gathering evidence that we face a major departure in rainfall regimes among other climatic patterns within the foreseeable future. Thus there is a growing premium on taking measures, especially preventative measures, while there is still a little time for manoeuver.

Two initiatives immediately arise for policymakers. The first is to generate mission-oriented research of a scope and scale that matches the issues. As tropical forests disappear, so humankind is witnessing an experiment with worldwide and presumably irreversible consequences. The repercussions for human welfare will be pervasive and profound. Yet we are conducting this superscale experiment in entirely unplanned manner. The amount of funding and other management resources that are allocated to the phenomenon is trifling as compared with some possible concealed costs of the experiment's outcome. Herein, then, lies a first and ultra-urgent recommendation: an emphatic expansion in research.

The second initiative available to policymakers is to adopt a risk-reduction approach toward the deforestation problem. In light of the broad-scope and adverse repercussions of tropical deforestation, the path of prudence dictates a safety-first response. That is to say, policymakers should pursue a strategy that seeks to respond to both the uncertainty inherent in the situation, and to the considerable risk factor (risk being defined here as the adverse outcome of a circumstance, multiplied by the likelihood of the unfortunate outcome actually coming to pass). Adopting a wait-and-see attitude--which is essentially what policymakers are doing right now--presumes that we shall sufficiently anticipate the amount and rate of tropical deforestation, plus its impacts, and that we are confident we can handle all the possible climatic repercussions.

The urgency of the issue is highlighted by recent evidence that a disruption of rainfall patterns on a hemispheric scale is already underway (Bradley _et al._, 1987). Over the past 20 to 30 years, a steady decline in rainfall over northern subtropical latitudes has been accompanied by a systematic increase in precipitation over higher latitudes. While the decline in rainfall has been most marked over northern Africa and the Sahel, it is also evident in Southeast Asia and around the Caribbean (Kelly, 1988).

In sum, even were the prospect of climatic dislocations to be considered slight, the repercussions could prove so disruptive, adverse and widespread that even a super-slight prospect of the dislocations actually occurring should be reckoned worth avoiding unless the costs of doing so prove prohibitively high. Under circumstances of risk associated with climatic dislocations (and as in other areas of public policy where the salient factor lies with uncertainty), it will be better for us to find we have been approximately right than precisely wrong.

References

Alfonso, P. 1987. Personal communication. Department of Zoology, University of the Philippines at Los Banos, Philippines.

Biswas, B.C. 1980. 'Effect of Forest Cover on Rainfall Distribution at Andaman and Nicobar Islands.' Mausan **31**: 178-180.

Bolle, H.-J., W. Seiler and B. Bolin. 1986. Other Greenhouse Gases and Aerosols: Assessing Their Role for Atmospheric Radiative Transfer. In B. Bolin, B.R. Doos, J. Jagel, and R.A. Warrick, eds. The Greenhouse Effect: Climatic Change and Ecosystems: 157-203. John Wiley and Sons, New York.

Bolin, B., B.R. Doos, J. Jagel, and R.A. Warrick, eds. 1986. The Greenhouse Effect: Climatic Change and Ecosystems. John Wiley and Sons, New York.

Bradley, R.S., H.F. Diaz, J.K. Eischeid, P.D. Jones, P.M. Kelly and C.M. Goodess. 1987. 'Precipitation Fluctuations Over Northern Hemisphere Land Areas Since the Mid-19th Century.' Science **237**: 171-175.

Chan, N.W. 1986. 'Drought Trends in Northwestern Peninsular Malaysia: Is Less Rain Falling?' Wallaceana **44**: 8-9.

Dickinson, R.E., ed. 1987. The Geophysiology of Amazonia: Vegetation and Climate Interactions. John Wiley and Sons, Chichester, U.K. and New York.

Dickinson, R.E. and A. Henderson-Sellers. 1988. 'Modelling Tropical Deforestation: A Study of GCM Land-Surface Parameterizations.' Quarterly Journal of Royal Meteorological Society **114**: 439-462.

Dickinson, R.E. and H. Virgi. 1987. Climate Change in the Humid Tropics, Expecially Amazonia Over the Last Twenty Thousand Years. In R.E. Dickinson, ed., The Geophysiology of Amazonia: Vegetation and Climate Interactions: 91-105. John Wiley and Sons, Chichester, U.K. and New York.

FAO and UNEP. 1982. Tropical Forest Resources. Food and Agriculture Organization, Rome, Italy, and United Nations Environment Programme, Nairobi, Kenya.

Fleming, T.H. 1986. 'Secular Changes in Costa Rican Rainfall: Correlation with Elevation.' Journal of Tropical Ecology **2**: 87-91.

Garrat, J.I. 1978. 'Flux Profile Relations Above Tall Vegetation.' Quarterly Journal of Royal Meteorological Society **104**: 109-211.

Goh, K.C. 1980. 'Rainfall Trends and Deficiencies in Peninsular Malaysia.' The Oriental Geographer **24**: 25-40.

Goreau, T.J. and W.Z. de Mello. 1988. 'Tropical Deforestation: Some Effects on Atmospheric Chemistry.' Ambio **17**: 275-281.

Green, S. and K. Minkowski. 1977. The Lion-Tailed Monkey and Its South Indian Rain Forest Habitat. In P. Rainie, ed., Primate Conservation: 289-337. Academic Press, New York.

Hansen, J.E. and T. Takahashi, eds. 1984. 'Climate Processes and Climate Sensitivity.' Geophysical Monograph No: **29**: American Geophysical Union, Washington D.C.

Hare, F.K. 1985. Climate Variations, Drought and Desertification. World Meteorological Organization, Geneva, Switzerland.

Harriss, R.C. and 11 others. 1988. 'The Amazon Boundary Layer Experiment: Dry Season, 1985.' Journal of Geophysical Research **93**: 1351-1360.

Henderson-Sellers, A. and V. Gornitz. 1984. 'Possible Climatic Impacts of Land Cover Transformations, With Particular Emphasis on Tropical Deforestation.' Climatic Change **6**: 231-257.

Henderson-Sellers, A. and P.J. Robinson. 1986. Contemporary Climatology. Longman Scientific and Technical Books Ltd., London, U.K., and John Wiley, New York.

Houghton, J.T., ed. 1984. The Global Climate. Cambridge University Press, New York.

Hougnton, R.A. and 11 others. 1987. 'The Flux of Carbon From Terrestrial Ecosystems to the Atmosphere in 1980 Due to Changes in Land Use: Geographic Distribution of the Global Flux.' Tellus **39B**: 122-139.

Hutchison, B.A. and B.B. Hicks, eds. 1985. The Forest-Atmosphere Interaction. D. Reidel Publishing Company, Dordrecht, Netherlands, and Boston, Mass.

Keller, M., W.A. Kaplan and S.C. Wofsy. 1986. 'Emissions of N2O, CH4 and CO2 from Tropical Forest Soils.' Journal of Geophysical Research **91**: 11,791-11,802.

Kelly, P.M. 1988. Personal communication, Climatic Research Unit, University of East Anglia, Norwich, U.K.

Lawson, T.L. 1986. Deforestion and Induced Changes in Meso/Microclimate. In R. Lal, P.A. Sanchez and R.W. Cummings, eds., Land Clearing and Development in the Tropics: 195-202. A.A. Balkema Publishers, Rotterdam, Netherlands.

Lee, R. 1978. Forest Microclimatology. Columbia University Press, New York.

Lettau, H., K. Lettau and L.C. Molion. 1979. 'Amazonia's Hydrologic Cycle and the Role of Atmospheric Recycling in Assessing Deforestation Effects.' Monthly Weather Review **107**: 227-238.

Lockwood, J.G. 1980. 'Some Problems of Humid Equatorial Climates.' Malaysian Journal of Tropical Geography **1**: 12-20.

McElroy, M.B. and S.C. Wofsy. 1986. Tropical Forests: Interactions with the Atmosphere. In G.T. Prance, ed., Tropical Rain Forests and the World Atmosphere: 33-60. Westview Press, Boulder, Colorado.

Meher-Homji, V.M. 1985. Disasters of Deforestation: Desertification or Deluge? In Proceedings of the Conference on Forest Resources Crisis in the Third World: 131-143. Sahabat Alam Malaysia, Penang, Malaysia.

Melillo, J.M., C.A. Palm, R.A. Houghton, G.M. Woodwell and N. Myers. 1985. 'A Comparison of Two Recent Estimates of Disturbance in Tropical Forests.' Environmental Conservation **12(1)**: 37-40.

Molofsky, J., C.A.S. Hall and N. Myers. 1986. A Comparison of Tropical Forest Surveys. U.S. Department of Energy, Washington D.C.

Mooney, H.A., P.M. Vitousek and P.A. Matson. 1987. 'Exchange of Materials Between Terrestrial Ecosystems and the Atmosphere.' Science **238**: 926-932.

Moore, P. 1986. 'What Makes Rainforests So Special?' New Scientist August: 38-40.

Myers, N. 1980. Conversion of Tropical Moist Forests (report to National Academy of Sciences). National Research Council, Washington D.C.

Myers, N. 1984. The Primary Source: Tropical Forests and Our Future. W.W. Norton, New York.

Myers, N. 1985. 'Tropical Deforestation and Species Extinctions: The Latest News.' Futures **17**: 451-463.

Oram, P.A. 1986. 'Sensitivity of Agricultural Production to Climatic Change.' Climatic Change **7**: 129-152.

Parry, M.L. and T.R. Carter. 1985. 'The Effect of Climatic Variations on Agricultural Risk.' Climatic Change **7**: 95-110.

Pereira, H.C. 1973. Land Use and Water Resources in Temperate and Tropical Climates. Cambridge University Press, Cambridge, Mass.

Prance, G.T. ed. 1986. Tropical Forests and the World Atmosphere. Westview Press, Boulder, Colorado.

Reynolds, E.R.C. and F.B. Thompson, eds. 1987. Vegetation Change and Regional Climate. United Nations University, Tokyo, Japan.

Rosenzweig, C. and R.E. Dickinson, eds. 1986. Climate-Vegetation Interactions. University Corporation for Atmospheric Research, Boulder, Colorado.

Rowntree, P.R., M.F. Wilson and A.B. Sangster. 1985. Impact of Land Surface Variations on African Rainfall in General Circulation Models. Meteorological Office, Bracknell, Berks., U.K.

Salati, E. and P.B. Vose. 1984. Amazon Basin: A System in Equilibrium. Science 225: 129-138.

Salati, E. A.E. de Oliveira, H.O.R. Schubart. F.C. Novaes, M.J. Dourojeanni and J.C. Umana. 1988. Changes in the Amazon Over the Last 300 Years. In B.T. Turner, ed. Earth Transformed by Human Action. Clark University, Worcester, Mass.

Sarmah, S.K. 1976. Some Observations on Variations of Monsoon Annual Rainfall in Assam. In Proceedings of Symposium on Tropical Monsoons: 458-464. Indian Institute of Tropical Meteorology, Puna, India.

Shukla, J. and Y. Mintz. 1982. 'Influence of Land-Surface Evapotranspiration on the Earth's Climate.' Science **215**: 1498-1500.

Shuttleworth, W.J., J. Gash, C. Lloyd, C. Moore and J. Roberts. 1984. 'Eddy Correlation Measurements of Energy Partition for Amazonian Forest.' Quarterly Journal of the Royal Meteorological Society **110**: 1143-1162.

Shuttleworth, W.J. 1988. 'Evaporation from Amazonian Rain Forest.' Proceedings of Royal Society of London B. 233: 321-346.

Spears, J.S. 1986. Personal communication. The World Bank, Washington DC.

Sud, Y.C., J. Shukla and Y. Mintz. 1985. Influence of Land-Surface Roughness on Circulation and Rainfall. Third Conference on Climate Variations: 93-94. American Meteorological Society, Los Angeles.

Sud, Y.C. and W.E. Smith. 1985. 'Influence of Local Land-Surface Processes on the Indian Monsoon: A Numerical Study.' Journal of Climate and Applied Meteorology **24**: 1015-1036.

Thompson, K. 1980. 'Forests and Climate Change in America: Some Early Views.' Climatic Change **3**: 47-64.

U.S. Department of Energy, 1985. State of the Art Reports on Carbon Dioxide, four books. Carbon Dioxide Research Program, U.S. Department of Energy, Washington D.C.

Warren, W.D.M. 1974. 'A Study of Climate and Forest in the Ranchi Plateau.' Indian Forester **100**: 291-314.

Windsor, D.M., A.S. Rand and W.M. Rand. 1986. Variation in Rainfall on Barro Colorado Island. Smithsonian Tropical Research Institute, Balboa, Panama.

Woodwell, G.M. and six others. 1983. 'Global Deforestation: Contribution to Atmospheric Carbon Dioxide.' Science **222**: 1081-1086.

Zhang, K. 1986. 'The Influence of Deforestation of Tropical Rainforest on Local Climate and Disaster in Xishuangbanna Region of China.' Climatological Notes **35**: 223-236.

THE ROLE OF THE HYDROLOGICAL CYCLE IN CLIMATE

Professor James C.I. Dooge
Centre for Water Resources Research
University College, Earlsfort Terrace, Dublin 2.

ABSTRACT

The wide range of scale involved in hydrological modelling is emphasised. The known analytical solutions for the problem of soil-controlled ponded infiltration corresponding to different assumptions on soil moisture characteristics are listed and the predicted times to ponding for a typical silt loam are compared. These are found to be remarkably similar inspite of the widely different shapes of profile appropriate to the different functional assumptions for the variation of soil moisture characteristics with moisture content. Empirical relationships for the dependance of actual evaporation on long term average precipation and potential evaporation are compared on a dimensionless basis and found to be similar in shape but to differ widely in their sensitivity to climate change under arid conditions. Brief reference is made to the stochastic-dynamic approach in macro-hydrology. Finally some suggestions are made in relation to future research.

Scale Effects and Feedback Mechanisms

To understand the role of hydrologic processes, it is necessary to appreciate the key importance of the scale at which the analysis is made. The very existence of a hydrological cycle (i.e. the occurrence of water in all three phases and its transformation from one phase to another) depends on the concurrence of two circumstances, one on the scale of the water molecule (10^{-10}m) and the other on the scale of the earth's orbit (10^{11}m) (Dooge 1983).

The main concern of hydrology is with the solution of the water balance equation of the hydrologic system under examination. In a recent review (Dooge 1988), the writer suggested that scale relationships in hydrology can be conveniently discussed on the basis of the nine-fold division of hydrologic scales shown on Table A.

On the micro-scale of physical hydrology, the differential equation of continuity at a continuum point is combined with some parametrized form of momentum equation and the latter is in nearly all cases non-linear. A solution must therefore be sought either by analysis based on some simplification of the dynamic equation or else by numerical computation of each particular case. At the scale of catchment hydrology, a lumped form of the continuity equation is combined with an assumed relationship between the relevant variables which is physically plausible rather than physically-based.

A major difficulty in providing a sound hydrological input to climate modelling is that the bulk of physically based analysis of hydrologic processes relate to the micro-scales of Table A whereas the scale of interest for general circulation models relates to the macro-scales of that table. Unfortunately the high degree of spatial variability involved in the system parameters and the inherent non-linearity of the hydrologic processes does not lend itself to the statistical mechanics approach to the study of very large aggregates of identical deterministic micro-processes. Just as the nature of liquid water is intermediate between the simple deterministic structure of ice and the unstructured stochastic

A. Berger et al. (eds.), Climate and Geo-Sciences, 355–366.

TABLE A
SPATIAL SCALES IN HYDROLOGY

CLASS	SYSTEM	Typical Length (Metres)
	PLANETARY	10^7
MACRO	CONTINENTAL	10^6
	LARGE CATCHMENT	10^5
	SMALL CATCHMENT	10^4
MESO	SUB-CATCHMENT	10^3
	MODULE	10^2
	REPRESENTATIVE ELEMENTARY VOLUME	10^{-2}
MICRO	CONTINUUM POINT	10^{-5}
	MOLECULAR CLUSTER	10^{-8}

TABLE B

ANALYTICAL SOLUTIONS FOR INFILTRATION

D \ $K/\sqrt{D}$ =	constant	linear	quadratic
delta-function	Green and Ampt (1911)		
	Mein and Larson (1973)		
constant	Carslaw & Jaeger (1946)	Philip (1966)	Philip (1974)
	Braester (1973)	Braester (1973)	Clothier et al. (1981)
Fujita-function	Fujita (1952)	(not solved)	(not solved)
	Knight &Phillip (1974)	Rogers et al. (1983)	Sander et al. (1987)

Upper name: solution for constant surface concentratition
Lower name: solution for constant surface flux

complexity of water vapour, so too is the study of hydrologic systems caught in the intermediate area between simple deterministic mechanisms and large unorganised aggregates that is described as the area of organised complexity on Figure 1 which is due to Weinberg (1975).

Apart from the difficulty of constructing a satisfactory hydrologic model at the GCM scale, there is the added problem of coupling together at the land surface an atmospheric GCM and a hydrologic model. The fluxes of energy, sensible heat and water vapour across the land surface serve as inputs to each of these coupled models and set up in each case a series of complex feedback mechanisms, some positive and some negative. Figure 2 due to O'Kane (Brandyk et al 1985) is a Forrester-type diagram showing the direction of some of the feedback mechanisms which arise from a change in the moisture content of the upper soil layers. Thus it can be seen that soil moisture and net radiation are linked by eight pathways, four of them involving overall positive feedback and four involving overall negative feedback. Apart from the difficulty of superimposing these various physical feedbacks to estimate quantitative effects, there is the added complication of bio-geophysical feedback mechanisms (e.g. soil moisture content - vegetation cover - runoff - soil moisture).

The Micro-hydrological Approach

If the soil moisture content were always to be at "field capacity", then the actual evaporation (AE) would be at the potential rate (PE) which is obtained as an output of the atmospheric GCM. Similarly unless the surface is ponded (i.e. the surface layers are saturated) the rate of infiltration into the soil will be equal to the rate of precipitation which is also given as an output of the atmospheric GCM. In each of these cases the atmospheric model and the hydrologic model are decoupled in regard to the flux of moisture across the land surface. If, however, the soil moisture falls below field capacity during a dry period or the surface layers become saturated during a wet period, then the flux at the surface is controlled by the soil moisture profile and the two models become interactive.

A basis for conceptual models of the soil moisture at the scale of a GCM grid aimed at simulating the coupling of atmospheric processes and subsurface processes can be sought from certain analytical results in physical hydrology. In the microphysical approach to the analysis of the vertical movement of water in the unsaturated upper layers of the soil, the equation of continuity at a continuum point is combined with a second equation relating moisture flux to soil moisture in order to produce a solution for given boundary conditions. For the purposes of the present discussion, it is convenient to combine the two equations in the diffusion form

$$\frac{\partial c}{\partial t} = \frac{\partial}{\partial z}\left[D(c)\frac{\partial c}{\partial z} \right] + \frac{\partial K}{\partial z} + r(z,t) \qquad (1)$$

where c(z,t) is the local moisture content of the soil at elevation z and time t, r(z,t) is the local rate of supply of water (e.g. by surface or subsurface irrigation) or abstraction of water (e.g. by plant roots), D(c) is the hydraulic diffusivity of the soil and K(c) its unsaturated conductivity. The cases for which this equation can be solved for a semi-infinite soil column of uniform initial moisture content are summarised in Table B (Kühnel et al 1989a).

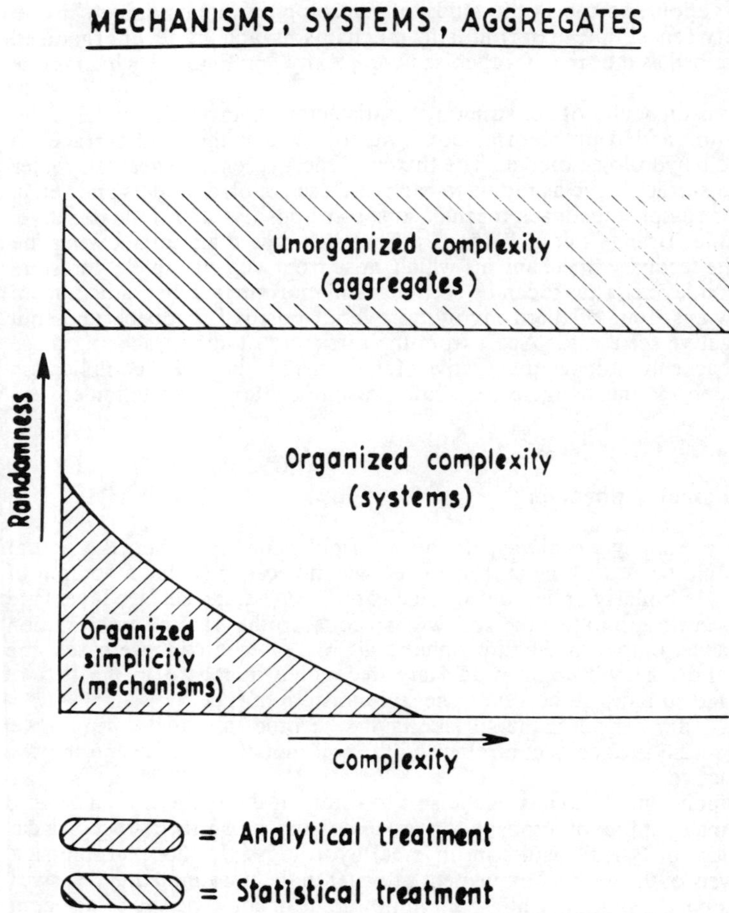

FIGURE 1 Mechanisms, systems, aggregates (Weinberg)

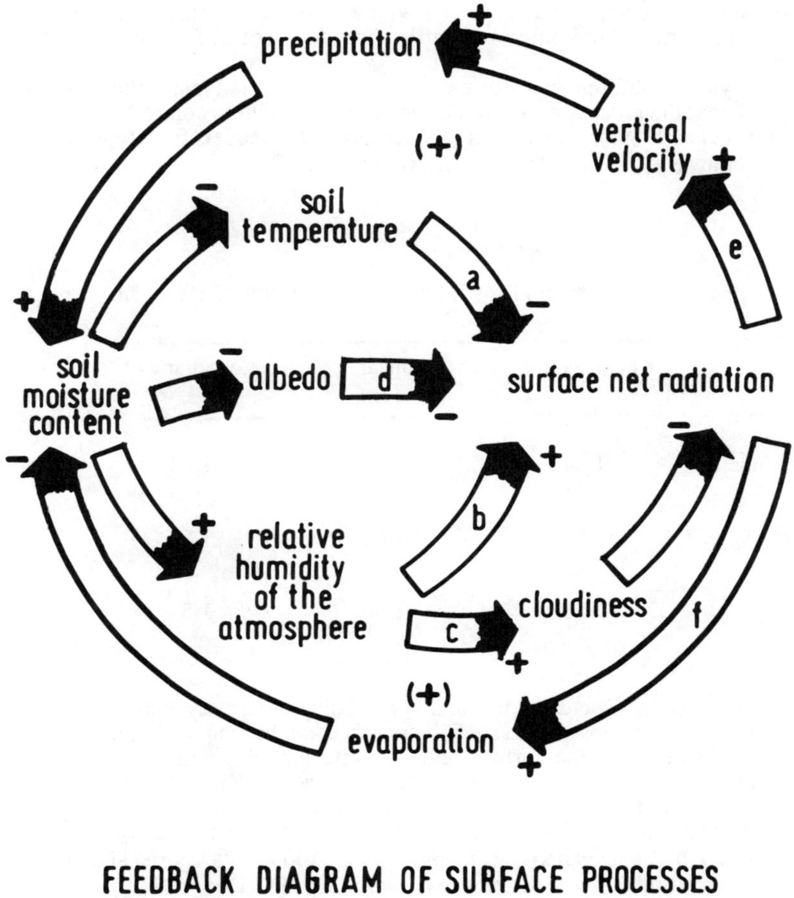

FIGURE 2 Feedback mechanisms of surface processes (O'Kane)

The shape of the soil moisture profile varies greatly with the assumption made but the rate (and consequently the cumulative volume) of infiltration does not differ much for the various solutions.

These analytical solutions can, for example, be used to calculate the time to ponding i.e. the time that the process of infiltration remains atmosphere controlled during constant rain (Kühnel et al 1989a). The results of a number of the models for a typical silt loam ($D = 200 cm^2/hour$, $K_{sat} = 0.1 cm/hr.$) are shown on Table C.

TABLE C

Time to ponding (hours) for silt loam with a rainfall of 10mm/hr.

D \ $D/\sqrt{K}$ =	constant	linear	quadratic
delta-function	1.34	1.34	1.34
constant	1.65	1.61	1.57
Fujita	1.58	1.56	1.53

It will be noted that this period of potential infiltration is relatively short compared with the length of inter-storm periods in most climates. The approach can be adapted in the case of upward fluxes of evaporation at the surface and the time for which the evaporation continues at the potential rate estimated for the different models (Kühnel et al 1988b). While microhydrology can offer some guidance, in these and other ways, to modelling the land surface processes at the GCM, the problem of parametrizing from local scale (lmm) to field scale (10 to 100 hectares) to catchment scale (100 to 1,000 km^2) and on to regional scale (10,000 to 100,000 km^2) remains extremely difficult (Dooge 1982).

Empirical approach to Macro-hydrology

In the hydrologic contribution to the modelling of land surface processes for climate purposes, the parametrization of hydrologic processes at the micro-scale must be complemented by an attempt to model the water balance at the macro-scale. The infinitisemal continuity equation is replaced by a lumped continuity equation relating to the total catchment area and to all forms of water storage, surface and subsurface. This may be written as

$$P(t) - E(t) - Q(t) = \frac{dS}{dt} \tag{2}$$

where P(t) is the precipitation, E(t) the actual evapotranspiration, Q(t) the total runoff (surface and subsurface), and S(t) the total storage. As in the micro-scale approach, a second relationship is required. In hydrologic modelling and in climate modelling empirical relationships are frequently invoked and only recently has a more systematic approach been attempted.

If a long term average relationship is sought, then equation (2) is simplified since the storage term on the right hand side of the equation becomes more and more insignificant compared to the other terms as the length of record considered is increased. The actual evapotranspiration is then given to the required accuracy by the difference between precipitation and runoff. The more arid the climate, the greater will be the proportion of soil-controlled evaporation and consequently the lower the ratio of actual evapotranspiration (AE) to potential evapotranspiration (PE). It is convenient to plot the results of long term measurements of catchment precipitation (P) and catchment runoff (Q) in terms of the variation of the ratio of actual to potential evapotranspiration (AE/PE) with the ratio of precipitation to potential evapotranspiration (P/PE). The latter ratio is in effect the reciprocal of the Budyko aridity index.

Typical empirical relationships which have been suggested are shown on Figure 3. Budyko and Zubenok (1961) examined the long term data for 1200 regions throughout the U.S.S.R. and found (a) that the data nearly all lay between the equation proposed by Schreiber in 1904 and that proposed by Ol'dekop in 1911, and (b) that the data fitted to geometrical mean of the two expressions within 10%. The simpler expression proposed by Turc (1954) and modified by Pike (1964) is very close to this geometric mean and much easier to handle. All these relationships reflect the asymptotic behaviour expected on physical grounds for the limiting cases of extreme aridity (AE = P) and extreme humidity (AE = PE) and are similar in general shape as shown on figure 3. However, this general similarity conceals some important differences that are of importance if such equations, based on long term equilibrium conditions, are to be applied to climate change. We can define a hydrological magnification factor for changes in runoff due to changes in precipitation (R_p) by

$$\frac{\Delta Q}{Q} = R_p \frac{\Delta P}{P} \tag{3}$$

and determine this factor analytically if the functional relationship between AE/PE and P/PE is given. When this is done, it is found (Dooge 1986) that, for increasingly humid conditions the hydrological magnification factors in all cases approach unity but that for increasingly arid conditions they diverge markedly. In the limit of extreme aridity (P/PE = 0), the Turc-Pike equation and the Ol'dekop equation give a magnification factor of 3 whereas the magnification factor for the Schreiber equation approaches infinity. It must be kept in mind however that the sensitivity of catchment runoff to changes in precipitation will be markedly affected by the assumption made in relation to direct effect of increased CO_2 on evapotranspiration (Wigley and Jones, 1985).

Models of dynamic catchment response with a plausible physical structure and lumped parameters are widely used in applied hydrology (Fleming 1975) but their accuracy does not always match their complexity (WMO 1975). Most of these catchment models involve a probability distribution of the variation of infiltration capacity over the catchment area which involves calibration of one or more extra parameters. Gleick (1987)

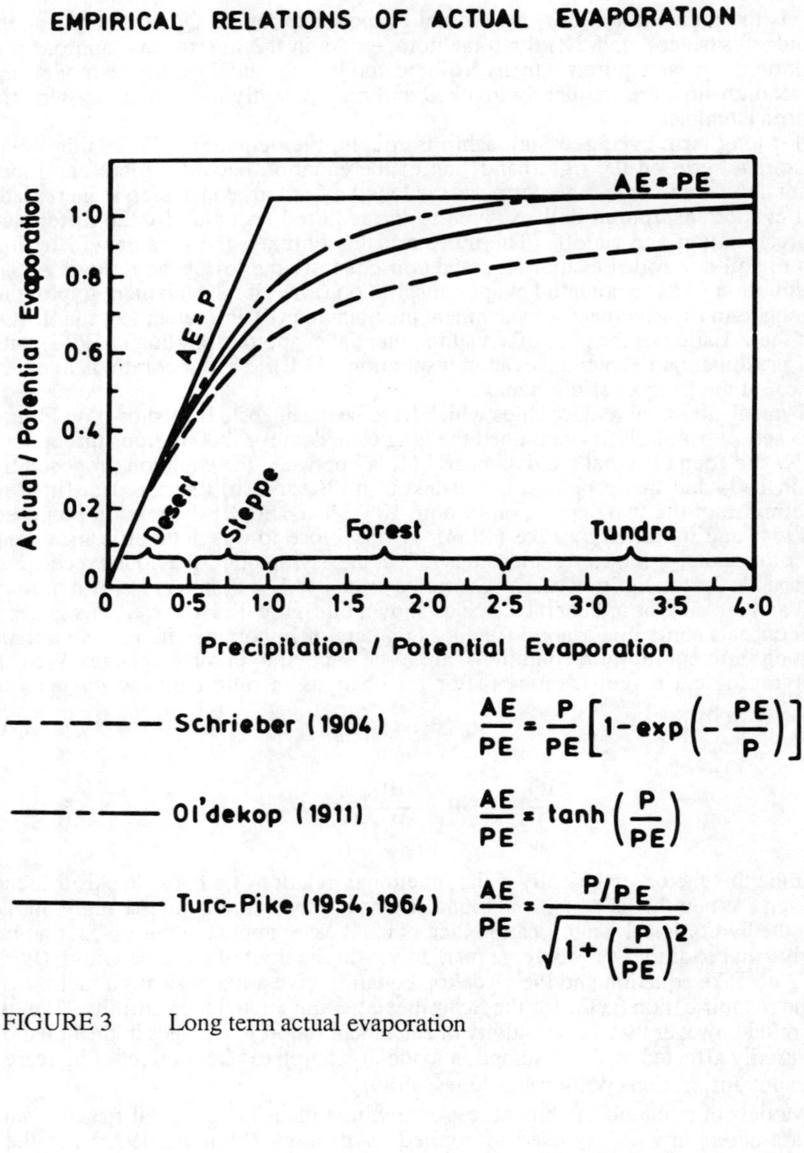

FIGURE 3 Long term actual evaporation

has demonstrated that for studying climate in a catchment area of 40,000 km^2 sufficient accuracy of the monthly water balance can be obtained by a careful division into two subcatchments based on differences in the form of precipitation and in soil conditions. The predictions from catchment models are also highly sensitive to the assumption about the direct effect of CO_2 on vegetation (Idso and Brazel, 1984).

The Stochastic-dynamic Approach

The most important attempt to provide a rational basis for the long term water balance is the work of Eagleson and his students at M.I.T. Eagleson (1978) used a stochastic dynamic approach which combined simplified models of the microscale hydrologic processes with representative probability density functions for the main characteristics of the precipitation and the potential evapotranspiration. In this way he derived probability density functions for actual infiltration during storms and for actual evapotranspiration between storms and thus provided the basis for predicting the ratio of actual to potential evapotranspiration as a function of a single soil climate factor formed from five parameters (three soil parameters and two vegetation parameters).

Eagleson (1982) extended this approach to study the relationship between the vegetal canopy density and the ratio of the unstressed rate of transpiration to the potential rate of evaporation from bare soil. An overall relationship was derived from the hypotheses that (a) in a water-limited region the canopy would adjust itself to produce minimum water demand stress in the root system, and (b) in a humid region any individual species would tend to maximize biomass productivity. In a further study directly related to climate variation and change, Eagleson modelled the relative density of trees and grassland (a savannah complex) on the basis of a two-layer soil model subjected to climatic frequency distributions (Eagleson and Segarra, 1985).

Future research needs

So that hydrology can contribute to an improvement in climate modelling, research is required in a great many areas to improve our understanding of the hydrological cycle on a macro-scale. Four areas appear to the writer as of particular importance. First, the attempt must be continued to parametrize from the micro-scale physical processes to the catchment scale. Secondly, modelling at the catchment scale must be critically studied from the point of view of simulation efficiency and physical basis rather than automatic calibration and data fitting. Thirdly, new laws of vegetation macrodynamics must be developed and tested. Fourthly, reliable data sets at the macro-scale need to be assembled, validated, analysed and disseminated. In all this work there will be an increasing need as time goes on to extend the analysis from consideration of mean flow to consideration of extremes and of the effect of extremes on storage requirement and reliability (Beran 1986, Dooge 1987).

A number of voices have been heard recently calling for a new departure in hydrology based on a clear understanding of the complex water balance relationship (Burges 1986, Beven 1987, Dooge 1988, Klemes 1988). If these voices are listened to, then the above research problems will be tackled in a spirit of deep analysis and of awareness of the geophysical and biological context of the problems involved.

REFERENCES

BERAN, M.A. (1986).
'The water resources impact of future climate changes and variability'.
In: Titus J.G. (ed.) Effects of changes in stratospheric ozone and Global Climate. EPA/UNEP. Washington. pp. 299-330.

BEVEN, K. (1987).
'Towards a new paradigm in hydrology'.
In: Water for the Future - edited by J.C. Rodda and N.C. Matalas.
Proceedings of IAHR/IAHS International Symposium.Rome. April 1987.
pp. 393 - 403.
IAHS Publication No. 164. Wallingford.

BRANDYK, T., DOOGE, J.C.I., and O'KANE, J.P.J. (1985).
Soil moisture in general circulation models.
Report No. 1 of EEC/NBST Research Project on - The role of soil moisture in climate modelling.
Centre for Water Resources Research, University College, Dublin.

BURGES, S.J. (editor) (1986)
'Trends and directions in hydrology'.
Water Resources Research
Vol. **22** August, 1986. No. 9 Supplement; pp. 1-14.

BUDYKO, M.I. and ZUBENOK, L.I. (1961)
'Opredelenie ispareniya s poverkhnosti sushi'.
(The determination of evaporation from the land surface)
Izv. Akad, Nauk SSSR. Ser. Geogr. No.6 pp. 3-17.

DOOGE, J.C.I. (1982)
'Parametrization of hydrologic processes'.
In: Land surface processes in atmospheric general circulation models. pp. 243-288.
Edited by P.S. Eagleson. Cambridge University Press.

DOOGE, J.C.I. (1983)
'On the study of water'.
Hydrol.Sci. J. **28**(1), pp. 23 - 48.

DOOGE, J.C.I. (1986)
Sensitivity of long term runoff to changes in precipitation and evaporation.
Technical note (20 pages)
Department of Engineering Hydrology, University College, Galway.

DOOGE, J.C.I. (1987)
'Climate change and water resources'.
In Water. WMO Publication No. 675. pp. 37-62.

DOOGE, J.C.I. (1988)
'Hydrology in perspective'.
Keynote lecture to IAHR/IAHS. International Symposium on Water for the Future, Rome, April, 1987.
Hydrological Sciences Journal. **33**(1). February, 1988.

EAGLESON, P.S. (1978)
'Climate, soil and vegetation'.
Water Resour. Res., **14**(5), pp. 705-776.

EAGLESON, P.S. (1982)
'Ecological optimality in water-limited natural soil-vegetation systems, 1, Theory and hypothesis'.
Water Resour. Res., **19**(2)
pp. 325-340.

EAGLESON, P.S. and SEGARRA, R.I. (1985)
'Water-limited equilibrium of savanna vegetation systems'.
Water Resources Research **21**(10): pp.1483 - 1493.

FLEMING, G. (1975)
Computer Simulation Techniques in Hydrology.
Elsevier. New York.

GLEICK, P.H. (1987)
'The development and testing of a water balance model for climate impact assessment - modelling the Sacramento Basin'.
Water Resources Research. **23**(6) : pp. 1049 - 1061.

IDSO, S.B. and BRAZEL, A.J. (1984)
'Rising atmospheric carbon dioxide concentrations may increase streamflow'.
Nature, London. **312**(5989) : pp.51-53.

KLEMES, V. (1988)
'A hydrological perspective'.
Journal of Hydrology **100** (1) : pp. 3-28.

KÜHNEL, V., DOOGE, J.C.I., SANDER, G.C., O'KANE, J.P.J. (1989a)
'Duration of atmosphere-controlled and of soil-controlled phases of infiltration for constant rainfall at a soil surface'.
Symposium on Moisture and Water in the Soil and Atmosphere.
European Geophysical Society. Bologna. March, 1988.
(To appear in Annales Geophysicae).

KÜHNEL, V., DOOGE, J.C.I., O'KANE, J.P.J. (1989b)
'Duration of atmosphere controlled and of soil controlled phases of evaporation for constant potential evapotranspiration at a soil surface'.
Symposium on Moisture and Water in the Soil and Atmosphere
European Geophysical Society. Bologna. March, 1988.
(To appear in Annales Geophysicae).

OL'DEKOP, E.M. (1911)
Sbornik trudov, ispol'nevnykh studentami pri Iur'ievskoj observatori (Tatu).
(On evaporation from the surface of river basins).
Tr. Meteorol. Observ. Iur'evskogo. Univ. Tartu, 4.

PIKE, J.G. (1964)
'Estimation of annual run-off from meteorological data in tropical climate'.
Journal of Hydrology, Amsterdam, **2**(2) : pp.116-123.

SCHREIBER, P. (1904)
'Ueber die Beziehungen zwischen dem Niederschlag und der Wasserführung der Flusse in mittel Europa'.
Z. Meteorol. **21**(10) : pp.441-452.

TURC, L. (1954 and 1955)
'Le bilan d'eau des sols, relation entre les précipitations, l'évaporation, et l'écoulement'.
Ann. Agron. **5** : pp.491-596 and Ann. Agron. **6** : pp. 5-131.

WIGLEY, T.M.L. and JONES, P.P. (1985)
'Influences of precipitation changes and direct CO_2 effects on stream flow'.
Nature. **314** : pp. 149-152.

WMO (1975)
Intercomparison of Conceptual Models used in operational Hydrological Forecasting.
Operational Report No. 7, W.M.O., Geneva.

CHANGES IN OCEAN TEMPERATURE AND THE SCIENTIFIC STRATEGY FOR INTERNATIONAL CLIMATE RESEARCH

P. MOREL
Director, WCRP
W.M.O.
CH-1211 Genève 20
Suisse

1. INTRODUCTION

The "Earth System" includes many components which can, on some time scale or other, force their own specific dynamics on the overall evolution of our environment. It is well known, for example, that the interplay between the accumulation of snow on the great land ice-sheets and the bending of the visco-elastic lithosphere helps to explain the character of glaciation-deglaciation cycles. The World Climate Research Programme (WCRP), sponsored by governmental agencies and weather services represented by WMO, as well as the scientific community, represented by ICSU, is concerned with faster climatic variations which may immediately affect our society and are of interest to managers and planners. On time-scales of seasons to several decades, the significant variable components of the Earth System are the atmosphere and the ocean, while land surface properties can either be taken for constant (e.g. elevation, soil properties and, to a large extent, land use) or varying under the immediate control of atmospheric fluxes.

Atmospheric forcing of the ocean circulation and heat content can take several forms, such as the wind stress on the surface, or fresh water input, or modulation of radiative fluxes, all of which are not straightforward to measure. In fact, the WCRP is only now beginning to make some progress toward the production of global climatologies of these fluxes, based on various combinations of conventional surface measurements, satellite observations and four-dimensional data assimilation by atmospheric general circulation models. Fortunately, the influence of the ocean on the atmosphere can be characterized by one single, easily measured parameter, sea surface temperature (SST). Thus, the hundred year long (or so), worldwide record of ship surface temperature data should provide us with a unique basis for diagnostics of natural climate variations on time scales of seasons to several years, and a guide for the WCRP scientific strategy. Now, of course, deforestation and the combustion of fossil fuel, as well as the release of various products of the chemical industry, are yet another cause of climate change, acting on relatively short time-scales, which must also be taken into account in the study of combined natural and anthropogenic climate variations.

A. Berger et al. (eds.), Climate and Geo-Sciences, 367–372.

2. VARIABILITY IN THE GLOBAL SST RECORD

The standardization of measurements by ship officers combined with the worlwide expansion of commercial shipping have allowed the collection of very long records of SST and, beginning at the turn of the century, the determination of quasi-global ocean temperature fields. Two global records : COADS compiled by NOAA and the (UK) Meteorological Office Ocean Surface Temperature Data Set (MOST) are known to be the most comprehensive. The information quoted here is based on a spectral analysis (in the spatial domain) of the interannual variability of seasonal mean SST averaged over 10° longitude by 10° latitude boxes, as given by MOST for the period 1901-1986 (Folland et al., 1986). Variability associated with the seasonal cycle is eliminated by the simple device of considering only data for one particular season, e.g. the three-month period december to February on each successive year. Spatial harmonics resulting from an empirical orthogonal function (EOF) analysis are ranked in order of decreasing importance.

The first EOF is in fact nearly constant everywhere and the time dependence of this mode hardly differs from the history of the mean global SST over the period. EOF-1 is basically the response of the upper ocean to global warming with very little indication of a specific large scale influence of ocean dynamics.

EOF-2 shows quite a large signal in the tropical Pacific Ocean with a distinct east-west contrast during the particular three-month period and fits almost ideally the accepted SST anomaly pattern for a mature "canonical" El Nino event. We know that the El Nino/Southern Oscillation (ENSO) phenomenon results from the interaction of the atmosphere with the tropical oceans, involving substantial zonal transport of heat by anomalous oceanic currents, and a displacement of the raising branch of the Walker circulation from Indonesia toward the date line or even eastward of the date line. It is interesting to see this confirmation of the significance of the Southern Oscillation, as the second-ranking mode of the variability of global air-sea interaction. We learn also that the strong signal occurs essentially, if not only, in the tropical Pacific Ocean. We do observe a relatively weak synchronous signature of the event in the Atlantic and Indian Oceans, which can only be caused by the instantaneous, essentially local, response of these oceans to changes in atmospheric circulation associated with ENSO.

Even more interesting is the consideration of the third ranking harmonic (EOF-3), which accounts for about 6% of the interannual variability of mean seasonal SST. The third harmonic does not evoke any localisation in the "tropical ocean waveguide" but rather a global seasaw between the North Atlantic-North Pacific one the one hand, and the South Atlantic and Indian Ocean, on the other, involving a change in the world ocean circulation and/or long-period variations in the annual mean heat storage by the world ocean. Nor does EOF-3 change much from year to year, as the dominant time-scale is of the order of one decade. We are thereby introduced to the poorly understood problem of <u>low-frequency</u>

climate fluctuations which we know are also manifest in aerological records.

3. GLOBAL ATMOSPHERIC PROCESSES : GEWEX

Understanding global temperature trends caused by changes in radiative transfer or mediated by changes in the frequency and geographical distribution of "meteors", calls for much refined quantitative knowledge of the interplay between a wide variety of atmospheric and surface processes, and the feedbacks resulting from their interaction. A typical example is the need to establish, on a quantitative basis, our knowledge of the cloud-radiation feedback, as clouds may, according to circumstances, cause a warming or a cooling of the surface. Almost the first initiative of WCRP was to launch the International Satellite Cloud Climatology Project (ISCCP), to begin assembling a global cloudiness data set.

Likewise, the abundance or depletion of soil moisture available to land vegetation, is another essential surface condition which can subsequently affect atmospheric moisture, weather and the general circulation regime. The sensitivity of sub-Saharan climate to changes in ground albedo, postulated by the late J. Charney, is one example of the response of global climate to changes in land surface processes and superficial hydrology. The Hydrological and Atmospheric Pilot Experiment (HAPEX) conducted in France during 1986 is the first major field study sponsored by WCRP to study these processes.

Obviously, reliable global statistics of rainfall, short- and long--wave radiation fluxes, latent and sensible heat fluxes from the land and ocean surface are essential information requirements. WCRP has undertaken to collect global estimates of these and other relevant climatological quantities derived from existing (inadequate) observations, e.g., the Global Precipitation Climatology Project, using infrared and microwave radiances measured from space, raingauge or meteorological radar data when available, and even, eventually, space-borne meteorological radar observations from the future Tropical Rain Measuring Mission (TRMM) or similar satellite projects.

None of these initial undertakings are completely adequate, however, and quite large data gaps exist in the most basic aerological fields such as wind in the tropics. A major step forward is needed to upgrade basic observational capabilities. The plans being made by the space agencies in Europe, USA and Japan, to develop a new generation of polar orbiting platforms and earth observation instruments give us an opportunity to realise the required experimental breakthrough in the period 1995-2000. For these reasons, the WCRP is launching the Global Energy and Water Cycle Experiment (GEWEX), with the objectives of observing, understanding and modelling the processes of the global atmospheric water cycle and energy budget, as well as interactions with the underlying land and ocean surface. GEWEX is envisaged as a comprehensive project, incorporating in a single co-ordinated programme, all relevant aspects from conceptual model development and data assimilation, to the actual preparation and operation of suitable land- and space-based observing systems (WCRP-5, 1988).

As atmospheric and near-surface hydrological processes have relatively short time scales, from hours to a few days or weeks, GEWEX will be intimately related to weather forecasting and the objectives of Stream 1 of WCRP : establishing the physical basis for long-range dynamical prediction of the general circulation. Nevertheless, GEWEX must be considered as an essential step toward the understanding of global climate change (Stream 3) since we can see now that further progress in the assessment of climatic impacts of greenhouse gases and other natural of artificial influences must be based on much improved quantitative knowledge of the relationship between atmospheric surface energy and water budgets and the basic parameters of the general circulation.

4. TROPICAL AIR-SEA INTERACTIONS AND TOGA

The meteorological insight of J. Bjerknes, the mathematical skill of A. Gill, the vast knwoledge of K. Wyrtki and many others gave us the basis for the conceptual model of the ENSO, resulting from almost linear forcing of the tropical ocean circulation by anomalous atmospheric winds and reverse forcing of the tropical atmosphere by the resulting ocean surface temperature anomaly, with longitudinal propagation of the oceanic signal along the "tropical wave guide" which allows relatively fast large-scale oceanic responses. This initial conceptual model was well borne-out by subsequent observations, and we are now approaching the stage where realistic seasonal prediction runs could be made with oceanic circulation models of the Pacific basin coupled with advanced GCMs, starting from initial ocean-atmosphere condition determined from observations.

Although the actual cause of successive ENSO cycles or the nature of the event(s) which trigger the onset of a new ENSO, are not known, it is clear that the <u>predictable</u> part of seasonal or interannual atmospheric variability, even outside the tropics, is linked to the second harmonic (EOF-2) of global SST variations in Folland's analysis. However, there is no indication of the existence of meaningful <u>recurring</u> variations of a similar nature involving intrinsic modes of variability of the tropical Atlantic (or tropical Indian Ocean) coupled with the global atmosphere. Whatever EOF-2 signal is found in the SST record outside the Pacific is only a consequence of variations in atmospheric forcing caused by ENSO. The specific propagation dynamics along the tropical Atlantic or Indian Ocean "wave-guides" do not appear to cause significant or frequent enough interannual changes to show as a distinct variation in the 85-year SST record. Thus, the scientific strategy of TOGA is shifting towards increasingly detailed attention to air-sea interaction in the region of the ascending branch of the (normal) Walker circulation, i.e. the warm water pool around Indonesia (the "maritime continent") and the western tropical Pacific. Plans are maturing for a "Coupled Ocean-Atmosphere Response Experiment" (COARE) to study specifically the dynamic and energetic response of the warm water pool to occasional westerly wind bursts generated by the atmospheric circulation.

5. LOW-FREQUENCY CLIMATE VARIABILITY AND WOCE

It has long been a prevailing view that the natural variability of weather and, by extension, seasonal or interannual anomalies in the general circulation due to the non-linear internal dynamics of the atmosphere or coupling with the tropical oceans, could all be lumped into some form of "meteorological noise", which the scientific community could forget to concentrate on the more exciting problem of global (climate) change in response to various real or hypothetical influences, i.e. to look for meaningful "climatological signals" in observational records and theoretical model simulations.

It is a basic contention in the scientific philosophy of WCRP that this distinction, although convenient at times, cannot be made in principle. In the first place, "climate" is no more than the cumulative effect of "weather" and climate change may result, for example, from a displacement of the mean "storm track" : there is no way to understand quantitatively the sensitivity of the mean track to various large-scale parameters without accounting realistically for the behaviour of individual storms. Many careful numerical experiments have shown explicitely that the differential sensitivity of a particular atmospheric model to a change in forcing parameters depends upon its own systematic biases. In other words, sensitivity studies can only be trusted inasmuch as the mathematical tool used is able to simulate realistically the present weather and climate. This condition places severe demands on the degree of detail and complexity with which physical processes need to be formulated and, of course, corresponding computing resource requirements.

Furthermore, there is no fundamentally valid way to separate the "signal", e.g. the long-term climatic trend whcih may be caused by sudden or progressive variations in forcing parameters, from the "noise" resulting from the internal dynamics of the Earth System on practically all time-scales accessible to our observation. Persistent changes in the relative strength of the mean zonal circulation compared to the total kinetic energy of the atmosphere (the so-called "index cycle") have been well-documented in the past. For example, persistent, yet reversible changes in the precipitation regime have caused some spectacular changes in hydrological resources, e.g. the decade of the "dust bowl" in north America, and recurrent Sahelian drought in the marginal belt between the West African Monsoon regime and the Saharan climatological subsidence.

Low frequency variability, similar in character to the north-south seesaw seen in Folland's EOF-3, has been detected in long integration runs with a global coupled atmosphere-ocean general circulation model designed by Manabe and Stouffer. They find transitions between two possible nearly-balanced states : one state features a strong thermohaline circulation of the north Atlantic Ocean and corresponding amount of deep water formation, the other state being consistent with no appreciable sinking in the north Atlantic or north Pacific (much like the inferred ocean circulation during the last glacial maximum). In truth, one must state that the cause or causes of low-frequency variability in the coupled atmosphere-ocean system are not known. They invoke global-scale changes in the world ocean circulation on time-scale of several years to several decades, which are inextricably mixed with detectable trends resulting from the changing greenhouse gases burden or land surface use or any other external factor. Because we are asked to come up with current assessments of possible climatic changes in the forthcoming decades, we cannot take the long-range view of historical or paleo-climatology and eliminate internal variability on interannual or decadal time-scales by suitable time-averaging : we must take the hard way of building-up a comprehensive, detailed and quantitative explanation of current variability as we can observe it. This is a formidable task, for which the world geo-science resources may be barely adequate. We shall test these capabilities in our next effort which will be to compile, for the first time in history, a global "synoptic" representation of the world ocean circulation, by means of in-situ oceanographic measurements along worldwide transects, global ocean surface observations from space and ocean circulation modelling in computing laboratories (see the Scientific Plan for the World Ocean Circulation Experiment, WCRP, 1986).

6. CONCLUDING REMARKS

A great range of processes are involved in the problem of global change, some fairly slow and others highly variable in space and time. It takes a special kind of organization to tackle effectively highly variable processes on the global scale, requiring dense operational or quasi-operational monitoring networks, real-time data flow and massive data processing facilities (including dedicated staff). The WCRP, having access to the resources of both WMO and the academic community (ICSU), and applying fairly rigid control on a few internationally co-ordinated programmes, such as GEWEX, TOGA and WOCE, is best suited to tackle such problems, involving the global atmospheric and oceanic circulation.

On the other hand, many outstanding issues in earth system sciences are either of a more exploratory nature or concern components which can be effectively studied without handling enormous data flows. The free associations and soft co-operation which can so conveniently be promoted by scientific unions or ICSU's International Geosphere-Biosphere Programme, would appear to be the best method to promote the development of science while avoiding cumbersome inter-governmental channels. It would be a mistake to oppose these two different organizational approaches, for both are equally well-suited to the kind of problems they are addressing, and both serve the progress of the geosciences.

SECTION 3 – SIMULATION OF THE GREENHOUSE WARMING

MODEL PROJECTIONS OF THE CLIMATIC CHANGES INDUCED BY INCREASED ATMOSPHERIC CO_2

MICHAEL E. SCHLESINGER
Department of Atmospheric Sciences and
Climatic Research Institute
Oregon State University
Corvallis, Oregon 97331
USA

ABSTRACT. The simulations of CO_2-induced equilibrium climatic change by energy-balance models, radiative-convective models and general circulation models are reviewed and characterized in terms of the direct radiative forcing of the increased CO_2, the response of the climate system in the absence of feedback processes, and the feedbacks of the climate system. Five recent simulations of CO_2-induced climatic change by atmospheric GCM/mixed-layer ocean models are contrasted in terms of their surface air temperature and soil moisture changes. The simulations of CO_2-induced transient climatic change by planetary energy-balance models, radiative-convective models and general circulation models are reviewed in terms of the e-folding time τ_e of the response of the climate system. Theoretical studies for a time-dependent CO_2 increase between 1850 and 1980 indicate that the sequestering of heat into the ocean's interior is responsible for the concomitant warming being only about half that which would have occurred in the absence of the ocean. These studies also indicate that the climate system will continue to warm towards its as-yet unrealized equilibrium temperature change, even if there is no further increase in the CO_2 concentration.

A. Berger et al. (eds.), Climate and Geo-Sciences, 375–415.

1. Introduction

If the Earth's atmosphere were composed of only its two major constituents, nitrogen (N_2, 78% by volume) and oxygen (O_2, 21%), the Earth's surface temperature would be close to the -18°C radiative-equilibrium value necessary to balance the approximately 240 Wm^{-2} of solar radiation absorbed by the Earth-atmosphere system. The fact that the Earth's global-mean surface temperature is a life-supporting 15°C is a consequence of the greenhouse effect of the atmosphere's minor constituents, mainly water vapor (H_2O, 0.2%) and carbon dioxide (CO_2, 0.03%). Measurements taken at Mauna Loa, Hawaii show that the CO_2 concentration has increased from about 315 ppmv in 1958 to 346 ppmv in 1985 (Keeling and Boden, 1986), a 10% increase in 28 years. Measurements of the air entrapped within the ice sheet of Antarctica indicate that the pre-industrial CO_2 concentration increased from about 280 ppmv in 1750 to 290 ppmv in 1880 (Siegenthaler and Oeschger, 1987). Rotty and Masters (1985) report that, with the exception of the periods of the Depression and World Wars I and II, the CO_2 concentration increased from 1860 to 1949 due to a 4.2% yr^{-1} growth in the consumption of fossil fuels (gas, oil, coal). Subsequently, the growth rate of CO_2 emissions was a steady 4.4% yr^{-1} from 1950 to 1973, and then decreased to 1.5% yr^{-1} from 1973 to 1982 as a result of the rise in the price of oil. A probablistic scenario analysis of the future usage of fossil fuels predicts about an 80% chance that the CO_2 concentration will reach twice the pre-industrial value by 2100 (Nordhaus and Yohe, 1983). Computer simulations of the equilibrium climatic change induced by a doubling of the CO_2 concentration have been made with a hierarchy of mathematical climate models, the most recent of which have given a warming as large as 5.2°C in the global-mean surface air temperature. Since such a global warming is comparable to that which is estimated to have occurred during the transition from the last ice age to the present interglacial (Gates, 1976a, b; Imbrie and Imbrie, 1979), there is considerable interest in the detection of a CO_2-induced climatic change, and in the potential impacts of such a change on the spectrum of human endeavors.

The majority of the simulations of CO_2-induced climatic change have been performed to determine the change in the equilibrium climate of the Earth resulting from an abrupt increase in CO_2 such as a doubling from 300 to 600 ppmv. The goal of these simulations has been to estimate the magnitude of the eventual climatic change which may occur as a consequence of the doubled CO_2 concentration projected to occur in the next century. These equilibrium climatic-change simulations have not been concerned with the time required for the climatic change to reach its equilibrium and, in fact, have generally tried to minimize that time in an effort to economize on computer time. More recently, concern has been focussed on the detection of a CO_2-induced climatic change. A simple scaling of the equilibrium $2xCO_2$-induced warming to the increase in CO_2 from 1861 to 1984 indicates an increase in the global-mean surface temperature of about 1.0°C, this for a sensitivity of 4°C for a CO_2 doubling. However, the observed surface temperature increase over this period is only about half this value (Jones *et al.*, 1986). This discrepancy suggests that either the equilibrium warming of our climate models is twice as large as that of nature, or that there is a lag in the response of the climate system. The latter possibility has been the focus of research on the transient response of climate to both abrupt and time-dependent increases in the CO_2 concentration. In the following, the studies of equilibrium and transient climatic change induced by increased CO_2 are reviewed in sections 2 and 3, respectively; a summary of these results is presented in section 4 and conclusions are given in section 5.

2. CO_2-Induced Equilibrium Climatic Change

Three different types of climate model have been used to simulate the change in the equilibrium climate resulting from an increase in the CO_2 concentration: energy-balance models (EBMs), radiative-convective models (RCMs), and general circulation models (GCMs). These models and their projections of CO_2-induced climatic change are described in turn below.

2.1. ENERGY-BALANCE MODELS

Energy-balance models predict the change in temperature at the Earth's surface, ΔT_s, from the requirement that $\Delta N = 0$, where N is the net energy flux expressed by $N = N(\mathbf{E}, T_s, \mathbf{I})$. Here **E** is a vector of quantities that can be regarded as "external" to the climate system, that is, quantities whose change can lead to a change in climate, but which are independent of climate. **I** is a vector of quantities that are internal to the climate system, that is, quantities that can change as the climate changes and, in so doing, feed back to modify the climatic change. The external quantities include, for example, the solar constant, the optically-active ejecta from volcanic eruptions and, for the present purpose, the CO_2 concentration (although eventually it may change as a result of climatic change). The internal quantities include all the variables of the climate system other than T_s. Since T_s is the only dependent variable in an EBM, the internal quantities must be represented therein by functions of T_s, that is, by $\mathbf{I} = \mathbf{I}(T_s)$.

A small change in the energy flux, ΔN, can be expressed as

$$\Delta N = \Delta Q - (G_o^{-1} - F)\,\Delta T_s\ ,$$

where

$$\Delta Q = \sum_i \frac{\partial N}{\partial E_i}\,\Delta E_i$$

is the change in N due to a change in one or more external quantity, ΔE_i,

$$-\,G_o^{-1}\,\Delta T_s = \frac{\partial N}{\partial T_s}\,\Delta T_s$$

is the change in N due to the change in T_s alone, and

$$F\,\Delta T_s = \sum_j \frac{\partial N}{\partial I_j}\,\frac{dI_j}{dT_s}\,\Delta T_s$$

is the change in N due to the change in the internal variables $\mathbf{I}$ through their dependence on T_s. When the equilibrium ΔT_s is reached in response to the forcing ΔQ, $\Delta N = 0$ and

$$\Delta T_s = \frac{G_o}{1 - G_oF}\,\Delta Q = \frac{G_o}{1 - f}\,\Delta Q = G_f\,\Delta Q\ . \tag{1}$$

This relation can be represented by a feedback block diagram for the climate system as shown in Fig. 1. It is seen that G_o is the gain (output/input) of the climate system when the feedback $f = G_oF = 0$. Thus G_o is the zero-feedback gain and $(\Delta T_s)_o = G_o\Delta Q$ is the zero-feedback temperature change; correspondingly, G_f and ΔT_s are the gain and the temperature change with

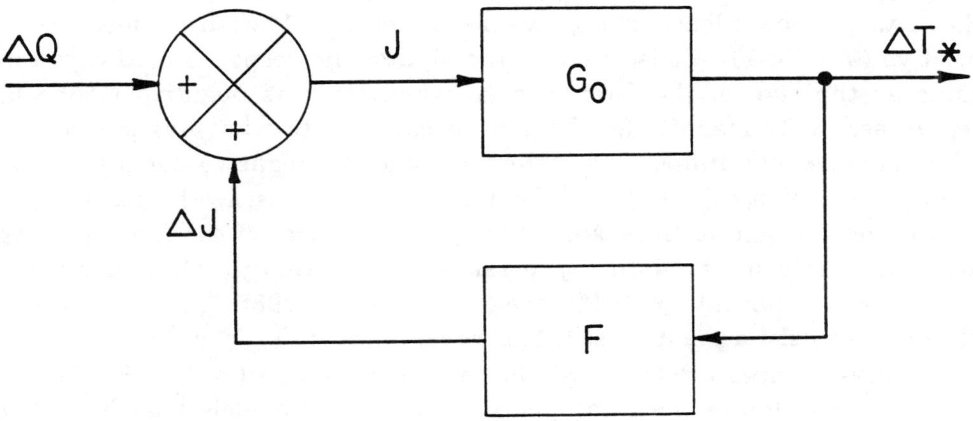

Fig. 1. Block diagram for the climate system. ΔQ and $\Delta T_s = G_0$ J are the forcing and surface-temperature response of the climate system, respectively, with G_0 the gain of the system in the absence of feedback and J = ΔQ + ΔJ, with $\Delta J = F\ \Delta T_s$. The feedback of the climate system is $f = G_0\ F$.

feedback Letting $R_f = \Delta T_s/(\Delta T_s)_0 = G_f/G_0$, it can be seen from Fig. 2 that $0 < R_f < 1$ when the feedback is negative ($f < 0$) and $R_f > 1$ when the feedback is positive ($0 < f < 1$). Consequently, the sign of the response is always the same as the sign of the forcing, even when there is negative feedback, regardless of its magnitude. This is in contrast to what has sometimes been erroneously inferred for the outcome of negative feedback (for example, see *Nature,* 1987). On the other hand, for positive feedback $f > 1$, $R_f < 0$ and the sign of the response is opposite to that of the forcing. This outcome, while mathematically possible (and actually obtained by one improperly formulated EBM; see Schlesinger, 1985, 1988a), is not physically-consistent and must therefore be rejected.

Energy-balance models impose the condition $\Delta N = 0$ at either the Earth's surface or the top of the atmosphere. The former models may be called surface energy-balance models (SEBMs) and the latter planetary energy-balance models (PEBMs). SEBMs have given a warming induced by a CO_2 doubling of about 0.2 to 10°C. This wide range is due in part to the inherent difficulty in specifying the behavior of the atmosphere in terms of the surface temperature in SEBMs, that is, $\mathbf{I}(T_s)$, and to the large sensitivity of ΔT_s in SEBMs to this specification (Schlesinger 1985, 1988a). Therefore, it is preferable to use models which calculate the atmosphere's behavior based on the fundamental laws of physics.

Before turning our attention to these physically-based models, it is useful to consider PEBMs for which

$$N = \frac{1 - \alpha_p}{4} S - \varepsilon \sigma T_s^4 ,$$

where S is the solar constant, α_p the planetary albedo, ε the effective emissivity of the Earth-atmosphere system, and σ the Stefan-Boltzmann constant. From this equation we can estimate the zero-feedback gain as

$$G_o = -\left(\frac{\partial N}{\partial T_s}\right)^{-1} = \frac{T_s}{(1 - \alpha_p) S} , \tag{2}$$

an expression which we will use subsequently. For a PEBM, ΔT_s is given by Eq. (1) with Eq. (2) and

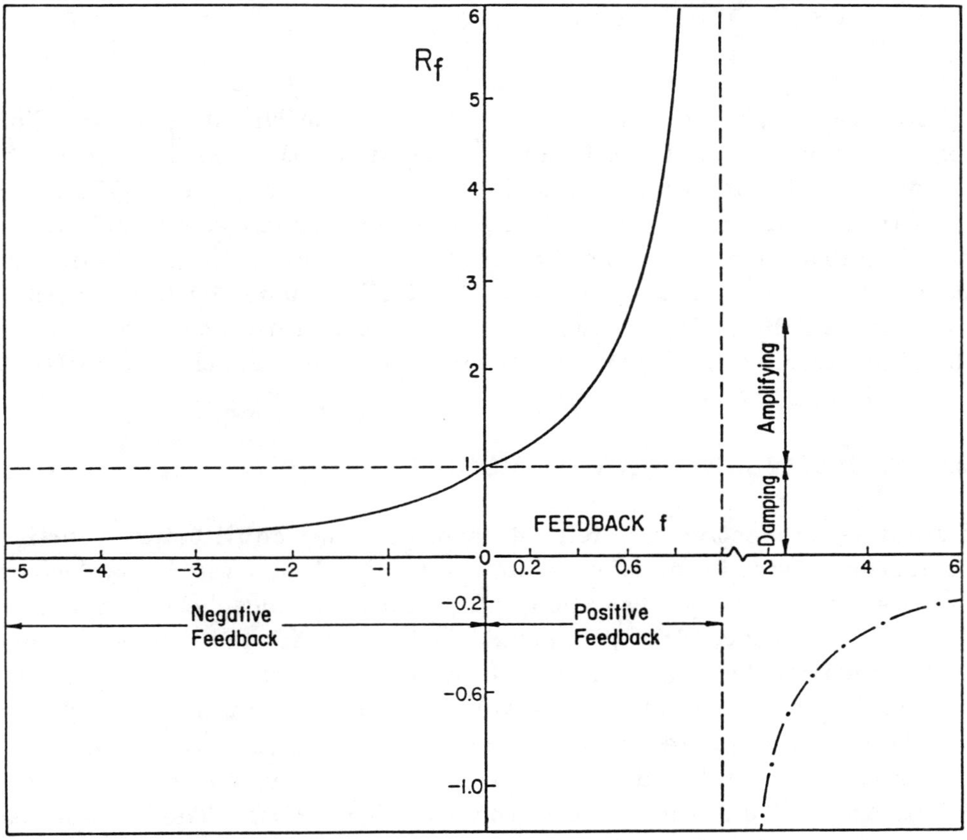

Fig. 2. The response ratio $R_f = \Delta T_s/(\Delta T_s)_o$ versus the feedback f, where ΔT_s is the surface-temperature response of the system with feedback and $(\Delta T_s)_o$ is surface-temperature response without feedback.

$$f = G_o \sum_j \frac{\partial N}{\partial I_j} \frac{dI_j}{dT_s} ,$$

where I_j are again the internal variables of the climate system. This expression means that the feedback depends on the specification of the behavior of the atmosphere and the Earth's surface. Thus PEBMs also have the same problem as SEBMs, namely, the need to treat the behavior of the climate system away from the energy-balance level. In PEBMs this has been done semi-empirically following the initial studies by Budyko (1969) and Sellers (1969). The equilibrium surface temperature change for a CO_2 doubling given by PEBMs ranges from 0.6°C (Rasool and Schneider, 1971) to 3.3°C (Ramanathan *et al.*, 1979).

2.2. RADIATIVE-CONVECTIVE MODELS

Radiative-convective models determine the equilibrium vertical temperature distribution for an atmospheric column and its underlying surface for given solar insolation and prescribed atmospheric composition and surface albedo. An RCM includes submodels for the transfer of solar and terrestrial (longwave) radiation, the turbulent heat transfer between the Earth's surface and atmosphere, the vertical redistribution of heat within the atmosphere by dry or moist convection, and the atmospheric water vapor content and clouds. The radiative transfer models used in RCMs are frequently identical to those used in GCMs. The surface heat exchange is treated either as an equivalent radiative exchange or is parameterized as a Newtonian exchange with a prescribed transfer coefficient. The vertical heat redistribution by convective atmospheric motions is modeled as an adjustment whereby the temperature lapse rate of the atmosphere is prevented from exceeding some given value. The amount of water vapor is determined in RCMs either by prescribing the absolute humidity or the relative humidity; in the latter case the amount of water vapor increases (decreases) with increasing (decreasing) temperature. Finally, the fractional cloudiness and the altitude of the clouds are either prescribed or predicted, the latter by some assumption about the behavior of clouds.

The equilibrium change in temperature simulated by RCMs for an increase in the CO_2 concentration shows a cooling in the stratosphere and a warming in the troposphere and at the Earth's surface, the latter with a range of 0.48 to 4.2°C. As indicated by Eq. (1), the change in the surface temperature induced by a doubling of the CO_2 concentration can be understood in terms of the direct radiative forcing ΔQ due to the CO_2 increase; the response of the climate system if the surface temperature alone changed, as characterized by the zero-feedback gain G_o; and the negative and positive feedbacks f which occur to reduce and enhance the zero-feedback response, respectively. Each of these is discussed below.

About 95% of the direct radiative forcing due to the increase in the CO_2 concentration occurs in the longwave radiation emitted by the Earth and 5% in the shortwave solar radiation (Ramanathan *et al.,* 1979). The direct radiative forcing of the troposphere-surface system ΔR_T has been found to be 4.0 Wm^{-2}, of which 2.7 Wm^{-2} is contributed by the reduction in upward flux from the troposphere, 1.55 Wm^{-2} by the increased downward flux from the stratospheric CO_2 increase, and -0.15 Wm^{-2} by the decreased solar flux at the tropopause (Lal and Ramanathan, 1984).

The response of the climate system to the direct radiative forcing of the increased CO_2 when only the surface temperature is allowed to change is given by Eq. (1), with f = 0 and $\Delta Q = \Delta R_T$, and by Eq. (2) as

$$(\Delta T_s)_o = G_o \, \Delta R_T = \frac{T_s(1xCO_2)}{(1 - \alpha_p) S} \, \Delta R_T \; .$$

Taking $S = 1370$ Wm^{-2}, $\alpha_p = 0.3$ and T_s as the observed surface air temperature $T_a = 288$ K, gives $G_o = 0.3°C/(Wm^{-2})$. Thus for $\Delta R_T = 4$ Wm^{-2}, $(\Delta T_s)_o = 1.2°C$. This value of $(\Delta T_s)_o$ is in agreement with what has been obtained by RCMs without feedbacks (see Schlesinger 1985, 1988a).

The feedback f of the CO_2-induced surface air temperature warming given by RCMs can be determined from Eq. (1) written as

$$f = 1 - \frac{(\Delta T_s)_o}{\Delta T_s} \; , \qquad (3)$$

together with the above estimate of $(\Delta T_s)_o$. Thus, for the $0.48 \leq \Delta T_s \leq 4.2°C$ range simulated by RCMs, $-1.5 \leq f \leq 0.7$. Several physical mechanisms are thought to be the cause of this wide range in the feedback of these models. For increasing T_s these mechanisms include: 1) the increase in the amount of water vapor in the atmosphere as a consequence of the near-constancy of the relative humidity, 2) the change in the temperature lapse rate, 3) the increase in the cloud altitude as the clouds maintain their temperature, 4) the change in cloud amount, 5) the change in the cloud optical depth, and 6) the decrease in surface albedo. The feedbacks of these physical processes, as determined by applying Eq. (3) to an ensemble of RCM simulations in which the feedbacks were added sequentially (see Schlesinger 1985, 1988a), are summarized in Table 1. This table shows that: 1) the water vapor, cloud altitude and surface albedo feedbacks are positive, with values that decrease in that order; 2) the cloud optical depth feedback is negative; 3) the temperature lapse rate feedback is either positive or negative, depending on whether the lapse rate is controlled by baroclinic adjustment (BADJ) or convective (MALR or PC) processes; and 4) the cloud cover feedback is unknown. However, as shown by Eq. (1) and Fig. 2, the influence of any of these feedbacks on the response of the climate system depends nonlinearly on the sum of the other feedbacks. For example, the addition of cloud altitude feedback with f = 0.2 would increase ΔT_s by 1.6°C if added to a system with an existing feedback of 0.5, but would increase ΔT_s by only 0.5°C if added to a system with no existing feedback.

2.3. GENERAL CIRCULATION MODELS

While EBMs and RCMs calculate only the surface temperature and the vertical temperature profile, respectively, GCMs calculate the geographical distributions of an ensemble of climatic quantities which includes the vertical profiles of atmospheric temperature, water vapor and velocity, and the surface pressure, precipitation, soil water and snow. In addition, GCMs used for climatic-change simulations must calculate the geographical distributions of the sea surface temperature (SST) and sea ice extent. (See Schlesinger 1984 and 1988b for further information on GCMs.) To study CO_2-induced equilibrium climatic changes with a GCM requires two simulations, a 1xCO_2 simulation with a CO_2 concentration generally taken to be 300-330 ppmv, and an NxCO_2 simulation with N generally taken to be 2 or 4. Each of the simulations is begun from some prescribed initial conditions and is run until these conditions are forgotten and the simulated climate reaches its statistically-stationary equilibrium state. The differences between the NxCO_2 and 1xCO_2 equilibrium climates are then taken to be the CO_2-induced equilibrium climatic changes.

The earliest GCM simulations of CO_2-induced climatic change were performed with so-called "swamp models" of the ocean. In such a swamp ocean model the ocean has zero heat capacity and no horizontal or vertical heat transports. Therefore, the swamp ocean is like perpetually wet land

Table 1. Summary of the feedbacks f in RCM and GCM simulations of CO_2-induced surface air temperature change.

Feedback Mechanism	RCM [a]	GCM GISS [b]	GCM GFDL [c]
Water Vapor	0.3 to 0.4	0.66	0.41
Lapse Rate [d]			
BADJ	0.1		
MALR	-0.25 to -0.4		
PC	-0.65		
Total		-0.26	0.05
Cloud			
Altitude	0.15 to 0.30		
Cover	Unknown		
Altitude & Cover		0.22	0.09
Optical Depth	0 to -1.32		
Surface Albedo	0.14 to 0.19	0.09	0.13
Total	-1.5 to 0.71	0.71	0.68

a. Based on the analysis of Schlesinger (1985, 1988a).

b. Based on the results of Hansen *et al.* (1984).

c. In this analysis a value of $\Delta R_T = 4.3\ Wm^{-2}$ is assumed. This value is based on the change in the net radiation at the top of the atmosphere given by Wetherald and Manabe (1988), $\Delta R_o = 2.28\ Wm^{-2}$, and the difference between the ΔR_T and ΔR_o results for this radiation model averaged over the five atmospheric profiles of the Intercomparison of Radiation Codes used in Climate Models (ICRCCM) study (Ellis, 1987, personal communication).

d. BADJ, MALR and PC denote baroclinic adjustment, moist adiabatic adjustment and penetrative convection, respectively.

and is always in thermodynamic equilibrium. Consequently, the time for the Earth-atmosphere system to reach its equilibrium climate with an atmospheric GCM/swamp ocean model depends only on the atmosphere and land surface, and generally requires about 300 days. While such models are economical of computer time, they must be run without the seasonal insolation cycle to prevent the freezing of the ocean in the latitudes of the polar night. Consequently, these models simulate the CO_2-induced change only for a surrogate of the annual-mean climate, namely, that which is obtained for the annual-mean solar insolation. More recently, atmospheric GCMs have been coupled to prescribed-depth mixed-layer ocean models which have heat capacity and sometimes a prescribed additional oceanic heating which includes the effects of oceanic heat transport. These atmospheric GCM/prescribed-depth mixed-layer ocean models are run with the seasonal insolation cycle and generally require about 20 simulated years to reach their equilibrium climate. An additional 10 years is generally simulated to obtain estimates of the means and other statistics of the equilibrium climate.

The annual cycles of the equilibrium climatic changes induced by a doubling of the CO_2 concentration have been simulated by the GCM/mixed-layer ocean models of: 1) the Geophysical Fluid Dynamics Laboratory (GFDL, Wetherald and Manabe, 1986), 2) the Goddard Institute for Space Studies (GISS, Hansen *et al.*, 1984), 3) the National Center for Atmospheric Research (NCAR, Washington and Meehl, 1984), 4) Oregon State University (OSU, Schlesinger and Zhao, 1989), and 5) the United Kingdom Meteorological Office (UKMO, Wilson and Mitchell, 1987). In the following two subsections the CO_2-induced changes in the surface air temperature and soil water simulated by these models are presented.

2.3.1. Surface Air Temperature Change. The equilibrium changes in the global-mean surface air temperature ΔT_s simulated by the five models for a doubled CO_2 concentration are presented in Table 2, together with the global-mean precipitation changes ΔP. This table shows that the simulated CO_2-induced changes in the annual-mean, global-mean surface air temperature range from 2.8 to 5.2°C, and the corresponding changes in precipitation from 7.1 to 15% of their $1xCO_2$ values. Furthermore, this table shows that the size of ΔP is positively correlated with the size of ΔT_s, this occurring as a result of the Clausius-Clapeyron relation between the saturation vapor pressure of water vapor and temperature.

A partial explanation for the range of ΔT_s, and thus of ΔP, is provided by Fig. 3 which displays the CO_2-induced warming plotted versus the annual-

Table 2. Changes in the global-mean surface air temperature (T_s) and precipitation rate (P) simulated by atmospheric GCM/mixed-layer ocean models for a CO_2 doubling.

Model/Study	ΔT_s (°C)	ΔP (% of $1xCO_2$ value)
GFDL/Wetherald & Manabe (1986)	4.0	8.7
GISS/Hansen *et al.* (1984)	4.2	11.0
NCAR/Washington & Meehl (1984)	3.5	7.1
OSU/Schlesinger & Zhao (1989)	2.8	7.8
UKMO/Wilson & Mitchell (1987)	5.2	15.0

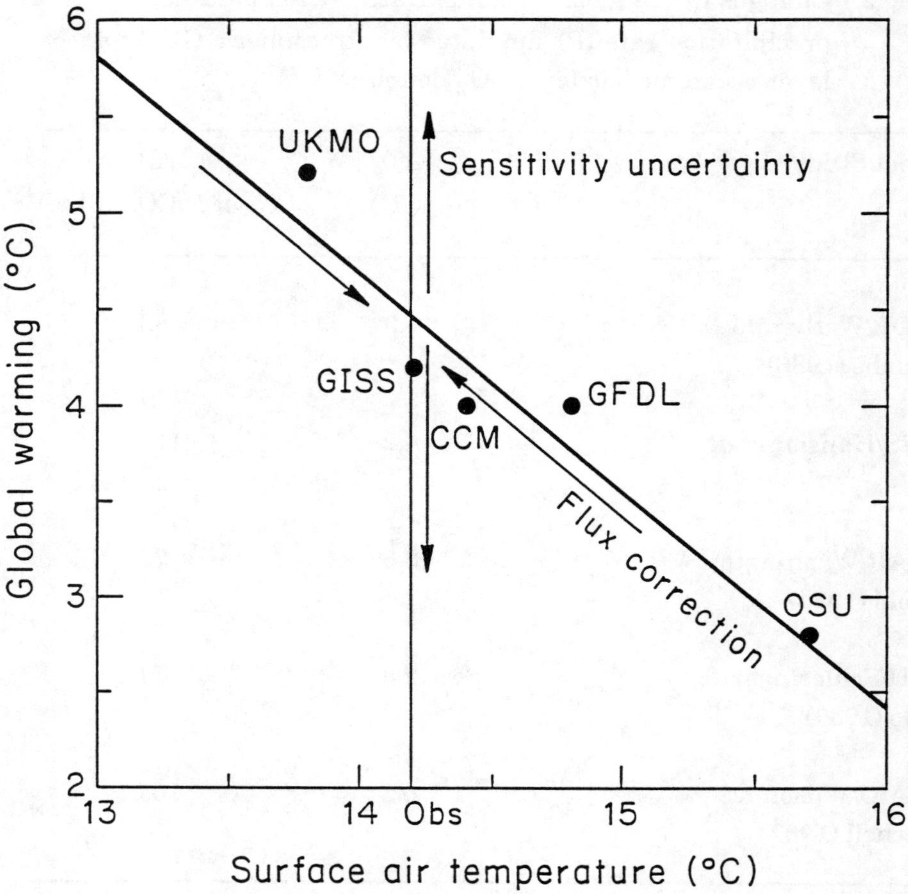

Fig. 3. The global-mean surface air temperature warming simulated for a CO_2 doubling by five GCMs [the GDFL model (Wetherald and Manabe, 1986), the GISS model (Hansen *et al.*, 1984), the NCAR model (Washington and Meehl, 1984), the OSU model (Schlesinger and Zhao, 1989), and the UKMO model (Wilson and Mitchell, 1987)] versus their simulated $1xCO_2$ global-mean surface air temperature. "Obs" indicates the observed global-mean surface air temperature based on the data of Jenne (1975). (Adapted from Cess and Potter, 1988).

mean, global-mean surface air temperature of the simulated $1xCO_2$ climate. This figure shows that the warmer the $1xCO_2$ control climate, the smaller the CO_2-induced warming. This is due in part to the decrease in the positive ice-albedo feedback with increasing $1xCO_2$ temperature as a result of there being less sea ice and snow. From this it appears that the CO_2-induced changes in temperature and precipitation simulated by these models would be in better agreement if their $1xCO_2$ global-mean surface air temperatures were in better agreement. Furthermore, it is tempting to conclude that the resulting common model temperature and precipitation sensitivities would be correct if the common $1xCO_2$ surface air temperature were in agreement with the observed temperature, perhaps through the use of the "Flux Correction Method."[1] However, while this is a necessary condition for the common sensitivities of the models to be in agreement with the sensitivities of nature, it is not a sufficient condition as is indicated by the vertical arrows labelled sensitivity uncertainty in Fig. 3. In fact, the potential for such a disparity is evidenced by Table 1 which shows that although the GISS and GFDL models simulate similar values of ΔT_s (Fig. 3), they do so with feedbacks that differ in both magnitude and sign, this despite the approximate agreement of their simulated values of T_s for the $1xCO_2$ climate. Consequently, to establish the correctness of the models' temperature and precipitation sensitivities when they simulate the present climate correctly requires the simulation of at least one climate different from that of the present, for example, that of Wisconsin Ice Age 18,000 years before the present, and the comparison of such a simulated paleoclimate with observations. This requirement for model validation is a fundamental and inherently difficult problem in climate modeling and simulation (Schlesinger and Mitchell, 1985, 1987). This notwithstanding, it is of interest here to present and compare further results of the CO_2-induced climatic changes simulated by these models.

The time-latitude distributions of the zonal-mean surface air

1. In this method, which was employed by the GISS and UKMO models, an oceanic heating is determined for each calendar month at each ocean grid point from a $1xCO_2$ simulation with the atmospheric GCM using the observed annual cycle of SST as its lower boundary condition over the ocean. This heating field is then prescribed in both the $1xCO_2$ and $2xCO_2$ simulations with the GCM/mixed-layer ocean model, thereby helping the model to reproduce the observed SST distribution in the $1xCO_2$ simulation. However this method does not allow the oceanic heat transport effects to change and contribute to the CO_2-induced climatic change.

temperature changes simulated by the five models for doubled CO_2 are presented in Fig. 4. This figure shows that the CO_2-induced temperature changes increase from the tropics, where the values range from about 2°C for the NCAR and OSU models to about 4°C for the GISS and UKMO models, toward the poles, and that the seasonal variations of the CO_2-induced surface temperature changes are small between 50°S and 30°N and are large in the regions poleward of 50° latitude in both hemispheres. In the Northern Hemisphere a warming minimum of about 2°C is simulated near the pole by all five models in summer. The models also simulate a warming maximum in fall with values that range from 8°C for the NCAR model to 16°C for the GFDL and UKMO models. This maximum extends into winter in all the simulations except that of NCAR which instead exhibits a warming minimum near the pole. The NCAR model also simulates another polar warming minimum in spring that is not found in the other simulations. In the Southern Hemisphere all five models simulate a maximum warming in winter and a minimum warming in summer. The summer warming maximum occurs near the Antarctic coast in all five simulations and ranges from 8°C in the GFDL and OSU simulations to 14°C in the NCAR and UKMO simulations.

The geographical distributions of the $2xCO_2$-$1xCO_2$ surface air temperature changes simulated for December-January-February (DJF) and June-July-August (JJA) by the five models are presented in Figs. 5 and 6, respectively. These figures show that all five models simulate a CO_2-induced surface air temperature warming virtually everywhere. In general, the warming is a minimum in the tropics during both seasons, at least over the ocean, and increases toward the winter pole. The tropical maritime warming minimum ranges from about 2°C in the NCAR and OSU simulations to about 4°C in the GFDL, GISS and UKMO simulations. Maximum warming in DJF occurs in the Arctic in all the simulations except that of NCAR which instead exhibits a warming maximum near 65°N. The maximum warming in JJA occurs around the Antarctic coast in all five simulations. The locations of the wintertime warming maxima in both hemispheres coincide with the locations where the $1xCO_2$ sea ice boundary retreats in the $2xCO_2$ simulation. The magnitude of the wintertime warming maxima in the Northern Hemisphere ranges from 10°C in the GISS and OSU simulations to 20°C in the UKMO simulation, and in the Southern Hemisphere it ranges from 10°C in the OSU simulation to 20°C in the UKMO simulation. In JJA there is a warming minimum in the Arctic of about 2°C in all five simulations.

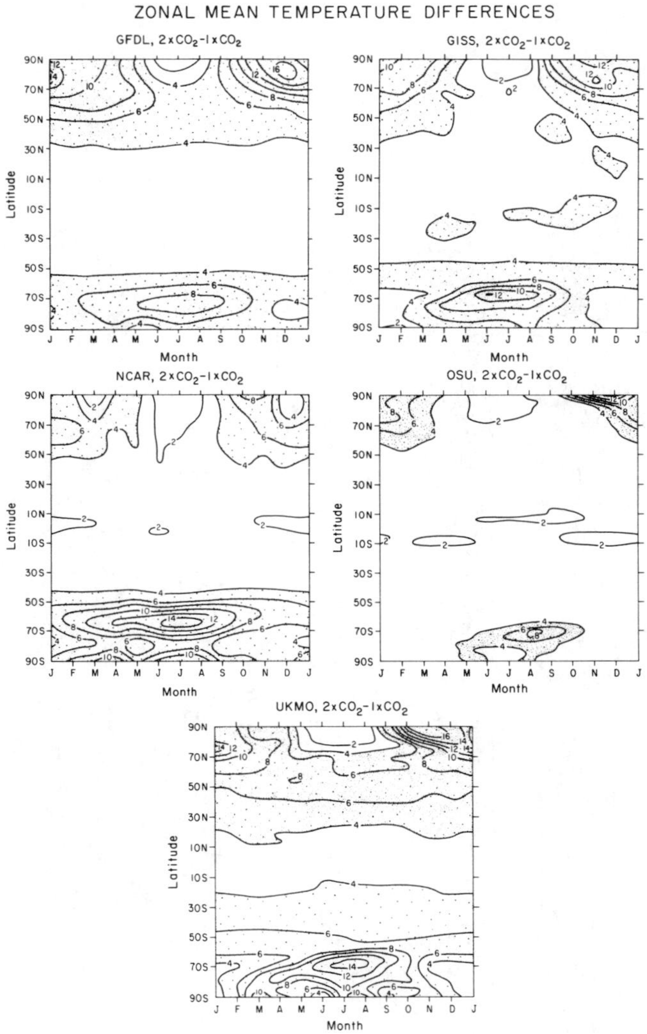

Fig. 4. Time-latitude distribution of the zonal-mean surface air temperature change (°C), $2xCO_2$-$1xCO_2$, simulated by: (a) the GDFL model by Wetherald and Manabe (1986); (b) the GISS model by Hansen *et al.* (1984); (c) the NCAR model by Washington and Meehl (1984); (d) the OSU model by Schlesinger and Zhao (1989); and (e) the UKMO model by Wilson and Mitchell (1987). Stipple indicates temperature increases larger than 4°C.

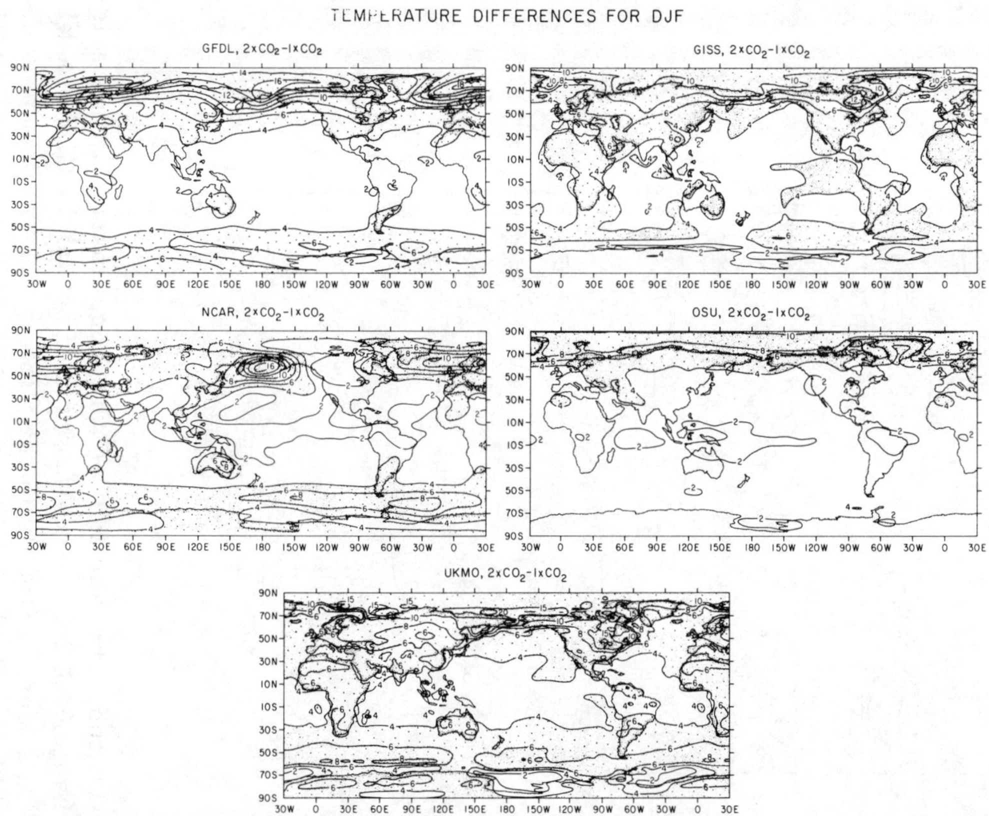

Fig. 5. Geographical distribution of the surface air temperature change (°C), $2xCO_2$-$1xCO_2$, for DJF simulated by: (a) the GDFL model by Wetherald and Manabe (1986); (b) the GISS model by Hansen *et al.* (1984); (c) the NCAR model by Washington and Meehl (1984); (d) the OSU model by Schlesinger and Zhao (1989); and (e) the UKMO model by Wilson and Mitchell (1987). Stipple indicates temperature increases larger than 4°C.

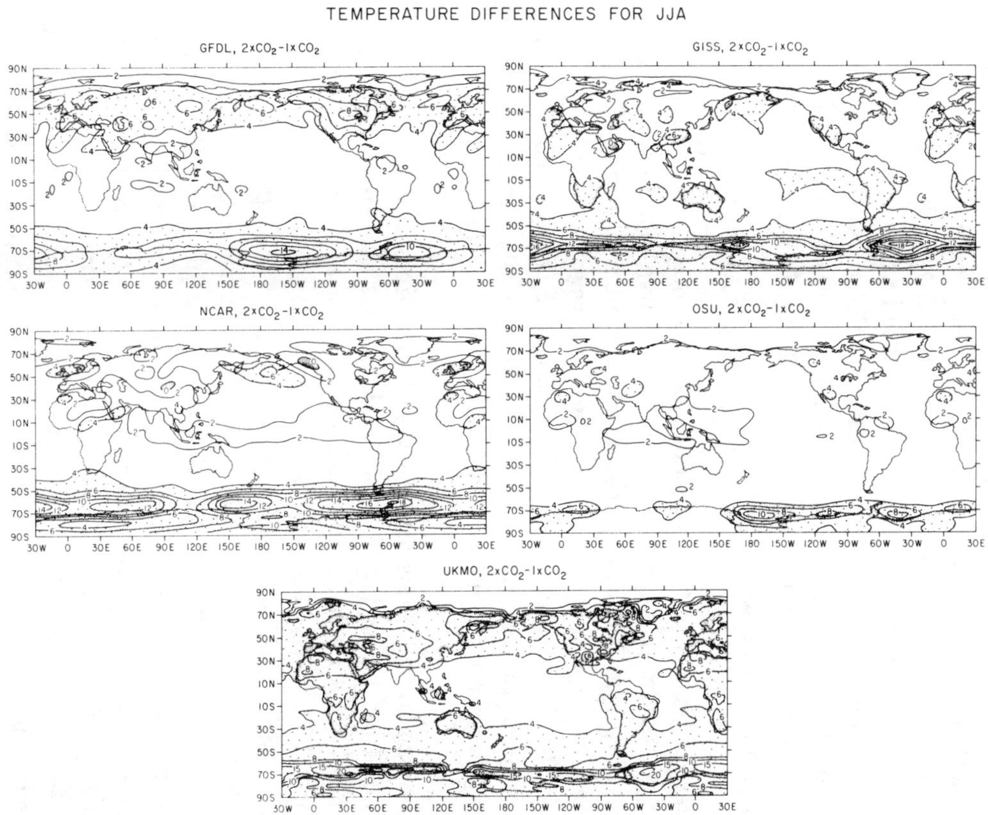

Fig. 6. Geographical distribution of the surface air temperature change (°C), $2xCO_2$-$1xCO_2$, for JJA simulated by: (a) the GDFL model by Wetherald and Manabe (1986); (b) the GISS model by Hansen *et al.* (1984); (c) the NCAR model by Washington and Meehl (1984); (d) the OSU model by Schlesinger and Zhao (1989); and (e) the UKMO model by Wilson and Mitchell (1987). Stipple indicates temperature increases larger than 4°C.

Figures 5 and 6 also show that although there are similarities in the CO_2-induced regional temperature changes simulated by the models, there are significant differences in both their magnitude and seasonality. For example, in North America the wintertime warming generally increases with latitude in the GDFL, GISS and UKMO simulations with values of 4°C in the south and 10°C in the north, while both the NCAR and OSU simulations exhibit a warming minimum of 2°C centered over Canada and over the Pacific Northwest, respectively. Also, the CO_2-induced warming in summer compared to that in winter is simulated to be smaller by the GISS and NCAR models, to be comparable by the GFDL, OSU and UKMO models, and to be larger by the NCAR model. Further similarities and differences can be seen for the CO_2-induced temperature changes simulated for the other continents.

2.3.2. Soil Water Change. The time-latitude distributions of the zonal-mean soil water change over ice-free land simulated by the five models for doubled CO_2 are presented in Fig. 7. This figure shows that all five models simulate an increased soil water during winter in the Northern Hemisphere from about 45°N to 70°N. In northern hemisphere summer the GISS and NCAR models simulate a minimum increase in soil water, while the GFDL, OSU and UKMO models simulate a decreased soil water from about 30°N to 70°N.

The geographical distributions of the CO_2-changes in soil water over ice-free land simulated by the models for DJF and JJA are presented in Figs. 8 and 9, respectively. As was evidenced by the zonal-mean soil water changes shown in Fig. 7, all five models simulate a moistening of the soil over much of Eurasia and North America in DJF. On the other hand, the GISS and NCAR models simulate both regions of increased and decreased soil water over Eurasia and North America in JJA, while the GFDL, OSU and UKMO models simulate a desiccation virtually everywhere in the Northern Hemisphere during summer. A similar desiccation of the northern hemisphere soil was first obtained in the "annual-mean" simulations with atmospheric GCM/swamp ocean models and with the first annual-cycle simulation with an atmospheric GCM/prescribed-depth mixed-layer ocean model (see Schlesinger and Mitchell 1985, 1987).

3. CO_2-Induced Transient Climatic Change

As discussed in the Introduction the CO_2 concentration has increased from

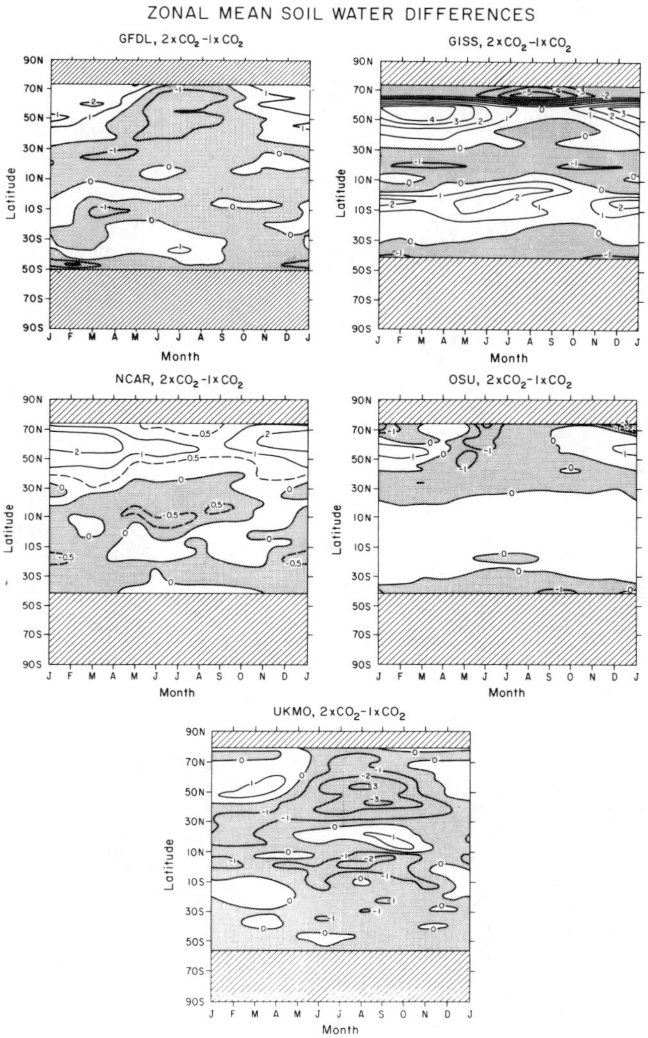

Fig. 7. Time-latitude distribution of the zonal-mean soil water change (cm), $2xCO_2$-$1xCO_2$, only over ice-free land simulated by: (a) the GDFL model by Wetherald and Manabe (1986); (b) the GISS model by Hansen *et al.* (1984); (c) the NCAR model by Washington and Meehl (1984); (d) the OSU model by Schlesinger and Zhao (1989); and (e) the UKMO model by Wilson and Mitchell (1987). Stipple indicates a decrease in soil water, hatching indicates latitudes where there is no ice-free land.

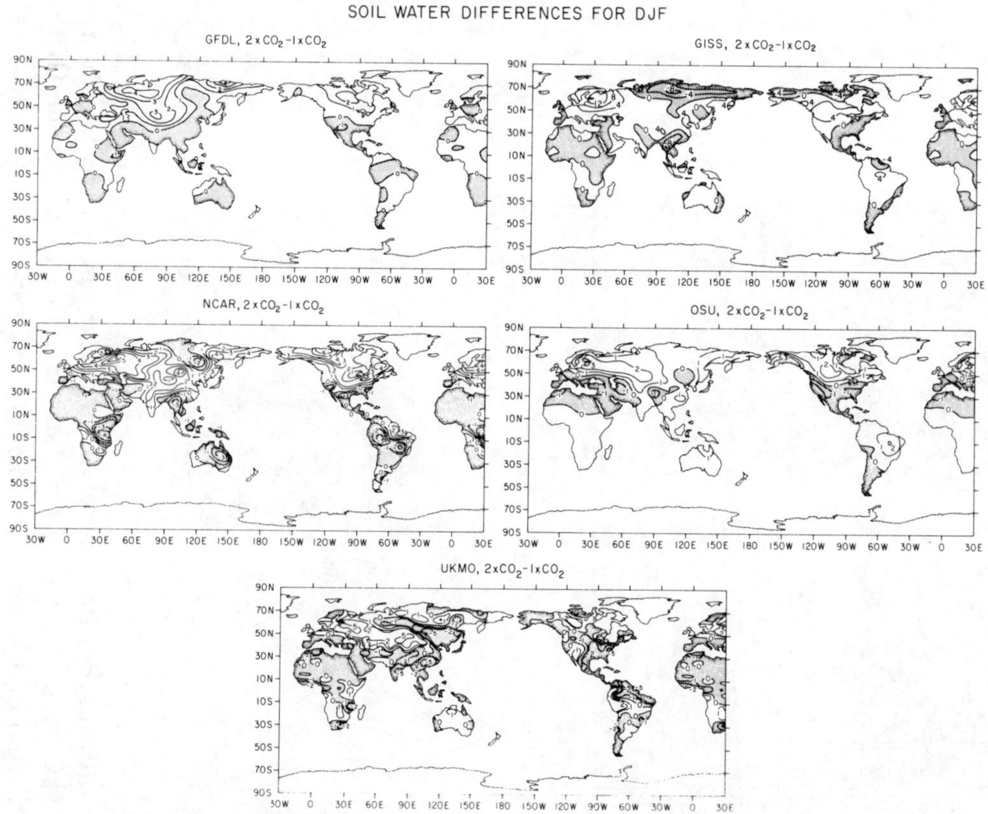

Fig. 8. Geographical distribution of the soil water change (cm), $2xCO_2$-$1xCO_2$, for DJF simulated by: (a) the GDFL model by Wetherald and Manabe (1986); (b) the GISS model by Hansen *et al.* (1984); (c) the NCAR model by Washington and Meehl (1984); (d) the OSU model by Schlesinger and Zhao (1989); and (e) the UKMO model by Wilson and Mitchell (1987). Stipple indicates a decrease in soil water.

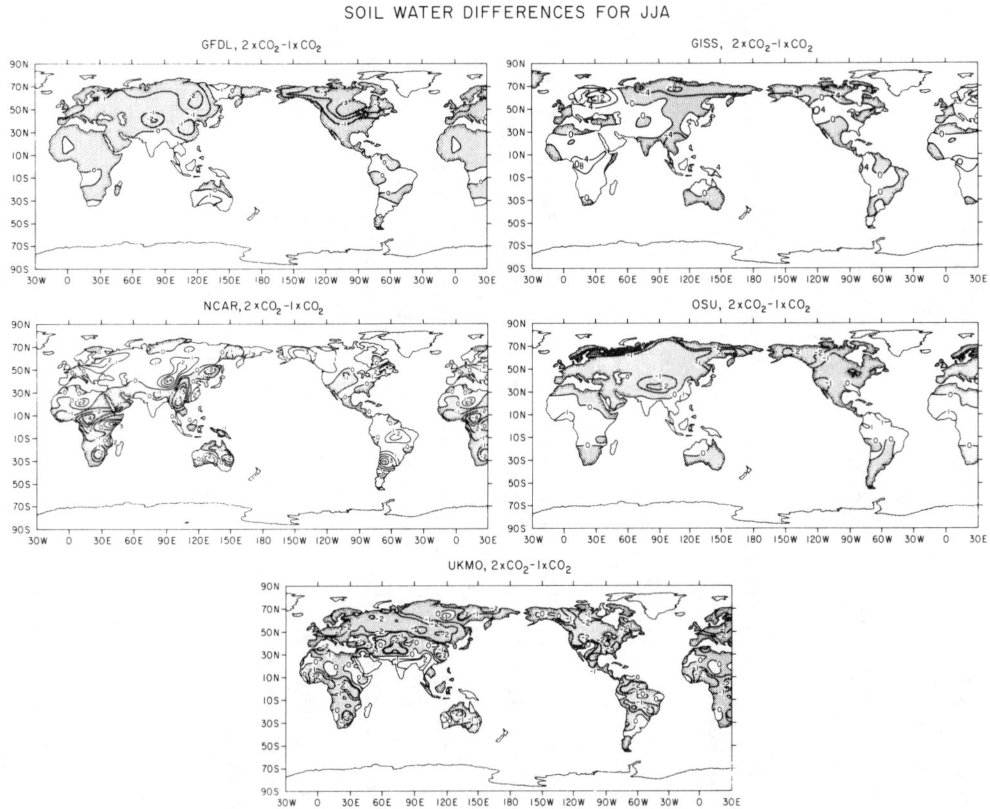

Fig. 9. Geographical distribution of the soil water change (cm), $2xCO_2$-$1xCO_2$, for JJA simulated by: (a) the GDFL model by Wetherald and Manabe (1986); (b) the GISS model by Hansen *et al.* (1984); (c) the NCAR model by Washington and Meehl (1984); (d) the OSU model by Schlesinger and Zhao (1989); and (e) the UKMO model by Wilson and Mitchell (1987). Stipple indicates a decrease in soil water.

about 288 ppmv in 1861 to 344 ppmv in 1984 (Siegenthaler and Oeschger, 1987; Keeling and Boden, 1986). The equilibrium surface temperature increase corresponding to this CO_2 increase can be written by Eq. (1) as

$$\Delta T_s = G_f \Delta Q ,$$

where the direct radiative forcing ΔQ can be scaled to that for a CO_2 doubling, ΔQ_{2x}, by (Augustsson and Ramanathan, 1977)

$$\Delta Q = \Delta Q_{2x} \left[\frac{\ln[C(1984)/C(1861)]}{\ln 2} \right] ,$$

where C is the CO_2 concentration. Combining these equations and noting that $(\Delta T_s)_{2x} = G_f \Delta Q_{2x}$ gives

$$\Delta T_s = (\Delta T_s)_{2x} \left[\frac{\ln[C(1984)/C(1861)]}{\ln 2} \right] . \tag{4}$$

If we assume that $(\Delta T_s)_{2x} = 4°C$ based on the results from the GFDL, GISS, NCAR, OSU and UKMO atmospheric GCM/mixed-layer ocean model equilibrium simulations, then for the CO_2 concentrations above, ΔT_s = 1.0°C. However, the reconstructed global-mean surface air temperature record (Jones *et al.*, 1986) indicates a warming from 1861 to 1984 of about 0.6°C. Does this difference mean that the gain G_f of our climate models is about twice as large as that of nature? The likely answer is no because the actual response of the climate system lags the equilibrium response because of the thermal inertia of the ocean. This can be illustrated by the energy-balance model

$$C_s \frac{d\Delta T_s}{dt} = \Delta Q - \frac{\Delta T_s}{G_f} ,$$

where C_s is the heat capacity of the upper ocean. When equilibrium is achieved, $d\Delta T_s/dt = 0$ and $(\Delta T_s)_{eq} = G_f \Delta Q$ as in Eq. (1). However, the transient solution

$$\Delta T_s(t) = (\Delta T_s)_{eq} \left(1 - e^{-t/\tau_e}\right)$$

shows that equilibrium is approached exponentially with a characteristic "e-folding" time $\tau_e = C_s G_f$, which is the time required to reach $1 - e^{-1}$ or 63% of the equilibrium response. In the following the studies of this lag of the climate system are reviewed.

3.1. RESULTS FOR AN ABRUPT CO_2 INCREASE

The transient response of the climate system to an abrupt CO_2 increase has been investigated with planetary energy-balance, radiative-convective and simplified atmospheric general circulation models in conjunction with box-diffusion, box-upwelling-diffusion and two-box ocean models. The box-diffusion ocean model consists of a fixed-depth mixed layer (the box) surmounting the thermocline and deep ocean in which vertical heat transport is treated as a diffusive process with a prescribed thermal diffusivity κ. The box-upwelling-diffusion ocean model is a box-diffusion model with a prescribed oceanic upwelling velocity W. The two-box ocean model is comprised of a fixed-depth mixed-layer box and an intermediate-water box which exchange heat vertically with a prescribed ventilation time. A transient simulation with a global, coupled atmosphere/ocean general circulation model has also been performed. In the following we summarize the results from these simple and more-comprehensive models.

3.1.1. Results From Simplified Models. The results from six studies of the transient response to abrupt heating are presented in Table 3. Hoffert *et al.* (1980) using a box-upwelling-diffusion ocean model, and Schneider and Thompson (1981) using a two-box ocean model, obtained e-folding times of about 10 to 20 years. A slightly larger e-folding time of 25 years was obtained by Bryan *et al.* (1982) and Spelman and Manabe (1984) with a simplified coupled atmosphere-ocean general circulation model in which the geographical domain was restricted to a 120°-longitude sector extending from equator to pole, with the western half of the sector occupied by land at zero elevation and the eastern half by ocean with a uniform depth of 5000 m. On the other hand, Hansen *et al.* (1984) used a box-diffusion ocean model and obtained 27-, 55- and 102-year e-folding times corresponding to assumed climate system gains G_f of 0.465, 0.698 and 0.977°C/(Wm^{-2}), respectively. Bryan *et al.* (1984) used an uncoupled global oceanic general circulation model and found an e-folding time of about 100 years in response to an imposed 0.5°C upper-ocean surface warming. Lastly, Siegenthaler and

Table 3. The e-folding time τ_e for abrupt heating from selected climate model studies.

Study	Model	τ_e (years)
Hoffert *et al.* (1980)	Planetary energy-balance climate model and a box-upwelling-diffusion ocean model	8-20
Schneider and Thompson (1981)	Planetary energy-balance model and a two-box ocean model	13
Bryan *et al.* (1982) Spelman and Manabe (1984)	Coupled atmosphere-ocean general circulation model with simplified geography and topography	25
Hansen *et al.* (1984)	Radiative-convective climate model and a box-diffusion ocean model	27 55 102
Bryan *et al.* (1984)	Global oceanic general circulation model	100
Siegenthaler and Oeschger (1984)	Planetary energy-balance climate model and a box-diffusion ocean model	60

Oeschger (1984) obtained a 60-year e-folding time in their study using a box-diffusion model.

The studies presented in Table 3 indicate that the e-folding time τ_e lies between 10 and 100 years. If $\tau_e \approx 10$ years, then the actual response of the climate system would be quite close to the equilibrium response and the disparity between the latter for the 1861-to-1984 warming and the corresponding observed warming would mean that the gain of our climate models is larger than that of nature. On the other hand, if $\tau_e \approx 100$ years, then the actual response of the climate system would be quite far from the equilibrium response, thus indicating that the gain of our climate models may be correct.

The factors that contribute to the wide range in τ_e have been discussed by Wigley and Schlesinger (1985) from their analytical solution for the energy-balance climate/box-diffusion ocean model. These authors found that τ_e depends quadratically on the climate system gain G_f and linearly on the thermal diffusivity κ. When a box-upwelling-diffusion model is used instead, the resultant e-folding time of the numerical solution can be expressed as shown in Fig. 10 by $\tau_e = 92.57\ G^{1.93}\ \kappa^{0.94}\ \exp(-0.15\ G^{0.97}\ \kappa^{0.09}\ W^{0.78})$. However, in view of this dependence of τ_e on G_f, κ and W, quantities which are prescribed in an energy-balance climate model, the determination of whether $\tau_e \approx 10$ or 100 years requires a global, coupled atmosphere-ocean general circulation model in which these quantities are self-determined. The results from a simulation with such a model are described below.

3.1.2. *Results From A Global, Coupled Atmosphere-Ocean GCM.* To investigate the transient response of the climate system, $1xCO_2$ and $2xCO_2$ simulations were performed with the OSU coupled atmosphere-ocean general circulation model (Schlesinger *et al.*, 1985; Schlesinger and Jiang, 1988). The atmospheric component of the coupled model is basically the two-layer AGCM described by Schlesinger and Gates (1980) and documented by Ghan *et al.* (1982). The oceanic component of the coupled model is basically the six-layer oceanic general circulation model (OGCM) developed by Han (1984 a, b) and extended by him to include the Arctic Ocean. The coupled model predicts the temperatures and velocities of the atmosphere and ocean, the atmospheric surface pressure and water vapor, the oceanic salinity, the land surface temperature and water content, the sea ice thickness, the snow mass and clouds, and includes both the diurnal and seasonal variations of solar radiation. The coupled model is global, has realistic continent/ocean geography and land and ocean-bottom

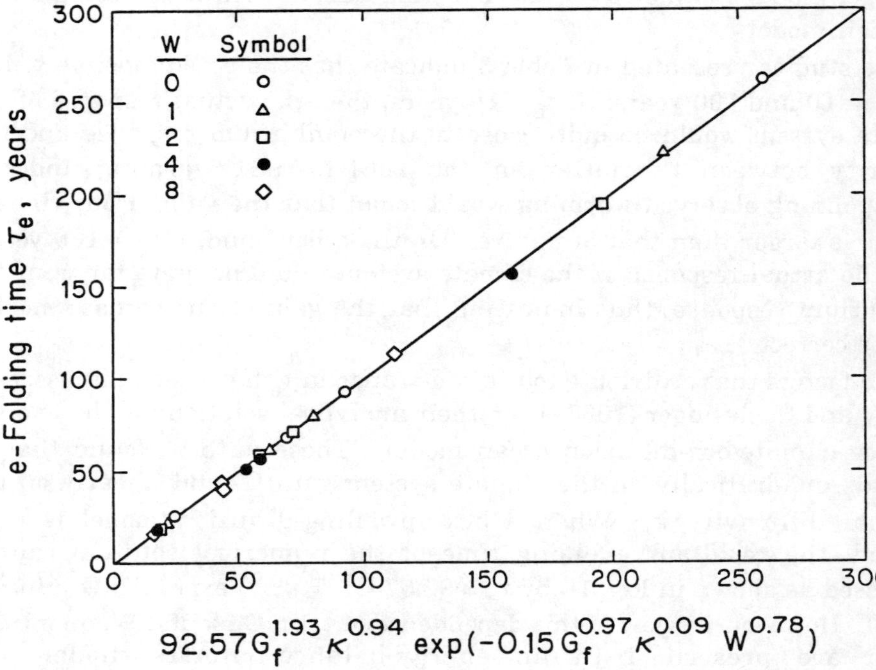

Fig. 10. The e-folding time τ_e for an abrupt heating perturbation versus $92.57\ G_f^{1.93}\ \kappa^{0.94}\ \exp(-0.15\ G_f^{0.97}\ \kappa^{0.09}\ W^{0.78})$. The data points were obtained from a numerical solution of a box-upwelling-diffusion model for upwelling velocities W = 0, 1, 2, 4 and 8 m yr^{-1}. The fit (straight line) was obtained by regression.

topographies, and is integrated synchronously, that is, both component models simulate the same period of time. The $1xCO_2$ and $2xCO_2$ simulations differed only in their CO_2 concentrations (326 and 652 ppmv, respectively), both were started from the same initial conditions, and each was integrated for 20 years. The evolution of the difference between the $2xCO_2$ and $1xCO_2$ simulations thus represents the transient climatic change induced by an abrupt CO_2 doubling.

The evolution of the change in global-mean temperature induced by the doubled CO_2 concentration is shown in Fig. 11 in terms of the vertical distribution of monthly-mean $2xCO_2$-$1xCO_2$ temperature differences for the atmosphere and ocean. The top panel of Fig. 11 shows an initially-rapid and vertically-uniform warming of the atmosphere followed by a progressively-slowing atmospheric warming. The bottom panel of Fig. 11 also shows an initially-rapid warming of the sea surface followed by a progressively-slowing warming. Figure 11 indicates that this decrease with time in the warming rate of the atmosphere and sea surface is the result of a downward transport and sequestering of heat into the interior of the ocean.

The temperature changes shown in Fig. 11, when normalized by the equilibrium temperature change, define the climate response function for the global-mean temperature. Clearly, these coupled atmosphere/ocean GCM simulations are not of sufficient duration for the equilibrium change to have been attained. However, estimates of the equilibrium temperature change and the vertical transport characteristics of the ocean have been obtained from a representation of the time evolution of Fig. 11 by a simple one-dimensional climate/ocean model (Schlesinger *et al.*, 1985). An energy-balance climate/multi-box ocean model gives an excellent representation of the evolution shown in Fig. 11 with self-determined parameters for which the estimated equilibrium temperature change is 2.8°C. The effective ocean thermal diffusivity κ, by which all vertical heat transport is parameterized in the simple multi-box ocean model representation, is 3.2 $cm^2\ s^{-1}$ at the 50 m level, 3.8 $cm^2\ s^{-1}$ at the 250 m level, and 1.5 $cm^2\ s^{-1}$ at the 750 m level. The mass-averaged effective ocean thermal diffusivity of $\kappa = 2.25\ cm^2\ s^{-1}$ is in agreement with the best estimate based on the value required by a box-diffusion ocean model to reproduce the observed penetration of bomb-produced radionuclides into the ocean (Broecker *et al.*, 1980; Siegenthaler, 1983). Moreover, an energy-balance climate/box-diffusion ocean model with κ = 2.25-2.50 $cm^2\ s^{-1}$ is successful in reproducing the evolution of the $2xCO_2$-$1xCO_2$ differences in the surface air

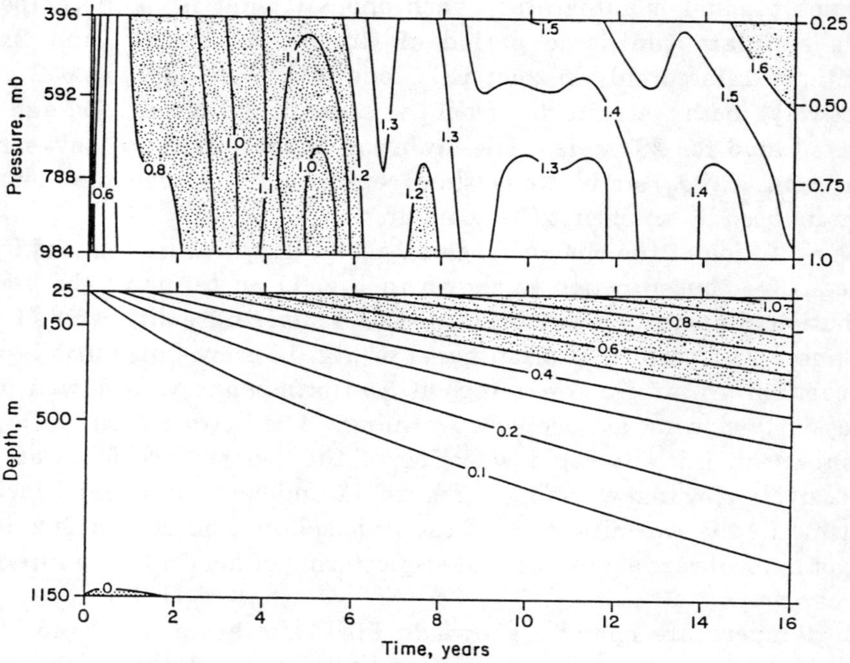

Fig. 11. Latitude-vertical distribution of the $2xCO_2$-$1xCO_2$ difference in the annual-mean, zonal-mean temperatures (°C) of the atmosphere (above) and ocean (below) for year 16. Atmospheric warming larger than 1.6°C is shown by light stipple, as is oceanic warming larger than 0.8°C, while oceanic cooling is indicated by heavy stipple. (From Schlesinger *et al.*, 1985).

and ocean surface layer temperatures simulated by the coupled atmosphere/ocean GCM. Consequently, it appears that the coupled model transports heat from the surface downward into the ocean at a rate which is commensurate with the rate observed for the downward transport of radionuclides.

The climate response function obtained from the energy-balance climate/multi-box ocean model representation of the coupled model $2xCO_2$-$1xCO_2$ temperature evolution is shown in Fig. 12 where a projection with the simple model has been made to year 200 after the abrupt CO_2 doubling. It is seen that the response function is not simply $1\text{-}\exp(-t/\tau_e)$, as it would be if only the oceanic mixed layer warmed without transporting heat to the thermocline and deeper ocean. Thus, the response function cannot be characterized by a single parameter such as τ_e. Nevertheless, the time to reach 63% of the equilibrium is useful for comparison with earlier studies (Table 3). Figure 12 shows that such an "e-folding time" is about 50-100 years.

3.2. RESULTS FOR A TIME-DEPENDENT CO_2 INCREASE

The results presented in the preceding section have been for the transient response of the climate system to an instantaneous doubling of the CO_2 concentration. However, the CO_2 concentration has not abruptly changed in the past, nor is it likely to in the future. Instead the CO_2 concentration has increased more or less continuously since the dawn of the industrial revolution.

Both Hansen *et al.* (1984) and Wigley and Schlesinger (1985) estimated the temperature change from 1850 to 1980 induced by the increasing CO_2 concentration during this 130-year period. Figure 13, based on the study of Wigley and Schlesinger (1985) in which the 1850 CO_2 concentration was taken as 270 ppmv, shows the 1850-1980 surface temperature change versus the equilibrium surface temperature change for a doubled CO_2 concentration. Again let us assume that the latter is 4°C. If the climate system had no thermal inertia, the 1850-1980 surface temperature change would be in equilibrium with the instantaneous 1980 CO_2 concentration, and the warming would be given by an equation similar to Eq. (4). As shown in Fig. 13, this instantaneous equilibrium warming would be 1.3°C. However, when the heat capacity and vertical heat transport of the ocean are considered, as shown by the curves for the oceanic thermal diffusivities $\kappa = 1$ and $\kappa = 3$ cm^2 s^{-1}, the 1850-1980 warming is reduced to about 0.5 to 0.7°C, a range which does not conflict with the observational record.

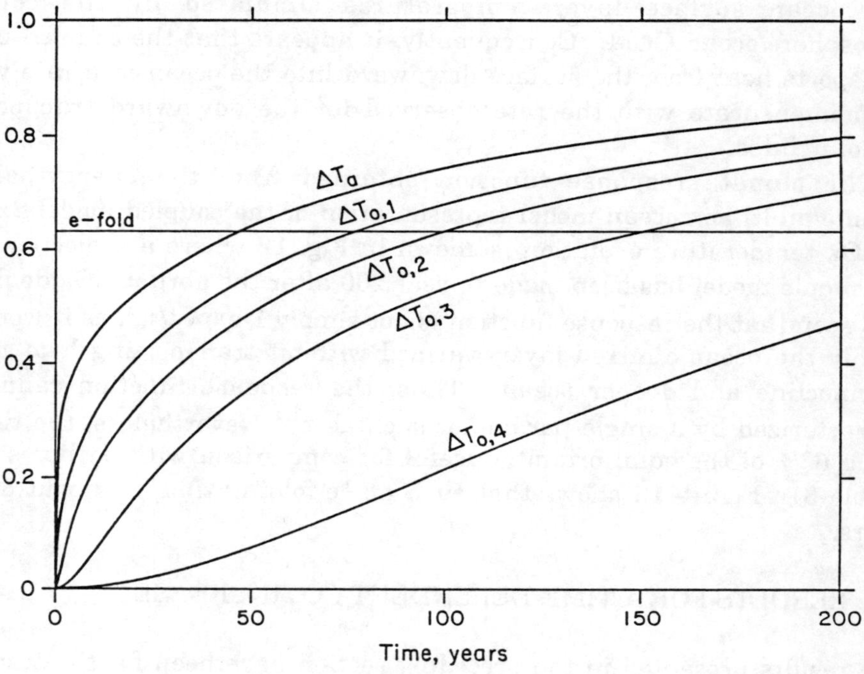

Fig. 12. Energy-balance climate/multi-box ocean model projection to year 200 of the evolution of the $2xCO_2$-$1xCO_2$ difference in global-mean surface air temperature (ΔT_a temperature ($\Delta T_{o,k}$, k = 1, 2, 3, 4) as a fraction of the equilibrium temperature difference (ΔT_{eq}). The horizontal line labelled e-fold indicates the level at which $\Delta T/\Delta T_{eq} = 1\text{-}1/e \approx 0.63$. (From Schlesinger *et al.*, 1985).

Consequently, it appears that the climate system gain of about 1°C/(Wm^{-2}) obtained by the most recent GCM equilibrium simulations may be reasonably correct.

Figure 13 also shows that even if the CO_2 concentration were to increase no further in the future, the Earth's surface temperature would continue to increase by about 0.7°C in its approach to its new equilibrium value. This demonstrates the "Catch-22" nature of the CO_2-induced climatic-change issue: The present warming is "small" and perhaps within the natural variation of climate because the ocean sequesters heat within its interior, but, because of this, when the warming becomes demonstrably evident, continued future warming is inevitable, even if the CO_2 concentration were prevented from increasing further.

4. Summary

In this paper we have reviewed the simulations of CO_2-induced equilibrium and transient climatic change made by energy-balance, radiative-convective and general circulation models. The equilibrium climatic-change simulations have been characterized in terms of the direct radiative forcing of the increased CO_2, the response of the climate system in the absence of feedback processes, and the feedbacks of the climate system. The transient climatic-change simulations have been characterized in terms of the e-folding time of the response of the climate system.

For a doubling of the CO_2 concentration the direct radiative forcing is about 4 Wm^{-2} for the surface-troposphere system. In the absence of feedbacks the gain (output/input) of the climate system is about 0.3°C/(W m^{-2}) and the change in the surface air temperature is therefore about 1.2°C. Surface energy-balance models (SEBMs) have given a warming induced by a CO_2 doubling of about 0.2 to 10°C. This wide range is the result of the inherent difficulty in specifying the behavior of the atmosphere in SEBMs. Radiative-convective models (RCMs) have given a warming of about 0.5 to 4.2°C for a CO_2 doubling. This range is the result of water vapor feedback, lapse rate feedback, surface albedo feedback, cloud altitude feedback, cloud cover feedback and cloud optical depth feedback. General circulation models (GCMs) have given warming of about 1.3 to 5.2°C for a CO_2 doubling. This also is the result of the feedbacks listed above, except that due to cloud optical depth feedback which has not yet been included in the GCM simulations of CO_2-induced climatic change.

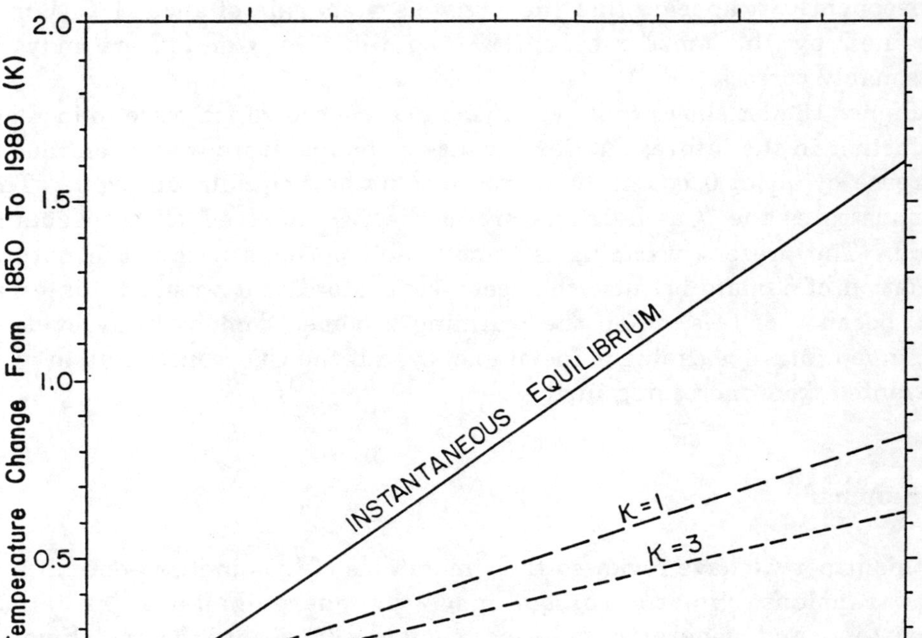

Fig. 13. The temperature change from 1850 to 1980 versus the equilibrium temperature change for doubled CO_2. The instantaneous equilibrium curve is given by Eq. (5). The curves $\kappa = 1$ and $k = 3$ cm^2 s^{-1} are from the results of Wigley and Schlesinger (1985).

The most recent simulations of CO_2-induced climatic change by the GFDL, GISS, NCAR, OSU and UKMO atmospheric GCM/mixed-layer ocean models have been presented. The changes in the annual-mean, global-mean surface air temperature simulated by these models for a CO_2 doubling range from 2.8 to 5.2°C, and the corresponding changes in precipitation from 7.1 to 15% of their $1xCO_2$ values. The CO_2-induced zonal-mean surface air temperature changes exhibit a seasonal variation which increases from the tropics toward the poles. The geographical distributions of the CO_2-induced temperature change display a warming virtually everywhere in both winter and summer. The warming is a maximum in the winter polar region where the $1xCO_2$ sea ice retreats poleward in the $2xCO_2$ simulation, and is a minimum in the summer polar region and in the tropics during both seasons. Over the continents the CO_2-induced warming simulated by the models exhibits qualitative differences in seasonality and quantitative differences in magnitude. The geographical distributions of the CO_2-induced soil water change simulated by all the models reveal a moistening of the northern hemisphere continents in winter. A drying of much of the northern hemisphere continents in summer is simulated by the GFDL, OSU and UKMO models, but not by the GISS and NCAR models.

The simulations of CO_2-induced transient climatic change by simplified models have given an e-folding time τ_e of about 10-100 years for the response of the climate system to an abrupt increase in the CO_2 concentration. Simplified analytical and numerical analyses show that this wide range in the estimates of τ_e is the result of the dependence of τ_e on three parameters of the climate system, namely, its gain and the oceanic upwelling velocity and effective thermal diffusivity. A simulation with a global coupled atmosphere/ocean general circulation model indicates that τ_e is 50-100 years as a result of the transport of the CO_2-induced surface heating into the interior of the ocean. Theoretical studies for a time-dependent CO_2 increase between 1850 and 1980 demonstrate that this sequestering of heat into the ocean's interior could reduce the concomitant warming to about half that which would have occurred in the absence of the ocean. These studies also reveal that, in the absence of other factors, the climate system will continue to warm towards its as-yet unrealized equilibrium temperature change, even if there is no further increase in the CO_2 concentration.

5. Conclusion

The five most recent simulations of the equilibrium climatic changes induced by a doubling of the CO_2 concentration have been made with atmospheric general circulation models coupled to prescribed-depth mixed layer models of the ocean. These ocean models do not include in a physical way the horizontal transports of heat, momentum and salinity (see footnote 1) which are fundamental to the maintenance and change of the general circulation of the ocean. Such models cannot therefore correctly simulate the sea surface temperature (SST) distribution which is of paramount importance for the general circulation of the atmosphere. Therefore, such atmospheric GCM/mixed-layer ocean models cannot correctly simulate the regional distribution of climate for either $1xCO_2$ or $2xCO_2$. Since only oceanic GCMs can simulate the horizontal and vertical transports upon which the SST and regional climates intimately depend, it is imperative that future simulations of CO_2-induced equilibrium climatic change be performed with coupled atmosphere/ocean GCMs.

A necessary condition for the acceptance of a coupled atmosphere/ocean model's simulation of a CO_2-induced equilibrium and/or transient climatic change is the validation of the model's simulation of the present ($1xCO_2$) climate by comparing it with the observed present climate. A rudimentary model validation has been performed by Schlesinger and Mitchell (1985) for the $1xCO_2$ simulations by the GFDL, GISS and NCAR models, and similar validations have been performed for the OSU and UKMO models by Schlesinger and Zhao (1989) and by Wilson and Mitchell (1987), respectively. These model validation studies show that, although the GCMs are the only type of model in the climate model hierarchy that can simulate the climate on the regional scale, they do so not without error, particularly for the components of the hydrological cycle such as precipitation. Therefore, further improvement in the simulation of CO_2-induced climate change requires the systematic analysis of the factors that contribute to the errors in the models' simulations of the present climate, and the subsequent correction of those errors. This will likely require the improvement of the methods used by the models for both the explicitly-resolved and parameterized physical processes, the former by a comparison of the solutions of the models' numerical methods with both exact solutions and the results of laboratory studies for simplified cases, and the latter by intercomparison studies among the models and observations following the approach of the Intercomparison of Radiation Codes in Climate Models (ICRCCM) program (Luther and Fouquart, 1984).

The successful validation of a model's simulation of the present climate, while a necessary condition for the acceptance of its simulated CO_2-induced climatic change, is not a sufficient condition for this purpose. While in all likelihood such a sufficient condition does not exist, there is a second necessary condition, namely, that the model correctly simulate a climate different from that of the present. Thus, to increase confidence that the CO_2-induced climatic changes simulated by the models are correct at the regional scale, the model validation methodology described above for the present climate must also be applied to at least one other climate such as that of the Wisconsin Ice Age 18,000 years before the present.

Acknowledgements. I would like to thank Syukuro Manabe and Richard Wetherald of the Geophysical Fluid Dynamics Laboratory; James Hansen, Gary Russell, Andrew Lacis and David Rind of the Goddard Institute for Space Studies; Warren Washington and Gerald Meehl of the National Center for Atmospheric Research; and John Mitchell of the United Kingdom Meteorological Office for making their results available to me. I express my gratitude to Dean Vickers for performing some of the calculations herein, and to John Stark, Larry Holcomb and Linda Haygarth for drafting. This study was supported by the National Science Foundation and the Carbon Dioxide Research Division, Office of Basic Energy Sciences of the Department of Energy under grant ATM 87-12033.

REFERENCES

Augustsson, T., and V. Ramanathan, 1977: 'A radiative-convective model study of the CO_2 climate problem.' *J. Atmos. Sci.,* **34**, 448-451.

Broecker, W. S., T.-H. Peng and R. Engh, 1980: 'Modeling the carbon system.' *Radiocarbon,* **22**, 565-598.

Bryan, K., F. G. Komro, S. Manabe and M. J. Spelman, 1982: 'Transient climate response to increasing atmospheric carbon dioxide.' *Science,* **215**, 56-58.

Bryan, K., F. G. Komro and C. Rooth, 1984: 'The ocean's transient response to global surface temperature anomalies.' In *Climate Processes and Climate Sensitivity*, Maurice Ewing Series, **5**, J. E. Hansen and T. Takahashi, Eds., American Geophysical Union, Washington, D.C., 29-38.

Budyko, M. I., 1969: 'The effect of solar radiation variations on the climate of the earth.' *Tellus,* **21**, 611-619.

Cess, R. D., and G. L. Potter, 1988: 'A methodology for understanding and intercomparing atmospheric climate feedback processes within general circulation models.' Submitted to *J. Geophys. Res.*

Gates, W. L., 1976a: 'Modeling the ice-age climate.' *Science,* **191**, 1138-1144.

Gates, W. L., 1976b: 'The numerical simulation of ice-age climate with a global general circulation model.' *J. Atmos. Sci.,* **33**, 1844-1873

Ghan, S. J., J. W. Lingaas, M. E. Schlesinger, R. L. Mobley and W. L. Gates, 1982: 'A documentation of the OSU two-level atmospheric general circulation model.' Report No. 35, Climatic Research Institute, Oregon State University, Corvallis, OR, 395 pp.

Han, Y.-J., 1984a: 'A numerical world ocean general circulation model, Part I. Basic design and barotropic experiment.' *Dyn. Atmos. Oceans,* **8**, 107-140.

Han, Y.-J., 1984b: 'A numerical world ocean general circulation model. Part II. A baroclinic experiment.' *Dyn. Atmos. Oceans,* **8**, 141-172.

Hansen, J., A. Lacis, D. Rind, G. Russell, P. Stone, I. Fung, R. Ruedy and J. Lerner, 1984: 'Climate sensitivity: Analysis of feedback mechanisms.' In *Climate Processes and Climate Sensitivity*, Maurice Ewing Series, **5**, J. E. Hansen and T. Takahashi, Eds., American Geophysical Union, Washington, D.C., 130-163.

Hoffert, M. I., A. J. Callegari and C.-T. Hsieh, 1980: 'The role of deep sea heat storage in the secular response to climate forcing.' *J. Geophys. Res.,* **85**, 6667-6679.

Imbrie, J., and K. P. Imbrie, 1979: *Ice Ages, Solving the Mystery*. Enslow Publishers, Short Hills, NJ, 224 pp.

Jenne, R., 1975: 'Data sets for meteorological research.' NCAR Tech. Note TN/1A-III, National Center for Atmospheric Research, Boulder, CO, 194 pp.

Jones, P. D., T. M. L. Wigley and P. B. Wright, 1986: 'Global temperature-variations between 1861 and 1984.' *Nature,* **322**, 430-434.

Keeling, C.D., and T.A. Boden, 1986: 'Atmospheric CO_2 Concentrations - Mauna Loa Observatory, Hawaii 1958-1986.' Carbon Dioxide Information Center (CDIC) Numeric Data Collection, NDP-001/R1, Oak Ridge National Laboratory, 17 pp with addenda.

Lal, M., and V. Ramanathan, 1984: 'The effects of moist convection and water vapor radiative processes on climate sensitivity.' *J. Atmos. Sci.,* **41**, 2238-2249.

Luther, F. M., and Y. Fouquart, 1984: 'The Intercomparison of Radiation Codes in Climate Models (ICRCCM): Longwave Clear-Sky Calculations.' World Meteorological Organization Report WCP-93, Geneva, Switzerland, 37pp.

Nature, 1987: Summary of letter to *Nature* by Roeckner *et al.,* v.

Nordhaus, W. D., and G. W. Yohe, 1983: 'Future paths of energy and carbon dioxide emissions.' In *Changing Climate*, National Academy of Sciences, Washington, D.C., 87-153.

Ramanathan, V., M. S. Lian and R. D. Cess, 1979: 'Increased atmospheric CO_2: Zonal and seasonal estimates of the effects on the radiation energy balance and surface temperature.' *J. Geophys. Res.,* **84**, 4949-4958.

Rasool, S. I., and S. H. Schneider, 1971: 'Atmospheric carbon dioxide and aerosols: Effects of large increases on global climate.' *Science,* **173**, 138-141.

Rotty, R.M., and C.D. Masters, 1985: 'Carbon dioxide from fossil fuel combustion: trends, resources, and technological implications.' In *Atmospheric Carbon Dioxide and the Global Carbon Cycle,* J.R. Trabalka, Ed., U.S. Department of Energy, DOE/ER-0239, 63-80.

Schlesinger, M. E., 1984: 'Climate model simulations of CO_2-induced climatic change.' In *Advances in Geophysics,* **26**, B. Saltzman, Ed., Academic Press, New York, 141-235.

Schlesinger, M. E., 1985: 'Feedback analysis of results from energy balance and radiative-convective models.' In *The Potential Climatic Effects of Increasing Carbon Dioxide,* M.C. MacCracken and F.M. Luther, Eds., U.S. Department of Energy, DOE/ER-0237, 280-319. (Available from NTIS, Springfield, Virginia.)

Schlesinger, M. E., 1988a: 'Quantitative analysis of feedbacks in climate model simulations of CO2-induced warming'. In *Physically-Based Modelling and Simulation of Climate and Climatic Change*, M. E. Schlesinger, Ed., NATO Advanced Study Institute Series, Kluwer, Dordrecht, 653-736.

Schlesinger, M. E., (Ed.), 1988b: *Physically-Based Modelling and Simulation of Climate and Climatic Change*, Parts 1 and 2, NATO Advanced Study Institute Series, Kluwer, Dordrecht, 1084 pp.

Schlesinger, M. E., and W. L. Gates, 1980: 'The January and July performance of the OSU two-level atmospheric general circulation model.' *J. Atmos. Sci.,* **37**, 1914-1943.

Schlesinger, M. E., and J. F. B. Mitchell, 1985: 'Model projections of the equilibrium climatic response to increased CO_2.' In *The Potential Climatic Effects of Increasing Carbon Dioxide,* eds., M. C. MacCracken and F. M. Luther, DOE/ER-0237, U.S. Department of Energy, Washington, D.C., 81-147. (Available from NTIS, Springfield, Virginia.)

Schlesinger, M. E., and J. F. B. Mitchell, 1987: 'Climate model simulations of the equilibrium climatic response to increased carbon dioxide.' *Rev. of Geophys.,* **25**, 760-798.

Schlesinger, M.E., and X. Jiang, 1988: 'The transport of CO_2-induced warming into the ocean: An analysis of simulations by the OSU coupled atmosphere-ocean general circulation model.' *Climate Dynamics*, **3**, 1-17.

Schlesinger, M.E., and Z.-C. Zhao, 1989: 'Seasonal climatic changes induced by doubled CO_2 as simulated by the OSU atmospheric GCM/mixed-layer ocean model.' *J. Climate,* **2**, (in press).

Schlesinger, M. E., W. L. Gates and Y.-J. Han, 1985: 'The role of the ocean in CO_2-induced climate warming: Preliminary results from the OSU coupled atmosphere-ocean GCM.' In *Coupled Ocean-Atmosphere Models,* J. C. J. Nihoul, Ed., Elsevier, Amsterdam, 447-478.

Schneider, S. H., and S. L. Thompson, 1981: 'Atmospheric CO_2 and climate: Importance of the transient response.' *J. Geophys. Res.,* **86**, 3135-3147.

Sellers, W. D., 1969: 'A global climate model based on the energy balance of the earth-atmosphere system.' *J. Appl. Meteor.,* **8**, 392-400.

Siegenthaler, U., 1983: 'Uptake of excess CO_2 by an outcrop-diffusion model of the ocean.' *J. Geophys. Res.,* **88**, 3599-3608.

Siegenthaler, U., and H. Oeschger, 1984: 'Transient temperature changes due to increasing CO_2 using simple models.' *Ann. of Glaciology,* **5**, 153-159.

Siegenthaler, U., and H. Oeschger, 1987: 'Biospheric CO_2 emissions during the past 200 years reconstructed by deconvolution of ice core data.' *Tellus*, **39B**, 140-154.

Spelman, M. J., and S. Manabe, 1984: 'Influence of oceanic heat transport upon the sensitivity of a model climate.' *J. Geophys. Res.,* **89**, 571-586.

Washington, W. M., and G. A. Meehl, 1984: 'Seasonal cycle experiment on the climate sensitivity due to a doubling of CO_2 with an atmospheric general circulation model coupled to a simple mixed-layer ocean model.' *J. Geophys. Res.,* **89**, 9475-9503.

Wetherald, R. T., and S. Manabe, 1986: 'An investigation of cloud cover change in response to thermal forcing.' *Climatic Change,* **8**, 5-23.

Wetherald, R. T., and S. Manabe, 1988: 'Cloud feedback processes in a general circulation model.' *J. Atmos. Sci.,* **45**, 1397-1415.

Wigley, T. M. L., and M. E. Schlesinger, 1985: 'Analytical solution for the effect of increasing CO_2 on global-mean temperature.' *Nature,* **315**, 649-652.

Wilson, C.A. and J.F.B. Mitchell, 1987: 'A doubled CO_2 climate sensitivity experiment with a global climate model including a simple ocean.' *J. Geophys. Res.,* **92**, 13,315-13,343.

Climate Sensitivity: Model dependence of results

John F. B. Mitchell

Meteorological Office

Bracknell, BERKS RG12 2SZ, ENGLAND

Abstract

There has been an increasing use of numerical models over the last 20 years to study climatic change. The simulated sensitivity of climate to a given perturbation is dependent on the model used. This dependence is illustrated using examples from numerical studies of the effect of increasing atmospheric carbon dioxide concentrations. The parametrization of cloud and the land surface and the modelling of the oceans, which give rise to the current major uncertainties in determining climate change, will be the focus of numerical modellers' attention over the next decade.

1. INTRODUCTION

Considerable interest in the understanding and prediction of climate change has been maintained over the last 15 years. There has been a gradual shift in emphasis from changes due to natural perturbations (for example due to variations in the atmosphere's turbidity due to volcanic eruptions, or changes in the solar constant) to potential changes due to man's activities (for example those leading to increased concentrations of radiatively active gases). The prediction of the climatic effects of increased carbon dioxide concentrations requires the solution of many of the problems inherent in predicting climate change in general. Thus, in this paper, examples of model dependence of climate sensitivity will be drawn from studies of the effect of increased CO_2 carried out over the last decade or so.

Attempts to predict the climatic effects of doubling CO_2 from globally averaged models have produced a wide range of results, giving surface temperature increases ranging from 0.2K (Newell and Dopplick, 1979) to 9.6K (Moller, 1963). Schneider (1975) attempted to clarify the reasons for the discrepancies between some of the earlier results, noting the dependence on the assumptions made and the models used. More recently, there have been a number of studies using 3-dimensional general circulation models (GCMs) producing estimates of equilibrium surface warming from 1.3K (Washington and Meehl, 1983) to 5.2K (Wilson and Mitchell, 1987) for doubling CO_2, and large variations in the simulated geographical changes in climate (Dickinson, 1986;

A. Berger et al. (eds.), Climate and Geo-Sciences, 417–433.

Schlesinger and Mitchell, 1987). The possible reasons for discrepancies between recent GCM studies will form the subject of this paper, since this is a topic in which the author has a special interest. See Schlesinger (1985) for a comprehensive discussion of studies of the effects of increased CO_2 using radiative-convective models.

We begin this review of GCM sensitivity studies in the following section by considering a hierarchy of experiments carried out by Manabe and his co-workers at the Geophysical Fluid Dynamics Laboratory Princeton (GFDL). The strength of selected feedback processes is assessed in Section 3, and the importance of using a model which produces a faithful representation of current climate is stressed in Section 4. In Section 5, the dependence of climate sensitivity on the choice of physical parametrization is illustrated using two examples, and the final section comprises a summary and some concluding remarks.

2. A HIERARCHY OF GENERAL CIRCULATION EXPERIMENTS ON THE EFFECT OF INCREASED CO_2

The first study of the effect of doubling atmospheric CO_2 using a GCM model was carried out at GFDL by Manabe and Wetherald (1975). This formed the first of a series of experiments in which increasingly realistic (and more complex) versions of the model were used in order to isolate the effect of various features on climate sensitivity. To enable a rough comparison of the different experiments, it will be assumed that the global mean temperature change on doubling atmospheric CO_2 concentrations is half that due to quadrupling CO_2 (see Augustsson and Ramanathan, 1977).

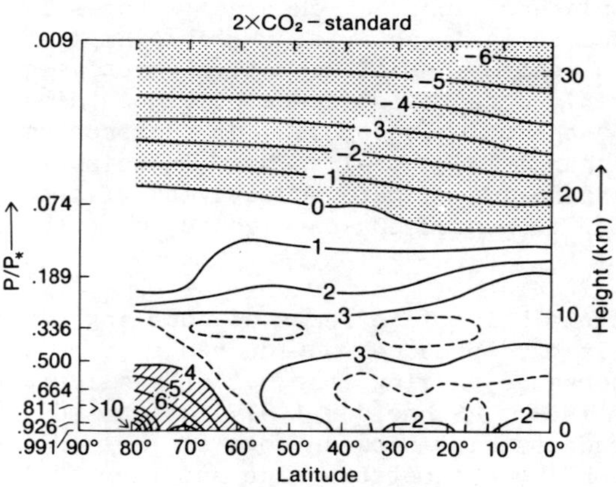

Figure 1 Height-latitude diagram of the zonally averaged temperature changes (K) due to doubling CO_2 in an idealized climate model (from Manabe and Wetherald, 1975).

Table 1

Global mean surface temperature change due to doubling atmospheric CO_2 amounts in GFDL experiments. Figures in brackets are obtained by halving the result for quadrupling CO_2. The second column of figures contains results using a revised radiation scheme.

Model. Change	Temperature	
1-D radiative convective	2.4	2.0
Annual mean sector	2.9	
Annual mean sector (variable cloud)		3.0
Mixed layer models		
Annual mean, sector		(3.0)
Seasonally varying, sector		(2.4)
Seasonally varying, realistic geography (1)		(2.0)
Seasonally varying, realistic geography (2)		3.2
Seasonally varying, realistic geography (variable cloud) (3)		4.0

Footnotes
(1) Manabe and Stouffer (1980), using climatological zonally averaged cloud.
(2) Wetherald and Manabe (1988), using prescribed cloud derived from the control simulation in (3). This differs from the cloud distribution in (1), giving different sea temperatures and sea-ice extents in the control simulation, and hence a different sensitivity to doubling CO_2.
(3) Wetherald and Manabe (1988), using the same model as in (2).

In the original GFDL experiment (Manabe and Wetherald, 1975), orography and the seasonal variation of insolation were ignored. The domain was restricted to a sector of the globe with an idealized land-sea distribution. The ocean was treated as a 'swamp', having no heat capacity or heat transport. Cloud amounts were prescribed. Doubling CO_2 concentrations produced a warming of 2.9K (Table 1).

This is somewhat larger than the 2.4K found using a radiative convective model (Manabe and Wetherald, 1967) presumably due to the large amplification of the warming in high latitudes obtained in the GCM (Figure 1).

Manabe and Wetherald attributed the polar amplification of the warming to temperature-albedo feedback, enhanced poleward transport of latent heat and the confinement of the warming to low levels under the strong polar inversion. The warming in the tropics is less than that found using the radiative convective model, which assumes a fixed 6.5K/km lapse rate. The moist convective parametrization used in the GCM adjusts the lapse rate towards the moist adiabatic, enhancing the warming in the upper troposphere and reducing it at the surface (see for example, Rowntree and Walker, 1978).

The 'swamp' model was modified to include a prognostic cloud scheme (Manabe and Wetherald, 1980). On increasing atmospheric CO_2 amounts there was a general decrease in middle and upper tropospheric cloud especially in convective regions, reducing reflected solar radiation and enhancing the radiative warming. However, as the mean cloud height in the tropics was also reduced, the effective emission temperature increased, enhancing thermal emission and almost compensating the solar effect. Low cloud also increased, particularly in high latitudes. However, because incoming insolation is small in these regions, the resulting increase in reflected insolation is again almost compensated by the corresponding reduction in thermal emission. Thus the global mean warming for a given increase in CO_2 concentrations was similar in the two models. It appeared from this experiment and a parallel study investigating the effect of changes in the solar constant (Wetherald and Manabe, 1980) that changes in cloud cover contributed little to climate sensitivity.

In order to include a seasonal cycle, Wetherald and Manabe (1981) replaced the 'swamp' model ocean with a 68 m 'mixed-layer' ocean in which cloud cover was prescribed and oceanic heat transport was still ignored. Two experiments were carried out in which CO_2 amounts were quadrupled - in the first, annual mean insolation was prescribed, and in the second, the full seasonal variation of insolation was included. The implied warming for a doubling of CO_2 was smaller in the case of the seasonal model (Table 1). The seasonal model produced a warmer control climate and hence less snow cover and a smaller temperature albedo feedback. In addition, the temperature albedo feedback in the seasonal model is probably smaller because in summer the albedo feedback is minimal since although insolation is at a maximum, snow and sea-ice extents are very small. When the sector is replaced by realistic geography and orography (Manabe and Stouffer, 1980), the global warming is reduced still further (Table 1). The lack of temperature albedo feedback over Antarctica, which arises as temperatures do not exceed freezing even in the enhanced CO_2 simulations, and the underestimation of southern hemisphere sea-ice extents both may contribute to the reduced sensitivity in this model.

A comparison of the earlier GFDL experiments with this study demonstrates that simplified models may give results which are qualitatively similar to those in a more detailed model, but are quantitatively inaccurate. Unfortunately, simplification may lead to qualitative differences, as a subsequent GFDL experiment showed.

Wetherald and Manabe (1988) studied the effect of doubling CO_2 amounts in a revised version of Manabe and Stouffer's (1980) model with both prescribed and predicted cloud. The changes in surface temperature were 3.2K and 4.0K respectively (Table 1), in contrast to the earlier "swamp" model experiments which indicated that changes in cloud did not affect climate sensitivity. In the more recent experiment, there were pronounced increases in high cloud in the tropics in addition to the decreases in middle and upper tropospheric cloud found with the "swamp" model. The associated increase in reflected insolation was small because high cloud has a low albedo, but the reduction in outgoing thermal emission was large, leading to a substantial positive feedback. In the earlier experiment, the model level nearest to the tropical tropopause was at 70 mb where changes in cloud were small, whereas in the later experiment, to the vertical distribution was changed, giving a level at 95 mb, corresponding to the maximum increase in cloud (Wetherald, 1988, personal communication). In addition, the increases in low cloud are smaller in the seasonal model, enhancing the overall positive feedback process of cloud cover (Wetherald and Manabe, 1988).

3. ANALYSIS OF FEEDBACKS

It is clear from the above experiments that changes in sea-ice and cloud amounts enhance the simulated response to increases in atmospheric CO_2. In order to quantify the contributions due to changes in individual parameters, Hansen et al. (1984) carried out an analysis of feedbacks in three climate change experiments. The feedback strengths, assumed to be linear, were estimated by inserting globally averaged changes from the three-dimensional climate model in a one-dimensional radiative-convective model. Following Hansen's analysis

$$\Delta T_{eq} = \Delta T_o \, (1-\sum g_i)^{-1} \qquad (1)$$

Where ΔT_o is the equilibrium temperature change due to doubling CO_2 in the absence of feedbacks, g_i are the 'gain factors' associated with individual feedbacks, and ΔT_{eq} is the equilibrium temperature change with all the feedbacks included. The gain factors associated with changes in water vapour, surface albedo and cloud (g_w, g_s and g_c respectively) from the Hansen et al. (1984) experiments are shown in Table 2.

Table 2

Gain factors from Hansen et al, (1984) derived from simulations with doubled CO_2 amounts

	Standard	Using model with enhanced sea-ice extents
Water vapour	0.40	0.37
Surface albedo	0.09	0.12
Cloud	0.22	0.26

The largest gain factor (and hence the strongest feedback) is due to changes in water vapour: the next strongest is due to changes in cloud cover. The second column of Table 2 shows results from 2 xCO_2 experiment carried out using a version of the model with a greater area of sea ice in the unperturbed simulations, leading to a stronger temperature-albedo feedback.

Although the 'gain factors' are additive, the associated amplification is highly non-linear (Table 3).

Table 3

Cumulative warming on doubling CO_2 amounts on adding selected feedback processes (Hansen et al, 1984).

No feedback	1.2
Water vapour	2.0
Water vapour, surface albedo	2.4
Water vapour, surface albedo, cloud	4.2

The estimated warming including water-vapour feedback along (Table 3) is compatible with the revised GFDL radiative-convective calculation (Table 1, top line), and that with water vapour, surface albedo and cloud feedback is compatible with the full GFDL model (Table 1, bottom line)

It is possible that a warmer climate would be accompanied by an increase in the water content of clouds, and hence a change in cloud radiative properties. Somerville and Remer (1984) speculated that this would produce a strong negative feedback, though this was not apparent in numerical simulations carried out by Roeckner et al, (1987). A further cloud feedback due to changes in cloud condensation nuclei has been postulated (Charlson et al. 1987). For detailed discussions of climate feedback processes see Dickinson (1986) and Schlesinger (1985).

4. IMPORTANCE OF THE SIMULATED CONTROL CLIMATE

We have seen in the preceding section that Hansen et al, (1984) found enhanced climate sensitivity when their model was altered to give greater sea-ice coverage in the control simulation. This change in sensitivity is particularly noticeable in results from a set of experiments reported by Spelman and Manabe (1984). Three versions of their model were used, differing only in the treatment of the oceans and the specification of the solar constant. The global mean temperature in the unperturbed simulation ranged from 278K to 288K and the sensitivity to quadrupling CO_2 varied by a factor of over 2 (Figure 2).

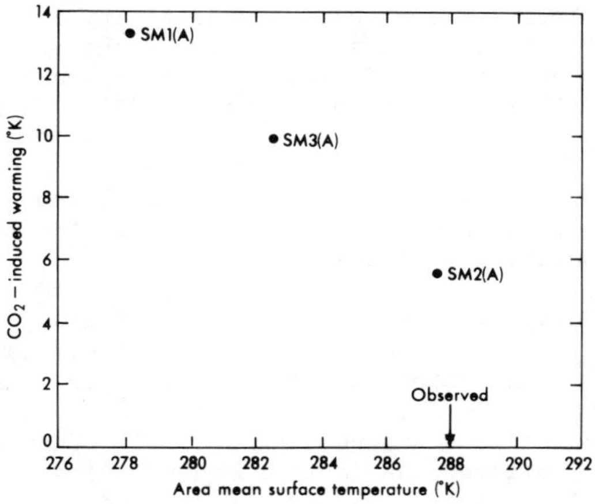

Figure 2 Relationship between an idealized 3-dimensional climate model's senstivity to a quadrupling of CO_2 and the simulated control temperature. (From Spelman and Manabe, 1984)

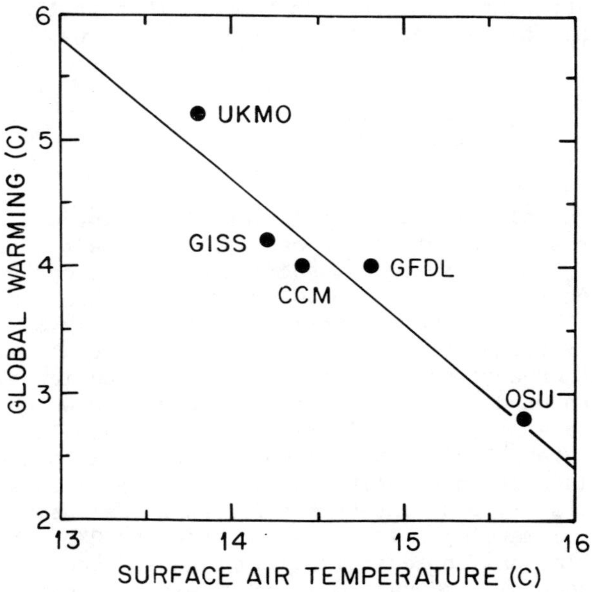

Figure 3 Global mean surface warming due to doubling CO_2 as a function of the mean surface temperature from five GCM experiments, UKMO, Wilson and Mitchell (1987); GISS, Hansen et al. (1984); NCAR, revised estimate based on Washington and Meehl (1984); GFDL, Manabe and Wetherald (1987); OSU, Schlesinger and Zao (1987).

Spelman and Manabe attributed the reduced sensitivity in the warmer models to a weaker albedo feedback as a result of the smaller area of sea-ice and snow. Other aspects of albedo feedback are discussed by Washington and Meehl (1986), Dickinson et al. (1987), Ingram et al. (1989).

Cess and Potter (1988) also note that the results from recent CO_2 studies suggest a decrease in climate sensitivity as the temperature of the unpertubed simulation is increased (Figure 3).

The models used in the derivation of the data in Figure 3 differ in many ways other than the temperature of the control simulation. This may be why the slope of the graph in Figure 3 (1.1) is much steeper than that deduced from Figure 2 (0.42), half the actual slope to allow for a doubling as opposed to a quadrupling of CO_2). They advocate various measures for isolating the factors contributing to climate sensitivity. First, instead of applying a radiative perturbation, such as that due to doubling CO_2 amounts, and using the resultant change in surface temperature as a means of determining climate sensitivity, they propose prescribing a surface temperature perturbation, and using the resulting change in radiative fluxes to determine climate sensitivity. This requires less computing power as the atmospheric response time is much shorter than that of the ocean mixed layer. This climate sensitivity λ is defined by

$$\lambda = \left(\frac{\partial F}{\partial T_s} - \frac{\partial Q}{\partial T_s}\right)^{-1}$$

where F is the outgoing longwave radiation at the top of the atmosphere, Q is the downward solar radiation at the top of the atmosphere and T_s is the surface temperature. (This can be related to Hansen's approach by noting $\Delta T_{eq} = \lambda G$, where $G = 4\ Wm^{-2}$ for doubling CO_2.) Secondly, they recommend carrying out fixed July simulations so that the albedo feedback (in the northern hemisphere at least) is minimised, using prescribed sea temperatures and sea-ice extents. This approach has been used in a series of experiments in which changes in sea surface temperature of -4, -2, +2 and +4K are applied in the version of the Meteorological Office model used by Wilson and Mitchell (1987). The values of the sensitivity λ derived from these experiments are fairly constant except for an anomalously high value on reducing the climatological sea temperatures by 2K. Thus, as noted earlier, the variations in sensitivity in Figure 3 are not due to differences in control temperature alone. Cess and Potter (1988), for example, find that the value of λ estimated from their model is very sensitive to the specification of cloud albedo.

Manabe and Bryan (1985), using a coupled ocean atmosphere model with seasonal mean insolation and realistic geography, demonstrated that changes in ocean circulation can also alter λ. With successive increases in CO_2 concentrations above present day values, they found little change in the oceanic meridional circulation or in climate sensitivity. However, with reductions in CO_2 amounts below the current level, there was a substantial weakening of the thermohaline circulation and a marked rise in climate sensitivity. This study emphasises the importance of including changes in oceanic circulation as well as using a model which produces an accurate simulation of present day climate.

In this section we have emphasised the dependence of global climate sensitivity on the control climate. Obviously, the prediction of regional climate change will depend on the accuracy of the simulation of present day regional climate, as has been demonstrated by Palmer and Mansfield (1986), and Mitchell et al. (1987).

5. DEPENDENCE ON PHYSICAL PARAMETRIZATIONS

We have seen in Section 3 that changes in cloud can enhance climate sensitivity. The use of alternative parametrizations may alter the simulation of present day climate which will also affect climate sensitivity, as seen in the previous section. Thus altering or adding a parametrization of a physical process can alter climate sensitivity directly (for example, by introducing an additional feedback) and indirectly through altering the climate of the control simulation. The indirect effects should be allowed for when analysing sensitivity experiments. Here, we consider two aspects of simulated climate change induced by enhancing atmospheric CO_2.

5.1 Amplification of the warming with height in low latitudes.

Numerical studies of the effect of doubling atmospheric CO_2 differ in the vertical profile of the warming in the tropics. (Schlesinger and Mitchell, 1987; Wilson and Mitchell, 1987). In studies using a moist convective adjustment scheme (Washington and Meehl, 1984; Wetherald, and Manabe 1988) the maximum warming in the tropical troposphere is 3 to 4K, less than twice the surface warming. In studies using a penetrative scheme (Hansen et al, 1984; Wilson and Mitchell, 1987), the upper tropospheric warming reaches 7 to 8K and is substantially greater than twice the surface warming. Thus it appears that the characteristics of the warming in the tropics are dependent on the convective parametrization. In order to isolate the effect of changing the convective parametrization, Cunnington and Mitchell (1989) have carried out two sensitivity studies in which CO_2 concentrations were doubled, and sea surface temperatures were everywhere increased by 2K. The first study used a penetrative convection scheme, the second a non-penetrative scheme (not moist convective adjustment). The upper tropospheric warming was about 1K greater in the experiment with the penetrative scheme (Figure 4) as this is more efficient in moving heat and moisture from the boundary layer to upper layers. The associated decrease in upper

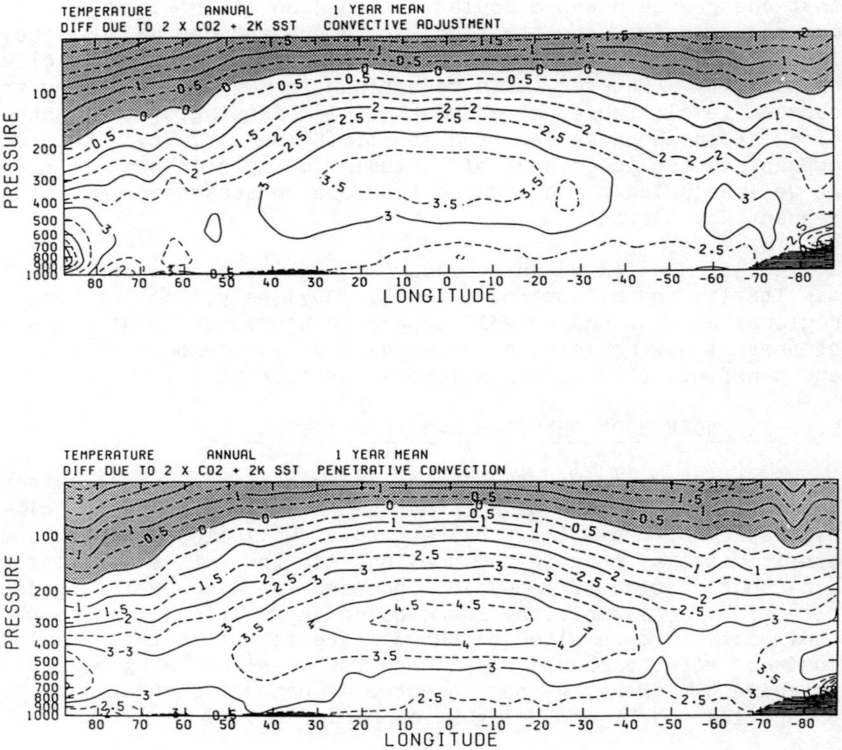

Figure 4 Changes in zonally averaged temperature (K) due to doubling CO_2 and enhancing sea surface temperatures by 2K.

(a) Using a non-penetrative convection scheme.

(b) Using a penetrative scheme.

(from Cunnington and Mitchell, 1988).

tropospheric cloud, and increase in cloud around the tropopause (not shown) is also more pronounced with the penetrative scheme as found in earlier CO_2 studies.

The surface warming in the tropics due to doubling CO_2 is also greater in models using a penetrative scheme (Hansen et al, 1984; Wilson and Mitchell, 1987) than in those using a non-penetrative convection scheme. The reverse is true in radiative-convective models using prescribed cloud amounts (for example, see Schlesinger 1985), where the penetrative schemes, being more efficient in transporting heat upwards, give a greater warming in the upper troposphere and a smaller warming of the surface on doubling CO_2. Note that in the GCM experiments, the increases in high cloud and reductions in middle and upper tropospheric cloud are larger in models with a penetrative scheme, indicating that the positive cloud feedback may be larger in these models and accounting for the apparent discrepancy with the radiative-convective model experiments.

5.2 Summer drying of northern mid-latitudes

Many equilibrium studies of the effect of increasing CO_2 suggest a drying of the northern extra tropical land surface in summer (Manabe et al, 1981; Mitchell, 1986; Manabe and Wetherald, 1987; Wilson and Mitchell, 1987; Schlesinger and Zao, 1987). On the other hand Hansen et al (1984) report substantial areas of increased soil wetness, and Washington and Meehl (1984) find a general moistening of the northern continents in summer with increased CO_2. Manabe et al. (1981) attributed the summer drying to enhanced evaporation, earlier snowmelt especially in more northerly latitudes and, in some areas, reduced precipitation. It also appears that in order to produce a drying,the soil should be close to saturation in early spring in the unperturbed simulation, so that the enhanced precipitation during the winter found on enhancing CO_2 concentrations is run off during snowmelt and not accumulated in the ground (see Figure 5, from Mitchell and Warrilow, 1987).

If the spring soil moisture in the unperturbed climate is reduced well below the runoff threshold, for example by running off snowmelt and rain if the sub-surface soil layers are frozen, then the enhanced rainfall in autumn and winter in the warmer enhanced CO_2 simulation is accumulated in the soil. Thus the drying out of the soil in spring starts from a higher level than in the control (Figure 5b) and the land surface is wetter for much of the summer with enhanced CO_2. This

may explain why the model of Hansen et al. (1984), which allows for enhanced runoff if the ground is frozen, does not produce a widespread summer drying. Meehl and Washington (1988) note that in their model (Washington and Meehl, 1984) the soil is below saturation at most grid points throughout the year in the control simulation, and so the land surface generally becomes wetter in northern mid latitudes in summer on doubling CO_2. They attribute the dryness of the land surface in their control simulation to unrealistically low values of surface albedo over land which elevate the rates of evaporation.

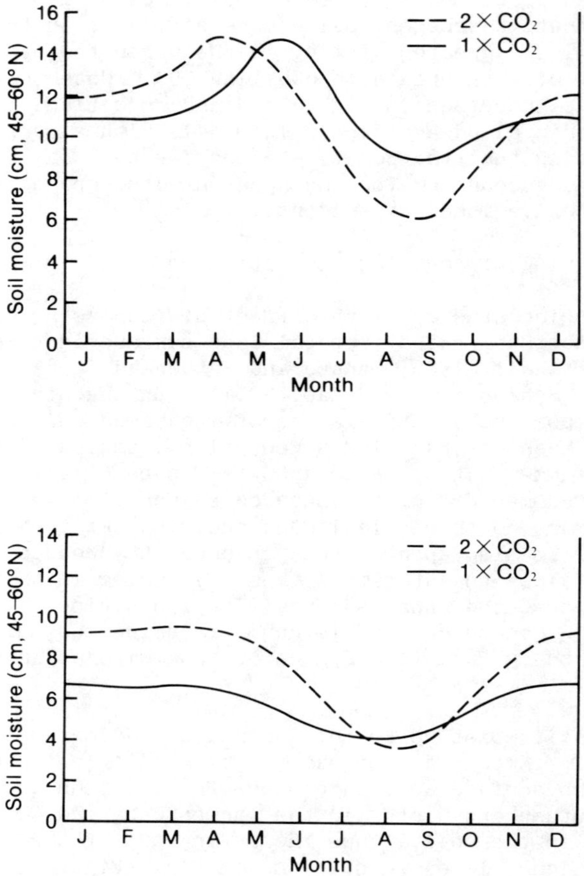

Figure 5 Soil moisture (cm, averaged over 45-60°N) in 1 and 2 x CO_2 simulations.

(a) With standard runoff parametrization

(b) With snowmelt and rain runoff over frozen ground.

(from Mitchell and Warrilow, 1987).

6. CONCLUDING REMARKS

In the preceding sections, we have identified various factors influencing climate sensitivity. These include

1) Climatic feedbacks including those involving cloud and sea-ice.

2) Dependence on the control simulation, especially on temperature.

3) The model domain (eg sector or realistic geography).

4) The physical parametrization (eg of convection or land surface processes).

There are still many differences between existing comparable studies of the effects of increasing CO_2 which are yet to be explained. For example, Rind (1987) has considered the influence of changes in the meridional gradient of sea surface temperature, but does not attempt to explain why the change varies from model to model. We are still at the stage of identifying mechanisms of climate change as opposed to providing quantitative estimates of regional changes.

Over the next few years, research will continue to be directed towards identifying those processes which are critical in determining climate sensitivity, and to removing current uncertainties. Topics which are of particular interest include:

1) The influence of clouds and climate. There is a requirement for observational studies of cloud to provide data on which to base the parametrization of cloud, including cloud radiative properties, in atmospheric models. This is at present an area of major uncertainty.

2) Little as yet is known about the likely response of the oceanic circulation to climatic perturbations. The continuing increase in computing power over the next decade will enable climate simulations of a century or more using detailed coupled dynamical models of the atmosphere and ocean. There is much to be learnt concerning the importance of the parametrization of deep convection, the seasonal mixed layer and sea-ice in determining the performance and sensitivity of such models. Coupled models will enable an evaluation of the transient as well as the equilibrium response to climate perturbations.

3) To date, most GCMs have not attempted to model the effects of vegetation. Sellers et al. (1986) and Dickinson (1984) have developed complex models of the land surface (including vegetation) which have been incorporated in GCMs but have not yet been used in climate studies. It remains to be seen whether or not vegetation changes are likely to modify climate change, and if the most recent parametrizations are adequate.

4) Most simulations of the effect of enhanced CO_2 to date have used relatively low horizontal resolution (typically 15 spectral waves or the finite difference equivalent). There is a need to use models with higher resolution in both the horizontal and the vertical to improve the simulation of present day mean climate and climatic variability.

5) There are many other aspects of climate models which require improvement. Although it is not always possible to distinguish which of two parametrizations is the more suitable (as for example with the case of non-penetrative and penetrative convection schemes), the increasing availability of satellite data and operational forecasting analyses will provide an opportunity for a more careful validation of simulations of present day climate.

As climate models become more complex, it is possible that the simulated response to imposed perturbations may become more diverse. The additional complexity will make it more difficult to unravel the reasons for the diversity in response. If one is to have confidence in numerical predictions of climate change, then the reasons for discrepancies in the simulated response must be understood, and if possible eliminated. Ideally, one would aim to perfect all aspects of the model, but in a world with limited resources, this is not possible. However, by carrying out carefully designed sensitivity experiments, it may be possible to identify those processes which are critical. Scientific effort can then be concentrated on developing accurate physically based representations of these critical processes which in turn will reduce the uncertainties in simulated climate change. This will be a continuing theme of climate research over the next decade.

7. ACKNOWLEDGEMENTS

I acknowledge constructive comments from an anonymous referee which have been incorporated in the text. Peter Rowntree reviewed an early version of the manuscript.

References

Augustsson T. and Ramanathan V., 1977. A radiative-convective model study of the CO_2 climate problem. J. Atmos. Sci. 34, 448-451.

Cess R.D. and Potter G.L., 1988. A methodology for understanding and intercomparing atmospheric feedback processes within general circulation models. J. Geophys. Res., 93, 8305-8314.

Charlson R.J., Lovelock J.E., Andrae M.O. and Warren S.G., 1987. Oceanic phytoplankton, atmospheric sulphur, cloud albedo and climate. Nature, 326, 655-662.

Cunnington W.M. and Mitchell J.F.B., 1989. On the dependence of climate sensitivity on convective parametrization. (Submitted to Climate Dynamics).

Dickinson R.E., 1984. Modelling evapotranspiration for three-dimensional climate models. In Climate Sensitivity (ed. J. Hansen and T. Takahashi) Geophysical Monograph 29, 130-163. American Geophysical Union, Washington DC.

Dickinson R.E., 1986. "How will climate change?" In "The Greenhouse Effect, Climate Change and Ecosystems", eds. B. Bolin, B.R. Doos, J. Jager and R.A. Warrick, SCOPE 13, Wiley, New York, 206-270.

Dickinson R.E., Meehl G.A., and Washington W.M., 1987. Ice-Albedo Feedback in a CO_2 - doubling simulation, Climatic Change 10, 241-248.

Hansen J, Lacis A, Rind D, Russell L, Stone P, Fung I, Ruedy R, and Lerner J., 1984. Climate Sensitivity and Analysis of Feedback Mechanisms. In Climate Processes and Climate Sensitivity (ed J. Hansen and T. Takahashi) Geophysical Monograph 29, 130-163. American Geophysical Union, Washington DC.

Ingram W.J. Wilson C.A. and Mitchell J.F.B., 1989. Modelling climate change: an assessment of sea-ice and surface albedo feedbacks. Submitted to J. Geophys. Res. (To appear).

Manabe S. and Bryan K. Jr., 1985. CO_2-Induced Change in a Coupled Ocean-Atmosphere Model and its Paleoclimatic Implications. J. Geophys. Res. 90, C11: 11,689-11,707.

Manabe S. and Stouffer R.J., 1980. Sensitivity of a global climate model to an increase in the CO_2 concentration in the atmosphere. J. Geophys. Res. 85, 5529-5554.

Manabe S. and Wetherald R.T., 1967. Thermal equilibrium of the atmosphere with a given distribution of relative humidity. J. Atmos. Sci. 24, 241-259.

Manabe S. and Wetherald R.T., 1975. The effects of doubling the CO_2 concentration on a general circulation model. J. Atmos. Sci. 32, 3-15.

Manabe S. and Wetherald R.T., 1980 On the distribution of climate change resulting from an increase in the CO_2 content of the atmosphere. J. Atmos. Sci. 37, 99-118.

Manabe S. and Wetherald R.T., 1987. Large scale changes of soil wetness induced by an increase in atmospheric carbon dioxide. J. Atmos. Sci. 44, 1211-1235.

Manabe S., Wetherald R.T. and Stouffer R.J., 1981. Summer dryness due to an increase of atmospheric CO_2 concentration. Climatic Change, 3, 347-385.

Meehl G.A. and Washington W.M., 1988. A comparison of soil moisture sensitivity in two global climate models. J. Atmos. Sci., 45, 1476-1492.

Mitchell J.F.B., 1986. On changes in soil moisture in CO_2 Climate Studies. In "ISLCSP: Parametrization of land surface characteristics; use of satellite data in climate studies; first results of ISLSCP ESA SP-248, 171-173.

Mitchell J.F.B. and Warrilow D.A., 1987. Summer dryness in northern mid-latitudes due to increased CO_2. Nature 330, 238-240.

Mitchell J.F.B., Wilson C.A. and Cunnington W.M., 1987. On CO_2 climate sensitivity and model dependence of results. Q.J.R. Meteorol. Soc., 113, 293-322.

Moller, F., 1963. On the influence of changes in CO_2 concentrations in air on the radiative balance of the earth's surface and on the climate. J. Geophys. Res., 68, 3877-3886.

Newell R.E. and Dopplick T.G., 1979. Questions concerning the possible influence of anthropogenic CO_2 on atmospheric temperature. J. Appl. Meteor. 18, 822-825.

Palmer T.N. and Mansfield D.A., 1986. Wintertime Circulation Anomalies during past El Nino Events, using a High Resolution General Circulation Model I. Influence of model climatology. Q.J.R. Meteorol. Soc., 112, 613-638.

Rind D., 1987. The Doubled CO_2 Climate: The impact of sea surface temperature gradient. J. Atmos. Sci., 44, 3235-3268.

Roeckner E., Schlese U., Biercamp J. and Loewe P., 1987. Cloud Optical depth feedbacks and climate modelling. Nature, 329, 138-139.

Rowntree P.R. and Walker J.M., 1978. The effects of doubling the CO_2 concentration on radiative-convective equilibrium. In Carbon Dioxide, Climate and Society, J. Williams (Ed.) 181-191, Pergammon Press, Oxford.

Schlesinger M.E., 1985. Feedback analysis of results from energy balance and radiative-convective models in "Projecting the climatic effects of increasing atmospheric carbon dioxide", edited by M.C.

MacCracken and F.M. Luther, pp 280-319, US Department of Energy, Washington DC, 1985 (Available as NTIS, DOE(ER-2037 from Nat. Tech. Inf. Serv. Springfield Va.)

Schlesinger M.E. and Mitchell J.F.B., 1987. Climate model simulations of the equilibrium climatic response to increased carbon dioxide. Review of Geophysics 25, 760-798.

Schlesinger M.E. and Zao Z., 1987. Seasonal climate changes induced by doubled CO_2 as simulated by the OSU atmospheric GCM/mixed layer model . Oregon State University Climatic Institute Report 70, 73 pp.

Schneider S.H., 1975. On the Carbon Dioxide Climate Confusion. J. Atmos. Sci. 32, 2060-2066.

Sellers P.J., Mintz Y., Sud Y.C. and Dalcher A., 1986. A simple biosphere model (SiB) for use within general circulation models. J. Atmos. Sci. 43, 505-531.

Somerville R.C.J. and Remer L.A., 1984. Cloud Optical Thickness Feedbacks in the CO_2 Climate Problem. J. Geophys Res. 89, 9668-9672.

Spelman M.J. and Manabe S., 1984. Influence of Oceanic heat Transport Upon the Sensitivity of a Model Climate. J. Geophys Res. 89, 571-586.

Washington W.M. and Meehl G.A., 1983. General Circulation Model experiments on the Climatic Effects due to a doubling and quadrupling of Carbon Dioxide Concentration. J. Geophys. Res. 88, 6600-6610.

Washington W.M. and Meehl G.A., 1984. Seasonal Cycle Experiment on the Climate Sensitivity Due to a Doubling of CO_2 with an Atmospheric General Circulation Model Coupled to a Simple Mixed Layer Ocean Model. J. Geophys.
Res. 89, 9475-9503.

Washington W.M. and Meehl G.A., 1986. General Circulation model CO_2 sensitivity experiments: Snow-ice albedo parametrizations and globally averaged surface air temperature climatic change, 8, 231-241.

Wetherald R.T. and Manabe S., 1980. Cloud Cover and Climate Sensitivity. J. Atmos. Sci. 37, 1485-1510.

Wetherald R.T. and Manabe S., 1981. Influence of seasonal variation upon the sensitivity of a climate model, J. Geophys. Res. 86, 1194-1204.

Wetherald R.T. and Manabe S., 1988. Cloud Feedback Processes in a General Circulation Model. J. Atmos. Sci, 45, 1397-1415.

Wilson C.A. and Mitchell J.F.B., 1987. A Doubled CO_2 Climate Sensitivity experiment with a GCM including a simple ocean. J. Geophys. Res., 92, 13315-13343.

CLIMATE RESPONSE TO GREENHOUSE WARMING: THE ROLE OF THE OCEAN

Kirk Bryan
GFDL/NOAA
Princeton, NJ 08542

1. Introduction

The scientific community is now confronted with unambiguous evidence that the abundance of radiatively important gases in the atmosphere is changing. The scientific tools exist to calculate the effect of these gases on the Earth's radiation balance. Is it then possible to draw firm conclusions with respect to projected climate change over the next half century? The answer is no, even if all processes internal to the atmosphere were completely understood. Climate change does not depend on the radiation balance of the atmosphere alone, but on the entire climate system, which includes the oceans, the biosphere polar ice masses and sea ice. Climate projections into the next century cannot be made with confidence until the role of the extra-atmospheric components of climate are better understood.

The purpose of this paper is to review the status of research on the ocean's role in climate change. For purposes of conceptualizing the role of the ocean in a period of rapid change in the Earth's environment we can imagine two extreme situations. In one case the World Ocean would simply be a shallow swamp, without heat capacity. In that situation the climate would respond rapidly to a change in the heat balance of the atmosphere, the only lags due to slow changes in the biosphere and the polar ice masses. The time scale of response of the atmosphere alone would only be of the order of few months (Hoffert et al., 1980). The other extreme situation would be one in which the ocean was well mixed on short time scales from top to bottom. Since the heat capacity of the ocean is more than three orders of magnitude greater than the atmosphere, the response time of the combined atmosphere ocean system would be a thousand times a few months, or several centuries. If surface waters were perpetually well mixed with deep waters, greenhouse warming of climate would not be appreciable for many decades.

Marine geochemistry provides some evidence to narrow down the range of possibilities between these two extremes. In the late 1950's and early 1960's bomb tests created an invasion of tritium and bomb-produced ^{14}C into the upper ocean. Tritium largely entered into

A. Berger et al. (eds.), Climate and Geo-Sciences, 435–446.

the Northern Hemisphere Oceans, but the invasion of bomb-produced ^{14}C was more uniform over the globe. The GEOSECS and TTO programs provide a partial description of how these tracers penetrate into the ocean. Recently, the invention of a device to measure chlorofluorocarbons (CFC's) in the oceans has opened up another source of information on vertical pathways. If it is assumed that a change in climate is sufficiently small so that the ocean circulation will not be significantly changed from its present state, then one can expect that warm and cold anomalies at the ocean surface will be carried downward in the same way as a passive tracer. In the following section we will show how this idea has been applied to project climate change.

2. Geochemical Modelling

A simple and elegant model for analyzing transient tracers was developed by Oeschger et al. (1975). The World Ocean is represented in a one-dimensional model, consisting of a deep diffusive layer capped by a well mixed slab at the surface. This basic model is frequently referred to as the "box-diffusion" model. It is a very useful conceptual tool for describing observations as well as the solutions of more complex models, (e.g. Schlesinger et al., 1985). In the box layer, $0 > z > -h$,

$$h \, \partial_t \, \alpha = F^s - F^h \tag{1}$$

where F^s is the flux of tracer with mixing ratio α through the surface and F^h is the flux of tracer through the base of the mixed layer. In the diffusive slab below, $-h > z > -H$,

$$\partial_t \, \alpha = K \, \partial_{zz} \, \alpha \tag{2}$$

where K is the vertical diffusivity and the boundary conditions are,

$$K \, \partial_z \, \alpha = F^h \qquad z = -h$$

and

$$\partial_z \, \alpha = 0 \qquad z = -H \, .$$

Observations taken during GEOSECS (Sarmiento, 1983) show that although tritium rapidly invaded the upper thermocline in the decade after the bomb tests, only a relatively small fraction (less than 10%) has penetrated into the North Atlantic deep water. The "box-diffusion" model is able to simulate the difference in time scales between the surface and deep water. In the model, tracer is rapidly taken up in the mixed layer and diffused rapidly into the upper part of the water column. As vertical gradients diminish with increasing depth, the downward penetration greatly slows down. However, Bacastow and Bjorkstrom (1981) point out that there are limitations to the quantitative application of the box-diffusion model to real data. The box diffusion model has essentially two adjustable parameters, h and K. These authors show that a careful calibration of these two parameters to bomb-produced ^{14}C data, does not give a good fit to the distribution of natural ^{14}C in the lower thermocline. They attribute this to the fact that the pathways to the deep water originate at very high latitudes, while the downward pathways to the main thermocline originate at the surface in middle latitudes. One could not expect that such very different processes could be precisely fitted by the same two parameter model with identical coefficients.

This has led to the development of more elaborate geochemical models in two and three dimensions. Of particular interest is the 12 box World Ocean model developed by Bolin, Bjorkstrom, Holmen and Moore (1983) which has recently been extended to an 84 box model of the Atlantic. The objective of this group has been to use geochemical data to construct the currents and mixing of the ocean by inverse methods. The calculated circulation is then used in a "forecast" model to predict the invasion of transient tracers such as tritium and bomb-produced ^{14}C. The only dynamical constraint introduced is the geostrophic vertical shear along east-west sections compatible with observed temperature and salinity distributions. The model provides a very reasonable picture of the total overturning circulation in the Atlantic and a good simulation of the transient tracers, although the horizontal resolution is not good enough for very detailed comparison with observations.

A pioneering simulation of transient tracers with a three-dimensional model was carried out by Sarmiento (1983). The ocean circulation used to transport the tracers was generated by the diagnostic model of Sarmiento and Bryan (1982). In this diagnostic calculation a 12 level numerical model of the North Atlantic is forced into a steady solution by a numerical integration with respect to time in which the predicted values of temperature and salinity are partially replaced by the observed values at each time step. The partial replacement parameter governs whether the emphasis is on a close fit to the data or to the model. Sarmiento's (1983) calculations demonstrate that winter-time convection is a much more important vertical pathway for tritium in the North Atlantic than is downward Ekman pumping.

Recently Maier-Reimer and Hasselmann (1988) have studied both the distribution of Carbon-14 and Tritium in a model of the World Ocean. In this case the ocean circulation model was time-stepped to equilibrium with only temperature and salinity specified at the surface. The tritium simulation for a north-south section in the Atlantic is compared to GEOSECS data in Fig. 1. Although the simulated penetration is not quite deep enough in the high latitudes of the North Atlantic and perhaps a little too deep near the equator, the observed pattern is captured by the model in remarkable detail. Fig. 1 illustrates how the transient tracer data are a powerful tool for validating ocean models. Similar validation experiments are being carried out which test ocean general circulation models with bomb-produced Carbon-14 and CFC (freon) data. It is a great advantage

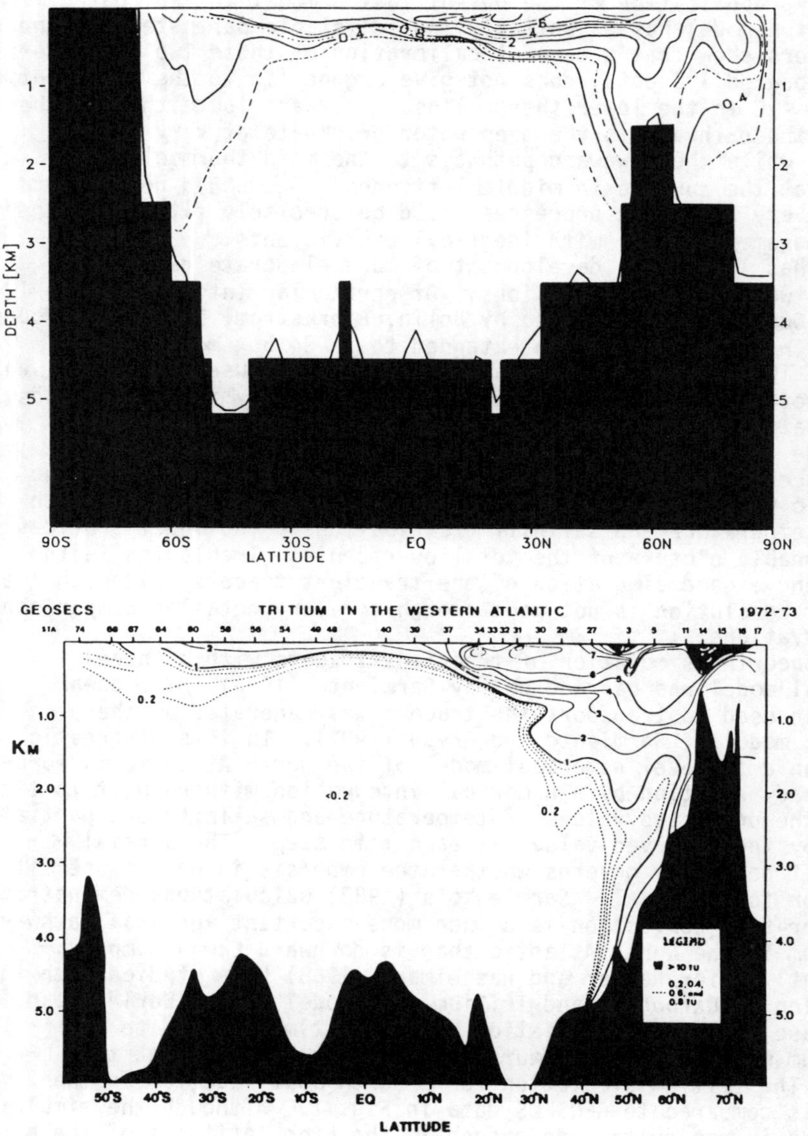

Figure 1. (a) A simulation of the penetration of Tritium into the Atlantic Ocean in a model of the World Ocean (Maier-Reimer and Hasselmann, 1988), (b) Tritium observations along a GEOSECS section made in 1972 in the Western Atlantic. tritium units.

to use several tracers, which have somewhat different source distributions at the ocean surface. A variety of different tracers will test the model's ability to simulate different pathways from the surface to the deep ocean, since pathways at all latitudes may be important as the ocean responds to greenhouse warming.

3. Energy Balance Models of the Ocean

The counterpart of the one-dimensional geochemical model of the ocean is the energy balance climate model (Hoffert and Flannery, 1985). The unmodified box-diffusion model is not satisfactory to represent the ocean's temperature structure, since diffusion would mix out the thermocline in an equilibrium state. A modification of the model adds a uniform upwelling, adding a third parameter. Hoffert et al. (1980), Cess and Goldenberg (1981) and Harvey and Schneider (1985) have used this type of model coupled to an energy balance model of the atmosphere to predict the effect of the ocean in the case of a carbon dioxide warming event. With a calibration based on transient tracers, these studies indicate that, because the warming of the ocean would be confined to the upper thermocline, climate warming would be delayed only a few decades by the ocean. This is quite different from a delay of centuries which would be expected if the entire ocean were well mixed.

Schneider and Thompson (1981) extended energy balance model studies of greenhouse warming to resolve differences of ocean mixing and land/ocean surface area at different latitudes. With a minimum of parameters their model takes into account the proportion of ocean at each latitude and the increase in depth of mixing by winter convection as one goes to higher latitudes. On the basis of this coupled ocean-atmosphere energy balance model Thompson and Schneider (1982) concluded that warming would take place much more rapidly in the Northern Hemisphere than the Southern Hemisphere. In a further refinement North et al. (1984) developed an energy balance model with a two-dimensional coverage of the entire globe. Applying this model to the actual ocean-land distribution, the authors confirm that temperature rise takes place much more slowly in the Southern Hemisphere. The results shown in Fig. 2 are for a "switch on" experiment in which the atmospheric carbon dioxide is suddenly doubled. The zonally averaged temperature normalized by the total expected change is shown as a function of latitude and time. Note that the parameterization of the ocean in this model implies, on the whole, a rather small impact of the ocean on climate change. In the Northern Hemisphere nearly 80% of the response has taken place within 5 years. In the Southern Hemisphere the delay is longer, as predicted by Schneider and Thompson (1981), but even there 80% of the response has taken place after 7 years.

As Hansen et al. (1984) point out, the speed of the response of a given model depends on how strong the feedback mechanisms are. The stronger the positive feedbacks the slower the approach to final

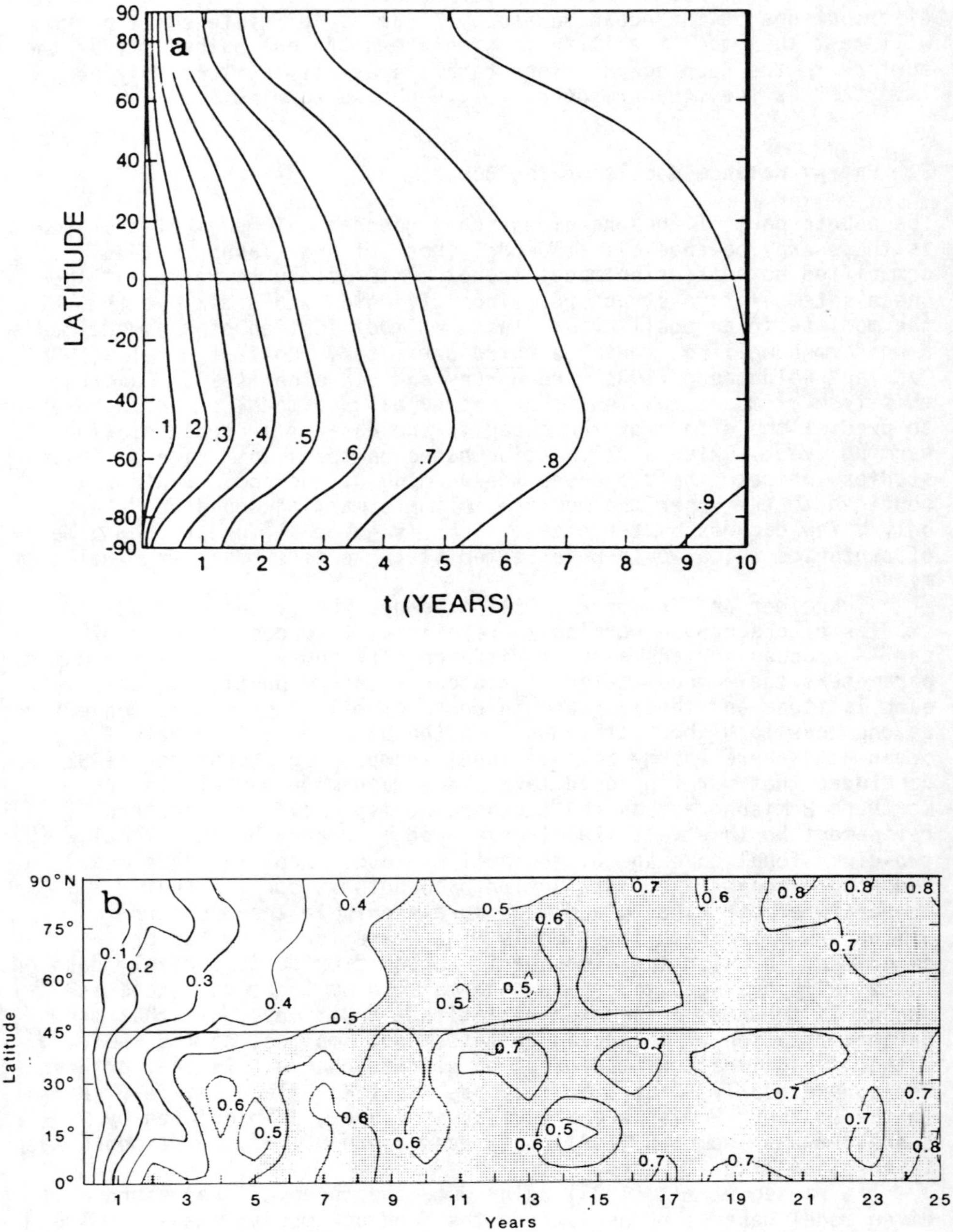
a
LATITUDE
90
80
60
40
20
0
-20
-40
-60
-80
-90
.1
.2
.3
.4
.5
.6
.7
.8
.9
1
2
3
4
5
6
7
8
9
10
t (YEARS)
b
Latitude
90° N
75°
60°
45°
30°
15°
0°
0.1
0.2
0.3
0.4
0.5
0.6
0.7
0.8
1
3
5
7
9
11
13
15
17
19
21
23
25
Years

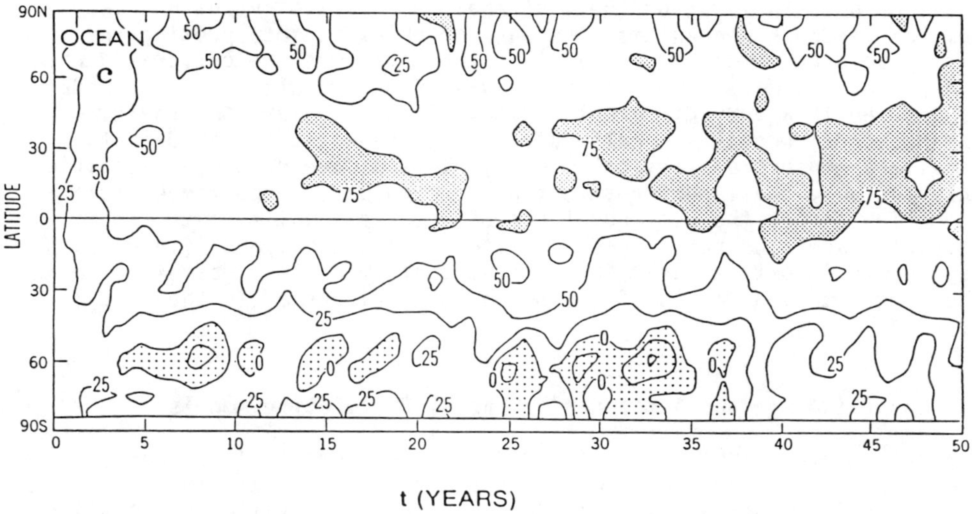

Figure 2. Air temperature response when atmospheric carbon dioxide is suddenly increased to a higher value. The zonally averaged and normalized temperature change is shown as a function of latitude and time after "switch on". Normalization is with respect to the temperature change expected for complete adjustment to the high atmospheric carbon dioxide. (a) Results of North et al. (1984) for two-dimensional, global energy balance model including a passive ocean. Carbon dioxide is doubled. (b) Results of the coupled ocean-atmosphere general circulation model of Bryan et al. (1982). Carbon dioxide in the atmosphere is increased four times. (c) A two hemisphere coupled atmosphere-ocean general circulation model (Bryan et al., 1988). The normalized response is multiplied by 100, and the atmospheric carbon dioxide is doubled. Note the delayed response high latitudes of the Southern Hemisphere.

equilibrium. In light of the fact that the climate system has many complex feedback mechanisms, the basic problem in interpreting the energy balance model results is that the model must be calibrated with the present climate. We cannot be sure of the results when the model is applied to a very different regime. In particular, the energy balance model oceans are calibrated using transient tracer data to provide a measure of vertical mixing. This procedure can be justified only for very small perturbations of temperature, since temperature anomalies affect the buoyancy field of the ocean while neutrally buoyant tracers do not. Feedback effects on the buoyancy field which may become important early in the next century can only be explored in the more detailed models of the ocean to be discussed in the next section.

4. Coupled Ocean and Atmospheric General Circulation Models

Recently the distinction between medium-range numerical weather prediction models and research climate models has become increasingly blurred. Since the various parameterizations of physical processes are finding their way into operational forecasting models, the main difference between operational medium-range models and climate research models is now in the horizontal resolution. As a result the atmospheric general circulation models are receiving a great deal of attention and have been extensively validated as forecast models. For obvious reasons equal attention has not been paid to the development and validation of ocean models, and climate research with coupled models of the atmosphere and the ocean is still in its infancy.

The basic dynamics of the ocean and the atmosphere are the same. One would think that ocean modelling would require only a straightforward adaptation of techniques used in atmospheric modelling. However, there are special features of the ocean that make this very difficult. One feature is the very great range of energy-containing spatial scales in the ocean relative to the atmosphere. Important currents are quite narrow, requiring detailed horizontal resolution for an accurate simulation. Another feature is the very long time scales of the ocean resulting from both the very slow propagation of important large-scale waves, and the ocean's enormous heat capacity. Thus the ocean is very wide banded with respect to frequency and spatial wave number relative to the atmosphere.

The oceanic equivalent of cyclones and anticylcones in the atmosphere are oceanic mesoscale eddies which are over an order of magnitude smaller than their atmospheric counterparts in middle latitudes, and are even smaller in polar regions. For example, a mesoscale eddy in the Gulf Stream has a scale of only a few hundred kilometers in comparison with a horizontal scale of thousands of kilometers for atmospheric disturbances. An important controversy among oceanographers concerns the role of mesoscale eddies. Can a model of the ocean be devised without explicitly resolving mesoscale eddies which will be adequate to represent the downward pathways for heat and tracers? Some insight on this question is available from a study by Cox (1985) in which the tracer penetration in eddy-resolving and non-eddy resolving models was carefully compared. It was found that the mixing by mesoscale eddies can be quite effectively parameterized, as suggested by Redi (1982), by rotating the mixing tensor so that it is aligned with density surfaces rather than with horizontal surfaces. Another approach to answering this question is strictly empirical, through comparison of model simulations with transient tracer data and water mass properties. An example is the comparison of GEOSECS tritium data with the simulation of the non-eddy resolving Hamburg model shown in Fig. 1. Given the uncertainties of the transient tracer input data, it is not clear whether an eddy-resolving model would provide a result which would be better in a statistically-significant sense.

The non-equilibrium problem of climate change can be studied only with fully coupled models. Since fully coupled atmosphere-ocean general circulation models are still under development, relatively little work has been done on this challenging problem. Response to greenhouse warming involves the exchange of heat between the surface and deeper water at both high and low latitudes, and thus requires a much more general model of the oceans than that required for modelling the El Niño-Southern Oscillation phenomenon. Two published studies of coupled ocean-atmosphere general circulation models by Bryan et al. (1982) and Schlesinger et al. (1985) have examined the effects of a sudden increase of atmospheric carbon dioxide. We will refer to these model studies as "switch on" numerical experiments. The first experiment (Bryan et al., 1982) was carried out for a very idealized geometry consisting of a 120° longitude sector bounded by two meridians running from the equator to the pole. Half of this area was covered by land and the other half by ocean. Mirror symmetry was assumed across the equator, and cyclic symmetry assummed at the meridional boundaries. Clouds were fixed and seasonal effects were not included in the model. Two equilibrium climates were determined, one corresponding to a normal level of CO_2 in the atmosphere, and one corresponding to four times the normal level. It would be impractical to find the climate equilibria by extended numerical integrations, because of the very long time scales of the deep ocean. To circumvent this problem special methods (Bryan, 1984) were utilized to accelerate convergence. The results in Fig. 2(b) show the zonally averaged surface air temperature as a function of the number of years after "switch on." The results can be compared directly with the global energy balance model results of North et al. (1984) shown in Fig. 2(a). It is reasonable to compare Fig. (2)b with the Northern Hemisphere results in Fig. 2(a) since the proportion of ocean to land is much greater than 1:1 in the Southern Hemisphere. It is evident that the ocean general circulation model takes up much more heat than the passive ocean in the North et al. (1984) model. After 25 years the normalized response is about 0.7 in the coupled general circulation model, while the same normalized response is reached in about 5 years in the coupled energy balance model. However, as pointed out earlier, the response time is very much a function of the feedbacks in the model. If cloud feedback had been included in these models, one might expect a much slower approach to the new equilibrium.

The first "switch on" calculation for a coupled general circulation model with true global geometry was carried out by Schlesinger et al. (1985). The ocean component has very much the same structure as that used by Bryan et al. (1982), while the atmospheric model was a finite difference two-layer model which has been extensively validated. Seasons and variable cloudiness were included in the atmospheric model. Rather than attempting to find a climate equilibrium for the coupled model, the "switch on" experiment was performed and the climate drift subtracted from the response. The sea surface temperature was predicted by the model, but the sea surface salinity was kept fixed as specified by observations. The results showed a nearly symmetric response in both hemispheres.

Recently, another "switch on" calculation (Bryan et al., 1988) has been completed which suggests that the "greenhouse" warming may not be symmetric in both hemispheres. The geometry of the ocean-to-land distribution is idealized, but care is taken to have the exact ratio of land to ocean at each latitude correspond to the Northern and Southern Hemisphere. Equilibria are obtained for climates corresponding to one and two times the carbon dioxide now in the atmosphere. Then the normal climate is perturbed by doubling the atmospheric CO_2 and the non-equilibrium response is monitored over a 50-year period. Both temperature and salinity in the oceans are free to adjust at the surface and deeper levels to the new fluxes of heat and water dictated by the atmospheric model. The normalized surface air response is shown in Fig. 2c. The predominantly land hemisphere, corresponding to the Northern Hemisphere, warms at about the same normalized rate predicted in the earlier calculation by Bryan et al. (1982) corresponding to a quadrupling of atmospheric carbon dioxide. On the other hand, the Southern Hemisphere warms at a much slower rate. Much slower, in fact, than predicted by the energy balance model results shown in Fig. 2(a). Analysis of the results show that there are two factors involved in the slow response of the predominantly ocean-covered hemisphere. One, the large extent of ocean allows greater heat storage. This effect was predicted by Thompson and Schneider (1982). Two, the very deep overturning of the ocean in the vicinity of the Antarctic Circumpolar Current, which is simulated in this model, sequesters heat very efficiently at deep levels in the ocean in the predominantly ocean-covered hemisphere. Thus the polar amplification of the greenhouse warming predicted by an equilibrium response calculation may not be seen in the Southern Hemisphere in a period of very rapid greenhouse warming.

5. Summary

The basic driving force for global greenhouse warming is the changing abundance of radiatively active gases in the atmosphere. The actual response in the near term of climate to a greenhouse warming event will depend on the large-scale interaction of the ocean and atmosphere. Attempts have been made to project climate change by assuming that the ocean responds passively to greenhouse warming as predicted by linear models calibrated from transient geochemical tracers. Coupled atmosphere-ocean experiments at this point are preliminary and do not provide consistent results. However, it is clear that the simpler models of the ocean's response are probably inadequate in the case of a rapid increase of greenhouse gases. A great deal of effort will be required to make reliable climate projections. The most obvious research directions are:

(a) Continued study of transient tracers in the ocean. The ocean component of coupled models must be validated using the transient tracer data.

(b) Historical climate records and paleoclimatic records must be carefully examined for information on the role of the oceans in modifying climate change.

(c) The equivalent of the effort that has gone into perfecting general circulation models of the atmosphere must now be devoted to developing models of the ocean and land surfaces.

References

Bacastrow. R., and A. Bjorkstrom, 1981: 'Comparison of ocean models for the carbon cycle.' In: Carbon Cycle Modelling, B. Bolin (Ed.), SCOPE, **16**, 29-80. John Wiley and Sons, New York.

Bolin, B., A. Bjorkstrom, K. Holmen, and B. Moore, 1983: 'The Simultaneous use of Tracers for Ocean Circulation Studies.' Tellus, **35B**, 206-236.

Bryan, K., 1984: 'Accelerating the convergence to equilibrium of ocean-climate models.' J. Phys. Oceanographer, **14**, 666-673.

Bryan, K., F.G. Komro, S. Manabe, and M.J. Spelman, 1982: 'Transient climate response to increasing atmospheric carbon dioxide.' Science, **215**, 56-58.

Bryan, K., S. Manabe, and M.J. Spelman, 1988: 'Interhemispheric asymmetry in the transient response of a coupled ocean-atmosphere model to a carbon dioxide forcing.' J. Phys. Oceangr., **18(6)**, 851-867.

Cess, R.D., and S.D. Goldenberg, 1981: 'The effect of ocean heat capacity upon global warming due to increased atmospheric carbon dioxide.' J. Geophys. Res., **86**, 498-502.

Cox, M.D., 1985: 'An eddy-resolving numerical model of the ventilated thermocline.' J. Phys. Oceanogr., **15**, 1312-1324.

Harvey, L.D.D., and S.H. Schneider, 1985: 'Transient climate response to external forcing on 10°-10^4 year time scales. Part II.' J. Geophys. Res., **90(D1)**, 2207-2222.

Hoffert, M.I., A.J. Callegari, and C.-T. Hsieh, 1980: 'The role of deep-sea storage in the secular response to climate forcing.' J. Geophys. Res., **85**, 6677-6679.

Hoffert, M.I., and B.P. Flannery, 1985: 'Model projections of the time-dependent response to increasing carbon dioxide.' In: The Potential Climatic Effects of Increasing Carbon Dioxide, edited by M.C. MacCracken and L.M. Luther, U.S. Department of Energy Report, ER-0237, National Tech. Information Service, Springfield, VA, USA.

Maier-Reimer, E., and K. Hasselmann, 1988: 'Transport and storage of carbon dioxide in the ocean an inorganic ocean-circulation carbon cycle model.' (Climate Dynamics)

Manabe, S., K. Bryan, and M.J. Spelman, 1989: 'Transient response of a global ocean-atmosphere model to a doubling of atmospheric carbon dioxide.' Manuscript, Geophys. Fluid Dyn. Lab., NOAA, Princeton, N.J.

North, G.R., J.G. Mengel, and D.A. Short, 1984: On the transient response patterns of climate to time dependent concentrations of atmospheric CO_2, in Climate Processes and Climate Sensitivity, J. Hansen and T. Takahashi, Eds. Geophysical Monograph 29, Amer. Geophys. Union, Washington, D.C. 164-170.

Oeschger, H., H. Siegenthaler, U. Schotterer, and A. Gugelmann, 1975: 'A box-diffusion model to study the carbon dioxide exchange in nature.' Tellus, **27**, 168-192.

Redi, M.H., 1982: 'Oceanic isopycnal mixing by coordinate rotation.' J. Phys. Oceanogr., **12(10)**, 1154-1158.

Sarmiento, J.L., 1983: 'A simulation of bomb tritium entry into the Atlantic Ocean.' J. Phys. Oceanogr., **13(10)**, 1924-1939.

Sarmiento, J.L., and K. Bryan, 1982: 'An ocean transport model for the North Atlantic.' J. Geophys. Res., **87**, 394-408.

Schlesinger, M.E., W.L. Gates, and Y.J. Han, 1985: 'The role of the ocean in carbon dioxide-induced climate warming: Preliminary results from the OSU coupled atmosphere-ocean GCM.' In Coupled Ocean-Atmosphere Models, (J.C.J. Nihoul, editor), pp. 447-478, Elsevier Press.

Schneider, S.H., and S.L. Thompson, 1981: 'Atmospheric carbon dioxide and climate: Importance of the transient response.' J. Geophys. Res., **86(C4)**, 3135-3147.

Thompson, S.L., and S.H. Schneider, 1982: 'Carbon dioxide and climate: The importance of realistic geography in estimating the transient temperature response.' Science, **217**, 1031-1033.

Unresolved issues and research needs pertaining to the transient climatic response to a CO_2 increase

L.D. Danny Harvey
Department of Geography
University of Toronto
100 St. George Street
Toronto, Canada, M5S 1A1

ABSTRACT. This paper explains the advantages and weaknesses of two different modelling approaches used to date to study the transient climatic response to a CO_2 increase, based on globally averaged advection-diffusion models and three-dimensional coupled atmosphere ocean models. Some key uncertainties and areas needing further research are identified.

1. Introduction

An important climatic problem is the computation of the time dependent or transient climatic response to a time dependent CO_2 increase. The transient climatic response to a CO_2 increase will be governed both by the equilibrium climatic sensitivity and by mixing of heat into the oceans. The oceans, however, are a heterogeneous and in most places horizontally stratified fluid which would respond to a change in surface heating in a complex, nonuniform manner with multiple time scales.

Our current knowledge of oceanic circulation and mixing processes is extremely scanty (Wunsch, 1984). Direct observation of the subsurface oceanic circulation is impossible in most parts of the ocean, so oceanic circulation patterns and sometimes velocities have been inferred from the three-dimensional distribution of such oceanic properties such as salinity, dissolved oxygen, nutrient content, and radioisotopes such as ^{14}C and tritium (Broecker, 1981; Bolin et al 1983).

2. Modelling approaches

Two modelling approaches have been largely used to investigate the transient climatic response to a CO_2 increase: one involving globally averaged one dimensional models with vertical resolution, and the other involving three-dimensional coupled atmosphere-ocean general circulation models (A/O GCM's). The 1-D models parameterize the net effect of all oceanic mixing processes either as a diffusion process, or as a combination of upwelling advection and diffusion (i.e.: Wigley and Schlesinger, 1985;

A. Berger et al. (eds.), Climate and Geo-Sciences, 447–451.

Harvey and Schneider, 1985a,b). These models thus greatly simplify reality and there are grounds for questioning whether they can adequately capture certain transient effects (Bryan and Spelman, 1985). On the other hand, these models have used realistic scenarios of CO_2 buildup, in contrast to work with A/O GCM's.

Ocean GCM's (OGCM's) used for global scale climatic simulation have a horizontal resolution on the order of 200 km or larger, whereas most of the eddy energy in the oceans occurs at scales smaller than 100 km (Woods, 1985). Ocean models used for climatic studies thus cannot resolve the important energy containing scales of motion and instead have to rely on parameterizations of eddy effects. The only eddy resolving ocean models built so far, with resolutions of 20-40 km, have been of single ocean basins, usually with idealized geometry and highly simplified physics (i.e.: Holland, 1985). Computer speed and memory limitations currently prevent construction of global, eddy-resolving OGCM's. There is disagreement at present as to whether or not non-eddy resolving OGCM's can capture the behaviour of eddy resolving OGCM's by parameterizing the effects of unresolved eddies. Other simplifications of OGCM's relevant to their transient response is the use of a vertical-horizontal coordinate system for diffusion rather than an isopycnal coordinate system (McDougall and Church, 1985), and modelling bottom water formation through convective overturning rather than as a boundary layer plume, as would be appropriate in some areas (Killworth, 1977). Even with the many simplifications and coarse resolution of global A/O GCM's, they are very costly to run and only a very small number of simulations have been performed with these models.

3. Unresolved issues

Of the few simulations of the transient response to a CO_2 increase with coupled A/O GCM's which have been published to date, all have been for the idealized case of a step function CO_2 increase (i.e.: Bryan et al 1982; Schlesinger et al 1985). Bryan and Spelman (1985) examined the penetration of positive heat anomalies into an OGCM following a step function 4-fold CO_2 increase. They found that warming of the surface layers of the ocean suppressed convective mixing, and since convection generally carries heat upwards, this allowed the heat anomaly to penetrate faster into the ocean than would otherwise have been the case. This is turn would tend to slow the mixed layer and hence atmospheric temperature response, and suggests that effective vertical diffusion coefficients (used in 1-D models) derived for equilibrium conditions may not be applicable during the transient. They also found the pattern of ocean circulation to be similar for normal and four times present CO_2, but to be distinctly different during the transient. However, Bryan and Spelman's results may be dependent on the imposition of a large, step function increase of CO_2; their results could very well be different for a gradual CO_2 increase.

The effect of the oceans on the transient climatic response to a time dependent CO_2 increase can be viewed either in terms of a lag of the actual climatic response behind the instantaneous equilibrium climatic response, or in terms of the ratio of the climatic response at a given time to the instantaneous equilibrium response. Wigley and Schlesinger (1985) computed both of these quantities for a wide range of model parameters using a 1-D diffusion model. They obtained lags ranging from 11 to 35 years and response ratios of 0.74 to 0.36. The response ratio is particularly important, since it indicates how much further climatic warming could be expected from the CO_2 increase that will have already occurred at any given time in the future.

Schneider and Thompson (1981) suggested that, as a result of latitudinal variation of the oceanic mixed layer depth and ocean fraction, the time evolution of horizontal temperature gradients during the transient could be significantly different from the equilibrium changes in horizontal temperature gradients predicted by climate models. Bryan et al (1982) found, for their A/O GCM, that the zonally averaged fractional temperature response (actual/equilibrium) following a fourfold, step-function CO_2 increase, was initially faster at high northern hemisphere latitudes than at lower latitudes, but became latitudinally uniform after about 25 years. Bryan et al (1982) thus concluded that equilibrium climate model responses can be used as a guide to the pattern of transient climatic response.

However, this issue is not settled as it is still possible that regional patterns of climatic change could in places be significantly different during the transient than in equilibrium, even if zonally averaged responses are similar. If the CO_2 increase is fast enough and the lag of the mixed layer behind its equilibrium temperature response long enough, then it is possible that the temperature difference in summer between continents and oceans in areas subject to monsoonal regimes could increase because of the faster temperature response of the continents. This in turn could enhance monsoonal rainfall during the transient, even if this does not occur in equilibrium. It is most likely, however, that the slower the CO_2 increase the smaller the differences between the normalized land and ocean responses, and the closer the patterns of climatic change during the transient will be to the equilibrium response. On the other hand, changes in precipitation pattern and soil moisture could respond quite nonlinearly to continuously changing temperatures and temperature gradients. A full resolution of this problem must await integrations of sufficiently long duration with coupled A/O GCM's using a realistic, time-dependent CO_2 increase.

4. Research needs

Among the important recommended areas of future research are

(1) Intensified work with limited domain eddy resolving OGCM's so as to fully understand their behaviour prior to global integrations with these models once computer hardware permits.

(2) Simulations of 100 years duration or longer of the climatic response to realistic CO_2 increase scenarios using existing coupled A/O GCM's.

(3) Continuing development of efficient yet accurate methods for asynchronously coupling A/O GCM's (see Schneider and Harvey, 1986), which will facilitate the long integrations required for gradual CO_2 increase scenarios.

5. References

Bryan, K., F.G. Komro, S. Manabe, and M.J. Spelman, 1982: 'Transient climate response to increasing atmospheric carbon dioxide', Science, **215**, 56-58.

Bryan, K., and M.J. Spelman, 1985: 'The ocean's response to a CO_2-induced warming', J. Geophys. Res., **90**, 11679-11688.

Bolin, B., A. Bjorkstrom, and K. Holmen, 1983: 'The simultaneous use of tracers for ocean circulation studies', Tellus, **35B**, 206-236.

Broecker, W.S., 1981: 'Geochemical tracers and ocean circulation', in B.A. Warren and C. Wunsch (eds.), Evolution of physical oceanography, MIT Press, Cambridge, Mass., 434-460.

Harvey, L.D.D., and S.H. Schneider, 1985a: 'Transient climate response to external forcing on 10^0-10^4 year time scales, 1, Experiments with globally averaged, coupled, atmosphere and ocean energy balance models', J. Geophys. Res., **90**, 2191-2205.

Harvey, L.D.D., and S.H. Schneider, 1985b: 'Transient climate response to external forcing on 10^0-10^4 year time scales, 2, Sensitivity experiments with a seasonal, hemispherically averaged coupled atmosphere, land, and ocean energy balance models', J. Geophys. Res., **90**, 2207-2222.

Holland, W.R., 1985: 'Simulation of mesoscale ocean variability in mid-latitude gyres', Advances in Geophysics, **28A**, 479-523.

Killworth, P.D., 1977: 'Mixing on the Weddell Sea continental slope', Deep Sea Res., 24, 427-448.

McDougall, T.J., and J.A. Church, 1985: 'Pitfalls with the numerical representation of isopycnal and diapycnal mixing', J. Phys. Oceanogr. **16**, 196-199.

Schlesinger, M.E., W.L. Gates, and Y.-J. Han, 1985: 'The role of the ocean in CO_2-induced climate change: Preliminary results from the OSU coupled atmosphere-ocean general circulation model', in J.C.J. Nihoul (ed.), Coupled Ocean-Atmosphere Models, Elsevier, Amsterdam, 447-478.

Schneider, S.H., and L.D.D. Harvey, 1986: 'Computational efficiency and accuracy of methods for asynchronously coupling atmosphere-ocean models. Part I: Testing with a mean annual model', J. Phys. Oceanogr. **16**, 3-10.

Schneider, S.H., and S.L. Thompson, 1981: 'Atmospheric CO_2 and climate: Importance of the transient response', J. Geophys. Res., **86**, 3135-3147.

Wigley, T.M.L., and M.E. Schlesinger, 1985: 'Analytical solution for the effect of increasing CO_2 on global mean temperature', Nature, **315**, 649-652.

Woods, J.D., 1985: 'The world ocean circulation experiment', Nature, **314**, 501-511.

Wunsch, C., 1984: 'The ocean circulation in climate', in J.T. Houghton (ed.), The global climate, Cambridge University Press, Cambridge, 189-203.

GLOBAL BIOGEOCHEMICAL CYCLES AND CLIMATE

R. WOLLAST[1] and F.T. MACKENZIE[2]

[1]*University of Brussels*
Laboratory of Oceanography,
Campus Plaine CP 208,
1050 Brussels, Belgium

[2]*Department of Oceanography*
and Hawaii Institute of Geophysics
University of Hawaii
Honolulu, HI 96822 U.S.A.

ABSTRACT. This paper highlights some important questions and problems involving greenhouse gas global cycles and climate response. Simplified biogeochemical cycles of the greenhouse gases CO_2, CH_4 and N_2O, and of the atmospheric compounds of sulfur are presented with particular emphasis on exchange fluxes involving Earth's surface and the atmosphere, where possible long-term natural fluxes are compared to gas fluxes resulting from human activities. It is demonstrated that the geological cycling behavior of these gases has been perturbed significantly by anthropogenic fluxes.

The inability of the atmosphere and other sinks to react rapidly enough to cope with increasing fluxes from societal activities has led to continuous change in the greenhouse gas composition of the atmosphere. This change in composition may lead to future climate change, including changes in temperature and precipitation patterns. The coastal and nearshore oceanic realm is particularly sensitive to material fluxes from societal activities, and its role in the exchange of gases between the atmosphere and Earth's surface is poorly known.

1. Introduction

During the last two decades, considerable effort has been invested in the formulation and analysis of the global biogeochemical cycles of elements. This effort has included, for example, studies of the cycling behavior of the major elements (e.g. Garrels and Mackenzie, 1971), the biologically-important elements of C, N, P, and S (e.g. Bolin and Cook, 1983), and trace elements (e.g. Mackenzie et al, 1979; Buat-Menard, 1986), as well as innumerable publications dealing with the carbon cycle at a variety of spatial and temporal scales (e.g. Sundquist and Broecker, 1985). Interest in the carbon cycle was certainly heightened by concern with emissions of the "greenhouse" gas CO_2 to the atmosphere, because of the burning of fossil fuels and land use practices, and its potential for changing climate.

A. Berger et al. (eds.), Climate and Geo-Sciences, 453–473.

The proposition that gases like CO_2 in the atmosphere act to retain reradiated heat energy and consequently increase the Earth's surface temperature above that found in an atmosphere without these gases (about -20°C) was first formulated in the late 1700's by Jean-Baptiste Fourier. It was not until 1896, however, that Svant Arrhenius attempted to calculate the effect of a doubling of atmospheric CO_2 content on Earth's average surface temperature. He concluded that a doubling would raise the temperature 5-6°C, an estimate that is slightly higher than but similar to that obtained today from climate model calculations (see Schlesinger, this volume). It is Arrhenius's work that was, perhaps, the first attempt to link quantitatively the biogeochemical cycle of an element with potential climatic change.

In this paper, we present simplified biogeochemical cycles of the "greenhouse" gases CO_2, CH_4 and N_2O, and of the atmospheric compounds of sulfur. Particular emphasis is placed on the relative magnitudes of the exchange fluxes between Earth's surface and the atmosphere owing to natural and human activities. It is demonstrated by comparing the long-term geological cycling behavior of these atmospheric compounds with their present-day cycles that the geological cycles are perturbed significantly by anthropogenic fluxes. These perturbations lead to atmospheric compositional changes, as well as other effects, and may be responsible for future climatic change. The coastal and nearshore oceanic realm is shown to be particularly sensitive to material fluxes from activities of society. Because of this sensitivity, the role of this environment in the exchange of gases between the atmosphere and the sea has been, and probably will continue to be, significantly altered.

In the models of the biogeochemical cycles of the compounds presented, we have culled and evaluated a large data base. Our reservoir and flux estimates were obtained from evaluation of various estimates given in the references, and by the authors listed under "Data Base References", and references therein. Commonly the data used by a particular author to derive a biogeochemical cycle are "cannibalistic"; that is, they are obtained from previous authors, and modified to fit newly acquired knowledge. A general trend, however, is that with time as knowledge of a particular element cycle improves, the number of reservoirs and fluxes necessary to describe cyclic interactions increases. Emphasis here is on the relative magnitudes of natural and anthropogenic flux estimates, the modification of atmospheric chemical composition, and potential climatic change caused by these fluxes.

2. The carbon dioxide cycle and climate

2.1 LONG-TERM AND MODERN CO_2 BIOGEOCHEMICAL CYCLES

The long-term cycling behavior of carbon dioxide is shown in Figure 1; fluxes owing to anthropogenic sources and their perturbation of the global, long-term, CO_2 cycle are illustrated in Figure 2.

By long-term we imply the behavior of carbon dioxide prior to the industrial revolution and prior to extensive land use activities that led to deforestation and increased erosion rates. It is not the representation shown in the multi-million year averaged cycles of authors like Garrels and Mackenzie, 1972; Berner, et al., 1983; and Oslon, et al., 1985. In these cycles most fluxes are "averaged" for post-Precambrian time (the last 600 million years), and thus, the values of the flux estimates are generally lower.

Some important features of Figures 1 and 2 are :

1. Prior to the extensive fossil fuel burning, land use and cement manufacturing activities of human society, the atmosphere contained about 595,000 million tons of CO_2, or 280 ppmv (cf. Siegenthaler and Oeschger, 1987; Barnola, et al., 1987). The CO_2 concentration value of the mid-1980's of 345 ppmv implies a 23 % increase in atmospheric CO_2 since the middle of the 18th century.

2. This increase in atmospheric CO_2 concentration for the past 200 years is primarily a result of the increased flux of carbon gases (CO_2, CO, CH_4) to the atmosphere from the land because of the burning of fossil fuels and deforestation and cultivation practices of society. It is likely that land use fluxes were more important than fossil fuel sources in the 19th and early 20th century, but during much of the latter century fossil fuel CO_2 releases appear to have been greater than net biospheric fluxes. The relative magnitudes of the fossil fuel and land use fluxes, and their historical trends, are hotly debated topics today (cf, for instance, Houghton et al., 1983, 1987; Siegenthaler and Oeschger, 1987). In our present-day CO_2 cycle, the anthropogenic flux of 6580 MTC/Y is the summation of 5370 MTC/Y (CH_4, CO, CO_2) from fossil fuel burning and 1210 MTC/Y from land use activities (including CH_4 from cultivation of rice paddies, biomass burning, coal mining, and ruminants).

3. Prior to human activities substantially affecting the CO_2 cycle, there was a net flux of CO_2 from the ocean through the atmosphere to the land, where the CO_2 was used in net production of organic carbon and weathering of minerals in continental rocks. Because of fossil fuel burning and land use activities, the net transfer of CO_2 from the ocean to the land has been reversed and the ocean is now an important sink of CO_2. In our model of the present-day CO_2 cycle, of the total carbon released to the atmosphere by activities of society, an important portion is taken up by stimulation of plant productivity because of elevated atmospheric CO_2 levels, and increases in forest growth and humus accumulation, particularly in northern hemispheric temperate regions. However, the temperature increase may enhance respiration of detrital organic matter causing a decrease of C storage (Woodwell, 1983). This partitioning of carbon is in accord with the early model of Revelle and Munk (1977), and the more recent arguments of Kohlmaier, et al. (1987). Thus, tropical and sub-tropical forests are apparent sources of carbon because of deforestation and cultivation activities (e.g. Woodwell, 1983), and the higher latitude biosphere a sink of anthropogenic carbon gases. A major problem for the modern CO_2 cycle is the relative future role of land use activities

releasing carbon gases to the atmosphere and potential CO_2 fertilization removing these gases from the atmosphere.

4. The amount of CO_2 accumulating in the atmosphere from anthropogenic activities during the mid-1980's is about 3000 MTC, or 1.4 ppmv, per year. This accumulation represents 56 % of the carbon released by fossil fuel burning, a value in accord with the data and arguments of Keeling, et al. (1979, 1980) and Siegenthaler and Oeschger (1987).

5. In the long-term cycle of CO_2, weathering of rock minerals and organic production in the terrestrial realm produced inorganic and organic carbon, respectively, which were transported to the ocean as dissolved inorganic carbon (DIC) and dissolved and particulate organic carbon (DOC, POC). Some detrital carbonate, about 17 MTC/Y, was also transported by rivers. The long-term weathering and organic production fluxes on land were largely supported by fluxes related to CO_2 evasion from the ocean, oxidation of the reduced carbon gases CH_4 and CO, derived mostly from biogenic sources on land, volcanic and diagenetic processes, and a small flux from weathering by atmospheric O_2 of fossil organic matter in sedimentary rocks exposed at Earth's surface.

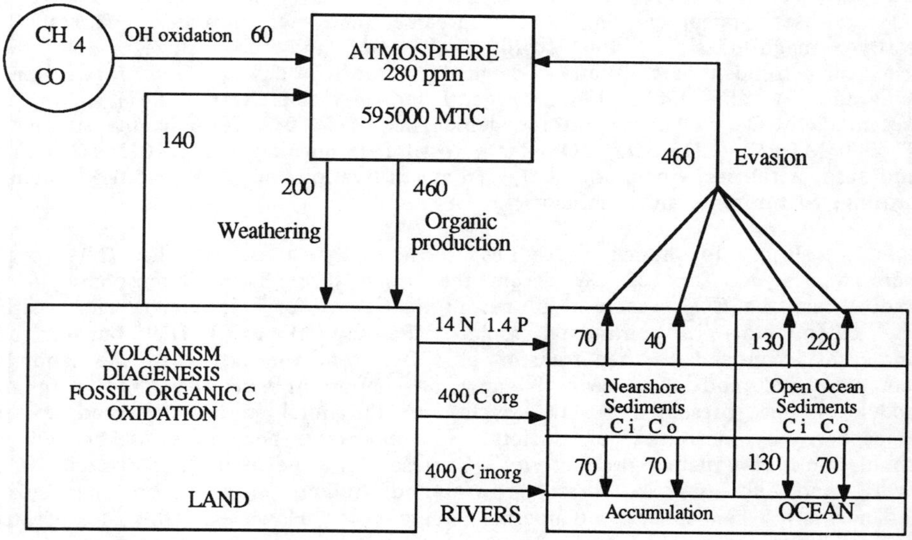

Figure 1. Long-term geological cycle of carbon. Fluxes in MTC/Y.

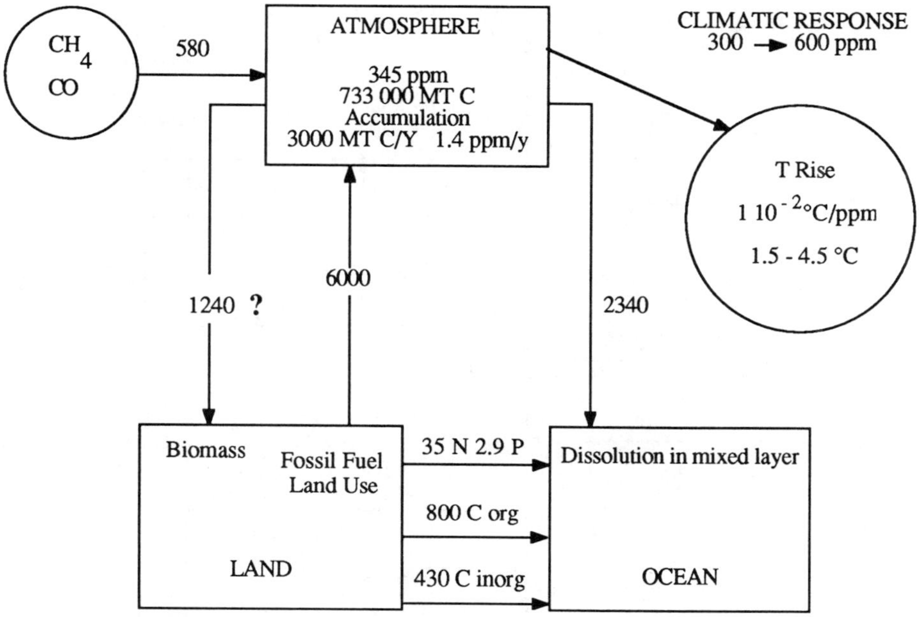

Figure 2. Anthropogenic perturbation of the carbon cycle. Fluxes in MTC/Y.

In the oceans half of the DIC riverine input accumulated in nearshore (coastal zones, oceanic banks and atolls, shelves) and open ocean (hemi-pelagic and pelagic) sediments as calcium and magnesium carbonate minerals (calcite, aragonite, magnesian calcite), and half eventually returned to the atmosphere owing to precipitation (e.g.) :

$$Ca^{2+} + 2HCO_3^- \quad = \quad CaCO_3 + CO_2 + H_2O.$$

Of the organic carbon entering the ocean, about 35 % accumulated in marine sediments as organic carbon with a molar C/N/P ratio of about 250/20/1 (Mackenzie, 1981); the rest was respired and oxidized to CO_2 and evaded the ocean. The nitrogen and phosphorus riverine fluxes of 14 and 1.4 MT/Y, respectively, represented the nutrient fluxes necessary to support the organic matter accumulation of 140 MTC/Y.

6. In the present-day CO_2 cycle, the interactions between land and ocean via rivers appear to have been affected substantially by the activities of humankind. Herein lies an important link between the carbon cycle and those of the nutrients N and P. This link is shown in Figure 3, which also demonstrates, when compared to the long-term carbon cycle of Figure 1, the effects of anthropogenic carbon on land-ocean-atmosphere CO_2 coupling.

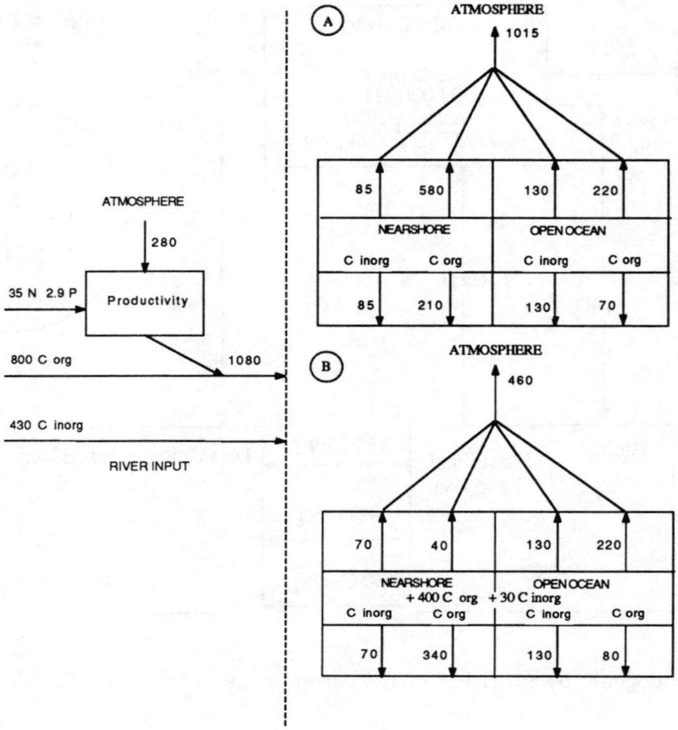

Figure 3. Effects of the perturbation on the marine system. Fluxes in MTC/Y.

First the application of fertilizers to the land surface has led to increased runoff of N and P from the land surface. Discharge of human wastes has augmented this flux. It appears that the total global nutrient flux today is about 2.5 times greater than the long-term geologic flux, and results in excess <u>accumulation</u> of organic carbon in the ocean (Meybeck, 1982; Wollast, 1983). Furthermore, because of land use activities, the flux of POC and DOC (Likens, et al., 1981) and DIC (Meybeck, 1982) from land via rivers to the ocean has been enhanced (Figures 2 and 3).

It is difficult to say what happens to these enhanced carbon fluxes today. In Figure 3, we present two opposing scenarios for the ocean. In one case (Figure 3A), most of the excess carbon brought by rivers is respired and oxidized to CO_2, and the excess organic carbon, supported by excess nutrient loading, is accumulated, primarily in nearshore environments. Also, all the riverine DIC is partitioned equally between deposition of carbonate minerals and CO_2 evasion to the atmosphere.

In the second case (Figure 3B), the excess carbon is stored in the ocean, either in the water column or in sediments, and the CO_2 evasion flux is maintained at the geologic rate. In this case, the oceanic reservoir gains 710 MTC/Y, an increase in storage capacity of 11 % of the anthropogenic flux of carbon to the atmosphere. Much of this increase takes place in the nearshore realm, although potential transport of DOC, POC, and DIC to the open ocean would also increase the loading of this part of the ocean.

2.2. CLIMATIC IMPLICATIONS AND QUESTIONS

Much has been written about the climatic implications of the observed increase of atmospheric CO_2 levels. The material is well summarized in this volume. Present modeling of the carbon cycle points to a doubling of atmospheric CO_2 concentration over the late 1800's value by the year 2030-2050. Global climate models indicate a range of global mean tropospheric temperature response to this doubling of 1.5 ° to about 5 °C. If there were a simple linear relationship between atmospheric CO_2 increase and temperature, the rate of change of equilibrium temperature would be about 1×10^{-2}°C/ppmv.

In this section we explore briefly some questions concerning the carbon cycle that bear on the rate of atmospheric CO_2 change induced by activities of humankind, and consequently, future climatic change. These questions and some comments about them are provided below.

1. *What may be the future net terrestrial biospheric response to additions of CO_2 to the atmosphere by activities of humankind ?* There seems to be emerging some consensus that land use practices in the late 19th-early 20th centuries were a more important anthropogenic source of atmospheric CO_2 than fossil fuel burning. Ecological renconstructions of the history of release of terrestrial biospheric CO_2 (Houghton, et al, 1985, 1987), although at odds with interpretations of the past 30-years record of atmospheric CO_2 increase and interpretation of ice core data on past levels of atmospheric CO_2, may be reconcilable with these data (Siegenthaler and Oeschger, 1987). The future role of the terrestrial biosphere, however, as a net source or sink of CO_2 is extremely important in model scenarios that attempt to ascertain the rate of rise of atmospheric CO_2 and potential consequent temperature change. This conclusion is evident from inspection of Figure 1; even without fossil fuel emissions, a net annual destruction of the terrestrial living biomass of 600,000 MTC of less than one percent would result in a flux comparable to the present-day anthropogenic flux.

2. *What are the chemical and biological processes and fluxes controlling the distribution of organic and inorganic carbon between the land and open ocean via rivers and the coastal realm ?* To our knowledge, all climate models today that link ocean and atmosphere CO_2 exchanges do not take into account carbon transfers in the nearshore realm, much less the coastal zone. Figure 2B demonstrates that increased euthrophication of nearshore marine environments, particularly bays, estuaries and seas, and increased riverine fluxes of organic and inorganic carbon could give rise to an

increased storage capacity of the oceans. The increase of the amount of C that can be stored annually in the coastal zone corresponds to 11 % of the anthropogenic flux of carbon to the atmosphere, a potentially significant sink of fossil fuel CO_2. However, much of this additional carbon could be respired and oxidized, and evade the ocean (Figure 3A).

For an assessment of the future climatic response to increasing atmospheric CO_2, it appears to us that we need a more quantitative picture of the fate of carbon in the coastal zone. Certainly we know that in some coastal marine environments, there has been increased accumulation of organic carbon derived from anthropogenic activities during the past century (e.g. Erlenkeuser, et al., 1974; Erlenkeuser, 1978). Also, we know that coastal zone waters may contain dissolved CO_2 concentrations greater than atmospheric, implying evasion of CO_2 to the atmosphere. The waters of the Bay of Calvi are a typical example of a system with high values of the partial pressure of CO_2 owing to organic matter respiration and upwelling (Figure 4).

On a global scale the relative importance of the potentially increased evasion and accumulation, as well as oceanic storage of carbon require further investigation. The balance in the coastal zone is of particular importance to climate models.

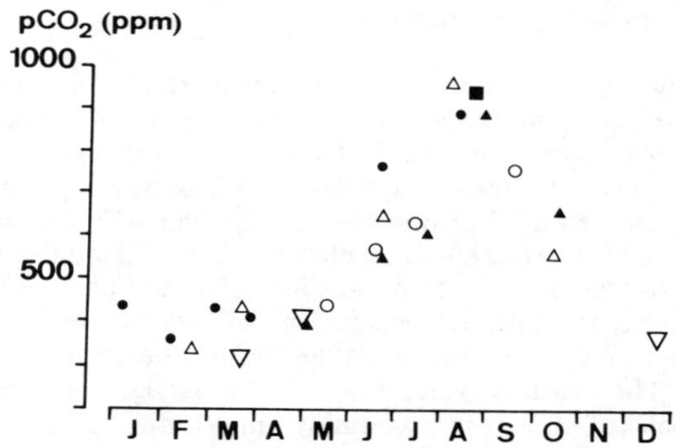

Figure 4. Annual change in the partial pressure of dissolved CO_2 in the waters of the Bay of Calvi (after Frankignoul, 1988).

3. The N_2O cycle and climate

3.1 N_2O BIOCHEMICAL CYCLE

The biogeochemical cycle of nitrogen, and the role of N_2O in the cycle, are poorly known today, particularly the estimates of fluxes involved in transfer of gases between the Earth's surface and atmosphere. In Figure 5, our interpretation of the N_2O cycle is given; a major difference between this cycle and others portrayed here is that the gas N_2O when released from Earth's surface is chemically inert in the troposphere. However, once in the stratosphere, N_2O reacts photochemically to nitric oxide (NO), which catalytically destroys ozone. The reaction scheme is :

NO formation (20-30 km)

$$N_2O + hv = NO + N \quad (<250\ \text{nm})$$
$$N_2O + O = 2NO$$

Ozone destruction

$$NO + O_3 = NO_2 + O_2$$
$$O_3 + hv = O_2 + O \quad (<650\ \text{nm})$$
$$NO_2 + O = NO + O_2$$

$$2O_3 + hv = 3O_2$$

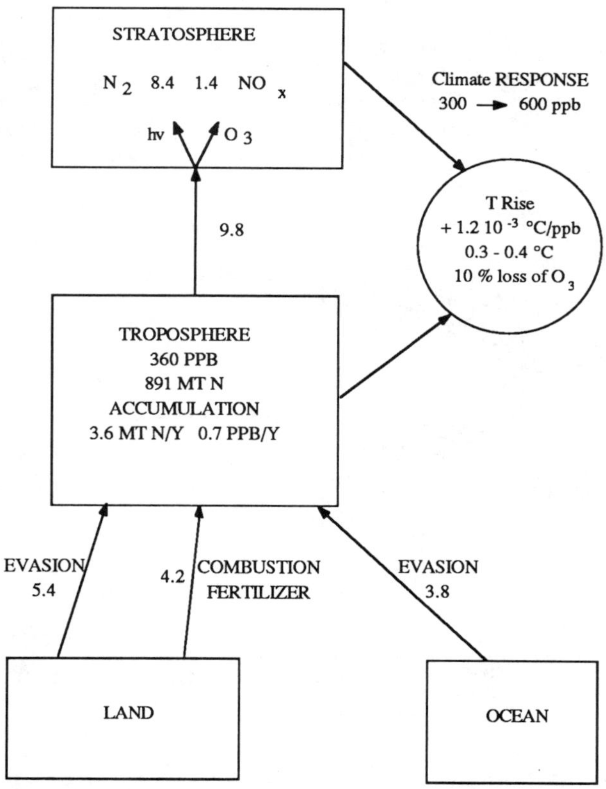

Figure 5. Global present-day biogeochemical cycle of N_2O. Fluxes in MTN/Y.

It is this series of reactions that has been the major regulator of stratospheric ozone concentrations through recent geological time. The chlorofluorocarbon gases ("freons") produced by anthropogenic activities are obviously new and synthetic chemical compounds that today may be augmenting the natural destruction rate of O_3 by N_2O released from Earth's surface.

The cycle of Figure 5 is constructed on the basis of (1) Liss' (1983) recent estimate of the N_2O evasion flux from the sea surface, (2) an estimate that about 7 % of the summation of the total nitrogen fixed to make fertilizer and the total nitrogen fixed by combustion sources annually (Mackenzie, 1987) is released to the atmosphere as N_2O, (3) an escape flux of N_2O to the stratosphere from the troposphere of about 1×10^{13}g N/year (Crutzen, 1983), and (4) an increase of tropospheric N_2O of 0.4 % per year, in reasonable agreement with the estimate of Weiss (1981) of 0.2 % for the late 1970's. The evasion flux from land, a poorly known estimate, is calculated by difference. The diagram implies a slight transport of anthropogenic N_2O from the troposphere to the stratosphere. The argument, which is controversial, for this conclusion is as follows.

The major perturbations of the N_2O cycle are due to application of urea and ammonium nitrate fertilizers to the land surface and the release of N_2O by fossil fuel combustion processes. Figure 6 shows for western European conditions of nitrogen fertilizer and manure application in agricultural areas that may be representative of world-wide conditions.

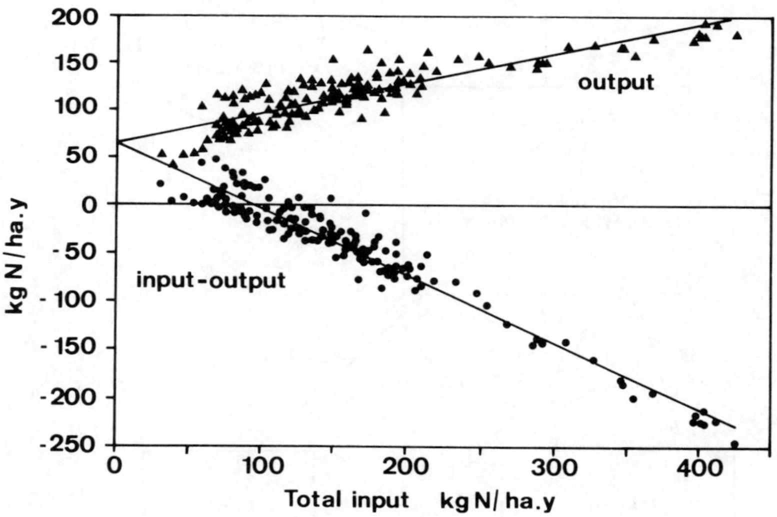

Figure 6. Relationship between nitrogen input by fertilizers and manure, nitrogen output by crops, and the difference between input and output in Western European agricultural regions during the time period 1945-1985 (after Behrendt, 1988).

Prior to 1955 the amount of nitrogen applied to cropland in this region was nearly exactly that removed in plant and animal protein. After 1955, however, there has been a nearly linear increase in the amount of nitrogen lost from fertilizers by the processes of leaching and denitrification; this latter process involves both N_2 and N_2O loss to the atmosphere. The N_2O loss from fertilizers coupled with a growing flux of N_2O to the atmosphere because of combustion processes has led to the mid-1980's flux estimate of anthropogenic N_2O to the troposphere of 4.2 MTN/year. This value is very close to the tropospheric N_2O accumulation estimate derived from actual measurements of N_2O concentrations in the atmosphere, and their trend with time (Weiss, 1981). Thus, the anthropogenic input to the troposphere may be totally accumulating there, with no stratospheric increased flux because of activities of humankind. However, we have followed the arguments of Crutzen (1983), and show that the stratospheric N_2O flux has increased by about 6 % because of combustion and fertilizer application.

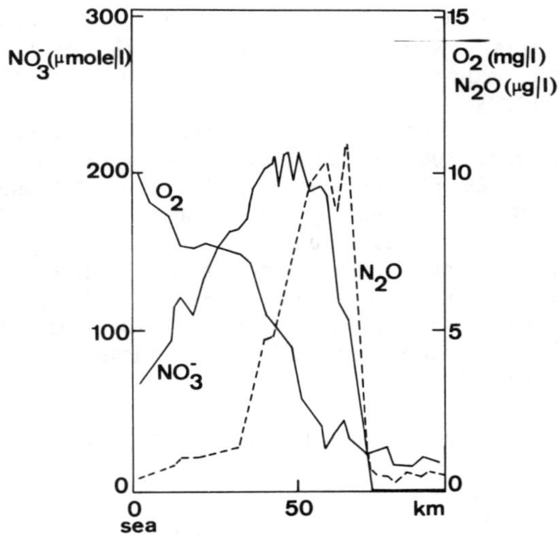

Figure 7. Longitudinal profile of O_2, NO_3^- and N_2O in the waters of the Scheldt Estuary.

The land is an obvious source of increased N_2O fluxes because of anthropogenic activities. The flux of N_2O from the sea surface, particularly in coastal regions, however, may also have increased. The scenario involves another link between the carbon and nitrogen cycles. The rate of denitrification in coastal marine systems may have increased because of increased riverine organic carbon fluxes to these systems, and because of increased eutrophication rates in these systems owing to increased fluxes of nutrients from anthropogenic sources. As shown in Figure 7, N_2O is an important by-product of denitrification (and nitrification). The production of N_2O, however, occurs within a narrow range of Eh-pH conditions in an aquatic system like the Scheldt estuary (Wollast, 1983). Maximum production rates for coastal systems vary by an order of magnitude : Scheldt Estuary, 0.1; Potomac Estuary, 0.1; and Hudson Estuary, 0.02 $mg/m^3/sec$ (McElroy et al., 1978; Deck, 1981). Also some of the N_2O produced may be utilized by bacteria and not escape the sea surface. This increased N_2O flux is poorly known because of these factors, but based on indirect arguments, Deck (1981) estimates a global value of this flux of 1.5 - 7 x 10^4 TN/year, or about 1 % of the global ocean evasion flux.

3.2 CLIMATIC IMPLICATIONS AND QUESTIONS

The two major responses of the climate system because of increased fluxes of N_2O to the atmosphere coming from anthropogenic sources are shown in Figure 5. First, N_2O is a trace "greenhouse" gas, and if its concentration in the atmosphere increases, there should be a response of heating of Earth's surface. Second, N_2O fluxes to the stratosphere are partially responsible for natural destruction of stratospheric ozone production from reaction of O_2 with radiant energy. Increased fluxes of N_2O to the stratosphere from anthropogenic activities could increase the rate of ozone destruction and lower stratospheric ozone levels. A doubling of N_2O levels in the atmosphere from 300 to 600 ppbv could give rise to an atmospheric temperature change at Earth's surface of + 0.3 - 0.4 °C, and a 10 % loss of stratospheric ozone (Crutzen, 1983). The linear rate of change for equilibrium temperature and ozone would be + 1.2 x 10^{-3} °C/ppbv and -0.03 %/ppbv, respectively. These rates are given mainly to illustrate the direction of climatic response. Actual rates depend on many complicating factors, including rate of change of N_2O loss from land to the atmosphere from processes of combustion and fertilizer application, and rate of change of the stratospheric flux.

There are at least three important questions involved with the cycle of N_2O and its climatic link :

1. *What were the pre-industrial fluxes of N_2O from Earth's surface to the atmosphere ?* This question is difficult to answer because the nitrogen cycle has already been severely perturbed. N_2O measurements of air trapped in glacial ice would help, as would better estimates of present-day fluxes. The magnitude of the denitrification flux of N_2O from the land to the atmosphere is especially poorly known, both from perturbed and natural ecosystems.

2. *What is the magnitude of the increased flux of N_2O from coastal regions because of increased organic carbon loadings and eutrophication in these environments ?* All estimates of this flux come from aquatic systems of developed countries. Large coastal regions of developing nations like India, China, and Brazil are influenced by increased organic carbon fluxes and eutrophication. These areas need be kept in mind when assessing global changes in the N_2O cycle.

3. *Has there been a loss of anthropogenic N_2O to the stratosphere ?* A corollary of this question is : what are the likely scenarios for partitioning of anthropogenic N_2O between the troposphere and stratosphere ? This partitioning is the controlling factor in the N_2O cycle governing climatic response, resulting in temperature change and/or change in stratospheric ozone levels.

4. The sulfur cycle and climate

The global biogeochemical cycle of sulfur emphasizing Earth's surface-atmosphere exchange is shown in Figure 8. Only some brief comments on the link between this cycle and climate response will be given here.

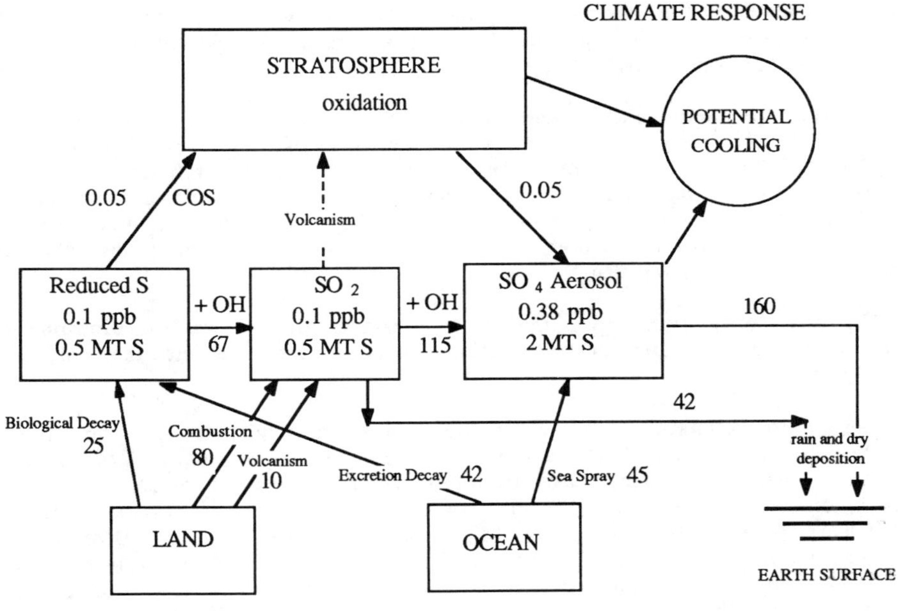

Figure 8. Global present-day biogeochemical cycle of sulfur. Fluxes in MTS/Y.

In the ocean and on the land biological processes result in the release of reduced sulfur gases to the atmosphere. The sulfur gases released from the two environments differ, mainly dimethylsulfide (DMS), and carbonyl sulfide (COS) characterize emissions from the ocean, whereas hydrogen sulfide (H_2S) is released from terrestrial soils and plants. COS is the most abundant reduced atmospheric sulfur species in the remote atmosphere, and because of its inert chemical behavior in the troposphere, and consequently its long residence time, it can enter the stratosphere

(Andreae, 1986). In the stratosphere, COS is oxidized photochemically to sulfate; thus, this flux is responsible for regulation of the stratospheric sulfate aerosol layer during periods of volcanic quiescense. Volcanism can inject sulfate aerosol directly into the stratosphere.

The reduced sulfur gases of DMS and H_2S are oxidized to sulfate in the troposphere. DMS produced at the sea surface appears to be an important source of cloud condensation nuclei (CCN) in the remote marine atmosphere. Charlson, et al. (1987) have pointed out that because the reflectance of clouds is sensitive to CCN density, there is a link in the tropospheric sulfur cycle between marine phytoplankton production and climate. They argue that a warming of Earth's climate, say by anthropogenic CO_2 emissions, could lead to enhanced phytoplankton growth, excretion, and emission of DMS from the sea surface. The increased DMS flux could result in increased production of CCN in the remote marine atmosphere and increased cloud albedo, giving rise to cooling of the troposphere. An interesting scenario but still somewhat controversial.

Volcanic processes and oxidation of reduced sulfur gases lead to SO_2 emissions to the atmosphere; however, combustion of fossil fuels constitutes about half of the SO_2 flux, or about 40 % of the total exchange of sulfur between Earth's surface and the atmosphere. This is a significant perturbation of the biogeochemical cycle of sulfur. The impact of this perturbation is primarily regional in nature leading to the acid rain problems of the northeastern U.S.A. and western Europe.

The sea spray flux from the ocean surface giving rise to atmospheric aerosol, and the rain and dry deposition sulfur fluxes complete the balanced cycle of Figure 8.

The major climatic response of potential future changes in the sulfur cycle because of activities of society may be one of cooling. DMS fluxes in coastal marine areas may have been enhanced because of eutrophication of these regions, and COS fluxes from biomass burning may be growing; currently this flux is estimated to be 0.1 MTS/year (Andreae, 1986). DMS fluxes may increase because of the release of "greenhouse" gases from anthropogenic sources, warming of the climate, and increased marine productivity. However, the feedback might operate in the opposite sense if, as many models suggest, a decrease of primary productivity results from reduced oceanic upwelling in relation with climatic warming.

Whatever the case, anthropogenic COS emissions escaping to the stratosphere could lead to a cooling of this region by increasing the stratospheric sulfate aerosol veil; volcanic SO_2 inputs to the stratosphere and generation of sulfate aerosol have been observed to lead to short-lived cooling responses of the stratosphere. Presumably, if the arguments of Charlson, et al (1987) are correct, then increased DMS fluxes from the sea surface can lead to a cooling of the troposphere.

5. The methane cycle and climate

We conclude the discussion of the link between biogeochemical cycles and climate with a brief look at methane; recent summaries of the details of methane sources and sinks are given in Senum and Gaffney, 1985; Bolle, et al., 1986; and Brasseur, this volume. A survey of the literature shows that global source and sink fluxes are controversial, and probably not known within a factor of two. Bearing this point in mind, we present in Figure 9 a tentative cycle of methane.

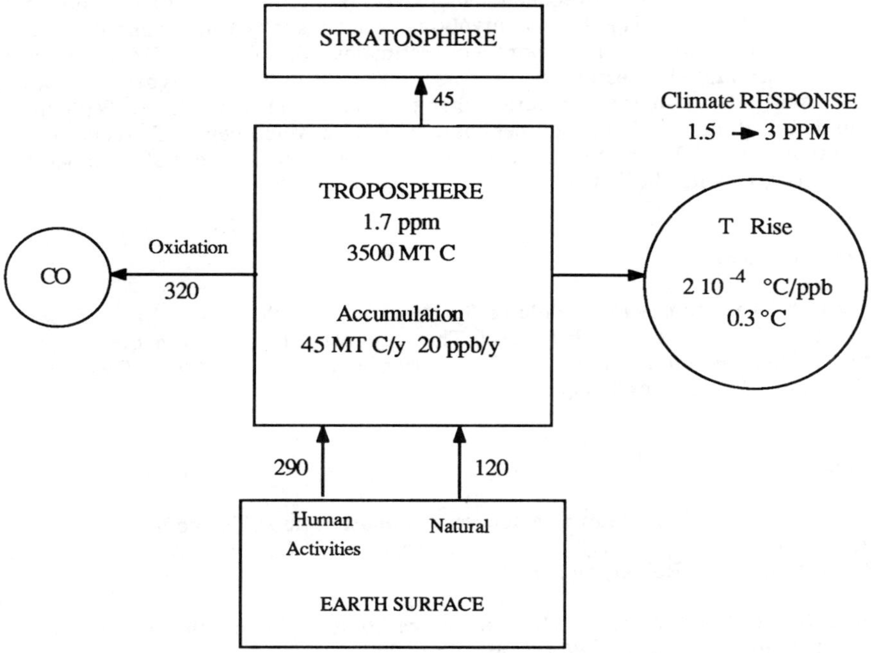

Figure 9. Global present-day biogeochemical cycle of CH_4. Fluxes in MTC/Y.

The cycle is based on the stratospheric loss estimate of Bolle, et al. (1986); the hydroxyl radical oxidation sink estimate of McElroy and Wofsy, 1985; an accumulation rate of CH_4 in the present troposphere of about 20 ppb per year (Blake and Rowland, 1988); and a total flux of CH_4 from the Earth's surface to the atmosphere from human activities (livestock rearing, farming of rice paddies, biomass burning, natural gas utilization, and coal mining) of 290 MTC/Y (Bolle, et al., 1986; values projected to 1990). The natural flux of CH_4 to the atmosphere is computed by difference. The important point is that CH_4 has accumulated in the atmosphere from a pre-1700's level of 0.7 ppmv to a concentration in 1987 of 1.68 ppmv (Blake and Rowland, 1988). Herein lies the important link between the CH_4 cycle and climate. The oxidation of methane by OH radical in the troposphere, the major atmospheric sink of CH_4, is not rapid enough to keep up with the increased

CH_4 flux from human activities. The resulting increase in atmospheric CH_4 concentrations is an important direct contribution to the "greenhouse effect". For comparison, one incremental CH_4 molecule in the atmosphere is about 20 times more effective in increasing temperature than one molecule of CO_2 (Blake and Rowland, 1988). Furthermore, oxidation of CH_4 escaping to the stratosphere generates H_2O vapor, thus enhancing the "greenhouse effect" by increasing stratospheric temperatures.

In conclusion, we indentify two major problems involving interactions between the CH_4 global biogeochemical cycle and climate. First, the large range of estimates of CH_4 fluxes involving Earth surface-atmosphere transfer needs to be narrowed, and improved estimates of future CH_4 emissions obtained. Second, the estimate of the sink strength of hydroxyl radical oxidation and its potential future change need refinement. Without a better understanding of CH_4 sources and sinks, development of scenarios of future trends in CH_4 atmospheric concentrations and potential temperature change are fraught with difficulty.

6. Acknowledgements

This research was completed while F.T. Mackenzie held the *Chaire Francqui* at the Université Libre de Bruxelles. The support of the Francqui Foundation and ACS-PRF Grant 17371-AC2 is gratefully acknowledged. Contribution no. 2049, Hawaii Institute of Geophysics.

7. References

Andreae, M.O., 1986, 'The ocean as a source of atmospheric sulfur com pounds.' *In : The Role of Air-Sea Exchange in Geochemical Cycling*, ed. P. Buat-Menard, D. Reidel, Dordrecht, 331-362.

Arrhenius, S., 1896, 'On the influence of carbonic acid in the air upon the temperature of the ground.' *Philos. Mag.* **41**, 237.

Barnola, J.M., Raynaud, D., Korotkevich, Y.S., and Lorius, C., 1987, 'Vostok ice core provides 160,000 year record of atmospheric CO_2.' *Nature*, **329**, 408-414.

Behrendt, H., 1988, 'Changes in nonpoint nutrient loading into European freshwaters : Trends and consequences since 1950 and not-impossible changes until 2080.' Working Paper, Internatl. Inst. Applied Systems Analysis, Laxenburg, Austria, 28 pp.

Berner, R.A., Lasaga, A.C., and Garrels, R.M., 1983, 'The carbonate-silicate geochemical cycle and its effect on atmospheric carbon dioxide over the past 100 million years.' *Am. J. Sci.*, **283**, 641-683.

Blake, D.R., and Rowland, F.S., 1988, 'Continuing worldwide increase in tropospheric methane, 1978 to 1987.' *Science*, **239**, 1129-1131.

Bolin, B., and Cook, R.B., 1983, *The Major Biogeochemical Cycles and Their Interactions*. John Wiley and Sons, New York, 532 pp.

Bolle, H.-J., Seiler, W., and Bolin, B., 1986, 'Other greenhouse gases and aerosol : Assessing their role for atmospheric radiative transfer.' *In : The Greenhouse Effect, Climatic Change and Ecosystems*, eds. B. Bolin, R. Doos, J. Jager, and R.A. Warrick, John Wiley and Sons, New York, 157-203.

Brasseur, G., this volume.

Buat-Menard, P., 1986, 'Air to sea transfer of anthropogenic trace metals.' *In : The Role of Air-Sea Exchange in Geochemical Cycling*, ed. P. Buat-Menard, D. Reidel, Dordrecht, 477-496.

Charlson, R.J., Lovelock, J.E., Andreae, M.O., and Warren, S.G., 1987, 'Oceanic phytoplankton, atmospheric sulfur, cloud albedo, and climate.' *Nature*, **326**, 655-661.

Crutzen, P.J., 1983, 'Atmospheric interactions - Homogeneous gas reactions of C, N, and S containing compounds'. *In : The Major Biogeochemical Cycles and Their Interactions*, eds. B. Bolin and R.B. Cook, John Wiley and Sons, New York, 67-114.

Deck, B.L., 1981, 'Nutrient-Element Distributions in the Hudson Estuary.' Unpubl. Ph. D. Thesis, Columbia Univ., 396 pp.

Erlenkeuser, H., Suess E., and Willkomm, H., 1974, 'Industrialization affects heavy metal and carbon isotope concentrations in recent Baltic Sea sediments'. *Geochim. Cosmochim. Acta*, **38**, 823-842.

Erlenkeuser, H., 1978, 'The use of radiocarbon in estuarine research'. *In : Biogeochemistry of Estuarine Sediments*, UNESCO, Paris, 140-153.

Frankignoulle, M., in press, 'Field measurements of air-sea CO_2 exchange'. *Limnol. Oceanogr.*

Garrels, R.M., and Mackenzie, F.T., 1971, 'Evolution of Sedimentary Rocks'. W.W. Norton, New York, 397 pp.

Garrels, R.M., and Mackenzie, F.T., 1972, 'A quantitative model of the sedimentary rock cycle'. *Mar. Chem.*, **1**, 27-41.

Houghton, R.A., Hobbie, J.E., Melillo, J.M., Moore, B., Peterson, B.J., Shaver, G.R., and Woodwell, G.M., 1983, 'Changes in the carbon content of terrestrial biota and soils between 1860 and 1980 : A net release of CO_2 to the atmosphere'. *Ecolog. Monogr.*, **53**, 235-262.

Houghton, R.A., Boone, R.D., Fruci, J.R., Hobbie, J.E., Melillo, J. M., Palm, C.A., Peterson, B.J., Shaver, G.R., and Woodwell, G.M., 1987, 'The flux of carbon from terrestrial ecosystems to the atmosphere in

1980 due to changes in land use : Geographical distribution of the global flux'. *Tellus*, **39**, 122-139.

Keeling, C.D., Mook, W.G., and Tans, P.P., 1979, '$^{13}C/^{12}C$ ratio of atmospheric carbon dioxide'. *Nature*, **217**, 121-123.

Keeling, C.D., Bacastrow, R.D., and Tans, P.P., 1980, 'Predicted shift in the $^{13}C/^{12}C$ ratio of atmospheric carbon dioxide'. *Geophys. Res. Lett.*, 7, 505-508.

Kohlmaier, G.H., Bröhl, H., Sire, E.P., Plöchl, M., and Revelle, R., 1987, 'Modelling stimulation of plants and ecosystem response of present levels of excess atmospheric CO_2'. *Tellus*, **39**, 155-170.

Likens, G.E., Mackenzie, F.T., Richey, J., Sedwell, J.R., and Turekian, K.K., 1981, 'Flux of Organic Carbon by Rivers to the Sea'. U.S.D.O.E. Conf. Rept. 8009140, Washington, D.C., 397 pp.

Liss, P.S., 1983, 'Gas transfer : Experiments and geochemical implications'. *In : Air-Sea Exchange of Gases and Particles*, eds. P.S. Liss and G.N. Slinn, D. Reidel, Dordrecht, 241-298.

Mackenzie, F.T., Lantzy, R.J., and Paterson, V., 1979, 'Global trace metal cycles and predictions'. *Mathematical Geology*, **2**, 99-142.

Mackenzie, F.T., 'Global carbon cycle : Some minor sinks for CO_2'. *In : Flux of Organic Carbon by Rivers to the Sea*, eds. G.E. Likens, F.T. Mackenzie, J.R. Sedwell, and K.K. Turekian, U.S.D.O.E. Conf. Rept. 8009140, Washington, D.C., 360-384.

Mackenzie, F.T., 1987, 'Global mixed-layer natural and anthropogenic fluxes'. *In : Dynamics of the Ocean Surface Mixed Layer*, ed. P. Muller and D. Henderson, Hawaii Inst. Geophys. Spec. Publ., Univ. Hawaii, Honolulu, 291-310.

McElroy, M.B., Elkins, J.W., Wofsy, S.C., Kolb, C.E., Duran, A.P., and Kaplan, W.A., 1978, 'Production and release of N_2O from the Potomac Estuary'. *Limnol. Oceanogr.*, **23**, 1168-1182.

McElroy, M.B., and Wofsy, S.C., 1985, 'Table 3-9 given in Atmospheric Ozone - 1985'. World Meteorol. Organiz., Rept. No. 16, 93.

Meybeck, M., 1982, 'Carbon, nitrogen, and phosphorus transport by world's rivers'. *Am. J.Sci*, **282**, 401-450.

Olson, J.S., Garrels, R.M., Berner, R.A., Armentano, T.V., Dyer, M.J., and Yaalon, D.H., 1985, 'The natural carbon cycle'. *In : Atmospheric Carbon Dioxide and The Global Carbon Cycle*, ed. J.R. Trabalka, U.S.D.O.E./ER000239 Rept., NTIS, Springfield, VA., 175-213.

Revelle, R., and Munk, W., 1977, 'The global carbon dioxide cycle and the biosphere'. *In : Energy and Climate*, Natl. Acad. Press, Washington, D.C., 243-280.

Schlesinger, M. E., this volume.

Senum, G.L., and Gaffney, J.S., 1985, 'A reexamination of the tropospheric methane cycle : Geophysical implications'. *In : The Carbon Cycle and Atmospheric CO_2 : Natural Variations Archean to Present*, eds. E.T. Sundquist and W.S. Broecker, A.G.U. Geophys. Monograph 32, Washington, D.C., 61-69.

Siegenthaler, U., and Oeschger, H., 1987, 'Biospheric CO_2 emissions during the past 200 years reconstructed by deconvolution of ice core data'. *Tellus*, **39**, 140-154.

Sundquist, E.G., and Broecker, W.S., 1985, 'The Carbon Cycle and Atmospheric CO_2 : Natural Variations Archean to Present'. *A.G.U. Geophys. Monograph* **32**, Washington, D.C., 627 pp.

Weiss, R.F., 1981, 'The temporal and spatial distribution of tropospheric nitrous oxide'. *J. Geophys. Res.*, **86**, 7185-7195.

Wollast, R., 1983, 'Interactions in estuaries and coastal waters'. *In : The Major Biogeochemical Cycles and Their Interactions*, eds. B. Bolin and R.B. Cook, J. Wiley & Sons, New York, 385-410.

Woodwell, G.M., 1983, 'Biotic effects on atmospheric carbon dioxide : A review and projection'. *In : Changing Climate*, Natl. Acad. Press, Washington, D.C., 216-241.

8. Data Base References

Andreae, M.O., and Raemdonck, H., 1983, 'Dimethyl sulfide in the surface ocean and the marine atmosphere : A global view'. *Science*, **221**, 744-747.

Berrensheim, H., and Jaeschke, W., 1983, 'The contribution of volcanoes to the global atmospheric sulfur budget'. *J. Geophy. Res.*, **88**, 3732-3740.

Blake, D.R., and Rowland, F.S., 1988, 'Continuing worldwide increase in tropospheric methane, 1978 to 1987'. *Science*, **239**, 1129-1131.

Bolin, B., and Charlson, R., 1976, 'On the role of the tropospheric sulfur cycle in the shortwave radiation of the Earth'. *Ambio*, **5**, 27.

Bolin, B., 1981, 'Carbon Cycle Modelling'. J. Wiley & Sons, New York. 390 pp.

Bolin, B., Degens, E.T., Kempe, S., and Ketner, P., 1979, *The Global Carbon Cycle*. J. Wiley and Sons, New York, 491 pp.

Bolin, B., Doos, B.R., Jager, J., and Warrick, R.A., 1986, *The Greenhouse Effect, Climatic Change, and Ecosystems*. J. Wiley and Sons, New York, 541 pp.

Broecker, W., Takahashi, T., Simpson, H., and Peng, T.-H., 1979, 'Fate of fossil fuel carbon dioxide and the global carbon budget'. *Science*, **206**, 409-418.

Carbon Dioxide and Climate, 1982, National Academy of Sciences Press, Washington, D.C., 72 pp.

Carbon Monoxide, 1977, National Academy of Sciences Press, Washington, D.C., 239 pp.

Chameides, W., and Davis D., 1982, 'Chemistry in the troposphere'. *Chem. Eng. News*, Oct. 4, 38-52.

Cohen, Y. and Gordon L., 1979, 'Nitrous oxide production in the ocean.' *J. Geophy. Res.*, **84**, 347-353.

Conrad, R., Seiler, W., and Bunse, G., 1983, 'Factors influencing the loss of fertilizer nitrogen into the atmosphere as N_2O'. *J. Geophy. Res.*, **88**, 6709-6718.

Crutzen, P., 1974, 'Estimates of possible variations in total ozone due to natural causes and human activities'. *Ambio*, **3**, 201-210.

Crutzen, P., 1976, 'Upper limits of atmospheric ozone reductions following increased application of fixed nitrogen to the soil'. *Geophy. Res. Lett.*, **3**, 169-172.

Crutzen, P., and Ehhalt, D., 1977, 'Effects of nitrogen fertilizers and combustion on the stratospheric ozone layer'. *Ambio*, **6**, 112-117.

Garrels, R.M., Mackenzie, F.T., and Hunt, C., 1975, 'Chemical Cycles and the Global Environment'. W. Kaufmann, Inc., Los Altos, CA, 206 pp.

'Global 2000 Report to the President', 1980, *The Technical Report*, Vol. **2**. Government Printing Office, Washington, D.C., 766 pp.

Graham, W., and Duce, R., 1979, 'Atmospheric pathways of the phosphorus cycle'. *Geochim. Cosmochim. Acta*, **78**, 1195-1208.

'International Symposium on Trace Gases', 1974. *Tellus*, **26**, 1-297.

Ivanov, M.D., and Freney J.R., 1983, 'The Global Biogeochemical Sulfur Cycle'. John Wiley and Sons, New York, 470 pp.

Khalil, M., and Rasmussen R., 1983, 'Sources, sinks and seasonal cycles of atmospheric methane'. *J. Geophy. Res.*, **88**, 5131-5144.

Likens, G., 1981, 'Some Perspectives of the Major Biogeochemical Cycles'. J. Wiley & Sons, New York, 175 pp.

Liss, S., and Slinn, W.G.N., 1983, 'Air-Sea Exchange of Gases and Particles'. D. Reidel, Dordrecht, 510 pp.

The Role of Air-Sea Exchange in Geochemical Cycles. 1986, Buat-Menard, P., ed, R. Reidel, Dordrecht, 549 pp.

Schlesinger, W., and Melack J., 1981, 'Transport of organic carbon in the world's rivers'. *Tellus*, **33**, 172-187.

Smil, V., 1985, 'Carbon, Nitrogen, Sulfur : Human Interference in Grand Biospheric Cycles'. *Plenum Press*, New York, 459 pp.

Stumm, W., 1977, 'Global Chemical Cycles and their Alterations by Man.' Dahlem Konferenzen, Berlin, 346 pp.

Svensson, B., and Söderlund, R., 1975, 'Nitrogen, Phosphorus and Sulphur Global Cycles'. *Ecol. Bull.*, Stockholm, 192 pp.

Varhelyi, G., and Gravenhorst, G., 1983, 'Production rate of airborne sea-salt sulfur deduced from chemical analysis of marine aerosols and precipitation'. *J. Geophy. Res.*, **88**, 6737-6751.

Woodwell, G.M., and Houghton, R.A., 1977, 'Biotic influences on the world carbon budget'. *In : Global Chemical Cycles and Their Alteration by Man*, ed., W. Stumm, Dahlem Konferenzen, Berlin, 61-72.

Zimmerman, P., Greenberg, J., Wandiga, S., and Crutzen, P., 1982, 'Termites : A potentially large source of atmospheric methane, carbon dioxide, and molecular hydrogen'. *Science*, **218**, 563-565.

ANTHROPOGENIC AND NATURAL PERTURBATIONS OF THE CARBON CYCLE

Ulrich Siegenthaler
Physics Institute, University of Bern
Bern, Switzerland

A summary is given of recent results and unsolved problems regarding the carbon cycle. There is a lack of knowledge concerning the amounts of CO_2 released by deforestation and land use, the response of terrestrial vegetation to increasing CO_2, gas exchange air-sea and the water exchange between surface and deep ocean. Analyses of polar ice cores have demonstrated that atmospheric CO_2 was lower during glacial than during interglacial times. These natural variations must be due to a change in operation of the oceans, but the precise mechanism is not yet known. Specific recommendations for experimental and modelling research regarding the global carbon cycle are given.

Introduction

It has become evident that atmospheric carbon dioxide plays an important role for the global climate via the greenhouse effect. On the one hand the anthropogenic CO_2 increase will with high probability lead to a modification of the climate in the near future. On the other hand, ice core measurements have shown that the atmospheric CO_2 concentration varied parallel to climate during the glacial-interglacial cycles, and its variations contributed, according to climate model results, significantly to the climatic changes. Ideally, the carbon cycle should therefore be treated as an internal part of the global climatic system. Today, however, the carbon cycle is still being modelled as a separate system, which is certainly justified in view of the limited knowledge and numerous open questions. In the following, a short summary of recent achievements and unsolved problems regarding the carbon cycle is given. Climatic effects of CO_2 are not discussed here.

Anthropogenic CO_2 increase

For modelling this problem, it is necessary to consider the three reservoirs atmosphere, ocean and land biosphere; reviews are given in Trabalka (1985), Bolin (1986) or Siegenthaler (1986). A part (ca. 10 - 40 percent) of the total man-made CO_2 emmissions stem from the land

A. Berger et al. (eds.), Climate and Geo-Sciences, 475–480.

biota, due to deforestation and agriculture. Estimates of this contributions have generally decreased in the past few years; they are difficult because of the heterogeneity of the biosphere and lack of sufficient data on areal extent of land modification, carbon loss per unit area and regrowth of abandoned areas. On the other hand, the unperturbed parts of the land biota may have responded to the higher CO_2 levels by increased carbon storage, but the efficiency of this CO_2 fertilization for natural ecosystems is rather uncertain (cf. Strain and Cure, 1985). There is some indirect evidence for such a response: the seasonal CO_2 amplitude at Mauna Loa, which reflects biospheric activity in the Northern Hemisphere, increased by ca. 18 % from 1958 to 1982 (Bacastow et al., 1985). It appears plausible that this is due to increased biospheric uptake and release of CO_2; it should, however, be noted that this is not equivalent to increased carbon storage.

As only about half of the man-made CO_2 remained in the atmosphere, the rest must mainly have been taken up by the oceans. Two steps control this uptake: (1) air-sea exchange, (2) transport from the surface to deeper ocean layers. The chemistry of the air-sea CO_2 exchange is well understood, but there are uncertainties regarding the gas transfer rate. The gas transfer velocity is controlled by the turbulence conditions near the interface. Its dependence on wind speed etc. has been studies in the field by means of the radium-radon disequilibrium method, but the results were not very conclusive (Roether and Kromer, 1984). New methods for investigating interfacial turbulence, e.g. by studying heat transfer (as an analogue) locally and with high time resolution using remote sensing methods, promise to give more insight (Libner et al., 1987). Synoptic large-scale wind fields obtained from satellite scatterometer data will become available in the future, so that, given an wind - transfer velocity relationship, it will be possible to compute gas exchange fluxes with high spatial and temporal resolution (although not necessarily with high accuracy).

The rate of water exchange between surface and deeper layer is a most important factor for the uptake of anthropogenic CO_2 as well as for natural CO_2 variations. Information on the vertical exchange has mainly come from oceanic ^{14}C data and from other isotopic and geochemical data (see e.g. Broecker and Peng, 1982). The progressive penetration of transient tracers (CFMs, bomb produced ^{14}C and ^{3}H, ^{3}He, ^{85}Kr) into the ocean provides a unique observational opportunity.

The oceanic uptake of CO_2 has so far been studied mainly by means of relatively simple, box-type exchange models, in which the ocean dynamics are lumped into a few parameters which have to be calibrated e.g. from the distribution of natural or of bomb-produced ^{14}C. An often used example is the (one-dimensional) box-diffusion model of Oeschger et al. (1975). A purely one-dimensional scheme does not account for the fact that vertical exchange rates vary strongly e.g. with latitude. In order to estimate the influence of rapid deep-sea ventilation in high latitudes, Siegenthaler (1983) modified the box-diffusion model by including a deep-sea outcrop. The calculated airborne fraction of fossil

CO_2 was for the same calibration method (using bomb-produced ^{14}C) significantly lower for the model with outcrop than without, illustrating the possible importance of the high latitude ocean as a sink for excess CO_2. This must, however, be studied by more realistic models.

Most modelling work concerning the past and future anthropogenic CO_2 increase has been based on the assumption that (1) there was a constant pre-industrial CO_2 concentration, and (2) generally the carbon cycle processes, especially those in the ocean, continue to work in a steady state. Assumption (1) can be justified by the result from ice core studies that any concentration variations in the past 10,000 years were small compared to the increase by 70 ppm since 1800 to the present value of ca. 350 ppm. However, for exploring possible changes in ocean circulation due to man-made climatic change and also their feedback on atmospheric CO_2, dynamic ocean models are required. Such models are also needed for analyzing the existing and expected large sets of geochemical and tracer data in an optimal way. Experiments with three-dimensional models of the oceanic carbon cycle have begun (Maier-Reimer and Hasselmann, 1987; Sarmiento et al., 1988); they allow the inclusion of phenomena of ocean dynamics and to study the pathways of CO_2 into the ocean in detail.

Natural variations

In different ice cores from Greenland and from Antarctica, CO_2 levels of 180 - 200 ppm during glacial time have been observed, compared to 280 ppm during the Holocene; the longest record goes 160,000 years back, i.e. well into the penultimate glaciation (Barnola et al., 1987). In addition to the relatively slow changes, rapid CO_2 variations have been observed in Greenland ice, with amplitudes of 50 ppm within less than 100 yr (Stauffer et al., 1984). It has not been possible until now to find corresponding rapid variations in ice cores from Antarctica, so that the question is still open whether they are an artefact (e.g. due to irregular stratigraphy of the ice) or do represent real atmospheric changes.

The reason for these variations must be looked for in the oceans, since on time scales of $\leq 10^5$yr, the atmospheric CO_2 concentration is essentially determined by an interplay of biological, chemical and physical processes in the ocean. In contrast to the perturbation models used for estimating the excess CO_2 uptake, models suited for studying natural CO_2 variations must therefore include the whole spectrum of these processes. Organisms fix carbon by photosynthesis in surface waters; after their death they sink and are oxidized by microbial action in deeper waters. In this way biological activity acts as a carbon pump that creates a CO_2 deficit at the surface. The activity of this pump is steered amongst other factors by the availability of the nutrients phosphorus and nitrogen. Broecker (1982) was the first to point out the link between nutrients in the ocean and atmospheric CO_2. He suggested that deposition and erosion of organic-rich shelf sediments during deglaciation and glacial growth was involved; cadmium measurements (Cd varies proportionally to phosphorus) in foraminifera have, however, not confirmed the required higher oceanic phosphorus content during glacial time (Boyle, 1988). As shown by three groups (Knox and McElroy, 1984; Sarmiento and Toggweiler, 1984; Siegenthaler and Wenk, 1984) the high-latitude oceans play an essential role for the atmospheric CO_2 level: there, in contrast to lower latitudes, the biological carbon pump is not regulated by the availability of the nutrients nitrogen and phosphorus. Therefore, changes in ocean circulation, insolation, ice cover etc. in high latitudes may affect the efficiency of the biological pump and consequently atmospheric CO_2. There is indeed evidence from deep-sea sediments that surface conditions and ocean circulation did change between glacial and interglacial time. Another possible explanation invokes changes in the ocean-sediment interaction and therefore in the average alkalinity of the ocean. These processes are relatively slow (time scale several 1000 yr) and could not have caused the rapid CO_2 variations mentioned above.

At this stage it is not possible to put the different pieces of evidence together to a coherent picture that could explain the detailed mechanism behind the CO_2 variations in convincing way. In order to get more insight into the cause of CO_2 variations, it will be necessary to consider ecological processes (Redfield ratios, factors regulating biological productivity) and physical oceanography (deep water formation, general ocean circulation), to obtain more information from deep-sea sediments (^{13}C and ^{14}C in planktonic and benthic forams, cadmium, organic carbon) and, not least, to perform more measurements of CO_2 and other trace gases in polar ice cores from glacial and Holocene time. An important, unresolved question is how the circulation of the ice-age ocean worked in detail, and how the changes came about.

Recommendations for future research

OBSERVATIONS:

- Studies on oceanographic processes as well as on the large scale circulation, especially on vertical exchange processes, by physical and by tracer methods.
- Data on present-day modifications of tropical and other terrestrial ecosystems.
- Laboratory and field studies on the response of biota to increased CO_2 levels and to changes of temperature and humidity.

- Ice core measurements of concentration and isotopic composition of CO_2 in glacial and Holocene time: size of Holocene CO_2 variations? Do the rapid variations observed in Greenland ice represent atmospheric variations or not? How did $^{13}C/^{12}C$ of CO_2 vary (relevant with respect to cause of CO_2 changes)?

- Observations from deep-sea sediments on the history of ocean circulation and chemistry.

MODELLING:

- Model studies on the role of the high-latitude oceans for anthropogenic as well as natural perturbations, especially vertical exchange processes and ocean-atmosphere gas transfer.

- 3D General Ocean Circulation Studies of pathways of anthropogenic CO_2 into the sea.

- Terrestrial ecosystem modelling of the reaction to CO_2 increase and to changed climatic conditions.

- Models of the biological, chemical and physical processes affecting the natural oceanic carbon cycle. Here, a hierarchy, ranging from simple few-box models to 3D models including dynamics is needed, and submodels e.g. of the marine biota have to be developed.

Acknowledgment:

Thanks are due to L.D.D. Harvey for constructive comments on the manuscript.

References

Bacastow, R.B., C.D. Keeling, and T.P. Whorf. 'Seasonal amplitude increase in atmospheric CO_2 concentration of Mauna Loa, Hawaii, 1959 - 1982'. J. Geophys. Res., **90**, 10, 529 - 10, 540, 1985.

Barnola, J.M., D. Raynaud, Y.S. Korotkevich, and C. Lorius. 'Vostok ice core provides 160,000-year record of atmospheric CO_2'.Nature, **329**, 408 - 414, 1987.

Bolin, B. 'How much CO_2 will remain in the atmosphere?' In: The Greenhouse Effect, Climatic Change, and Ecosystems, **SCOPE 29** (eds. B. Bolin et al.), pp. 93 - 156, J. Wiley & Sons, 1986.

Boyle, E.A. 'Cadmium: chemical tracer of deepwater paleoceanography'. Paleoceanography, **3**, 471 - 490, 1988.

Broecker, W.S. 'Ocean chemistry during glacial times'. Geochim. Cosmochim. Acta, **46**, 1689 - 1705, 1982.

Broecker, W.S. and T.-H. Peng. Tracers in the Sea. Eldigio Press, Lamont-Doherty Geological Observatory, Palisades, N.Y., 1982.

Knox, F. and M. McElroy. 'Changes in atmospheric CO_2: Influence of marine biota in high latitudes'. J. Geophys. Res., **89**, 4629 - 4637, 1984.

Libner, P., B. Jähne, and E. Plate. 'Eine neue Methode zur lokalen und momentanen Bestimmung der Wiederbelüftungsraten von Gewässern (A new method measuring water reaeration locally and fast)'. Wasserwirtschaft, **77**, no. 5, 1 - 6, 1987.

Maier-Reimer, E., and K. Hasselmann. 'Transport and storage of CO_2 in the ocean'. Climate Dynamics, **2**, 63 - 90, 1987.

Oeschger, H., U. Siegenthaler, U. Schotterer, and A. Gugelmann. 'A box-diffusion model to study the carbon dioxide exchange in nature'. Tellus, **27**, 168 - 192, 1975.

Roether, W., and B. Kromer. 'Optimum application of the radon deficit method to obtain air-sea gas exchange rates'. In: Gas Transfer at Water Surfaces (eds. W. Brutsaert and G.H. Jirka), pp. 447 - 457, D. Reidel, 1984.

Sarmiento, J.L., J.R. Toggweiler, and R. Najjar. 'Ocean carbon cycle dynamics and atmospheric pCO_2'. Phil. Trans. R. Soc., **A**, in press, 1988.

Siegenthaler, U. 'Uptake of excess CO_2 by an outcrop-diffusion model of the ocean'. J. Geophys. Res., **88**, 3599 - 3608, 1983.

Siegenthaler, U. 'Carbon dioxide: its natural cycle and anthropogenic perturbation'. In: The Role of Air-Sea Exchange in Geochemical Cycling (ed. P. Buat-Ménard), pp. 209 - 247, D. Reidel, 1986.

Siegenthaler, U., and T. Wenk. 'Rapid atmospheric CO_2 variations and ocean circulation'. Nature, **308**, 624 - 625, 1984.

Stauffer, B., H. Hofer, H. Oeschger, J. Schwander, and U. Siegenthaler. 'Atmospheric CO_2 concentration during the last glaciation'. Annals of Glaciology, **5**, 160 - 164, 1984.

Strain, B.R., and J.D. Cure (eds.). 'Direct effects of increasing carbon dioxide on vegetation'. U.S. Dept. of Energy, **DOE/ER-0238**, Washington D.C., 1985.

Trabalka, J.R. (ed.). 'Atmospheric Carbon Dioxide and the Global Carbon Cycle'. U.S. Dept. of Energy, **DOE/ER-0239**, Washington D.C., 1985.

Modelling biospheric control of carbon fluxes between atmosphere, ocean and land in view of climatic change

J.Goudriaan
Agricultural University
P.O. Box 430
6708PD Wageningen
The Netherlands

ABSTRACT. The amount of carbon in the atmosphere is less than in the terrestrial biosphere and much less than in the ocean. The level of atmospheric CO_2 is the result of a delicate balance in exchange fluxes with ocean and biosphere. Climatic change has the potential to alter this balance.

A model for the global carbon cycle was used for assessment of the importance of a number of biosphere and ocean characteristics that control atmospheric CO_2. This model included CO_2 feed back on the exchange fluxes, but no temperature feed back.

According to this model covering previously bare land with vegetation would lower atmospheric CO_2 by only 1.5 ppm for each 10^{12} m^2(100 Mha) of land additionally covered. At the present size of the terrestrial biosphere a global doubling of longevity of soil carbon would cause a decrease of 65 ppm of atmospheric CO_2.

The accumulated net biospheric release of carbon to the atmosphere during the last 120 years is estimated at about 30 Gt C. This figure is consistent with measured data of carbon isotopes and with data of atmospheric CO_2.

Marine photosynthesis is responsible for building up high phosphate and carbonate levels in the deep sea by precipitation of organic material. Marine photosynthesis is mostly limited by phosphate in the surface waters, and therefore an increase in total phosphate in the sea would lower atmospheric CO_2. Precipitated fraction and phosphate content of organic material are equally important.

The present difference in chemical composition between the Atlantic ocean on one hand and the Indian and Pacific oceans on the other can be explained by a much larger mixing rate between deep and surface water in the Atlantic ocean. Reducing this mixing rate in the Atlantic to that in the Pacific, would cause atmospheric CO_2 to drop by 25 ppm within a few hundred years time. Similarly, a sudden increase in mixing rate at the glacial-interglacial transitions could explain the strong and fast rise in atmospheric CO_2.

A. Berger et al. (eds.), Climate and Geo-Sciences, 481–499.

1. INTRODUCTION

1.1 Reservoirs and fluxes

Three major reservoirs of carbon, the atmosphere (700 Gt C), the oceans (39000 Gt C) and the terrestrial biosphere (2000 Gt C, including soil) constantly exchange carbon in the form of CO_2. The biosphere acts as a powerful driving force in the exchange of carbon between atmosphere, ocean and land. Modelling these fluxes is necessary for a quantitative assessment of the effects of the present and continuing injections of CO_2 into the atmosphere. It should also lead to a better understanding of the large and fast changes of atmospheric CO_2 at glacial-interglacial transitions(Oescher and Stauffer,1986).

Broecker(1984) pointed at the importance of nutrient levels in ocean water for marine photosynthesis, and for atmospheric CO_2, but he also noted that the rapidity of the glacial-interglacial transitions dismisses total nutrient contents as the prime factor. Ocean ventilation rate is more likely a candidate.

Terrestrial photosynthesis is also potentially able to sequester large amounts of carbon, but compared to the enormous reservoir of carbon in the oceans the role of the terrestrial biosphere is modest. In absence of other disturbances, an external injection of carbon will eventually be partitioned over atmosphere, ocean and terrestrial biosphere in the ratio's 0.11,0.71 and 0.18 respectively.

With industrialization, fossil carbon is more and more being injected into the global carbon cycle. At present the oceans absorb a fraction of only 0.4 of this man-made addition and not of 0.71, due to lack of time for redistribution.

Reclamation of land for agriculture and for other human utilization also causes release fluxes of CO_2. These fluxes are due to oxidation of soil carbon, and to decomposition of long-lived biomass such as wood. Simultaneous regrowth of biomass in other regions, and a global stimulus of growth by increased CO_2 largely restore the balance.

1.2 Model development

In this paper a model for the global carbon cycle(Goudriaan and Ketner,1984) will be used and further developed to investigate the controlling potential of some factors. The model performance for the present situation will be discussed and the importance of some of the model parameters for control of atmospheric CO_2 will be investigated. The model will be validated with published data sets such as depth profiles of chemical constituents in the oceans, time series of carbon isotopes and of atmospheric CO_2. The effect of climatic change will not be explicitly modelled, but some potential points of impact will be indicated.

2. MODELLING THE TERRESTRIAL BIOSPHERE

2.1 Disaggregation with respect to residence time

The average residence time of carbon in a simple one box model would be about 40 years. A more realistic model consists of a parallel-serial circuitry of components with various residence times. For humus and dinert carbon this type of structure can be derived from the work of Kortleven (1963) and of Olson (1963), who both showed that the response of humus level to litter input is of a first order character. Schlesinger (1986) reviewed several data sources, leading to support of first order dynamics of humus. He mentioned considerable losses in soil carbon upon reclamation of virgin land (from an equilibrium of 20 kg C m^{-2} to a new equilibrium of 15 kg C m^{-2}, to be reached after several decades).

In simulation models such losses are reproduced, by virtue of a shorter residence time of humus in agricultural soil (20 yr) than in grassland or in forest (50 yr). Resistant carbon (residence time 500 years) which amounts to about 10 kg C m^{-2}, is included in total soil carbon, and is much less affected by these land use changes.

2.2 Geographical disaggregation

The potential distribution of ecosystems on earth is primarily governed by climate. The Holdridge classification scheme (Holdridge,1967) uses climate variables to map vegetation, and can be exploited for an analysis of the effects of climatic change (Warrick et al.,1986a). Amounts of carbon stored in soils can be mapped in the Holdridge diagram (Fig 6.2 in Houghton et al.,1985) which enables a direct estimation of potential carbon sequestering in soils upon a change in climate.

A much higher degree of complexity of disaggregation was realized in the description by Matthews(1983) who compiled a high resolution data base of the geographical distribution of ecosystems. Remote sensing data have contributed considerably to the mapping of the distribution of green leaf area on earth (Tucker et al.,1986).

In the model of Goudriaan and Ketner(1984) only 6 ecosystems were distinguished, to represent the major features of the geographical distribution of biotic terrestrial carbon.

2.3 Human disturbance of ecosystems

Human activities such as agriculture, forest logging, pollution and urbanization have strongly altered vegetation and will continue to do so. They are locally clearly visible and tend to mask the physiological and climatic effects of rising atmospheric CO_2 on ecosystems.

Forest logging can be done with two purposes that must be clearly distinguished. One possibility is that the land is going to be permanently used as agricultural land, the other one is just removal of wood, perhaps with the intention of a temporary agricultural utilization. Both types of disruption can be modelled by a transition matrix that represents the annual area to be transferred from one ecosystem to another.

If no change in land destination occurs, the annually treated land area occurs on the diagonal of this transition matrix. Its effect was here modelled as a sudden release of most of the above ground carbon to the atmosphere (burning). A small fraction of the biomass, however, becomes long-lasting charcoal, and is so excluded from atmospheric circulation for a long time. This effect reflects the twofold consequence of burning: it not only immediately releases CO_2 into the atmosphere, but it also fixes a small fraction of the carbon of the material burnt into the highly inert charcoal pool (Seiler and Crutzen, 1980). Therefore repeated burning of forests, agricultural land and savanna's increases the carbon content of soils, at the expense of the atmospheric and oceanic reservoirs.

When the land use destination is changed as well, the contents of the soil carbon reservoirs are transferred to the corresponding reservoirs of the new ecosystem type, usually agricultural land. The residence time of humus in agricultural land is shorter, resulting in a considerable loss of soil carbon during the next decades. For instance, in tropical forests the mean areal density of carbon is 140 ton C ha^{-1} in contrast with only 50 ton C ha^{-1} in agricultural land. A rate of transfer of 12 Mha yr^{-1} therefore means that about one Gt C will be released in total for each year that this deforestation occurs.

However, this release is not immediate but distributed in time. With a residence time of humus in agricultural soil at 20 years, the release rate is overestimated if immediate release is assumed. With a dynamic simulation model such a delay can be easily taken into account.

2.4 Terrestrial photosynthesis as affected by CO_2

The weight fraction of carbon in dry weight of plant material varies only between 40 and 50%, so that there is a close connection between CO_2 assimilation and dry weight gain. Plant physiological and agricultural research has shown the importance of environmental factors such as radiation, temperature, water and nutrient supply for primary production (De Wit et al.,1979; Johnson,1981). Aereal CO_2 itself has been proven to be important as well (Lemon,1984; Warrick et al.,1986b).

In general, the normal ambient CO_2 partial pressure of about 300-330 μbar is suboptimal (Strain and Cure,1985) for C_3-plants which form 95% of the biomass. Stimulation of seasonal dry weight gain by increasing ambient CO_2 continues up to about 1000 μbar partial pressure. Over a large range of CO_2(200- 1000 μbar) the response of dry weight gain to CO_2 is logarithmic:

$$NPP = NPP_o \, (1 + \beta \log(CO_2/CO_{2,o})) \qquad (1)$$

where : NPP net annual primary production (g m^{-2} yr^{-1})
NPP_o NPP in the reference situation at 300 μbar.

The value of the response factor β is about 0.7 under good growth conditions otherwise(Goudriaan et al.,1985), but declines with increasing nutrient shortage (Goudriaan and De Ruiter,1983). Under water shortage (Gifford,1979; Rosenberg,1981) the growth stimulating effect of atmospheric CO_2 is not reduced, but may even be enhanced. This interaction with water shortage is caused by partial stomatal closure under increased CO_2. In this situation C_4-plants benefit from increased CO_2 just as much as C_3-plants.

Ideally, the value of β should be separately determined for each type of ecosystem, but up to now in modelling the global carbon cycle a single value is used, usually between 0.2 and 0.5 . Goudriaan and Ketner(1984) whose model will be used here, adopted 0.5. This relatively high value was chosen to allow for eutrophication as well.

2.5 Terrestrial photosynthesis as affected by biomass

The logarithm of the growth rate of a free standing individual plant is often linearly related to the logarithm of plant size itself: an allometric relationship. This relationship has been used in a number of models as reviewed by Bolin(1986). However, this relationship cannot be used on the scale of a field, and even less on that of an ecosystem,. Because of competition between plants, the effect of biomass on growth should then rather be described in terms of fraction of radiation absorbed by green leaf area (Milthorpe and Moorby,1979). Due to mutual shading of leaves, the fraction of absorbed radiation is soon saturated with increasing leaf area. Above a leaf area index (LAI) of 3 m^2(leaf area) m^{-2}(land area), which corresponds with a leaf dry weight of 150 g m^{-2}(land area) no further increase of NPP can be expected with increasing leaf biomass. Positive feedback of biomass on NPP is further weakened by increased partitioning of growth to non-photosynthesizing organs such as stems and branches, after the soil has been sufficiently covered with green leaf area. Therefore models that use an allometric relationship between NPP and biomass on a global scale tend to overestimate the positive feedback with biomass. Only in case of clustering of vegetation in bushes, as often occurs in dryland vegetation, does the photosynthesis-biomass relationship continue much longer, but it does so at the expense of growth rate per unit land area. In this situation bush coverage and green leaf coverage within the bushes must be separately modelled (Goudriaan,1986). For a global scale model however such a refinement is not warranted.

Expansion of vegetation onto previously bare land is something else and should be explicitly modelled as area growth of ecosystem coverage. It is clear that such processes can be important when the climate changes (see 5.2.1).

Thanks to pioneer vegetation, carbon fixation is not much reduced after removal of biomass. As far as modelling of carbon is concerned, it does not matter what the type of vegetation is that takes up the carbon from the atmosphere. It will take decades however, to restore carbon pools with long residence times, such as stem wood. The time course of net ecosystem production (NEP) after deforestation typically shows an overshoot. After some decades the losses of slowly decaying dry matter increase in reservoirs with long residence times (stems in particular) and they finally consume almost all production.

Both the cyclic (diagonal component of the transition matrix) and permanent (triangular component of transition matrix) forms of land use change grow at a global scale. Thanks to regrowth the cyclic components of deforestation and of other forms of biomass removal from ecosystems do not have much influence on atmospheric CO_2 (see 4.3)

3. MODELLING THE OCEAN

3.1 Compartmentalization

Reviews of some of the existing models for carbon exchange between ocean and atmosphere can be found in Björkström(1986), Sarmiento(1986), Broecker and Peng(1982),Baes et al.(1985). As Björkström(1986) pointed out, some of these models have presently been compartmentalized to the extent that they need as many as 600 parameters. The work presented here aims at compromising the needs for high resolution on one hand and for simplicity on the other. In the model presented by Goudriaan and Ketner(1984) the ocean was compartmentalized into 10 strata, of which the top one was further split into three zones : a high latitude zone (after Viecelli et al.,1981) and a low latitude zone consisting of two strata. Diffusion was permitted between all strata (K=4000 m^2 yr^{-1}), and the Gordon circulation was simulated by a mass flow sinking from the high latitude surface zone. This mass flow was equally distributed over the nine deep strata, where it forced an upwelling into the low latitude surface zone. The circulation was closed by continuous mass flow from the low to the high latitude surface zone.

In modification of the Goudriaan and Ketner(1984) model, the surface zones were considered to be internally well-mixed (K infinite), but a resistance of 0.25 yr m^{-1} was modelled for exchange with the atmosphere. The size of this resistance was chosen by calibration with the rate of decline of the ^{14}C-peak after the nuclear tests in the early sixties (Fig 6). It means that an annual molar flux of CO_2 can pass through the surface, equivalent to the total carbon content of 4 meters of sea water. Because of the small size of the chemical capacitance, however, an approximately ten times(Revelle factor) larger depth is annually involved in chemical equilibration.

In the 1984 model marine photosynthesis was added as a fixed driving force. The decomposition of precipitated organic material could account for observed features such as high partial pressure of CO_2 in deep sea water, followed by CO_2 release in the upwelling zone and net uptake in the high latitude downwelling zone.

Since this model considered the ocean as one stratified unit, it could not account for the large observed differences between the Atlantic ocean on one hand and the Pacific and Indian on the other. In glacial periods these oceans may have been more alike (Broecker and Peng,1982). To investigate the effect of realization of this possibility by simulation, it was necessary to separate the two ocean systems in the model, and characterize them with their own parameters.

In the modified model the two ocean systems differed mainly by their rate of internal turning over of deep and surface water, which is much higher in the Atlantic than in the Pacific&Indian ($2\ 10^{15}$ m^3 yr^{-1} versus $0.2\ 10^{15}$ m^3 yr^{-1}) (Baes et al.,1985).

The two ocean systems were connected, via the atmosphere for carbon dioxide and for oxygen, and via the Antarctic Sea for phosphate, oxygen

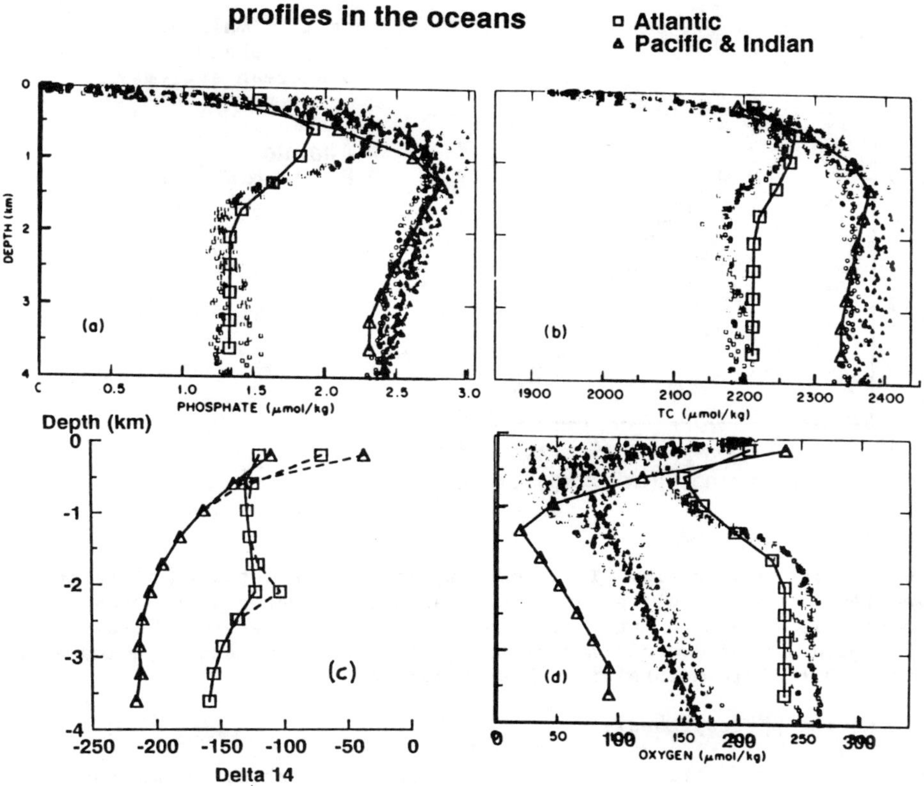

Figure 1 Measured and simulated (large connected symbols) depth profiles of phosphate(a), of total carbon (b) and of oxygen(d) in the Atlantic and in the Pacific and Indian. These simulation results are presented as an overlay on Fig 5.6 in Baes et al.(1985). For $\delta^{14}C$ (c) simulated profiles are given before the nuclear tests and for the year 1972 (dashed lines)

and total carbon. The exchange via the atmosphere is obvious, but the one via the Antarctic Sea needs some explanation. A strong circumpolar flow of surface water around Antarctica was modelled, connecting both oceanic systems. Some exchange of deep sea water may occur as well, but was not included in the model. At the 45.2 °/oo isopycnal a strong gradient is observed of many characteristics (GEOSECS data in Broecker and Peng,1982), which means that there is little exchange between the deep waters of the Antarctic and the oceans.

The downward water flow from the Antarctic Sea towards the deep sea of both ocean systems stems from a common source of cold surface water

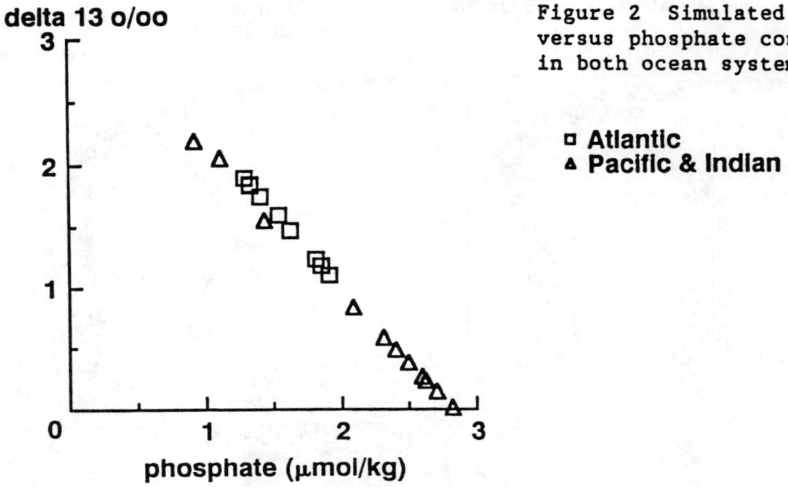

Figure 2 Simulated δ^{13} (°/oo) versus phosphate concentration in both ocean systems.

in the circumpolar flow. This feature, in combination with different ventilation rate could generate the large differences in chemical compositions in both ocean systems, as are also observed in reality. The precipitation of organic material, originating from marine photosynthesis was essential in maintaining this difference.

3.2 Marine photosynthesis

Since nutrients strongly affect marine photosynthesis, it was necessary to include nutrient transport. Phosphate is considered as the main limiting factor. The total amount of ocean phosphate was initialized as an input parameter (92 Gt P, on basis of Baes et al,1985), but the distribution of phosphate was established by simulation. The ratio between downward precipitation of organic material, fixed in marine photosynthesis, and phosphate concentration of the upper layer was given as a model parameter (278 g C g P^{-1} m yr^{-1}). This parameterization leads to a downward flux of approximately 3 Gt C yr^{-1} in total, sinking through the 400 m depth level. The phosphorus content of organic material, and its oxygen demand upon decomposition was taken according to the Redfield equation (Baes et al.,1985;pp 88), leading to 0.024 g P g^{-1}C and 3.47 g O g^{-1}C respectively.

Surface layers received phosphate from the deeper layers by upwelling and by diffusion. The precipitating organic material was allowed to decompose between 400 and 1500 m depth, thereby raising the phosphate and carbonate levels and consuming oxygen (Fig 1). The profiles generated in this way are similar to observed profiles (Baes et al,1985).

This biotic pump is independent of atmospheric CO_2, since carbon is not a limiting factor for marine photosynthesis. Assuming that the ocean profiles have been in equilibrium with this process before atmospheric

CO_2 started to rise, marine photosynthesis plays no role in enhancing the rate of carbon uptake by the oceans. However, if the rate of marine photosynthesis is increased by eutrophication or for other reasons, its role can be significant.

3.3 Carbon isotopes

Exchange of isotopes within the sea was modelled analogous to other passive admixtures. Between sea and air a second flux component was modelled, driven by the difference between isotope ratio's in water and air, and by their discrimination upon solution. This flux component faces the total amount of carbon for isotopic equilibration, and is therefore about ten times(Revelle factor) slower than chemical equilibration. As a result of this method the prenuclear equilibrium of $\delta^{14}C$ in the surface ocean layers was simulated to be about -100 °/oo (Fig 1c). The long residence time of water in the deep Pacific shows up in low values of $\delta^{14}C$. The simplified model structure for massflow generated a tongue of downwelled water in the Atlantic at a depth of about 2000 m.

The assumed discrimination of ^{13}C in photosynthesis (Mook,1986) caused a correlation between $\delta^{13}C$ and phosphate (Fig 2), practically the same as observed in reality(See Fig 6-12 in Broecker and Peng,1982).

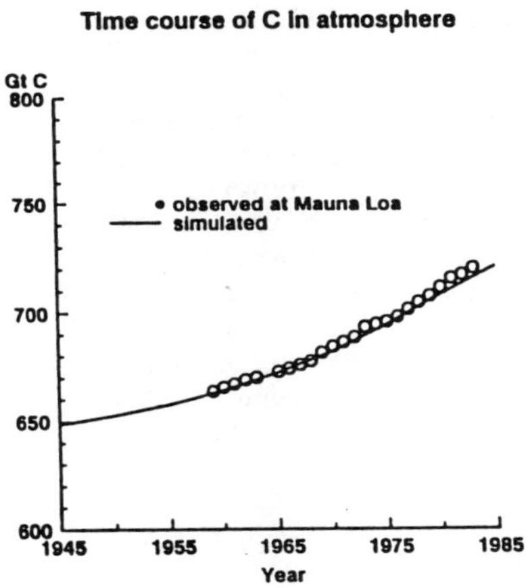

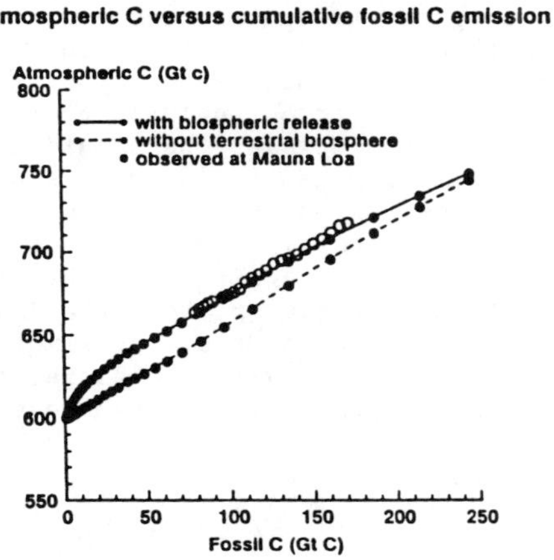

Figure 3. Simulated and measured time courses of atmospheric CO_2.

Figure 4. Simulated and measured atmospheric CO_2 versus cumulative injection of fossil carbon, every five years, the last one in 1995. The simulated results are given with and without biospheric exchange.

4 MODEL BEHAVIOUR FOR THE MODERN TIME

4.1 Atmospheric CO_2 concentration and airborne fraction

Observed and simulated atmospheric CO_2 are given in Fig 3. The fossil emission to the atmosphere was based on the data of Rotty and Masters (1985), with extrapolation to the near future. The simulation was started in 1780 at 285 μbar (599.6 Gt C).

Atmospheric CO_2 is highly correlated with cumulative fossil fuel emission (Fig 4). The slope of this graph is the so-called marginal airborne fraction α_m (Bolin,1986). Over the period 1960-1980 it has an observed value of 0.576, whereas values of 0.564 and 0.634 were simulated with and without terrestrial biosphere respectively (Fig 4). The difference was caused by a net positive biospheric uptake in the last decades.

To be able to compare models with and without biosphere, Bolin introduced the term "total airborne fraction" which stands for atmospheric increase divided by the sum of biospheric and fossil fuel emissions. Of course, net biospheric emission should be used in this calculation and not the gross emission. The difference can be considerable, if one considers that the simulated gross biospheric emission between 1860 and 1980 was 122 Gt C, in contrast to 30 Gt C net emission. If the atmospheric increase over this period is divided by the sum of gross biospheric and fossil fuel emission, α can be found to be less than 0.3 . Such a low value of α seems to indicate an unrealistically high capacity of the biosphere to absorb carbon, which was a reason for Bolin(1986) to criticize the model of Goudriaan and Ketner(1984). However, he should have used net emission, not gross. Gross emission generates ample ecological space for regrowth.

If net biospheric emission is used, α becomes a characteristic of ocean uptake only. Between 1960 and 1980 the net biospheric emission was simulated to be negative, so that the simulated total airborne fraction was higher, not lower, than the "marginal" one and became 0.595 . This value is not the same as was simulated without biosphere, as one might have thought. However, the different time path of atmospheric CO_2 in both simulation runs affected ocean uptake.

4.2 Carbon isotopes

Due to an assumed $\delta^{13}C$ of -25°/oo for fossil carbon and a fractionation of -19°/oo in photosynthesis, the simulated atmospheric $\delta^{13}C$ decreased with 1.47 °/oo between 1860 and 1980, with a continuing rate of decrease in 1980 of 0.03 per year. These figures practically coincide with the data recorded by Freyer(1986) (Fig 5a), but they show a faster rate of decrease than Stuiver(1986) has found. For atmospheric ^{14}C a similar, but much stronger, decrease (the Suess effect) was simulated, from -23.5 °/oo in 1860 down to -46 °/oo in 1950. These model simulations fall in line with the data as given in Broecker and Peng (1982) and Bolin(1986) (Fig 5b). The modelled decline of atmospheric radiocarbon after the nuclear tests is presented in Fig 6.

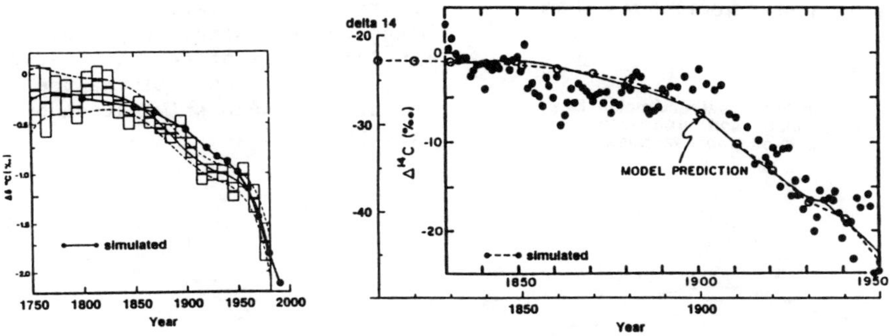

Figure 5. Simulated and measured time courses of carbon isotope ratio's. The simulation results are presented as overlays.
a) For $\Delta^{13}C$ on the data given by Freyer (1986) in his Figure 7.5.
b) For $\Delta^{14}C$ on the data given by Broecker and Peng (1982) in their Figure 10-23. Their "MODEL PREDICTION" and my "simulated" results practically coincide.

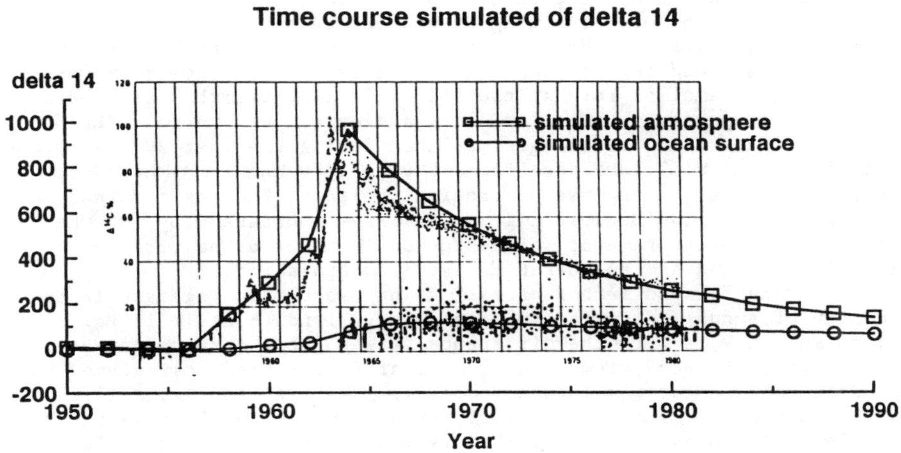

Figure 6. Simulated and observed time courses of $\Delta^{14}C$ in the atmosphere and in the sea surface water at low latitudes, for the period after the nuclear tests. The simulation results are presented as an overlay on the data given by Bolin (1986) in his Figure 3.6

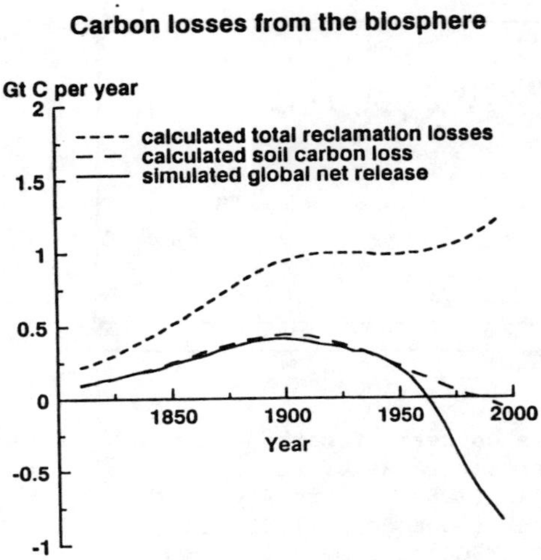

Figure 7. Calculated losses are based on statistics only. The simulated rate of net biospheric release of carbon includes the effect of growth stimulation by atmospheric CO_2.

4.3 Biospheric release

A detailed review of the subject was given by Houghton(1986). It highlights the uncertainties of current estimates, both of current carbon contents and of the timing of the releases.

In modification to the model given by Goudriaan and Ketner(1984), a faster rate of land reclamation around the turn of this century was assumed, and a smaller rate at present. This larger emphasis on early reclamation represents the colonial expansion in that period. The net effect of these land use changes was that during the last 200 years the area of agricultural land rose from 1000 to 1800 Mha (100 Mha=10^{12} mPw2), mainly at the expense of tropical forests. Temperate forests were also reduced (by 100 Mha), but grassland area increased (by 140 Mha). The assumed present rate of deforestation is about 6 Mha yr^{-1} to agricultural land and another 6 Mha yr^{-1} to grassland.

A straighforward method of estimating biospheric releases is to use the rate of annual land transfer from virgin land to agricultural soil, and to multiply it with the difference between soil carbon content typical for both ecosystems. The result of this method, calculated for the data used in our simulation model, is given in Fig 7 as "calculated soil carbon loss". The high rate of reclamation around 1900 is clearly visible. The cross-over around 1980 is caused by increasing reclamation from tropical forest land, which has a low soil carbon content.

Similarly the rate of removal of above ground biomass can be calculated by multiplying carbon areal density with the areal rate of permanent deforestation. Added to "calculated soil carbon loss" this rate gives the "calculated total reclamation losses" in Fig 7. The strong increase in recent times is caused by much faster deforestation in the tropics. In this procedure the estimates of Brown and Lugo (1984) were used, which are lower (about half) than assumed before. Naturally the estimates for biomass releases are smaller as well.

However, for a global estimate of net carbon release from the terrestrial biosphere to the atmosphere these type of calculations are unsufficient. First, a considerable portion of deforested land becomes grassland, with a much higher soil carbon content than arable land. Second, the soil carbon is not released immediately, but it is gradually decomposed over a period of decades. Third, partial humification and charcoal formation reduce the losses by 10 to 20% . Fourth, elsewhere reclamation of dry land occurs by irrigation. In these places soil carbon is increased, not decreased. Fifth, and most important, a global stimulus of NPP may be assumed due to rising atmospheric CO_2(see 2.4). This slowly changing global background does not produce dramatic sights, but it still has large consequences. Without it, the terrestrial biosphere (including soil carbon) would now contain about 50 Gt C less. In other words, the photosynthetic stimulus has delayed the rising curve of atmospheric CO_2 by about ten years time.

It is central to this model result that the longevity of the various biospheric components is not decreased. As long as longevity is the same, carbon storage will be proportional to increased NPP. This assumption is not unrealistic, and it may even be to conservative. The observed lowered nitrogen contents connected with increased CO_2 may even lead to larger longevities.

Together, these factors explain the difference between "calculated total reclamation losses" and "simulated global net release" in Fig 7. The difference is the more striking as the same rates of land transfer were used.

Beside permanent transfer of land between ecosystems, biomass is also temporarily removed by litter burning and by deforestation for shifting cultivation. An assumed rate of annual burning of biomass (mainly litter) on grassland and agricultural land may seem large (800 Mha yr^{-1}) as it causes a release of about 5 Gt C per year. However, when this litter burning was set at zero, the simulated atmospheric CO_2 concentration in 1980 was very little affected (6 ppm lower). This small effect on net biospheric release is not amazing, considering that the litter would have decomposed anyway during the year to come. Whether or not shifting cultivation is considered did not have a large effect either: a simulated concentration that is only 1 ppm lower in 1980.

Between 1860 and 1980 the accumulated loss of biospheric carbon was simulated to be no more than 30 Gt C. This figure is much lower than others have found on basis of isotopic data. However, the simulation results presented here give no reason to assume larger releases(Fig 5).

Several reasons may exist for the deviation between my rather low estimate and earlier larger ones. First, differences of differences ($\Delta^{13}C$ versus $\Delta^{14}C$) are inaccurate. Second, in my estimate of the resistance of the atmosphere-ocean interface, based on the rate of decline of $\Delta^{14}C$, I allowed for the simultaneous dilution in the biosphere. My resistance estimate was therefore larger than without allowance for this dilution. A larger resistance leads to a faster decline of $\Delta^{13}C$, and requires a smaller release of biospheric carbon for compatibility with observed rates of decline of $\Delta^{13}C$. Several other small model differences may also have contributed.

5 POTENTIAL GLACIAL-INTERGLACIAL PARAMETER CHANGES

5.1 Redistribution of CO_2 released into the atmosphere

Total terrestrial biotic carbon can be estimated at about 2000 Gt C for a prehuman situation. Since the cascade of biospheric components driven by terrestrial NPP was described by linear equations, the equilibrium level of their contents will follow the logarithmic response of Eqn 1 as well. The carbon content of the oceans is about 39000 Gt C, responding with a Revelle factor of about 10 to atmospheric CO_2, which reduces their effective content to 3900 Gt C. Starting with an atmosphere with 600 Gt C, a quick calculation shows that CO_2 emission into the atmosphere will eventually be partitioned as 11% to the atmosphere, 18% (for β=0.5) to the terrestrial biota and 71% to the ocean.

Therefore the ocean must be included in even a simple model for the effect of biospheric changes, but on the other hand the biosphere can be omitted if just first order effects of ocean parameter changes are studied. This consideration was followed below.

5.2 Terrestrial biosphere

5.2.1 Change in vegetated area

A dramatic increase of the vegetated area by 1000 Mha ($10. 10^{12}$ m^2) from bare land to grassland and forest did not have a large impact on atmospheric CO_2. Such an increase would cause a carbon sequestering of some 230 Gt C (23 kg C m^{-2}). Growth of stemwood and accumulation of soil carbon being a rather slow process, atmospheric CO_2 dropped never more than 25 ppm below the starting level of 285 ppm. It reached an equilibrium after one thousand years, which was only 15 ppm lower than the initial level. This result means that about 200 Gt of the required 230 Gt C must have been released by the ocean in response to the decreased partial pressure of CO_2 in the atmosphere.

5.2.2 Change in longevity of soil carbon

Longevity of soil carbon (humus and charcoal) is of large importance. When its value was instantaneously doubled, atmospheric CO_2 dropped from 285 to around 220 ppm within a few decades, and remained at that level. About 1000 Gt C was sequestered in the soil, of which 840 Gt C was extracted from the sea, and 140 Gt C from the atmosphere.

The wetness of climate has a large effect on the amount of soil carbon. Doubling the precipitation/evaporation ratio also practically doubles soil carbon (Fig 6.2 in Houghton et al,1985). A globally mean wetter climate could therefore induce a large drop in atmospheric CO_2.

5.3 Ocean

5.3.1 Changes in marine photosynthesis

Two model characteristics dominate marine photosynthesis: the phosphate content of organic material (Redfield ratio) and the effect of phosphate concentration in the surface waters on precipitation of organic material. When this effect was tripled, precipitation instantaneously tripled as well. However, phosphate was then withdrawn at a much faster rate, and soon the phosphate in the water became more depleted. Because of this feedback mechanism the effect of the precipation-phosphate ratio was much less than proportional. A similar mechanism operated when the Redfield ratio was reduced: decreased withdrawal of phosphate caused some accumulation of phosphate in the surface layers, thereby stimulating photosynthesis and carbon precipitation to the deep sea. Only when the Redfield ratio and precipitation-phosphate ratio were simultaneously altered, could a lasting strong effect on atmospheric CO_2 be observed (Table 1).

Of course, a direct change in the total amount of phosphate in the ocean had also a large effect: about 1 ppm decline of atmospheric CO_2 per percent of total ocean phosphate.

Table 1 Simulated effect of parameters affecting marine photosynthesis on atmospheric CO_2(control 285 ppm)

		Precipitation/phosphate ratio (g C g P^{-1} m yr^{-1})	
		278	834
Redfield ratio	0.024	285 ppm	236 ppm
Redfield ratio	0.01	247 ppm	160 ppm
Total phosphate	92 Gt P	285 ppm	
Total phosphate	101 Gt P	277 ppm	

5.3.2 Changes in ocean ventilation

The model behaviour was remarkably insensitive to the size of the Circumpolar flow, connecting both ocean systems. If it was decreased from 2 to 0.1 10^{15} m^3 yr^{-1}, the difference between the ocean systems was largely maintained, and atmospheric CO_2 dropped from 285 to 278 ppm.

A much stronger effect resulted from changes in the massflow ventilation within each ocean system. If the Gordon flow in the Atlantic was reduced from 2. to 1. 10^{15} m^3 yr^{-1}, atmospheric CO_2 dropped from 285 to 276 ppm and a further reduction to 0.2 10^{15} m^3 yr^{-1} resulted in an atmospheric CO_2 of 260 ppm.

This effect can be explained by decreased upwelling, a lowering of the phosphate concentration in the surface layer of the Atlantic and increase of both phosphate and total carbon in the deep Atlantic. In the Pacific very little changed.

Similarly, if the ventilation in the Pacific was increased from 0.2 10^{15} m^3 yr^{-1} in the control run to 2. 10^{15} m^3 yr^{-1}, the atmospheric concentration was increased from 285 up to 344 ppm in 500 years time.

The relaxation time of these changes was about 200 years.

Changing the depth at which the downwelling water of the Atlantic enters the deep sea from 1500 to 3500 m reduced the residence time of the deep water (as signified by ^{14}C) but almost nothing otherwise. This lack of importance of the reentering depth is connected with the assumed complete decomposition of precipitated organic material in the upper 1500 m, below which almost no gradient remained.

6. CONCLUSION AND DISCUSSION

During the glacial periods atmospheric CO_2 was much lower than during interglacials, whereas the vegetated area was presumably much smaller. Although soil carbon may have been higher in the tropics than at present, it seems unlikely that it could have compensated for the loss of terrestrial carbon in the high latitudes. Climatic wetness stimulates accumulation of soil carbon, but during the glacials climate was probably dryer, not wetter. From a physiological point of view the dryness was accentuated by low atmospheric CO_2 which stimulates stomatal opening and transpiration. Dry and CO_2 poor conditions definitely favoured C_4-species, and have probably stimulated their evolution. In view of these considerations it is most likely that the terrestrial carbon pool was much smaller during the glacials than in between. In addition to going the wrong way, it is hard to see how the terrestrial biosphere could have caused the rapidity of the changes of atmospheric CO_2.

Marine photosynthesis itself can probably not be altered much in its response to the limiting nutrient phosphate. However, it is the precipitated fraction that matters for carbon accumulation in the deep sea. Also the mean depth at which it decomposes is important. These factors are probably biology-linked and controlled by species composition. If release of phosphate upon decomposition of precipitating organic material could occur faster than of carbon, the remaining material would be depleted in phosphate. Such a shift would cause a further drawing down of atmospheric CO_2.

The main actor on atmospheric CO_2 must have been the ocean circulation. Changes in ocean ventilation have a very large and quick effect. The question remains what has initiated them to change their pattern or magnitude. The temperature feedback on solubility of CO_2 in sea surface water was estimated to amplify an externally caused change in atmospheric CO_2 by a factor 1.5 .

According to the ocean model presented here the differences between the major ocean systems can be explained without deep sea currents from the Atlantic to the Pacific. Such currents further strengthen the nutrient-enriched character of the deep Pacific as compared to the waters of the Atlantic (Broecker and Peng,1982).

Model parameters are always uncertain. For instance, the assumed value of the biotic growth factor β of 0.5 may be a little too high, which is also suggested by Fig 4. The eutrophication effect, which was included in the value of β, should be treated separately.

The simulation results of carbon isotopes give no reason to believe that the net accumulated carbon release from the biosphere during the last century has been larger than about 30 Gt C. Both the isotopic data and the CO_2 data themselves are consistent with this limited net release of biotic carbon. Even a slight imbalance in the growth of terrestrial ecosystems on a global scale is sufficient to absorb the carbon released by deforestation (Lugo and Brown,1986). Such an imbalance may be due to increasing atmospheric CO_2, as assumed here, but other environmental factors may be involved as well.

7. REFERENCES

Baes, C.F., A. Björkström and P.J. Mulholland, 1985. 'Uptake of carbon dioxide by the oceans'. In Trabalka (Ed)

Björkström, A., 1986. 'One-dimensional and two-dimensional ocean models for predicting the distribution of CO_2 between the ocean and the atmosphere'. In : Bolin et al. (Eds)

Bolin, B, B.R. Döös, J. Jäger and R.A. Warrick (Eds), 1986. The greenhouse effect, Climatic change, and ecosystems. SCOPE 29. Wiley &Sons, Chichester.

Bolin, B.,1986. 'How much CO_2 will remain in the atmosphere?'. In : Bolin et al.(Eds)

Broecker, W.S. and T.H. Peng, 1982. Tracers in the sea. Eldigio Press, Palisades, New York.

Broecker, W.S., 1984. 'Carbon dioxide circulation through ocean and atmosphere'. Nature 308 :602

Brown, S. and A.E. Lugo, 1984. 'Biomass of tropical forests: a new estimate based on forest volumes'. Science 223 : 1290-1293

Day W. and R.K. Atkin (Eds), Wheat growth and modeling. NATO ASI Series, Series A: Life Sciences Vol. 86

Freyer, H.D., 1986. 'Interpretation of the northern hemisphere record of $^{13}C/^{12}C$ trends of atmospheric CO_2 in tree rings.' In: Trabalka and Reichle (Eds).

Gifford, R.M., 1979. 'Growth and yield of CO_2-enriched wheat under water-limited conditions'. Australian J. of Plant Phys. 6: 367-378

Goudriaan, J. and H.E. de Ruiter, 1983. 'Plant growth in response to CO_2 enrichment, at two levels of nitrogen and phosphorus supply'. Netherlands Journal of agricultural Science 31 : 157-169

Goudriaan, J. and P. Ketner, 1984. 'A simulation study for the global carbon cycle, including man's impact on the biosphere'. Climatic Change 6: 167:192

Goudriaan, J., H.H. van Laar, H. van Keulen and W. Louwerse, 1985. 'Photosynthesis, CO_2 and plant production.' In: Day and Atkin (Eds)

Goudriaan, J., 1986. 'Simulation of ecosystem response to rising CO_2, with special attention to interfacing with the atmosphere.' In: Rosenzweig and Dickinson (Eds)

Holdridge, R., 1967. Life Zone Ecology. Tropical Science Center. San Jose, Ca. USA.

Houghton, R.A., W.H. Schlesinger, S. Brown, J.F. Richards, 1985. 'Carbon dioxide exchange between the atmosphere and terrestrial ecosystems.' In : Trabalka (Ed)

Houghton, R.A., 1986. 'Estimating changes in the carbon content of terrestrial ecosystems from historical data'. In: Trabalka and Reichle (Eds).

Johnson, C.B., 1981. Physiological processes limiting plant productivity. Butterworths. London.

Kohlmaier,G.H., H. Bröhl, E.O. Siré, M. Plöchl and R. Revelle, 1987. 'Modelling stimulation of plants and ecosystem response to present levels of excess atmospheric CO_2.' Tellus 39B: 155-170

Kortleven, J., 1963. Kwantitatieve aspecten van humusopbouw en humusafbraak. Verslagen Landbouwkundige Onderzoekingen 69-1. Pudoc. Wageningen, Netherlands

Lemon, E.R. (Ed), 1984. CO_2 and plants. Westview Press, Colorado, USA

Lugo, A.E. and S. Brown, 1986. 'Steady state terrestrial ecosystems and the global carbon cycle'. Vegetatio 68: 83-90

Matthews, E., 1983. 'Global vegetation and land use: new high resolution data bases for climate studies.' J. of Climate and appl. Met. 22: 474-487.

Milthorpe, F.L. and J. Moorby, 1979. An introduction to crop physiology. Cambridge University Press. Cambridge. UK.

Monteith, J.L., 1981. 'Does light limit crop production?.' In: Johnson(Ed)

Mook, W., 1986. '^{13}C in atmospheric CO_2'. Netherlands Journal of Sea Research 20: 211-223

Oeschger,H. and B. Stauffer,1986. 'Review of the History of Atmospheric CO_2 recorded in Ice Cores.' In: Trabalka and Reichle(Eds).

Olson, J.S., 1963. 'Energy storage and the balance of producers and decomposers in ecological systems.' Ecology 44 : 322-331.

Rosenberg, N.J., 1981. 'The increasing CO_2 concentration in the atmosphere and its implication for agricultural productivity.' Clim. Change 3 : 265-279

Rosenzweig C. and R. Dickinson (Eds), Climate Vegetation Interactions. NASA Conference Publication 2440. Greenbelt Md, USA

Rotty, R.M. and C.D. Masters, 1985. 'Carbon dioxide from fossil fuel combustion: trends, resources, and technological implications.' In: Trabalka(Ed)

Sarmiento, J.L., 1986. Three-dimensional ocean models for predicting the distribution of CO_2 between the ocean and atmosphere. In : Bolin et al.(Eds)

Schlesinger, W.H., 1986. 'Changes in soil carbon storage and associated properties with disturbance and recovery.' In : Trabalka and Reichle (Eds)

Seiler, W. and P.J. Crutzen, 1980. ' Estimates of gross and net fluxes of carbon between the biosphere and the atmosphere from biomass burning'. Climatic Change 2:207-247

Sneep, J. and A.J.T. Hendriksen, 1979. Plant breeding perspectives. Pudoc, Wageningen, Netherlands.

Strain, B.R. and J.D. Cure, 1985. Direct effects of increasing carbon dioxide on vegetation. DOE/ER-0238, US Dept. of Energy, Washington DC USA

Stuiver, M.,1986. 'Ancient carbon cycle changes derived from tree-ring ^{13}C and ^{14}C.' In Trabalka and Reichle (Eds).

Trabalka, J.R., 1985. Atmospheric carbon dioxide and the global carbon cycle. DOE/ER-0239, US Dept. of Energy, Washington DC USA

Trabalka J.R. and D.E. Reichle,1986. The Changing Carbon Cycle, a Global Analysis. Springer Verlag.

Tucker, C.J., I.Y. Fung, C.D. Keeling and R.H. Gammon, 1986. 'Relationship between atmospheric CO_2 variations and a satellite-derived vegetation index'. Nature 319 : 195-199

Viecelli, J.A., H.W. Ellsaesser and J.E. Burt, 1981. 'A carbon cycle model with latitude dependence'. Climatic Change 3 : 281-302

Warrick, R.A., H.H. Shugart, M.Ja. Antonovsky, with J.R. Tarrant and C.J. Tucker, 1986a. 'The effects of increased CO_2 and climatic change on terrestrial ecosystems'. In : Bolin et al. (Eds)

Warrick, R.A. and R.M. Gifford, with M.L. Parry, 1986b. ' CO_2, clmatic change and agriculture.' In : Bolin et al (Eds)

Wit, C.T. de, H.H. van Laar and H. van Keulen, 1979. 'Physiological potential of crop production.' In: Sneep and Hendriksen (Eds).

SECTION 4 – TECHNOLOGY FOR CLIMATE STUDIES

SUPERCOMPUTING AND MASSIVE PARALLELISM

BILL BUZBEE
Scientific Computing Division
National Center for Atmospheric Research
P. O. Box 3000
Boulder, CO 80307

ABSTRACT: The geosciences community is dealing with problems which will tax the capability of even the most powerful supercomputers that are available in state-of-the-art equipment. Achieving further performance increase will necessitate use of parallel processing. In this paper we use the Ware model of parallel processing to show that shared-memory supercomputers with 16 processors can provide about an order of magnitude performance provided the models are highly concurrent and engender low overheads from task management. We also briefly describe the Connection Machine which is a massively parallel array and outline our plans to explore its potential to provide even higher levels of performance relative to state-of-the-art equipment.

1. Introduction

Historically, the atmospheric sciences have made extensive use of high performance computers for modeling of weather, climate, oceans, and related phenomena. Further, and as noted by Holland and McWilliams,[1] the capability of a model is often paced, or limited, by the power of the computer that is available to execute it. That is, a more powerful computer makes possible more realistic models. For example, 15 years ago the CDC 7600 was used for climate simulation and one hour of CPU time was required per 24 hours of simulated time. Consequently, models were typically formulated in the perpetual-January mode (i.e., with constant astronomical, chemical, and ocean forcing appropriate for the month of January) and at most, a few months were simulated. With the availability at the National Center for Atmospheric Research (NCAR) of the CRAY-1 and its high performance in vector mode, less than two minutes of CPU time were required per 24 hours of simulated time. This permitted climate researchers to venture beyond the perpetual-January mode and investigate such areas as the interannual variability of time-averaged atmospheric states. Nevertheless, as much as 200 CPU hours could be required for a twenty-year simulation. Availability at NCAR of the CRAY X-MP/48 with a 256 million word Solid State Storage Device (SSD) has likewise extended the scientific capability of models. For example, scientists are using the X-MP to study potential

A. Berger et al. (eds.), Climate and Geo-Sciences, 503–511.

relations between variations in the earth's orbit and the recent ice ages of 18,000 years ago.[2] Each simulation requires 30 to 40 CPU hours on one processor of an X-MP. Raymond Roble[3] and coworkers are using the X-MP to study the response of the thermosphere to variations in the sun's energy output. Each simulation requires one hour or more of CPU time and at least one million words of memory. Eventually, they hope to gain a better understanding of radio signal propagation, satellite drag, and chemical transport in the upper atmosphere. Terry Clark[4] and coworkers are using the X-MP for cloud simulation. Their cumulus entrainment calculations use two million words of memory per variable and there are 20 to 30 variables. Typical simulations require on the order of 20 hours of CPU time. Robert Chervin,[5] and Albert Semtner,[6] are using the parallel processing capability of the CRAY X-MP/48 in their respective climate and ocean models. Although both of these models executed over 400 million floating point operations per second (Mflops), typical simulations require 100 or more wall-clock hours of computation. Thus, throughout the era of electronic computation, the atmospheric science community has consistently acquired the most powerful computers available and has exploited their highest performance levels.

2. Supercomputers Versus Massive Parallelism

The architectural design space for parallel processors is multi-dimensional. Two prominent classes of parallel architectures are shared-memory supercomputer systems (SMSC) and distributed-memory microprocessor systems (DMMP). Many state-of-the-art supercomputers are SMSC (see Table 1), wherein a few fast and expensive processors share a large memory. Massively parallel processors are often DMMP systems, wherein thousands of relatively slow and low cost processors each have local memory and are interconnected via some communication geometry. At present, SMSC and DMMP systems appear to offer the greatest potential for successful application of parallel processing to large simulations. Thus, in the remainder of this paper we will review Ware's model for estimating performance of a parallel processor, then we will estimate performance of supercomputers in 1990, and finally, discuss the potential of massively parallel systems.

3. The Ware Model of Parallel Performance

To estimate increased speed, Ware[7] developed a simple but effective model of performance of parallel systems. Throughout these discussions we use the notion of speedup (S_p) which we define as

$$S_p = \frac{\textit{execution time} \text{ for } \textit{best known serial implementation}}{\textit{execution time} \text{ for } \textit{parallel implementation using } p \textit{ processors}} \tag{1}$$

For a specific calculation, we define α to be the fraction of computational work in an application that can be done in parallel. Ware's model assumes that at any instant in time either all processors are operating or only one of them is operating, i.e., it assumes a two-state machine. If the execution time for serial execution is normalized to unity, then

$$S_p = \frac{1}{(1 - \alpha) + \dfrac{\alpha}{p}} \tag{2}$$

Note that the first term in the denominator is the execution time for that part of the calculation that cannot be parallel processed; the second term is the time for that part that can be parallel-processed. How does S_p vary with α? Differentiating S_p yields

$$\frac{\partial S_p}{\partial \alpha}\Big|_{\alpha = 1} = p^2 - p \tag{3}$$

Thus the rate of change of S_p near $\alpha = 1$ is quadratic in p. This behavior is shown in Figure 1 for two choices of p as a function of α. The conclusion is immediate--high performance necessitates a high degree of concurrency. Thus development of highly concurrent algorithms is a critical path to success in parallel processing.

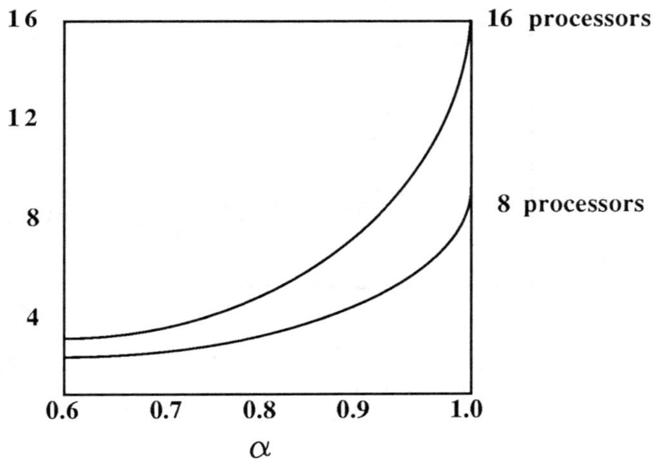

Figure 1. Speedup as a fraction of work (α) that can be done in parallel; results are shown for $p = 8$ and 16.

4. The Modified Ware Model

Though useful, Ware's model is inadequate because it assumes that the parallel and serial implementations contain exactly the same amount of work; in other words, identical instruction streams. This is never the case because parallel processing requires additional instructions for interprocessor communication and synchronization. Furthermore, a highly concurrent algorithm may involve more arithmetic than the best-known serial algorithm. Thus, Ware's model can be improved by adding a third term to the parallel execution time. This term called σ, accounts for the additional work, or "overhead" required in parallel computation.

$$S_p = \frac{1}{(1-\alpha) + \frac{\alpha}{p} + \sigma} \tag{4}$$

Note that as p increases, the term α/p becomes insignificant while σ may increase. In that case, S_p will behave as illustrated in Figure 2 and, indeed, instances of "speeddown" have been observed in practice. For further discussion of performance models of parallel computation, work by Flatt and Kennedy[8] should be reviewed.

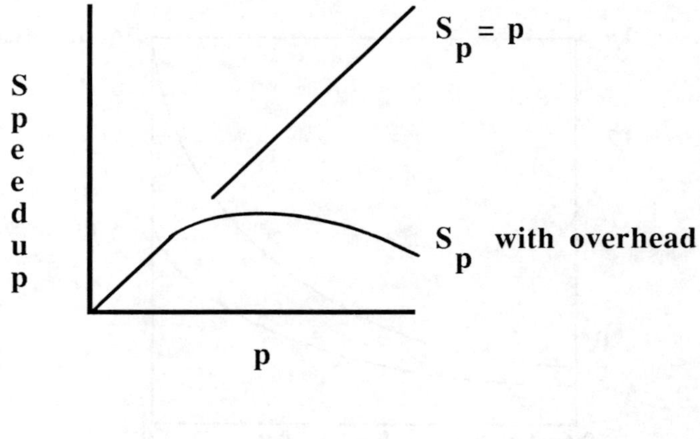

Figure 2. Speedup as a function of the number of processors, two cases are shown, one with overhead costs and one with $S_p = p$.

Thus, to achieve high performance in parallel processing, models must meet stringent "boundary conditions" associated with concurrency and overhead costs of task management. This has far-reaching implications for problem formulation, algorithms, and system software.

5. Trends in Shared-Memory Supercomputer Performance

What will the performance of supercomputers be in 1990? On the basis of current performance, we will attempt a brief and cautious answer to this question. It is axiomatic today that supercomputer performance is a function of application, implementation, and machine architecture. Thus, measuring and evaluating supercomputer performance is complex; in fact, it is a research area of growing importance. Of course, supercomputers are often characterized by their peak performance, but as Jack Dongarra of Argonne National Laboratory has observed, "Peak performance is that level of performance that the manufacturer guarantees cannot be exceeded."

The CRAY X-MP has hardware facilities for monitoring various aspects of its operation, and the machine has been in use long enough for substantial performance data to accumulate. For example, Richard Sato[9], NCAR Scientific Computing Division (SCD) has made measurements over periods of time that span several days. His measurements indicate that a single X-MP processor typically averages about 50 Mflops across a broad spectrum of NCAR applications. We emphasize that this is average performance of a single processor over long periods of time. Such should not be confused with optimal performance because a single processor of the X-MP often executes numerical linear algebra software at 140 to 200 Mflops.[10]

Comparative studies, such as those by Lubeck, Moore, and Mendez,[11] and by Kampe and Nguyen,[12] show that the single-processor performance of other supercomputers is within a factor of two of the X-MP. Thus we will use the X-MP as our frame of reference for estimating 1990 performance.

Two pacing factors in the performance of a single processor are signal propagation and packaging. Photons travel at about one foot per nanosecond, while electrons travel somewhat slower. That simple fact, combined with the cycle times shown in Table I, provides insight into the packaging problem for supercomputers. To achieve high speed, electronic componentry must be driven at high power, most of which is then released as heat. Thus the design challenge for supercomputer processors is how to "shorten the longest wire while powering and cooling the beast." As a result, it will be very difficult to substantially increase the performance of a single processor over the next five years. So in estimating supercomputer performance in 1990, we will take 50 Mflops per processor as representative of today's equipment. Conservative estimates are that by 1990 the performance of a single processor will at most double, i.e., 100 Mflops.

Table I. State-of-the-art Supercomputers

Machine	Cycle Time ns	Maximum Rate In Mflops	Memory In MBytes	Extended Memory In MBytes	Number of Processors
Amdahl 1400 (Fujitsu VP-400)	15 (7.5)	1066	256	--	1
CRAY X-MP	8.5	235/proc	128	4096	1,2,4
CRAY Y-MP	6	333/proc	256	4096	8
CRAY-2	4.1	488/proc	2000	--	4
CRAY-3	2	1000/proc	16000	--	16
ETA-10	7-10	1250/proc	2048*	--	2,4,6,8
Hitachi S-810/20	14	630	256	1024	1
Hitachi S-820/80	4	3000	512	12000	1
IBM 3090/600	17.2	696	128	256	6
NEC SX-2	6	1300	256	2048	1

*Also 32 Mwords of local memory with each processor.

SMSC systems are already in use and some organizations are beginning to use these systems to parallel process large applications, e.g., NCAR and ECMWF. For example, at NCAR a major climate model incorporates a three-dimensional grid and a typical simulation contains about four million mesh points; five variables are carried per mesh point.[5] Thus, 20 million numbers must be computed to advance the model from one point in time to another and about 300 MBytes of information must be stored whenever a history file of the model variables is made. The model has been optimized for vector and parallel processing and averages over 450 Mflops on the NCAR CRAY X-MP/48. Nevertheless, in excess of 100 hours of wallclock time are required to run problems of the type desired by scientists. Parallel processing will likely be commonplace on multiprocessor supercomputers by 1990, so we are led to ask what performance gains can be expected from it.

Over the past five years, a lot of data has been acquired on parallel processing of scientific applications on four-processor SMSC systems. On the average, use of four processors tends to increase performance over a single processor by a factor about 3.75. By 1990, systems with up to 16 processors sharing a common memory should be available. At least one study, Williams and Bobrowicz[13] (1985), indicates that use of 16 processors will increase performance over use of a single processor by a factor of 10 to 15. This is easily confirmed via Ware's model. For example, a speedup at 3.75 using four processors implies $\alpha = .98$. Sixteen processors and $\alpha = .98$ yields a speedup of about 12. In view of the modified Ware model, we will assume a factor of 10 and hope that it is conservative. To achieve a factor of 10 from parallel processing, the modified Ware model shows that we will have to give careful attention to parallelization and overheads from task management.

The combination of a factor of two in single processor performance and a factor of 10 from parallel processing gives a potential increase of a factor of 20 over the next five years. Such an increase in processing capability, together with tremendously large memories, will make solvable a number of important problems that are beyond computation today. Thus, the decade of the nineties should be an exciting time for geoscience provided we are willing to use parallel processing on SMSC systems.

6. Massively Parallel Systems

Massively parallel systems have the potential to offer high performance at relatively low cost. High performance can be achieved by incorporating thousands of processors into a single system. Low cost can be achieved by use of high volume, microprocessor and memory componentry. One of the most powerful DMMP systems available today is the Connection Machine (CM) from Thinking Machines Corporation. We will use the CM to exemplify the potential of massive parallelism.

The CM can be configured with up to 64K (65,536) processors with .5 Gbytes of memory. It can also contain up to 80 Gbytes in its secondary storage system which is known as "the data vault." Aggregate transfer rates to the data vault can be as high as 320 MBytes per second. The interprocessor communication geometry is designed to minimize the overhead that we discussed in the modified Ware model. The CM is a single-instruction machine in that all processors are synchronized and execute the same instruction stream. Actually, on each cycle a single instruction is broadcast from a front-end computer to each CM processor. If a processor is active (enabled), then it applies the instruction to data in its local memory. If the processor is inactive (disabled), the operation is not performed. The set of enabled/disabled processors can be changed dynamically as a function of both declarative control structures and computed results. In computer parlance, the CM is a single-instruction, multiple-data (SIMD) stream architecture. Interestingly, systems with multiple, independent processors (MIMD) are often programmed such that the same program is loaded into each of the processor memories. This is known as the single-program, multiple-data (SPMD) mode of parallel computation. While SPMD offers more flexibility than SIMD, SIMD is easier to debug.

The technology underlying the CM is both innovative and conservative. Sixteen processors are fabricated per chip. Two sets of sixteen processors share a floating-point microprocessor. Each processor has 8K bytes of local memory. Because the CM is SIMD, all memory is available for data. Total power dissipation is 28K watts, so the machine is air cooled. A full array of 65,536 processors fits in a five-foot cube. Peak performance of such an array is 32 billion floating point operations per second (32 bit precision) and a full array costs about six million dollars. Thus the CM offers:

High performance
Relatively low cost
Large memory, and
Programability

But it follows from our discussion of the Ware model that high performance requires a high percentage of concurrency in the associated model.

As noted earlier, the NCAR X-MP averages 50 Mflops per processor and models that use multitasking drive the machine at over 400 Mflops. These levels of performance show that a high degree of vectorization and parallelization has been achieved within some models from the geoscience community. And, models with a high degree of vectorization are good candidates for implementation on a DMMP machine such as the CM. To test this thesis we have implemented a 2-D shallow water model on the CM.[14] We estimate a performance level of 1.7 Gflops using 65,536 processors. Thus, personnel at NCAR and the University of Colorado are attempting to establish a Center for Applied Parallel Processing (CAPP) that will include a CM. Our research will include development of a sequence of 3-D ocean models of increasing complexity. In particular, we will start with a simple, "stacked slab" model. Our long-term objective is to determine the capability and suitability of the CM for support of computationally intensive models from the geoscience community.

REFERENCES

1. Holland, W. R. and McWilliams, J. C., "Computer modeling in physical oceanography from the global circulation to turbulence," *Physics Today*, October, 1987.

2. Kerr, R. A., "Climate since the ice began to melt," *Science*, October 19, 1984.

3. Roble, R., "Modeling the general circulation of the earth's thermosphere and ionosphere," paper presented at the Third International Conference on Supercomputing, Boston, May 19, 1988.

4. Clark, T., Hull, W. D., and Smolarkiewicz, P., "3-D cumulus entrainment studies," NCAR, Boulder, CO (unpublished).

5. Chervin, R. M., "Super problems, supercomputers and super models: Meeting tomorrow's climate modeling challenges...today," paper presented at the Third International Conference on Supercomputing, Boston, May 19, 1988.

6. Semtner, A., "An ocean modeling system for problem solving in the next decade using supercomputing architectures," paper presented at the Sixth Annual SCD Users Conference, NCAR, Boulder, January 12, 1988.

7. Ware, W., "The ultimate computer," *IEEE Spectrum*, **Vol. B9**, No. 3, March 1972, p. 89.

8. Flatt, H. and Kennedy, K., "Performance of parallel processors," *Journal of Parallel Computing*, North Holland, (to appear).

9. Sato, R., "Preliminary analysis of the performance monitor statistics for the NCAR job mix," *Supercomputer Applications*, IEEE Newsletter, (to appear).

10. Dongarra, J., "Performance of various computers using standard linear equations software in a Fortran environment," Tech. Memo. Chicago: Argonne National Laboratory, April 23, 1986.

11. Lubeck, O., Moore, J., and Mendez, R., "A benchmark comparison of three supercomputers: Fujitsu VP-2000, Hitachi S810/20, and CRAY X-MP/2," IEEE *Comput.* December, **Vol. 18**, No. 12: 10-24, 1985.

12. Kampe, F. and Nguyen, T., "Performance comparison of the CRAY-2 and CRAY X-MP on a class of seismic data processing algorithms," in press.

13. Williams, E. and Bobrowicz, F., "Speedup predictions for large scientific parallel programs on CRAY X-MP-like architectures." *Procedures 1985 International Conference on Parallel Processing.*

14. McBryan, O., Sato, R., and Swarztrauber, P., "Trip report," visit to Thinking Machines Corporation, NCAR, Boulder, January 26-29, 1988.

DATA MANAGEMENT METHODS; DATA FOR EUROPE

Roy L. JENNE
National Center for Atmospheric Research (NCAR)
Data Support Section
P.O. Box 3000
Boulder, CO 80307
U.S.A

"Data Management Methods; Data For Europe"

KEYWORDS/ABSTRACT: data management / data handling / binary-packed data / formats / data availability / climate data for Europe / archival storage costs

There has been good progress in making climate data available. Selected data management strategies and trends as well as the changing cost of archival storage since 1960, are given. Some data organizations emphasize preparing data and making it available, others spend most time on the interactive display of data. Our activity at NCAR emphasizes the former. There is still a problem in obtaining climate datasets for all of Europe for long periods of time.

- -

Many of the scientific disciplines face similar problems in the preparation and archival of datasets. Data are needed for both relatively small-scale problems and for all the earth. Some datasets have large enough volumes that they are difficult and expensive to use. Therefore, careful attention must be given to the way the data are structured, and to data subsets to minimize problems of access. The strategies for managing and accessing the data will be considered.

At the request of reviewers, we will first give a summary of our data activities at NCAR. Then examples of progress in data support over the last 20 years will be given. The Data Support Section, of the Scientific Computing Division at NCAR, maintains a large archive of computer-accessible research data that generally is made available to scientists around the world. The archive represents an irreplaceable store of observational data and analyses, and is used for major national and international atmospheric and oceanic research projects. There are now over 300 distinct datasets in the archive; each may be stored on one to several hundred tapes. NCAR has millions of atmospheric analyses, each usually with 1,000 to 10,000 grid points. There are even more observations. One dataset contains a total of about 20 million reports each year from 8,000 major weather stations around the world. These data are decoded from world telecommunications bulletins. Another has over 72 million three- or six-hourly observations taken from ships at sea, after 1854.

A. Berger et al. (eds.), Climate and Geo-Sciences, 513–524.

NCAR also archives delayed datasets such as all daily upper air raob data prepared by Asheville/NCDC (National Climatic Data Center) from 1946 onwards for U.S. sources. Similar data from a few other countries such as Canada, Australia, and France are available. There are datasets that have monthly surface and upper air data for many stations in the world for long periods of time. Selected oceanographic data are similarly obtained.

Climatological grid point data (Global) are available for sea level pressure, surface temperature and dew point, and for variables on the 1000 to 100 mb pressure levels, plus the northern stratosphere. A variety of fields such as ice age climatology, modern vegetation cover and the cloud climatology are also present. There are year-monthly data for a number of variables including SST, long-wave radiation, tropical winds, and selected means of daily analyses listed above. Selected data from a few climate models are also available.

More detailed information about some types of data will be given. There are daily analyses from NMC (National Meteorological Center, U.S.A.), ECMWF (European Centre for Medium-range Weather Forecasts), U.S. Navy, Australia, etc. Daily global sea surface temperature (SST) analyses, daily winds at the lowest NMC boundary layer, and many analyses at different atmospheric levels are available. Daily analyses for N. hemisphere sea level pressure are available from 1899. Higher levels started between 1946 and 1962. The University of Washington put a selection of these analyses (through 1985) on a compact disc (CDROM), (Mass, et al, 1987). For precipitation, there are daily and monthly HRC data (Highly Reflection Clouds) for the tropics, selected geosynchronous satellite data from the Climate Analysis Center and daily and monthly data from selected observing stations for decades. The NCAR archives are described in more detail in Jenne, 1975 and later updates.

International projects are usually focussed on a limited period of time. Research on these projects often requires comparisons with previous data for a long time period (such as El Niño over a 40 or 100-year period). NCAR tries to have the data that permits global-scale studies for very long periods of time. Since only a few people have been available at NCAR to do this work, data were gathered from many sources; selected cooperative agreements with others were made, to help prepare major datasets (such as the COADS project for surface marine data). The total volume is about 12 x 10^{12} bits of satellite data and 2 x 10^{12} bits of other data.

NCAR routinely adds recent data to many of the 300 datasets (nine sets are updated each one to three months, 18 each year, and 57 each one to five years). About 20 new datasets are added each year. The data are described in Jenne, 1975 and in a short 'Data Availability at NCAR,' that is updated each year.

The Data Support Section at NCAR helps scientists in locating data appropriate to their research needs. The users may obtain copies of data on tape or use it on-line to the NCAR supercomputers. The Section also assists scientists by providing data access programs (to read in data and unpack it) and other software for data manipulation, as well as a variety of written material. During a year, about 650 different people from universities use the NCAR computers. There are another 400 users at NCAR. A portion of the computing tasks require the use of observed data, analyses, climatology or physical data such as elevation data for the earth. The on-line data banks at NCAR enable users to quickly access data without having to obtain it elsewhere.

1. Progress in Data Availability

Often we think so much about the existing data problems that the successes of previous years are not considered. For example, in 1967, the NCDC in Asheville had about 600,000,000 punched cards (80 characters each). They filled the building, including many hallways. NCAR had to pay $4.50 per 1,000 cards (put onto tape) to obtain data in 1967–70. By 1978, NCDC, at Asheville, had put the data on tape; NCAR could buy 500,000 cards on a 1600 BPI tape for about $60. This is a cost of about $0.12 per 1,000 cards. The progress has continued. Now, NCAR can easily afford to have all U.S. daily rawinsonde data since 1948, readily available to the main computers at NCAR. Universities around the country can obtain a copy of what they need or use the data on-line by using the computers at NCAR.

NCAR selected NMC analyses from 2,000 tapes for 1958–73. Now, the data are on only 50 tapes. A set of world surface synoptic data first was prepared on about 107,000,000 cards by the Air Force. Then it was put onto 1,414 tapes. It has been moved to 350 reels; it could be put onto about 150 tapes, still in character format. With binary packing, it would require even fewer tapes.

Until 1982, the world's surface marine synoptic data was on several hundred tapes, in a few formats, many duplicates, and a number of problems. Between 1982–83, the data (100,000,000 observations) was all reprocessed. Now 40 tapes (binary packed) have all of the resulting data: 72,000,000 observations for 1854–1979. Also, many statistics were prepared that describe the marine climate each month of each year. These have been requested by over 80 research groups around the world. Any statistic for each month, for all years, can be obtained from NCAR on only two tapes.

Let us consider an example for the case of satellite data. All primary atmospheric sounder data (eight infrared channels) from NOAA polar orbiting satellites for Dec 1972 through Feb 1979, were on nearly 1,200 old tapes. The data was in danger of being lost. In 1986, it was copied to 48 tapes (6250 BPI). A similar history for many other types of data could be given. The progress has been large, but the appetite for data has also been increasing.

All progress is not involved with making large volumes of data available, as above. There are other activities like preparing a climatology on tape, gathering rainfall data from Africa, improving the world's holdings of monthly surface data, preparing tapes that give vegetation cover, and soil type, etc. Here, the time needed to prepare the data is large even though the volume is small.

If data are on many tapes or if the cost is high, individual scientists will not use it very much. The limit for individual investigators is often in the range of 3 tapes to 20. As the cost of data and handling problems are reduced, the data become available for much more research. A central archive will not see all of this new use of data. Large datasets that once were located only in central archives could be copied for main research archives as costs came down. Now, a number of datasets are distributed down to the level of the university department, where several people use the same data. A key role of the larger archive is to accomplish all of the tasks of data gathering and preparation so that the data can be readily used at the archive or distributed.

The key-entry of old data is a large task. A 6250 BPI tape holds about 125 megabytes of data, if properly blocked. It will hold about 1,600,000 of the old 80-column punched cards. Each card usually represented one "observation" that contained a number of variables. Let us assume that such an observation usually has about 60 data digits (key strokes), and that the cost of manual key-entry and initial processing is about $.10 per card. The cost to key-enter the 1,600,000 cards that will fit onto a tape is then about $160,000. Considering the size of the archives above, we have inherited a large legacy of data from past generations. We should make sure that it is preserved, and that a selection of other old data is digitized.

In early 1986 NCAR started using IBM 3480 cartridges for mass storage of data. A cartridge is small, yet it holds 200 MB. This technology has been excellent. The error rate has been very low. The stated data rate is three MB/sec. NCAR achieves an average rate almost this high when moving data directly from a cartridge to a fast computer.

Optical disk technology is becoming more interesting as time passes. The CDROM is based on standards and new disks can be "stamped" like a music disk and sent to a large number of people. One CDROM holds about 600 MB. Other optical storage devices still suffer from lack of standards, high media costs compared to the best magnetic tape options, and low data rates. However, the optical read devices are relatively inexpensive. The CDROM readers are under $1,000. One optical unit reads 12-inch, 3.2 Gbyte disks at an observed rate of about 220 KB/sec. and costs about $7,000. A controller that can serve several drives costs about the same. The write rate is about 120 KB/sec. The density will soon double to 6.4 Gbytes per 12-inch platter. Some types of optical media will probably have a shelf life of 30 to 100 years. Other types are not as good. Hardware to read a given type of optical media will probably only be available for 10 to 20 years.

Technical progress in computer capability and in storage costs (Table I) have made our advances in data management possible. This progress is continuing at a rapid rate.

Table I: Changes in Archival Storage Costs, With Time

The term Bytes is used below to mean 8 bit quantities, even though characters were 6 bits long in earlier years.

			---- Media Costs ----	
		Effective Cap.(Gbit)	Per Gbyte	Per Tbit
1960	556 BPI tape ($14), 10 Mbytes	.08	$1,400	$175,000
1972	1600 BPI tape ($14), 42 Mbytes, holds 37.5 Mbytes	.30	374	46,700
1980	6250 BPI tape ($13), 150 Mbytes, holds 125 Mbytes	1.00	104	13,000
Feb 1986	IBM 3480 tape cart ($14), 200 Mbytes, holds 174 Mbytes	1.39	81	10,070
1986-88	Sony Optical 3.2 GB, holds 2.9 ($270)	23.2	94	11,720
Feb 1987	3480, $10.50 each	1.39	60.5	7,560
June 1987	3480, $7.50 each	1.39	43.2	5,400
1988	Honeywell VHS Cassette, 5200 MB, holds 4900 MB, costs $10.50	39.2	2.14	268

2. What Governs the Use of Data?

For many types of research data, we should not expect multitudes of users. For example, in the case of ozone, there are national and international reasons why we should continue to spend significant amounts of money for both satellite sensors and ground stations to monitor the amount of ozone and the changes. But, the world probably only needs 20 to 50 people who actively analyze each piece of evidence. They will use the basic datasets. Some other people will use data that are prepared in arrays by the first group. The rest of us will probably obtain most of our information about ozone from reports and figures that are prepared by the most active scientists in this area. The use of data depends on such factors as interest in the given type of data, ability to find that the data exists, cost, ease of use, etc.

In the case of atmospheric soundings of Mars, only a few people will use the data and report their findings to the scientific community. Similarly, most of us will look at pictures of Mars that have been prepared by someone else, rather than going back to the basic satellite scanner data.

Some people say we should just archive the results and throw away the basic data. This usually is a bad idea because for complicated data, the algorithms to derive information are usually improved after the complexities are better understood. Also, many types of data have multiple uses that are not understood at first. Archives are often under pressure to cut costs and save space. In a recent example, an archive threw away a one-tape dataset of basic climatological data because it had not been requested in three years. Fortunately another archive still had it. If a dataset contributes a reasonable amount to the basic knowledge of mankind, and can't be readily regenerated, it should be saved indefinitely. But it certainly doesn't make sense to use significant extra resources to offer millisecond access time and to prepare quantities of special software for a dataset that will be used only once a year or once a decade.

It is instructive to work directly with users of data and observe what their problems are. The frustrations often come from rather fundamental sources, such as:

- No one has prepared the data they need into a useful dataset;
- The volume or cost of the data is too large to cope with;
- Sometimes the data are not easy to use;
- They get parity errors when reading the tape;
- They struggle with their systems people just to properly read a tape from another computer;
- The time delay to obtain the data is long;
- Contents and format of data are not adequately described or wrong;
- The dates given on portions of the data are wrong, or the data are out of order;
- They can't find out if given data are available.

This list of problems is given in more detail in Jenne and Joseph, 1978. It often is surprising that meetings about data management rarely talk about such fundamental problems. Often, these meetings cover such topics as new storage technology, display systems, common formats, communications, and catalogs. These are useful topics, but they often do not address our most serious problems.

How much does it cost to store a dataset for one year? For most datasets, the big cost is in preparing data and its documentation, not in storing it. To achieve low storage costs, one must make sure that the data and its documentation can be easily moved to new data storage systems. Consider the cost of storing a relatively small dataset of only 30 megabytes. Compare this volume to a familiar book, the King James Bible, with about 4,000,000 characters. Nearly 200 megabytes of data can be put onto a new IBM cartridge tape that costs $8. The tape will have to be recopied in about 8 years. Even if a backup tape is kept, the storage costs for 30 Mbytes can be kept down to $1 to $4 per year, if the system is organized so that little time of people is needed for datasets already in storage. Consider data that should be saved 1,000 years. Books must be moved to new libraries to survive; with digital data, it must be moved to new media about each eight years. Copies of important data should exist in more than one archive.

2.1 WHY ARE SOME DATA STORED IN BINARY FORM INSTEAD OF CHARACTERS?

Suppose that we need to store numbers that range in size from 0 to 4000. It would take four characters to store them (usually requiring 32 bits), or they could be stored in 12 binary bits. Usually, the difference in volume between character and binary data is a factor of about 2.5. For small datasets, the volume doesn't really matter; they are often stored using characters because more people are familiar with this form of data.

In addition to data volume, an advantage of binary-packed data is that it saves a lot of computing time. Typically, it takes 10 to 30 times as much computer time to unpack a number stored in characters compared to the time to unpack a binary number. This becomes very significant when there are many numbers, Jenne, 1984b.

Many will want to know if a lot of effort is needed to unpack binary numbers. If users had to do machine-language programming to unpack the binary data, it is clear that binary packing would not be practical. However, the user can handle all of the unpacking using a routine that we call GBYTES that can be called from a Fortran program. One call will unpack one number or 600 numbers from an array. The only problem is that to be efficient, the subroutine GBYTES needs to be written in machine language for a given class of computer. It is available for CRAYS, CYBERS, IBM, VAXES, and some others. The unpacking software that uses GBYTES can be written in such a way that it is portable across computers, even when the word length changes. However, there would be problems if the packed numbers were 20 bits long and one wanted to use them on a 16-bit computer.

If the subroutine GBYTES is available on your computer, it is possible to handle any form of data that people might send you, just as long as you can read the data into the computer. The routine GBYTES was defined at NCAR in 1965. It has spread around. In 1986, a scientist in India had obtained a copy from a university in England, to use in India, to unpack meteorological analyses from ECMWF in Europe. A tape having GBYTE subroutines and documentation is available from NCAR. The document 'Techniques for the Processing, Storage, and Exchange of Data' (Jenne and Joseph, 1974) describes how to manipulate binary data and accomplish other tasks.

3. Issues in Data Management

Over the last 20 years, different issues about data management have received varying degrees of emphasis. A manager needs to maintain balance in the methods used. Several principles for data management were summarized in 'Planning Guidance for the

World Climate Data System', (Jenne, 1982). It covers a host of issues including data requirements, formats, and quality control. Another publication considers the overall strategies to use in managing scientific data, Jenne, 1980. Over the years, different issues have been discussed at length, in the data community. We will briefly review several of these and then make a few comments on simplicity and on the future. Some of the popular discussion items for data have been as follows:

3.1 COMMON FORMATS

It has been said that all of these many data formats are wasting our time and resources. What we need to do is to define standard formats for everything. This emphasis was most common about 1971–1976. There are legitimate concerns here. But, complete agreement on formats simply is not necessary in most cases. In some cases, the agreement on a common format merely creates one more format, but doesn't get rid of previous ones. In a few cases, a useful observed variable was not included in the data because it was not considered when the format was defined. On one ocean project, a scientist told me that the data people wanted them to put the data into a complicated format structure. The scientists "revolted" and said that they couldn't do that much extra work. The scientists said they would prepare the data in their own format and that it could be changed later, if necessary. Probably the format requested was much more complicated than necessary. For one type of data, it is, in fact, difficult for users if the data are received in many different formats. Also, note that there is no reason at all to try to force rawinsonde data into a format for analysis data.

Various factors to consider when defining formats are briefly described in Jenne, 1981. The factors include data blocking, data packing, and whether variable length data blocking is needed.

3.2 DATA BASE MANAGEMENT SYSTEMS

From at least the late 1960s, formal database management systems (DBMS) have been under development. In the approximate period 1977–81, there were claims that the use of such systems was the only way to make progress — no matter what the problem and no matter what the size of the dataset. People assumed that what worked well on a 20 Mbyte dataset would work just as well on a 10,000 Mbyte set. They did not think about the time taken if there is a 0.1 or 0.2 second delay to obtain each piece of information. One major climate data center experimented (~1979-81) with the use of DBMS systems on very large datasets. In one case, they found that just to load the system with about 100,000,000 observations (about 10 numbers in each) would take over 10 years (actually, the quote was 26 years, but I've never understood why it would be that long). Now, it is better understood when such systems are useful, and when they are not. They have applications especially for control information such as data catalogs. Also, they have been applied to some data browse tasks.

To attain good processing speeds, one needs to keep the data moving rapidly by delivering relatively large blocks of data at once. When proper data blocking is not used, the rate of data transfer becomes very slow, as described in Jenne and Joseph, 1978. The data flow in supercomputers is discussed in 'Scientific Data Processing,' Jenne, 1982b.

3.3 ALL DATA MUST BE ON-LINE

I heard this many times from about 1975–85. It has never made any sense. If data are used often, on-line storage is desirable. A given tape in a large archive often is used only once each one to five years. If a person has waited a year to use a dataset, surely he/she can wait two or three minutes to bring the data on-line. The point is that we still don't want users to have to wait for days. The cost of off-line storage is much cheaper than on-line storage for large datasets; therefore, it should usually be used. There are also cases of frequent data use where it is less expensive to keep the data on-line, especially for small datasets.

3.4 ALL DATA CAN BE SENT BY RAPID COMMUNICATIONS WHEN NEEDED

The almost exclusive use of communications for data shipment was often pushed, from about 1984–86. Others have been quick to point out that large amounts of data can usually be sent more easily, with better error control, at less cost, and often faster by airmail. The half-inch tape standard has made it easy to exchange data worldwide. We will continue to need a few types of exchange media. There are, however, many time-critical data flows where communication systems are essential. Also, for gathering small pieces of a large dataset and for exchanging small datasets, communications are often most practical.

3.5 WE DON'T HAVE TO GATHER DATASETS; COMMUNICATIONS CAN GET THE COMPONENTS WHENEVER NEEDED

A given type of data such as station precipitation, or aircraft weather reports, usually needs to be gathered together from many sources in order to have a useful dataset. We have found that the gathering process also serves another purpose. It enforces a discipline on everyone to **really** prepare their data into a useful form, and to make certain that it reaches a central archive. A given archive often specializes in a certain category of data. Thus, one archive can focus on climate and another on astronomy.

3.6 ALL DATA ARE EQUALLY VALUABLE AND NEED EQUAL ARCHIVE CARE

Actually, data vary widely in both importance and in the amount of use they receive. It makes sense to offer more software services, for those datasets where the need is strongest.

3.7 WE NEED INTEGRATED DATASETS, NOT SEPARATE DATA FOR EACH SENSOR

This saying was most common from about 1974–82. There is a certain amount of truth to the saying, but not much. Datasets of a similar type such as rawinsondes or daily station data must be gathered together from many stations. But, when too many types of data are put into one dataset, it becomes harder to use any one type of data. This happened in the Global Weather Experiment (GWE or FGGE, 1979). To study drifting buoys, one needed to go through about 100 tapes in order to study data that would fit onto one or two tapes. For this reason, the FGGE data was again separated into broad categories by data type. If a user needs data from two datasets, it is not difficult to simultaneously input data from the two sets.

3.8 INTEROPERABLE DATA SYSTEMS

The goal to prepare data systems that work smoothly, even though different computers, communications, and storage media are involved, is an important part of present planning. It seems to me that the goal is good. One tendency of planning that will not work well, is to demand too many standards in order to make the system work. The problems of media standards and formats have to be addressed separately. An overall design of data systems should permit much individual initiative, and other actions that do not have to be completely coordinated in order to achieve good results. In this regard, we can learn from the experiences of national economic systems. That experience tells us that centralizing decisions at high organizational levels, and trying to plan and control most aspects of a society, leads to bad results. The same applies to data management. Good systems can cope with a reasonable amount of diversity, and are better because of it. In this regard, I like the statements by Guerard in 1959-60:

He first noted that mankind craves verbal idols. The great virtue of verbal idols is that by appeasing our hunger and thirst for righteousness, they soften our harsh duty to think clearly and act consistently. In our context, the examples could be some of the headings listed above. In a later discussion, he added:

"We are led to realize that the pursuit of excellence implies a fight on two fronts:

- One against chaos which paralyzes;
- The other against conformity which deadens.

We have to shift our forces from one front to the other."

4. How to Make Powerful Systems That Are Simple

Many aspects of computing and data processing are necessarily complicated, but good designs can achieve an overall simplification. For example, the methods used to quickly and accurately calculate the sine function of an angle are complicated, but no one notices this fact because the complexity is all hidden in a subroutine which accomplishes this specific task for a user. Brody, 1986, wrote that the common creed in Japanese factories is "simplify, simplify." When a toolmaker planned to automate a factory, the staff laid out every part and every tool being used on two large tables. Plant engineers analyzed the hardware and tried to justify why each item was needed. The result was that the number of tools was reduced from 672 to 46, all but six of which had multiple uses. Prasad, who manages industrial control systems for Ford Motor Company, said "An American company would probably try to build a sophisticated database to manage 672 tools." I think he is right, but I want to add a work of caution here. It is often possible to find simplifications such as these, but if carried too far, one ends up with tools (or programs) that are too general, too complicated, too hard to improve, and too costly.

Simplicity usually is achieved by finding an organization of a problem in which you don't have to worry about lower levels of detail until they become entirely necessary. For example, when managing a library, a book is just an item with a number, title, author and perhaps an abstract. It can be kept track of without any real personal knowledge about its contents. A volume of digital data on a mass storage system is handled in the same way. The system knows its name and location; it doesn't know anything about the format of the data within the dataset or what the data means. This is a great simplification. When, for example, all of the data must be moved from an older type of storage device to a new one, the move can be made without knowing about the contents of the data or the structures of datasets. A user knows that he or she can get back the same data that was put into the system. The user doesn't have to know what storage media a given dataset is on. The philosophy of data communications is similar. A user typically doesn't want to know how his data are sent. He only wants the assurance that the data will not be altered while being sent.

The problems of achieving compatibility from one computing system to another are discussed in 'Software Compatibility and Data Communications', an Appendix in NAS, 1984. There are some things that a programmer can do to help achieve easier portability of programs. One example is: do not directly call the I/O routines of a particular machine. Instead, call a subroutine which can be rewritten for the new computer. The text of the report also suggests a few steps that computer companies could take to make portability easier. An example is that the subroutine GBYTES that allows users to easily handle binary-packed data (or any form of data from another computer), should be provided to users as part of the standard software.

5. What Will the Future Bring?

There will be higher-density storage media for low costs, so that small research and applied groups can have large data libraries on-site. The coupling of good mass storage systems with powerful computers makes it possible to process large amounts of data without having to personally handle and keep track of the media. People will receive data on tape, disk, CDROM or by communications. There will gradually be more software that is sent with data, to unpack the archive format and put the data into arrays that can be used for computing. Data manipulation software can also take data from the arrays to make various displays. In order to permit this to happen, we need to make a significant effort to enhance the quality and extent of present datasets.

6. Problems in Data Availability for Europe

There is a major problem in obtaining observed data for climatic research for Europe. If a person only wants to study the climate in one European country, he or she can often obtain the necessary data by writing to the archives of that country. However, many studies need to have data to study climate on continental or global scales. Suppose that the person in Europe needs daily digital data for 400 stations, covering 15 European countries, including surface and upper-air data for 10 to 50 years. The scientist will probably soon discover that it is hopeless to obtain the data with any reasonable amount of effort by going to each national archive. The scientist will probably abandon the study or try to see if the data may be available from the U.S.. Roger Barry, director of World Data Center A for Ice and Snow, tried to gather a set of snow fall and snow depth data for Europe. First, he got a nice dataset from Austria and was quite encouraged. Then, he started asking for data from countries that had very high prices and which had restrictions on giving the data to anyone else. He finally had to give up on the idea of a dataset for Europe — so the possible research is limited. A European scientist helped me to obtain some data from his national archive. With some frustration he wrote to me: "You know these data people. They would rather preserve their precious data than make them available for use." This was humorous and hopefully, not very true.

I have another recent example of the difficulty of people finding the European data and advice that they need. In February 1988, an investigator, from the Wildfowl Trust in the UK, wrote to me at NCAR, for help with European data. In order to study the cold weather movements of waterfowl in Western Europe, they needed daily data for air temperature, wind speed and direction, state of the ground (frozen or snow covered) and other details. Could NCAR "help with the difficult problem of selecting sufficient meteorological stations to enable a regional analysis of weather conditions in any one winter? Would you have or be able to supply a suitable list of weather stations with continuous records from about the 1940s to the present?"

The Wildfowl Trust research staff had written to WMO (World Meteorological Organization) for information. WMO suggested that they write to NCAR. Actually, NCAR has a considerable amount of the desired digital data, but the record is not continuous, nor do we have the regional expertise that would be desirable. The main point is that the request for data was reasonable and there should have been a dataset in Europe that could have quickly met their needs for digital data (probably for 75 stations), at a cost of a few hundred dollars.

6.1 A PROPOSED DATABASE FOR EUROPE

Selected datasets should be prepared for Europe that would support regional applied studies, and also support regional and global climate research. The high resolution data such as rainfall each 2 km is not needed in the regional archives.

What data would such a regional archive contain? In Jenne, 1982, we summarized various data requirements for climate studies. Some types of data such as total ozone measurements are already being gathered together under international arrangements that are working well. Some of the datasets suggested for Europe are as follows:

- Daily precipitation, snowfall, and snow depth for stations each 30 to 50 km;
- Daily max-min temperature, precipitation, etc. for stations each 75 km;
- Data from surface synoptic stations each three hours for many years, about each 100 to 150 km;
- Daily sunshine and radiation data, from stations about each 100 to 150 km, as available; It is desirable to have some of these data from the same sites as the surface synoptic data.
- Monthly surface station data for very long periods for both the earliest city locations, and for undisturbed rural sites;
- Library data about station moves, etc.;
- Daily and monthly river discharge data;
- Daily upper-air winds and radiosonde temperatures for very long records.

It would not be necessary for any one European laboratory to gather all of the above data. One institution could accept the task of gathering only one type of data. Then many laboratories would probably choose to have a copy of the data. A copy should also be available from the World Data Centers.

The cost of gathering data is a strong function of how often a dataset is updated. Since global GTS real-time data can supply most of the needs for recent data, it is suggested that an update cycle of each one or two years would be adequate.

The U.S. has considerable amounts of older European surface and upper air data that are on many magnetic tapes. The volume is large and the contents are not easy to describe as yet. The data were prepared under massive projects by the U.S. Air Force in the 1950s and 60s. They probably punched over 200,000,000 cards of data. Data for a number of years in the period 1925 – 1965 are often available. Some of these data could be made available to help establish a European archive, if Europe would fill gaps, add recent years, and make the data openly available at a reasonable cost.

7. Summary

Significant progress toward better data availability has been made over the last 20 years. During the same time, there have been evolving ideas about what methods of data management help us accomplish the goals. Some of these have been reviewed. One problem is that meteorological data and climate data from past years in Europe is not easy to access. An archive is proposed in which individual countries could volunteer to help gather one or two types of data.

REFERENCES

Guerard, A., 1959-60: 'The Quest of Excellence, a Humanistic Utopia.' *The American Scholar*, Winter 1959-60, Vol **29**, No. 1. pp 31-42.

Jenne, R.L., and D.H. Joseph, 1974: 'Techniques for the Processing, Storage, and Exchange of Data.' NCAR-TN/**IA-93**, Boulder, CO., U.S.A, 46 pp.

Jenne, R.L., 1975: 'Data Sets for Meteorological Research.' NCAR-TN/**IA-111.** National Center for Atmospheric Research, Boulder, Colorado, 80307, 194 pp.

Jenne, R.L. and D.H. Joseph, 1978: 'Management of Atmospheric Data.' From NASA Conference Publication **2055**: Engineering and Scientific Data Management, 261 pp

Jenne, R.L., 1980: 'Strategies to Develop and Access Large Sets of Scientific Data.' Workshop: Frontiers in Data Storage Retrieval and Display. Conf. Publ. by NGSDC, NOAA, Boulder, Colorado.

Jenne, R.L., 1982: 'Planning Guidance for the World Climate Data System.' WCP-19, WMO, Geneva. 136 pp.

Jenne, R.L., 1982b: 'Scientific Data Processing.' Paper in *Scientific Justification for an Advanced Vector Computer*, University Corporation for Atmospheric Research, Boulder, Colorado. pp 63-73.

NAS, 1984: 'Solar-Terrestrial Data Access, Distribution, and Archiving', A National Academy of Science Panel Report, National Academy Press, Wash, DC 1984.

Jenne, R.L., 1984b: 'Data Packing, Data Flow and Storage Costs for Different Memory Levels.' Prepared for Remote Sensing Workshop, Nov. 1984. 23 pp. In book, *Advances in Remote Sensing Retrieval Methods*, Ed: Deepak, et al. 1985.

Brody, Herb, 1986: 'Good is Never Enough.' *High Technology*, Aug 86, pp 20-21

Jenne, R.L., March 1987: 'Data Availability at NCAR,' P.O. Box 3000, Boulder, Colorado, 80307. 18 pp. (unpublished; updated each 12 to 18 months)

Mass, C.F., H. Edmon, H. Friedman, N. Cheney, and E. Recker, 1987: 'The use of compact disks for the storage of large Meteorological and Oceanographic datasets.' *Bul. AMS*, Vol **68**, No 12, pp 1556-1558.

SATELLITES AND CLIMATE

E. C. Barrett
Remote Sensing Unit
Department of Geography
University of Bristol
Bristol BS8 1SS

ABSTRACT. Attention is paid to the urgent and growing need for climate data from satellites to supplement the highly inadequate supplies from in situ observing networks. The special attributes and disattributes of satellites as atmospheric monitoring platforms are summarised, and progress with the use of satellite data in climatology reviewed, firstly for the first 15 years after 1960, and secondly for the subsequent 15 years to the present. Areas meriting further, or new, attention, are noted. It is generally concluded that increased cooperation amongst space powers would have great benefits for future satellite/sensor system design and use for climate applications.

1. INTRODUCTION

Climatology is specially demanding in its data requirements. Not only does it, like many environmental sciences, call for data which are continuous, homogeneous, closely and evenly spread, but unlike many other such sciences, it necessitates data sets of long duration. Although climate studies may range over periods from a minimum of 5 days and upwards, periods of 30-35 years are commonly held to be appropriate for the compilation of standard "climatological normals".

Satellites are naturally well-suited to provide data sets which are spatially much more even and continuous than sets which any conceivable network of "conventional" observation platforms could facilitate. Furthermore, as the third decade of meteorological satellite operations draws to a close in 1989, the capability of satellite systems to provide long-term climate data sets might not appear to be in question. Unfortunately, meteorological satellite observing systems have been not only many, but also varied, and few homogeneous data sets therefrom have covered periods more than a few years in length. Furthermore, the establishment of full and readily accessible satellite climate data archives has generally been accorded a priority too low to ensure adequate funding for the same.

Thus the use of satellites in support of climatology has been orientated more towards research than routine climate monitoring.

A. Berger et al. (eds.), Climate and Geo-Sciences, 525–545.

Much has been achieved, but, in the context of our small, worn planet whose natural resources are being placed under ever-increasing pressures resulting from simultaneous rises in both populations and expectations of living standards, climatic research alone is not enough. Whereas studies of past periods and even recent extreme events have great intrinsic values, the needs for near real-time climate monitoring, and at least short-term climate forecasting, are now quite overwhelming. It is clear that today's requirements for climate data are so great as to totally defy any effort to meet them from *in situ* sources and it is vital that satellite data gathering, data collection (from remotely-controlled *in situ* sensors), and related data preprocessing, processing, analysis, interpretation and archiving should be planned, provided, and managed in ways which will best serve the whole user community - within which climatologists should rank neither last nor least.

2. FACTORS AFFECTING THE USE OF SATELLITE DATA IN ATMOSPHERIC SCIENCE

Unquestionably satellite systems and the data they provide carry many strongly positive *attributes* for both climatological research and operational monitoring programes when compared with conventional data-gathering systems. First listed by Barrett & Martin (1981), the most important of these positive attributes may be summarised as follows:

(a) Satellite systems can provide convenient, unified data sets for the entire globe, or large parts of it.
(b) Satellite data can be spatially continuous (imagery) or relatively regularly distributed (e.g. soundings) for any area(s) of interest.
(c) Single sensors or sensor systems can provide homogeneous data sets over very wide areas, even over the globe as a whole.
(d) Some satellites yield data with a very high temporal frequency: from geostationary satellites half-hourly data are commonplace.
(e) Satellites afford new views of the atmosphere, monitoring it from above rather than within. Thus integrations of parameters (e.g. radiation fluxes) along lines, over areas, or through volumes of the atmosphere can be obtained readily.
(f) All data from satellites, being remotely sensed, have physical meanings more or less different from, and therefore usually complementary to, those from *in situ* sources.
(g) Satellite data can be obtained for very wide areas in or near real time, in forms which are commensurate with automatic data reception, handling, and analysis computing systems.

Unfortunately, however, satellite data also carry a range of disattributes which hinder their use, or even preclude it in some circumstances &/or for some potential applications. The most notable of these neutral or negative attributes may be summarised as follows, again following Barrett & Martin (1981):

(a) Data from satellites are rarely obtainable for synoptic reporting hours, raising difficulties over their merger with conventional data sets.

(b) Elaborate transformation procedures may be required to convert satellite measurements (antennae temperatures) into observed radiances, and, thereafter, into evaluations of climatologically useful parameters. As a consequence, satellite algorithm development has been patchy.

(c) Users are distant from the data source(s), and most may have little or no influence on sensor system design and operation.

(d) The types and resolutions of the data yielded are often less than optimal for many classes of actual &/or potential users.

(e) Information extraction can be difficult &/or inefficient for, not uncommonly, the data content of interest to an individual user is but a tiny fraction of the total.

(f) Archiving of data has left, and continues to leave, much to be desired. In particular, most runs of weather satellite data are still much too short for classical climatological research.

(g) Rapid improvements in satellite systems, and related new procedures for data analysis and application, can be hard to achieve, because of international political problems, and operational inertia respectively.

These attributes and disattributes of satellite systems and data comprise the frame within which the picture of satellite climatology has been rapidly but unevenly developing. It seems unlikely that the characteristics of this frame will change significantly in the foreseeable future, though the scale and significance of the individual points may be expected to vary, especially those over which human control can be exercised.

3. SATELLITES AND CLIMATE, 1960-1974.

3.1. General Remarks

We have been reminded that the first specialised atmospheric satellite was launched in 1960. Mid way between then and now the text entitled "Climatology from Satellites" (Barrett, 1974) was published. It is convenient, therefore, to organise the present review of Satellites and Climate accordingly, briefly summarising first the general "half-life" points made by the 1974 text, then following these with a more detailed summary of advances from the mid-1970s to the present.

3.2. Concepts of Climatology from Satellites

In the mid-1970s both the need for, and the possibility of, climatology from satellites were relatively new concepts, the cases for which had to be argued.

3.3. Satellite/Sensor Systems

The USA & USSR were the only countries to have placed weather satellites in orbit around the Earth. Although both the operational system of NOAA polar-orbiting weather satellites and the research and development family of Nimbus satellites were already well-established, relatively little geostationary-based research had yet been published. Atmospheric sounders were in their infancy, but had already begun to revolutionise upper atmosphere climatology (see, e.g. Barnett et al., 1970). The first passive microwave radiometer (ESMR) had been orbitted on Nimbus 5, but the first results and their implications were not generally known.

3.4. Satellite Data Analysis Systems and Techniques

Techniques of data analysis were dominantly manual, and therefore subjective, though objective techniques were being actively developed, e.g. for cloud characteristics (especially brightness, temperature, cover, and type), wind speeds and directions (based on temperature thresholding, e.g. Endlich et al. 1971), and, most notably, for temperature mapping and radiation budget monitoring (e.g. by Vonder Haar, 1968).

3.5. Global-scale Studies of Selected Atmospheric Elements

Reports of the first satellite-based global or quasi-global mapping and short-term climatological monitoring exercises had been published for a rather limited range of parameters, most notably for radiation and energy budget components (see Winston, 1967, and Vonder Haar, 1968, respectively), global cloud cover (see Winston, 1970, and Miller, 1971), and tropospheric relatively humidity (see Bandeen et al. 1965). Stratospheric temperature patterns had been mapped, first from Tiros radiometer data by Nordberg (1965), then with additional attention to the observed changes and seasonal marches of temperature, and patterns of deviations from the latitudinal norms by Fritz & Soules (1970).

3.6. Synoptic & Sub-synoptic Scale Studies

Numerous studies of synoptic and sub-synoptic scale weather systems, principally evidenced by satellite-imaged clouds, had already been undertaken, adding considerably to knowledge and understanding of atmospheric patterns and processes primarily in more conventional data-sparse regions of the tropics and mid-latitude oceans particularly in the Southern Hemisphere. Worthy of special mention is the work of Hubert et al. (1969) on the Inter-Tropical Cloud Band; Bjerknes (1969), who up-dated the Walker Circulation theory of the 1920s and so stimulated the present era of interest in the El Nino effect centred on the South Pacific, but seemingly influential on a global scale through related atmospheric teleconnections; Sikdar & Suomi (1971) who, from early geostationary satellite data, began to model scale interactions between convective building blocks of tropical disturbances and the structure, circulation, and significance of the tropical atmosphere as a whole; Carlson (1969) & Chang (1970) through their synoptic climatological studies of easterly waves over the Atlantic and Pacific Oceans respectively; Oliver (1969) with his pioneering satellite-based classification of tropical storms and hurricanes; and Streten (1968a & b & Streten & Troup, 1972) with studies both of the range of types of Southern Hemisphere mid-latitude frontal depressions and their geographical distributions. New light was thrown on key regional climatological patterns and processes, e.g. through monthly maps of infrared radiances over Southern Asian and Indian Ocean region, e.g. by Kreins & Allison (1969); and on newly-identified weather systems, e.g. the burst bands of the equatorial Eastern Pacific, discovered and modelled by Fujita et al. (1969).

4 SATELLITES AND CLIMATE, 1974-1988

4.1. General Remarks

If the period from 1960-74 was the period of primary satellite research and development and satellite climatological pioneering, that from 1974 to the present can be epitomised as a period of rapid algorithm development and application to much more complete ranges of regions, atmospheric systems and scales of activity: the earlier pioneer exploration of the satellite climatological field by a few has given way to widespread and increasingly enthusiastic colonisation by many. Consequently, the literature linking satellites and climate has mushroomed - presenting much larger problems to those who seek to summarise it. Therefore in the account which follows, emphasis must once again be placed only on the selected highlights of this period.

4.2. Concepts of Climatology from Satellites

The inclusion of this paper in the present Workshop is one of many evidences of the now widespread acceptance of satellite climatology as an active and profitable sphere of atmospheric science.

4.3. Satellite/Sensor Systems

Since the mid-1970s particular progress with satellite observation platforms for atmospheric science has been made in 3 directions. First, in operational spheres the long-standing system of U.S. polar-orbiting satellites (the ESSA/NOAA series), which has functioned almost continuously since February 1966, has been supplemented by U.S. operational geostationary satellites (of the SMS/GOES families) since May 1974, stimulating a big upsurge in interest in mesoscale patterns and processes. Closely associated advances have been made in the design and deployment of atmospheric sounders, most notably the Tiros Operational Vertical Sounders (TOVS) of recent polar-orbiting NOAA satellites, and the Vertical Atmospheric Sounders (VAS) on present geostationary GOES satellites. (See Table I). Meanwhile, operational satellite advances in the USSR, the other major space power during this period, have been similarly impressive, although less information

TABLE I: Climate-related systems on representative recent satellites

Payloads	Missions
NOAA 6, 7, 8(1978 through 1983)	
-Advanced Very High-resolution Radiometers (AVHRR)	Global visible and infrared imaging. Medium resolution direct readout, and stored, imagery. High-resolution direct readout, and limited stored, imagery. Improved sea-surface temperature mapping, and snow and ice discrimination.
-Tiros Operational Vertical Sounder (TOVS)	Direct readout, and stored global, water vapor and temperature soundings.
-Data Collection System (DCS)	Reception and processing of environmental data from fixed or moving platforms.
-Solar Environment Monitor	Measurements of proton, electron and alpha particle densities for solar disturbance prediction.
NIMBUS 7 (1978 in part through present)	
-Earth Radiation Budget (ERB)	Monitoring of Earth and solar radiation fluxes.
-Limb Infrared Monitoring of the Stratosphere (LIMS)	Measurement of gas concentrations, and retrieval of stratospheric temperature profiles.
-Stratospheric Aerosol Measurement-II (SAM-II)	Retrieval of aerosol backscatter profiles, and measurements of optical properties of stratospheric aerosols.
-Stratospheric And Mesospheric Sounder (SAMS)	Measurement of gas concentrations and temperature profiles in stratosphere and mesosphere.
-Solar Backscatter Ultra-Violet Energy & Total Ozone Mapping (SBUV)	Ozone profile retrieval, and evaluations of total atmospheric ozone, solar and terrestrial irradiances.
-Scanning Multichannel Microwave Radiometer (SMMR)	Monitoring sea ice, ocean and land surface conditions, and precipitation. (Recently superceded by the DMSP SSM/l)
-Total Ozone Mapping Spectrometer (TOMS)	
GOES 4,5(1982-1989)	
-Visible and Infrared Spin Radiometer (VISSR)	Visible disc, visible and infrared Scan imaging and readout.
-Atmospheric Sounder (VAS)	Direct readout of water vapor and temperature soundings.
-Space Environment Monitor (SEM)	Monitoring of Earth's magnetic field and energy particle flux, and solar X-rays.

is available on them, and resulting data are generally unavailable for broad-scale climate studies.

Second, the range of research satellites and sensors developed and demonstrated with climatological potential has continued to broaden, especially through the continuation of the excellent Nimbus satellite series until its culmination and completion with Nimbus 7 (see Table I), but also through the U.S. Department of Defense Meteorological Satellite program (DMSP), and new-concept satellites such as Seasat with its active microwave sensors, and satellites built and operated by other countries or groups of countries, most notably the European Space Agency's Meteosat family with its atmospheric water vapour imaging channel, and geostationary satellites from India (Insat) and Japan (GMS). Special mention must be made also of the Earth Radiation Budget Satellite (ERBS) orbitted in 1985 to make observations in wavebands selected specifically for radiation budget studies, including those of a climatological nature, following the great success of the ERB experiment sustained by Nimbus 7.

Third the increasing attempts at co-ordinated, international planning, design and development of satellites and sensor systems for the next 15 years and more, including several capable of sustaining research and operations in climatological time frames. Of these efforts the most notable have been those organised under the U.S. Earth Observation System (EOS) umbrella, now focussing primarily on the system of Space Station Polar Platforms. These are being conceived a megasatellites, much larger than any previously used for Earth observation. (See e.g. McElroy & Schneider, 1985). We will return to these related questions in Section 5.1.

4.4. Satellite Data Analysis Systems & Techniques

The past 15 years has been marked by the rise of both automatic (objective), and interactive (man-machine), methodologies in satellite climatology, these relegating manual methods to relatively minor roles mainly in support of synoptic weather system evaluation and tracking (e.g. hurricanes, based on the detailed type classification of WMO 1979), and low-cost, hard-copy based regional climatological studies (e.g. the preparation of a 10-year Atlas of Potential Rain Clouds over the Sultanate of Oman and surrounding regions by Barrett, Hamilton & Power, 1986). Not surprisingly it is at the larger space scales, and longer time scales, that the objective techniques have become practically universal; inter-active techniques linger on in some satellite cloud motion vector monitoring programmes - and have spread in some medium to small area studies, especially those involving precipitation, but mainly in support of techniques focussing on special and/or extreme events. Examples are the Bristol/NOAA InterActive Scheme (BIAS) for satellite-improved rainfall monitoring, especially of advectional rain systems (see Barrett, et al., 1986); and the Scofield-Oliver technique for severe convectional storm evaluation, now the basis of the U.S. Weather Bureau's IFFA (Interactive Flash Flood Analysis) system (Scofield, 1984).

In many areas the rapid growth and increase in availability of digital processing facilities has encouraged the development of "total systems" approaches to the analysis and interpretation of satellite data, permitting ranges of (output) data products to be derived simultaneously from the original (input) data sets. An exciting example has been the derivation of sets of various quasi-global climatological products from the NOAA multispectral HIRS/AMSU instruments, as exemplified by Fig. 1 (Chahine, pers. comm.).

4.5. Recent Satellite Science & Results

4.5.1. Radiation & temperature. Mapping and monitoring of components of the Earth/atmosphere radiation and energy budgets has advanced rapidly, involving both global-scale inventories (e.g. the 5-year analysis of NOAA data prepared by Winston et al., 1979), and local regional studies (e.g. by Ohring & Gruber, 1983., see Fig. 2). Further, geostationary studies (e.g. by Saunders & Hunt, 1980) have begun to add finer temporal detail than was possible before. From such studies it has emerged that the tropics are generally less reflective, and warmer, than previously believed; significant variations may occur from year to year in radiation budget components, e.g. the global annual average albedo may vary as much as 2-3%; local climates can be typed by their satellite-derived radiation budget and cloud characteristics; and marked diurnal fluctuations may be seen in budget components especially over land. Currently, efforts are being made to intercompare data sets from different satellites and sensors so that these may be used more interchangeably than hitherto.

4.5.2 Atmospheric moisture. Numerous satellite-based studies of global cloud cover have been undertaken, well reviewed by Henderson-Sellers & Hughes (1982). At present interest is being focussed on multispectral classifications, and how such computer-intensive methods might be compressed for broad-scale use. Rainfall evaluation and monitoring from satellites is attracting increasing attention, as the review by Atlas & Thiele (1981) first indicated. Just as importantly, much understanding has been accumulated as to how and why some methods perform better than others (see Barrett & Martin, 1981). The choice of methods and output products is now very wide (Barrett & D'Souza, 1986), and, whilst the current choice for most operations uses visible &/or infrared imagery, passive microwave data, especially from the DMSP-SSM/I (see Table I), carry most promise for future advances. (See, e.g. Spencer, 1986, & Barrett & Kidd, 1987). In the meantime, water vapour data from both the Meteosat imager and the NOAA-Tiros Operational Vertical Sounder have stimulated new interest in the spatial distribution of atmospheric water vapour, and its temporal variations.

In middle and high latitudes, significant progress is being made in the discrimination between cloud, and ice and snow, using either multispectral imagery from NOAA satellites, or microwave data (active or passive). These are already promising improved snow-area monitoring procedures, including some for short-term climatological time periods (e.g. for 7 days; see Harrison & Lucas, in press).

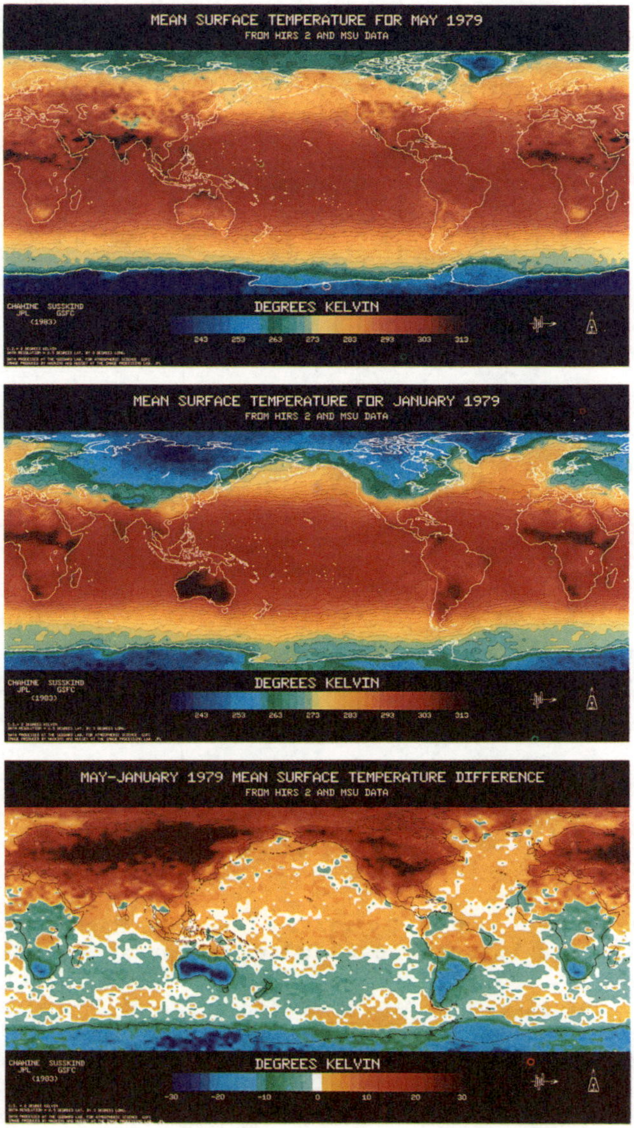

Figure 1: Mean surface ("skin") temperature distributions for May & January, 1979, & the differences from January to May, derived from the High Resolution Infrared Sounders & the Microwave Sounding Units on NOAA satellites. The data analyses method removed the effects of clouds, atmosphere and solar radiation. These are examples of a wide range of global climate products now possible using satellite data. (Courtesy, P.Chahine, JPL, Pasadena, Ca.).

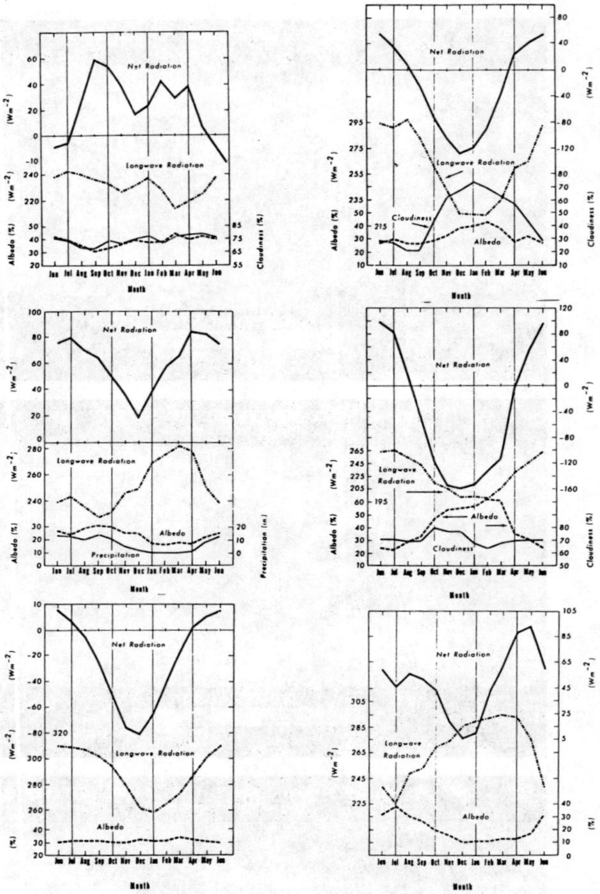

Figure 2: The annual variation of the Earth-atmosphere system radiation budget and cloudiness for an equatorial location (equator, 67.5°W - Sao Gabriel, Brazil), (top left); a subtropical continental location 40°N, 55°E - Aidin, USSR), (top right); an equatorial monsoon location (10°N, 107.5°E - Saigon, Vietnam), (middle left); a midlatitude continental location (52.5°N, 82.5°E - Barnaul, USSR), (middle right); a tropical continental location (25°N, 32.51E - Aswan, Egypt), (bottom left); an equatorial monsoon location (15°N, 70°E - Indian Ocean), (bottom right). Net radiation (-), longwave radiation (-.-.). and albedo (---). (From Ohring & Gruber, 1983).

4.5.3. Winds and air circulation patterns. Considerable progress has been made with the refinement of wind flow and air circulation algorithms and procedures, despite the fact that assessment of these parameters from satellite evidence is often relatively difficult and the results more sporadic. Indeed, wind speed evaluation through objective identification of geostationary satellite cloud motion vectors is now routine; so too is the evaluation of maximum sustained wind speeds in tropical storms and hurricanes; tracking a wide range of synoptic weather systems from satellite evidence has become standard practice in many weather centres.

4.6. New Emphases in Satellite Climatology

During the second half of the lifetime of satellite climatology at least three new areas of emphasis have opened up in the interface between satellite operations and climate studies.

4.6.1. Applied Satellite Climatology. Whereas, understandably, most early studies in all areas of satellite climatology were experimental, there is now a growing emphasis on practical applications, not only in respect of centralised meteorological and climatological services, but also increasingly in support of other weather and climate-affected operations. These include water resources assessment and monitoring, crop prediction, pest control, ice and snow surveys, atmospheric pollution monitoring etc. etc. (See Barrett, 1986). This is a sphere of activity which may be expected to continue to expand rapidly.

4.6.2. Middle & Upper Atmosphere Studies. As our understanding of the lower atmosphere (troposphere and lower stratosphere) has improved, and new satellite sensing systems have been developed to monitor higher levels of the atmosphere it has been recognised increasingly that the atmosphere is, in some senses at least, an integrated whole. Whilst important advances had been made even before the mid-1970s in respect of middle and upper atmospheric circulations (see Section 3.5), more recently attention has been focussed on atmospheric chemistry, and its relations with radiative and dynamical processes. (See NASA, 1987a). Interesting satellite-based evaluations of upper level constituents have already included carbon monoxide mixing ratios in the upper troposphere from Space Shuttle MAPS data, nitric acid distributions in the stratosphere from the LIMS experiment, and ozone distributions in the stratosphere from Nimbus-7 TOMS data. The immediate challenge is to extend such studies, and to formulate a coupled global view of the upper atmosphere as an integral extension of the lower regions.

4.6.3. Integrated Programmes. Since the mid-1970s another new type of activity has begun to open up involving, and with consequences for, satellite climatology which entails levels and degrees of variety and integrated far in excess of any in the previous half-life period. This is in the realm of international scientific programmes, the range of which has been growing since the First GARP Global Experiment

(FGGE), which took place from 1 December 1978 to 30 November 1979. Described at the time as "The greatest single undertaking in atmospheric sciences" (Tanczer et al., 1981), it featured meteorological satellites in a key data-collection role, and brought about the still-unique fully-integrated use of a set of 5 geostationary satellites approximately equally spaced around the Equator. Subsequent programmes drawing data from very extensive &/or widely-separated regions are now under way, including some with broad objectives, e.g. the International Satellite Land Surface Climatology Project (ISLSCP), a wide-ranging but rather loosely coordinated scheme to further the detection and quantitative assessment of climate using satellite-derived information; and others with narrower foci, e.g. the International Satellite Cloud Climatology Project (ISCCP) which is intended to use normalised data from both geostationary and polar-orbiter satellites to provide improved cloud estimates for climate research for 5 years beginning in 1983; and the Global Precipitation Climatology Project (GCPC) to produce a 10-year set of estimates of large-scale, coarse-resolution, 'global' rainfall from geostationary satellite infrared imagery. Now complementary projects are being organised to focus in much more closely on detailed parameters and system component interactions, e.g. the First ISLSCP Field Experiment (FIFE), which is centred on a 15 km^2 area near Manhattan, Kansas. Such integrated projects have captured environmental scientific imagination, and are likely to proliferate. The results will reflect the lively interchanges which characterise the project meetings.

5. PRESENT CHALLENGES AND FUTURE OPPORTUNITIES FOR SATELLITE CLIMATOLOGY

Whilst many opportunities and challenges may be identified in the area of overlap between environmental satellite activities and climate monitoring and research, several seem to merit special mention in the present context. The first involves the so-called 'space segment', the second the 'ground segment', and the third and fourth, dependent science. Some of the key issues involved are as follows.

5.1. The Space Segment

Earlier (see Section 4.3) reference was made to the Space Station project, led by the USA, with anticipated support from a long list of other nations. Although very detailed planning has already been applied to this project (See Table II), numerous problems remain to be resolved. In particular, it remains to be seen whether this system finally emerges as the fully-coordinated, multinational system currently envisaged, with U.S., Canadian, European & Japanese contributions to each of a set of three such platforms. This is because associated political problems have yet to be finally resolved, involving not only the magnitudes of the contributions from individual nations and groups of nations, but also conflicts between military and

TABLE II. Proposed EOS Instruments

Instrument	Measurement
Advanced Data Collection & Location System (ADCLS)	Data and command relay and location of remotely sited measurement devices.
SISP - Surface Imaging and Sounding Package	
Moderate-Resolution Imaging Spectrometer (MODIS)	Surface and cloud imaging in the visible and infrared 0.4-2.2 nm, 8-14 μm. resolution varying from 10 nm to 0.5 μm.
High-Resolution Imaging Spectrometer (HIRIS)	Surface imaging 0.4-2.2 nm; 10-20 nm spectral resolution.
High-Resolution Multi-frequency Microwave Radiometer (HMMR)	1-94 GHz passive microwave images in several bands.
Lidar Atmospheric Sounder and Altimeter (LASA)	Visible and near-infrared laser back-scattering to measure atmospheric water vapor, surface topography, atmospheric scattering properties.
SAM - Sensing with Active Microwaves	
Synthetic Aperture Radar (SAR)	L-, C-, & E-band radar images of land, ocean, and ice surfaces at multiple incidence angles and polarizations.
Radar Altimeter	Surface topography of oceans and ice, significant wave height.
Scatterometer	Sea surface wind stress to 1 m/s 10° in direction K_u band radar.
APACM - Atmospheric Physical and Chemical Monitor	
Doppler Lidar	Tropospheric winds to 1 m/s doppler shift in laser backscatter.
Upper Atmosphere Wind Interferometers	Upper atmospheric winds to 5 m/s doppler shift in O_2 thermal emissions.
Tropospheric Composition Monitors	Trace chemical composition passive emission detectors at wavelengths from UV to microwave.
Energy and Particle Monitors	Solar emissions from 150-400 nm, 1 nm spectral resolution, Earth radiation budget, total solar irradiance, particles and fields environment.

civilian aspirations. An associated problem is that there has been some, though probably not enough, debate as to whether the most effective Earth observation systems of the future should mainly involve relatively large numbers of satellites dedicated to particular sciences, or a small number of much larger multi-mission platforms. Since the latter are the more likely to be configured after inter-disciplinary competition, they are the more likely to end up as compromises. Unfortunately, climatology has been mostly conceived not as a primary, but a secondary, customer for satellite data. As such it seems unlikely that its counsel will prevail in satellite system specification studies. Therefore, with important exceptions including the planned Tropical Rainfall Monitoring Mission (TRMM) (see NASA, 1987b), the future is less bright for satellite climatology than it might be. On the other hand, the recent switch of research interest within meteorology as a whole, away from numerical modelling towards integrated climate studies across interdisciplinary boundaries is strengthening the case for improved satellite climate monitoring. Perhaps the dramatic discovery of the Antarctic "ozone hole", and the convenience with which this can be both demonstrated and monitored by satellites, will be seen retrospectively to have been the most important single factor bringing respect for, and respectability to, climatology from satellites during its first three decades of life.

5.2. The Ground Segment

In addition to the above prospects and problems, the past 15 years have witnessed an exciting new trend in the development of satellite data analysis systems and related techniques. Whereas earlier satellite data reception and analysis facilities were very expensive, and therefore centralised in a very few locations, the costs of satellite ground station and data-processing equipment have fallen sharply. Today there is a growing tendency for satellite receivers supported by modest but effective processing equipment to come within the budgets of smaller nations, regional offices in larger nations, and individual research laboratories and institutes. This trend is sure to increase as micro-computers become more capable, and as packaged applications software becomes increasingly extensive. Already, to cite just one example, it is now possible to purchase and instal a Meteosat Primary Data User's Station (a reception facility able to collect full-resolution digital image data from the Meteosat satellite(s)), supported by two IBM or IBM-compatible PCs (one to manage the data reception and preprocessing, the other to handle data processing and analysis) for approximately $US 35K. (See Fig. 3). Accordingly, the race is on amongst the more established satellite applications scientists, in conjunction with software experts, to prepare, market, and extend packages of programmes at least for regional data analyses.

Figure 4 shows a rainfall-oriented PC programme concept developed recently for a customer in Italy. It features the production of Meteosat-based rainfall products for a wide area using an automatic, bispectral thresholding scheme (code-named ADMIT) for 10, 30, 90 and

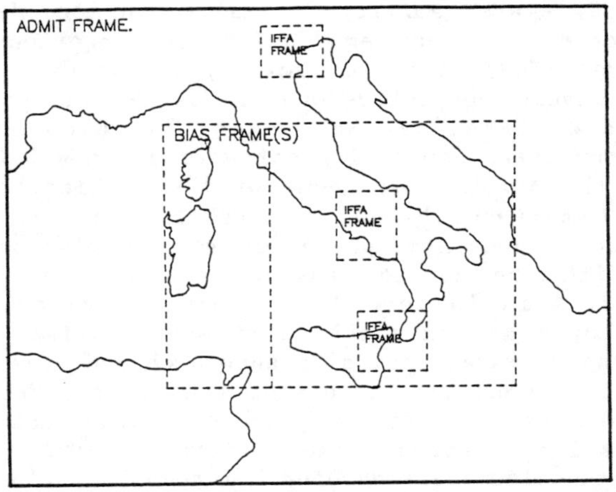

Figure 3: Configuration of a proposed satellite reception facility (Meteosat PDUS), and associated personal computers (PCs) for local satellite weather/climate data reception and processing, and dependent output production.

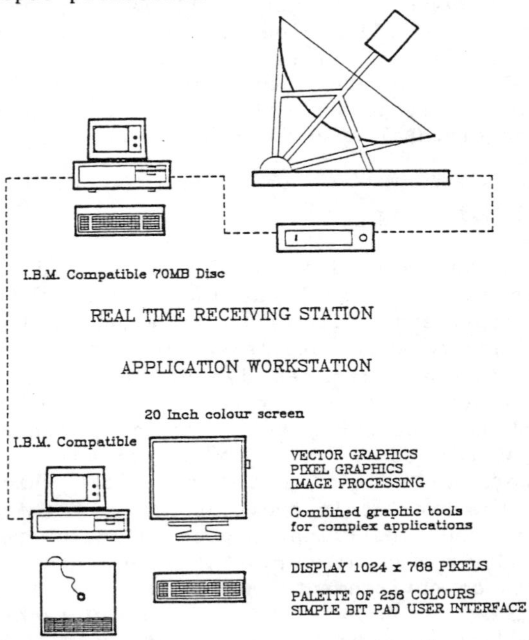

Figure 4: Conceptualisation of a regional satellite/conventional rainfall monitoring operation based on PCs to provide (a) broad-area, climate rainfall products; (b) analyses of local regional rain systems; & (c) forecasts of flash floods. Windows for (b) & (c) can be scaled and located as required.

365 day periods; a related ability to analyse in more detail smaller areas and selected rain systems using the previously-mentioned interactive scheme (BIAS) plus the facility to evaluate in even more temporal detail severe convective storms using further interactive scheme (modelled on IFFA, see Section 4.4). Even more broadly, packages of environmental monitoring software are now being developed for example for the AUTOSAT programme of the British Meteorological office, and the GEM (General Environmental Monitoring) programme of SUPARCO, the Space & Upper Atmosphere Research Commission of Pakistan in conjunction with the Remote Sensing Unit in the University of Bristol, England. This is intended to monitor climate variables including cloud cover and rainfall, plus other variables influenced by, and influencing climate, including snow area and vegetation.

If such advances continue, involving the design, development and operation of satellite/sensor systems, and of ground station reception and data processing systems, the future growth of satellite climatology at all scales seems assured. Indeed, a rapid acceleration of it may be expected. Perhaps the main challenge will be to ensure at least some degrees of international uniformity into what is done and how it is done. Enabled by national donors, the World Meteorological Organisation is currently organising and helping to implement a PC-based system code-named 'CLICOM' for the management and archiving of conventional climatological data in an increasing number of Developing Countries. A similar scheme for satellite data and products would help facilitate future projects seeking to evaluate climate-scale patterns, processes and related problems over regions up to and including continental or oceanic scales.

5.3. Environmental Modelling

The third area of new activity likely to expand rapidly in the foreseeable future is, potentially at least, not only multinational, but also <u>multidisciplinary</u>. It is to be expected that significant progress on climate-related problems such as the increase in atmospheric CO_2, the anticipated depletion of the O_3 layer, El Nino-related modifications of weather patterns, and acid precipitation must raise questions concerning the integrated functioning of the Earth/atmosphere system as a whole (NASA, 1984). The search for satisfactory answers will involve the processes which govern this system, including the hydrologic cycle and biogeochemical cycles, all of which involve weather and climate, plus climatic processes acting, as it were, on their own account. Consequently there will be increasing emphasis on environmental modelling. Much of this will be at the larger scales, which satellites alone can rather conveniently address, but satellites are expected to contribute at smaller scales also, as they have done recently down to the mesoscale in studies of tropical rainstorms, one feature of which has been the 'rediscovery' of stratiform rain in the tropics. (See, e.g. Houze, in press).

5.4. Climatic Change

Other than a few quite modest, short-term, investigations, studies of climatic change based on, or involving, satellite data have been most notable by their absence. This is largely because of the primary prerequisite for such studies: homogeneous data sets of adequate length to support them - and the lack of suitable satellite data sets of such kinds. Until present archiving problems, and questions of the compatability of records from different sources, have been resolved, substantive satellite studies of climatic change will not be possible. In many ways climatic change would seem to be the most elusive challenge for satellite climatologists.

6. CONCLUSION

One of the doyens, and chief visionaries, of satellite meteorology and climatology has been Professor Verner E. Suomi of the University of Wisconsin, USA. He has drawn attention to three distinctive aspects of future satellite programmes; these may be expressed, and qualified in respect of atmospheric satellite activities as follows:

6.1. What <u>should</u> be done

Determination of this in the context of satellites and climate can be made by experts in these fields. Generalising broadly, it seems likely that the provision of data suitable for the improved initialisation and validation of climate models, whether relatively local, regional, or global, represents the highest presently-identifiable objective.

6.2. What <u>could</u> be done

This is limited by presently available observing techniques, plus those which are possible in the light of recent advances in both science and technology. The best approach in the future will be to operate proven instruments in parallel with new observing systems, and, of special importance for satellite climatology, to ensure that provision is made for long-term, reasonably homogeneous, data sets to be built up.

6.3. What <u>will</u> be done

This will be constrained by a combination of scientific requirements, technical feasibility, and the availability of financial resources. Unfortunately, politics may be the over-riding influence. Today unfortunately, the fragmentation of the global effort into a number of international and, increasingly, individual national, initiatives results in much duplication of effort (e.g. in the provision of meteorological satellites to support operational weather forecasting), whilst other aspects of atmospheric science go underprovided. It is a sobering thought that, <u>were the currently available total global</u>

budget for atmospheric satellites and supporting systems to be spent in a fully-coordinated global way, all reasonable scientific objectives which present technology can support would probably be possible. A much improved and better balanced climatological data set would be one, very desirable, result.

Hopefully one of the results of meetings such as this Workshop will be to help ensure that what is done is much more nearly the same as what could and should be done. Politicians often do not deliver to their constituents all they might reasonably expect; it is one proper function of scientists to convince them that they could, and of technologists to show them how. In the case of climate monitoring and forecasting it has to be made clear that the need is great, the problem is pressing - and there are no suitable alternatives to the increased use of satellites for these purposes.

7 REFERENCES

Atlas, D. & Thiele, O.W. (eds.) (1981): Precipitation Measurements from Space, NASA, Goddard Space Flight Center, Greenbelt, Md., 220pp.

Bandeen, W.R., Halev, M. & Strange, I. (1965): A radiation climatology in the visble and infra-red from the Tiros meteorological satellites, NASA Technical Note No. TN D-2534, Goddard Space Flight Center, Greenbelt, Maryland, 30 pp.

Barnett, J. et al. (1972): 'The first year of the selective chopper radiometer on Nimbus IV', Quarterly Journal of the Royal Meteorological Society 98, pp. 17-37.

Barrett, E.C. & D'Souza, G. (1986): The Development of an Objective Range of Dry Day Products for Drought Monitoring by Meteosat over Africa, Final Report, Contract No. 6137/84/D/JS9SC), ESA, Paris, 68 pp.

Barrett, E.C. & Kidd, C. (1987): 'The use of SMMR data in support of a VIS/IR satellite rainfall monitoring technique in highly-contrasting climatic environments', in NOAA Technical Report, NESDIS 35, NOAA, Washington D.C., pp.109-123.

Barrett, E.C. & Martin, D.W. (1981): The Use of Satellites in Rainfall Monitoring, London: Academic Press, 340pp.

Barrett, E.C. Beaumont, M.J., Harrison, A.R. & Richards, T.S. (1985): I^{2}S Users Guide, Final Report, US Dept. of Commerce, Cooperative Agreement No. NA85AA-H-RA010, 72pp.

Barrett, E.C. (1974): Climatology from Satellites, Methuen, London, 418pp.

Barrett, E.C. (1986): 'Satellite Remote Sensing Applied to Climatology', In Report of Study Week on: Remote Sensing and its impact on developing countries, June 16-21, (Eds. C.Chagas & V.Canuto), Vatican Academy of Sciences, pp. 479-502.

Barrett, E.C., Hamilton, M.G. & Power, C.H. (1986): A Satellite Climatology of Potential Rain Clouds in the Region of the Sultanate of Oman, Final Report (Part 1) to the PAWR, Sultanate of Oman, Contract No. C/Z91985, 98pp.

Bjerknes, J. (1969): 'Atmospheric teleconnections from the equatorial Pacific', Monthly Weather Review 97, pp. 163-72.

Carlson, T.N. (1969): 'Synoptic histories of three African disturbances that developed into Atlantic hurricanes', Monthly Weather Review 97, pp.256-76.

Chang, C-P. (1970): 'Westward propagating cloud patterns in the tropical Pacific as seen from time-composite satellite photographs', Journal of the Atmospheric Sciences 27, pp. 133-8.

Endlich, R.M., Wolf, D.E., Hall, D.J. & Brain, A.E. (1971): 'Use of a pattern recognition technique for determining cloud motions from sequences of satellite photographs', Journal of Applied Meteorology 10, pp. 105-17.n

Fritz, S. & Soules, S.D. (1970): 'Large-scale temperature changes in the stratosphere observed from Nimbus III', Journal of the Atmospheric Sciences 27, pp. 1091-7.

Fujita, T.T., Watanabe, K. & Izawa, T. (1969): 'Formation and structure of equatorial anticyclones caused by large-scale cross-equatorial flows determined by ATS-1 photographs', Journal of Applied Meteorology 8, pp. 649-67.

Harrison, A.R. & Lucas, R.M. (in press): 'Multispectral Classification of Snow Using NOAA AVHRR Imagery', International Journal of Remote Sensing, in press.

Henderson-Sellers, A. & Hughes, N.A. (1982): 'Albedo and its importance in climatic theory', Prog. Phys. Geog. 6, pp. 125-163.

Houze, R.A. (in press): 'Convective & stratiform precipitation in the tropics', in Proceedings of the International Symposium on Tropical Precipitation Measurements, Tokai University, Tokyo, Japan, NASA/NASDA, in press.

Hubert, L.F., Krueger, A.F. & Winston, J.S. (1969): 'The double intertropical convergence zone - fact or fiction?', Journal of the Atmospheric Sciences 26, pp. 771-3.

Kreins, E.R. & Allison, L.J. (1969): An atlas of Tiros VII monthly maps of emitted rdiation in the 8-12 micron atmospheric window over the Indian Ocean area, NASA Technical Note, TN D-5101, Goddard Space Flight Center, Greenbelt, Maryland, 25 pp.

McElroy, J.H. & Schneider, S.R. (1985): The Space Station Polar Platform: NOAA Systems Considerations & Requirements, NOAA Technical Report, NESDIS 22, NOAA, Washington, D.C., 155pp.

Miller, D.B. (1971): Global atlas of relative cloud cover, 1967070, based on data from meteorological satellites, U.S. Dept. of Commerce, U.S. Air Force, Washington, D.C., 237 pp.

NASA (1984): Earth Observing System, NASA Technical Memorandum 86129 I & II, NASA, Greenbelt, Md., 117pp.

NASA (1987a): From Pattern to Process: the Strategy of the Earth Observing System, EOS Science Steering Committee Report, 2, NASA, Greenbelt, Md., 140pp.

NASA (1987b): On Requirements for a Satellite Mission to Measure Tropical Rainfall, NASA Reference Publication 1183, NASA Scientific & Technical Information Branch, Washington, D.C., 55pp.

Nordberg, W. (1965): 'Geophysical observations from Nimbus I', Science 150, pp. 559-72.

Ohring, C. & Gruber, A. (1983): 'Satellite radiance observations and climate theory', Advances in Geophysics 25, pp. 237-304.

Oliver, V.J. (1969): 'Tropical storm classification system', in Satellite Meteorology, Bureau of Meteorology, Melbourne, Australia, pp.27-19.

Saunders, R.W. & Hunt, G.E. (1980): 'Meteosat observation of diurnal variation of radiation budget parameters', Nature, pp.283-47.

Scofield, R.A. (1984): 'The NESDIS Operational Convection Precipitation Estimation Technique", in Preprint Volume, 10th Conference on Weather Forecasting & Analyses, 25-29 June 1984, Clearwater Beach, Fla., American Meteorological Society, Boston, Mass., pp.171-180.

Sikdar, D.N. & Suomi, V.E. (1971): 'Time variations of tropical energetics as viewed from a geostationary altitude', Journal of the Atmospheric Sciences 28, pp. 1700-80.

Spencer, R.W. (1986): 'A satellite passive 37 GHz scattering based method for measuring oceanic rain rates', Journal of Climate & Applied Meteorology, 25, pp.754-766.

Streten, N.A. & Troup, A.J. (1973): 'A synoptic climatology of satellite observed cloud vortices over the Southern hemisphere', _Quarterly Journal of the Royal Meteorological Society_ **99**, pp. 56-72.

Streten, N.A. (1968): 'A note on multiple image photo-mosaics for the southern hemisphere', _Australian Meteorological Magazine_ **16**, pp. 127-36.

Tanczer, T., Gotz, G. & Major, G. (1981): _First FGGE Results from Satellites_, Pergamon Press, 332 pp.

Vonder Haar, T.H. (1968): 'Variations of the Earth's radiation budget', in _Meteorological Satellite Instrumentation an Data Processing_, Final Report, Contract NASw-65, 1958-68, Dept. of Meteorology, University of Wisconsin, 179 pp.

Winston, J.S. (1967): 'Planetary-scale characteristics of monthly mean long-wave radiation and albedo and some year-to-year variation', _Monthly Weather Review_ 95, pp. 235-56.

Winston, J.S. et al. (1979): _Earth-Atmosphere Radiation Budget Analyses Derived from NOAA Satellite Data June 1974-February 1978_ 1 & 2, US Department of Commerce, NOAA, NESS, Washington, D.C., 220pp.

World Meteorological Organization (1979): _Operational techniques for forecasting tropical cyclone intensity and movement_, WMO No. 528, Geneva, 78pp.

RADIATION BUDGET AND CLOUDS

EHRHARD RASCHKE
Institut für Geophysik und Meteorologie
Universität zu Köln, Kerpener Straße 13
D-5000 Köln 41
Federal Republic of Germany

ABSTRACT. We briefly review the instrumental and sampling requirements to measure and/or provide adequate data sets for global determination of the planetary radiation budget at the top of the atmosphere, the surface radiation budget parameters, cloud parameters and of the radiative flux divergence within the atmosphere. This paper does not stress other important climate parameters, such as precipitation, wind, water vapour etc. Some consideration is also given to ground-based and other validation measurements.

1. Introduction

Undoubtless, the atmospheric dynamics and consequently an even wider range of dynamical processes within the ocean and at the continents are driven by gradients in radiative energy supply. Without repetition of other introductory remarks we will simply state that, at the present state of our capability to model the climate system, a thorough knowledge of radiation budget parameters at the top of the atmosphere and at ground, and also within the atmosphere, is of crucial importance. A major step forward might be possible once we are able to understand and model cloud-radiation feedback processes correctly !

What quantities do we need to derive ? These are the following:

- at the top of the atmosphere: solar irradiance, reflected and emitted radiation flux densities.
- at the surface: downward and reflected solar radiation, or surface albedo; downward and upward terrestrial (longwave) radiation;
- within the atmosphere: vertical radiation flux divergences and, in particular, cloud field properties: areal coverage, height distribution, liquid and ice water content, possibly particle size distribution, and radiative transfer properties: emittance, reflectance, transmittance and absorptance.

It has been stated in several WCRP and COSPAR reports or elsewhere that the space resolution should be adequate to meet the need for model verification; the time resolution should be good enough to avoid typical sampling errors in annual, monthly and even daily averages. Such observations should also allow analyzing typical diurnal variations which are associated with or even force dynamical systems. Further error sources, due to instrumental and orbital limitations of the observing system, must be controlled.

A. Berger et al. (eds.), Climate and Geo-Sciences, 547–549.

2. Quantities to be measured

In this connection we can only briefly discuss the extent to which present and future satellite and instrumental systems could meet all these requirements.

2.1 RADIATION BUDGET AT THE TOP OF THE ATMOSPHERE

The instruments provided by the Earth Radiation Budget Experiment (ERBE), flown onboard two polar-orbiting (NOAA 9 and 10) and one low-inclination (ERBS) satellites are at present the best one might obtain. But studies have shown that they - in particular the scanning radiometers - need improvements. Inversion errors are still large due to insufficient knowledge in angular dependence of the outgoing radiation fields. Temporal sampling errors might be overcome by the use of properly corrected measurements of geostationary satellites. In general, one would whish operational geostationary satellites to carry well-calibrated "broad-band" radiometers. A new radiometer type, the <u>ScaRaB</u>, is being developed by French, Soviet and German groups and should be ready for launch in 1991.

2.2 RADIATION BUDGET AT GROUND

Radiation quantities at the surface can only be derived, not measured from span snice in particular the longwave parameters are completely decoupled from outgoing radiances in the case of a cloudy atmosphere. Of critical importance are the optical properties and altitude range of clouds, but also the surface temperature and near surface air temperature. We would expect considerable improvements, once passive spectral imaging measurements and imaging laser data are available. But, one also urgently requires a well-balanced ground-based radiometer-network, with due consideration to sampling and maintenance requirements. No serious measurement system has been conceived yet for the vast oceanic areas or the climatically effective polar sea-ice sheets.

2.3 RADIATION BUDGET WITHIN THE ATMOSPHERE/CLOUDS

The vertical flux divergence in clean air can be computed from data which are operationally available now. One important parameter is still required: aerosol optical density and disperties which might be obtained in the mid-nineties by imaging lidar data.

Clouds are again the dominating factor controlling radiation energy gains and losses. One needs to measure: areal coverage and altitude range, radiative transfer properties for the solar and terrestrial radiation (the "absorption paradox" is still making its way through the literature), liquid and solid water contents - but also the life cycle and variations of these parameters in relation with the large-scale circulation systems.

The present operational observing systems, based on passive measurements only, cannot provide all these parameters, although sophisticated algorithms - such as the operational ISCCP cloud-retrieval algorithm - have already been developed to estimate cloud and radiation properties. A major step forward may be expected once active <u>imaging</u> optical (lidar) and microwave (radar) systems become available.

But the validation of such inferred products is an extremely difficult task due to extremely high spatial (and also temporal) variability of most cloud field properties. Multiple-aircraft experiments - which are very difficult and subject to many measurement errors - are urgently required in many parts of the world.

3. Concluding Remarks

We find that observing systems, which will be available in the next years - with some instrumental improvements - provide already most of the basic data to derive the above mentioned climatic parameters. A significant improvement will be obtained with imaging lidar and radar systems (the latter will also map better than before precipitations) and adequate computing power (and research funds) for analysis of comprehensive data sets from multiple sensors.

One must not overlook the need for a new and reliable ground-based radiometer network, complemented by sporadic airborne surveys.

4. Literature

Major references are given in relevant publications in the WCP and WCRP report series, especially:

WCP-95 ISCCP Catalogue of Data and Products (WMO, 1985)

WCP-115 Surface Radiation Budget for Climate Applications (WMO, 1986)

WCP-136 Radiative Flux Measurements (WMO, 1987)

WCRP-10 Radiation and Climate, JSC Working Group on Radiative Fluxes (WMO, 1988)

SECTION 5 – CLIMATE ENVIRONMENT AND INTERNATIONAL SECURITY

THE GLOBAL COMMONS AND NATIONAL SECURITY

Paul R. Ehrlich
Department of Biological Sciences
Stanford University
Stanford, CA 94305
U.S.A.

ABSTRACT. Deterioration of the global environment, in part a result of runaway human population growth, is a major threat to the security of all nations. Rapid change in the atmospheric commons that (among other things) will alter climate and decrease agricultural productivity, is a key component of that threat. Gaining control of that commons will involve some degree of surrender of national sovereignty, an insinuation of a form of world government from below, rather than (as in the standard model) an imposition of authority from above. A substantial diversion of funds from military enterprises that decrease security to attempting to maintain the habitability of Earth is urgently needed.

World government has been a common dream of those who wish to end the constant wars that plague humanity and seek to avoid thermonuclear Armageddon. At the end of the Second World War, when the United Nations was launched with 51 member states, many hoped that it would evolve into such a government. Today that hope is largely gone, even though the U.N. has grown to over 150 member states and the positive roles played by that organization are much more substantial than it usually gets credit for. Nations have not shown the slightest interest in surrendering any sovereignty; if anything nationalism has become more widespread -- and in many cases more virulent.

Most human beings quite clearly desire to live in peace. Nonetheless, distrust between nations is such that the establishment of a world federation empowered to settle international disputes is not possible at the moment. The threat of nuclear war is not sufficient stimulus; however real the prospects, such an event remains almost unimaginable. Government officials themselves are a barrier to federation, since it is <u>their</u> power that would, in part, be

A. Berger et al. (eds.), Climate and Geo-Sciences, 553–562.

given up. Worse yet, times when the international situation seem most threatening are exactly those that bring chauvinism to the surface in both leaders and populaces. The very circumstance that should elicit a strong movement towards federation, a palpable increase in the chances of war, instead elicit responses exemplified by slogans such as "better dead than red."

A nuclear exchange between the superpowers that exhausted a substantional portion of their nuclear arsenals could destroy civilization in the Northern Hemisphere in a very short time. Prompt deaths could number 500 million or more (WHO, 1984, using what is probably an unrealistically pessimistic scenario, estimated 1000 million). Environmental effects, mediated primarily through atmospheric changes, could multiply that number severalfold (Ehrlich et al., 1983). There has been controversy over the exact magnitude of climatic change that would follow a large-scale nuclear conflict --that is, the severity of the "nuclear winter." But results of the most sophisticated recent atmospheric studies (e.g., Covey et al., 1985; Thompson, 1985) indicate that biological damage would be extremely severe even if the most optimistic models are correct. It makes little difference to corn or wheat, for example, whether the sky is pitch black and temperatures fall to -40° C in July in Iowa, Bordeaux, and Kazakhstan or whether there simply are a few weeks of twilight and scattered frosts. Whatever agricultural potential remained in the absence of inputs and social organization would be destroyed in either case by cold, low light levels, high levels of ionizing radiation, increased UV-B flux, and other less widely discussed atmospheric sequelae of the war. Civilization clearly could "end with a bang."

Destruction from war is not, however, the only catastrophic threat to national security as the world moves toward the twenty-first century. The other is arguably as serious, but less widely perceived. Today the human population is over five billion people, and the world is badly overpopulated. That can be seen from the simple fact that the present population can only be supported (with billions living in poverty) by consuming a one-time inheritance from Earth (Ehrlich and Ehrlich, 1986, 1987). We are destroying our capital -- not just petroleum and other mineral resources but rich agricultural soils, "fossil" groundwater, and biodiversity (populations and species of other organisms). These resources are not renewable on a time-scale of interest to society. They have accumulated in periods ranging from millennia to hundreds of millions of years, and are being dispersed and destroyed in decades to centuries.

Even if there is no third world war, if current trends in human population growth and per-capita impact on the global environment continue, then civilization will probably end much as it would following a nuclear winter. The only difference would be that, rather than collapsing in a matter of months, it will take decades. Civilization can also "end with a whimper."

Changes in the atmosphere would virtually certainly play a major role in such a collapse, especially by reducing agricultural productivity and accelerating the deterioration of natural ecological systems. But in all likelihood, epidemic diseases, terrorism, revolution, and local wars would play a substantial role. A key point to be made here is that there are many dimensions to national security other than the military one. While the nations of Earth spend roughly a trillion dollars annually on their armed forces, the overall <u>security</u> of every country is rapidly declining.

Civilization ending with a whimper would amount to a gigantic "tragedy of the commons" (Hardin, 1968). People and nations acting independently in ways that are in their short-term best interest create situations that in the long term destroy them. As Hardin said, "Ruin is the destination toward which all men rush, each pursuing his own best interest in a society that believes in the freedom of the commons. Freedom in a commons brings ruin to all." Our ruin will stem both from the human treatment of the biosphere as a commons with almost no international controls, and the organization of human societies in ways that make it difficult to recognize or act on that fact. But in the potential tragedy lie the seeds of salvation -- a strategy to achieve a degree of world government.

No feature of Earth is more a commons, a feature of the planet used by all and owned by none, than the atmosphere. The atmosphere interacts intimately with both oceans and soils. It connects all nations and people, past and present. You have breathed oxygen atoms that Julius Caesar and Chairman Mao breathed; the temperature of the atmosphere over the Arctic can influence grain yields in Kansas, and driving automobiles, generating power from fossil fuels, and eating fast-food hamburgers in the United States affects the temperature of the Antarctic atmosphere. By traveling through the atmosphere, people can quickly move from one place on Earth to another; by altering the atmosphere, nations can assault one another, even if they don't intend to.

Of course, changes in the atmospheric commons concern us only to the degree that they influence biological systems, which includes human societies. The major route of such influence is, of course, recognized in the concept of an ecosystem -- the biological community in an area plus the physical environment with which it interacts. The atmosphere is a major component of the global ecosystem, being greatly influenced by the organisms of the planet. Photosynthesizing organisms, for example, are thought to have generated virtually all of the oxygen in the atmosphere, largely in the past; and plants, animals, and microorganisms play a major role today in the maintenance of its gaseous composition and a lesser role in the patterns in which those gases circulate. Reciprocally, the atmosphere influences the organismal component of ecosystems, since its makeup helps create conditions that make life itself possible.

Ecosystems provide human society with an array of free "ecosystem services" without which civilization would soon collapse. Some of these centrally involve the atmosphere -- keeping the air breatheable, ameliorating the climate, supplying fresh water. In others, the atmosphere is involved largely to the degree that its qualities are supportive of the lives of the organisms providing the services. Forests are a key to flood control, but if the atmosphere continues to be loaded with oxides of nitrogen and sulphur, that ecosystem service may ultimately be lost. The waste disposal and nutrient recycling services of ecosystems, and their provision of food from fresh water, may also be compromised if atmospheric influences lead to increased acidity of soils, lakes, and streams. The ability of ecosystems to provide pollination and pest control services and to maintain nature's vast "genetic library" can be reduced by climate change. If change is very rapid, it carries the risk of damage to ecosystems regardless of the direction of change.

The last service is especially important. Humanity has already withdrawn from that library the very basis of civilization in the form of domestic plants and animals, industrial products, and medicines -- and yet the potential of the library has barely been tapped.

All of this makes clear why there is a close connection between the state of the atmospheric commons and international security. An increase of a few degrees in global temperature, as is expected to follow the buildup of greenhouse gases, could change patterns of atmospheric circulation, altering both the climate and composition of the atmosphere (e.g., load of dust or pollutants) over large areas.

Consider only some of the most obvious possible effects of such an event. The agricultural productivity of the great plains of North America might be severely reduced by heat stress while the already shaky grain production system of the Soviet Union might fall apart. The economic situation of both nations would deteriorate seriously, and the U.S. might be unwilling or unable to sell grain to the Soviets. Tension between the superpowers could increase greatly, and those tensions could well be fueled by agricultural problems in client states, which could no longer depend on food imports. For instance, over 100 nations now depend to one degree or another on grain imported from North America to feed their populations (FAO, 1986).

Changing rainfall patterns, virtually certain to accompany a warming, could exacerbate tensions over increasingly scarce water supplies in places as disparate as the Middle East and western North America. If rainfall becomes sparser in Israel and its neighbors, an already explosive situation would become much worse. Egypt is already tottering on the brink of chaos, and further reduction of the flow of the Nile could push it over the brink. The water in the Jordan, Yarmuk, and Litani rivers already are a central source of tension, and Israel may be very short of water in the 1990s (Cooley, 1984). The hardnosed Israeli stand on the West Bank in 1988 may just be a small taste of what is to come. Similarly, a significant decline in the dependable supply of water available from the Colorado could lead to serious trouble between Mexico and the United States. Conflicts could be greatly increased if more and more Mexicans seek to cross into the United States in order to make a living (Ehrlich et al., 1979).

The flow of ecological refugees in some parts of the world seems bound to increase as agricultural conditions change. In some cases, of course, climatic change may relax tensions. But even where the transformation is beneficial in the long run, disruption associated with change itself may cause serious problems, especially when that change is rapid (Schneider, 1988). Thus increased rainfall in semi-desert areas may manifest itself primarily as flooding, decrease food supplies in the short term, and lead downstream nations into conflict with upstream ones that open dams and pass floodwaters downstream.

Other places where international tensions could be heightened by reduced river flows are between Pakistan and India (over the Indus) and India and Bangladesh (Ganges). Disputes over water have already occurred in those places and in the basins of the Tigris/Euphrates, Amur, Mekong, Rio de La Plata, and Nile. The water of 120 of the world's 200 major river systems is shared by two or more countries (Myers, 1985); with the entire planet overpopulated and serious regional water shortages already widespread, the potential for drought-induced conflict is enormous.

Transnational air pollution problems also will be altered by global warming, but in ways that are virtually impossible to predict. Forest systems, already stressed by air pollution, are not likely to react well to rather sudden changes in temperature and rainfall regimes. Indeed, there is evidence that some forests are already being degraded by climate change (e.g., Hamburg and Cogbill, 1988). And in the longer term, atmosphere-ocean coupling and changed freshwater flows could change the dynamics of oceanic fish populations and at least temporarily reduce productivity. Per capita fisheries productivity has been declining for almost two decades, a trend that is likely to continue even without changes in oceanic environments that are (in terms of evolutionary time) very rapid. Shots have been fired over fishing rights; that may only be a hint of things to come unless fisheries management aspects of the Law of the Sea can be kept in effect.

Climate change obviously will seriously interfere with the crucial "genetic library" maintenance function of natural ecosystems. Biological reserves increasingly are islands of relatively natural habitat in seas of human disturbance. Even if those reserves were flawlessly guarded against human invasion and development, they cannot be protected against climate change. Terrestrial organisms have evolved in response to climatic regimes. Most populations are relatively intolerant to substantial and rapid changes in those regimes. Throughout Earth's history, species have responded to climate change by gradually changing their geographic distributions and by evolving tolerances to new conditions. But today migration is often not possible, since it would have to occur through impassible barriers of disturbance. And the rate of climate change now envisioned for the near future will be much too rapid for all but the most rapidly reproducing organisms (such as bacteria and some insects) to have a chance of surviving by genetic adaptation.

Of course, climatic change is not always undesirable. Even in the short run, some agricultural areas of marginal rainfall might receive somewhat increased precipitation and produce larger or more stable yields. In the long run, warmer temperatures or increased precipitation might open new areas to agriculture, although problems with soils may often have to be overcome. But overall, one must assume that rapid climatic change will be harmful simply because social systems, agricultural systems, and natural ecosystems are adapted to current conditions and tend to change slowly.

In A.E. Housman's words, we must "train for ill and not for good." We cannot risk the future of civilization on the Pollyanna-ish hope that climate change will improve the lot of humanity as it struggles in the next few decades with the most serious problems it has ever faced.

The world is already suffering from overpopulation, shortages of resources, and increasing environmental deterioration. Humanity is suffering from an increasing maldistribution of wealth, including a growing gap between rich and poor nations, and rampant racism, sexism, and religious prejudice -- all of which make finding solutions to pressing global problems more difficult.

The human epidemiological environment grows ever more precarious as human numbers increase, the average susceptibility to disease increases (with malnutrition on the rise), and the speed with which disease carriers can move around the globe also increases. Malaria is resurgent, and AIDS has the potential for destroying a large portion of humanity. The AIDS virus is the third lethal virus to threaten a large-scale epidemic since 1967. Whether or not it can be contained will depend primarily on how rapidly education can slow its spread, when <u>and if</u> the medical community can find satisfactory preventatives or treatments, and, above all, luck. A virus that infects millions of people may well evolve new transmission characteristics, and if the ability of AIDS to be transmitted heterosexually increases or it evolves the ability to concentrate in sputum, for instance, an already catastrophic situation could worsen.

Climatic change induced by global warming therefore is just one of many gigantic problems facing humanity, and not even the only one that is primarily atmospheric. After all, there is also the problem of increased UV-B flux as a consequence of ozone depletion (ozone depletion will, of course, also contribute to climate change), acid deposition, and other forms of air pollution. But climate change alone has great potential both for for contributing to international conflict and for contributing to a decline in the economic security of all nations. Aside from large-scale war and possibly epidemic disease, no single facet of the human predicament is more frightening.

There is, however, one positive aspect to the prospect of climate change. It clearly is an issue that must be dealt with globally. Leaders may still be convinced that their nations can avoid war if they keep themselves armed to the teeth. Superpowers may imagine that they can impose peace on the world. Even if peace-keeping fails, a few countries such as New Zealand might be able to ride out the effects of a large-scale nuclear war with its society battered but intact. By use of quarantines and isolation, some communities might conceivably be able to escape the ravages of AIDS or other worldwide epidemic diseases. But it can be made clear to everyone that there is no practical way for any segment of humanity to avoid the consequences of a rapid change in global climate, no national policies that can lead to national security.

If the rate of change is to be slowed, as it should be, and its effects are to be minimized, as they must be, it will require worldwide cooperation. In essence, we must bring world government in through the back door, by establishing an Global Commons Regime with power to regulate human interactions with the atmosphere and oceans. A little reflection shows, of course, that such a regime would have to regulate to some degree such diverse activities as agriculture, the clearing of tropical forests, the generation of power, the use of internal combustion engines, the manufacture of plastics and hair sprays, and the disposal of toxic wastes. In order for a Commons Regime to accomplish anything, nations would have to surrender considerable sovereignty to it -- something they have been unwilling to do in the hope of achieving peace.

There is no guarantee that an effective Global Commons Regime could pull humanity through the coming crises, but it seems certain that, without such a transnational effort to deal with global environmental problems, civilization will collapse. One must hope that approaching the surrendering of sovereignty in an area where sovereignty is obviously meaningless, that gradually and cooperatively introducing much more international control of key activities, will be simpler than attempting to impose world government from the top down. The task is barely begun, and what needs to be done will require effort that is nothing less than monumental. But the benefits for all of humanity if the commons can be regulated would also be monumental.

A central problem facing us at this conference therefore is finding ways to convince decision makers and the general public that opportunities for action are fast slipping away. Actions to try and slow climatic change should be taken now, ranging from much more intensive campaigns to increase the efficiency of energy use (to decrease the production of CO_2) and place greater restrictions on chlorofluorocarbon manufacture to massive subsidization of poor tropical nations for environmental preservation and restoration. Indeed, regrowth of forests worldwide should move near the top of the human agenda. The pioneering Guanacaste project in Costa Rica, regrowing tropical seasonal forests, can provide a model for similar projects worldwide. The project has been created through the energetic determination of one brilliant ecologist, Daniel Janzen, but it involves the full cooperation of local people. Research and development of energy-mobilizing technologies that do not deposit CO_2 in the atmosphere should be accelerated, including the development of nuclear power reactors that have safety as a primary characteristic.

Above all, humane programs to bring population growth to a halt and start a slow decline are desperately needed. Overpopulation both contributes directly to the factors causing climatic change, and makes its potential consequences more dire. We must move as rapidly as possible toward a number of people that Earth can support in comfort primarily on income (Ehrlich and Ehrlich, 1986, 1987). The prospects of trying to deal with the dislocations caused by climatic change alone (to say nothing of the other elements of the human predicament) in a world with 8-12 billion people are daunting indeed. Consider, for example, what would happen if there were a birth control miracle in India, and that nation achieved replacement reproduction around 2025. In that case India's population would continue to grow until almost the end of the next century, and when it stopped India would have about two billion people -- the number on the entire planet in the year of my birth, 1932. Picture what vagaries in the monsoon would mean to two billion Indians!

Since, regardless of any actions started now, climatic change will occur while Earth is vastly overpopulated, planning should be started now for dealing intelligently and humanely with the inevitable dislocations. First and foremost, part of the money now being wasted on weapons should be diverted into enterprises that will increase the security of nations. Establishing many more agricultural experiment stations and extension services worldwide to help farmers improve their food production and respond to climate change is an obvious requirement. So is improvement of long-term food storage and transport facilities to help insure against catastrophic famines. You can think of others. The things that need to be done to help humanity get through the crisis decades ahead are fairly obvious; generating the social and political will to get them done will be horrendously difficult. Atmospheric scientists have special leverage since they deal professionally with the most important global commons. It is a heavy responsbility, and one that I hope you will discharge with vigor.

REFERENCES

Cooley, J. 1984. The war over water. Foreign Policy 54:3-26.

Covey, C., S.L. Thompson, and S.H. Schneider, 1985. "Nuclear winter": a diagnosis of atmospheric general circulation model simulations. J. Geophys. Res. 90:5615-5628.

Ehrlich, P.R., 1986. Ecology of nuclear war: population, resources, environment. In A. Cohen and S. Lee (eds.), Nuclear Weapons and the Future of Humanity: The Fundamental Questions, Rowman and Allanheld, Totowa, N.J., pp. 85-101.

Ehrlich, P.R., 1988. The ecology of nuclear war. In P. Ehrlich and J. Holdren (eds.), The Cassandra Conference, Texas A & M Press, College Station, pp. 181-199.

Ehrlich, P. and A. Ehrlich, 1986. World population crisis. Bull. Atomic Sci., April:13-19.

Ehrlich, P. and A. Ehrlich, 1987b. Earth. Franklin Watts, New York.

FAO, 1986. Production Yearbook, 1985. Rome.

Hamburg, S.P. and C.V. Cogbill, 1988. Historical decline of red spruce populations and climatic warming. Nature 331:428-430.

Hardin, G. 1968. The tragedy of the commons. Science 162:1243-1248.

Myers, N. 1985. The environmental dimensions of security issues. Mimeo

Schneider, S.H. 1988. The greenhouse effect: what we can or should do about it. Proc. First North Amer. Conf. on Preparing for Climate Change, in press.

Thompson, S.L. 1985. Global interactive transport simulations of nuclear war smoke. Nature 317:35-39.

WHO, 1984. Effects of Nuclear War on Health and Health Services. Geneva.

THE GLOBAL ENVIRONMENTAL ISSUES - AN APPROACH TO THEIR SOLUTION

P. USHER
UNEP
P.O. Box 30552
Nairobi
Kenya

ABSTRACT. The paper identifies present environmental issues affecting, or predicted to affect, the world. Using the issue of ozone layer depletion as an example, the paper shows how a solution can be developed and applied. It draws attention to the risks and inadequacies of uncoordinated response and recommends an internationally orchestrated programme as a means to preparing an effective and rapid solution.
The paper identifies scientific research and assessment as the basis for political response and suggests that advantage should be taken of the international organizations as a suitable mechanism for catalyzing the intergovernmental action. The predicted global climate change is discussed in the light of experience gained in addressing stratospheric ozone depletion and an outline is given of the ongoing programme addressing the issue.

INTRODUCTION

The Global environmental issue is a relatively recently recognised consequence of human misuse of the earth's resources. There are several unique aspects associated with global environmental issues which require their being addressed in ways for which there are no precedents.

While recent times have thrown up several issues which could be categorized as global in scope, most are closely related, requiring as they do a transport mechanism to deliver the pollutant or introduce the second order impacts to a region geographically distanced from the source. The atmosphere, therefore, as a consequence of its efficient global circulation of air both horizontally, vertically and interhemispherically, ensures the global nature of the most pressing of the environmental issues under review :

The risk to the stratospheric ozone shield.
Climatic Change caused by greenhouse gas emissions.
Long range transport of pollutants - acid rain.
Atomic radiation resulting from nuclear power related accidents.

A. Berger et al. (eds.), Climate and Geo-Sciences, 563–575.

Another problem currently generating major international attention is the spread of communicable diseases such as AIDS.

Most future global environmental issues are likely to have considerable similarities with those present problems listed above, generating some hope that if effective solutions for current issues can be identified and implemented then problems of the future are not likely to be insoluble.

This paper will deal only with the issues of risks to the ozone layer and of greenhouse gas emissions. A probable solution to the former issue has been agreed and is currently being implemented. The second issue has sufficient similarities with the ozone layer problem to suggest it might be addressed in a similar way. However, it has greater complexities which, despite the experience of addressing ozone depletion, will make finding the appropriate solution for climatic change extremely difficult.

The cause of environmental concern not infrequently conflicts with commercial interest. The environment receives short shrift when in competition with industrial profitability. Near term advantage invariably triumphs over long term resource sustainability. The voice of caution, no matter how authoritative, is seldom heeded until environmental degradation reaches proportions compelling action against the polluter. The first theories on ozone depletion were published in the 1970s yet it will be the closing decade of the century before corrective action is initiated. The lesson here is obvious, protection of the environment will not automatically follow expressions of concern. A minimum of convincing scientific argument is a prerequisite for governmental directives against the polluter, a situation made doubly difficult in the case of ozone depletion and climatic change where quantitative proof will be synonymous with irreversible environmental change. Theoretical prediction of ozone destruction and climatic warming must have the soundest scientific basis if it is to be accepted as an indicator of future environmental damage.

RISKS TO THE OZONE LAYER

Global problems require global solutions. National response no matter how drastic are palliatives rather than remedies. In the mid 1970s, the United States and some other countries applied strict governmental control on the use of chlorofluorocarbons as aerosol propelelants, restricting their use to essential purposes only - the delivery of medicinal preparations for example - and effectively cutting CFC use in this sector by 90% nationally. The effect on global CFC production was immediate but short lived. By 1984 estimated global production of CFCs had once more risen by 16% in the two previous years bringing it to only slightly below (-13%) the 810 kt produced by CMA reporting companies in 1974, the peak year[1] (Figure 1). The reason for the recovery lies in changing use patterns of CFC 11 and 12 with demand moving away from the aerosol sector into other applications - refrigeration, air conditioning, open and closed cell foam production. The increasing use of other CFCs, such as CFC 113, used extensively as a solvent in the electronics industry and the development of Halons, bromine containing compounds similar to CFCs, used extensively as fire retardants despite their high ozone depleting potential - estimated at ten or twenty times that of CFC 11 and 12[2] were further factors threatening the ozone layer.

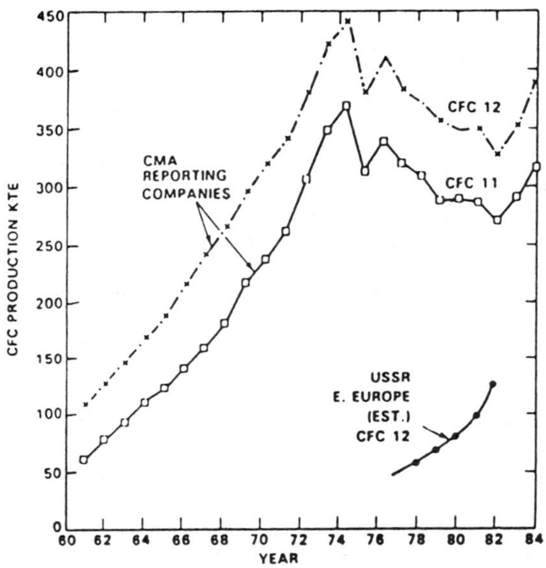

Figure 1. Chlorofluorocarbon production history for CFC-11 and CFC-12, 1961-84 for CMA reporting companies. Soviet Union/Eastern European production for CFC-12 is estimated separately, USSR production is based on Borisenkov and Kazadov, 1977 with estimates continued by CMA to include Eastern Europe from 1976 through 1982 (CMA, 1984).

It is possible to conclude that unilateral action in defence of the ozone layer had not only limited global effect but might also have compromised some national negotiating positions when global controls were being determined. Nations with previous little or no production and use control, now obliged to limitations under international law, could at least make their consumption cuts in the previously unregulated or partially controlled non-essential use sectors. Unlike the United States, Canada and the Scandinavian countries which had already trimmed that particular fat and would be obliged to cutback elsewhere. Nor would unilateral action defuse the time bomb of current non-users becoming the major consumers and producers of the future. regulatory controls, as will be seen later, would have to take account of genuine development needs, particularly of the emerging nations, and at the same time ensure a degree of control consistent with a worldwide reduction of chemical emissions sufficient for environmental safety.

However, to agree and effect a global pollution control process there is need for an international process and forum for debate. In the case of ozone layer depletion, the United Nations Environment Programme, with its wealth of experience in environmental protection and in formulating agreements among nations for pollution control served as the vehicle for international action[3].

THE CONTROL OF OZONE DEPETING SUBSTANCES

Ozone, the triatomic form of oxygen, exists naturally in the atmosphere, particularly in the stratosphere between 10 and 40 kilometers above the earth's surface. It is produced by photochemical action and destroyed by chemical decomposition. Although exhibiting annual, interannual and spatial variability, under normal conditions, it exits in dynamic balance protecting biological life on the planet through its ability to filter out damaging solar ultraviolet radiation of wavelength below 320nm. Theoretical work by Rowland, Molina and others suggested that the balance of ozone production and loss could be upset by the effect of certain trace gases, particularly the manmade chlorofluorocarbons, leading to significant ozone depletion and consequent risk to human health and biological productivity. According to theory, CFCs, although stable and non-creative in the troposphere (properties that commend their use in a variety of industrial processes and domestic appliances) could, in the high atmosphere, be broken down by sunlight to release chlorine atoms, which are catalytically reactive with ozone. Each chlorine atom, not being removed itself during the reaction process, would continue to exist in the stratosphere for long periods destroying many thousands of ozone molecules.

Based on current production rates it was predicted that a 5% reduction in total column ozone would eventually occur and, if production increased by as much as 3% per year then by the middle of the next century ozone depletion might exceed 25%[4]. As 1% ozone depletion might increase human skin cancer incidence by as much as 4-6% the threat of ozone destruction of this magnitude was alarming[5] (Figure 2).

But ozone is not only a filter for UV-B; by absorbing ultraviolet radiation it heats the surrounding atmosphere. Thus, a change in the abundance of ozone may change the temperature structure of the atmosphere and thereby alter the atmospheric circulation[6].

Ozone destruction is also predicted to be accompanied by changes in its vertical distribution in the atmosphere. Unnatural amounts of CFCs and ozone in the troposphere will contribute significantly to the greenhouse effect. Yet, despite this threat, in general undisputed by the scientific community, the global consumption of CFCs continued to rise notwithstanding the application of national controls by some countries. There is little reason to doubt that without global ragulation production would continue to rise, despite the predictions of risk, to the point where serious and virtually unrepairable damage could be suffered by the environment.

The classical UNEP approach to a potential environmental problem is assessment of the nature and scale of the issue, identification of the

GROWTH SCENARIO

(Impossible Under Protocol)

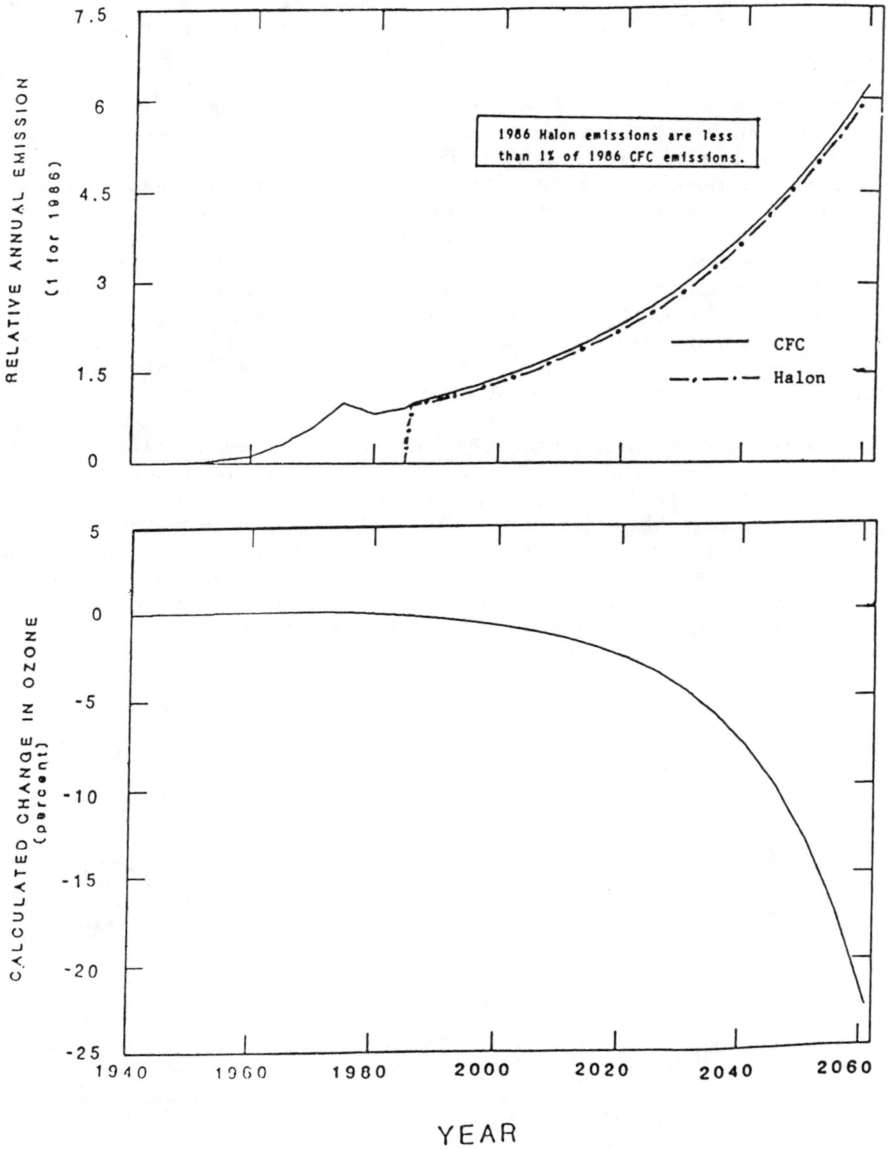

most appropriate response and assistance in its application for rational environmental management. The issue of ozone layer depletion was drawn to the attention of UNEP by its Governing Council in 1976. The following year UNEP organized an international meeting on the problem which recommended a World Plan of Action for a research programme and the Assessment of Ozone Layer Modification to be implemented by Governments and Specialised Agencies of the United Nations Systems[7]. UNEP's role would be as co-ordinator of the programme. Assisting UNEP in this task was a newly convened Co-ordinating Committee on the Ozone Layer made up of those Organizations implementing the plan, governments with active research programmes concerned with the ozone layer, and non-governmental organizations such as the International Council of Scientific Unions and the Chemical Manufacturers Organization.

The requirement of a convincing scientific basis for action was convincingly demonstrated to UNEP at an early stage. Its first assessments were widely criticized following major variations in the annual estimates of future ozone depletion - from 7% in 1977, 15 to 20% in 1978 and 1979 and 2 to 3% by 1982[8] (Figure 3). The fact that these re-evaluations owed less to discovered errors in the theory of ozone destruction and more to a growing realisation that there was a greater number of substances and chemical reactions involved in ozone modification than was previously suspected and that many of the reactions were interdependent and often behaved differently in the troposphere from the way they did in the stratosphere did not satisfy those resistant to CFC control. From 1981, better understanding of ozone layer chemistry and dynamics resulted in depletion estimates in a more narrow band and pressure was mounting both within and outside UNEP for action to reduce the environmental threat posed by the new escalation in CFC emissions.

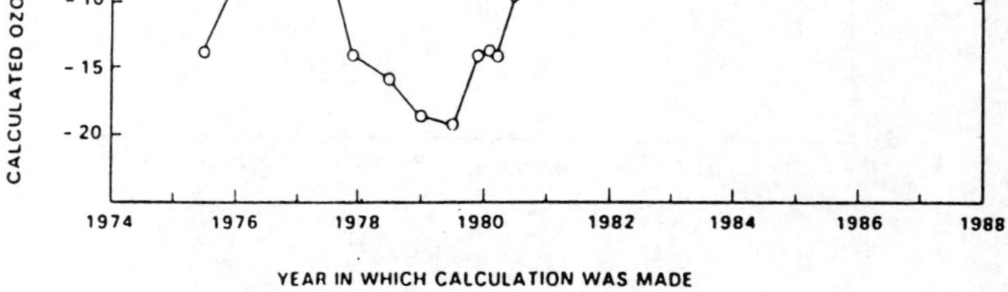

Figure 3. Calculated ozone-column change to steady state CFC-11 and CFC-12 emitted continuously at 1974 rate.

In accordance with its Governing Council decision[9] UNEP began a programme aimed at developing a global framework convention for the protection of the ozone layer. Governmental legal and technical experts met regularly over a four year period to translate the Council's decision into a legal text which would provide a basis for scientific co-operation, a programme of monitoring and research and information exchange on scientific, technical, socio-economic and legal matters.

The Vienna Convention for the Protection of the Ozone Layer was agreed and signed on 22 March 1985[10]. Attempts to append a protocol to the Convention outlining specific regulatory measures to limit chlorofluorocarbon emissions failed as a consequence of States being unable to agree a text of a protocol at the time of adoption of the Convention. Further work was necessary, and, over the next two years UNEP organized two scientific meetings to update the assessment of ozone depletion and its effects. It also joined WMO, NASA and four other bodies in preparing a massive three volume scientific document "Atmospheric Ozone 1985, Assessment of our understanding of the processes controlling its present distribution and change". UNEP further organized two intenational economic workshops detailed the current production, emissions, use by category, imports and exports of chlorofluorocarbon substances and the national and regional legislation in place or contemplated which had a bearing on the control of CFCs[11]. The criteria for possible international control measures were also carefully elaborated and suggestions for a regulatory formula were examined for possible inclusion in a protocol to the Vienna Convention[12]. Armed with this information and increasingly worried by news of the discovery of an unexpected and major springtime ozone loss over Antarctica, UNEP reconvened the Vienna Group of Ad Hoc Legal and Technical Experts to elaborate a protocol on the control of ozone depleting substances. The "ozone hole" as it became popularly known, was a major issue during the negotiations. It was unpredicted and unexplained although theories variously attributed the phenomenon to increasing levels of chlorine in the Antarctic stratosphere, dynamic effects related to unusually cold stratospheric temperatures and the unique meteorology of Antarctica characterized as the Polar Vortex[13]. What the appearance of the hole did do was to point up the inadequancy of our knowledge of atmospheric processes and to emphasize the risk involved in deliberately altering the composition of the atmosphere when the consequences of such experimentation are, at best, speculative.

Despite the scientific consensus that there was compelling observational evidence for increases in the concentrations of most atmospheric trace gases which control ozone and that such increases were likely due to human activities; that the emissions of potential ozone depleting substances were annually increasing and cataclysmic ozone depletion was being measured over Antarctica which could probably be, and subsequently was, attributed to manmade chemicals, it still proved a major task to agree a protocol which would limit chemical emissions to the atmosphere.

Points at issue included whether regulations should be limited to CFC 11 and 12, the most widely use chemicals or should they include a wider range of chlorine and bromine containing substances. Another, was how the developing and low-consuming countries should be treated. A freeze at current consumption levels would be likely to disadvantage those yet to realise their development plans which would include requirements for refrigeration and materials containing or using CFCs in the manufacturing process. Yet another issue concerned the impact of stringent regulations on the economic feasibility of producing allowable levels of CFCs and on the long term commitments for CFC production and export locked into already operational economic programmes. How exports and imports would be handled in the control formula and how trade would be regulated between compliers and non-parties to the future agreement were other problems to be solved. Deep division also existed between those countries claiming a freeze in production was sufficient and others demanding production cutbacks of 85 percent and more. UNEP's primary concern was to ensure effective protection of the environment and it was able to demonstrate at a specially convened meeting in Wurzburg involving atmospheric modellers from both sides of the Atlantic, that to protect the ozone layer the full basket of ozone depleting substances had to be regulated and that the simple freeze in production advocated by some negotiators would be insufficient to prevent ozone layer depletion without further significant reduction in halocarbon emissions[14].

The rest is history; common sense prevailed and on 16 September 1987 the Montreal Protocol on Substances that Deplete the Ozone Layer was signed by more than twenty countries[15]. In the following months further signatories to the Act have doubled the original number and preparations are in hand for the ratification and coming into force of the protocol by 1989.

Concessions were made in agreeing the final formula for control. Temporary production increases were allowed in the interests of industrial rationalisation, concessions were given to low consuming countries allowing their production to rise up to an agreed maximum and additional time given to them before application of the strict regulations for reductions in consumption of the controlled chemicals. Accommodation was also made for Eastern Bloc countries with centrally planned economies and agreed development plans extending beyond the date requiring CFC consumption controls.

The question is whether or not application of the protocol limits will be sufficient to protect the ozone layer. Based on current knowledge the answer is yes. Adherence to the reduction formula will maintain the ozone layer within the limits of natural variability with a projected depletion held to within 2% of its average concentration[16].

Not an ideal solution but a pragmatic one balancing adroitly both environmental and economic wellbeing. The protocol provides a mechanism whereby the stringency of the controls can be quickly tuned should evidence emerge changing the degree of risk. The carrying out of scientific research, global monitoring and international information exchange is an integral part of the legal agreement centrally co-ordinated by a Secretariat which will be operated by an International Organization probably UNEP.

What has been achieved is an effective, internationally accepted and practical measure to protect the global environment. From the earliest examination of ozone destruction theory to adoption of the

Final Act of the Montreal Protocol took ten years, only half of which was necessary for development of the legislation. Given the fact that the threat remains theoretical, even though the reality of it is becoming increasingly certain, the marshalling of the global response has been relatively swift particularly when it is appreciated that the treaty is unprecendented in international law.

THE CLIMATE CHANGE ISSUE

Can the ozone layer agreement be a model for addressing other global issues ? Climate change is a similar atmospheric problem of a magnitude calculated to impact on the social fabric of all the earth's inhabitants. As for ozone, there is scientific consensus that the continuing emissions of greenhouse gases to the atmosphere will lead to an equivalent doubling of carbon dioxide by as early as 2030 (Figure 4) and will result in an average increase in global temperate of 1.5 to 4.5°C with accompanying sea level rise of between 20 cm and 140 cm[17].

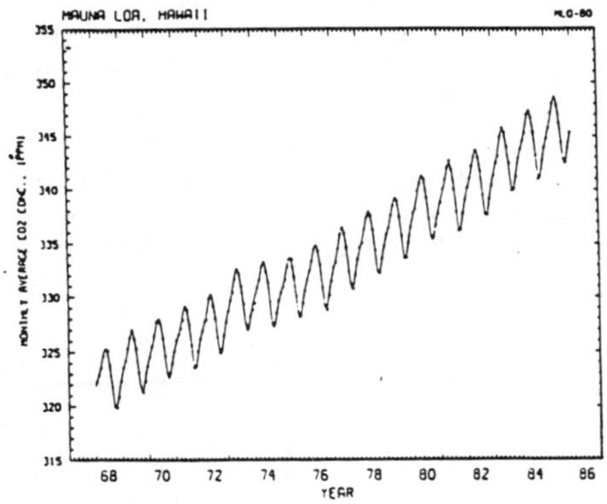

Figure 4. Concentration of atmospheric carbon dioxide at Mauna Loa Observatory, Hawaï (The Greenhouse Effect Climatic Change and Ecosystems - SCOPE 29 ed. Bolin Doos Warrick Jager).

Chlorofluorocarbons are Greenhouse gases as well as being modifiers of the ozone layer. The Montreal Protocol, when put into effect, will slow the progress of climate change caused by CFCs but not enough to be considered as anything more than a small part of the solution. An effective solution is impossible without a global co-operation to address the problem, it being patently obvious that unilateral control will at best only modify and not prevent the change from our present climate regime to another outside of human experience (the projected change is equivalent to that which occur between the last Ice Age and the present Interglacial). The problem is global in origin; the major human-induced impact is from industrial emissions but deforestation and unwise land-use practices in the developing world are other factors promoting climate instability.

However, the climate problem diverges from that of ozone in that not all the projected impacts are unwelcome. Potential agricultural land presently unused through too long winters and excessive cold, and ice and permafrost limits to northward extension of shipping routes and human habitation could be different in a warmer world.

The probable response to global change in the absence of international agreement would be regional acceptance and adaptation, although it is hard to contemplate the sort of adaptation some island nations would make when condemned by current predictions to become ocean shallows.

Attention is moving, if not away from the adaptation strategy, at least in parallel contemplation of avoidance, reduction and delaying devices. Is the value of fossil fuels to economic development greater than of environmental stability ? Is it possible to internationally limit the consumption of fossil fuels when direct benefits are unlikely to accrue equally to all nations (indeed some might be particularly disadvantaged by such altruism) ?

At the present time, we do not know enough about the problem particularly of the likely regional impacts and of the extent of the controls on greenhouse gases sufficient to contain climate change within a familiar range, say, 1°C temperature and 12 cm sea level rise per century (Figure 5). Yet it is possible to find out and, unless there is a climatic surprise equivalent to the ozone hole about to spring unexpectedly upon us there is time to develop appropriate responses that could ensure irreversible change is avoided or kept within acceptable bounds.

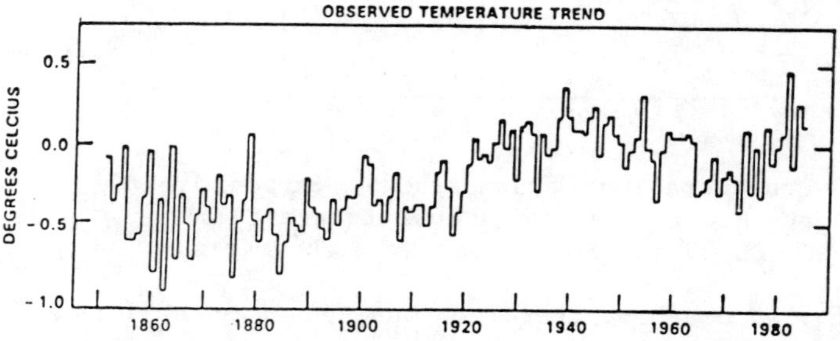

Figure 5. Observed surface air temperature trends for land masses of the Northern Hemisphere. In the past one hundred years or so average temperatures have risen by about 0.5°C (Report of the Eighth Session of the United Nations Environment Programme Co-ordinating Committee on the Ozone Layer - February 1986).

Based on its experience with the ozone layer issue, UNEP is initiating an action oriented programme that will identify the range of national, regional and global responses to climate change and to establish the process through which the best practical solutions can be put into effect.

The basis for this programme must be scientific understanding and in addition to national research programme the activities of the World Climate Research Programme and the International Geosphere/Biosphere Programme are already eliminating the uncertainties relative to man's impact on climate and the interactive relationship among the earth, the atmosphere and the oceans.

Emphasis is being placed on better quantifying regional climate as, it is only when this is better understood, can regional impacts of climate variability and change be refined to the extent that response options can be identified.

These actions would preferably occur sequentially but of necessity will develop in parallel.

Dialogue between the scientist and decision maker must also be a continuous process in order to translate technical alternatives into practical choices. A series of conferences, seminars and workshops have occurred or are planned with the object of refining the decision making process. Significant among future examinations of the problem and responses to it will be the Canadian Conference on the Changing Atmosphere in June 1988 and the Second World Climate Conference in 1991.

The development of effective energy control strategies and the identification of the range of possible solutions to the problem requires the establishment of a co-ordinated research programme for that purpose.

Regional analyses of the socio-economic impact of climate change, sea level rise and the elaboration of policy response options need be undertaken immediately even though it is necessary to consider all possible and even the not impossible scenarios of climate change rather than the probable one still to be identified.

Informing and sensitizing public opinion on the issue will promote the early acceptance of mitigative and preventative strategies no matter how unpalatable in the short term.

The likely political choice that will be confirmed by scientific examination and advice could be some form of international agreement on energy control requiring perhaps greenhouse gas emission limits of 50% of today's values (the provisional estimate for control necessary for atmospheric stabilization).

The agenda for action proposed by UNEP is an aggressive one. Early emphasis on research - physical, biological and socio-economic will be supplemented by identification of policy alternatives and the choice and implementation of the most practicable. Internationally co-ordinated assessments of climate change and its impacts will be a continuing process during the programme. If international legislation is necessary to ensure control of greenhouse gas emissions then intergovernmental legal and technical experts will be brought together to elaborate the appropriate legal instruments either individually for the particular problem of climate change or as an all-encompassing Law of the Air to address existing and future problems of the atmosphere. A provisional estimate suggests this process could be well in hand by 1992. Hopefully, the world can enter the twentyfirst century not as victims of disturbed

Nature but in control of our environment and well prepared to reap its advantages.

There may be other global environmental problems of the future to supplement or replace those of the present. They can be solved or reduced to an acceptable level by a process of international scientific and political co-operation. The mechanism for it is already in place and is demonstrably effective.

REFERENCES

1. Atmospheric Ozone, 1985 - World Meteorological Organization - Global Ozone Research and Monitoring Project - Report n°16.
2. UNEP - Report of the _ad hoc_ Working Group of Legal and Technical experts for the Preparatory of a Protocol to the Vienna Convention for the Protection of the Ozone Layer, 1986.
3. Register of International Treaties and Other Agreements in the Field of the Environment - UNEP/GC/INFORMATION/11/REV.1 - UNEP May 1985 and UNEP/GC/INFORMATION/11/REV.1 Supplement 1 - UNEP April 1987.
4. Report of the UNEP Co-ordinating Committee on the Ozone Layer - UNEP/CCOL/VIII 28 February 1986.
5. UNEP/EPA Effects of Changes in Stratospheric Ozone and Global Climate - August 1986.
6. Stratospheric Ozone - United Kingdom Stratospheric Ozone Review Group H M Stationery Office - August 1987.
7. Report of a UNEP Meeting of experts designated by Governments and Intergovernmental and Nongovernmental Organizations - Washington, DC March 1977 - UNEP/WG 7/25/Rev. 1.
8. Present State of knwoledge of the upper atmosphere - an assessment report - United States National Aeronatics and Space Administration. January 1984.
9. United Nations Environment Programme. Report of the Governing Council on the work of its Ninth Session - U.N. General Assembly Official Records thirtysixth session - Supplement N° 25(A/36/25) New York 1981.
10. Vienna Convention for the Protection of the Ozone Layer Final Act UNEP Nairobi 1985.
11. Report of the first part of the Workshop on the control of chlorofluorocarbons - Rome - UNEP/WG.148/2, May 1986.
12. Report of the second part of the Workshop on the control of chlorofluorocarbons, Leouburg, UNEP/WG.148/3, September 1986.
13. The Antarctic Ozone Hole Richard S. Stolarski - Scientific American, January 1988, volume 258 number 1.
14. Report of Ad Hoc Scientific Meeting to compare model generated Assessments of ozone layer change for various strategies for CFC Control - Wurzburg 9-10 April 1987 - UNEP/WG.167/INF.1 and UNEP/WG.-167/INF.1/Add.1 - 10 April 1987.
15. Montreal Protocol on Substances that deplete the ozone layer Final Act - UNEP Nairobi 1987.
16. Stratospheric Ozone Protection in the light of the Montreal Protocol Chemical Association Washington DC. October 1987.

17. Report of the International Conference on the Assessment of the role of Carbon Dioxide and of other greenhouse gases in climate variations and Associated Impacts - WMO N°661, 1986.

HUMAN IMPACT ON THE ENVIRONMENT

Sir Crispin TICKELL
United Kingdom Mission
to the United Nations
845 Third Avenue
New York, NY 10022
U.S.A.

All human societies are subject to environmental constraints, and some are the products of them. Land resources, natural life (and its concomitant disease), the vagaries of climate including water, and not least the rich variety of our own species have together given each society its special characteristics, successes, failures and possibilities. Many wise things will already have been contributed to this symposium about the climatic aspect, and I shall not attempt to repeat them.

For those who look back in order to look forward, a recurring consolation is that human society is about as unpredictable in its development as the environment from which each has sprung. All our history is of course very short. Our minds are not genetically programmed to understand the enormity of time and even the tiny span of a human society. The concept of 10,000 years ago, when the last Ice Age came to an end, can no more be grasped as a reality than that of 160,000 years ago when a blue giant star blew up in the Greater Magellan Cloud and began to send the light which reached us on 23 February 1987. In its present form the human species did not then exist. It probably existed for 100,000 years before organized societies began to come together. When at last they did, they learnt about environmental constraints the hard way. The Harappa civilization of the Indus Valley around 4,000 years ago probably collapsed for a combination of environmental reasons. More recently Greeks, Nabataeans, and Romans abused and destroyed the fertility of the land around them. The shores of the Mediterranean have never recovered. The same happened to the mountain plateaux and tropical lowlands of pre Columbian Mexico. These societies grew in very different climatic circumstances. There was no determinism about their rise and fall. It was just that none understood the constraints, or was able to see the consequences of its actions.

Where are we now ? Are we ready to learn the lessons of the past ? Few, I suspect, have grasped the peculiar state of the world today. It is not fate which governs our actions but we ourselves. Reading our familiar history books which usually tell the story of nations or countries, as if the rest of the world scarcely existed, we can all too

A. Berger et al. (eds.), Climate and Geo-Sciences, 577–582.

easily miss the central event of our time : the industrial revolution of barely 200 years ago which has changed the face of the earth as nothing since the glaciers made their last major advance and retreat.

The underlying reason for the industrial revolution was the growth of science and technology in societies disposed to use them. It grew out of European history and the agricultural revolution which immediately preceded it. The physical environment of the temperate parts of the earth where industrialization took place was resilient, with big natural resources, varied soils, neither too hot nor too cold and with regular rainfall.

But for the rest of the world it has not been so easy. Anyone who has visited poor countries can report on the kind of shell shock which the impact of the West has had on their cultures and way of life. For many of them industrialization cannot grow naturally out of their past; their agriculture remains primitive; and in any case the process is telescoped in time. Their environment and resources are also different. Many live in desert or tropical areas of high vulnerability. The complex eco-systems which sustain life on the edge of the desert or in the rain forests are extremely fragile, and once destroyed cannot be replaced. Most of their soils are poor. Those who have oil do not have water. Those who have water see irrigation schemes silt and salt up. Rapid change in the customs of those who live in this environment, and the application of methods useful elsewhere, can too often create gross economic and social disruption.

Many of our problems are getting worse. You will have heard them described already, but I run quickly through the list : desertification in all its aspects; depletion of the world's tropical forests; mass extinction of species and associated loss of genetic resources; rapid population and urban growth; mismanagement and shortage of fresh water resources; over-fishing and marine pollution; threats to health arising from mismanagement of pesticides, chemicals and new pathogens; atmospheric pollution of all kinds, including damage to the ozone layer; and mismanagement of fossil and other fuels, in particular wood, with all its consequences for land use.

I come to climatic change last. In a nutshell the activities of our species in changing the chemistry of the atmosphere and the land surface of the earth could lead to a measurable increase in global surface temperatures within the next few decades. Here our failure to grasp causes and effects over time is of critical importance. Few governments, like few people, think in terms of decades. If governments are to look at this problem as seriously as most at this symposium think they should, then we must put ourselve into the minds of decision-makers. The first question a government will ask is : can you prove it ? The second is : what will happen and when ? Will it be good or bad ? and the third and last question is : what should be done about it ? It look at these questions in turn.

Could we prove it ? The answer must still be No. We still cannot demonstrate an irrefutable linkage between the undoubted increase in the green-house gases and climatic warming.

Next : what will happen and when ? Will it be good or bad ? Here also clear answers are impossible. We have to reckon with the enormous complexity and multifarious character of any major change in the natural environment. There are the virtually unknowable chains of causation in the loops of negative and positive feed-back. Nor are the effects of the current accelerating changes in the world's land surface on climate yet clear.

Would warming be good or bad ? In the crude sense that some areas now arid or semi-arid would get more rain and that monsoons would become more regular, the countries concerned might reckon that warming would be beneficial to them. Likewise an enhanced concentration of carbon dioxide in the atmosphere should, at least theoretically, improve certain crop yields and thus agricultural production. Laboratory experiments suggest that, depending on growing conditions, and excluding climatic change, a doubling of the concentration of carbon dioxide would cause an increase in growth and yield of up to 10 % in maize, sorghum and sugar cane, and an increase of up to 50 % in wheat, soya bean and rice. All would depend on the combination of such factors as temperature, precipitation, radiation and soil nutrients, and the potential benefits would vary widely from region to region.

But if some were advantaged, more would be disadvantaged. Again in a crude sense areas which now enjoy climatic benefits might find them drastically reduced. Precipitation is not a zero sum game and might on average be enhanced; but according to most forecasts, many regions including those with the most productive soils might face increasing aridity.

In short geopolitics - the balance of wealth and power between different regions and countries in the world - would certainly be changed. For all the biggest problem would be change itself.

In earlier times people could - and did - react to change with their feet. They just moved on to where conditions were better. But with increasing population density and resources which are exploited to the maximum, large movements of people would scarcely be possible. The more perfectly adapted a society - like a species - to its particular environment, the less easy for that society - or species - to adapt itself. This is as true of societies with limited resources which suddenly receive a bonanza of rain as it is for societies to whom rain is cut off and water becomes precious. So in our overstretched human world all change means dislocation. Time - the rate of change - is the key question we are least able to answer.

Then there is a related factor arising from the loss of species and genetic resources. We are used to looking at natural problems in a myopically human way. But climatic change would affect the myriad interdependent networks of all living things - animals, fish, insects, plants, funguses - of which we are in truth so tiny a part. Age old relationships and characteristics would be upset, and new ecosystems would emerge. The role of adapted bacteria and viruses would be important, not only for our own species, but for those other species of animal and plant on which we depend.

The example familiar to all is the fate of the river Nile. This great river, which has nourished civilization for more than 7,000 years, is in trouble for both natural and man-made reasons. Most of the water which reaches the Mediterranean comes from the Blue Nile from Ethiopia.

Yet Ethiopia is now in its eighth year of severe drought. Water from the White Nile now fails to compensate. So the flow of water into Lake Nasser above the Aswan Dam has fallen year by year. Reserves are now lower than in living memory and the Aswan Dam itself has proved an uncertain benefit. It has brought greater security, electricity for domestic and industrial use (around 28 % of total Egyptian requirements), and extension of irrigation schemes.

At the same time more water has evaporated, water-borne disease have spread, and the supply of nutrient silts has been so reduced that the Delta is beginning to wash away into the sea and the East mediterranean fishing industry has been virtually ruined. The other riverine states - Ethiopia, the Sudan, Uganda and Kenya, all of which take water from the river - are now talking about taking more to meet the needs of their increasing populations. Egypt itself, with an annual rate of increase of 3 % - or 1 million people every 8 months - has the biggest needs of all.

Now supposing that the effect of warming was permanently to reduce the rainfall which reaches the Nile, the effects on the Egyptian economy could be disastrous within less than a human generation. Far from expanding its irrigation schemes, it would not be able to maintain existing ones. Egypt may be a special case. Few countries are so dependent on a single resource. But droughts across Africa, recurrent flooding in Bangladesh, and the disruptions caused by El Nino have served to show the vulnerability of many countries to what are no more than blips in the operation of the weather machine.

Warming would also bring a rise in sea levels, perhaps somewhere between 20 cms and 1.4 m. The effect on coastlines and archipelagoes can be calculated. Less obvious would be the effect on estuaries, river systems, and even underground aquefers of fresh water. Preliminary work has been done in respect of the Netherlands, the United States, Bangladesh and Egypt. Taking the case of one of them - again Egypt - a rise of between 1 and 3 metres could erode up to 20 % of the nation's arable land, affecting some 10 million people. The cost of work to protect lowlying areas worldwide, including ports and industrial plants, would be astronomical.

Still more dramatic would be the effect on the Maldive Islands in the Indian Ocean, with their population of almost 200,000 people. As the Maldives President said in the debate on the Brundtland Commission Report in the United Nations General Assembly on 19 October "a mean sea level rise of 2 metres would suffice virtually to submerge the entire country of 1,190 small islands. With a mean 1 metre rise, a storm surge would be catastrophic and possibly fatal."

I come to the last of the questions which Governments must ask themselves. Faced with the prospect of a global warming, what should be done about it ? Obviously the first requirement is for more knowledge. On the one hand governments and the international community generally must acquire an understanding on which future decisions could be taken; and on the other public opinion must be aware of the nature and scope of a problem which will affect the pattern of life itself.

Certain governments and their agencies have long had research programmes of their own. For the international community we have the World

Climate Programme with its four research programmes, the World Meteorological organization, the International Council of Scientific Unions and the United Nations Environment Programme. It sounds a lot. But most would agree that the funds devoted to research on this subject are not enough.

We need a clear identification of signs of greenhouse gas induced warming; a better idea of the consequences in key areas; improved understanding of the sensitivity of different agricultural systems; and precise evaluation of the effects of a rise in sea levels.

Public awareness has a direct effect on the readiness of governments to finance research and eventually take decisions. Conferences, parliamentary and congressional hearings, symposiums, reports in the press and on television are all to the good. Last years a big impact was made by the report of the Brundtland Commission on Environment and Development. It was debated in the United Nations General Assembly on 19 and 20 October. The participation of such speakers as Rajiv Gandhi of India and Robert Mugabe of Zimbabwe should help dispose of the vulgar illusion that concern for the environment is a luxury of the rich. In December Ronald Reagan and Mikhail Gorbachev agreed at their summit meeting to cooperate in work on how to deal with any climatic warming.

More important has been the public debate on the twin issues of acid deposition and ozone depletion. In the case of acid deposition, the dying of millions of trees and the poisoning of hundred of lakes was clear for people to see. Enough is now known to persuade governments and industries to take steps to mitigate and reduce the problem to manageable levels.

More remarkable, because the effects are invisible, has been the reaction to the news of the depletion of the ozone layer. Ever since the discovery of seasonal ozone holes over the Antarctic by the British Antarctic Survey in 1985, concern has been intense. Again the science is incomplete. But the role of chlorofluorocarbons is certain, and the effects of increased ultra violet radiation equally so. Vested interests did not favour regulation, either national or international. But the result was the Montreal Agreement of last September to limit production and use of chlorofluorocarbons throughout the world.

With greater knowledge and sustained by public concern, governments should be better equipped to make a difficult choice : whether to do the easy thing and let mankind - as well as nature - take its course, thereby permitting an experiment of formidable dimensions; or whether, as with ozone, to work towards global arrangements involving some measure of regulation. The implications of warming reach into every aspect of life. They are correspondingly difficult to grasp, let alone manage.

Obviously the main approach must be international with national governments in a sustaining role. There are existing international institutions, some private, others supported by governments, which could take coordinating responsibility. The World Climate Programme was created for this very purpose. Yet in this case, covering so many imponderables, I believe something else is required, at least at the beginning. If government leaders are to be persuaded to take action, then some catalyst in the form of a commission or panel of people distinguished in political as well as scientific life might be created. There are many precedents for groups of this kind. The Brandt Commission of the 1970s and the Burndtland Commission of the 1980s both had an impact which existing institutions, usually composed of experts, could never have hoped to achieve.

The functions of such a body could be three-fold. First the members could, with the help of experts, draw up an inventory of the causes of likely warming and the corresponding remedies. Next they should draw up a code of good international behaviour to act as a framework. Such a code would not be mandatory, but would amount to a reference system on which consensus could be built. Lastly they should examine the possibility of a network of agreements on the points defined in the code, which might, sooner or later, have more binding characteristics. Each stage represents an ascending order of difficulty. It would also represent progression in time so that each could reflect the needs of the problem as it - and public awareness of it - developed.

The remedies also seem to be three-fold. First there is the better and proper use of energy. We need better conservation of energy; we need to make less use of coal and oil and more of natural gas; and we need to look into other renewable sources of energy of a kind still described as unconventional : they range from nuclear energy (especially fusion) to solar, tidal and wind generated energy. Secondly we have to look at substitutes for such industrial chemicals as chlorofluorocarbons, and apply the right inducements and penalties to industry. Thirdly we have to examine the problem of land use, and the deforestation which has so contributed to atmospheric pollution.

Of course the main responsability for carrying out any plan would rest with individual governments. To them also would fall the responsability for coping with whatever may happen. Governments cannot do more than indicative planning over periods of decades, and rely heavily on individuals and natural market forces. But public investment on the scale likely to be necessary, at least in some countries, has long lead times, and work on what to do, and on what to avoid, may already be desirable. Planners should look at all land uses, particularly food supplies; at the management of fresh water, whether for agricultural, industrial or energy purposes; at the directions of industrial growth; at measures necessary to protect ports, cities, industries and agricultural land liable to flooding; and at the effects of what could be a major redistribution of population in changing economies.

The geopolitics of the world will anyway be very different by 2088. Power is as fluid as mercury. But the time to recognize the vulnerability of our planet - and of ourselves - is now. The multiplication of our own kind, and the way we have changed the land, sea and air around us, require us to management the consequences. It concerns us all.

SOCIETY AND FUTURE CLIMATE CHANGE

Evan Vlachos
Colorado State University
Fort Collins, Colorado 80523
USA

ABSTRACT. The paper outlines the increasing concern with climatic instability and the potential major dislocations of human societies. An initial presentation of the interaction of climate and society is followed by a range of impacts and alternative responses. The catastrophic preoccupation allows for a concluding speculative section utilizing alternative images of the future and broad strategies for prudent responses to a transforming environment.

INTRODUCTION

From around 1960 on, both in the popular mind in the professional literature as well as there seems to be an increasing recognition that the former stability of climate tends to disappear and significant oscillations have increased the concern as to present and future impacts of long-term climatic changes on the surrounding environment. Newspaper articles, special sections in Time magazine, dramatic presentations in the evening news as to ozone depletion, greenhouse effect, or new glacial ages, are all manifestations of a free-floating anxiety and of a preoccupation with new apocalyptic scenarios of impending doom.

The present paper does not attempt to recapitulate the perennial arguments and debates as to long-term climatic changes. Instead, it attempts to concentrate on the importance of the interaction of climate and society, especially by relating how changes in climatic conditions seem to impact on the surrounding social environment. More important, the paper summarizes responses and adjustments to such changes, and attempts to speculate on future challenges from a social policy point of view.

The pages that follow have a two-prong emphasis. First, there is a pragmatic concern as to how we can approach the question of how societies can adapt to changes, particularly changes that tend to have a long-term horizon. Second, there is an effort to utilize some speculative efforts and imaginary approaches concerned with visions of things to come, alternative scenarios if you wish, related to long-term climatic changes.

A. Berger et al. (eds.), Climate and Geo-Sciences, 583–602.

The question of climate can be viewed as a "megaproblem." Tonn uses climate as part of a class of environmental problems which pose significant long-term risks to the world's societies. He refers to such problems as "500-year" problems. These types of problems, according to Tonn, share three significant attributes. First they take centuries to either develop or to show their effects. Second, they take centuries to solve or require centuries of effort to forestall. And, finally, they entail causal loops that run from society to the physical/natural system back to society.

Yet, significant as such problems may be for the long-run survival of society, climatic changes occur too slowly to provide politically salient support for policies designed to prevent, retard, or alter long-term transformations. Since climatic change seems to lack for many people a compelling environmental rationale, it is so far characterized by reactive approaches and and a prevailing "crisis" management style of operation. Finally, a megaproblem like climatic change, although global in potential effects and consequences, lacks clearly identifiable constituencies or parties-at-interest which could mobilize resources and personnel and, thus, become a rallying dramatic point for massive demands for concerted responses and policy changes.

The diffusiveness of the topic, the contrasting and competing interpretive models, and the relative paucity of data and knowledge (although they have been increasing in recent years) make concern with climatic changes not the type of environmental problem that one usually associates with significant environmental movements or demands for drastic policy changes. But, as the increasing barrage of negative news and the expressed social and ecological concerns come to dominate the popular press, they also tend to increase a sense of tension and a mood of impending catastrophe. Again, however, even though there might be agreement that significant climatic changes may have important global ramifications in the years to come, it is equally difficult to transform scientific confirmations to political act in particular situations and for particular regions of the planet.

As it will be related through the paper the concern with both an overall cooling as well as with the warming resulting from a "greenhouse effect" are not antithetical interpretations of on-going climatic changes. They represent different time periods and are both considered reasonable interpretations. The missing point in all of these interpretations is a clear understanding of the human induced changes in what are otherwise "natural" climatic trends. The human presence tends to exacerbate or reinforce climatic variations. Anthropogenic changes become difficult to understand and equally difficult to determine in terms of far-reaching social upheavals and dislocations

The literature abounds in recent assessments of the concern with changing climate and its impact on society. There is no need to relate again competently conducted assessments or studies of various academies and other august bodies of knowledge which have tried to understand both short- as well as long-term effects of climatic change on society. The selected bibliography at the end summarizes works particularly useful to the flow of the present argument. When all is said and done, the emphasis of the present paper is to speculate a little bit further on future developments and on the challenge for societal mobilization and socio-cultural adaptations if climatic changes acquire the importance and characteristics that various studies tend to predict for the years to come.

1. Interaction of Climate and Society

The increasing range of environmental and ecological studies of the last 20 years has promoted significantly our understanding of the effect of human activities upon this planet. We begin to have an increased appreciation of the far-reaching impacts and consequences of human activity on land, water, and the atmosphere. Furthermore, we have come to recognize a problem of much more expanded scale, i.e., the long-term, persistent effects of human activities on the entire planetary life support system.

In approaching the climate-society interaction most of the ongoing research has pointed out to four critical views of climate, i.e., climate as setting, as determinant, as hazard, and, finally, climate as natural resource (Riebsame, 1985). All such views tend to coalesce to expressions of human vulnerability and to concerns towards a natural resource base that can be mismanaged or abused. The findings that periods of gentle climates between ice ages are becoming much shorter than had ever been thought before, have reinforced the preoccupation with the doom and gloom of rapid changes and the proximity of potential catastrophe. The evidence is increasing in terms of such items as the duration of the Arctic snow cover, animal migrations, sea-surface temperatures, microfossils on the ocean floor, ozone holes, more icebergs breaking away from Antartica, and so on. All of them, taken as a whole, point toward what the New Scientist described as "an atmosphere in convulsions."

When large scale climate fluctuations take place they tend to be attributed to three basic and interrelated categories of concern: a) events associated with the earth's relationship to the solar system; b) events associated with the processes occurring within the earth's system; c) events associated with human activities.

The puzzle of trying to understand the interaction of this vast array of forces that govern climate become even more complex when the changes in the earth's movements and orbits around the sun are also affected by the presence or absence of vegetable and animal life, by cycles of feedback, and by astronomical cycles of the type suggested by Milankovitch in the 1920s. All such elements provide a very complex equation as to the type of solar energy that the earth receives and, at the same time, seem to contribute to the earth's periodic major ice ages as well as short-term cold and warm spells.

In terms of human-induced changes there are two prevailing views as to how such activities may affect the climate. First, we have the concern with the increasing dust content of the atmosphere through pollution as well as wind blown soil from agricultural land (what has been referred to by Hardy as "the human volcano"). The second view or major theory of climatic changes induced by human activities revolve around the build up of carbon dioxide in the atmosphere due to burning of fossil fuels and the simultaneous destruction of forests. The list of human-induced changes can be expanded if one also adds such human alterations of the atmosphere as radioactivity in rainfall, the depletion of the stratospheric ozone layer from microbial transformations of nitrogenous fertilizers and the use of chlorofluorocarbons, chlorinated hydrocarbons, the injection of fine particulate pollutants, and acid rain. All these are not only symptoms of the cumulative impacts that natural and human-induced changes bring about, but also part of a much more significant transformation of the earth's ecosystem.

These vast changes must be seen within the context of long-term weather or climatic cycles. The question that arises in the literature has to do with what one may consider "normal" weather versus "freak" climatic occurences. Indeed, the extremes that have been observed in recent years tend to be declared by competent researchers as more or less the normal circumstances rather than unique years. The fluctuations of the 1970s and the record-breaking droughts and blizzards all over the planet are not really "unseasonable" events but more or less historically consistent climatic conditions. What has changed is increased global awareness and the ability of societies to make efficient use of resources under adverse circumstances.

The changing context of both population and societal activities that has significantly changed from previous historical periods has also affected our perception of climatic changes. On the one hand, the earth is much more densely populated than earlier years and, therefore, the disruptions attributable to climate can be much more pronounced and much more severely felt by larger segments of population. On the other hand, the intensification of the exploitation of the planet by increasing numbers of people and the developmental efforts of many countries have created an agricultural system that is much more delicately tuned to climatic changes. Climatic variations can affect now a much larger pool of population and be more severely felt by large scale mechanized agriculture of the scale found in the Great Plains.

The last observation brings forward also an important argument when one discusses the context of change, namely the centrality of a diachronic perspective. The presence of short time oscillations implies also that dislocations can produce severe local effects as both societal activities and populations try to adapt on a long-term scale. There may be little "social time" left to develop long-term coherent and global policies to alleviate the immense transformations anticipated in the next 40 to 50 years.

The data on all such changes and transformations are really impressive and disturbing. As one looks at the increasing knowledge on both the nature and context of climatic changes, one must recognize that there are significant conflicts, disagreements and debates as to the nature, causes and consequences of perceived changes. We may classify all such debates in three major categories. The first source of uncertainty and conflict can be labeled as <u>cognitive</u> or <u>methodological conflicts</u> and reflect disagreements as to the state-of-the-art. Here we have all the types of conflicting data and information as to on-going trends and the difficulties associated with the modeling of such a complex system. Such debates become particularly acute since they are driven by a central assumption of many scientists monitoring climate, namely that the surrounding atmosphere will respond to human influence in a steady gradual way. Others argue that all these surprising climatic transformations provide for rather sharp variations such as changes of ozone which affect in a rather short period and in a dramatic fashion the circulation of the upper atmosphere. Some even go as far as to argue that the climate can suddenly (even within a century or less) change into an entirely different state. The recognition of climate as a system of highly complex interlocking parts implies also that the present simple computer models do not seem to be particularly sensitive to interactive, synergistic changes and impacts.

A second major conflict and area of disagreement has to do with what one may call stakeholder conflicts. This area of disagreements reflects the fact that as in any other use of natural resources populations tend to coalesce into power groups. Such groups or parties-at-interest have different perspectives as to questions of what is at stake, of who pays and who benefits, as well as to preferences for particular interventions or policies for correcting disequilibria in the surrounding environment. There are powerful national and transnational interests that would oppose measures restricting the use of fossil fuels or chemicals or who believe that certain intervention measures could severely affect local economies.

Finally, a third major area of disagreements has to do with ideological conflicts or modeling disagreements. Here the various paradigms, models or theories of climatic change reflect preferences and values as well as visions of appropriate social structure, postures towards development and towards the use of resources. Such conflicts represent not only points of view but also important mechanisms for collecting data, for presenting solutions, for estimating risks, and for proposing strategies for solutions. In the literature of climatic change there seem to be three basic categories of opinion in modeling. The first is represented by the pessimists who consider climatic doom as imminent and, therefore, propose very little for solution other than perhaps the total reconstruction of society following an apocalyptic change. The second group is composed of those who maintain a "middling" position and who believe that while there may be some serious short-term dislocations, perhaps a hundred years from now, the deterioration of the climate at the present time can be met by some concerted attention from responsible people and existing institutions. This group also believes that we should proceed in some careful manner as to the future by noticing that while some climatic change may be important there are other problems that need much more immediate attention such as questions of overpopulation, lack of food, and shortages of energy. Finally, at the other end are the extreme optimists who believe that climate really doesn not change very much in any time scale relevant to society and a number of on-going changes actually can counteract each other. An example of the last emanates from the observation that the heating of the atmosphere may counteract a natural cooling trend. The optimists usually include people who believe in technological solutions and in the basic premise that when societies are confronted with serious challenges they tend to respond appropriately through technological and social changes.

In any case, information and knowlege on climate have been increasing as to cycles and weather changes and recent research has amassed evidence as to patterns of rainfall especially during warm and cold years since 1925. Past records and more accurate data point out that the transition from a natural cooling trend to an unnatural warming is not likely to be a smooth one. Whatever one may want to describe as the weather machine, it will take some time to adjust to changing conditions with the first response being increasing variability around the globe and with more extreme conditions recurring from year to year. This means that there will be more droughts and floods and more severe winters and scorching summers. This pattern has already been identified with the events of the 1970s, especially in the "typical" year of 1972 which could be the benchmark and the crude guideline for planning for many years to come for both agriculture and industry.

In the interaction between climate and society one conclusion should stand particularly out, namely the well-known fact among atmospheric scientists that they are indeed broad historical cycles of global warming and cooling. Among widely accepted conclusions one should point out are that the earth's surface gradually began warming after the last Ice Age peaked 18,000 years ago; the variability of the natural climate will continue during the next hundred years with the possibility of a more extreme and probably cooler climate; man is already influencing to a significant extent climate on a local scale, but with no clearly established role as to the effect of anthropogenic changes on climatic cycles; and, that the effect of man's interventions will probably be a warmer climate with significant changes of the world's rainfall patterns.

The most debatable argument in this context has to do with how anthropogenic causes could significantly alter climate patterns that have lasted for millenia. Three major areas of concern really summarize the kinds of discussions and reports that appear, especially in the popular press, and constitute the areas of mass concern. One has to do with the greenhouse effect, namely with the extent to which unusual global warming has indeed begun and that as an emerging natural phenomenon would have both negative and positive consequences. A second area of concern is the perennial preoccupation with potential ice ages and the discernable natural trend towards cooling, a trend which seems to reflect the widest agreement among experts. Finally, the third area has to do with stratospheric ozone depletion which has become the focal point, especially in terms of public health and the potential for widespread skin cancer as well as with the vulnerability of all life on earth by damaging ultraviolet radiation.

What all of these point out is that the stability that has characterized the earth's climate together with the well known periodic cycles and correcting mechanisms seem now to have been upset and the balance of temperature and moisture sustaining collective life seem to have become more precarious. As research is piling up evidence on the natural cycles and rhythms of the climate, there is a concerted concern as to the extent to which humans are upsetting, perhaps irrevocably or irreversibly, this precarious balance. Furthermore, fears are expressed as to the responsiveness of humans to altered conditions, since societies tend to be particularly tuned to the present (rather favorable) climate. And, not only are temperature changes looked upon carefully, but also precipitation patterns and the balance between rainfall and evaporation tend to become extremely important.

What all of these changes and potential impacts of warming and cooling forces imply is instability and long-term vulnerability. For historical examples of the vast effects that significant climatic change can have one can use the years of 1816-1817 when volcanic eruptions affected the weather in a way that produced far-reaching harvest failures, famine, social dislocations, disease, demographic upheavals and socio-political consequences (Post, 1977; Stommel and Stommel, 1983).

As one looks at this instability and high uncertainty concerning climate history and contemporary events, it is easy to understand the consternation and also confusion about decisions that must be made in order to affect societal adaptations to climatic changes. Conflicting advise about possible and probable effects is really adding up to the disagreements pointed out earlier, i.e., in terms of knowledge of facts, stakeholder interests and value differences included in modeling attempts. (See, for example, the dramatic juxtapositions between Fred Hoyle's Ice: The Ultimate Human Catastrophe, NY: Continnuum, 1981, and Howard A. Wilcox's Hot-House Earth, NY: Praeger, 1975.) One thing seems to be sure, however, as various climate conferences have pointed out, namely that this variability of the natural climate will continue in the foreseeable future and that we must deal in terms of both anticipation and hedging against the most possible, if not plausible, catastrophes that present knowledge can estimate (Schneider, 1977). We are asked upon to act in a combination of knowledge, experience and common sense, keeping in mind a cautious approach which emphasizes that it is better to be approximately right rather than precisely wrong. In order to adapt better to future variabilities in the climate, broad societal changes must take place that demand not only the capability of forecasting potential changes but also of accomodating new far-reaching patterns of social interactions.

The nature and extent of climatic changes, then, have been heavily debated in recent years. From a sociological point of view the key questions are not the very pragmatic concerns with the characteristics, causes and impacts of the changes of the physical environment, but most important such questions as:

a. How much time is needed for societal adaptation?
b. What strategies can be undertaken especially for a worst case scenario of truly climatic upheavals?
c. What information and monitoring is needed that would allow us to develop relevant strategic interventions in the foreseeable future?
d. How can we understand better global interdependencies and develop much more sensitive models and transnational

responses?

However one may phrase uncertainties involved in the examination of climate, it seems that there is increasing agreement in the literature based on evidence already available that the climatic picture is disturbing enough to merit at least prompt attention to restricting atmospheric emission from fossil fuels and other large sources. Strong measures are already suggested if we are to prevent further degradation of the natural ecosystem, especially in view of the time lag involved, e.g., in the case of the ozone between the release of the gases and their reaching of the upper atmosphere. While the immediate future does not seem to be quite as bleak as some scientists have suggested (notably Bryson) the views of most atmospheric scientists do not allow for complacency, but do suggest that we must at least investigate and model better the interaction between possible climatic changes and solutions to long-term problems.

When one looks at the prophets both of warming and cooling scenarios, one is struck by the noticeable agreement on one particular item, namely that climatic changes can come quickly, even within decades, and can have devastating consequences for human life. This urgency is the one that seems to drive the concern of future climate conditions as the signs of possible vast climatic disruptions on society and on the economy become either visible or better understood. Together with the vulnerability resulting from the previously mentioned population increases and in increasing demands for food, human-induced activities at work seem to be upsetting the stability of the present climate and together they will have a most significant future influence even though the estimates of the magnitude of such changes tend to differ widely. How to live with these changes, how extreme they are and how long they will last are the urgent questions that are increasingly raised in any discussion and analysis of on-going climatic changes.

2. Changes of Climatic Conditions and Their Impacts

We can see potential effects of climatic transformations by making a distinction between natural and human-induced changes as well as between near-term (temporary) and long-term (permanent) changes. Such a distinction can helps us separate temporary perturbations from radical climatic and "evolutionary" changes. Obviously this is only but one way of classifying climatic changes. One could also introduce such additional dimensions such as inadvertent vs. deliberate changes; of natural vs. artificial changes; pervasiveness vs. boundedness; transitoriness vs. chronicity; global vs. regional or local impacts; high vs. low variability; etc. However one may want to classify the different ways in which climatic impacts may occur, it is very difficult to directly trace climatic events and their influence upon society on both a global and a local scale. This is particularly true in view of the difficulty of providing a clear line separating short-term weather stresses against long-term, gradual climatic changes.

The literature on the impacts of climatic changes points out to a broad concensus as to the sensitive interrelationship between climatic variations and agriculture, water resources, building, transportation, energy activities, and health. In addition, there is continuous concern about impacts on insurance, governmental expenditures, and recreation (see here the detailed discussion of Maunder, 1986). At the same time, impact studies have not only examined general impacts in terms of a series of social and economic consequences but have also concentrated on specific geographic regions particularly susceptible to climatic fluctuations, such as drought-prone arid areas.

Climatic changes impinge on human affairs in diversified ways but most pronounced in the literature tend to be effects associated with water supplies especially as they affect groundwater levels and soil moisture, river and lake levels and demands for hydropower. Other significant effects are prevailing temperatures and their direct impact on human health, fuel demands and crop growth; and effects on trade, travel, communication, tourism, forestry, agriculture and horticulture, plant, animal and human health and diseases. They are all part of concepts of disasters, disruptions, and disolocations contributing to social upheaval and social disorganization.

We are talking, of course, here about gradual, long-term climatic changes and not such temporary disasters as droughts and floods, although they in turn are caused by long-term fluctuations. But such recurring climatic changes contribute also to failures of food supplies which may lower the resistence of human populations to sickness and disease.

All in all, climatic changes tend to have significant effects on human affairs by acting both as destabilizing influences and also as catalysts for far-reaching social changes. This is particularly true in less developed countries where a combination of rapid population growth, inadequate protection, and accelerating competition for limited natural resources compound an already precarious situation. The literature abounds in examples of how human societies have paid the price for significant weather and climatic changes and how the effects of such changes have severely affected the course of civilization. History is replete with examples of international migrations, political rebellions, social upheavals and pandemic diseases associated with severe cold and harvest difficulties, large scale desertification and similar disturbances of local and global climates.

Most difficult to trace is the movement from first-order, direct, traceable impacts of climatic and weather changes to secondary, indirect, cumulative consequences. As one moves from direct to indirect impacts, the causal linkages become much more complicated and difficult to trace. Yet, climatic events are no less important in the sense of far-reaching consequences on social organization, human health, economy, and social institutions. The important thing to notice is that the less defined social impacts tend to become really crucial in understanding the challenges of long-term climatic fluctuations. Not only there is high variability as to how societal impacts come about as a result of climatic changes; but also as to how societies differ in their ability to cope and take advantage of such changes and shifts in both economic and social activities.

Useful in this context is also the distinction that Barhum draws of Homeostatic disasters which result from natural catastrophes which reflect the rhythms and limits of nature and assume return to equilibrium. Metastatic disasters as artificial catastrophes caused by human behavior and whose unclear spatial and temporal boundaries make return to equilibrium problematic. Finally, Hyperstatic disasters include artificial catastrophes as well as large-scale ecological imbanances intensified to the point of completely obliterating discernible spatial and temporal boundaries through global extension and system-destroying properties. Hyperstatic disasters are becoming now the dominant form in modern complex societies which are now more rather than less vulnerable. The conclusion of Barhum and other writers is that the world is becoming more threatening, not only because absolute levels of possible destruction have risen, but because the capacity to absorb destruction has not risen with comparable speed. In other words, as disaster shifts to hyperstatic, society-wide forms, and as the limits of resource elasticity set in, adaptation becomes more difficult and rational responses less prevalent.

The complexity of the climatic system that has been repeatedly referred to in the literature provides a tremendous uncertainty as to not only the range and intensity of impacts but also as to potential responses. An example here from the current literature can exemplify the importance of what can really happen when climatic changes are used as an explanation of current events. As reported in The Economist of February 27, 1988, Egypt is passing through a drought period which many have argued (although tentatively) that is the result of gradual changes in the world's climate. These climatic changes are traced to the greenhouse effect, which may be exaggerating the temperature differentials between the northern and southern hemispheres, warming up southern oceans more. As a result Africa's rainfall patterns can be seen as changing in the long term so that the rainfall belt which should provide rain for the Sahel, in West Africa, is not penetrating as far north as it once did. If, then, this greenhouse effect is playing a similar role in East Africa the Ethiopian sources of the Blue Nile could be dry for a long time to come with devastating consequences on Egypt as its water supply is exhausted while its population continues to dramatically increase.

One important item in any consideration of changes of climatic conditions and their impacts on society is the extent to which one can answer questions of differential "vulnerability." Here the literature distinguishes between environmental or ecological vulnerability in terms of the degree to which humans violate ecological principles which maintain stability within the system; technological vulnerability in terms of how technological systems violate ecological principles; economic vulnerability in terms of climatic impacts resulting for problems of externality, equitability, and development; and, finally, social vulnerability in terms of the organization of society being capable of sustaining shocks of climatic fluctuations. The conclusions seem to be that the most vulnerable societies tend to be those labelled as "underdeveloped" or "developing", i.e., the poorest and socio-politically unstable. Developed countries are particularly vulnerable to high technological developments (see in this regard the discussion by Warrick and Riebsame in Chen et al. 1983).

Together with vulnerability any relevant discussion must also consider the resilience of socio-economic systems and their sensitivity to climate-society interface. This is why previous studies such as the one by the SCOPE Workshop (1978) has pointed out that there is a stratification of climatic impacts dependent on climatic regime; biosphere and economy (agriculture, energy use, etc.); and social and political considerations. The exposure to risk and the level of reaction against natural hazards vary greatly from community to community and from nation to nation.

Before leaving this brief discussion on impacts of altered rainfall and temperature on society, it is interesting to notice that there are also a number of researchers who point out that despite predictions of calamities from increasing releases of carbon dioxide in the atmosphere there are potentially positive effects particularly in the boosting of agricultural production as a result of the stimulation of plant photosynthesis and increased food production. Indeed, such researchers point out that global warming would likely open to intense crop production what are now considered marginal farming areas. (UPI in January 6, 1980 quoted the Director of the Michigan State Agricultural Experimental Station as saying: "People love to talk about calamities. Climate changes don't have to be bad. They might be good. On balance, we may be better off than we are now." Gribbin also supports the notion that, on the whole, the earth will have a climate more favorable for feeding the increasing population. Similarly, Bruce Kimball, a soil scientist with the Departmant of Agriculture has shown that higher concentrations of carbon dioxide can act as a natural fertilizer for crops --American Scientist, March-April 1988, p. 130).

3. Responses and Adjustments to Changes

The concern with responses to on-going or perceived climatic changes can be more broadly related to the challenge of adapting to a transforming social environment. These larger concerns point out that the preoccupation with both the impacts and the responses to climatic changes are part of larger issues as to what can be done to minimize present socio-cultural disruptions and to develop national and international mechanisms for responding to a complex, crisis environment. Since all estimates and the best models point out that the rate of warming would be appreciably larger in the years to come, a number of immediate responses come to mind as the simplistic question, "What can be done?" is raised. They include: a) the reduction of the rate at which fossil fuels are burned and other greenhouse gases are produced; b) the disposal of gases and fluorocarbons elsewhere than in the atmosphere; and c) the acceptance of a changing climate and the adaptation to it.

The sensible option in terms of responses to either warming or cooling would be to hedge and to develop strategies of graduated responses. The fear is that the current preoccupation with the threat of the greenhouse effect may lead to expensive research and search for technological fixes and for expensive solutions, rather flexible strategies accommodating a wider spectrum of climatic variations.

Perhaps one way to summarize the responses and adjustments to climatic changes is by developing a global vigilance strategy. The concrete outcome would be a contingency planning posture involving three basic premises. The first would be expanded knowledge particularly by understanding and forecasting climatic changes, developing basic science, improving measurement, and remote sensing data processing and information storage and retrieval technologies. This is a highly complex scientific problem requiring vast improvements in quantitative physical theory and estimates about changes that are likely to be brought about by different causes. The second premise is continued vigilance in terms of systems of monitoring and of assessing consequences of different types of natural or human-induced climatic changes, especially through sensitive warning systems. The third premise of such a broad strategy is to increase our respect for nature by learning to live within a highly complex interrelationship of humans, biosphere and climate and by avoiding further deterioration of the ecosystem as, e.g. through a slowing of the increse of carbon dioxide, and through strategies that increase resilience to climate change (agricultural practices, wise water management, adequate global food reserves, etc.). Such a respect of nature requires concerted efforts for planning and implementation and both national and transnational cooperation among all inhabitants of the planet.

As we look back at the literature in terms of larger options for adapting to a changing climate, various authors have pointed out once again that the choices tend to be rather stark. A first choice would be that of no action, a laissez-faire approach that would emphasize that despite oscillations and rather bizzare weather the climate is not really significantly changing. Another choice is that of societal adaptation to the changing climate especially in view of a potential of "Little Ice Age" similar to the one that was observed in Europe and North American between 1350 and 1850. However, increasing populations and expanding developmental demands make it difficult to maintain a level of adaptation that would restrict national demands (especially by developing countries) for expanding and bettering their economies. A third choice would be concerted efforts for manipulating and controlling global climate. Weather modification as well as any other type of environmental modification, however, present us with significant problems of international cooperation and with intricate changes which may have other unintended catastrophic consequences. Past experience with weather modification has not been encouraging on a global scale and the gains expected tend to be relatively marginal.

The end result of all such alternatives is that each particular choice tends to involve a variety of risks and potentially disastrous consequences. The most wise course is a mix of options involving such larger strategies as the storing up of food reserves (the "Genesis Strategy"); interbasin transfers of water to accommodate ecological inadequacies; continuous conservation measures; and, certainly, restructuring of both national and international institutions. Obviously the strategies or mix of policies we may want to adopt in dealing with the carbon dioxide problem, the ozone depletion challenge or the long-term cooling of our earth depends to a large extent on the adequacy of national and international decision-making processes. Even if knowledge exists and more precise estimates can be made as to climatically induced damages, there are extreme difficulties in mobilizing and implementing world-wide social changes of introducing unpopular alternatives. Entrenched and powerful vested interests, national differences, and differently conceived forms of adaptation limit global mobilization.

The question behind all such discussions is that while prevention and some temporary adjustments may work, long-term adaptation provides us with the challenge of accommodating massive population migrations into regions which are favored by relevantly benign climate; controlling the rate of industrialization in places where the development cycle may not have been completed as yet; and dissipating impacts through strongly enforced regulatory measures.

We may now summarize our potential responses to climatic changes by distinguishing between three processes, i.e., adjustment, accommodation and adaptation. These three terms often encountered in the literature of adjusting to natural hazards and changes reflect an idealized continuum between cumulative long-term responses to natural hazards (adaptation), short-term responses described as adjustments, with accommodation lying somewhere inbetween the two.

Material on the types of adjustment, accommodation and adaptation comes from a variety of sources. A major source is the "disaster" literature, a rich vein of research and theory concerning community adaptation to catastrophic events (as, e.g., the insightful work of Barton, 1969 and the vast inventory of findings for short-term disasters in Drabek, 1986). However, the studies of short-duration localized disasters have little to say on long-enduring stress such as the ones resulting from extended climatic changes (especially when people and societies can do little about them). Other sources for potential guidance here include works on the threat of impending disasters (e.g. preoccupation and fears with nuclear holocaust); fictional accounts where authors present extreme scenarios and plausible sequence of events; examples from severe weather (killer weather) accounts; long-term consequences of civilizational upheavals such as the Black Death and similar epidemics; even writings speculating on adaptation to hostile environments, including confinement in space stations, lunar colonies, etc.

When looked at in a broader perspective, all such responses and mechanisms of adjustment and adaptation tend to emphasize that because of uncertainties associated with such large scale transformations, it is important to exercise restraint particularly with our present resource intensive and highly consumptive life style. At the same time, the literature points out there is need to mitigate and where possible the ill effects of air pollutants in the atmosphere. In this connection the larger strategies outlined above also point out that while climate change cannot be prevented altogether, it may be possible to limit its extent, delay its onset or reduce potential effects through the increased conservation and shifts to alternative energy soucres. Other authors also maintain that there are indeed a host of relatively sound strategies available currently to make human societies more resilient to climatic fluctuations. Such policies include the use of marginal lands, countervailing modifications in hydrology, closed environment agriculture, reforestation so that more carbon dioxide can be absorbed, altered genetics and so on.

4. Speculating on Future Developments and Challenges

If it has not become apparent so far, there seems to exist a pervasive gloom and doom mood that has society pinned between the two extremes of climate changes, i.e., from the scorching consequences of the greenhouse effect to the shivering anticipations of a new ice age. On the side of optimism, an intriguing idea has been recently forwards as an explanation of how life can survive despite enormous odds. Known as the Gaia hypothesis, it can serve as a model for planetary and cellular dynamics and it is part of a new paradigm of wholeness (together with the Santiago school of cognitive biology, the Parisian school of self-organizing systems biology and the epistemological work of Gregory Bateson).

According to this hypothesis, Earth can be viewed as a self-regulating system of the biota and the environment, with the capacity to maintain both the climate and the chemical composition of the atmosphere in a state favorable to life. The originators of this "theory", James Lovelock and Lynn Margulis, have proposed that the atmosphere appears as peculiar as it does because it is a part of a living entity. Thus, we must regard the biosphere plus the atmosphere as a unity where humans are but only a small element in this living organism that covers the surface of the earth. If we were to adopt such a remarkable hypothesis which suggests that life on earth with the atmosphere represent a single entity, we can conclude that, taken as a whole, the entire planet surface including the very thin layer that we call the atmosphere keeps us alive despite outside influences.

Reaction to the Gaia hypothesis has been quite emphatic not because scientists do not agree with the paradigm of wholeness -- a paradigm widely accepted by ecosystem scientists -- but more because it became a new symbol for both those who search for unity with nature and for those polluters who could use this organic metaphor as a justification for current practices. While the Gaia hypothesis is a profound concept, many atmospheric scientists have rejected the idea that life somehow optimizes its environment. There are many who agree with the very important role of feedback mechanisms between organic and inorganic components of the earth, but they reject the proposition that such an interaction is always self-regulatory for life. More important, the idea of a planetary-scale homeostasis allows us to take unnecessary risks whatever foolish human activities. But the Gaia hypothesis remains the countervailing metaphor for an optimism describing the co-evolution of the organic and inorganic components of the Earth and a source of hope against the abyssmal pessimism of a threatened atmosphere.

While researchers search for valid and reliable knowledge as to the impacts of climatic changes on the biosphere and society, and while intriguing hypotheses attempt to view Earth as a self-regulating system of life and environment for their mutual benefit, we have rather limited material on both the consequences and responses to long-term significant alterations. In this last speculative part we may use fictional accounts as a substitute source for exploring fears and apprehensions associated with the dramatic shifts in social structure and cultural life, when climate stops being the accepted familiar context.

Without entering into a long discussion as to the themes and expressions of social strain in novels, we may point out the abundant presence of a diversified number of extrapolative fiction concerning the general theme of the end of the world (especially through assorted ecocatastrophes) and of the presence of bleak future social systems. Such efforts provide us with additional manifestations as to present shared apprehensions concerning anticipated social breakdowns and the prospects for human survival following significant climatic alterations.

There is an inescapable conclusion in reading such works as well as in referring to the current vast number of books on the general environmental plight: despite our preoccupation with technological solutions, it becomes more than obvious that the ultimate solution will be based on a transformation of our life style. Such a change, however, requires major social transformations that a number of authors cannot perceive happening in a rather orderly manner. Only some major disaster, a cataclysm, an apocalypse, or a major revolutionary upheaval, may bring about the "new world."

Perhaps we should not push the point of the so-called "social" novel and its apocalyptic emphasis, but it is interesting to speculate as to the avalanche of books that have come about describing nuclear holocaust, Malthusian nightmares, and assorted ecocatastrophies culminating in "new civilizations." Two major tendencies tend to describe the dystopian end: one may be called the "bang theory" where all will end up on some fiery crush or some apocalyptic cataclysm resulting usually from the backlash of a widely abused nature. The second may be described as the "whimper theory," an approach that places more emphasis on gradual erosion, small but cumulative changes, a loss of elan vitale, and of the will to continue.

There is little to say here about the images of the future and the many faces of the apocalypse as they come upon us through fictional accounts. The point that needs only to be reiterated here is that speculative fiction tends to express rather dramatically the strange ideological position of an oscillation between the poles of hope and despair. If nothing else, these writings are another manifestation of the free-floating anxiety and part of the background that may help explain how the Gaia hypothesis can serve as an amorphous rationalization for millennial expectations of salvation and as a long-term solution to a decaying environment. A potentially rich vein, then, is the perusal of fictional accounts relating potential futures and adaptations to perceived climatic upheavals. Especially, if as Krieger points out a good doomsday prediction combines scientific, moral and political considerations; the story it tells is plausible, probable and consequential; allows for the knowledge of an oncoming catastrophe; and the prediction is seen as reliable and authoritative.

Jay Christopher's The Long Winter (1962) outlines the apparent arrival of a devastating ice age which closes down civilization in Europe and England driving the survivors south into Africa. We have here an interesting reversal of roles as the tropical countries become the major world powers. The book of Arnold Federbush, Ice (1967) shows through the eyes of a young woman her tragic fate in the terrible weather of an approaching ice age precipitated by nuclear testing.

One of the most competent works in this category is the book by Douglas Orgill done in collaboration with the well-known atmospheric scientist John Gribbin (The Sixth Winter, 1970). The technicalities of meteorology are well explained and the context is quite realistic when the plot follows the inescapable conclusion that the earth is headed into the "normal" condition of a glacial age. The authors' main thesis is that an ice age can come with astounding rapidity and at this point it would only take six extremely severe winters in close succession to start the irreversible chain of events which would cover much of the world with a glacier a mile thick. Governments try to maintain their state ideologies even as geographical boundaries are swept away. When cooperation is finally forced upon a crumbling Soviet Union, the northern hemisphere ceases to exist as a political entity. Depleted populations move southwards across national boundaries, feeding on microbiological rations. The international group which becomes snowbound in the North of Siberia and is forced to continue East over the Bering Straights to Alaska remind us of earlier human and animal migrations. City after city is buried by tons of ice and snow and the death toll rises into the tens of millions. The authors provide answers to such questions as to what happens in a world where Chicago is a mile or so under ice? What happens to the world's food supply? How can people survive such conditions? And throughout there are continuous warnings at us, the Interglacial Generations, that we are complacent and unprepared for the doom to come.

Finally, such other books as George Stone's Blizzard (1977) replete with unprecedented amounts of snow besieging the Eastern seaboard underscores the notion that the predominant emphasis on cooling as is also true in Robert Walker's Sub-zero (1979), David Axton Prison of Ice (1976), James Follett Ice (1978), Killian Crawford Icequake (1980), Terry R. Setler After the Snow (1981), Ian Weekley The Moving Snow (1974) and Edwin Woodward and Heather Woodward Bishoff Storehouses of the Snow (1980).

A few books such as Arthur Herzog's Heat 1977) point out the consequences of dramatic increases in carbon dioxide. All over America, and elsewhere, violent storms erupt, the ecological balance is disrupted, streets buckle and city parks turn into tropical swamps as the earth heads towards global asphyxiation and lethal heat. Of such gasping endings are also the nightmarish scenarios of Jon Hartridge's

Earthjacket (1970) and Trevor Hoyle's The Last Gasp (1983) where the ocean's plankton is dying and our oxygen is steadily depleted.

Dakota James, in his Greenhouse (1984) sets up a date -- 1997 -- for the dreaded Greenhouse Effect, when continental masses are burned to desert and coastlines are inundated with the melting of the polar icecaps. In a Chicago, nine years hence, one can see a society committing suicide through unrestrained development. How do people react to such a catastrophe? In the voice of the author: "The people of the United States reacted to the crisis the way a great people could be expected to react, with a mixture of bewilderment, name-calling, catatonic aplomb within the conservative community, sarcastic wit within the liberal community, and in the high-tech community, the morbid conceit of a haughty aristocrat about to blow his brains out with an heirloom dueling pistol."

These are but a few examples of speculative novels of an increasing number of "ecofiction" reinforcing a pervasive pessimistic mood as to the capacity of societies to respond in time to impending catastrophes. In turn, such novels reinforce some underlying themes of the scientific literature, although they tend to differ markedly from the generally optimistic conclusions of disaster studies. But whether with logarithmic graphs, rigorous scientific accounts, or with the imaginative thrusts of speculative fiction the conclusion must be that hysteria, panic or overreaction are not the proper responses to our shared concern with changing climate (unless used as a strong literary device for counteracting complacency). There can be remedies for more prudent response, including interdisciplinary research, expanded knowledge, sophisticated models that can make forecasts over decades, holistic vision, and broader new paradigms in science; protection of the world's tropical forests; strong conservation measures; detailed monitoring of the atmosphere; and, finally, institutional changes and international agreements enhancing the flow of information and the sharing of knowledge, establishing long-term environmental scanning, and mobilizing people and organizations for concerted institutional responses to the challenges of collective survival.

SELECTED BIBLIOGRAPHY

Ausubel, Jesse H. 1983. 'Can We Assess the Impacts of Climatic Changes?' Climatic Change 5:7-14.

Ausubel, Jesse and Asit K. Biswas (eds.). 1980. Climatic Constraints and Human Activities. New York: Pergamon Press.

Barkun, Michael. 1977. 'Disaster in History.' Mass Emergencies 2:219-231.

Barton, Allen H. 1969. Communities in Disaster: A Sociological Analysis of Collective Stress Situations. Garden City, NY: Doubleday.

Butzer, Karl W. 1980. 'Adaptation to Global Environmental Change.' Professional Geographer 32:269-278.

Butzer, Karl W. 1983. 'Human Response to Environmental Change in the Perspective of Future, Global Climate.' Quaternary Research 19:279-292.

Bryson, Reid A. and Thomas J. Murray. 1977. Climates of Hunger. Madison: The University of Wisconsin Press.

Chen, R.S., E. Boulding, and S.H. Schneider (eds.). 1983. Social Science Research and Climate Change. Boston: D. Reidel Publishing Company.

Claiborne, Robert. 1970. Climate, Man, and History. New York: W.W. Norton & Co., Inc.

Cooper, Charles F. 1978. 'What Might Man-Induced Climate Change Mean?' Foreign Affairs, 56(3):500-520.

d'Arge, R.C., William D. Schulze, and David S. Brookshire. 1982. 'Carbon Dioxide and Intergenerational Choice.' American Economic Association Papers and Proceedings 72:251-256.

Don, Frank. 1981. Earth Changes Ahead: The Coming Great Catastrophes. NY: Warner Destiny Books.

Drabek, Thomas E. 1987. Human System Responses to Disaster: An Inventory of Sociological Findings. NY: Springer-Verlag.

Federov, E.K. 1979. 'Climatic Change and Human Strategy.' Environment 21(4): 25-31.

Flohn Hermann. 1977. 'Climate and Energy: A Scenario to a 21st Century Problem.' Climatic Change 1:5-20.

Flohn, Hermann. 1980. Possible Climatic Consequences of Man-made Global Warming. Laxenburg, Austria: International Institute for Applied Systems Analysis.

Flohn, Hermann, and Roberto Fantechi. 1984. The Climate of Europe: Past, Present and Future. Boston: D. Reidel Publishing Company.

Graham, W.R. and M. Warshaw. 1971. 'A Model of Global Climate and Ecology.' In George A. Bekey (ed.). Mathematical Models of Public Systems. Simulation Councils Proceedings Series. 1(1):25-31.

Green, Fitshugh. 1977. A Change in the Weather. New York: W.W. Norton & Co., Inc.

Gribbin, John. 1976. Forecasts, Famines, and Freezes: Climate and Man's Future. NY: Walker and Company.

Gribbin, John (ed.). 1978. Climatic Change. Cambridge: Cambridge University Press.

Gribbin, John. 1979. What's Wrong With Our Weather? The Climatic Threat of the 21st Century. New York: Charles Scribner's Sons.

Gribbin, John. 1982. Future Weather and the Greenhouse Effect. New York: Delacorte Press.

Haas, J.E. 1975. 'Social Impact of Induced Climate Change.' In d'Arge, R. et al. (eds.) Climate Impact Assessment Program CIAP), Monograph 6: Economic and Social measures of Biologic and Climatic Change. Washington, D.C.: U.S. Department of Transportation.

Halacy, D.S., Jr. 1978. Ice or Fire? Surviving Climatic Change. New York: Harper & Row.

Hardy, Ralph, Peter Wright, John Kington, and John Gribbin. 1982. The Weather Book. Boston: Little, Brown and Company.

Hare, F. Kenneth. 1984. 'Changing Climate and Human Response: The Impact of Recent Events on Climatology.' Geoforum 3:383-394.

Hare, F. Kenneth. 1985. 'Future Environments: Can They be Predicted?' Institute of British Geographers Transactions New Series 10:131-137.

Hare, F. Kenneth and W.R. Derrick Sewell. 1986. 'Awareness of Climate.' In R.W. Kates and I. Burton (eds.), Geography, Resources, and Environment. Vol. II, Themes from the Work of Gilbert F. White. Chicago: University of Chicago Press, pp. 107-139.

Hillary, Sir Edmund (ed.). 1984. Ecology 2000: The Changing Face of Earth. New York: Beaufort Books, Inc. (esp. Normal Myers, 'Will the Climate Change?' Chapter 10, pp. 171-188).

Kates, Robert W. 1980. 'Climate and Society: Lessons From Recent Events.' Weather 35:17-25.

Kates, Robert W., Jesse H. Ausubel, and Mili Berberian (eds.). 1985. Climate Impact Assessment. Chichester: John Wiley & Sons.

Kellogg, W.W. 1980. 'Modeling Future Climate: It is Time to Plan for a Climate Change and a Warmer Earth' Ambio. 9:216-221.

Kellogg, W.W. 1987. 'Mankind's Impact on Climate: The Evolution of an Awareness.' Climatic Change 10:113-136.

Kellogg, W.W. and Robert Schware. 1981. Climate Change and Society: Consequences of Increasing Atmospheric Carbon Dioxide. Boulder: Westview Press.

Kellogg, W.W. and Robert Schware. 1982. 'Society, Science and Climate Change.' Foreign Affairs 60(5):1076-1109.

Krieger, Martin H. 1987. 'The Possibility of Doom.' Technology in Society 9:181-190.

Ladurie, Emannuel Le Roy. 1971. Times of Feast, Times of Famine, A History of Climate Since the Year 1000. Garden City, NY: Doubleday & Company.

Lamb, H.H. 1982. 'Climatic Changes in our Own Times and Future Threats.' Geography 67, Pt. 3, 296:203-220.

Lamb, H.H. 1982. Climate History and the Modern World. London: Methuen.

Mather, John R., Richard T. Field, Laurence S. Kalkstein, Cort J. Willmott, and W.J. Maunder. 1981. 'Climatology: The Impact of the Seventies and the Challenge of the Eighties.' Weather and Climate 1:69-76.

Maunder, W.J. 1986. The Uncertainty Business; Risks and Opportunities in Weather and Climate. London: Methuen.

Meyer-Abich, Klaus M. 1980. 'Socioeconomic Impacts of CO2-Induced Climatic Changes and the Comparative Chances of Alternative Political Responses: Prevention, Compensation, and Adaptation.' Climatic Change 2:373-385.

Mintzer, Irving M. 1987. A Matter of Degrees: The Potential for Controlling the Greenhouse Effect. Washington: World Resources Institute (Research Report #15).

Morrisette, Peter M. (ed.). 1987. A Selected Annotated Bibliography of Climate and Society Research. Washington, D.C., NOAA Technical Report NCPO 002.

Nanda, Ved P. (ed.). 1983. World Climate Change: The Role of International Law and Institutions. Boulder: Westview Press.

National Defense University. 1978. Climate Change to the Year 2000. Washington, D.C.: National Defense University.

Parry, Martin L. 1986. 'Some Implications of Climatic Change for Human Development.' In W.C. Clark and R.E. Munn (eds.), Sustainable Development of the Biosphere. Cambridge: Cambridge University

Press, pp. 378-407.
Ponte, Lowell. 1976. The Cooling: Has the Next Ice Age Already Begun? Can We Survive It?. Englewood Cliffs, NJ: Prentice-Hall.
Post, John D. 1977. The Last Great Subsistence Crisis in the Western World. Baltimore: The Johns Hopkins University Press.
Riebsame, William E. 1985. 'Climate Hazards: Lessons from Recent Events in the United States.' Disasters. 9:295-300.
Riebsame, William E., Henry F. Diaz, Tod Moses, and Martin Price. 1986. 'The Social Burden of Weather and Climate Hazards.' Bulletin of the American Meteorological Society 67:1378-1388.
Roberts, Walter Orr, and Henry Lansford. 1979. The Climate Mandate. San Francisco: W.H. Freeman and Company.
Schelling, Thomas C. 1984. 'Anticipating Climate Change: Implications for Welfare and Policy.' Environment 26:6-9.
Schneider, Stephen H. 1977. 'Climate Change and the World Predicament: A Case Study for Interdisciplinary Research.' Climatic Change 1:21-43.
Schneider, Stephen H. with Lynne E. Mesirow. 1976. The Genesis Strategy: Climate and Global Survival. New York: Plenum Press.
Schneider, Stephen H. and Richard L. Temkin. 1978. 'Climatic Changes and Human Affairs.' In John Gribbin (ed.), Climatic Change. Cambridge: Cambridge University Press, pp. 228-246.
Smith, V. Kerry. 1982. 'Economic Impact Analysis and Climate Change: A Conceptual Introduction.' Climatic Change 4(1):5-22.
Sonka, Steven T. and Peter J. Lamb. 1987. 'On Climate Change and Economic Analysis.' Climatic Change. 11(3):291-311.
Stommel, Henry and E. Stommel. 1983. Volcano Weather: The Story of 1816, The Year Without a Summer. Newport, R.I.: Seven Seas Press.
Timmerman, Peter. 1981. Vulnerability, Resilience and the Collapse of Society: A Review of Models and Possible Climatic Applications. Toronto: University of Toronto, Institute for Environmental Studies, Environmental Monograph No. 1.
Tonn, Bruce E. 1986. 'Using Possibility Functions for Long-Term Environmental Planning.' Futures 18(6):795-807.
Verschuur, Gerrit L. 1978. Cosmic Catastrophes. Reading, MA: Addison-Wesley.
Warshofsky, Fred. 1977. Doomsday: The Science of Catastrophe. NY: Reader's Digest Press.
Wigley, T.M.L., M.J. Ingram, and G. Farmer (eds.). 1981. Climate and History: Studies in Part Climates and Their Impact on Man. Cambridge: Cambridge University Press.
Wilkinson, John. 1976. 'Climate as a State Variable: What Should be a Rational Policy?' Technological Forecasting and Social Change 9:301-308.
Wilson, L.S. 1981. 'The World's Changing Climate -- Some Issues for Planners.' Long Range Planning 14(5):83-89.

GLOBAL CLIMATIC CHANGES AND GEOPOLITICS: PRESSURES ON DEVELOPED AND DEVELOPING COUNTRIES

Peter H. Gleick [1]
Energy and Resources Group
Building T-4, Room 100
University of California
Berkeley, California 94720

ABSTRACT. Large-scale climatic changes may lead to international frictions and resource competition between both developed and developing countries for three principal reasons: the major responsibility for the production of greenhouse gases lies with the industrialized countries; the socioeconomic and environmental impacts of climatic changes will be far more equally distributed; and there are major differences in the ability of countries to respond or adapt to climatic impacts. These inequities may complicate the possibility of negotiating international agreements to prevent or slow the greenhouse effect. This paper discusses these three issues, the principal vulnerabilities of both developed countries and developing countries to climatic changes, and mechanisms for improving international cooperation.

1. INTRODUCTION

Global climatic changes will lead to a variety of major socioeconomic and political consequences for both developed and developing countries. Among these consequences is the possibility that international frictions between East and West or between North and South will be worsened by changes in the access to or the quality of various resources, or that political agendas will complicate the possibility of an international agreement to prevent or slow the greenhouse effect. Given the extent and severity of the possible climatic changes, we must begin to ask how such changes will affect international relationships, economics, behavior, and security. [2]

---

[1] SSRC-MacArthur Foundation Fellow in International Peace and Security; Visiting Research Scholar, Energy and Resources Group, University of California, Berkeley, California, 94720.

[2] Threats to security can be defined to include actions that (1) threaten to drastically and quickly degrade the quality of life for the inhabitants of a state, or (2) threaten to significantly narrow the range of policy choices available to governments or non-governmental entities (Ullman 1983). Climatic changes that lead to the deterioration of environmental quality or limit access to resources fall into both categories.

A. Berger et al. (eds.), Climate and Geo-Sciences, 603–621.

There are three fundamental characteristics of the climate issue that will affect both perceptions of climate impacts and the actions taken to prevent or adapt to climatic changes. First, the responsibility for the production of greenhouse gases lies primarily with the industrialized countries. In contrast, the impacts of climatic changes will be far more widely distributed among both developed and developing countries. And finally, there are major differences in the ability of countries or regions to react and adapt to climatic changes. These characteristics may play as much of a role in determining the extent to which tensions and conflicts develop over climatic changes as the details of the impacts themselves. This paper explores these three factors for both developed and developing countries and highlights those vulnerabilities most likely to affect international politics.

2. THE SOURCE OF THE PROBLEM: THE PRODUCTION OF GREENHOUSE GASES

Climatic changes are being driven by increasing concentrations of atmospheric trace gases that are produced by human activities, such as the combustion of fossil fuels, deforestation and forest burning, the manufacture of semiconductors, foam production, and leakage of industrial coolants. Table I lists the principal "greenhouse" gases and their estimated annual rates of increase.

The prime responsibility for the production of these gases lies with the industrialized world. Table II shows the high percentage of carbon dioxide and chlorofluorocarbons (CFC) produced by the major industrialized countries in the mid-1980s. Approximately half of the expected climatic change from greenhouse gases is due to carbon dioxide. As Table II shows, the northern hemisphere industrialized countries are currently responsible for about 70 percent of the global production of carbon dioxide. The United States alone is responsible for nearly a quarter of all CO_2 production and 20 to 30 percent of CFC production. When these data are coupled with population data, the discrepancy between the industrialized and developing worlds becomes even more apparent. Figure 1 plots the per-capita CO_2 production as a function of population for many of the world's countries and regions.

The principal responsibility for carbon dioxide and chlorofluorocarbon production is slowly shifting from the major industrialized powers to developing countries with growing populations and a growing industrial base. By the mid-1980s, developing countries produced as much carbon dioxide from fossil-fuel combustion as did the United States, or the Soviet Union and other centrally-planned economies (Rotty 1987b). Similarly, under the Montreal Protocol on Substances That Deplete the Ozone, the share of consumption of chlorofluorocarbons by developing countries will rise considerably, even if some key producers fail to abide by the treaty (United States Congress 1988). Nevertheless, the industrialized world will continue to be responsible for a disproportionate share of greenhouse gases for many years to come.

Table I

Principal Greenhouse Gases

Greenhouse Gas	Atmospheric Concentration (ppm)	Annual Rate of Increase (%)
Carbon Dioxide	350	0.4
Nitrous Oxide (N_2O)	1.7	1.0
Methane (CH_4)	0.3	0.25
CFC-11	0.23	5
CFC-12	0.4	5

Table II

Percent of World Production of Carbon Dioxide and Chlorofluorocarbons

	CO_2 (1)	CFC-11 (2)	CFC-12
United States	24	23	30
Western Europe & Canada	16	49	33
Japan, Australia, New Zealand	6	20	15
Soviet Union and other CPEs	26	6	19
All others	28	2	3

Notes.

1. These figures include carbon dioxide from fossil fuel combustion and industrial fuel use, but do not include the contribution from biomass burning (Rotty 1987a).

2. Chlorofluorocarbon data are from Mintzer (1987).

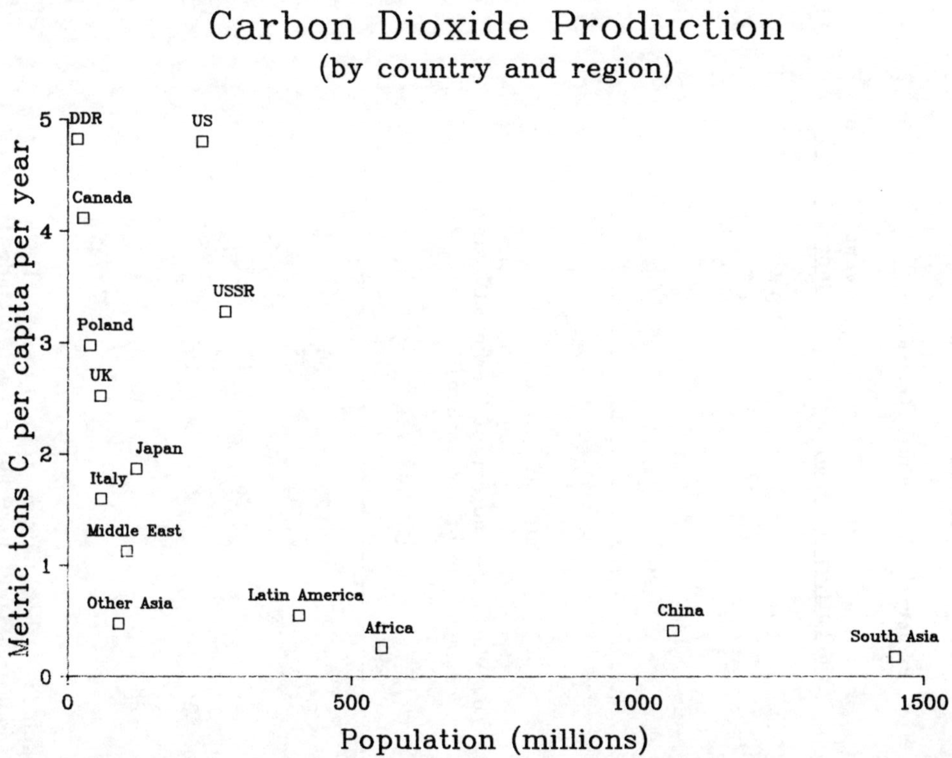

Figure 1. Per capita production of carbon dioxide (in metric tons of carbon per capita) by country or region for 1983 versus population (in millions). Note the high per capita annual CO_2 production for the industrialized countries with low populations, while China, Latin America, Africa, and other developing countries and regions produce little carbon dioxide per capita.

3. THE DISTRIBUTION OF THE IMPACTS OF CLIMATIC CHANGES

Unlike the distribution of responsibility for greenhouse gases, the distribution of impacts from greenhouse warming will be global in extent and delayed in time. Among the victims of climatic change will be those nations (and generations) least responsible for the production of greenhouse gases, least able to either adapt to or mitigate the changes, and with little international economic or political clout. As a result, there is a distinct possibility that climatic changes will worsen international frictions and disputes.

Given the current state-of-the-art in climate modeling, accurate estimates of the regional impacts of climatic changes may not be available until the impacts are upon us. For this reason, it is important to look at the existing vulnerabilities to climatic variability in order to understand future vulnerabilities to more permanent future changes.

Five areas stand out as particularly important for national and international relationships, behavior, and policy: agricultural productivity, the availability and quality of freshwater resources, access to energy resources, sea-level rise, and ocean/ice conditions in the high latitudes. These areas are listed in Table III, with an estimate of their relative importance. Table IV lists separately the likely routes of conflict between developed countries, between developing countries, and between developing and developed countries.

4. POINTS OF VULNERABILITY FOR DEVELOPED AND DEVELOPING COUNTRIES

4.1 Agriculture

Threats to the basic food supplies of a country are cause for frictions and tensions between nations (Wallensteen 1986, Gleick 1987). Possible mechanisms for such threats include trade embargoes or other forms of political manipulation of access to food, environmental degradation such as loss of soil fertility, or competition among conflicting land uses. Because scarcity is a fundamental condition for a good to become a political tool, the disparity in food needs and food resources between the developing and the developed world has long hinted at the possibility of future conflict over access to food resources.

Food availability depends on a complex array of factors, including patterns of production, purchasing ability, and the operation of food distribution systems. The vulnerability of political behavior to the availability and quality of agricultural resources has been highlighted in recent years by internal conflicts and riots over food shortages throughout the Sahel in the 1970s, in Sudan in 1981 and 1985, in Poland in 1980, in Tunisia in 1983 and 1984, and in Morocco in 1984. These internal events often serve to increase external tensions as well (Schneider 1983, Wallensteen 1986).

In many developing countries, agricultural productivity is acutely vulnerable to natural climatic variability that may cripple their own food production or substantially reduce the supply and raise the price

Table III

Sensitivities of Developed and Developing Countries to Climatic Changes (1)

	Developing Countries	Developed Countries
Agriculture		
Irrigated	M	M
Rainfed	H	L
International Transfers	M	M
Water Resources		
Quality	M	L
Quantity	M	M
Access to Energy Resources	L	M
Sea-Level Rise	M	M
Arctic and Antarctic Ocean/Ice Conditions	L	L

Notes.

1. Specific impacts depend on regional details of climatic changes and socioeconomic effects. These indicators are meant to give a sense of the types of impacts that have implications for international tensions and conflicts and to highlight relative vulnerabilities to climatic changes.

L = low impact
M = moderate impact
H = high impact

Table IV

Possible Routes of Conflicts

Conflict Between Developed Countries

Agricultural Productivity and Trade
Ice-Free Arctic Ocean/Open Northern Passages
Access to and Sales of Arctic and Antarctic Resources
Patterns of Energy Use

Conflict Between Developing Countries

Access to Freshwater Resources
Agricultural Productivity and Trade
Sea-Level Rise and Population Migration

Conflict Between Developing and Developed Countries

Access to Freshwater Resources
Patterns of Energy Use
Agricultural Productivity and Trade
Access to and Sales of Arctic and Antarctic Resources

Table V

Rivers with Five or More Nations Forming Part of the Basin

	#	Area (km^2)
Danube	12	817,000
Niger	10	2,200,000
Nile	9	3,030,700
Zaire	9	3,720,000
Rhine	8	168,757
Zambezi	8	1,419,960
Amazon	7	5,870,000
Mekong	6	786,000
Lake Chad	6	1,910,000
Volta	6	379,000
Ganges-Brahmaputra	5	1,600,400
Elbe	5	144,500
La Plata	5	3,200,000

Source. United Nations (1978).

of foodstuffs on the world market. Rainfed agriculture, in particular, is sensitive to the vagaries of weather, especially the timing and length of the rainy seasons and the total amount of rain received. Under conditions of higher temperatures, changing rainfall patterns, and growing population, the situation for some developing countries may grow more precarious. The inability to accurately predict which regions will receive more or less rainfall further complicates analysis and planning.

The situation for large grain producers is equally uncertain. As temperatures increase, irrigated agricultural production could expand into northern regions of the United States, the Soviet Union, China, and Canada only if soil conditions, water availability, and other factors (such as infrastructure) permit. But output in regions that are currently productive, such as the Central Plains of the United States, the Ukraine, and Kazakhstan, could be reduced by higher temperatures and changes in water availability. Similarly, a near-term improvement in yield due to higher temperatures or greater water availability could be followed by a longer-term deterioration of yield as temperatures continue to rise or as beneficial rainfall patterns disappear. Figure 2 demonstrates this effect.

International tensions and conflict may be provoked by changes in the productivity of major grain importers, by the demand for grain on the international market, and by the ability of present grain suppliers to continue to generate surpluses. For example, climate plays an important role in limiting the productivity of Soviet agriculture, and periodic crop failures have been a distinct political danger. The direct effect of Soviet grain productivity problems is to force them to international markets to make up deficiencies. Yet the dependence on international markets makes them vulnerable to the political risks of international grain embargoes. In 1980, the United States imposed such an embargo on the Soviet Union in response to the Soviet intervention in Afghanistan.

The principal effect of the 1980 embargo was not economic or political. Rather it gave the Soviet leadership a sharp reminder of the risks of depending on world markets for such a critical commodity. The willingness of the Soviets to continue to purchase foreign grain results from their ability as the single largest grain purchaser to find favorable terms from a mix of suppliers, from their ability to continue to generate the foreign currency needed for grain purchases through the sale of other natural resources (notably fossil fuels, gold, and certain minerals), and from their belief that reliance on foreign grain sources is only temporary. In fact, the vulnerabilities arising from international trade in grain cut both ways. In recent years, drought in Russia and Kazakhstan has been one of the major uncertainties for Western grain markets, and sudden and massive purchases by the Soviet Union have played havoc with world prices.

Analysis of the net effect (both regionally and globally) of climatic changes on food production is further complicated by the difficulties of estimating the effect of changes in yields on world agricultural markets. Short-term reductions in yields alone are not necessarily bad for overall long-term productivity and food availability. Confounding factors include the size of stocks, subsequent investments in other regions, planting patterns,

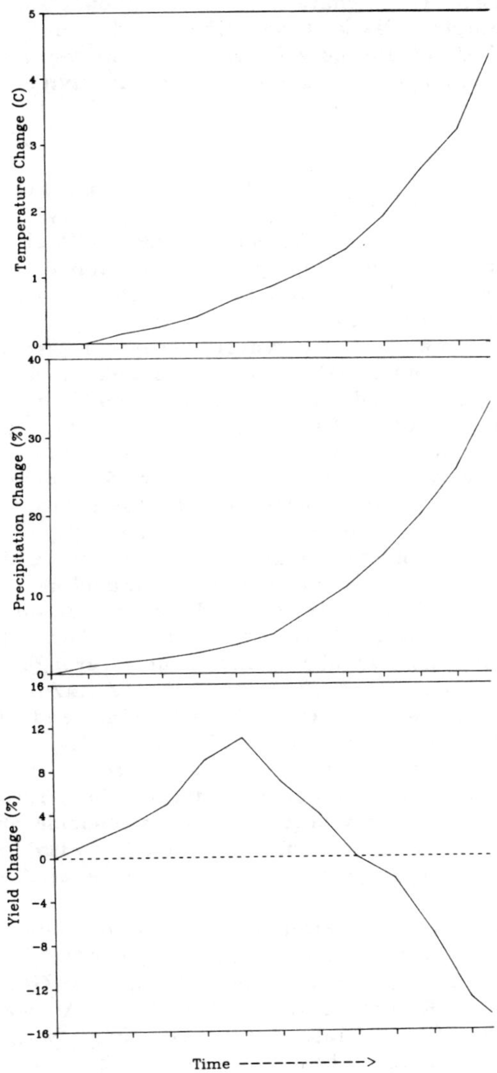

Figure 2. Possible transient changes in agricultural yield with time. The bottom graph shows yields initially increasing as mean temperatures (top graph) and mean precipitation (middle graph) increase. Further increases in yields, however, are prevented by additional increases in temperature and precipitation, which may cause crop damage when they reach certain levels. This type of effect complicates the equilibrium analysis of climatic impacts.

international prices, and the character and mechanisms of trading agreements (see, for example, Clark 1985, Chen and Parry 1987). Thus changes in the comparative advantage of the different actors can play a greater role than absolute changes in agricultural productivity.

4.2 Water Resources

Historically, international political frictions and tensions have arisen over the control of, access to, or the quality of freshwater resources (Naff and Matson 1984, Falkenmark 1986, Gleick 1987, 1988). These conflicts occur in both developing and developed countries and depend to a greater extent on the nature of the water resources than the level of industrial or economic development. Even in the absence of climatic changes, pressures on existing water resources are growing due to increases in population, industrial water demand, and development. Where water resources are shared, as in international river basins or bodies of water bordering more than one country, the possibility of friction and conflicting demands exists.

Freshwater resources are widely shared. Nearly 50 countries on four continents have more than three-quarters of their total land area falling within international river basins, while globally 47 percent of all land area falls within international river basins (United Nations 1978). Over 200 river basins are multinational, including 57 in Africa and 48 in Europe. Table V lists thirteen different major rivers with five or more nations forming part of the watershed. Examples from the different continents include the Danube, which runs through 12 different nations; the Nile, which runs through nine; the Amazon, which runs through seven; the Mekong, which runs through six; and the Ganges-Brahmaputra, which runs through five (United Nations 1978).

As mentioned earlier, the identification of regional problems is limited by our ability to model regional details of future climatic changes. In the case of water resources, there is considerable value to simply identifying where existing competition over water resources is already a problem for reasons of limited water availability or for reasons of political dispute.

Regions with a history of international tensions or competition over water resources include the Jordan and Euphrates Rivers in the Middle East, the Nile, Zambesi, and Niger Rivers in Africa, the Ganges in Asia, the Colorado and Rio Grande Rivers in North America. As water demands increase, the probability of conflict over remaining water resources will also increase. In these cases, mechanisms for allocating shortages must be designed and agreed upon before climatic changes begin to alter the hydrologic regime. International water law, though immature in its development and application, must play a role (International Law Association 1967, 1986, Gleick 1988). By codifying these allocation schemes, mechanisms can be put into place to resolve conflicts before they become acute.

4.3 Northern Energy Resources

Significant energy resources underlie Arctic continental and offshore regions and these resources play a vital role in national economies and world economic markets. Any change in climate that affects the ability to extract these resources will be a factor in the response of international actors to initiatives to control climatic change. Access to northern oil and gas is already constrained in some regions by climatic conditions and it is dependent on expensive and vulnerable methods and materials.

The oil and gas potential of the northern Arctic regions is large. Despite limited exploration, over 20 percent of the "proved" reserves of the United States are in Alaska, and nearly 40 percent of Soviet proved reserves are in northern and western Siberia. Table VI shows the Arctic proven reserves of the countries of the region. These figures should be compared with estimated global proved oil reserves of 95 billion tonnes (700 billion barrels) of oil, and gas reserves of 100 trillion cubic meters (3600 trillion cubic feet).

Given the importance of northern energy resources, climatic constraints are unlikely to prevent future development. Even today, despite the difficulties in exploiting northern minerals, the Western Siberian oil fields produce about half of all Soviet oil production (Mote 1983). More than 50 percent of all Soviet hard currency income during the 1980s has been derived from oil.

The more relevant question, however, is whether or not future climatic changes will significantly increase or decrease the difficulty--and hence the expense--of that production. Current analyses of Soviet economic conditions suggest that the maintenance or increase of oil exports is a key to sustaining import levels (including grain) essential for the Soviet Union's economic plans and political priorities (Bergesen et al. 1987). Similarly, U.S. dependence on Middle Eastern oil is claimed to threaten U.S. security interests and development of the Alaskan/Arctic oil reserves is often promoted as a way of reducing this dependence. The uncertainties posed by future climatic changes will complicate these problems.

Although problems of access to northern energy resources do not directly affect developing countries, there are a number of indirect effects that need to be explored. How, for example, will world energy markets be affected by easier (or more difficult) access to northern oil and gas? Will the world market price of fossil fuels be affected? Certainly, if exploiting these resources becomes easier, their cost will decrease and their use may increase. Perversely, this could further exacerbate the greenhouse effect by encouraging more combustion of carbon-based fuels.

4.4 Sea-Level Rise

One of the most pressing direct effects of greenhouse warming will be rising sea-level due to both thermal expansion of the oceans and melting of glaciers and land ice. While the rise in sea-level will most likely be incremental, the damages caused by this rise will occur during

Table VI

Arctic Oil Resources

	Arctic Potentially Recoverable[1] (10^9 bbl)	Arctic Proved Reserves (10^9 bbl)	National Proved Reserves[3] (10^9 bbl)	Arctic Proved Reserves as a Percent of National Reserves
United States	17 - 55	7	33	21
Soviet Union	50 - 80	22	59	37
Canada	15	1 - 2	8	12 - 24
Norway	10 - 15	(2)	11	(2)

Notes.

Sources. Garrett 1984, Bergesen et al. 1987, British Petroleum 1987.

(1) Undiscovered potential. The mean U.S. value is 17 to 24 billion barrels. The high-side estimate for the United States is 39 to 55 billion barrels (Garrett 1984). Gas resources could be as great as 60 trillion cubic meters (2000 trillion cubic feet). These estimates are highly uncertain.

(2) Accurate Norwegian oil reserves data are unavailable, but estimates range between 100 and 300 million tonnes (or 0.7 to 2 billion barrels).

(3) Estimates of world 1986 proven oil reserves total 700 billion barrels, of which 480 billion are in the Middle East.

Table VII

Major Storm Disasters in Coastal/Delta Regions

Year	Location	Estimated Death Toll
1963	Bangladesh	22,000
1965	Bangladesh	17,000
1965	Bangladesh	30,000
1965	Bangladesh	10,000
1970	Bangladesh	250,000 - 500,000
1971	India	10,000 - 25,000
1977	India	10,000
1985	Bangladesh	10,000

(Source: Bardach 1987)

specific events such as storm surges, hurricanes, and typhoons. Although the impacts of sea-level rise will vary depending on the geophysical characteristics of the region, some observations can be made about the differences between developing and developed countries.

In developing countries, coastal populations and coastal agricultural developments are particularly vulnerable to storms. Other than coastal cities, sea-level rise will be most severely felt in deltaic areas, especially in countries where rivers are large and deltas extensive and highly populated. The highly-populated coastal plains of India and Bangladesh, for example, are already extremely vulnerable to storms. Since 1960, India and Bangladesh have been struck by at least eight tropical cyclones that each killed more than 10,000 people (see Table VII). In late 1970, storm surges killed approximately 300,000 people in Bangladesh and reached over 150 kilometers inland over the lowlands. Recent estimates show that a climatically-induced one-meter sea-level rise would cover land presently occupied by nearly 10 percent of the Bangladesh's population (Broadus et al. 1986). A far greater fraction of the population would be put at risk by the increased consequences of storms.

The delta in Bangladesh is also vulnerable to the activities of upstream states: India, Nepal, China, and Bhutan. Disputes already exist between India and Bangladesh over the construction of dams on the Ganges, and over the use of major upstream tributaries. Upstream activities can affect both water availability in Bangladesh and the formation and ecology of the delta, which offers some protection from storm surges and sea-level rise. One solution, population migration, offers a way of reducing the human vulnerability in some areas, but at the high societal cost of relocation and absorption of those displaced.

Of greater concern in the developed world is the protection of coastal cities and industrial sites by building structures to protect against sea-level rise. In the Netherlands--a small, but vulnerable country--over half the country would be uninhabitable if not protected against the sea (Goemans 1986) and 8 million people would be displaced. Indeed, the Dutch see greenhouse-induced sea-level rise as a growth industry capable of providing a new global market for their contractors, hydraulic engineers, and consultants. In the United States, attention is being given to threats to cities, with a mixed strategy of abandonment and protection being considered. In almost all of these studies, no consideration other than "natural adaptation" is given to the risks to natural ecosystems.

4.5 Arctic and Antarctic Ocean and Ice Conditions

Arctic and Antarctic conditions are beginning to play an important role in both strategic and commercial fields. In the Antarctic, questions about access to mineral resources are beginning to re-emerge as we near the 30th anniversary of the Antarctic Treaty of 1959, and as capabilities for minerals extraction under severe conditions improve. Despite several years of discussion, no treaty that would regulate the exploitation of Antarctic mineral resources has yet been worked out (Westing 1986). As climatic conditions begin to alter extraction conditions in the higher latitudes, disputes could become more prevalent, especially since both developed and developing countries maintain unresolved sovereignty claims in the Antarctic.

In the Arctic, commercial sea routes in the Northeast Passage by the Soviet Union are kept open at considerable expense by Soviet icebreakers. There is a possibility that as the world warms, the Arctic will become ice-free for a large part of the year. The implications for the commercial balance are unclear, but one important consequences could be the reliable opening of the Northeast Passage over the Soviet Union and the possible opening of a satisfactory Northwest Passage over Canada.

There are also implications for the strategic balance between northern hemisphere superpowers. The Arctic is considered to be an area of strategic importance to the United States (Johnson et al. 1984) and to the Soviet Union, yet activities there are severely hindered by environmental conditions. This ice-covered region provides cover for the actions of nuclear-powered submarines of the United States and the Soviet Union, over the occasional objections of the other nations--particularly Canada--that border the area. While it is difficult to predict the importance of submarines (or even the relations between the United States and the Soviet Union) in 25 to 40 years, the possibility of much less severe ice conditions in the Arctic could have important security ramifications.

5. THE ABILITY TO RESPOND OR ADAPT TO CLIMATIC IMPACTS

There is a fundamental discrepancy between the ability of developed and developing countries to adapt or respond to the impacts of climatic change. In some cases, climate impacts may be more severe in countries that are the least able to adapt. For these countries, prompt preventative measures are far preferable. Yet the burden for prevention rests with those countries that bear the greatest responsibility for the production of greenhouse gases. Where the severity of the impacts are uncertain, developed countries are likely to lean toward adaptation. One risk of this strategy, of course, is that unexpectedly severe or rapid impacts may overwhelm the ability to adapt, or that there are thresholds or nonlinearities in societal capabilities for adaptation (Clark 1985).

One of the impetuses for a treaty to protect stratospheric ozone by controlling the production of chlorofluorocarbons was the discovery of a hole in the ozone layer over Antarctica. This discovery was a "discontinuity"--an unexpected environmental effect with possible global consequences. In the absence of similarly prompt and severe climatic impacts associated with greenhouse gases, industrialized nations are likely to prefer the "wait and see" strategy of adaptation, which produces the fewest near-term costs to society. The corollary to this, however, is that adaptation may produce the greatest long-run costs from climatic impacts. Developing countries with limited resources for responding to major climatic changes, such as sea-level rise, loss of agricultural productivity, and so on, have the most to gain by forcing greenhouse gas producers to cut back on carbon dioxide production.

A further complication may be a desire of certain actors (alliances, nations, sub-national groups, corporations) to capitalize on perceived regional advantages of climatic changes. Those actors who believe--rightly or wrongly--that they will benefit from a warmer earth will have no direct incentive to cooperate in any international agreement to prevent climatic change. If these actors are among those most responsible for greenhouse gas production--and hence most able to affect the outcome--others may be forced into an adaptation strategy.

6. DISCUSSION

Although developing countries are increasing their production of carbon dioxide and other greenhouse gases, a disproportionate share of the responsibility for climatic changes will continue to lie with the industrialized countries for many years to come. The societal impacts of these climatic changes, however, will be far more equally distributed, posing important, and still unresolved, questions about equity and fairness. Finally, the ability to adapt to severe climatic changes--to build sea-walls or relocate populations, to alter agricultural infrastructures and cropping techniques, or to provide adequate freshwater resources, is highly variable around the world. Yet no region or country can expect to benefit from rapid climatic changes that would overwhelm the capacity of even wealthy countries to adapt.

With few exceptions, developing countries are least able to mobilize the resources for adapting to severe impacts, while developed countries are more likely to prefer to adapt to climatic changes than to drastically alter their energy and industrial infrastructures. This fundamental dichotomy is likely to lead to international frictions and disputes between developed and developing countries, unless mechanisms can be worked out to more equitably share the costs of climatic impacts and the costs of preventing them. Disputes may also arise between developed countries over access to northern energy resources, over changes in international agricultural trade, and over patterns of energy use.

International laws and agreements are tools for cooperative efforts on international resource problems. Such tools are supposed to provide the structures and mechanisms through which nation-states manage their interactions (Bilder 1980). Unfortunately, international law presently lacks the necessary depth to resolve the complications posed by climatic change and impacts. There are some broad guidelines--such as the U.N. Declaration on the Environment (1976),[3] but these have never been

---

[3] Principle 21 of the UN Declaration on the Environment, adopted in 1976, says: "States have, in accordance with the Charter of the United Nations and the principles of international law, the sovereign right to exploit their own resources pursuant to their own environmental

satisfactorily applied to environmental problems that are global in scope.

International laws are only one factor among many--including political, economic, social, scientific, and technological factors--that influence national policymaking and decisions. Indeed, energy policy appears to be the most important factor in the climate issue, because of the role of fossil fuels in producing carbon dioxide and the role of fossil fuels in economic and industrial development. Unless actions on energy policy begin soon to reduce the emissions of carbon dioxide, the earth will be irreversibly committed to substantial warming.

Finally, differing perceptions about the severity of global climatic changes must not be allowed to stop comprehensive international discussions about this problem and negotiations on ways of preventing the worst consequences. Many actions that would prevent or delay climatic change are appropriate in their own right, such as energy conservation, a reduction of dependence on fossil fuels, reforestation and alternatives to fuelwood use, and a ban on the use of certain trace gases with long-term atmospheric effects. These cooperative strategies can reduce the rate of climatic change and give us time to both improve our understanding of climatic impacts and to reflect on appropriate international responses.

---

policies, <u>and the responsibility to ensure that activities within their jurisdiction or control do not cause damage to the environment of other States or of areas beyond the limits of national jurisdiction".</u> (emphasis added.)

REFERENCES

Bardach, J.E. 1987. 'Global Warming and the Coastal Zone: Some Effects on Sites and Activities'. For the Conference on Developing Policies for Responding to Future Climatic Changes. September 28 to October 2. Villach, Austria. 80 pp.

Bergesen, H.O., Moe, A., and W. Ostreng. 1987. Soviet Oil and Security Interests in the Barents Sea. (Frances Pinter, Publishers, London, England.) 144 pp.

Bilder, R.B. 1980. 'International Law and Natural Resource Policies'. In Dorner, P. and M.A. El-Shafie (editors), Resources and Development. (The University of Wisconsin Press, Croom Helm, London.) pp. 385-421.

British Petroleum (BP). 1987. BP Statistical Review of World Energy. British Petroleum Company. (Britannic House, Moor Lane, London.) 36 pp.

Broadus, J., Milliman, J., Edwards, S., Aubrey, D., and F. Gable. 1986. 'Rising Sea Level and Damming of Rivers: Possible Effects in Egypt and Bangladesh'. In J. Titus (editor), Effects of Changes in Stratospheric Ozone and Global Climate: Volume 4, Sea Level Rise. U.S. Environmental Protection Agency and United Nations Environment Programme. (Washington, D.C.) pp. 165-189.

Chen, R.S. and M.L. Parry (editors). 1987. 'Policy-Oriented Impact Assessment of Climatic Variations'. International Institute for Applied Systems Analysis, Laxenburg, Austria. IIASA-RR-87-7. 54 pp.

Clark, W.C. 1985. 'On the Practical Implications of the Carbon Dioxide Question'. International Institute for Applied Systems Analysis, Laxenburg, Austria. IIASA-WP-85-43. 84 pp.

Falkenmark, M. 1986. 'Fresh Waters as a Factor in Strategic Policy and Action'. In A.H. Westing (editor), Global Resources and International Conflict: Environmental Factors in Strategic Policy and Action. (Oxford University Press, New York.) pp. 85-113.

Garrett, J.N. 1984. 'Conventional Hydrocarbons in the United States Arctic: An Industry Appraisal'. in Westernmeyer, W.E. and K.M. Shusterich (editors), United States Arctic Interests: The 1980s and 1990s. (Springer-Verlag, Inc., New York). pp. 39-58.

Gleick, P.H. 1987. 'The Implications of Global Climatic Changes for International Security'. Background Paper No. 14 for the Conference on Developing Policies for Responding to Future Climatic Change, Villach, Austria, September 28-October 2, 1987.

Gleick, P.H. 1988. 'The Effects of Future Climatic Changes on International Water Resources: The Colorado River, The United States, and Mexico'. In press, Policy Sciences.

Goemans. T. 1986. 'The Sea Also Rises: The Ongoing Dialogue of the Dutch with the Sea'. In J. Titus (editor), Effects of Changes in Stratospheric Ozone and Global Climate, Volume 4: Sea Level Rise. United States Environmental Protection Agency and United Nations Environment Programme. (Washington, D.C.) pp. 47-56

International Law Association. 1967, (revised 1986). 'The Helsinki Rules on the Uses of the Waters of International Rivers.' 52nd Conference of the International Law Association, Helsinki, Finland. (London) 56 pp.

Johnson, G.L., Bradley, D. and R.S. Winokur. 1984. 'United States Security Interests in the Arctic'. in W.E. Westermeyer and K.M. Shusterich (editors), United States Arctic Interests: The 1980s and 1990s. (Springer-Verlag, New York.) pp. 268-294.

Mintzer, I.M. 1987. A Matter of Degrees: The Potential for Controlling the Greenhouse Effect. World Resources Institute. (Washington, D.C.) 60 pp.

Mote, V.L. 1983. 'Environmental Constraints to the Economic Development of Siberia'. In Jensen, R.G., Shabad, T. and A.W. Wright (editors), Soviet Natural Resources in the World Economy. (The University of Chicago Press, Chicago). pp. 15-71.

Naff, T. and R.C. Matson. (editors). 1984. Water in the Middle East: Conflict or Cooperation. (Westview Press, Boulder, Colorado). 236 pp.

Rotty, R.M. 1987a. 'Estimates of Seasonal Variation in Fossil Fuel CO_2 Emissions'. Tellus 39B, No. 1-2, pp. 184-202.

Rotty, R.M. 1987b. 'A Look at 1983 CO_2 Emissions from Fossil Fuel (With Preliminary Data for 1984)'. Tellus 39B, No. 1-2, pp. 203-208.

Schneider, S.H. 1983. 'Food and Climate: Basic Issues and Some Policy Implications'. In V.P. Nanda (editor), World Climate Change: The Role of International Law and Institutions. (Westview Press, Boulder, Colorado.) pp. 46-63.

Ullman, R.H. 1983. 'Redefining Security'. International Security 8, No. 1, pp.129-153.

United Nations. 1978. Register of International Rivers. (Pergamon Press, Oxford, England.) 58 pp.

United States Congress. 1988. 'An Analysis of the Montreal Protocol on Substances that Deplete the Ozone Layer'. Office of Technology Assessment Staff Paper. (Washington D.C.) 41 pp.

Wallensteen, P. 1986. 'Food Crops as a Factor in Strategic Policy and Action'. In A.H. Westing (editor), Global Resources and International Conflict: Environmental Factors in Strategic Policy and Action. Stockholm International Peace Research Institute, United Nations Environment Programme. (Oxford University Press, Oxford.) pp.143-158.

Westing, A.H. (editor). 1986. Global Resources and International Conflict: Environmental Factors in Strategic Policy and Action, Stockholm International Peace Research Institute, United Nations Environment Programme. (Oxford University Press, Oxford.) 280 pp.

CLIMATE CHANGE IN THE CONTEXT OF MULTIPLE THREATS

ROBERT S. CHEN
Brown University
Alan Shawn Feinstein World Hunger Program
Box 1831
Providence, Rhode Island 02912
USA

ABSTRACT. Climate change is only one of a number of interlinked environmental threats that could affect human welfare. Simple differential calculus illustrates that, from a policy viewpoint, many different types of interactions among human activities and environmental processes must be considered. Unfortunately, our understanding of such interactions is limited, and the uncertainties tend to multiply. An alternative approach, drawn from the medical experience with multiple organ system failure, is to develop ways of buying time for the biosphere until underlying problems can be resolved.

1. Introduction

A changing climate is one major environmental threat facing the world over the next century, but is by no means the only one. Soil erosion, air and water pollution, heavy-metal contamination, and hazardous wastes are just a few of the many other large-scale environmental problems that threaten human welfare—and that may have important interactions with climatic changes. Nor will the impacts of or responses to climatic changes occur in a vacuum—in the coming century, significant changes in population, technology, and social and economic organization are also likely to occur.

Unfortunately, most if not all previous efforts to assess the impacts of climate change (or of any other environmental threat) have tended to focus narrowly on the direct effects of human activities on climate and the immediate implications of any consequent climate changes for human welfare. Few climate impact assessments have considered, let alone quantified, possible indirect effects of human activities, climate changes, or climate impacts and possible interactions with non-climatic trends. In particular, few such assessments have considered the possibilities that *a)* disruption of one environmental sub-system could increase the potential for disruption in other environmental sub-systems and *b)* disruption of multiple environmental sub-systems could increase the potential for catastrophic collapse of the entire system.

2. Environmental Threats in a Simple World

Let us assume a simple world of one activity $\mathbf{a_1}$, one environmental variable $\mathbf{e_1}$, and one impact variable $\mathbf{I_1}$ of concern. In general, we are concerned with net changes in $\mathbf{I_1}$ as the result of changes in the activity $\mathbf{a_1}$. Using the notation of calculus, this can be represented

A. Berger et al. (eds.), Climate and Geo-Sciences, 623–631.

a function of both $\mathbf{a_1}$ and $\mathbf{e_1}$, i.e., there may be both direct and indirect effects of activity $\mathbf{a_1}$. The basic chain rule for the total derivative then yields

$$\frac{d\mathbf{I_1}}{d\mathbf{a_1}} = \frac{\partial \mathbf{I_1}}{\partial \mathbf{a_1}} + \frac{\partial \mathbf{I_1}}{\partial \mathbf{e_1}} \frac{d\mathbf{e_1}}{d\mathbf{a_1}} \tag{1}$$

The first term on the right-hand side of equation 1, the partial derivative of $\mathbf{I_1}$ with respect to $\mathbf{a_1}$, represents the direct effect of $\mathbf{a_1}$ on $\mathbf{I_1}$. The second term, the product of the partial derivative of $\mathbf{I_1}$ with respect to $\mathbf{e_1}$ and the total derivative of $\mathbf{e_1}$ with respect to $\mathbf{a_1}$, represents the indirect effect of $\mathbf{a_1}$ on $\mathbf{I_1}$ exerted through $\mathbf{e_1}$. An example is given in Figure 1a.

Equation 1 immediately highlights one limitation of many impact assessment studies, i.e., the tendency to focus only on impacts that involve the environmental variable of interest (the second term) and to consider "all else equal"—thereby overlooking the direct effects of an activity on the impact variables (the first term). For example, the direct effects of changing energy use on agricultural yield are often ignored, even though the historical record of energy use in mechanized agriculture clearly shows yield impacts as large or larger in magnitude than the potential yield impacts of climate. If the direct and indirect effects are of opposite sign, e.g., the first term is positive and the second term is negative, then such an omission can be particularly important: activity $\mathbf{a_1}$ can still have a net positive effect on $\mathbf{I_1}$ even if its effect exerted through $\mathbf{e_1}$ is negative (so long as the direct effect is larger in magnitude). Thus, the direct benefits of fossil-fuel use in agriculture may far outweigh the indirect costs to agriculture due to the incremental climate warming caused by that fossil-fuel consumption. This highlights the danger of making policy judgements only on the basis of environmental considerations (i.e., the second term of equation 1) rather than on the total derivative itself, which includes both environmental and non-environmental influences.

Equation 1 also underscores the importance of understanding *both* the effect $\frac{d\mathbf{e_1}}{d\mathbf{a_1}}$ of an activity on the environment *and* the effect $\frac{\partial \mathbf{I_1}}{\partial \mathbf{e_1}}$ of the resulting environmental change on the impact variable. If $\frac{d\mathbf{e_1}}{d\mathbf{a_1}}$ is defined as positive in sign (e.g., an incremental increase in fossil-fuel consumption leading to an incremental increase in temperature), then $\frac{\partial \mathbf{I_1}}{\partial \mathbf{e_1}}$ must be large and negative (i.e., adverse) in order for the product term $\frac{\partial \mathbf{I_1}}{\partial \mathbf{e_1}} \frac{d\mathbf{e_1}}{d\mathbf{a_1}}$ to make $\frac{d\mathbf{I_1}}{d\mathbf{a_1}}$ differ significantly from $\frac{\partial \mathbf{I_1}}{\partial \mathbf{a_1}}$. In other words, environmental considerations are presumably only of interest from a policy viewpoint if the environmental changes themselves have substantial adverse effects on the impact variable of concern. If $\frac{\partial \mathbf{I_1}}{\partial \mathbf{e_1}}$ is zero, then even a large physical response of the environment to an activity would not affect the total derivative. This implies that as much attention should be devoted to quantifying and reducing uncertainties in $\frac{\partial \mathbf{I_1}}{\partial \mathbf{e_1}}$ as is generally given to improving estimates of $\frac{d\mathbf{e_1}}{d\mathbf{a_1}}$.

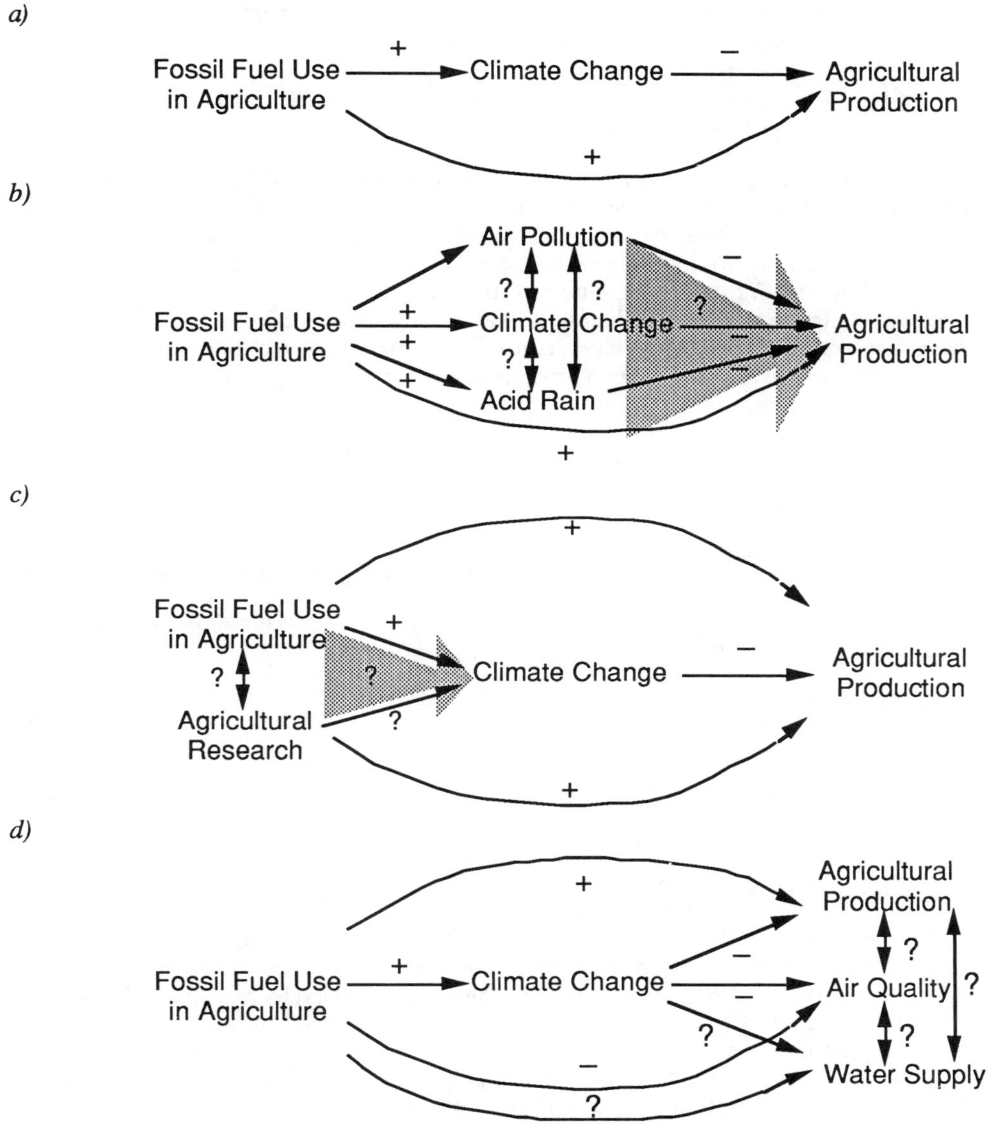

Figure 1. Schematic relationships between human activities, environmental variables, and impact variables: *a)* example of one activity, one environmental variable, and one impact variable; *b)* example of one activity, multiple environmental variables, and one impact variable; *c)* example of multiple activities, one environmental variable, and one impact variable; *d)* example of one activity, one environmental variable, and multiple impact variables. Shaded arrows represent possible synergistic effects.

If we expand our simple world to include additional environmental (or other) variables $\mathbf{e_i}$ which are also only functions of $\mathbf{a_1}$, then the chain rule yields

$$\frac{d\mathbf{I_1}}{d\mathbf{a_1}} = \frac{\partial \mathbf{I_1}}{\partial \mathbf{a_1}} + \frac{\partial \mathbf{I_1}}{\partial \mathbf{e_1}}\frac{d\mathbf{e_1}}{d\mathbf{a_1}} + \frac{\partial \mathbf{I_1}}{\partial \mathbf{e_2}}\frac{d\mathbf{e_2}}{d\mathbf{a_1}} + \ldots + \frac{\partial \mathbf{I_1}}{\partial \mathbf{e_i}}\frac{d\mathbf{e_i}}{d\mathbf{a_1}} \tag{2}$$

The impact of changes in $\mathbf{a_1}$ on $\mathbf{I_1}$ includes all indirect effects via other intermediate variables. For example, changes in fossil-fuel use may not only affect the climate, but also acid precipitation, air and water pollution, agricultural land use, and, of course, the atmospheric carbon dioxide concentration itself—all of which could in turn affect agricultural yields (Figure 1b). It is noteworthy that the atmospheric carbon dioxide concentration is clearly not independent of climate change, implying that the assumption of independence among the $\mathbf{e_i}$'s implicit in equation 2 would not be met in this case.

Alternatively, there may be another activity $\mathbf{a_2}$ that affects $\mathbf{e_1}$, so that the total change in $\mathbf{I_1}$ is

$$d\mathbf{I_1} = \left(\frac{\partial \mathbf{I_1}}{\partial \mathbf{a_1}}\right)_{\mathbf{a_2}} d\mathbf{a_1} + \left(\frac{\partial \mathbf{I_1}}{\partial \mathbf{a_2}}\right)_{\mathbf{a_1}} d\mathbf{a_2} \tag{3a}$$

where $\left(\frac{\partial \mathbf{I_1}}{\partial \mathbf{a_1}}\right)_{\mathbf{a_2}}$ and $\left(\frac{\partial \mathbf{I_1}}{\partial \mathbf{a_2}}\right)_{\mathbf{a_1}}$ are the partial derivatives of $\mathbf{I_1}$ with respect to $\mathbf{a_1}$ (holding $\mathbf{a_2}$ constant) and $\mathbf{a_2}$ (holding $\mathbf{a_1}$ constant), respectively. If we assume that $\mathbf{a_1}$ and $\mathbf{a_2}$ are independent, the chain rule yields

$$\left(\frac{\partial \mathbf{I_1}}{\partial \mathbf{a_1}}\right)_{\mathbf{a_2}} = \frac{\partial \mathbf{I_1}}{\partial \mathbf{a_1}} + \frac{\partial \mathbf{I_1}}{\partial \mathbf{e_1}}\frac{\partial \mathbf{e_1}}{\partial \mathbf{a_1}} \tag{3b}$$

$$\left(\frac{\partial \mathbf{I_1}}{\partial \mathbf{a_2}}\right)_{\mathbf{a_1}} = \frac{\partial \mathbf{I_1}}{\partial \mathbf{a_2}} + \frac{\partial \mathbf{I_1}}{\partial \mathbf{e_1}}\frac{\partial \mathbf{e_1}}{\partial \mathbf{a_2}} \tag{3c}$$

As in equation 1, the terms $\frac{\partial \mathbf{I_1}}{\partial \mathbf{a_1}}$ and $\frac{\partial \mathbf{I_1}}{\partial \mathbf{a_2}}$ represent the direct effect on $\mathbf{I_1}$ of variations in $\mathbf{a_1}$ and $\mathbf{a_2}$, respectively, but in this case holding both $\mathbf{e_1}$ and the other activity constant. The second terms represent the indirect effects of each activity on $\mathbf{I_1}$ exerted through $\mathbf{e_1}$, holding the other activity constant. These relationships are illustrated by example in Figure 1c.

Note that, even if the indirect effect of an activity exerted through $\mathbf{e_1}$ is small, the direct effect need not be. For example, changes in agricultural techniques are not likely to have large impacts on climate, but may of course affect agricultural yields significantly. However, few climate impact assessments look beyond those activities that bear directly on the climatic variable of immediate concern to include all of the other activities that may directly affect the selected impact variable. These activities include such "background" trends as changes in demography, technology, and social and economic organization. This type of omission can be especially significant if activities are not independent of each other, since strong effects on an impact variable could be exerted through omitted activities. For instance, changes in energy use may influence not only the degree of climate change, but also a whole range of agricultural activities—all of which could affect agricultural yield.

Another extension to our simple world is to consider additional impacts $\mathbf{I_i}$ (Figure 1d). We could, for example, write

$$\frac{\mathbf{dI}}{\mathbf{da_1}} = \frac{\mathbf{dI_1}}{\mathbf{da_1}} + \frac{\mathbf{dI_2}}{\mathbf{da_1}} + \ldots + \frac{\mathbf{dI_i}}{\mathbf{da_1}} \tag{4}$$

This assumes that impacts are independent and can be translated into equivalent terms (e.g., monetary values or another "utility" function). In a sense, by taking a sectoral approach, by ignoring all but the most "climate sensitive" sectors, and by expressing results in terms of monetary costs and benefits, most climate impact assessment studies to date have implicitly treated the overall impact of a change in activity as this type of "sum" of the individual impacts in a few key sectors. We know, however, that there are likely to be important interactions between impact variables and sectors, e.g., between agricultural yield, energy consumption, and water demand. And few studies have examined the degree to which gains or losses in different sectors can be equated realistically—is, for example, a loss of **x** dollars in one key sector due to climatic change equivalent to losses of **x/n** dollars in **n** different sectors?

Relaxing the assumptions of independence among the activities, environmental variables, and impacts vastly increases the complexity of the problem—and the information needed to solve it. Clearly, it can be a monumental task to specify the functional relationships (the partial and total derivatives) between all important $\mathbf{a_i}$, $\mathbf{e_i}$, and $\mathbf{I_i}$. In our "simple" world, a system with only 3 activities, 3 environmental variables, and 3 impact variables would require specification of 36 relationships between pairs of elements—72 if relationships are not reversible and even more if they are conditional on specific pathways. Any realistic system of activities, environmental variables, and impacts would likely involve many more than 9 elements. Table I, for example, lists more than twenty major long-term threats to agriculture, loosely grouped by the principal domains they affect. This list is by no means comprehensive or consistently defined, but it does serve to highlight the varied environmental pathways that exist, any of which could constrain agricultural production at regional or global scales.

Table I. Some possible environmental threats to agriculture.

Land/Water	Atmosphere	Biota
1) Desertification	*10)* Temperature change	*19)* Bioaccumulation
2) Soil erosion	*11)* Precipitation change	*20)* Pest/disease outbreaks
3) Water logging	*12)* Change in growing seasons	*21)* Reduction in diversity
4) Salinization	*13)* Change in weather extremes	*22)* Change in fire frequency
5) Groundwater depletion	*14)* Volcanic dust	
6) Heavy metal contamination	*15)* Nuclear winter/fall	
7) Radioactive contamination	*16)* Ozone depletion	
8) Coastal flooding	*17)* Acid deposition	
9) Land subsidence	*18)* Air pollution	

Further complications arise from possible feedbacks and discontinuous relationships among variables and from widely varying response times. For example, efforts to reduce acid precipitation are likely to include some reductions in fossil-fuel use on time scales of less than a decade, thereby affecting longer-term trends in energy use. Responses to the recent evidence for diminishing ozone concentrations in the stratosphere may well affect the impact and response functions for other environmental variables. These complexities suggest that simple matrix-like approaches to any realistic set of activities, environmental threats, and impacts, although potentially useful as a qualitative framework (e.g., Crutzen and Graedel, 1986), may not be the most fruitful for more realistic quantitative analysis.

3. Cascading Uncertainty

A key element that is not well represented in the simple world and by the simple mathematics discussed above is the high degree of uncertainty in the probability, magnitude, distribution, and timing of most if not all environmental threats. Many environmental threats, such as a possible collapse of the West Antarctic Ice Sheet or a major volcanic eruption (e.g., Krakatoa or Tambora), are events of low probability but potentially catastrophic consequences (e.g., Schneider and Chen, 1980). Others like soil erosion and deforestation are already important in many localized areas, but the threat they pose at the global level is less certain. Still others, most notably the depletion of stratospheric ozone, are global threats for which there is significant evidence, but the timing and distribution of impacts remain unclear.

Uncertainty surrounds not only the probability of each threat vis-à-vis one or more key activities, but also the probability of other potentially important interactions among activities, threats, and impacts. For example, we know little about how long-term changes in rainfall patterns due to a warming climate might affect soil erosion, groundwater depletion, salinization, heavy metal contamination, acid deposition, and fire frequency. We know of cases in which temperature and weather extremes have affected the frequency and severity of pest and disease outbreaks (e.g., Waggoner, 1983), but predicting such effects under changing climatic regimes is difficult. Possible synergistic effects on forests subject both to climatic changes and to pollution and acidic deposition have been recognized, but not quantitatively assessed (e.g., Chen and Parry, 1987). Shifts from fossil fuels to nuclear power could increase the potential for widespread radioactive contamination of land areas and water bodies, but controversy continues as to how likely such accidental releases are, even after the example of Chernobyl. In general, the high degree of uncertainty has meant that these sorts of interactions have received little serious attention.

4. Towards a Multiple Threat Approach

Although it is theoretically possible to develop a mathematical model that can take into account the complexities and uncertainties described above in detail, the practical difficulties are enormous. Many environmental threats are essentially unprecedented, making it difficult to find appropriate and sufficient data on which to base realistic probabilistic estimates. Too few of the underlying and interacting physical, biological, and ecological processes are understood sufficiently to be certain that all significant pathways and interactions have been

Table II. Mortality as a function of the number and duration of organ system failures (OSFs). Length of failure is measured in terms of the number of days in an intensive care unit since the first OSF. After a day of failure, a patient may die, get better, have an additional OSF, or remain in the same condition. Data are based on 2,719 admissions to 13 hospitals. The number of patients in each category is given in parentheses. Source: Knaus *et al.* (1985).

No. of OSFs	Percent Mortality Duration of OSF (days) 1st	2nd	3rd	4th	5th	6th	7th
One	22% (2070)	31% (847)	34% (607)	35% (455)	40% (356)	42% (279)	41% (195)
Two	52% (458)	67% (219)	66% (156)	62% (191)	56% (171)	64% (122)	68% (82)
≥ Three	80% (191)	95% (74)	93% (54)	96% (52)	100% (38)	100% (33)	100% (32)

included. On time scales of years, decades, or longer, human interventions must be taken into account.

An alternative approach that my colleague Myron Fiering of Harvard University and I have begun to explore is suggested by the medical literature on "multiple organ system failure." In the human body, acute organ system failures (OSFs) can occur for at least five inter-linked functions: the cardiovascular, respiratory, renal, hematologic, and neurological systems. Current medical technology can keep many patients alive for a few days or more if one of these system fails, but the survival rate falls dramatically with the duration of failure (defined by the number of days in an intensive care unit with at least one OSF) and with the number of different OSFs (Table II). With only one OSF, mortality rates increased in one study (Knaus *et al.*, 1985) from 1 in 5 patients on the first day of failure, to 1 in 3 patients on the second to fourth days, and to 2 in 5 patients on the fifth to seventh days. But with two OSFs, mortality rates jumped from 1 out of every 2 patients on the first day of failure to 2 out of 3 patients on subsequent days. With three or more OSFs, mortality rates climbed from 4 in 5 patients on the first day of failure to every patient by the fifth day of failure. The ability of a patient to survive three OSF's for any length of time appeared to depend greatly on the patient's age and health prior to admission to the hospital.

To what degree is this analogy applicable to environmental processes? Clearly, there are strong linkages among thermodynamic, hydrologic, atmospheric, biogeochemical, ecological, and other environmental sub-systems. Extreme variation or disruption in one can propagate quickly through other sub-systems. Although managed and unmanaged ecosystems may be able to survive or adapt to one insult or set of changes, their ability to deal successfully with more than one challenge at a time may be lower (see Holling, 1986 for several examples). The length of time under stress may also be a factor. Of course, whether environmental systems at the global level are connected by feedback and control processes analogous to those that operate in the human organism is a controversial issue (e.g., Lovelock, 1986; Clark and Munn, 1986; Kerr, 1988).

The medical experience in treating patients with multiple OSFs is also instructive. Therapies such as artificial ventilation, hemodialysis, and vasoactive drugs can temporarily maintain organ system function, but Knaus *et al.* (1985: 685) point out that

> [i]n the mid-1970's, it was recognized that, for some patients, support of multiple organ system failure (OSF) did not result in improved long-term survival but merely delayed death. At the same time, it was reported that patients with a variety of traumatic and nontraumatic conditions were dying in multiple OSFs. The exact cause of this syndrome remains elusive, but as our experience with acute OSF grows, it is becoming clear that the major role of mechanical and pharmacological support of OSF is to buy time—time during which the primary disease process can be identified and effectively treated. Used in this way, organ system support is life saving.

The lesson here is clear. It may be necessary to develop a *dual* strategy for dealing with multiple environmental threats. The first challenge is to improve understanding of basic environmental processes and the consequences of environmental change—i.e., the "primary disease processes." This is an objective that is being addressed in the context of the planned International Geosphere-Biosphere Programme (International Council of Scientific Unions, 1986). But a second, equally important challenge is to develop techniques for maintaining or indeed *resuscitating* stressed and damaged environmental systems—i.e., ways to "buy time" until the underlying causes can be "identified and effectively treated."

This need is especially sobering in the light of recent evidence of significant hemispheric-scale depletions in stratospheric ozone concentrations—we may already be in a situation where one critical environmental system is starting to degrade. Many managed and unmanaged ecosystems are likely to be severely stressed by increased ultraviolet radiation and may, as a result, be even more susceptible to other environmental stresses including other climate changes. Our planet may, in other words, be approaching the situation of a patient with one organ system failure, whose risk of death increases greatly the longer the failure continues and the more other organ systems are damaged. Will we be prepared to provide the "intensive care" needed to stave off death until a permanent "cure" can be found?

5. References

Chen, R.S. and Parry, M.L. (eds.): 1987, *Policy-Oriented Impact Assessment of Climatic Variations*, Research Rept. RR-87-7, International Institute for Applied Systems Analysis, Laxenburg, Austria, 54 pp. Reprinted as *Climate Impacts and Public Policy*, United Nations Environment Programme and International Institute for Applied Systems Analysis.

Clark, W.C. and Munn, R.E. (eds.): 1986, *Sustainable Development of the Biosphere*, Cambridge: Cambridge University Press.

Crutzen, P.J. and Graedel, T.E.: 1986, 'The Role of Atmospheric Chemistry in Environment-Development Interactions', in W.C. Clark and R.E. Munn (eds.), *Sustainable Development of the Biosphere*, pp. 213-250. Cambridge: Cambridge University Press.

Holling, C.S.: 1986, 'The Resilience of Terrestrial Ecosystems: Local Surprise and Global Change', in W.C. Clark and R.E. Munn (eds.), *Sustainable Development of the Biosphere*, pp. 292-317. Cambridge: Cambridge University Press.

International Council of Scientific Unions: 1986, *The International Geosphere-Biosphere Programme: A Study of Global Change*, International Council of Scientific Unions, August 4, 1986, 21 pp.

Kerr, R.A.: 1988, 'No Longer Willful, Gaia Becomes Respectable' (Research News), *Science* **240**(4851), 393-395.

Knaus, W.A., Draper, E.A., Wagner, D.P. and Zimmerman, J.E.: 1985, 'Prognosis in Acute Organ-System failure', *Ann. Surg.* **202**(6), 685-693.

Lovelock, J.E.: 1986, 'Geophysiology: A New Look at Earth Science', *Bull. Am. Meteorol. Soc.* **67**(4), 392-397.

Schneider, S.H. and Chen, R.S.: 1980, 'Carbon Dioxide Warming and Coastline Flooding: Physical Factors and Climatic Impact', *Ann. Rev. Energy* **5**, 107-149.

Waggoner, P.E.: 1983, 'Agriculture and a Climate Changed by More Carbon Dioxide', in *Changing Climate: Report of the Carbon Dioxide Assessment Committee*, pp. 383-432. Washington DC: National Academy Press.

CLIMATE AND PUBLIC POLICY IN THE UNITED STATES CONGRESS

ALAN ROBOCK
Department of Meteorology
University of Maryland
College Park, Maryland 20742
USA

ABSTRACT. While there is no particular expertise in the United States Congress in meteorology, several atmospheric issues are receiving much attention. These include ozone depletion, global warming, acid rain, nuclear winter, and earth system science. The status of each of these is briefly discussed.

The ability of the governments of the United States and other countries to deal with each of these issues depends on both the spatial scale of the environmental impact and the societal costs and disruptions involved in the solutions that are being proposed to deal with the problems. Ozone depletion is a much simpler problem to deal with in this regard than acid rain or global warming.

"Politics is harder than physics." – Albert Einstein

1. Introduction

Of the 540 Members of the United States (U.S.) Congress and the staffs of their personal offices and committees, there are very few scientists and no meteorologists. Nevertheless, several issues relating to the atmosphere are receiving considerable attention in national public policy debates. By holding hearings, utilizing the Congressional Research Service and Office of Technology Assessment, and consulting with staffers, Members of Congress are making policy decisions involving every resident of the United States and the rest of the world, and involving the spending of billions of dollars.

I spent one year, 1 September 1986 through 31 August 1987, working in Congress, on the personal staff of Congressman Bill Green from New York City, and on the staff of the Environmental and Energy Study Conference (EESC), a bipartisan Congressional caucus. Details of my experience can be found in Robock (1988). It became clear that several issues related to the atmosphere are receiving much attention in Congress. In this paper I briefly discuss the important issues currently being considered by Congress that are related to atmospheric science, and the problems involved in coming to solutions. These include ozone depletion, global warming, acid rain, nuclear winter, and earth system science.

2. Atmospheric Issues

In testimony at a recent hearing, Senator Albert Gore (D-TN) stated that, except for the threat of nuclear weapons, the issues of ozone depletion and global warming were the

A. Berger et al. (eds.), Climate and Geo-Sciences, 633–637.

most important ones for the future of mankind. This demonstrates the high level of interest in Congress about a variety of issues related to the atmosphere. Each of the issues discussed below is very complex, and each is in a different stage of action. A brief description of the status of each as of Fall, 1987, is given.

2.1. OZONE DEPLETION

The recent discovery of the Antarctic "ozone hole" has catalyzed ongoing national and international activities. The possibility of anthropogenic destruction of stratospheric ozone has been recognized for over a decade, and resulted in a ban on chlorofluorocarbon (CFC) spray can propellants in the U.S., Canada, and Scandinavia, and international negotiations to require the reduction of CFC emissions. In Montreal in September, 1987, an international agreement was signed to require nations to reduce CFC production by 50% over the next decade. Congressional hearings during the past year heard testimony from the State Department negotiators, scientists studying the problem, industry groups that would be hurt by the ban, and environmental groups urging quicker and larger reductions. At the same time, environmental groups were suing the Environmental Protection Agency to require CFC reductions already mandated by current laws. The many hearings educated Congress and the public about the issues, and accelerated the ongoing processes, but did not require new legislation.

Because the nations of the world perceived this to be a global problem, with everyone losing if it is not solved, and because there seem to be relatively inexpensive CFC substitutes that are much less damaging to stratospheric ozone, it was possible to come to an international agreement to help solve the problem. Various compromises were included in the Montreal agreement which took into account regional economic and industrial situations, and the mandated reduction in CFC production is not as fast as some environmentalists and scientists would like, but the fact is that this was the first international agreement to solve a global environmental problem. Furthermore, the problem is currently only theoretical, and the governments of the world have shown enough faith in scientists to act now rather than wait for definitive observations. The discovery of the "ozone hole" has certainly affected the psychology of the negotiating climate, but it was not explained by existing theories at the time of the agreement, and the agreement was in response to theories of lower latitude ozone depletion. This agreement may become a model for the much more difficult negotiations in the future on global warming and acid rain. And by limiting CFC production, it already helps to slow greenhouse warming.

2.2. GLOBAL WARMING

Several hearings on this topic were held recently in the U.S. Congress. Testimony was given by scientists, science agencies and environmental groups describing the potential problems that could result decades in the future. The EESC also started a special study group for Members of Congress on the subject. Green (1987) and Robock (1987) discuss details of the public policy aspects of global warming. This is a rare instance of concern in Congress about a potential problem that is only theoretical and will not manifest itself during the terms of most of the Members. It demonstrates the growing trust, confidence, and respect that Congress has for scientific research.

It turns out that there are many things Congress can do to slow the predicted rate of greenhouse warming caused by CO_2 and trace gas emissions. CO_2 emissions can be reduced by energy conservation, including more public transportation, lower speed limits, better building insulation, and more efficient appliances. Shifting to gas or oil from coal produces less CO_2 per unit heat produced, and renewable energy sources, such as solar, geothermal, hydropower and windpower produce no CO_2. Nuclear power also does not produce CO_2, but has its own environmental problems. Discouraging deforestation can lower this important source of atmospheric CO_2. Reducing other trace gas emissions will address about one half of the problem. These issues are addressed in much more detail in Robock (1987).

Any international agreement on limiting CO_2 emissions will have to deal with the various levels of economic development in the world. It would be difficult for the industrialized countries to argue that the global good prevents the other countries from following the same path they took to become industrialized, namely use of cheap fossil fuels and extensive deforestation. For instance, how can the U.S., which deforested its eastern third during the past century, or Western Europe, which was deforested even earlier, argue that tropical countries should not chop down their trees, despite a current pressing need for farmland and wood fuel? How can Eastern Europe be influenced to substitute expensive, but less-polluting, oil for cheap "brown coal?" How can China be urged to start building diesel locomotives rather than steam engines which burn coal for their railroads, when coal is much cheaper? Massive economic incentives will be necessary to seriously attack this problem on a global basis.

2.3. GLOBAL CHANGE

A related issue is the new Earth System Science (or Global Change, or International Geosphere-Biosphere Program) proposals emerging from several groups. They propose a new integrated approach to studying the earth that links studies of the atmosphere, ocean, cryosphere, lithosphere and biosphere and their interactions. Several agencies are proposing such programs and the reaction in Congress has been quite favorable at hearings. The national investment in this program would be quite cheap, on a relative basis, and the potential societal benefit is enormous. The prospect of more federal money for research in these areas seems good. Since large climatic warmings are predicted in spite of any immediate policy shifts, the argument that society must learn how to predict and adapt to these changes has been well received. It will take a new Administration, however, with different overall objectives, to produce a massive infusion of new funds.

2.4. ACID RAIN

This is a problem that has been recognized for years, but has resisted easy solutions. Unlike ozone depletion, which represents a problem for the whole world, or global climate change, which may result in benefits and losses distributed on a regional or national basis, acid rain pits states against states in the U.S. In the first place, much controversy exists about the extent of the problem and the need for an immediate remedy. Suggested reductions of SO_2 and NO_x emissions by half would harm the economy in states that are the main sources, such as in the Ohio River valley and help states downwind that have the problem. States in the West, that do not perceive themselves to have a problem, resist helping to pay for the cleanup. Solutions do not appear to be imminent, especially since the Senate Majority Leader, Senator Robert Byrd from West Virginia, is from the state that is the greatest source of SO_2, and the Chair of the relevant House committee, Rep. John Dingle from Michigan, resists any further restrictions on NO_x emissions, which he perceives would hurt the automobile industry in his state.

On an international basis, this problem among the nations of Europe has difficulties similar to those among the U.S. states. But with more discoveries of the effects of acid rain in places such as China, and the development of new technologies to restrict SO_2 and NO_x emissions, regional benefits may force nations and states to come to some agreements.

2.5. NUCLEAR WINTER

Several hearings on this topic were held in the past few years, but none in the past year. They resulted in a requirement in the Defense Department FY87 authorization that a new comprehensive report be prepared and submitted to Congress by 1 November 1987 that describes "the atmospheric, climatic, biological, health, and environmental consequences of nuclear explosions and nuclear exchanges and the implications that such consequences have for the nuclear weapons, arms control, and civil defense policies of the United States." In January 1988, the Defense Department wrote a letter to Congress refusing to do the study and stating that their earlier, brief, inadequate report (Weinberger, 1985) was reflective of the current status of research. Meanwhile, international cooperative research goes on, which refines and basically confirms the validity of earlier work. Congress may again hold hearings and issue legislation depending on the reaction to the refusal of the Defense Department to prepare the report. The policy implications of this theory have still not been adequately addressed.

Initial alarmist publicity about some of the worst case results from the first studies has since been shown to have been exaggerated. This has resulted in a public perception that the theory of nuclear winter has been disproven. In fact, we have learned that although climate changes will not be as extreme as initially forecast, biological systems are more sensitive than initially thought, and that there would be serious disruptions to the global food supply as a result of a large-scale nuclear war. This should be a lesson to scientists about premature "Chicken Little" expressions to the public before theories have had a chance to be examined in the normal scientific process.

3. Conclusions

Policy makers are now beginning to take seriously the theoretical warnings of scientists about environmental dangers. The eventual policy remedies will have to take into account the economic costs of solving the problems as compared to the perceived benefits. In the case of ozone depletion, this calculation was relatively easy. In the cases of global warming and acid rain, the solution will be much slower and more difficult. Scientists can aid this process by expressing their results in terms non-scientists can understand, and by being as honest and unbiased as they can in their assessments. Alarmist expressions can result in immediate attention, but in the longer run can come back to discredit not only the work of the alarmists, but also that of honest scientists. When scientists have come to some kind of consensus, however, it is our responsibility to inform, or even warn, the public in as straight-forward a way as we can.

4. References

Green, Bill, 1987: Policies on global warming and ozone depletion. *Environment*, **28**, No. 3, pp. 5, 45. [ghostwritten by Alan Robock]

Robock, Alan, 1987: *The greenhouse effect: Global warming raises fundamental issues.* (Environmental and Energy Study Conference, U.S. Congress, Washington), 8 pp.

Robock, Alan, 1988: Atmospheric science issues in Congress. *Proceedings of the Twelfth Annual Climate Diagnostics Workshop,* (NOAA, Washington), 335-341.

Weinberger, C. W., 1985: *The Potential Effects of Nuclear War on the Climate,* A report to the United States Congress. (Department of Defense, Washington), 17 pp.

CLIMATIC CHANGES AND WATER RESOURCES DEVELOPMENT

L.V. da Cunha
Scientific Affairs Division, NATO
1110 Brussels
Belgium

ABSTRACT: Due to the social, economic and environmental importance of water resources, the consequences of climatic changes can be quite relevant and may require changes in the current criteria for water resources management and planning. The possible consequences of climatic change on water resources are analyzed, considering separately the impacts on the inland waters and on the coastal regions. The use of general circulation climate models and of regional hydrologic models to assess the impacts of climatic change are revised and some recommendations formulated, in particular as regards research needs related to these problems.

1. CLIMATE AND WATER RESOURCES

1.1 Introduction

It is presently accepted that significant man-induced climatic change may occur in the next decades, in particular related to the increase of CO_2 concentration in the atmosphere, due mainly to the intensive consumption of fossil fuels. A doubling of CO_2 in the atmosphere would have as a consequence an increase in temperature of about 3^oC which would imply considerable changes in the precipitation and evaporation regimes and thus also in runoff. Important consequences for water resources availability are thus to be expected with significant social, economic and environmental impacts. In fact, through the workings of the hydrologic cycle, climate plays an essential role in the quantity and quality of the available water. Thus, climatic changes will have direct consequences on water resources availability and characteristics, as well as indirect consequences on the vegetation, the soil and the environment in general.

From a practical standpoint, it is important to try and evaluate the magnitude of the consequences of climatic changes on water resources, identify the possible difficulties that they will raise for water resources managers and, finally, establish if and how the

A. Berger et al. (eds.), Climate and Geo-Sciences, 639–660.

current water resources management and planning strategies should be modified. The objective would be to assess the impact of the foreseen climatic changes on water resources in order to help water resource managers and planners to take the necessary measures in due time to alleviate the adverse impacts and to take the best advantage of favourable impacts.

Figure 1 gives a view of the relation between climate, water resources, water management and economic and social development. Not only are the main relations (continuous line arrows) shown, but also the feedbacks (dotted line arrows), the whole giving a simple perspective of the impact of climate on water. From this diagram it is easy to appreciate the complexity of the problems under discussion, which involve an interdisciplinary approach with numerous professional skills, as well as an intersectoral approach which results from multiple water uses.

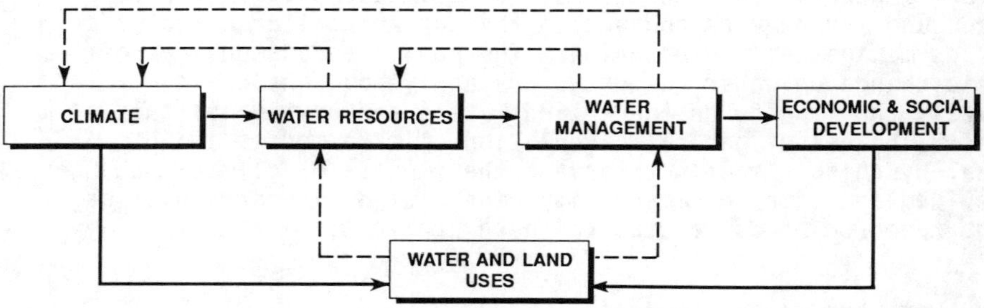

Fig.1 - Relation between climate, water resources and development

It is currently recognized that social and economic development can be strongly affected by the quantity and the quality of available water resources. Water, unlike other natural resources, cannot be replaced in most of its uses, e.g. urban, rural and industrial water supply, irrigation, power production, navigation and recreation. Therefore, water plays a major role in economic development and social welfare. Moreover, a significant part of the water used is returned to the environment with a lower quality, and this may affect the existing water reserves causing, for instance, the pollution of surface or groundwater, the eutrophication of lakes and reservoirs or the salinization of aquifers.

The water resources that can actually be used depend not only on the existing quantity of water, but indeed very much on the quality of the available water. The situation, of course, varies considerably from region to region, but even an analysis in overall terms gives evidence of the present and especially the future difficulties. LVOVICH 1979 refers to the fact that the world annual water withdrawal in 1970 was of the order of 600 km^3, which corresponds to only approximately 4% of the available resources (14,000 km^3/year).

Consumption was estimated to be about one quarter of the withdrawal, i.e. 150 km^3/year, thus corresponding to approximately 1% of the existing supply. Even if these figures, as presented, do not give evidence of any reason for special concern, it is important to note that the annual volume of waste water rejected into the water bodies was 450 km^3 in 1970 and that these rejections, using the present treatment technologies, mobilize approximately 6000 km^3 of river and lake waters, i.e. approximately 40% of the available resources, in absorbing the resulting pollution. This is indeed of concern, particularly because due to marked differences in world geographical distribution of water, the situation can become extremely serious in certain regions.

In the future, an increase in population and also in the standards of living will, of course, lead to an increase in water demand. Urbanization, industrialization and increased irrigated agriculture will also determine a substantial increase in water demand. Water pollution will likewise increase in accordance with the evolution referred to above. Processes, such as deforestation, will also contribute in certain regions to creating serious water deficits and a degradation of water quality as a consequence of erosion. LVOVICH 1979 also provides a scenario for the year 2000 in which the volume of water withdrawal would increase up to 7000 km^3/year and the waste water up to 6000 km^3. This would require approximately 38,000 km^3 of water to absorb the rejected pollutants. If this figure is compared to the foreseen available river runoff of 22,000 km^3 in the year 2000, it becomes obvious that an extremely difficult situation may arise. The above estimates are, of course, only approximate. However, they illustrate a very unfavourable trend and call attention to the absolute need to establish new water management strategies for the future (FALKENMARK, DA CUNHA and DAVID 1987).

Climatic changes may still aggravate this situation by creating an imbalance between the water resources available and the demand, and tend to cause new problems or to aggravate existing ones. This is, for example, the case with extreme phenomena, such as floods and droughts, and of erosion and sedimentation, which may be aggravated by climatic changes.

Impacts of climatic variations on water resources depend not only on climatic change mean values but also on changes in climatic variability. KELLOG and SCHWARE 1981, for example, have concluded that the impact is much greater when changes in variability are also present.

Also, several human activities can be responsible for changes in the hydrological regime. Urbanization, changes in agricultural practices, deforestation and afforestation, as well as the construction of civil engineering works interfering with surface flow may also change the hydrologic regime considerably. These types of man-induced changes are, however, better known than future climatic changes, though this does not imply that they are always taken into consideration as they should be.

The impact of climatic changes on water resources can be identified only if substantial in comparison to the impact of changes related to human activities and if these two types of impacts can be reasonably separated. For this reason, as NOVAKY et al. 1985 point out the assessment of human-caused hydrologic changes should always go hand-in-hand with the assessment of water-related climatic impacts. In fact, because the conditions of water use have been changing recently, as in the case of water supply to large metropolitan areas or large industrial settlements, the increase of irrigated agriculture or the urbanization of river plains subject to flooding, the consequences of possible climatic modifications will be different than they would have been in the past before these modifications occurred.

1.2 Consequences of the Greenhouse Effect on Water Resources

The main factor responsible for the greenhouse effect is CO_2, it being currently assumed that its concentration in the atmosphere would approximately double that of pre-industrial times by the middle of the next century (WMO 1986). CO_2 alone would be responsible for about 50% of the rise in atmospheric temperature, the remainder being due to the increase in concentration of other gases, in particular ozone, CH_4, N_2O and the CFCs (chlorofluorocarbons or freons). The results of numerous climatic models predict an increase in atmospheric temperature of 1.5^oC to 4.5^oC (NAS 1985), a good average for the global increase of temperature being 3^oC.

The rise in temperature induced by the greenhouse effect would be associated with an increase in evapotranspiration and, consequently, an increase in humidity of the air, precipitation and cloudiness. This increase in cloudiness would, as a matter of fact, somewhat reduce the consequences of the greenhouse effect, because clouds would reduce the solar radiation reaching the Earth's surface.

As mentioned, the increase in temperature should, in principle, determine drier conditions, implying an increase in evaporation. However, ASTON 1984 claims that increased evaporation would not occur, because the increase of the CO_2 would raise the stomatal resistance of vegetation, decreasing evapotranspiration and consequently increasing runoff. Along the same lines, WIGLEY and JONES 1985 consider that the magnitude of evapotranspiration changes due to an increase in CO_2 is unknown. This is due to the fact that evapotranspiration changes are influenced by several factors, namely higher temperatures, changes in the growing season, changes in wind velocities and changes in net radiation. The combination of these factors could lead to increased or decreased evapotranspiration.

The first comprehensive analysis intended to establish the magnitude of the effects of climatic changes on water resources availabilities was made by NAS 1977 and is not very conclusive as, in general terms, it states that the magnitude of these effects cannot be adequately estimated. In this study, several US scientists examined the problem of water shortage, which may or may not have resulted from climatic change and made an assessment of the probable effects of climatic variability and climatic change on water supply needs, policy and design. The different studies presented in NAS 1977 agree on the need for further research.

Other comprehensive and more recent analyses deserving special reference are those by GLEICK 1986, CHANGNON 1987 and LEWIS 1987. The first two have a general scope and the last one centres mainly on the case of Canada, but nevertheless provides a good general view of the problems involved.

1.3 Current Design Procedures for Water Resources Works

When designing dams and other hydraulic works, engineers assume the climate to be static, i.e. they assume that the climate during the next decades (50 to 100 years, sometimes more), corresponding to the anticipated lifetime of the works, will not be significantly different from the climate recorded in the last 5 or 10 decades. However, engineers also take into consideration an element of uncertainty related to the stochastic nature of the hydrologic variables and, in particular, runoff.

In general, engineers and hydrologists are happy when they have 25 years of historical data on precipitation and/or streamflow and based on these data, they try to estimate future extreme situations of floods and droughts that are likely to occur. Works are usually designed seeking for a compromise between the minimization of the impact of the extreme conditions likely to occur and the need to ensure acceptable economic costs.

Decisions related to the design of dams and other hydraulic works normally involve huge expenses and thus possible errors in determining the size of the work are not only expensive but also difficult to rectify, as hydraulic works once completed are usually difficult to change. Large hydraulic works usually involve a long period of conception, design and construction, often lasting 25 years or more, and this makes long term planning for water resources most crucial.

Hydraulic works are designed for a certain range of climatic fluctuations. It is usual to estimate a maximum probable flood which the hydraulic works should be able to withstand, and also take into consideration the lower discharges during a dry period in which the works are expected to ensure a minimum water discharge. It is expected that the works will function under certain conditions of reliability which are accepted, considering, on one hand, the economic costs associated with the construction of the works and, on the other hand, the economic, social and environmental impacts that may result from the work being under-designed.

One aspect to consider when dealing with climatic changes and water resources is the estimation of the range of the expected changes in runoff and the comparison of these changes with those associated with fluctuations which always occur under the assumption of a static climate. In the normal water resources management and planning process, a certain degree of climatic variability is always taken into account and safety margins are considered when designing hydraulic works, namely in relation to extreme phenomena such as floods. Difficulties will be created for the water resources planners only if the variations resulting from the anticipated climatic changes are important when compared to the the random variations associated with the stationary stochastic process.

Large reservoirs are currently the most effective means of regulating natural irregular streamflow, normally ensuring that water is available during the drier periods and preventing floods in the wetter periods. Water resources management strategies can involve both planning for new projects or improved management of existing ones.

Apart from those so-called structural measures to cope with irregularity of streamflow, such as dams and river dykes, one can also resort to non-structural measures such as, for instance, taking steps to prevent or slow down climatic changes, reduce water demand, resort to functional zoning in alluvial valleys subject to floods, implement incentives for a rational water resources use (such as pricing schemes, pollution standards and pollution changes), or just abandon certain areas affected, e.g. those flooded due to changes in river regime or sea-level rise.

Water resources planners and managers have, thus, a large number of means available to increase the buffering capacity of hydraulic systems to absorb the effects of climatic changes. This capacity concerns both the ability to absorb surprises and the ability to recover from failure. Using the terminology used by MATALAS and FIERING 1977, one can say that the hydraulic systems as currently designed by engineers have some kind of provision for "robustness" and "resilience" when faced with climatic changes.

1.4 Regional Climatic Changes

From a practical standpoint the most interesting aspect is the prediction of the consequences of climatic changes for regional water availability, it being especially important to forecast the timing and magnitude of surface runoff and soil moisture fluctuations. This implies that we should be able to determine not only global, but also regional climatic changes and it is still impossible to do this adequately. One cannot yet simulate accurately the details of regional climate and consequently one is unable to predict with confidence the locations and magnitudes of regional climatic changes.

When passing from the analysis made at a global level to that made at a regional level, one should try to answer questions such as the following:

- What are the regional climatic scenarios developed for the regions under consideration and are these scenarios consistent?
- How do these scenarios affect the components of the hydrologic cycle and, in particular, precipitation and evapotranspiration?
- What are the consequences of these changes in precipitation and evapotranspiration in the various water resources uses?

- How should the present strategies of water resources management and planning be changed in order to face the consequences of future regional climatic changes?
- What are the short and long term R&D needs to solve the existing problems in the areas of regional climatic changes, hydrology and water resources management?

It should also be noted that the practical consequences of climatic changes in relation to water resources uses depend not only on the actual climatic changes that may take place, but also on the idea that planners may have of expected climatic changes. In other words, uncertainty also plays an important role. As referred to by NAS 1977 the uncertainty is due to the "threat of climatic change, real or imagined, which can alter what farmers will plant, as well as what economists think they will plant".

2. POSSIBLE CONSEQUENCES OF CLIMATIC CHANGES ON WATER RESOURCES DEVELOPMENT

2.1 Introduction

As referred to in Section 1, the greenhouse effect would determine a global increase in evapotranspiration as well as an increase in precipitation, expressing a kind of activation of the hydrologic cycle. According to WIGLEY and JONES 1985, doubling of CO_2 would cause an increase in precipitation varying between 3 and 11%, depending on the models used. However, the incidence of these variations would not be the same in all regions due to transport of humidity in the atmosphere from some regions to others. For example, it appears to be expected that the warming effects will be the greatest in the northern hemisphere's polar and mid-latitude regions, especially during the autumn and winter. The increase in temperature would increase the runoff in the high latitude and cause drier summers in the northern mid-latitude regions.

NEMEC and SCHAAKE 1982 have also shown that moderate fluctuations in precipitation may be much amplified in changes of runoff. In the case studied by the authors, an increase in precipitation of 25% resulted in a corresponding increase in runoff of 50 to 70%. BISWAS 1984a also notes that it has been observed that fluctuations in precipitation in certain regions tend, in the dry years, to be associated with greater fluctuations in runoff than in precipitation. This is so because evapotranspiration requirements which are relatively constant from one year to another have to be met prior to aquifer recharge and surface runoff.

Precipitation and potential evapotranspiration are essentially the characteristics of climate which condition surface flow. An aridity coefficient obtained by dividing potential evapotranspiration by precipitation can be used together with long-term average temperatures to specify nine different types of climate as done by SZESZTAY 1965 (Fig. 2).

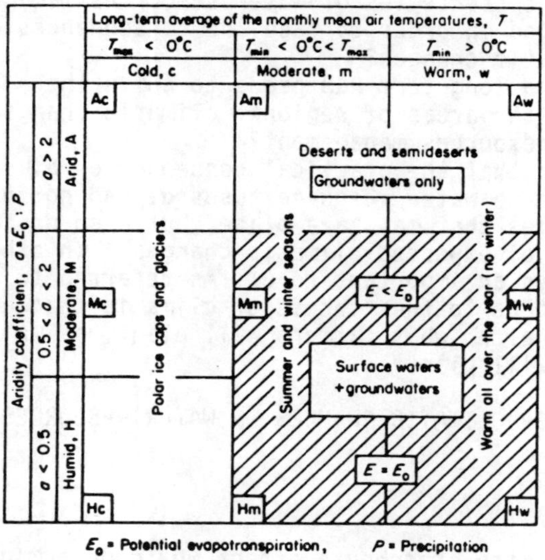

Fig. 2 - Climate Classification based on aridity coefficient and temperature

As indicated in the figure, for five of these nine types of climate there is no streamflow for two different reasons. Three of these types, covering about 12% of total land area, correspond to polar ice and glaciers, and there is neither streamflow nor groundwater; the two other types (covering 26% of the land area) correspond to deserts and semi-deserts: here, there is no streamflow but groundwater may be present.

For the four other types, corresponding to 62% of the land area, there is streamflow: for two of these types there are well-defined summer and winter seasons and, for the other two, weather is warm year-round.

Climatic changes can determine that regions having a certain type of climate can shift to another type. For instance, an increase in temperature will imply a reduction of the land area associated with the climatic types in the left column of the figure, implying reduction of ice caps and consequent increase of sea level; or, the increase of evapotranspiration would imply an increase of the land area associated with climatic types in the first row implying an increased desertification.

One important aspect of the relationship between climate and water resources is that available water being determined by climate is also in itself an essential ingredient of climate. Water is an intrinsic factor of the climatological processes which is not only extremely important but also, as NOVAKY 1985 points out, "almost ubiquitous because it stores, transports, delivers and redistributes energy in many ways and on many scales."

Climatic change can affect both inland water resources (influencing the amount of water resources available and its geographical, seasonal or interannual distribution) and coastal waters (mainly in relation to the problems caused by sea-level rise). The main possible consequences of climatic change on these two types of water resources are reviewed subsequently.

2.2. Consequences of Climatic Changes on Inland Water Resources

The main consequences of climatic changes related to inland waters are the following:

a) changes in the global amount of water resources available and in the spatial and temporal distribution of these resources;
b) changes in soil moisture;
c) changes in extreme phenomena related to water resources, i.e. floods and droughts;
d) changes in water quality;
e) changes in sedimentation processes;
f) changes in water demand.

Reference is made below to each one of these possible consequences.

a) Changes in the global amount of water resources available and in the spacial and temporal distribution of these water resources. The changes in the available amount of surface and/or groundwater are mainly related to the variations of precipitation and evapotranspiration as referred to in section 2.1. Climatic changes can also determine a modification of the regional and seasonal distribution of water resources with possible implications for crop production, for the capacity of reservoirs needed to store water for different uses, and in the impacts of floods and droughts.

b) Changes in soil moisture. If soil moisture changes occur in regions of the world where the impact on agricultural productivity is relevant, this may determine important economic and social impacts.

c) Changes in extreme phenomena related to water resources, i.e. floods and droughts. One important question is to try and find out how floods and droughts are modified when climatic changes occur. There are not many studies available covering this aspect as generally, during the periods to be studied, other man-induced changes, in particular those related to the construction of hydraulic works and/or modifications of land use, make the analysis of climatic change impacts difficult. Examples of not very conclusive studies are the one by MEKO and STOCKTON 1984 relative to the western part of the United States and another one by CHANGNON 1983 who studied a case in Illinois, USA.

d) Changes in water quality. The consequences of climatic change on water quality may be related to possible changes in precipitation regime or the occurrence of acid rain. Moreover, there may be direct consequences of climatic changes on water quality. First, temperature increases may determine decreases in levels of saturation dissolved oxygen in water. Second, biochemical oxygen demand (BOD) also increases with temperature. The result of these two effects could be a decrease in the dissolved oxygen concentration in water, which on the one hand determines the reduction of the capacity of absorption of pollutants by water, and on the other hand, may have negative impacts on aquatic ecosystems and, particularly, on fisheries. Finally, climatic changes are also likely to have indirect consequences on water quality, since the decrease of river discharges, particularly during the dry season, will increase the concentration of industrial pollutants in water bodies. NAS 1985 presents the results of an US-Canadian assessment of toxic substances in the Great Lakes and concludes that a change in precipitation is likely to have serious effects on the concentration of toxic contaminants in the Lakes.

e) Changes in sedimentation processes. Water is the cause of erosion, transport and deposition of sediments and thus climatic changes can affect these sediment processes as they affect the water processes. Sediment management should thus be considered as an important issue going hand in hand with water management. Conditions of soil erosion caused by water can be changed if the climate changes, with consequences for the forest and plant cover of the soil, and possibly for the stability of buildings and other structures. Sediment transport in rivers is associated with such important issues as navigation, flood control, and stability of river training works. Sediment deposition is crucial for reservoir silting (reducing thus the capacity of water storage and causing erosion downstream), estuary silting or delta behaviour. Finally, biological and ecological consequences may also be associated with sedimentation problems, such as the adverse impact of high sedimentation on the spawning of fish and the distribution of biota, reduction in photosynthesis due to increase in turbidity or loss of nutrients retained in reservoirs.

f) Changes in water demand. Climatic changes do not influence only the water availability but can also affect the water demand. For example, the water demand for irrigation is largely affected by climatic change which conditions evapotransporation. The water demand for domestic or industrial uses can also be affected by climatic change, namely as the result of temperature increases which influence consumption associated with uses such as water for cooling systems or water for bathing, washing and gardening.

2.3 Consequences of Climatic Changes on Sea Level Rise

Another effect of the greenhouse effect would be the rise of sea level due to the melting of glaciers. Several authors have predicted by the end of next century sea-level rises of 0.5 to 3.5m (e.g. NAS 1983, BARTH and TITUS 1984 and HOFFMAN 1984) due to the climatic modifications induced by an increase in CO_2.

The impact of a sea level rise would be quite different from one region to another, depending on factors such as land elevation, local morphologic and hydrologic conditions, degree of urban and economic occupation and degree of development of flood protection schemes and of water resources management in general. The main consequences of climatic change related to coastal waters are as follows:

a) flooding of urban, industrial and service areas;
b) rendering inoperable civil engineering works and other facilities;
c) intrusion of sea water in aquifers and estuaries;
d) changes in coastal sedimentation processes;
e) changes in the coastal environment.

These different consequences are analyzed subsequently.

a) Flooding of urban, industrial and service areas. The economic and social impact of the rise in sea level (possibly aggravated in some regions by land subsidence) would be very important, since many urban centres and expensive industrial and service facilities have developed and are still developing preferentially in the low lands along the maritime coasts, estuaries and deltas. These facilities could be affected in their operation, put completely out of use or even destroyed due to sea level rises. Destruction of land property in coastal or estuarine areas is another possible consequence.

b) Rendering inoperable civil engineering works and other facilities. Some water works, such as intakes and pumping stations, could be put out of use or dykes could be overrun and destroyed, determining the flooding of areas protected by these dykes. Sea level rises could also reduce clearance for boats under bridges, affecting fluvial transport. To counteract these impacts, the water works would have to be rebuilt at higher elevations and the dykes would have to be raised, implying heavy economic consequences. The increase in sea level could also seriously affect urban or agricultural drainage systems, rise above water supply pipe systems and destroy them by corrosion, affect the foundations of buildings, or flood underground transportation systems or underground cables. Also, the coastal and estuarine reservoirs which have been created in some parts of the world by damming of tidal lagoons or estuaries would be completely upset in their present operation, since gravity drainage to the sea would no longer be possible.

c) Sea water intrusion in aquifers and estuaries. Sea level rise would also cause the intrusion of sea water, contaminating fresh water aquifers which could thus no longer be used for water supply. A rise in sea level would also be the cause of sea water moving further upstream in estuaries, determining important modifications in the water quality of present fresh water and the need to shift in upstream direction existing water intakes for urban or industrial water supply, or for irrigation. This would imply expenses in building new water withdrawal works and extending the pipes or canals for water transportation. The environmental impact of this increased intrusion of salt water in estuaries will also in many cases be considerable.

d) Changes in coastal sedimentation processes. A rise in sea level would determine changes in coastal sediment transport affecting coastal erosion, silting of harbours and stability of beaches. These changes would also be dependent upon, and mutually influenced by, the climate-induced changes in the sediment transport regime in rivers and estuaries. Sediments are a resource that must be appropriately managed in any case and especially in the presence of climatic change.

e) Changes in the coastal environment. Sea level changes due to climatic change can determine important modifications of the coastline, beaches and barrier islands, affect coastal recreational areas or wetlands and cause changes in fish habitat.

To face an expected sea level rise one can consider various alternative measures:

- to evacuate the present coastal areas, where flooding would be allowed and retreat to the new future coastal areas, abandonning the facilities submerged and/or destroyed.
- to defend the present coastal areas, by protecting the shoreline with levees, sea walls or dykes.
- to nourish the present coastal areas, by feeding them with sand transferred from other areas, e.g. with sand dragged from estuary entrances to facilitate navigation.
- to modify the hydrologic cycle by intersecting river runoff to the sea, building dams to create reservoirs or directing the water to natural surface or groundwater storage areas.

The first alternative corresponds to just letting nature do its work and may be, in some cases, the most economical option. The second and third alternatives correspond to options currently adopted by engineers; finally, the last alternative is less conventional but has been defended by NEWMAN and FAIRBRIDGE 1986 who conjecture that an increase in water storage of 500 km^3/yr would absorb an annual sea level rise of 1.25 mm.

In practice one should resort to a combination of these alternatives, as recommended by the conditions of each specific case. And perhaps, due to the slow evolution of the situation, the problem to be faced is not too serious. This seems to be the conclusion of a study by NAS 1987. This study considered three possible scenarios for the year 2085 corresponding to a sea level rise of 0.5m, 1.0m and 1.5m and reviewed the practical implications for coastal planners, engineers and government agencies. The conclusion was that although the economic consequences can be important, particularly due to the loss of land property in coastal areas, there will no cause for alarm or emergency action since changes will occur gradually, allowing for ensuing changes and protection of major facilities in the course of normal maintenance schemes.

3. OVERVIEW OF THE MAIN ATTEMPTS TO ASSESS THE IMPACT OF CLIMATIC CHANGES IN WATER RESOURCES

3.1 Introduction

The approaches followed by meteorologists and hydrologists to deal with the interaction between climatic changes and water resources are quite different.

Meteorologists tend to prefer large scale tools such as general circulation climate models or climatic analogue scenarios. General circulation models have the disadvantage of not ensuring enough resolution to provide detailed climatic information at regional level and/or seasonal level. Climatic analogue scenarios can be constructed by three different methods, namely by reconstructing particularly warm historic periods, by using statistical models of atmospheric dynamics to estimate climatic changes and by using past instrumented climatic data (PITTOCK and SALINGER 1982 and WIGLEY et al. 1986). This third method has the advantage of incorporating the variability of the climate and allowing the definition of regional and seasonal scenarios.

Hydrologists, in contrast to climatologists, like to work with models on the scale of a river basin. As LEWIS 1987 points out "the details of the hydrologic processes are much better described in hydrologic models than in climatic models, but in the first case the climate is held static. The hydrologic models, being quite detailed in reproducing the physical phenomena at the drainage basin levels, do not include important climatic-hydrologic feedback mechanisms which are so necessary in describing the highly non-linear climatic system".

The determination of the impacts of climatic changes on water resources availability can be made by combining general circulation models with regional hydrologic models, either conceptual models or water-balance models. GLEICK 1986, after comparing different models, has concluded that the water-balance models have important advantages over other methods for assessing the impact of climatic changes on water resources.

Climate-water resources relationships are established resorting to transfer functions which can be classified in three categories: statistical, analytical and numerical. NOVAKY et al. 1985 deal with these categories of transfer functions stating that "statistical transfer functions are, for example, the empirical relationships between proxy information on fluctuations of climate (such as tree-ring indices or glacial activity) and water resources; analytical transfer functions are based on simplified physical principles, such as the balance between climatic elements (for example, precipitation or evaporation) and water resources (runoff, change of storage in soil moisture etc.); and numerical transfer functions are based on conceptual hydrologic models, which allow for more detailed physical considerations than the analytical functions, but also require digital computers for their application".

The normal procedure when making use of hydrologic modelling to estimate the changes in water resources due to climatic changes starts by fitting an adequate conceptual hydrologic model to existing historic hydrologic data relating to a river basin. These data refer, in particular, to precipitation, temperature and streamflow. Then, a certain climatic change is assumed, the corresponding conditions of precipitation and evapotranspiration are derived and subsequently taken as an input to the hydrological model which simulates a streamflow series. This series is then taken as an input to study the consequences in the performance of different water use systems. One typical case is the analysis of the modified output of a storage reservoir when the climatic change occurs. This could, for instance, lead the water resources planners to increase the size of a reservoir in order to ensure the same water output with the same reliability.

One difficulty with this method is the fact that existing conceptual hydrologic models are not usually geographically or climatically transferable, in the sense that they do not perform satisfactorily under conditions different from those for which they have been calibrated. So, the use of models in the way described has a number of limitations and it appears that they can be more useful to perform a sensitivity analysis, rather than to provide a quantitative information on the modified water resources situation.

It should also be pointed out that the hydrologic models based on the water balance equation involve a major difficulty resulting from the fact that the structure of the water balance equation expresses a linear relationship among the variables involved, and this is not at all the case. KLEMES 1985 has performed a comprehensive review of several aspects of these questions in a project on the sensitivity of water-resources systems to climatic variations, developed within the World Climate Change Programme of the World Meteorological Organization, and presents recommendations and proposals for further work of WMO in this area.

3.2 Use of General Circulation Climate Models

As mentioned before, water resources changes induced by climatic changes have been studied, in some cases resorting to general circulation climatic models. Examples of these studies are those carried out by MANABE et al. 1981, MITCHELL 1983, MANABE and WETHERALD 1986, WASHINGTON and MEEHL 1984, RIND and LEBEDEFF 1984, and COHEN 1986 and 1987. These studies are briefly referred to below.

As a rule, it can be stated that the results of general circulation models indicate that there will be changes in precipitation and evapotranspiration that can have an important impact on water resources availability. For example, MANABE et al. 1981, using the general circulation model of the Geophysical Fluid Dynamics Laboratory and considering an unlikely assumption of a four-fold increase in CO_2, identified in two regions, at middle and high latitudes, an increase in temperature and evaporation and a reduction in mean soil moisture, with particular incidence in the summer months. They also noted a positive feedback effect of dryness creating a self-amplified process.

Following similar approaches, MITCHELL 1983 and MANABE and WETHERALD 1986 have also predicted a shift to the north of the mid-latitude rainbelt, as well as changes in timing of snowmelt which would determine significant reductions in summer soil moisture patterns in the middle latitude regions.

WASHINGTON and MEEHL 1984 found different results, indicating that changes in timing of precipitation and positive feedbacks among soil moisture, precipitation and low cloud formation would determine an increase in soil moisture in summer months.

RIND and LEBEDEFF 1984, at the U.S. Environmental Protection Agency, using the general circulation model of the Goddard Institute for Space Studies and considering a two-fold increase in CO_2, have made an analysis of hydrologic changes over North America. The result of the study shows variation of precipitation between +24% and -7%, of evaporation between +37% and -11%, according to the region, defined as a grid element of 8 degrees latitude by 10 degrees longitude. The study shows that decreases in precipitation and evaporation both take place roughly in the middle and low latitudes of North America. Rind and Lebedeff have also derived the change in runoff referring to the same grid, as between +42% and -27% depending on the region. Regions with a decrease in runoff have been identified at all latitudes but with a greater incidence in the higher and lower latitudes.

COHEN 1986 and 1987 has predicted changes in Great Lakes basin water resources by using general circulation models produced by the Goddard Institute for Space Studies and the Geophysical Fluid Dynamics Laboratory, and by combining three scenarios of CO_2-induced climatic change (assuming different wind speeds) and projections of population and consumptive use of water to the year 2035. The decrease in net supply ranged from 29% to 12% for the various scenarios.

3.3 Use of Regional Hydrologic Models

Most of the studies on the impact of climatic change on water resources availability make use of regional hydrologic models to investigate the consequences of changes of average temperature and precipitation in water resources. Examples of these are the studies by STOCKTON and BOGGESS 1979, NEMEC and SCHAAKE 1982, QUINN and CROLEY 1983, REVELLE and WAGGONER 1983, IDSO and BRAZEL 1984, and ASTON 1984. A brief reference is made below to these studies.

STOCKTON and BOGGESS 1979 have considered four climate scenarios involving changes in temperature of $\pm$ 2°C and in annual precipitation of $\pm$ 10% to evaluate the impact on runoff in each of the 18 river basins of conterminous United States. They concluded that a change toward a warmer and drier climate would have the greater negative impact at the national level, whereas a change toward a cooler and wetter climate would be in general beneficial, except for some increased flooding in some rivers. The study by Stockton and Boggess was completed by another study, carried out by KNOX 1979, intended to ascertain geomorphologic impacts in the same basin.

NEMEC and SCHAAKE 1982 used the Sacramento hydrologic model, established by BURNASH et al. 1973, to simulate streamflow sequences corresponding to changed climatic conditions in two different river basins in the United States - one in a humid region in Mississippi approximately 2000 km^2, and the other in an arid region in Texas approximately 5 times larger - and then used these sequences to assess the future performance characteristics of a system of reservoirs to store water. Nemec and Schaake have investigated a wide range of possibilities by considering different climatic scenarios characterized by combinations of four possible changes of precipitation (increase and decrease of 10% or 25%) and three changes in potential evaporation (increase of 12% or 4% and decrease of 4%).

One conclusion of the study was that the changes of streamflow were much more conditioned by changes of precipitation than of evapotranspiration. The study also determined how the climatic changes could be compensated by changes in reservoir storage. However, the authors themselves note that the conclusions reached are somewhat limited by the sampling uncertainties arising from the shortness of the historical record used.

A propos of the work performed by Nemec and Schaake 1982, KLEMES 1985 expresses some criticisms about the rationale of developing scenarios in the way as by those two authors. He claims that a high sampling variability in climate and runoff exists due to the small size of data samples, corresponding only to 10 to 20 years. Thus changes due to climatic variability could be considerably higher than those due to climatic change.

QUINN and CROLEY 1983 have used a water balance model to study the changes in runoff for Lakes Superior and Erie, assuming an increase of 3°C in monthly temperatures and an increase of 6.5% in monthly precipitation for Lake Superior basin. They concluded that increases of 17% and 12% for Lakes Superior and Erie respectively could occur and they anticipated long-term decreases in water availability of 10% and 33%, also for the 2 basins.

REVELLE and WAGGONER 1983 have studied the impact of an increase in temperature of 2°C and an increase or decrease of 10% in precipitation in the western United States, particularly in the Colorado River basin. They discovered that there was a decrease in runoff even when precipitation is increasing due to a rise in evapotranspiration.

IDSO and BRAZEL 1984 have made an analysis similar to that performed by Revelle and Waggoner but adding the effect of CO_2 in vegetation by increasing the stomatal resistance and consequently decreasing evapotranspiration. They found that a 40% decrease in runoff, calculated without considering the effect of stomatal resistance, became a 40% increase when considering this effect.

ASTON 1984 has also considered the effect of stomatal resistance of vegetation. He studied small river basins in Australia and, assuming a doubling of atmospheric CO_2, his model would indicate 40% to 90% increase in runoff. This result is explained by Aston because, in spite of the fact that the 2° to 4°C increase in temperature due to a rise in CO_2 should determine an increase in evaporation, what happens is that the increase of CO_2 raises stomatal resistance of vegetation, decreasing evapotranspiration leading to an increase in runoff.

As an overall conclusion of the analysis made of these studies, it can be stated that regional-scale modelling techniques, incorporating all the complexities of the climatic and hydrologic processes, seem to be the solution to allow for some progress. Among the factors to be considered in this modelling, besides the information on future climatic changes, one should consider such hydrologically relevant aspects as temperature, precipitation, topography, vegetation cover, snowmelt, soil characteristics, surface and groundwater storage and seasonal variability.

4. CONCLUSIONS AND RECOMMENDATIONS

As mentioned in section 2, when considering the effects of climatic change on water resources, one should distinguish between the effects on inland waters and the effects on coastal waters. The effects of climatic change on inland water resources are somewhat uncertain, as some studies predict a decrease in the available water resources, while others predict an increase, depending on the climatic change scenario and the geographical regions under consideration. The effects of climatic change on sea level are certain in the sense that all studies agree that a rise in sea level will occur, only the magnitude of this being uncertain.

There are a number of possible alternatives to cope with the consequences of climatic change on water resources including: (i) prevention strategies; (ii) changes in water resources management and planning; (iii) abandonment; and (iv) compensation.

Progress in studying impacts of climatic change on water resources requires an interdisciplinary approach involving atmospheric scientists, oceanographers, geochemists, climatologists, meteorologists, hydrologists, engineers, water planners, economists and social scientists. The development of coordinated interdisciplinary efforts of these groups of professionals as well as of the research centres working in those areas would indeed be a condition for a better mutual understanding and for progress to be achieved. In particular, there is a need to investigate further the linkages between climate, meteorology and hydrology, especially by getting better models which simulate the physical reality in the river basins and which take into consideration the regional aspects of climatic change.

Conceptual hydrologic models - which reconstitute the hydrologic response of a basin under changing weather conditions at short time-steps and use the data corresponding to periods of years or decades - are particularly useful in determining the impact of a change in precipitation and temperature on soil moisture, streamflow and groundwater. The social and economic impact of these changes in hydrologic variables should then be assessed as required. However, the hydrologic models currently used have some limitations which may strongly condition the results obtained. Those limitations are related to such aspects as: (i) limited historical verification of models; (ii) non-consideration of non-linear relationships, in particular between precipitation and runoff; (iii) exclusion of snowfall and snowmelt; and (iv) non-consideration of all the pertinent physical factors involved and, in particular, some important characteristics of the basins.

The application of a certain hydrologic model to a given region should always be made following a careful verification of the climatic transferability of the model from the conditions in the region where it has first been applied.

The current practice of setting some scenarios based on the results of global circulation models but without being able to assign probabilities to these scenarios is somewhat limited. This limitation will only be overcome when it will become possible to establish a direct relationship between the climatic modelling and the hydrologic modelling. What is needed is that the hydrologic phenomena are incorporated into global circulation models with an adequate resolution. Linking global circulation modelling and regional water-balance modelling is likely to be the best way to obtain new and better information on the impact of global climatic change on regional water resources.

The impacts of climatic change on water resources should be presented to the policy makers with appropriate detail, as regards regional distribution and timescale, and should be associated with appropriate information on statistical likelihood.

In the future, water resources planning should always include the consideration of plausible climatic change scenarios when doing the sensitivity analysis and deriving social, economic and environmental impacts which are part of the planning process.

The timescale for the prediction of the water resources impact of climatic change should be adequate for the specific needs of short, mean, and long-term water resources planning.

While waiting for a more precise definition of the impact of climatic change on water resources, one way to face the adverse consequences is to honour the well-recognized engineers' practice of keeping a safety margin that accounts for uncertainty in general, including that associated with climatic change.

To conclude, one can try and identify the main research needs, to ensure a better understanding of the impact of climatic change on water resources development. These research needs can be described as follows:

- the quantity and quality of the available data should be improved, in particular, by creating better data networks in the regions not adequately served, namely in relation to solar radiation, precipitation, evapotranspiration and soil moisture.
- historical analysis of climatic impact on water resources should be improved as the sampling uncertainties due to insufficient historical records themselves impose limitations on the usefulness of very elaborate modelling efforts. This is particularly important in relation to extreme phenomena, i.e. floods and droughts for whose study stationarity of climate over time is usually assumed.

- in particular, increased attention should be given to the study of drought situations, aiming to determine how the current water resources management strategies ought to be changed when drought conditions are present. In fact, in the past greater attention has been devoted to coping with floods than to coping with droughts.
- the climatic scenarios should be developed at a regional level for the different regions of interest for prediction of water resources availablity, particularly those where the social, economic and environmental impacts are likely to be more relevant.
- the development and sophistication of the linkage between large scale geographical circulation models and the small scale river basin hydrologic models seems essential.
- the follow-up of the effort launched by WMO 1975 in order to compare several hydrologic modelling approaches is highly recommended. At the same time the establishment of an adequate data base which could be used to test the various models would also be recommended.
- the relationship of climatic change and water quality should be further investigated as these may correspond to a measured impact of climatic change.
- as regards the sea level rise, efforts have to be made in order to reach a more accurate prediction of magnitude and timescale, since the current average tidal levels may show considerable variations due to wind patterns and temperature distributions in the sea water.

REFERENCES

ASTON, A.R. 1984 - The Effect of Doubling Atmospheric CO_2 on Streamflow: A Simulation. Journal of Hydrology, 67, pp. 273-280.

BARTH, M.C.; TITUS, J.G. (Eds.) 1984 - Greenhouse Effect and Sea Level Rise. Van Nostrand Reinhold, New York.

BISWAS, A.K. 1984a - Climate and Development. Tycooly International Publishing Limited, Dublin.

BISWAS, A.K. 1984b - Climate and Water Resources. In BISWAS 1984a.

BOLIN, B.; DOOS, B.R.; JAGER, J.; WARRICK, R.A. (Eds.) 1986 - The Greenhouse Effect, Climatic Change and Ecosystems. John Wiley & Sons, Chichester, Sussex.

BURNASH, R.C.; FERRAL, R.L.; McGUIRE, R.A. 1973 - A Generalized Streamflow Simulation System: Conceptual Modelling for Digital Computers. U.S. National Weather Service and California Department of Water Resources, Joint Federal-State River Forecast Center, Sacramento, California.

CHANGNON, S.A. 1983 - Trends in Floods and Related Climate Conditions in Illinois. Climatic Change, 5, pp. 341-362.

CHANGNON, S.A. 1987 - An Assessment of Climate Change, Water Resources and Policy Research. Water International, 12, pp. 69-76.

COHEN, S.J. 1986 - Impacts of CO_2-Induced Climatic Change on Water Resources in the Great Lakes Basin. Climatic Change, 8, pp. 135-153.

COHEN, S.J. 1987 - Sensitivity of Water Resources in the Great Lakes Region to Changes in Temperature, Precipitation, Humidity and Wind Speed. IUGG/IAHS Symposium, Vancouver.

FALKENMARK, M.; DA CUNHA, L.; DAVID, L. 1987 - New Water Management Strategies Needed for the 21st Century. Water International, 12, pp. 94-101.

GLEICK, P.H. 1986 - Methods for Evaluating the Regional Hydrologic Impacts of Global Climatic Changes. Journal of Hydrology, 88, Amsterdam, pp. 97-116.

HOFFMAN, J.S. 1984 - Estimates of Future Sea Level Rise. In BARTH and TITUS 1984.

IDSO, S.B.; BRAZEL, A.J. 1984 - Rising Atmospheric Carbon Dioxide Concentrations May Increase Streamflow. Nature, 312, pp. 51-53.

KATES, R.W.; AUSUBEL, J.H.; BERBIAN, M. (Eds.) 1985 - Climate Impact Assessment. John Willey & Sons, New York.

KELLOGG, W.; SCHWARE, R. 1981 - Climate Change in Society. Westview Press, Boulder, Colorado.

KNOX, J. 1979 - Hydrogeomorphic Implications of Climatic Change. US Army Institute for Water Resources, Fort Belvoir, Virginia.

KLEMES, V. 1985 - Sensitivity of Water Resource Systems to Climate Variations. World Climate Programme, World Meteorological Organization.

LEWIS, J. 1987 - Climatic Change and its Effects for Water Resources in Canada: A Review. Science Council of Canada, Ottawa.

LVOVICH, M.I. 1979 - World Water Resources, Present and Future. Geo-Journal, 3(5), Wiesbaden.

MANABE, S. et al. 1981 - Summer Dryness Due to an Increase of Atmospheric CO_2 Concentration. Climatic Change, 3, pp. 347-388.

MANABE, S.; WETHERALD, R.T. 1986 - Reduction in Summer Soil Wetness Induced by an Increase in Atmospheric Carbon Dioxide, Science, 232, pp. 626-628.

MATALAS, N.; FIERING, M. 1977 - Water Resources Planning. In NAS 1977.

MEKO, D.M.; STOCKTON, C.W. 1984 - Secular Variations in Streamflow in the Western United States. Journal of Climate and Applied Meteorology, 23, pp. 889-897.

MITCHELL, J.M. 1983 - An Empirical Modeling Assessment of Volcanic and Carbon Dioxide Effects on Global Scale Temperature. Second Conference on Climate Variations, American Meteorological Society, New Orleans, Louisiana.

NAS 1977 - Climate, Climatic Change and Water Supply. National Research Council, Studies in Geophysics, National Academy of Sciences, Washington, D.C.

NAS 1983 - Changing Climate. National Academy of Sciences, National Academy Press, Washington, D.C.

NAS 1985 - The Great Lakes Water Quality Agreement: An Evolving Instrument for Ecosystem Management. National Academy of Sciences, Washington, D.C.

NAS 1987 - Responding to Changes in Sea Level: Engineering Implications. Committee on Engineering Implications of Changes in Relative Mean Sea Level, Marine Board, Commission on Engineering and Technical Systems, National Academy Press, Washington, D.C.

NEMEC, J.; SCHAAKE, J. 1982 - Sensitivity of Water Resources Systems to Climate Variations. Journal of Hydrological Sciences, 27(3), pp. 327-343.

NEWMAN, W.S.; FAIRBRIDGE, R.W. 1986 - The Management of Sea Level Rise. Nature, 320 (6060), March, pp. 319-321.

NOVAKY, B.; PACHNER, C.; SZESTAY, K.; MILLER, D. 1985 - Water Resources, in KATES, R.W. et al. 1985.

PITTOCK, A.B.; SALINGER, M.J. 1982 - Towards Regional Scenarios for a CO_2-Warmed Earth. Climatic Change, 4, pp. 23-40.

QUINN, F.D.; CROLEY, T.E. 1983 - Climatic Basin Water Balance Models for Great Lakes Forecasting and Simulation. Fifth Conference on Hydrometeorology, American Meteorological Society, Boston.

REVELLE, R.R.; WAGGONER, P.E. 1983 - Effects of a Carbon-Dioxide Induced Climatic Change on Water Supplies in Western United States. In NAS 1983, pp. 419-432.

RIND, D.; LEBEDEFF, S. 1984 - Potential Climate Impacts of Increasing Atmospheric CO_2 with Emphasis on Water Availability and Hydrology in the United States. U.S. Environmental Protection Agency, Washington, D.C.

STOCKTON, C.W.; BOGGESS, W.R. 1979 - Geohydrological Implications of Climate Change on Water Resources Development. US Army Institute for Water Resources, Fort Belvoir, Virginia .

SZESZTAY, K. 1965 - Some Aspects of Hydrological Network Design with Special Regard to Mountainous Areas. Publication no. 68, International Association of Scientific Hydrology, Quebec.

WASHINGTON, V.M.; MEEHL, G.A. 1984 - Seasonal Cycle Experiment on the Climate Sensitivity due to a Doubling in CO_2 with an Atmospheric General Circulation Model Coupled to a Simple Mixed-Layer Oceans Model. Journal of Geophysical Research, 89(D6), pp. 9475-9503.

WIGLEY, T.M.; JONES, P.D. 1985 - Influences of Precipitation Changes and Direct CO_2 Effects on Streamflow. Nature, 314, pp. 149-152.

WIGLEY, T.M.; JONES, P.D.; KELLEY, P.M. 1986 - Empirical Climate Studies. In BOLIN et al. 1986.

WMO 1975 - Inter-comparison of Conceptual Models Used in Operational Hydrological Forecasting. Operational Hydrology Report No. 7, World Meteorological Organization, Geneva.

WMO 1986 - International Conference on the Assessment of the Role of Carbon Dioxide and of other Greenhouse Gases in Climate Variations and Associated Impacts. Report on the Villach Conference (October 1985), World Meteorological Organization, Geneva.

HYDROLOGICAL IMPACTS - WHAT IF WE ARE RIGHT?

MAX BERAN
Institute of Hydrology
Wallingford
Oxon OX10 8BB
UK

ABSTRACT. Water supply policy makers are expected to take the greenhouse effect into account in forward planning. Their advice must come from the research community who need therefore to consider the information needs of this sector. A checklist is offered to aid this task. While accustomed to uncertainty there are severe mismatches between the climate model outputs and the climatic inputs to hydrological transfer functions. These relate to spatial, temporal and statistical discrimination. Research is needed to firm up hydrological prediction and to bridge these gaps.

1. Introduction

The question posed in the title is deceptively simple. As climatologists searching for evidence of past and future change our problem is classic: hypotheses are framed, tested and either rejected or improved, and little opprobrium attaches to an incorrect forecast. The international community, through the 1985 Villach statement, urged scientists to reduce uncertainty in forecasting the effect of CO_2-induced change. That same statement also enunciates the need to take account of climate change when framing policies for economic development and environmental control. While the first call is directed at researchers, the second concerns policy makers which, in the current context, means those who plan and manage water supplies. This latter group lacks specialist knowledge, so responsibility for advising them reverts to scientists.

This note reviews the difficulties in translating the current knowledge about the future climate into "actionable" guidance for those whose role it is to provide the world with water. It also aims to inform climate specialists about how their information will be employed in a checklist of tests, and considers supporting research needs.

2. Water In The Scheme Of Things

The framework for scientific enquiry in the hydrological and water resources field has been set, at the international level, by the World Meteorological Organization through WCP(Water) with a range of pertinent project areas and contributing research programmes. The World Climate Programme highlights the importance of land surface processes within the general circulation and here hydrological science has been productive in improving understanding and modelling. The role of water permeates many natural processes, biological and physical, and the International Geosphere Biosphere and GEWEX programmes will continue to involve scientific hydrologists in the scientific enquiry.

However this note is more concerned with water supply practice than hydrological science

A. Berger et al. (eds.), Climate and Geo-Sciences, 661–665.

though, even at this level, water's pervasive role means that the concern about change ranges wider than just water supply. This note focuses on this area, in particular to discuss the form of the scientific advice and the supporting research.

3. Water Resource Impact Assessment

Three stages can be recognised in the process of quantifying the water resources impact of future climate change. Firstly it is necessary to construct a climatic change scenario. Secondly climate variables have to be translated into hydrological variables. In the third stage the water resources implications of these altered hydrological conditions must be evaluated (see Beran, 1986 for bibliography).

As with other impacts of climatic change the root problem is what assumptions to make about the future climate. The convention adopted elsewhere of a doubling of the greenhouse gases is normally taken. Practitioners are learning the jargon of GCMs and scenarios; the latter is an especially useful concept conveying the idea of plausibility and internal consistency, but not burdened by the unattainable accuracy implicit in the word "forecast".

The well-publicised uncertainties in GCM-based estimates have lead hydrologists to consider scenario construction in its broadest sense. This encompasses the use of palaeoclimate and recent historical warmings as analogues for future warming, and also entirely hypothetical scenarios have also been used. In the latter approach the input climate variables are perturbed by preset and arbitrary amounts across a range, and the sensitivity of response variables is studied. Perhaps the most quoted example of this approach is Nemec and Schaake (1982) who were able to demonstrate important amplifications in certain key hydrological variables relevant to reservoir storage. GCM-based scenarios have also been popular; a recent example is due to Bultot et al (1988) who have traced impacts to contrasting Belgian catchments.

Concerning the climate:hydrology transfer function two problems are paramount stemming from the mismatch between GCM results and hydrological models: (i) the paucity of information from GCMs on temporal and statistical variability and (ii) their overlarge spatial unit. Meaningful hydrological analysis requires knowledge of interannual and seasonal variation about individual catchments, river basins and aquifers. Much of the "skill" in water impact studies lies in the reconstitution of this necessary temporal and spatial detail. Several studies (eg Cohen, 1986) have adopted unadjusted data from nearby GCM grid points. Other procedures may be envisaged in which more explicit use is made of information on extremes and to use them to choose spatial analogues (ie regions remote from the study area whose current climate approximates the study area's altered climate). This technique is being applied in an ongoing project at the Institute of Hydrology into reservoir performance in the UK using the maritime region of S.W. France.

When considering the hydrology:water resource transfer hydrologists are on much firmer ground. However there is a need to consider a wider range of hydrological responses. Most impact studies have focussed on surface water resources whereas groundwater, soil moisture, water quality, glaciers, physical and chemical properties of water and the integrity of the watercourses have been passed over relatively unnoticed. Also there has been a geographical bias to studies; a current study for FAO within my Institute in which we are considering irrigation scheme integrity in Southern Africa is one of very few examples where the problems of developing countries have been considered.

Nor has as much consideration been given to the myriad other areas where water is an important determining factor in decisions or design. Here one may list weather sensitive activities such as building design, and machinery subject to cold, damp, or potentially corrosive environments. Demand patterns appropriate to an altered climate also need to be incorporated into studies. Issues relating to agriculture and the natural environment have received more attention than those concerned with the man-made environment.

4. Advising Water Managers

We come now to the central issue of how to frame advice to aid water management which, it must be appreciated, is not an amorphous topic in the hands of a homogeneous population. The nature of the climate, hydrological regime and state of development stamp very different characters on technical and institutional structures for supplying water. Also the infrastructure that is appropriate for dealing with matters of land drainage and flood control may differ from those relevant to domestic supply, energy production or drought management.

In order to solve individual design or operational problems those concerned with water resources have already to weigh many interacting factors. In addition to this complexity, uncertainty impinges at almost every point: political and security issues of international rivers; economic uncertainties affecting demand patterns; demographic, food and health issues; technological change; inadvertent man-induced side effects; as well as natural disasters which may affect supply and demand. A worrying problem, encountered worldwide but particularly acute in developing countries, is that decision makers have often failed to take action even when presented with accurate information on impending disaster (Glantz, 1977).

The low awareness in the water resource community of climate change and its impacts had, during the 1970s, been justified by the conflicting views then broadcast by scientists about the climate future, and by the lack of climate information tailored to its particular objectives. With broad consensus established behind the greenhouse effect and the near certainty of some consequential warming it is no longer legitimate for them to opt for the "do-nothing" approach. Nor is uncertainty a legitimate reason for taking this line - water managers are accustomed to operating within an uncertain environment.

5. An Action Checklist For Water Management

The climate is changing - this needs to be preached as an unequivocal fact - and it was ever so. Most of the traditional analytical procedures are underpinned by a world view which sees the water cycle as a global experiment in which a fixed target is aimed for but always missed; ie a stationary process.

Given the present state of the art - where the impact of climate change can be quantified only as a very broad band which probably intersects the band reflecting current uncertainties and already embodied in design criteria - few would wish physically to modify existing schemes. However major reviews and new schemes merit a report analogous to an Environmental Impact Statement which considers scheme resilience to a climate shift and which encompasses the following:

a) A search for evidence for non-stationarity in regional climate and hydrological data sets.

b) Demonstration of the degree of resistance to hypothetical changes to input hydrological series perhaps constructed from the more disadvantageous periods from within the historic record.

c) A review of the overall performance of the project against more general criteria including environmental and social impacts within the existing climate regime.

d) Consideration of the sensitivity of components of the project to first order effects of climate change - ie higher temperature, reduced snow-pack, glacier runoff, and higher sea levels - about which uncertainty is less.

e) Informal inspection of GCM-based scenario for 2 x CO_2 anomalies and evaluation of project design criteria if rainfall and temperature time series were adjusted to accord with the revised mean values.

Items (a), (b) and (e) will reveal sensitive aspects of a scheme and allow adjustments to be made to optimise the resource under a more severe climatic regime. Item (c) reflects a conclusion, noted uniformally across all impact areas, that projects which are tuned to their environment - ecologically, socially, technologically - are better able to withstand the pressures of change. Item (d) acts as a reminder that water resources impacts are often linked to other impact areas. Obvious examples are direct temperature effects and sea-level rise.

This list can be used as the touchstone for providing climatological advice and data. The main problem areas are likely to be, not the uncertainty, but the absence of statistical information on extreme events. Given that such information is not available from current models it will remain a task for hydrologists to infill basing the scenario on control climate variability.

6. Research Needs

On the scientific front programmes should be encouraged that allow macro-scale description of hydrological processes to improve the description of land phase processes within General Circulation Models. Past and current examples of these are the HAPEX and ISLSCP experiments but local desk and field studies in which processes at the atmosphere land interface remain pertinent.

Researches into palaeohydrology will have increased importance for linking past hydrological regimes to their ambient climates. The quantification of regional flood and drought corresponding to times of earlier warming episodes will be most useful. Quantitative geomorphology and lake ecology are associated topics that should be encouraged.

New methods of analysing site and regional long-term records should be devised which recognise explicitly the "fluctuating" nature of climatic processes. A start in this direction has been made within the WCP(Water) in which a global long hydrological dataset is being assembled and its properties evaluated.

The main difficulty in quantifying the impact on water resources of future climatic change will remain the problems of modelling the processes, especially those in the unsaturated zone, and handling the complex social and economic interactions. Therefore studies into analytical techniques for solving the problems of design and rational operation should continue. Particular attention needs to be given to the problems of developing countries at all levels; this includes those hydrological processes that are characteristic of arid areas, and the problems of information dissemination so that the threat of the greenhouse effect can be countered in these fragile environments.

Water resources exist to cope with departures from mean values and spatial variability. Research is needed into techniques for bridging this disparity between what climatologists can provide and what water resource modellers require.

Another area where attention is needed concerns the construction of internally coherent scenarios for water demand. Past climate sensitivities might give some guidance on future demand but the effect of a sustained change would obviously give rise to a more profound adjustment. The problem is compounded by the possibility of a direct CO_2 impact on plant and crop water use.

7. Summary

Planners and engineers who make practical decisions about designing and operating water supply facilities require clear statements to the effect that the climate is innately variable, has always been, and with high probability will respond to the warming action of increasing concentrations of greenhouse gases. Uncertainties exist in particularising the change but this needs not stand in the way of estimating the risks to water supply arrangements.

It may be several decades before sufficient skill and data have accumulated to permit a quantitative estimate of, for example, revised reservoir dimensions to maintain an existing demand, or levee height to protect against inundation. Meanwhile steps should be taken to increase the robustness of systems and test their resilience by reference to scenarios constructed from severe periods within the data of past decades and GCM based "guesstimates". A checklist has been offered but much research remains to be done.

8. References

Beran, M.A. 1986. 'The water resources impact of future climate change and variability'. In Titus, J.G. (ed) *Effects of changes in stratospheric ozone and global climate. USEPA.*

Bultot, F., Dupriez, G.L. and Gellens, D. 1988. 'Estimated annual regime of energy-balance components, evapotranspiration and soil moisture for a drainage basin in the case of a CO_2 doubling'. *Climatic Change* **12** 39-56.

Cohen, S.J. 1986. 'Impacts of CO_2-induced climatic change on water resources in the Great Lakes basin'. *Climatic Change* **8** 135-153.

Glantz, M.H. 1977. 'The value of a long-range weather forecast for the West African Sahel'. *Bulletin of the American Meteorological Society*, **58**, 150-158.

Nemec, J. and Schaake, J.C. 1982. 'Sensitivity of water resource systems to climate variation'. *Hydrological Sciences Journal*, **27**, 327-343.

CLIMATE CHANGE: EFFECTS ON BIOLOGICAL SYSTEMS

Anne H. Ehrlich
Department of Biological Sciences
Stanford University
Stanford, CA 94305
U.S.A.

Abstract

Three major classes of human-induced changes in the atmosphere pose serious global problems for humanity's future. These classes of problems are: the increasing atmospheric concentrations of gases, including CO_2 and methane, leading to a "greenhouse" warming; the addition of chemicals that cause depletion of the stratospheric ozone layer and admission of increased UV-B radiation; and pollutants in the troposphere that damage life and property both directly and indirectly, as acid precipitation. The build-up of greenhouse gases in particular seems likely to lead to worldwide changes in climates, with the potential for seriously disrupting both agriculture and the functioning of already impaired natural ecosystems. The large and still rapidly growing human population is dependent on the health of both kinds of ecosystems for its sustenance and can little afford serious disruptions of them. Policies to address all these global problems must soon be developed and implemented worldwide if the consequences of atmospheric changes, especially climate changes, are to be averted or ameliorated.

Research in the past ten to twenty years has revealed not only many details of Earth's climatic system and the circulation patterns and chemistry of the atmosphere, but also much about the atmosphere's susceptibility to human-induced change and, in turn, the biosphere's sensitivity to atmospheric changes. Findings about atmospheric and climatic changes during Earth's long history are interesting in themselves and for their influence on the history and evolution of life (Schneider and Londer, 1984). The history of climatic change also illuminates changes that may occur or already are occurring and their effects on biota (COHMAP, 1988, Bryson, 1988). As a biologist, I am mainly concerned about the potential effects of current and anticipated atmospheric changes on life, including human life.

A. Berger et al. (eds.), Climate and Geo-Sciences, 667–679.

Introduction

In my perception, the atmospheric changes now occurring that should most concern civilization fall into three general classes. The first is the buildup of CO_2 and other "greenhouse" gases, warming the atmosphere (Bolin et al, 1987; Rasmussen and Khalil, 1986; Dickenson and Cicerone, 1986; Blake and Rowland, 1988; McElroy, 1988; Lemonick, 1989). In the second category of problems are chemical reactions that are causing destruction of the ozone layer and admitting increased UV-B radiation (Lemonick, 1987a, 1987b; Stolarski, 1988; Bowman, 1988; Rowland, 1988), and the consequent effects on both biota and climate (Scotto et al, 1988). The third category includes industrial pollutants emitted to the lower atmosphere, which are directly damaging to life and property or, after undergoing chemical changes in the atmosphere, become components of acid precipitation (NRC, 1986; Harte, 1988).

In addition to the current chronic anthropogenic additions of gases to the atmosphere, the potential exists for catastrophic consequences arising from a sudden injection into the atmosphere of massive amounts of particulates such as smoke and dust, which might be caused by a violent volcanic explosion, a collision of Earth with an extra-terrestrial body, or a large-scale nuclear war (Ehrlich et al, 1984; Pittock et al, 1985; Schneider and Thompson, 1988). The first two events obviously are beyond human control, and at least the second is highly unlikely to occur within a time-scale of interest to society. The possiblity of a nuclear war certainly cannot be discounted, although recent progress toward reductions in the nuclear arsenals of the two superpowers may have reduced its likelihood.

Climate and Life in Earth's History

Changes in climate over the ages undoubtedly have had profound effects on Earth's complement of life-forms, and they in turn evidently have influenced the gaseous composition of the atmosphere -- notably the oxygen deposited in the atmosphere through the action of photosynthesizing organisms while carbon was sequestered by tectonic and other geologic action. That climate has substantially varied through time is very clear. The past two million years or so have seen massive glacial advances and retreats in the northern hemisphere, whereas earlier epochs seem to have been characterized by a warm, equable climate extending well into arctic regions.

The apparently stable warm climate of the Cretaceous may have been due to a higher atmospheric concentration of carbon dioxide and/or other greenhouse gases (possibly including water vapor), to a slightly different inclination of the planet's axis, or to some combination of these factors. An upset of that benign state of affairs about 65 million years ago was accompanied by a catastrophic extinction event; along with the dinosaurs, thousands of species of diverse major taxonomic groups, including numerous terrestrial and marine plants and animals, disappeared in a relatively short time, geologically speaking. Regardless of the cause of that event, which is still a matter of controversy, global climate seems to have been somewhat changed afterward; and life evolved accordingly, with mammals replacing reptiles as the most diverse large land animals.

While life eventually restored itself to new levels of abundance and productivity, a long period of recovery from the biotic impoverishment imposed by the extinction event no doubt was required. And the relatively large swings in mid-latitude climates during the past 50 million years or so may well have stimulated development of today's extremely diverse flora and fauna. Yet the changes, rapid glacial advances especially, also must have caused local and regional extinction of many kinds of organisms and at least temporary reductions in terrestrial biotic productivity -- the photosynthetic fixation of solar energy into carbohydrates by green plants and some other organisms minus the energy those organisms use for their own life processes -- which biologists call net primary production (NPP). During periods of maximum glacial coverage, it is even possible that global NPP was significantly diminished, even though sea level was lower, exposing more land in unglaciated regions.

During the Pleistocene, as glaciers advanced and retreated and climates warmed and cooled, forests or grasslands stretched unbroken across continents, interrupted only by rivers, mountain chains, or other natural discontinuities. Plants, animals, entire ecosystems could respond to a change in climate occurring over decades to centuries by evolving tolerance for new conditions or by gradually shifting their distributions to more favorable areas as temperatures rose or fell or as weather became drier or wetter.

HUMAN IMPACTS

Today, because of human expansion, continuous natural habitats no longer exist in temperate or subtropical continental areas or the seasonal tropics, and they are rapidly disappearing in both moist tropical and subarctic regions (CEQ, 1980; World Resources Institute, 1988). Most natural areas on land have been replaced by human-controlled ecosystems -- cropland, tree farms and managed forests, pastures and grazed rangeland, or urban-industrial development. Even the continuity of many riverine ecosystems has been disrupted by extensive damming, diversion of water, and river channeling.

In the United States, for instance, the formerly vast eastern deciduous forests have been reduced to isolated patches of differing sizes, while the natural grasslands of the central states have all but disappeared. Similar, often even more extensive changes have occurred throughout the world. And the remaining natural areas are far from undisturbed; ecosystems and their component species everywhere are subjected to stresses from pollutants in air, water, and soils.

Atmospheric Changes and Consequences

Thus, even as human activities are causing climate to change with unprecedented speed, human occupation and use of land interferes with the capacity of flora and fauna to respond to the changes (Peters and Darling, 1985; Peters, 1988; Pain, 1988; Peters et al, 1989). Evolutionary adjustments are inhibited because of genetic impoverishment of reduced numbers and sizes of populations of numerous organisms (Peters, 1988; Murphy and Weiss, 1989); migration is effectively blocked by huge stretches of human-controlled systems and urban development, including highways (EPA, 1988; Roberts, 1988). In addition, neither evolution nor migration seems capable of providing an adequate response to changes as rapid as are anticipated with the greenhouse warming.

A major consequence therefore of global warming and changed precipitation patterns caused by the atmospheric buildup of greenhouse gases will be a rise in rates of extinction -- in addition to the cascade of extinctions already being caused by humanity through habitat conversion or disruption (CEQ, 1980; Ehrlich and Ehrlich, 1981). Biologists have only begun to study the probable effects of such large-scale and rapid changes on individual species and on natural communities.

ACID PRECIPITATION

An important factor in habitat disruption throughout north temperate regions, of course, is acid precipitation (Hordijk, 1988). In concert with other components of air pollution and, possibly, stresses from climatic warming, acid precipitation appears to be literally killing forests in Europe and North America (Harte, 1988; World Resources Institute, 1986), in addition to their better-known lethal effects on lakes and streams. Increased acidity is damaging or lethal to many forms of life. In forest ecosystems, the acidity seems to act both directly (through deposits on leaves) and indirectly (by effecting complex changes in soil chemistry and soil organisms) on trees and other plants, weakening them and increasing susceptibility to diseases, insect attack, and extreme weather conditions. When forest trees die, so do numerous other species of plants, animals, fungi, and microbes that depend on each species of tree one way or another (Ehrlich and Ehrlich, 1981).

EFFECTS OF OZONE DEPLETION AND INCREASED UV-B

Releases by humanity to the atmosphere of chemicals that cause depletion of stratospheric ozone (chlorofluorocarbons and oxides of nitrogen), thereby allowing increased ultraviolet radiation from the sun to reach Earth's surface, pose another set of problems for civilization in coming decades. Certain wavelengths of ultraviolet radiation are known to have potentially damaging effects on virtually all forms of life, particularly to genetic mechanisms, but also affecting a wide variety of other physiological processes (Ehrlich et al, 1984; Harwell and Hutchinson, 1985). UV-B radiation is known to inhibit photosynthesis and retard growth and reproductive processes in many kinds of plants, including important crops (Begley et al, 1988). It can kill many small organisms, such as bacteria and planktonic species. Known effects on human beings include an increased incidence of skin cancers and cataracts and impairment of immune responses.

While damaging effects of enhanced UV-B radiation have been observed in a broad spectrum of organisms, and sensitivity varies among species and among different life-stages of many organisms, these have not been studied exhaustively. Exactly how enhanced UV-B exposure resulting from depletion of the ozone layer would affect particular species or ecosystems therefore cannot now be predicted, but it is reasonably safe to say that widespread disruptive effects on both agricultural and natural ecosystems are likely.

In addition, CFCs and oxides of nitrogen also contribute to greenhouse warming (Gribbin, 1988). The need to find benign substitutes for their uses is therefore doubly important.

HUMAN IMPACTS ON BIOTIC PRODUCTIVITY

Virtually all animals and other non-photosynthesizing organisms (including human beings) are completely dependent on the energy produced by photosynthesis for life. But human activities, in displacing and/or damaging natural ecosystems worldwide (through deforestation and desertification as well as by replacement of natural systems with less productive human-controlled ones such as croplands, have diminished worldwide biotic productivity on land by an estimated 13 percent just since about 1950. Moreover, nearly 30 percent of all net primary production (NPP) on land, including that directly consumed by people and livestock, has been diverted into human-controlled systems (Vitousek et al, 1986). Thus, a total of nearly 40 percent of the <u>potential</u> NPP on land, plus a few percent of that in the oceans, is now accounted for by humanity.

Vulnerability of Living Systems to Climate Change

The implication of this massive degree of human displacement of natural biota is that very little room is left for mistakes that could lead to further reductions in NPP -- mistakes such as inadvertent significant climate changes on a global scale. Not only are other organisms dependent on that source of food, so are people. The human population is still expanding rapidly, and demands for food and other materials are rising, indicating increased pressure on Earth's NPP in future decades to meet human needs. As those demands continue to rise, the natural systems supplying vital ecosystem services that underpin the entire human enterprise will be still further squeezed (Ehrlich and Ehrlich, 1981).

AGRICULTURE AND CLIMATE CHANGE

Agriculture, which supplies most of humanity's food, is highly vulnerable to climate changes. Crops and livestock are closely adapted to their native local climates, and many modern high-yield crops are even more sensitive to weather deviations than traditional varieties. Sudden and extreme shifts in weather patterns can lead, as they did in summer 1988, to substantial declines in harvests (Begley et al, 1988; Brown, 1988).

If farmers understand that a change is permanent, not merely a short-lived stretch of "bad weather," they can switch to using different varieties of crops and livestock -- assuming that genetic stocks of alternative varieties have been preserved and plant geneticists can produce appropriate new ones. The difficulty lies in predicting exactly what the changes will be in any given place and how stable the new climatic regime will be. In view of the lack of predictability and the likely instability of new regimes in many key agricultural regions, substantial fluctuations and significant drops in yields and harvests may be a regular feature of world agriculture in the years ahead.

NATURAL ECOSYSTEMS AND CLIMATE CHANGE

Much of the biota of natural ecosystems is just as sensitive to local climates as crops are. Even if continous habitat still existed, time would be required for an adjustment of individual species through shifts in ranges or evolutionary changes in physical tolerances (Peters, 1988; Peters et al, 1989). Furthermore, since different species are likely to be differentially sensitive to temperature and moisture regimes, coevolved relationships between species might be upset, causing further disruption of ecosystems. Ample time for adjustments both by individual species and biotic communities has normally been available during natural shifts in climate, but the expected anthropogenic changes will come too swiftly. Thus, unless the remaining fragmentary ecosystems are carefully managed by humanity, there is little hope of their surviving major rapid climate changes without substantial losses of species diversity and, quite probably, significant temporary drops in productivity (Peters et al, 1989; EPA, 1988).

A large change in global climate now therefore may cause a far greater extinction event than the quite drastic shifts of the Pleistocene, for several reasons: first, greenhouse and ozone-induced changes will occur too fast for evolutionary adjustments and possibly even too quickly for adequate cultural adaptation in human systems; second, evolutionary adjustments in natural systems through shifts in range will be thwarted by the discontinuity of habitat; third, the sensitivity of individual species to climate change will very likely be enhanced by other environmental assaults to which they are already being subjected, such as exposure to pesticides and other toxic substances, air pollutants, and acid precipitation.

The Challenge of Global Policy

Designing policies to address this complex of atmospheric problems will not be easy, for several reasons. For one, just assessing and monitoring the atmospheric changes and their climatic consequences is a difficult technological challenge in itself (Jones et al, 1986; Kerr, 1988a, 1988b). The atmosphere is a dynamic system having numerous complex chemical interactions with the ecosphere (Mooney et al, 1987) that often are not easily distinguished from anthropogenic changes (Schneider, 1988a, 1988b). Reaching agreement on the dimensions and sources of the problems is hard enough; achieving a consensus on appropriate policies both to mitigate the consequences and to adjust to the inevitable changes will be even more difficult (Gleick, 1988; Ramirez, 1988). And implementing effective policies will require international cooperation on an unprecedented scale.

Excluding the threat of nuclear war, whose atmospheric and biological consequences would dwarf those described here (Ehrlich et al, 1984; Pittock et al, 1985; Harwell and Hutchinson, 1986), all these atmospheric problems must be attacked simultaneously (Kellogg and Scwhare, 1982; Tangley, 1988), in part because, to some extent, they have synergistic effects on biotic systems.

So far, an international agreement has been negotiated for only one of the three major sets of problems -- phasing out of production some of the chemicals responsible for ozone depletion (Garelick, 1987). And, in light of the latest findings, that first agreement appears to be inadequate (Pearce, 1988; Kerr, 1988b, 1988c, 1988d; Crawford, 1988; MacKenzie, 1988). Efforts to constrain the pollutants that cause acid precipitation have been far too little too late for a process that, once entrained, is largely irreversible (Harte, 1988).

CONTROLLING GREENHOUSE GASES

No effort at all has been made to slow or halt the injection of greenhouse gases into the atmosphere, although this may be the most important of the three problems. The most practical way to reduce atmospheric inputs of carbon dioxide is to burn less fossil fuels and wood (Rose et al, 1984). Some atmospheric CO_2 will eventually be taken up by the oceans and in vegetative growth.

The buildup of CO_2 could be substantially slowed through some important changes in policy and behavior, allowing more time for the oceans to absorb CO_2 and for both natural and agricultural ecosystems to adjust to higher temperatures and altered rainfall patterns. A key change would be to reduce the burning of fossil fuels and to switch from using coal, which produces the most CO_2 per unit of energy generated, to oil and natural gas, which produce less. Serious energy conservation and substitution of solar and other alternative energy sources for fossil fuels should be major components of such a strategy. In addition, an aggressive worldwide program of reforestation and revegetation, which is urgently needed for other reasons besides, could be launched (Postel, 1988; Booth, 1988; Goreau and de Mello, 1988) to help take up CO_2 from the atmosphere.

Buildups of other greenhouse gases also need to be addressed; methane emissions (Blake and Rowland, 1988) might be curbed by reducing cattle production (such a reduction is also desirable for other reasons) or finding ways to reduce their methane production, and by changing methods of rice cultivation.

Policy development is only beginning on these atmospheric problems, although the urgency of establishing new policies to address them is being more widely recognized by decision makers. Bills have been introduced into both houses of the United States Congress and hearings have been held. Several international conferences have been held, and the international community is now grappling with them.

PUBLIC INFORMATION AND THE ROLE OF SCIENCE

Public awareness, which is necessary for successful policy implementation in such economically central arenas as energy use and agriculture, has risen dramatically in the past year -- thanks partly to the practical lesson provided by the severe drought in North America, China, and the Soviet Union, as well as by several extreme weather events such as the inundation of Bangladesh, unprecedentedly powerful hurricanes in the Caribbean, and out of season tornados in the United States. That these events may never be traceable to greenhouse warming does not matter; that they were the kinds of event that greenhouse warming is expected to produce is sufficient for them to serve as object lessons.

Scientists have a vital role to play explaining to the public and to political leaders the significance for the future of humanity of these slow-developing and difficult-to-perceive problems, as well as the consequences of postponing action to mitigate their effects. Unless unified insistence that these problems receive prompt and serious attention comes from the worldwide scientific community, little is likely to happen before it is too late.

ACKNOWLEDGEMENTS

The author thanks Paul R. Ehrlich, Stephen H. Schneider, and an anonymous reviewer for their helpful comments on this paper.

REFERENCES

Begley, S., M. Miller, and M. Hager, 1988. 'The endless summer?' Newsweek, July 11.

Blake, D.R. and F.S. Rowland, 1988. 'Continuing worldwide increase in tropospheric methane, 1978 to 1987.' Science 239:1129-31.

Bolin, B., B.R. Doos, J. Jager, and R.A. Warrick (eds.), 1986. The Greenhouse Effect, Climatic Change, and Ecosystems: A Synthesis of the Present Knowledge. Wiley, New York.

Booth, W., 1988. 'Johnny Appleseed and the greenhouse.' Science 242:19-20.

Bowman, K.P., 1988. 'Global trends in total ozone.' Science 239:48-50.

Brown, L.R., 1988. 'Growing grain gap.' World Watch, 1:5 (Oct.).

Bryson, R.A., 1988. 'Civilization and rapid climate change.' Environmental Conservation, 15:1, pp. 7-15.

COHMAP Members, 1988. 'Climatic changes of the last 18,000 years: observations and model simulations.' Science 241:1043-1052.

Council on Environmental Quality (CEQ) and U.S. Department of State, 1980. The Global 2000 Report to the President U.S. Government Printing Office, Washington, D.C.

Crawford, M., 1988. 'EPA: Ozone treaty weak.' Science 242:25.

Dickenson, R.E. and R.J. Cicerone, 1986. 'Future global warming from atmospheric trace gases.' Nature 319:109-115.

Ehrlich, P.R. and A.H. Ehrlich, 1981. Extinction: The Causes and Consequences of the Disappearance of Species, Random House, New York.

Ehrlich, P.R. and J.P. Holdren (eds), 1989. The Cassandra Conference: Resources and the Human Predicament. Texas A & M University Press, College Station, Texas.

-----------, C. Sagan, D. Kennedy, and W.O. Roberts, 1984. The Cold and the Dark W.W. Norton, New York.

Environmental Protection Agency (EPA), 1988. The Potential Effects of Global Climate Change on the United States (draft). Office of Policy, Planning, and Evaluation, EPA, Washington, D.C.

Garelick, G., 1987. 'A breath of fresh air.' Time, Sept. 28.

Gleick, P.H., 1988. 'Climate change and California: past, present, and future vulnerabilities.' In Glantz, M.H., ed. Societal Responses to Regional Climate Change: Forecasting by Analogy, Westview Press, Boulder, Colorado.

Goreau, T.J. and W.Z. de Mello, 1988. 'Tropical deforestation: some effects on atmospheric chemistry.' Ambio 17:4.

Gribbin, J., 1988. 'CFCs could alter world's heat balance.' New Scientist, 19 May.

Harte, J., 1988. 'Acid rain.' In Ehrlich, P.R. and J.P. Holdren (eds), The Cassandra Conference: Resources and the Human Predicament. Texas A & M University Press.

Harwell, M.A. and T.C. Hutchinson, et al. 1985. Environmental Consequences of Nuclear War, SCOPE 28, Vol II, Ecological and Agricultural Effects. John Wiley & Sons, New York.

Hordijk, L., 1988. 'A model approach to acid rain.' Environment 30:2. See also references therein.

Jager, J., 1988. 'Anticipating climate change.' Environment 30:7.

Joyce, C., 1988. 'America counts the cost of global warming.' New Scientist, 29 October.

Kellogg, W.L. and R. Schware, 1982. 'Society, science, and climate change.' Foreign Affairs 60:5, pp. 1076-1109.

Kerr, R.A., 1988a. 'Is the greenhouse here?' Science 239:559-561.

-----------,1988b. 'Stratospheric ozone is decreasing.' Science 239:1489-91.

-----------, 1988c. 'Evidence of Arctic ozone destruction.' Science 240:1144-1145.

-----------, 1988d. 'Ozone hole bodes ill for the globe.' Science 242:785-786.

-----------, 1988e. 'Report urges greenhouse action now.' Science 242:7-8.

Jones, P.D., T.M.L. Wigley, and P.B. Wright, 1986. 'Global temperature variations between 1861 and 1984.' Nature 322:430-434.

Lemonick, M.D., 1987a. 'Culprits of the stratosphere.' Time, Sept. 21.

-------------, 1987b. 'The heat is on.' Time, Oct. 19.

-------------, 1989. 'Feeling the heat; Deadly danger in a spray can.' Time, January 2.

MacKenzie, D., 1988. 'Coming soon: the next ozone hole.' New Scientist, 1 September.

McElroy, M., 1988. 'The challenge of global change.' New Scientist, 28 July.

Mooney, H.A., P.A. Vitousek, and P.A. Matson, 1987. 'Exchange of materials between terrestrial ecosystmes and the atmosphere.' Science 238:926-932.

Murphy, D.D. and S.B. Weiss, 1989. 'The effects of climate change on biological diversity in western North America: species losses and mechanisms.' In Peters, R.S., et al. Consequences of the Greenhouse Effect on Biological Diversity, Yale University Press, in press.

National Research Council (NRC), National Academy of Sciences, 1986. Acid Deposition: Long-Term Trends, NAS, Washington, D.C.

Pain, S., 1988. 'How the heat trap will wreak ecological havoc.' New Scientist, 15 October.

Pearce, F. 1988. 'Ozone threat spreads from the Arctic.' New Scientist, 24 March.

Peters, R.L., 1988. 'Effects of global warming on species and habitats: an overview.' Endangered Species Update 5:7.

Peters, R.L. and J.D.S. Darling, 1985. 'The greenhouse effect and nature reserves.' BioScience 35:11, pp.707-717.

Peters, R.S., et al. 1989. Consequences of the Greenhouse Effect on Biological Diversity, Yale University Press, in press.

Pittock, A.B., T.P. Ackerman, P.J. Crutzen, M.C. MacCracken, C.S. Shapiro, and R.P. Turco, 1985, Environmental Consequences of Nuclear War, SCOPE 28, Vol. I, Physical and Atmospheric Effects. John Wiley & Sons, New York.

Postel, S., 1988. 'A green fix to the global warm-up.' World Watch, 1:5 (Oct.).

Ramirez, A., 1988. 'A warming world: what it will mean.' Fortune, July 4.

Rasmussen, R.A. and M.A. Khalil, 1988. 'Atmospheric trace gases: Trends and distributions over the last decade.' Science 232:1623-24.

Roberts, L., 1988. 'Is there life after climate change?' Science 242:1010-1012.

Rose, D.J., M.M. Miller, and C. Agnew, 1984. 'Reducing the problem of global warming.' Technology Review, May/June, pp.49-58.

Rowland, F.S., 1988. 'Chlorofluorocarbons, stratospheric ozone, and the Antarctic "ozone hole".' Environmental Conservation 15:2, pp.101-115.

Schneider, S.H., 1988a. 'Climate and life.' In Ehrlich, P.R. and J.P. Holdren (eds), The Cassandra Conference: Resources and the Human Predicament. Texas A & M University Press.

---------------, 1988b. 'Doing something about the weather.' World Monitor, 1:3 (December).

---------- and R. Londer, 1984. The Coevolution of Climate and Life. Sierra Club Books, San Francisco.

---------- and S.L. Thompson, 1988. 'Simulating the climatic effects of nuclear war.' Nature (in press).

Scotto, J., G. Cotton, F. Urbach, D. Berger, and T. Fears, 1988. 'Biologically effective ultraviolet radiation: Surface measurements in the United States, 1974 to 1985.' Science 239:762-64.

Stolarski, R.S., 1988. 'The antarctic ozone hole.' Scientific American 258:1 (Jan.).

Tangley, L., 1988. 'Preparing for climate change.' BioScience 38:1, pp. 14-18.

Vitousek, P.M., P.R. Ehrlich, A.H. Ehrlich, and P.A. Matson, 1986. 'Human appropriation of the products of photosynthesis.' BioScience 36:6, 368-73.

World Resources Institute and International Institute for Environment and Development, 1986. World Resources 1986. Basic Books, New York.

World Resources Institute and International Institute for Environment and Development, 1988. World Resources 1988-89. Basic Books, New York.

FOREST PRODUCTIVITY AND HEALTH IN A CHANGING ATMOSPHERIC ENVIRONMENT

Michael A. Fosberg
U.S. Department of Agriculture, Forest Service
Washington, DC USA

INTRODUCTION

Historically, forest managers and researchers have believed that forest productivity and health would be assured by wise silviculture and by protection from insects, disease, and fire. Much of our knowledge and procedures for strategic planning and management are based on observations during the past 100 years. Because much of this knowledge is emperically based, such as growth and yield tables, it is now being questioned as to its validity under a changed climate and chemical environment. Our knowledge of risks associated with management options is also based on past observations. These, too, will undoubtedly change and reduce our ability to choose the best option. These changes will probably not occur uniformly over the globe, but instead will show regional variations in amount of change. Our traditional knowledge base may not fit the events and effects of a future climate and chemical environment.

A large number of future scenarios can be constructed. These scenarios range from climate shifts outpacing the rate at which vegetation can migrate and soils evolve, to increased insect and disease activity from disturbed host-predator relations. Increased ultraviolet radiation could decrease insect and disease activity through mortality and result in increases in forest productivity and vigor. It is also possible that the host plant will become more resistant to predators. In addition, carbon uptake by vegetation may increase plant vigor in these future scenarios.

While there is consistency between models on the predicted temperatures from increased carbon dioxide, there is still great uncertainty in regional precipitation predictions (Schneider, 1987). It should be noted that the lower range of the temperature predictions are very near the expected range of natural temperature variability (Bernabo, 1981). Long-term variability in drought (Stahle et al, 1988) will also make it difficult to detect the systematic change in precipitation resulting from carbon dioxide induced climate change. Perhaps the rate of change of temperature and precipitation should be of more concern than the magnitude of these changes.

Given the uncertainty in prediction of regional precipitation patterns, can other than highly speculative future scenarios be described? I will use four examples where some information exists which will allow us to make some estimates which are not totally speculative.

COTTONWOOD LEAF RUST

Cottonwood (Populus deltoides) is rapidly becoming an important source of pulp for the paper industry. Its use is expected to double in the next 50 years and become one third of the fiber used for paper in the southeastern United States (USDA Forest Service, 1982).

A. Berger et al. (eds.), Climate and Geo-Sciences, 681–688.

Eastern cottonwood is a widely distributed species requiring abundant moisture during the growing season. Lower slopes and river bottoms are preferred sites. Cottonwood can be readily cultivated in monoculture plantations, can stand temperature extremes, and is resistant to flood damage. It is very susceptible to fire damage and to the Melampsora Leaf Rust (Hepting, 1971). The rust can overwinter at high latitudes in an alternate host (Larix spp.) and spread to warmer regions during the growing season. It also overwinters at low latitudes where the summer spore stage, which is genetically stable, spreads northward into the major cottonwood areas. The normal overwinter range is the lower Rio Grande Valley (the southern border between the United States and Mexico), where the mean low temperature in the winter is approximately 4C. A warming of 5C in the minimum temperature would move its overwinter range approximately 600 kilometers north. In years when the minimum temperatures during winter have been higher than normal, this spatial range has been observed (McCracken, et al., 1984). The current variation in overwinter range of the rust is similar to the variations that might be expected under the climate change scenario. If this historical variability were superimposed on a temperature increase of 2C, the potential overwinter range of the rust would extend 1000 kilometers north. Use of a 2C increase here is arbitrary and represents a low estimate of change (Dickenson, 1982).

BARK BEETLES

Bark Beetles (Dendroctonus spp.) are major pests in most areas in which conifer trees grow. McCambridge and Knight (1972) observed that bark beetles in northern latitudes and high latitudes are capable of developing tolerance to extreme cold (-29C). Conifers which are under moisture stress are less able to resist attack, and in a recent drought bark beetles were responsible for 64 percent of the tree mortality in California (Smith et al., 1988).

In southern pine, winter hardiness of the beetles is not a factor, but a longer warm summer increases the number of generations in a given year. In the range of the southern pine beetle, the period of active production could be extended by 6 to 8 weeks with a temperature increase of 2C (climate data from Vishar, 1954). Trees killed by bark beetles and later salvaged for lumber produce a lower quality of wood, namely a 12 percent reduction in the modulus of elasticity, and a 1 to 2 percent reduction in compression strength (Becker, 1987).

RANGE OF FOREST TYPES

In this example, I use the intensively managed commercial species of loblolly pine (Pinus taeda L.) to describe the relationship between climate and species range. Loblolly pine grows on upland, well drained soils throughout the south and southeastern United States and is tolerant of a wide range of soil types. Because loblolly is intensively managed in plantations and natural stands, much is known about both natural and artificial regeneration (Brender, 1973). Seed production in the Piedmont is correlated with weather. Higher summer precipitation results in greater seed production 17 months later, and spring precipitation and temperature are correlated with greater seed production 7 months later (Lamb, et al., 1973). Using a mean spring temperature of 15C as an indicator of the potential change of range of loblolly, a northern migration of several hundred kilometers is possible (based on a warming of 2C) and is consistent with migration rates for *Pinus* (Shugart, *et al*., 1986; Warrick, *et al*., 1986) provided that precipitation during the spring and summer is over 80 cm and sites with well drained soils are available (climate data from Vishar, 1954).

FOREST FIRES

Some of the natural enemies of loblolly are the above mentioned bark beetles, fusiform rust, damage from ice storms, and fire.

Forest fire intensity and damage depend both on the vegetation amount and structure (the forest type) and on the moisture content of the fuels (Fosberg and Furman, 1973; Fosberg, *et al*., 1981). The relationship between the forest stand and fire is that fine fuels control fire spread (Rothermel, 1972) while large fuels dominate fire intensity (Deeming, *et al*., 1972). As noted above, climate, particularly precipitation, exerts a strong influence on the fuels as expressed in forest type. Short-term variations of weather as expressed in humidity, seasonal rainfall, and wind determine the frequency and intensity of forest fires. Because humidity, seasonal rainfall, and wind are poorly predicted in climate change scenarios, I resort here to a series of possible future states. Here I will assume that the distribution about a mean climate value remains unchanged but that the mean shifts. For the case of temperature, this is always to a higher mean. An increase of humidity and a decrease of wind results in lower fire danger, and conversely, a decrease of humidity and an increase of wind results in higher fire danger. The threshold for fire severity will remain constant. That is, the biophysical/biochemical relations will not have changed. Because the threshold for severe fires will remain constant, it is reasonably safe to assume that a change in mean values will result in a changed distribution of fire frequency and severity.

We can anticipate a change of fuel loading (forest type) over a period of time as the ecosystem responds to climate change. In much of central North America, which experiences a spring and fall fire problem, we might expect a major shift in forest fire suppression activities because of decreased summer precipitation. Using the scenario in Warrick, *et al*. (1986), and Shugart, *et al*. (1986), we might expect shrubs and herbs to replace the dominant hardwood forests. The original forests are likely to be fire tolerant by structure rather than genetics. Most arid zone shrub species survive fire, and may depend on fire for germination, or can produce new growth from roots protected by soil (Noble and Slatyer, 1977). The transition period, with a remnant species from the old climate, structurally safe from fire, and an invading species, genetically aggressive with fire occurrence, would likely lead to a more rapid forest change. Specifically, the lower canopy shrubs which are fire tolerant would provide a continuous fuel ladder to the earlier forest canopy. The effects of drought, or increased frequency of drought, would cause the pre-climate change deciduous forests to be under severe threat from fire and to be replaced by fire tolerant shrublands.

A UNIFYING THEME

In three of the four examples, moisture was the dominant factor in predicting the possible future scenarios, and in the fourth, leaf rust and moisture availability dominated the host species. This clearly places a high priority on improving the confidence in regional precipitation distributions in future climate scenarios. Also, because terrestrial ecosystems show high sensitivity to interannual variability, some effort needs to be made to capture this in the predictive process of climate change. While probability of adverse events (ice storms) will not be predictable, joint probabilities of current climate adverse events and systematic climate change will be needed to approximate future probabilities. If a higher temperature increase were assumed, then the problem is further complicated by uncertainty in regional precipitation distribution and interannual variability.

A PRIORITY RESEARCH PROGRAM FOR TERRESTRIAL ECOSYSTEMS

Because climate change will require a new knowledge base for forest management, the USDA Forest Service has begun planning and implementation of a priority research program. The priority research program on forest health and productivity in a changing atmospheric environment will use multiple-stress ecosystem approach. The broad scientific questions to be addressed are:

(1) What processes are involved with atmospheric effects on forests?
(2) How will atmospheric changes influence forest health and productivity?
(3) What effects will forest management practices have on the atmosphere?

The physical and chemical climate determine, to a large extent, the structure, function, and productivity of forest, range, and related ecosystems. Our current understanding of the biogeophysical processes is based on the historical data obtained during steady or slowly changing environments. This data tells us little about a future with changed physical and chemical climates. Our research programs will focus on four specific areas:

(1) Processes of atmospheric effects on forests;
(2) Effects of forest management on the atmosphere;
(3) Long-term monitoring of effects; and
(4) Development of ecosystem models to manage forests.

These four areas have been broken down into 21 specific problems that will be addressed by Forest Service scientists and their cooperators.

(1) Processes of Atmospheric Effects on Forests

The initial need is to determine what effects the changed atmospheric chemistry and increased variability of the physical and chemical climate have on the ecosystem. Fundamental problems in the biosphere-geosphere interaction are:

(a) Determine the simultaneous responses of carbon, water, and nutrient cycles at altered physical and chemical atmospheric conditions.

(b) Refine understanding of the processes by which the atmosphere transports energy and materials within forest canopies, and how these materials are utilized in the photosynthesis process.

(c) Develop methods to predict changes in species compositions and distributions resulting from global climate change.

(d) Develop methods to predict changes in water yields from forest and rangeland resulting from global climate change.

(e) Develop methods to predict changes in fire severity resulting from global climate change.

(f) Determine the sensitivity of aquatic ecosystems and fisheries to changes in the physical and chemical atmosphere.

(g) Determine the sensitivity to and impact on wildlife species to changes in the physical environment and climate change.

(h) Determine the effects of altered climate and chemical composition on the quality and quantity of wood.

(2) **Effects of Forest Management on the Atmosphere**

A second area of needed research is in the general area of determining what effects human activity in the forests will have on climate variability. These include:

(a) Refine estimates and mechanisms of emissions from prescribed fire and wildfire.

(b) Develop understanding of transport, dispersal and removal of pollutants from prescribed fire and wildfire.

(c) Determine effects of large area forest removal on carbon, oxygen, and water budgets and relative composition of the atmosphere.

(3) **Monitoring the Problem**

A third need is to develop an appropriate monitoring program which would relate the improved understanding of the processes in the two previous sections to the methods used in management strategies. Much of our current knowledge base was obtained on our Experimental Forests and Rangelands. We need to continue experiments on these sites in order to compare the actual changes against predicted changes. Some of these experimental reserves have been monitored for over 50 years.

(a) Support long-term measurements of forest, range, and related ecosystems, particularly on Experimental Forests and Rangelands and on Research Natural Areas.

(b) Identify critical variables necessary to assess ecosystem response to climate changes and develop methodologies to monitor these variables on a wide basis, including monitoring within Forest Plans.

(c) Identify ecological, physiological, and/or morphological characteristics that can serve as indicators of chemical and/or physical environmental changes.

(d) Identify sensitive ecosystem boundaries which are sensitive to chemical and/or physical environmental changes and where single or multiple stresses can be monitored.

(4) **Develop Management Options**

The final area of research needs is to use methodologies, the knowledge of the processes, and the database from the monitoring activities to develop management strategies.

(a) Develop and verify physiological model(s) to determine the carbon, water, and other chemical budgets in forest stands.

(b) Improve biophysical aspects of plant growth models and incorporate physical and chemical environmental changes in these model(s).

(c) Develop and/or improve models of water quantity which incorporate physical and chemical environmental changes.

(d) Develop and/or improve physiologically based stand process models which are linked to growth and yield models.

(e) Develop models on the impact of physical and chemical climate variations on wildlife and fish.

(f) Develop techniques to improve the interface between models at differing spatial and temporal scales.

(g) Develop socio economic and risk analysis techniques from projected physical and global climate change.

REFERENCES

Becker, Michael, 1987. Quality and utilization of wood from damaged trees in Impact of air pollution damage to forests for roundwood supply and forest product markets. United Nations, New York: pp. 1-18.

Bernabo, J. Christopher, 1981. Quantiative estimates of temperature changes over the last 2700 years in Michigan based on pollen data, Quartenary Research, 15: 143-159.

Brender, Ernest V., 1973. Silviculture of loblolly pine in the Georgia Piedmont. Georgia Forest Research Council Report No. 33, 74 pages.

Deeming, John E.; Lancaster, James; Fosberg, Michael A.; Furman, R. William; and Schroder, Mark J., 1972. National Fire Danger Rating System, USDA Forest Service Res. Paper, 165 Pages. RM-84. Ft. Collins, CO.

Dickenson, Robert G., 1982, Modeling climate changes due to carbon dioxide increases. IN: Carbon Dioxide Review, 1982. ed. William C. Clark, Clarendon Press New York: pp. 101-133.

Fosberg, Michael A. and Furman, R. William. 1973. Fire climates in the Southwest. Agri. Meteor. 12: 27-34.

Fosberg, Michael A.; Rothermel, Richard C.; and Andrews, Patricia, 1981. Moisture content calculations for the 1000-hour timelag fuels. Forest Science, 27: 19-26.

Hepting, George H., 1971 Diseases of forest and shade trees of the United States, USDA Forest Service Handbook 386: 658 pages.

Lamb, Robert C.; Waters II, Marshall P.; and Brender, Ernest V., 1973. Apparent influence of weather upon seed production of loblolly pine. Southeast Forest Experiment Station, USDA Forest Service, Res. Note, SE-183; 7 pages.

McCambridge, William F.; and Fred B. Knight, 1972. Factors affecting spruce beetles during a small outbreak. Ecology, 53: 830-839.

McCracken, Francis; Schipper, A.L.; and Widin, K.D.; 1988. Observations on occurrence of cottonwood leaf rust in central United States. European Journal of Forestry, 14: 226-233.

Noble, I.R.; Slatyer, R.O., 1977. Post-fire succession of plants in Mediterranean ecosystems in proceedings symposium on environmental consequences of fire and fuel management in Mediterranean ecosystems. USDA Forest Service General Technical Report WO-3: 27-36.

Rothermel, Richard C., 1972. A mathematical model for predicting fire spread in wildland fuels. USDA Forest Service, Research Paper INT-115, 40 pages.

Schneider, S.H., 1987. The Greenhouse Effect: What We Can or Should Do About It. IN: First North American Conference on Preparing for Climate Change Proceedings. Government Institutes Inc. Washington: pp. 18-34.

Shugart, H.H.; Antonovsky, M. Y; Jarvis, P.G.; and Sanford, A.P., 1986. CO_2, Climate change and forest ecosystems. B. Bolin, B.R. Doos, J. Jager and R. Warrick eds. IN: The Greenhouse Effect, Climate Change and Ecosystems: pp. 475-521.

Smith, Richard S. Jr.; Roettgering, Bruce E.; and Schultz, David E. 1988. Pest tree mortality in the 1976-1977 California drought. Forest Pest Management Report 86. USDA Forest Service, San Francisco.

Stahle, D.W.; Cleaveland, M.K.; and Hehr, J.G., 1988. North Carolina Climate Changes Reconstructed from Tree Rings: A.D. 372 to 1985. Science, 240: 1517-1519.

USDA Forest Service, 1982. An analysis of the timber situation in the United States, 1952-2030. Forest Resource Report 23, 499 pages.

Vishar, Stephen S., 1954. Climate atlas of the United States. Harvard University Press, Cambridge 403 pages.

Warrick, R.A.; Shugart, H.H.; Antonovsky, M. J; Tarant, J.R.; and Tucker, C.J., 1986. The effects of increased CO_2, and climate changes on terrestrial ecosystems. IN: B. Bolin, B.R. Doos, J. Jager and R.A. Warrick eds. The Greenhouse Effect, Climate Change and Ecosystems: pp. 363-392.

CLIMATE ENVIRONMENT IN INTERNATIONAL SECURITY : THE CASE OF DEFORESTATION IN THE BRAZILIAN AMAZON

Ph.M. FEARNSIDE
Department of Ecology
National Institute for Research in the Amazon (INPA)
C.P. 478
69011 Manaus-Amazonas, Brazil

ABSTRACT. Deforestation in Brazilian Amazonia can affect the security of other countries if it contributes to climatic changes such as those expected in a world warmed by the greenhouse effect. Conversion of Brazil's Legal Amazon to cattle pasture, a transformation now rapidly proceeding, would add significantly to atmospheric CO_2 and trace gas releases. It would also reduce rainfall in the region and in neighboring areas through reduction of evapotranspiration. Foreign countries can help slow deforestation by refraining from activities that speed the clearing, such as financing highways and similar developments, direct investments in enterprises that clear forest, importation of products produced unsustainably, and funding unproductive public works that fuel the inflation that drives land speculation and consequent deforestation. Positive contributions from foreign countries could include research on such subjects as the impacts of deforestation, sustainable uses of standing forest, and economic mechanisms that would encourage sustainable management and discourage competing nonsustainable uses. Foreign sources, including multilateral development banks, can play an important role in financing alternative means of support for the population now migrating to Amazonia, including industrial development to create urban employment in the migrant source areas.

1. INTRODUCTION

General agreement exists that climate change affects international security. The countries affected by climatic change are often not the same ones where the processes that provoke the changes are located. Means must be developed to channel international efforts such that the roots of unfavorable climate changes are effectively attacked. This must be done without offending the sovereignty of the countries involved, otherwise the measures will be ineffective and will create additional strains on the international security of all concerned.

A case in point is the link between deforestation in tropical areas, especially Amazonia, and the greenhouse effect provoked by carbon dioxide, methane, nitrous oxide and other products of forest combustion

A. Berger et al. (eds.), Climate and Geo-Sciences, 689–694.

and decomposition. Many of the most immediate climatic impacts are in polar and temperate regions. I have calculated that converting Brazil's 5×10^6 km^2 "Legal Amazon" region from its original vegetation to cattle pasture would release about 50 gigatons (G tons) of carbon; were this to occur over the next 50 years (a conservative assumption), one G ton would be released per year, or about 20% of the present global total release from fossil fuels and all other sources (Fearnside, 1985a, 1987a). This has been disputed by Lugo and Brown (1986) and defended by me (Fearnside, 1986a). Calculations based on the very low estimates for forest biomass made by Brown and Lugo (1984) result in smaller contributions from tropical deforestation (e.g. Detwiller and Hall, 1988), but examination of available data supports substantially higher values for average biomass in Amazonian forests (see Fearnside, 1987a). While far from the entire problem, the potential contribution of Brazil on the order of 20% is very substantial over the coming decades when forest is being felled. The measures that would be needed to restrain the trend of forest conversion to pasture in Brazil have a high financial and political cost. Many of these measures will have to be taken by the Brazilian government itself, but international sources can contribute significantly to give these changes the speed and effectiveness they merit.

2. INTERNATIONAL RESPONSABILITY FOR PREVENTING DEFORESTATION

A portion of the deforestation now occurring in Brazil is the direct or indirect result of actions by other countries. These activities constitute an appropriate starting place to focus international efforts aimed at slowing or halting forest loss. An additional set of international activities should be encouraged in order to provide financial and scientific support for measures to reduce forest clearing.

Activities that should not be continued by foreign countries, especially those in the industrialized portion of the Northern Hemisphere, include the financing of destruction through multilateral and private banks. Loans from such sources as the World Bank and the Interamerican Development Bank often fund the key component within the system of forces that drives deforestation : construction of highways into rainforest areas (Fearnside, 1987b). This is precisely the factor to which deforestation appears to be most sensitive, since the presence of roads sets in motion a cycle of immigration, land speculation and population turnover that leads to clearing and is beyond the control of government authorities (Fearnside, 1984, 1986b, 1987c).

A second set of activities from abroad that contributes to deforestation is direct investments by foreign firms. Volkswagen's 139,000 ha Rio Cristalino Ranch in Pará is the best known example (although its sale to Brazilian investors is under negotiation). Others include the 72,000 ha Armour-Swift/Brascan ranch in Pará and the 678,000 ha Suia-Missú Ranch in Mato Grosso of the Italian multinational Liquigas (also under negotiation for sale). While the thousands of Brazilian ranching operations account for most of the clearing by this sector, the presence of international firms is not insignificant (see Fearnside, 1983, 1986c).

Foreign countries also contribute to deforestation by importing many products that are produced in deforested land or are taken from the forest in unsustainable ways. The "hamburger connection" between Central American beef exporters and North American markets is the best known (Myers, 1981; Nations and Komer, 1983), although hoof-and-mouth disease (Afthosis) protects Amazonia from this scourge. Pig-iron exports, principally to Japan, will soon add a new and mighty force to deforestation in eastern Amazonia once production from the Grande Carajás area reaches its expected scale. Pig-iron is to be produced using charcoal which, were it produced from planted *Eucalyptus*, would require an area ten times that of the managed plantations at Jari (Fearnside, 1988). Since plantations on this scale are expensive and risky, the wood can be expected to come from native forest for as long as accessible stands remain in existence.

A fourth means by which foreign countries speed deforestation in Amazonia is indirectly through the contribution their activities make to Brazil's astronomical inflation. Protection from inflation is a major motivation for land speculation, which leads to deforestation in order to secure title to the land and to keep squatters or other speculators from usurping the claim. Inflation is driven by expenditures for large public works that do not contribute a proportional return to the economy. When people earn money in constructing such a public work they create demand for products in the marketplace; since supply remains the same, demand exceeds supply and prices increase. Examples of massive unproductive investments include the Angra nuclear power plants near Rio de Janeiro (with equipment sold and financed by Germany) and the Balbina hydroelectric project near Manaus (with equipment financing and purchase from France). Even when developments are more productive than these notorious examples, the return from the projects is often so slow that significant inflation can be expected in the meantime. The inducement of easy financing for such projects leads political leaders to initiate new projects as fast as they can. The direct and indirect impact on the forest is generally the last consideration in these decisions.

In addition to the above types of activities that foreign countries should avoid, many positive activities *should* be pursued. These include research on the climatic and other impacts of deforestation. One major impact with potentially severe consequences within Brazil itself is alteration of the water cycle (Marques et al., 1977; Salati et al., 1978, 1979; Villa Nova et al., 1976). The impacts on rainfall would extend to neighboring areas, such as south-central Brazil (Salati and Vose, 1984). Because of the high variability of rainfall in the region and the critical contribution of evapotranspiration to precipitation during the dry season, reductions could lead to burning within the standing forest as occurred in Borneo in 1982/1983 (see Malingreau et al., 1985), and/or set in motion a positive feedback loop leading to less rainfall and more open xerophytic vegetation (Fearnside, 1985b). Better studies of both the contributions of the forest to maintaining climatic equilibria and the potential impact of climate change on the forest are urgently needed. One major effort being planned to improve our ecological understanding tropical forest and its relation to global processes is the International Biosphere-Geosphere Program. Unfortunately, the program appears to intend to assess tropical forest from study sites in such politically attractive places as Puerto Rico. Because of Amazonia's great size relative to other tropical forest areas of the world, there can be no substitute for data collected in the region itself if globally-relevant conclusions are to be drawn. Results from studies carried out in Amazonia can also lend important weight to arguments aimed at convincing Brazilian decision-makers to make halting deforestion a top priority.

Research on substainable uses of the forest is also needed. This concludes forestry management studies and collection of the basic ecological information needed to give a secure founding to systems that maintain the structure of the forest intact. It also includes studies of non-wood products that can be obtained from the forest, especially the many pharmacological compounds contained in forest plants and animals. The only feasible means of surveying the vast store of potential compounds is to use the knowledge of indigenous tribes in the region. These tribes are rapidly disappearing; their protection is not only a moral but also an economic and environmental imperative.

Some of the principal barriers to sustainable use of Amazonian forest are economic and social rather than technological. Considering the importance of the task, very little is being done to identify the measures that the government might take to make sustainable systems profitable and the competing non-sustainable ones unprofitable. Some needed changes are obvious, such as an end to financing and tax incentives for pasture, heavy taxes on capital gains to impede land speculation, and disallowing cattle pasture as <u>benfeitoria</u> (improvement) for purposes of establishing land tenure claims. Many other changes that might be made are not yet sufficiently well understood to be translated into operational suggestions. For example, evaluation of forestry management plans must be based on criteria that give appropriate weight to long-term gains (rather than simple comparison with a banker's discount rate), and that include non-monetary environmental and social benefits (Fearnside, 1989). Foreign researchers and funding sources can contribute to filling these important lacunae.

Major funding from international sources will be needed to rechannel development in ways less damaging to the forest. Migration from other parts of Brazil, such as the south-central state of Parana, is a major factor in the exponential increase in deforested areas observed in the Amazonia states Mato Grosso, Rondonia and Acre. The migrants are being forced to leave small farms in Parana because of continuing concentration of land tenure and transformation of agriculture from labor-intensive crops to mechanized cultivation, especially of soybeans for export as cattle feed to North America and Europe. One government initiative has been land reform, now stalled due to strong opposition from large landholders. Clearly this would be very expensive if property owners are to be adequately compensated for land that is redistributed. Land redistribution is a prerequisite if large populations are to be maintained in the countryside. Whether or not land reform is implemented, population exodus is likely to continue : Brazil cannot expect to keep 32% of its population in the countryside (Brazil, IBGE, 1987:57) when its agriculture is transforming into systems similar to those in

North America where less than 5% of the population is rural. Many people leaving the countryside have moved to favelas (shantytowns) around large cities like Sao Paulo where their visible povery, crime, and tendency to support opposition political parties are unwelcome to national leaders. Diverting this population flow to Amazonia by paving highways and establishing colonization projects is a financially and environmentally costly, as well as temporary, solution. International funding sources can play a vital role in promoting industrial developments to provide urban employment as an alternative for these people. The international funding now used to build highways and other projects causing deforestation in Amazonia must be redirected to these ends.

3. REFERENCES

Brazil, IBGE, 1987. Anuário Estatístico do Brasil 1986, **47**, Instituto Brasileiro de Geografia e Estatistica (IBGE), Rio de Janeiro, 628pp.

Brown S. and A.E. Lugo, 1984. 'Biomass of tropical forests : A new estimate based on forest volumes'. Science, **223**, 1288-1293.

Detwiller R.P. and C.A.S. Hall, 1988. 'Tropical forests and the global carbon cycle'. Science, **239**, 42-47.

Fearnside P.M., 1983. 'Land use trends in the Brazilian Amazon Region as factors in accelerating deforestation'. Environmental Conservation, **10**(2), 141-148.

Fearnside P.M., 1984. 'Land clearing behaviour in small farmer settlement schemes in the Brazilian Amazon and its relation to human carrying capacity'. pp. 255-271 In : A.C. Chadwick and S.L. Sutton (eds), Tropical Rain Forest : The Leeds Symposium. Leeds Philosophical and Literary Society, Leeds, U.K., 335pp.

Fearnside P.M., 1985a. 'Brazil's Amazon forest and the global carbon problem'. Interciencia, **10**(4), 179-186.

Fearnside P.M., 1985b. 'Environmental Change and Deforestation in the Brazilian Amazon', pp. 70-89 In : J. Hemming (ed.) Change in the Amazon Basin : Man's Impact on Forests and Rivers. Manchester University Press, Manchester, U.K. 222pp.

Fearnside P.M., 1986a. 'Brazil's Amazon forest and the global carbon problem : Reply to Lugo and Brown'. Interciencia, **11**(2), 58-64.

Fearnside P.M., 1986b. 'Spatial concentration of deforestation in the Brazilian Amazon'. Ambio, **15**(2), 72-79.

Fearnside P.M., 1986c. Human Carrying Capacity of the Brazilian Rainforest. Columbia University Press, New York, 293pp.

Fearnside P.M., 1987a. 'Summary of progress in quantifying the potential contribution of Amazonian deforestation to the global carbon problem'. pp. 75-82 In : D. Athie, T.E. Lovejoy and P. de M. Oyens (eds), Proceedings of the Workshop on Biogeochemistry of Tropical Rain Forests : Problems for Research. Universidade de São Paulo, Centro de Energia Nuclear na Agricultura (CENA), Piracicaba, São Paulo, 85pp.

Fearnside P.M., 1987b. 'Deforestation and international economic development projects in Brazilian Amazonia'. Conservation Biology, **1**(3), 214-221.

Fearnside P.M., 1987c. 'Causes of deforestation in the Brazilian Amazon'. pp. 37-53 In : R.F. Dickinson (ed.) The Geophysiology of Amazonia : Vegetation and Climate Interactions. John Wiley & Sons, New York, 526pp.

Fearnside P.M., 1988. 'Jari at age 19 : Lessons for Brazil's silvicultural plans at Carajas'. Interciencia, **13**(1), 12-24.

Fearnside P.M., 1989. 'Forest management in Amazonia : the need for new criteria in evaluating development options'. Forest Ecology and Management, **26**, (in press).

Lugo A.E. and S. Brown, 1986. 'Brazil's Amazon forest and the global carbon problem'. Interciencia, **11**(2), 57-58.

Malingreau J.P., G. Stephens and L. Fellows, 1985. 'Remote sensing of forest fires : Kalimantan and North Borneo in 1982-83'. Ambio, **14**, 314-321.

Marques J., J.M. dos Santos, N.A. Villa Nova and E. Salati, 1977. 'Precipitable water and water vapor flux between Belém and Manaus'. Acta Amazonica, 7(3), 355-362.

Myers N., 1981. 'The hamburger connection : How Central America's forests become North America's hamburgers'. Ambio, **10**(1), 3-8.

Nations J.D. and D.I. Komer, 1983. 'Rainforests and the hamburger society'. Environment, **25**(3), 12-20.

Salati E.A. Dall'Olio, E. Matusi and J.R. Gat, 1979. 'Recycling of water in the Brazilian Amazon Basin : An isotopic study'. Water Resources Research, **15**, 1250-1258.

Salati E., J. Marques and L.C.B. Molion, 1978. 'Origem e distribuição das chuvas na Amazônia'. Interciencia, **3**(4), 200-206.

Salati E. and P.B. Vose, 1984. 'Amazon Basin : A system in equilibrium'. Science, **225**, 129-138.

Villa Nova N.A., E. Salati and E. Matusi, 1976. 'Estimativa da evapotranspiraçao na Bacia Amazônica'. Acta Amazônica, **6**(2), 215-228.

WHERE IS CLIMATE RESEARCH GOING ?

J. PERRY
National Academy of Sciences
National Research Council Board on
Atmospheric Sciences and Climate
Washington, DC 20418, USA

The broad range of research reviewed in this symposium reveals clearly the rapid evolution - no, revolution - in our understanding of climate over the past decades. When the older members of our number, such as I, began our scientific careers, climate was disposed of in our textbooks by means of a few statistics and charts, a rough-and-ready energy diagram, and a few words about the ice ages. A conference on climate would have been sparsely attended, unbearably dull, but mercifully brief.

Not much more than a decade ago, the study of climate began to emerge as an exciting scientific challenge, rather than a dull statistical chore. Many factors contributed to this evolution. Study of sedimentary evidence on land and the ocean, and analysis of glacial cores, coupled with new sophistication in theoretical investigation, revealed a richly varied tapestry of past climatic variability that provided an inescapable challenge to science. Space-based observing systems let us observe the global system responsible for these variations in unprecedented scope and detail. Electronic computers and clever mathematics allowed us to realize Richardson's dream of numerically simulating the atmosphere. Exploitation of these capabilities in the Global Atmosphere Research Programme enhanced our understanding of the atmosphere's behaviour. Finally, the observations of atmospheric composition carried out with dogged dedication by Charles David Keeling reminded us forcefully that we were conducting a great geophysical experiment in the tiny but precious test tube in which we live.

This evolution in understanding and concern set the stage for the World Climate Conference, organized in Geneva in 1979 by the World Meteorological Organization. Further motivation for the Conference was provided by a series of climatic fluctuations that stressed the world food production and distribution system in the early 1970's. The Conference assembled a group of scientists and scholars that would have been unthinkable a decade before to review all aspects of contemporary understanding of climate variability, change, and interaction with life. The review confirmed the growing appreciation of climate as a central

A. Berger et al. (eds.), Climate and Geo-Sciences, 697–701.

issue for science and society, and deepened global concern for the future.

The Conference set in motion the World Climate Program, which in turn stimulated a still-growing flood of scientific effort directed at improving our understanding of climate - past, present and future - and climate's interaction with society. It is a tribute to the success of this world-wide crusade that a Symposium such as this week's could be assembled so smoothly and effectively, and with minimal resources. Moreover, it now seemed natural - indeed inevitable - that a discussion of climate should involve not only meteorologists, but also astronomers, geologists, geochemists, biologists, oceanographers, mathematicians, computer scientists, hydrologists, and political scientists.

The reviews we heard during the week revealed how greatly our understanding of climate has been enhanced. We now know a great deal about the climates of the past, from the Archaean through the brief moments of our own species'time on Earth. The roles of astronomical forcing and of life processes in shaping climate seem even more firmly established than a decade ago. In modeling the behaviour of the climate system, it still appears that the ocean, clouds, and the land surface pose major challenges. We learned much about influences of chemistry, the problems of coupling media, the information gleaned from proxy records, climate modeling techniques, atmosheric chemical processes, and the details of modeled response and atmosphere-biosphere interactions.

However, discussions such as this between specialists tend to focus on details, uncertainties, and the many things we wish we knew better. Indeed, they tend to obscure the common core of consensus on what we do know, what we do believe. So it is perhaps worthwhile to state explicitly the consensus that implicitly underlies all our work :

. The atmosphere, and indeed the entire global environment, is changing;

. We humans are today the principal agents of change, and the rich nations of NATO are among the most prominent of these **agents provocateurs**;

. In particular, the climates of all the world's nations will change, rich and poor alike;

. Climate is an important resource, constraint, and perhaps hazard that must be taken into account in the prudent management of our little planet condominium.

But the question facing this terminal session is : "Where do we go from here ?" I believe that we can answer this question in at least three ways, according to the definition of "we". First, if "we" are the researchers concerned with the physical climate system, I believe that the papers presented during the week laid out a challenging range of scientific problems. The speakers this morning have identified many of these tasks. It is heartening to know that a well-organized international attack on this range of problems is being carried out by WMO and ICSU under the aegis of the World Climate Research Programme. It is clear that the considerable momentum of the many well-conceived projects developed under this Programme should be maintained.

If the "we" is the broader array of scientists concerned with all the interlinked changes going on in our complicated planet, then a companion international effort - the International Geosphere-Biosphere Programme (IGBP) - promises to play a similar catalytic and coordinating role. As described by Professor Dooge, this program seeks "to describe and understand the interactive physical, chemical and biological processes that regulate the total Earth system, the unique environment that it provides for life, the changes that are occurring in this system, and the manner in which they are influenced by human actions." The programme's content is being elaborated by a distinguished Special Committee of the International Council of Scientific Unions (ICSU), supported by the National Committee in many countries, including my own. Although its scope in principle encompasses that of the World Climate Research Programme (WCRP), I believe that in practice the IGBP will be complementary to, not competitive with, the WCRP, in particular by providing a context for the involvement of geochemists, biologists and ecologists in the effort to understand what's happening to our planet.

In fact, however, I suppose that the "we" in "Where do we go from here ?" is the "we" who participate in the NATO Science Programme. As Dr. da Cunha outlined at the beginning of the week, this effort seeks to advance the scientific and technological capabilities of the alliance by promoting international scientific cooperation in scientific disciplines relevant to the challenges facing modern society. These efforts in the form of Study Institutes, Research Workshops, research grants and fellowships are in part focused through Special Programmes - like the Global Transport Mechanisms in Geo-Sciences Programme of which this Symposium is the terminal activity. Has the work of this Programme, and more particularly the information brought forward in this Symposium, revealed a need for some sort of follow-on activity by NATO ?

I believe that our discussions this week indicate strongly that the problems posed by man-made influences on climate comprise a prime example of exactly the type of inscapably international, interdisciplinary issue in which activities such as NATO's have great and beneficial leverage. Certainly the potential importance of climatic change to the NATO nations needs little re-emphasis. Moreover, the NATO nations collectively form the major source of the emissions that we believe are changing our world. Thus, for both practical and moral reasons, man-made climatic change should remain high on the scientific agenda of NATO.

Moreover, this area of research is one that can especially benefit from activities such as NATO's that are directed to improving collaboration by enhancing the international mobility and intercommunication of scientists. The number of scientists actively contributing to this research is but a trace constituent in the body of science. Parenthetically, I would guess that the abundance of climate researchers in the population of scientists is in fact somewhat less than the abundance of carbon dioxide in dry air. Like CO_2, they are also distributed all over the face of the globe. In order to work effectively, they require access to resources of data and computing power that are available in only a few places, and they must be able to gather together from time to time in a critical intellectual mass.

Furthermore, as Paul Ehrlich so forcefully held, the ultimate goal of our research is to influence policies in the economic and political realms. Results of our research will influence policies only if politicians understand them and believe in them. To build this broad base of confidence and credence we must somehow involve the scientific establishments, government bureaucracies and diverse constituencies of many countries.

Thus, I believe that we have ample grounds to argue that some aspects of climate should be the focus of a continuing Special Programme of NATO. But what tasks should such a program address ?

First, I believe that we need enhanced interactions among the scientists who are trying to understand climate as a scientific problem.

Second, we need to mobilize and entrain the energies and talents of many relevant disciplines of the natural sciences that have as yet given only peripheral consideration to the problem of climate change. As I noted above, the history of climate research has been characterized by a steady growth in disciplinary scope. A broadly integrated attack on the problems of environmental change is the objective of the IGBP. The new NATO program should certainly foster scientific exchanges that will advance climate research and the goals of the IGBP. Thus, we must involve biologists, ecologists, hydrologists, soil scientists, and others.

Thirdly - and perhaps most urgently - we must move vigorously into the human dimension. Our current models - implicit or explicit - of the future influences of society on climate and the impact of climate on society are rudimentary at best. We urgently need to entrain first rate social scientists of many persuasions to work intimately with natural scientists to seek a joint understanding of the future of mankind's climate. As Diana Liverman pointed out, we need to involve people who understand the forces that will - with climate - be shaping our future world : technology, economics, population, and politics. As emphasized by the 1985 WMO (World Meteorological Organisation) - ICSU - UNEP (United Nations Environment Programme) Villach Conference, it is high time to involve policy-makers in the problem of climate change.

With these thoughts in mind, I believe that the central conclusion of the Symposium is that a NATO Special Programme in this area should be developed as a matter of urgency. A suitable title might be :

Climate and Society into the 21st Century

The Programme would comprise the usual mix of activities - ASI's, ARW's, CRG's, and fellowships - and would focus on three closely related objectives :

(1) Improve our understanding of the behaviour of the physical climate system;
(2) Improve our understanding of the interactions among the physical climate system, the natural and managed ecosystems, and the human economic, social, and technological systems;
(3) Explore the implications of this understanding for policy, and the possible options for amelioration, adaptation, and control.

The design of such a program is not a task for a Symposium, even one populated by such enlightened souls as we. A planning and steering group should be established in the near future. I feel confident that such a program would permit us to organize, say five years from now, an even better and more illuminating Symposium than we have enjoyed this week. If our hypotheses and projections are correct, the new improved and much broader insights we will then be able to present will be even more in demand by NATO's and the world's nations.

But in one respect at least, it will be difficult to improve on this week's program. We could hardly expect to hold a meeting in a better setting, or in a more beautiful and interesting country, or under better leadership and hospitality than provided by our gracious host, Prof. Berger, and his most helpful staff. So, with thanks to NATO and our hosts, let us go back home and back to work !

AUTHORS INDEX

SUBJECTS INDEX